Edited by
Kristin Bowman-James, Antonio Bianchi,
and Enrique García-España

Anion Coordination Chemistry

Related Titles

Sliwa, W., Kozlowski, C.

Calixarenes and Resorcinarenes

Synthesis, Properties and Applications

2009

ISBN: 978-3-527-32263-3

Ribas Gispert, J.

Coordination Chemistry

2008

ISBN: 978-3-527-31802-5

Balzani, V., Credi, A., Venturi, M.

Molecular Devices and Machines

Concepts and Perspectives for the Nanoworld

2008

ISBN: 978-3-527-31800-1

van Leeuwen, P. W. N. M. (ed.)

Supramolecular Catalysis

2008

ISBN: 978-3-527-32191-9

Diederich, F., Stang, P. J., Tykwinski, R. R. (eds.)

Modern Supramolecular Chemistry

Strategies for Macrocycle Synthesis

2008

ISBN: 978-3-527-31826-1

Edited by Kristin Bowman-James, Antonio Bianchi, and Enrique García-España

Anion Coordination Chemistry

WILEY-VCH

WILEY-VCH Verlag GmbH & Co. KGaA

The Editors

Prof. Dr. Kristin Bowman-James
Department of Chemistry
University of Kansas
1251 Wescoe Hall Drive
Lawrence, KS 66045
USA

Prof. Dr. Antonio Bianchi
University of Florence
Department of Chemistry
Via della Lastruccia 3
50019 Sesto Fiorentino
Italy

Prof. Dr. Enrique García-España
Instituto de Química Molecular
Departamento de Química Inorgánica
C/ Catedrático José Beltrán 2
46980 Paterna (Valencia)
Spain

Library of Congress Card No.: applied for

British Library Cataloguing-in-Publication Data
A catalogue record for this book is available from the British Library.

Bibliographic information published by the Deutsche Nationalbibliothek
The Deutsche Nationalbibliothek lists this publication in the Deutsche Nationalbibliografie; detailed bibliographic data are available on the Internet at <http://dnb.d-nb.de>.

Typesetting Laserwords Private Limited, Chennai, India
Printing and Binding Fabulous Printers Pte Ltd, Singapore
Cover Design Formgeber, Eppelheim

Printed in Singapore
Printed on acid-free paper

Print ISBN: 978-3-527-32370-8
ePDF ISBN: 978-3-527-63952-6
oBook ISBN: 978-3-527-63950-2
ePub ISBN: 978-3-527-63951-9
Mobi ISBN: 978-3-527-63953-3

Contents

Preface

While Park and Simmons provided the first seminal report of the supramolecular chemistry of anions in 1968, it was Jean-Marie Lehn who suggested in 1978 that it was truly a form of coordination chemistry. At that time supramolecular chemistry, which refers to the interactions of molecular and ionic species beyond the covalent bond, was in its formative years. The term supramolecular chemistry was built on the lock and key concept first proposed by Emil Fischer in 1894. The actual term, however, was coined by Jean-Marie Lehn at the early stages of the development of this field. In many respects this concept can be merged with another key chemical concept, that of coordination chemistry, also introduced in the late nineteenth century by Alfred Werner. All three men, Fischer, Lehn, and Werner, were recognized for their seminal contributions to science with Nobel Prizes.

As pointed out in Chapter 1, anions were of interest to chemists as early as the 1920s. Yet in the early years of supramolecular chemistry, the focus on anions began only as a small seedling that has now grown into a giant tree with many branches. Anion coordination chemistry now impinges on numerous fields of science, including medicine, environmental remediation, analytical sensing, as well as many aspects of the global field of nanotechnology. Scientists from all areas of chemistry and beyond have joined forces to explore this exciting new field.

By the early 1990s, there were a number of texts devoted to various aspects of supramolecular chemistry, but none that focused entirely on anions. At that time the three of us realized the need for such a text, and we gathered the expertise of anion researchers far and wide to contribute to the book that was published in 1997, *Supramolecular Chemistry of Anions*. Since that time a small number of excellent texts and many reviews have been published, focusing on anions and reporting advances in this rapidly evolving field. In this sequel to our earlier text, using the same strategy of enlisting the aid of noted scientists in the field, we have tried to incorporate some of the imagination and excitement that has gone into the science of anions in the last 15 years. The chapters are laid out in a manner similar to that in our first volume, covering basic topics in anion coordination. Chapter 1 approaches the historical development of anion chemistry from a slightly different viewpoint than usual, by covering both biological and supramolecular developments. It is followed by two chapters outlining what we consider to be the core foundations of anion coordination, thermodynamic and structural aspects. Synthetic aspects of some of the more commonplace receptors are reviewed in Chapter 4. The following two chapters explore some of the more recent and exciting aspects that illustrate the growth of the field: the use of anions as synthetic templates in Chapter 5 and anion-π interactions in Chapter 6. Chapters 7 and 8 focus on biological implications of anions and include an overall view of hosts for biologically relevant anions and receptors designed for membrane transport, respectively. The book concludes with a chapter exploring an important application of anion coordination, sensors for anions.

This book has been possible only because of the outstanding scientists who have contributed exceptionally well-written chapters. We extend our warm thanks for the time and effort that they have dedicated to this process. We would also like to thank the many funding agencies worldwide that have made this research possible. K.B.-J would like to express appreciation to the National Institutes of Health and the Department of Energy, and especially the National Science Foundation grant CHE CHE0809736 for the current funding. EGE thanks the Spanish Ministry of Science and Innovation and Science (MCINN), Projects CONSOLIDER CSD 2010-00065, CTQ 2009-14288-C04-01 and Generalidad Valenciana (GVA), project Prometeo 2011/008.

Last but not least, we would like to take this opportunity to acknowledge our families, research groups, and students. Our families have provided patience and encouragement throughout the making of this book. Our students and other researchers in our groups have made significant contributions to some of the science reported here. We would also like to thank the many researchers in the anion community who have conducted the outstanding science that has now become part of this book.

Lawrence, Kansas, USA — *Kristin Bowman-James*
Florence, Italy — *Antonio Bianchi*
Valencia, Spain — *Enrique García-España*

List of Contributors

Valeria Amendola
Università di Pavia
Dipartimento di Chimica
via Taramelli 12
27100 Pavia
Italy

Rowshan Ara Begum
University of Kansas
Department of Chemistry
Lawrence, KS 66045
USA

Andrea Bencini
Università di Firenze
Dipartimento di Chimica
"Ugo Schiff"
50019 Sesto Fiorentino (Florence)
Italy

Antonio Bianchi
Università di Firenze
Dipartimento di Chimica
"Ugo Schiff"
50019 Sesto Fiorentino (Florence)
Italy

Kristin Bowman-James
University of Kansas
Department of Chemistry
Lawrence, KS 66045
USA

Jack K. Clegg
University of Sydney
School of Chemistry
NSW 2006
Australia

Megan M. Daschbach
Washington University
Department of Chemistry
St. Louis, MO 63130
USA

Victor W. Day
University of Kansas
Department of Chemistry
Lawrence, KS 66045
USA

Pere M. Deyà
Universitat de les Illes Balears
Departament de Química
Crta de Valldemossa km 7.5
07122 Palma de Mallorca
(Baleares)
Spain

Luigi Fabbrizzi
Università di Pavia
Dipartimento di Chimica, via
Taramelli 12
27100 Pavia
Italy

Antonio Frontera
Universitat de les Illes Balears
Departament de Química
Crta de Valldemossa km 7.5
07122 Palma de Mallorca
(Baleares)
Spain

Enrique García-España
Instituto de Química Molecular
Departamento de Química
Inorgánica
C/ Catedrático José Beltrán 2
46980 Paterna (Valencia)
Spain

George W. Gokel
University of Missouri – St. Louis
Department of Chemistry and
Biochemistry
Center for Nanoscience
One University Blvd
St. Louis, MO 63121
USA

and

University of Missouri – St. Louis
Department of Biology
Center for Nanoscience
One University Blvd
St. Louis, MO 63121
USA

Sung Ok Kang
University of Kansas
Department of Chemistry
Lawrence, KS 66045
USA

and

Chemical Sciences Division
Oak Ridge National Laboratory
Oak Ridge, TN 37831
USA

Stefan Kubik
Technische Universität
Kaiserslautern
Fachbereich Chemie –
Organische Chemie
Erwin-Schrödinger-Straße
67663 Kaiserslautern
Germany

Maurizio Licchelli
Università di Pavia
Dipartimento di Chimica
via Taramelli 12
27100 Pavia
Italy

Leonard F. Lindoy
University of Sydney
School of Chemistry
NSW 2006
Australia

José M. Llinares
Universitat de Valéncia
Instituto de Ciencia Molecular
(ICMol)
Departamento de Química
Orgánica
C/ Catedrático José Beltrán n 2
46980 Paterna (Valencia)
Spain

David Quiñonero
Universitat de les Illes Balears
Departament de Química
Crta de Valldemossa km 7.5
07122 Palma de Mallorca
(Baleares)
Spain

Angelo Taglietti
Università di Pavia
Dipartimento di Chimica
via Taramelli 12
27100 Pavia
Italy

1
Aspects of Anion Coordination from Historical Perspectives

Antonio Bianchi, Kristin Bowman-James, and Enrique García-España

1.1
Introduction

Supramolecular chemistry, the chemistry beyond the molecule, gained its entry with the pioneering work of Pedersen, Lehn, and Cram in the decade 1960–1970 [1–5]. The concepts and language of this chemical discipline, which were in part borrowed from biology and coordination chemistry, can be to a large extent attributed to the scientific creativity of Lehn [6–8]. Recognition, translocation, catalysis, and self-organization are considered as the four cornerstones of supramolecular chemistry. Recognition includes not only the well-known transition metals (classical coordination chemistry) but also spherical metal ions, organic cations, and neutral and anionic species. Anions have a great relevance from a biological point of view since over 70% of all cofactors and substrates involved in biology are of anionic nature. Anion coordination chemistry also arose as a scientific topic with the conceptual development of supramolecular chemistry [8]. An initial reference book on this topic published in 1997 [9] has been followed by two more recent volumes [10, 11] and a number of review articles, many of them appearing in special journal issues dedicated to anion coordination. Some of these review articles are included in Refs [12–52]. Very recently, an entire issue of the journal *Chemical Society Reviews* was devoted to the supramolecular chemistry of anionic species [53]. Since our earlier book [9] the field has catapulted way beyond the early hosts and donor groups. Because covering the historical aspects of this highly evolved field would be impossible in the limited space here, a slightly different approach will be taken in this chapter. Rather than detail the entry of the newer structural strategies toward enhancing anion binding and the many classes of hydrogen bond donor groups that have come into the field, only the earlier development will be described. This will be linked with aspects of naturally occurring hosts, to provide a slightly different perspective on this exciting field.

Interestingly enough, the birth of the first-recognized synthetic halide receptors occurred practically at the same time as the discovery by Charles Pedersen of the alkali and alkaline-earth complexing agents, crown ethers. While Pedersen

Anion Coordination Chemistry, First Edition. Edited by Kristin Bowman-James, Antonio Bianchi, and Enrique García-España.

submitted to *JACS* (*Journal of the American Chemical Society*) his first paper on crown ethers in April 1967 entitled "Cyclic Polyethers and their Complexes with Metal Salts" [1], Park and Simmons, who were working in the same company as Pedersen, submitted their paper on the complexes formed by bicyclic diammonium receptors with chloride entitled "Macrobicyclic Amines. III. Encapsulation of Halide ions by in, in-1, (k + 2)-diazabicyclo[k.l.m]alkane-ammonium ions" also to *JACS* in November of the same year [54].

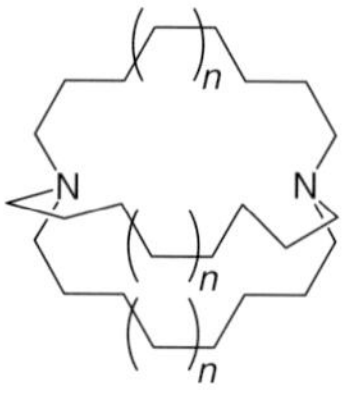

$n = 1$ (**1**)
$n = 2$ (**2**)
$n = 3$ (**3**)
$n = 4$ (**4**)

These cage-type receptors (**1-4**) were called *katapinands*, after the Greek term describing the swallowing up of the anionic species toward the interior of the cavity (Figure 1.1). The engulfing of the chloride anion inside the katapinand cavity was confirmed years later by the X-ray analysis of the structure of Cl^- included in the [9.9.9] bicyclic katapinad [55]. However, while investigations on crown ethers rapidly evolved and many of these compounds were prepared and their chemistry widely explored, studies on anion coordination chemistry remained at the initial stage. Further development waited until Lehn and his group revisited this point in the late 1970s and beginning of the 1980s [56–62].

Nevertheless, evidence that anions interact with charged species, modifying their properties, in particular their acid–base behavior, was known from the early times of the development of speciation techniques in solution, when it was noted that

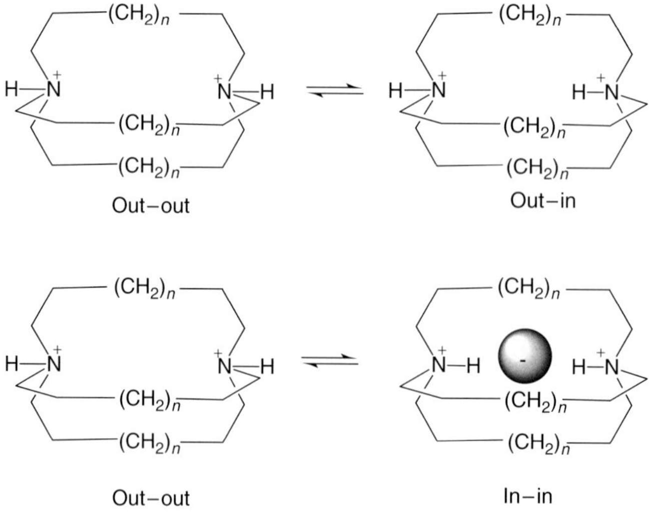

Figure 1.1 In–in and out-out equilibria, and halide complexation in katapinand receptors.

protonation constants were strongly influenced by the background salt used to keep the ionic strength constant [63]. Following these initial developments, Sanmartano and coworkers did extensive work on the determination of protonation constants in water with and without using ionic strength. In this way, this research group was able to measure interaction constants of polyammonium receptors with different anionic species [64, 65]. Along this line, Martell, Lehn, and coworkers reported an interesting study in which the basicity constants of the polyaza tricycle (**5**) were determined by pH-metric titrations using different salts to keep the ionic strength constant [66]. The authors observed that while the use of $KClO_4$ did not produce significant differences in the constants with respect to the supposedly innocent trimethylbenzene sulfonate anion (TMBS), the use of KNO_3 and KCl led to higher pK_a values, particularly as more acidic conditions were reached. From these titrations, binding constants of nitrate and chloride with hexaprotonated **5** were determined to be 2.93 and 2.26 logarithmic units, respectively.

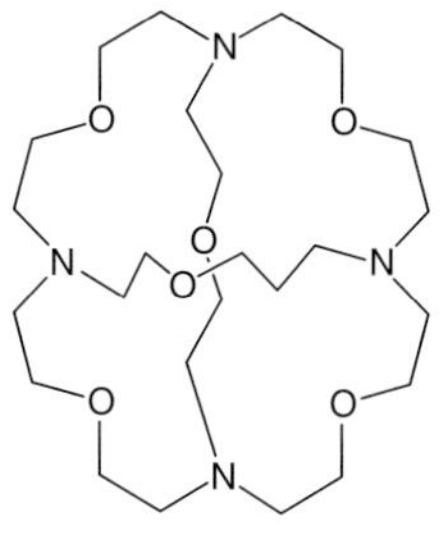

5

Similar events were observed in the biological world many years ago. The well-known Hofmeister series or lyotropic series [67] was postulated at the end of the nineteenth century to rank the relative influence of ions on the physical behavior of a wide variety of processes ranging from colloidal assembly to protein folding. The Hofmeister series, which is more pronounced for anions than for cations, orders anions in the way shown in Figure 1.2. The species to the left of Cl^- are called *kosmotropes*, "water structure makers," and those to the right of

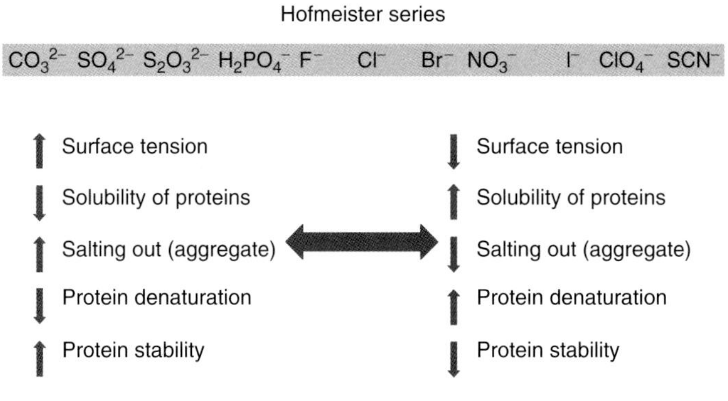

Figure 1.2 Representation of the Hofmeister series.

chloride are termed *chaotropes*, "water structure breakers." While the kosmotropes are strongly hydrated and have stabilizing and salting-out effects on proteins and macromolecules, the chaotropes destabilize folded proteins and have a salting-in behavior.

Although originally these ion effects were attributed to making or breaking bulk water structure, more recent spectroscopic and thermodynamic studies pointed out that water structure is not central to the Hofmeister series and that macromolecule–anion interactions as well as interactions with water molecules in the first hydration shell seem to be the key point for explaining this behavior [68–72].

In this respect, as early as in the 1940s and 1950s, researchers sought to address the evidence and interpret the nature of the binding of anions to proteins [73]. Colvin, in 1952 [74], studying the interaction of a number of anions with the lysozyme, *calf thymus* histone sulfate, and protamine sulfate proteins using equilibrium dialysis techniques, concluded that although electrostatic charge–charge interactions may be chiefly responsible for the negative free energy of binding, there were other contributions such as van der Waals and solvation energies that can equal or even exceed the charge to charge component.

More recently, the use of X-ray diffraction techniques for unraveling the structure of proteins and enzymes has provided many illustrative examples of key functional groups involved in anion binding. In this respect, the phosphate-binding protein (PBP) is a periplasmic protein that acts as an efficient transport system for phosphate in bacteria. The selection of phosphate over sulfate is achieved taking advantage of the fact that phosphate anion is protonated at physiological pH and can thus behave as both a hydrogen bond donor and an acceptor. The strong binding of phosphate (dissociation constant, $K_d = 0.31 \times 10^{-6}$ M) is achieved through the formation of 12 hydrogen bonds to a fully desolvated HPO_4^{2-} residing inside a deep cleft of the protein (Figure 1.3a) [75]. One of these hydrogen bonds, which is crucial for phosphate over sulfate selectivity, involves the OH group of phosphate as a donor and one aspartate residue as the acceptor (Asp141 in Figure 1.3a). Analogous to PBP, the sulfate-binding protein (SBP) is a bacterial protein responsible for

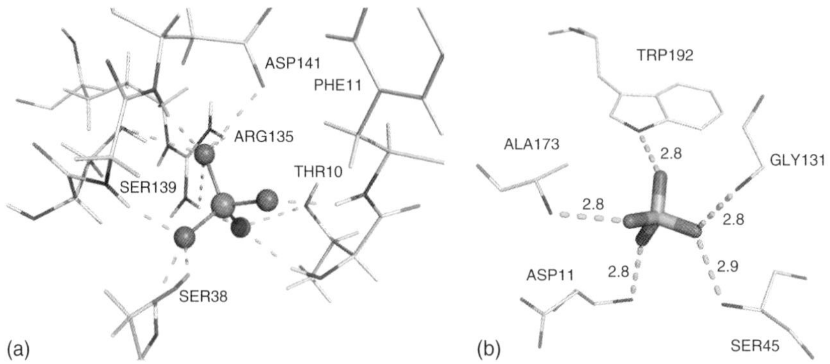

Figure 1.3 Scheme of the active sites of PBP (a) and SBP (b).

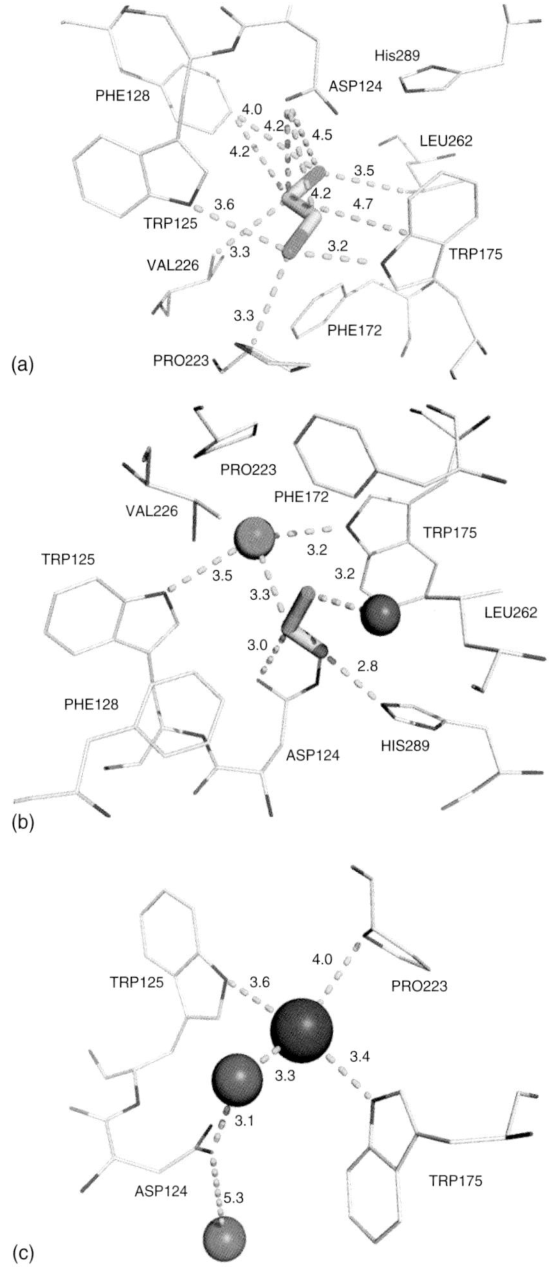

Figure 1.4 Schematic view of the interactions occurring in the active site of dehalogenase: (a) with the substrate before the start of the reaction, (b) with the alkyl intermediate and the chloride ion during the reaction, and (c) with the chloride ion and water molecules after hydrolysis.

the selective transport of this anion. Sulfate binding relies on the formation of a hydrogen bond network in which sulfate accepts seven hydrogen bonds, most coming from NH groups of the protein backbone (Figure 1.3b). The selectivity for sulfate over phosphate is about 50 000-fold in this protein [76].

Another bacterial protein whose crystal structure has revealed interesting binding motifs to anions is haloalkane dehydrogenase, which converts 1-haloalkanes or α,ω-haloalkanes into primary alcohols and a halide ion by hydrolytic cleavage of the carbon–halogen bond with water as a cosubstrate and without any need for oxygen or cofactors [77]. The crystal structure of the dehalogenase with chloride as the product of the reaction shows that the halide is bound in the active site through four hydrogen bonds involving the Nε of the indole moieties of two tryptophan residues, the Cα of a proline, and a water molecule (Figure 1.4).

One of the most important characteristics of anions is their Lewis base character. Therefore, compounds possessing suitable Lewis acid centers can be appropriate anion receptors. Several families of boranes, organotin, organogermanium, mercuroborands, acidic silica macrocycles, and a number of metallomacrocycles have been shown to display interesting binding properties with anions. Examples of this chemistry are included in Figures 1.5 and 1.6 and Refs [78–94].

Anion coordination chemistry and classical metal coordination chemistry have an interface in mixed metal complexes with exogen anionic ligands. Indeed, most of the ligands are anionic species belonging to groups 15–17 of the periodic table. Metal complexes can express their Lewis acid characteristics if they are coordinatively unsaturated or if they have coordination positions occupied by labile ligands that can be easily replaced. If this occurs, metal complexes are well suited for interacting with additional Lewis bases, which are very often anionic in nature, giving rise to mixed complexes. Mixed complexes in which the anionic ligand bridges between two or more different metal centers have been termed, in the new times of supramolecular chemistry, "*cascade complexes*" [95].

Formation of mixed complexes is the strategy of choice of many metalloenzymes dealing with the fixation and activation of small substrates. A classic example is

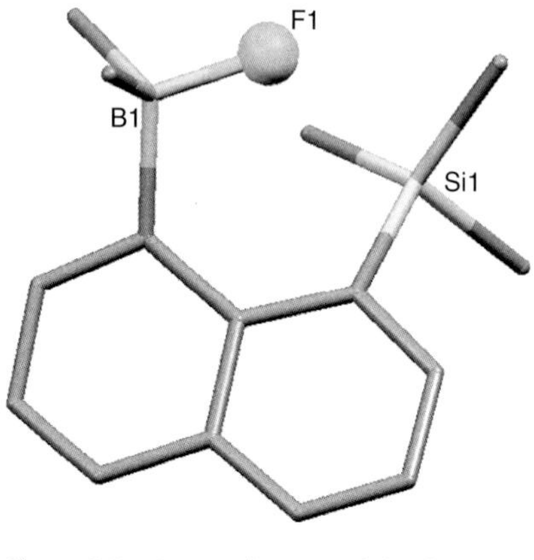

Figure 1.5 ORTEP diagram of the fluoride complex of a boron–silicon receptor. Taken from Ref. [85].

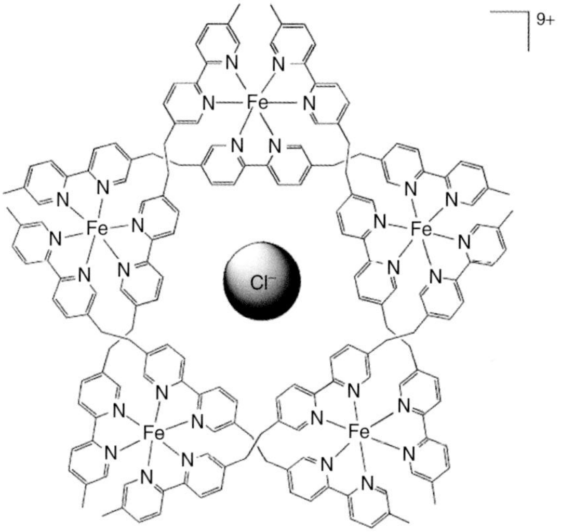

Figure 1.6 Reaction of $FeCl_2$ and a tris-bipyridine ligand gives rise to a double helix with the chloride as a template [94].

the family of enzymes called *carbonic anhydrases* [96–98]. Carbonic anhydrases are ubiquitous enzymes that catalyze the hydration reaction of carbon dioxide and play roles in processes such as photosynthesis, respiration, calcification and decalcification, and pH buffering of fluids. Human carbonic anhydrase II (HCA II) is located in the erythrocytes and is the fastest isoenzyme accelerating CO_2 hydrolysis by a factor of 10^7. Therefore, it is considered to be a perfectly evolved system, its rate being controlled just by diffusion. The active site of HCA II is formed by a Zn^{2+} cation coordinated to three nitrogen atoms from histidine residues and to a water molecule that is hydrogen bonded to a threonine residue and to a "relay" of water molecules that interconnects the coordination site with histidine 64 (Figure 1.7). The pKa of the coordinated water molecule in this environment is circa 7, so that at this pH, 50% is hydroxylated as Zn-OH^-, thus generating a nucleophile that will attack CO_2 to give the HCO_3^- form.

The rate-determining step is precisely the deprotonation of the coordinated water molecule and the transfer of the proton through the chain of water molecules to His64, which assists the process.

Phosphatases are the enzymes in charge of the hydrolysis of phosphate monoesters. Metallophosphatases contain either Zn^{2+} or Fe^{3+} or both; one of their characteristics is the presence of at least two metal ions in the active site. *Escherichia coli* alkaline phosphatase contains two Zn^{2+} and one Mg^{2+} metal ions in the active center. In the first step of the catalytic mechanism, the phosphate group of the substrate interacts as a bridging η,η'-bis(monodentate) ligand through two of its oxygen atoms with the two Zn^{2+} ions, while its other two oxygen atoms form hydrogen bonds with an arginine residue rightly disposed in the polypeptide chain (Figure 1.8).

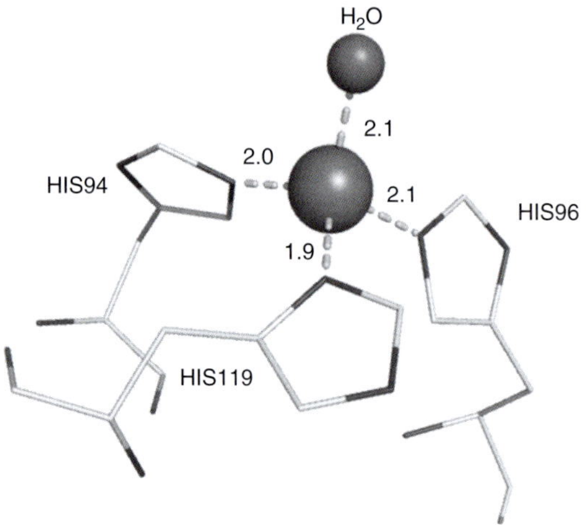

Figure 1.7 Schematic representation of the active site of HCA II showing the tetrahedral arrangement of three histidine residues and a water molecule.

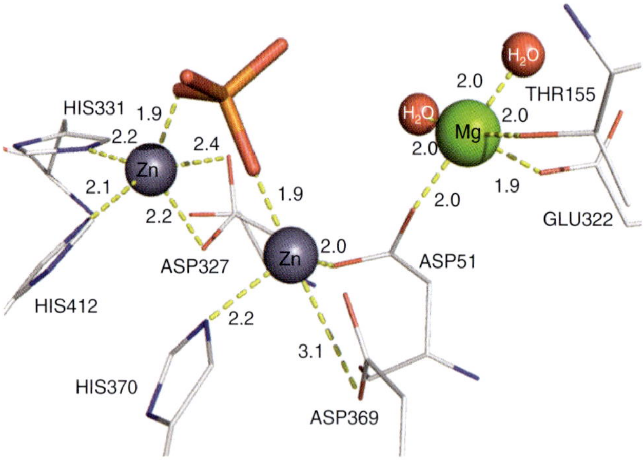

Figure 1.8 Active site of alkaline phosphatases. Adapted from Ref. [99, 100].

A last example that we would like to recall is ribulose-1,5-bisphosphate carboxylase/oxygenase (*rubisco*), which is the most abundant enzyme in nature [101]. *Rubisco* is a magnesium protein that is present in all the photosynthetic organisms participating in the first stage of the Calvin cycle. A lysine residue interacts with CO_2, forming an elusive carbamate bond, which is stabilized by interaction with the Mg^{2+} ion and by a hydrogen bond network with other groups of the polypeptidic chain (Figure 1.9). The ternary complex formed interacts with the substrate, which is subsequently carboxylated.

In all these examples, anionic substrates bind (coordinate) to a metal ion in key steps of their catalytic cycles, which assists the process as a Lewis acid.

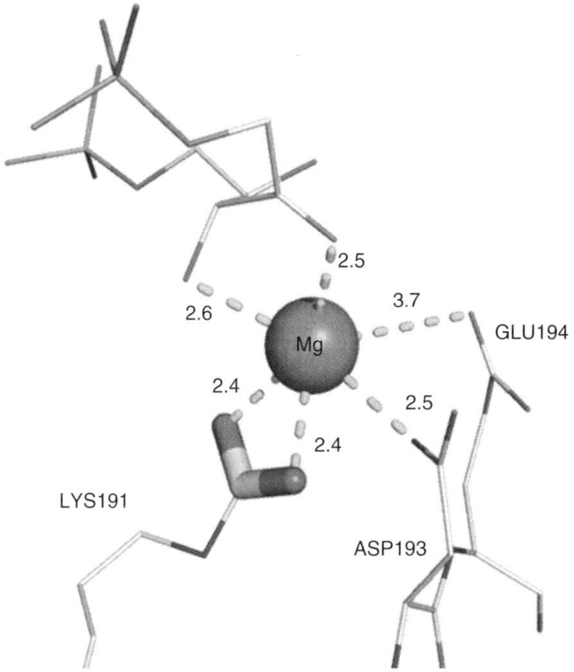

Figure 1.9 Active site of the enzyme *rubisco*. Adapted from Ref. [100].

1.2 Halide and Pseudohalide Anions

Having all these points in mind, there is no doubt that the birth of supramolecular anion coordination chemistry as an organized scientific discipline can be traced back to the work started by Lehn and coworkers in the mid-1970s. The first seminal paper of Lehn's group dealt with the encapsulation of halide anions within tricyclic macrocycles **5–7** [56]. The parent compound of the series **5**, already mentioned in the previous section, which is known as the *soccer ball ligand* in the jargon of the field, had been synthesized one year in advance by the same authors [102].

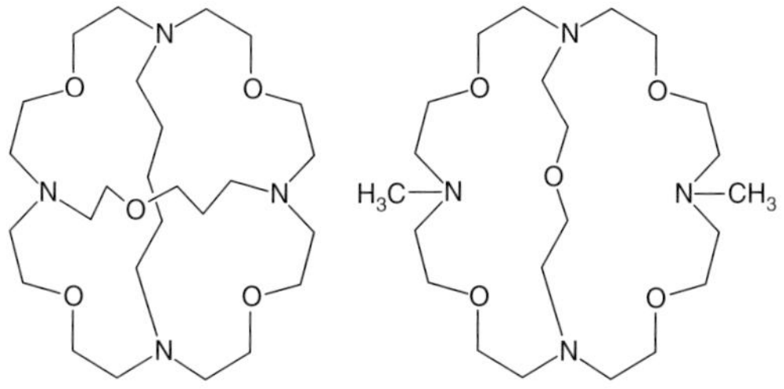

6 **7**

The authors started this paper stating that "Whereas very many metal cation complexes are known, stable anion complexes of organic ligands are very rare indeed." By means of ^{13}C NMR, the authors proved the inclusion of F^-, Cl^-, and Br^- within the macrotricyclic cavity at the time when they found a remarkable Cl^-/Br^- selectivity in water of circa 1000.

No interaction was observed with the larger I^- and with the monovalent anions NO_3^-, CF_3COO^-, and ClO_4^-. The crystal structure of [$Cl^- \subset H_4(\mathbf{1})^{4+}$], where the mathematical symbol $\subset$ stands for inclusive binding, shows that chloride was held within the tetraprotonated macrocycle by an array of four hydrogen bonds with the ammonium groups [103]. Years later, Lehn and Kintzinger, in collaboration with Dye and other scientists from the Michigan State University, used ^{35}Cl NMR to study the interaction of halide anions with **5**, **6**, and several related polycycles [61].

This premier study on spherical anion recognition was followed by the work performed in Munich by Schmidtchen, who described the synthesis of a quaternized analog of **5** (receptor **8**) [104]. In the same paper, similar macrocycles with hexamethylene and octamethylene bridges connecting the quaternary ammonium groups placed at the corners of the polycycles were also reported (**9** and **10**).

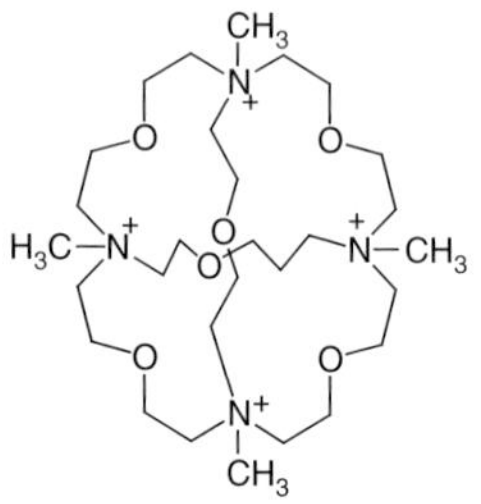

8

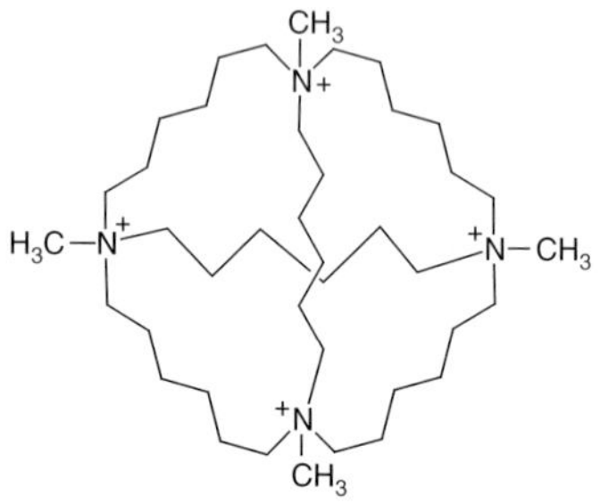

9

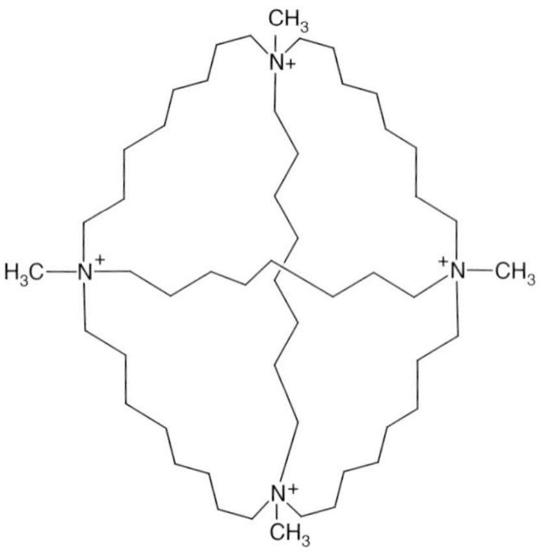

10

These azamacropolycycles, whose binding ability does not depend on pH, show modest affinity for halide anions in water. In the case of **9** and **10**, selectivity for bromide and iodide over chloride was found. However, binding is clearly weaker

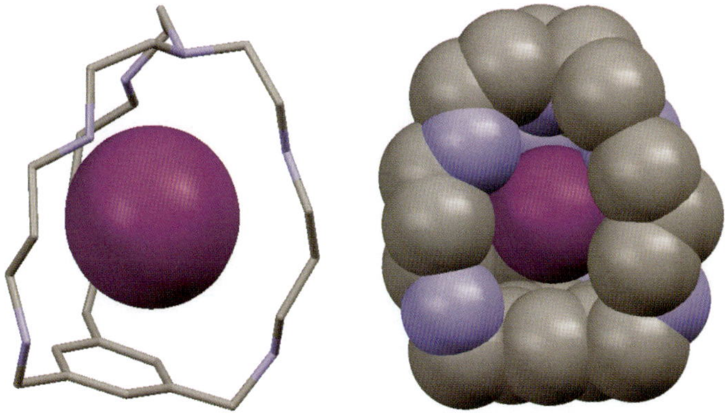

Figure 1.10 Views of the inclusion complex of I^- into the cavity of **9**.

than when auxiliary hydrogen bonding can occur. The crystal structure of an iodide complex with **9**, having hexamethylene bridges, confirmed the inclusion of the anion in the macrotricyclic cavity [105] (Figure 1.10). This series expanded over a wide range of studies illustrating the conceptual utility of these systems for understanding the kinds of binding forces involved in anion coordination [106–118].

Recognition of fluoride came up a little bit later, probably because of the higher difficulties in binding this anion in aqueous solution, which are associated with its high hydration energy in comparison to the other halides. In this respect, it has to be emphasized that most of the pioneering studies in anion coordination were carried out in water. The first stable fluoride complex was obtained with the bicyclic cage nicknamed O-BISTREN (**11**) [119].

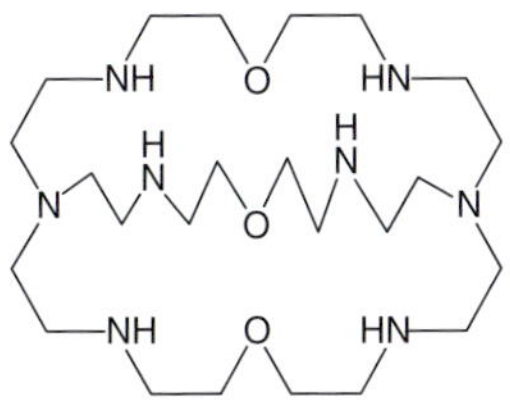

11

However, as illustrated in Figure 1.11 [120], the fitting of fluoride within the cavity was not very snug. The anion sits off-center, forming hydrogen bonds with just four of the six ammonium groups of the macrocycle. Consequently, although higher constants were found for the interaction of fluoride with $[H_6(\mathbf{11})]^{6+}$, the selectivity over the other halides, Cl^-, Br^-, and I^-, was not very large (log K 4.1, 3.0, 2.6, and 2.1 for F^-, Cl^-, Br^-, and I^-, respectively). In Figure 1.12, it can be seen that chloride fits more tightly into the cavity of **11**. In this case, hydrogen bonds are formed between the encapsulated anion and all six ammonium groups of the cryptand, although some of them are relatively weak.

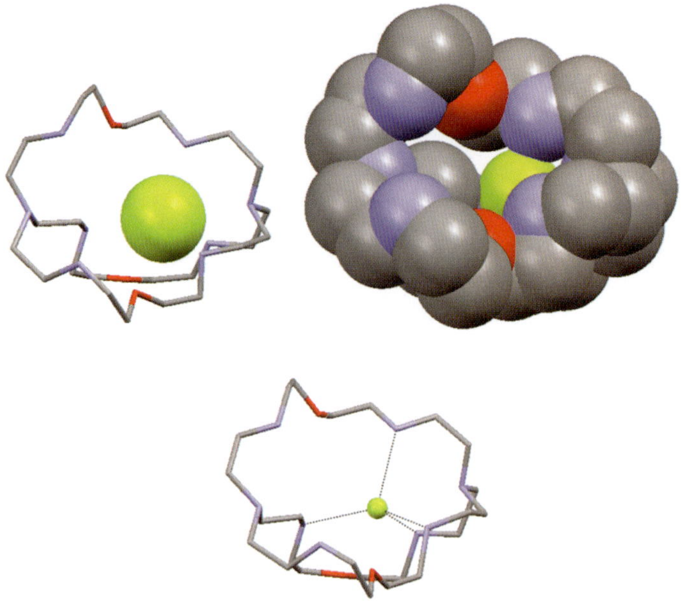

Figure 1.11 Views of F^- included in the molecular cavity of hexaprotonated **11** showing the mismatch in size. Hydrogen atoms have been omitted.

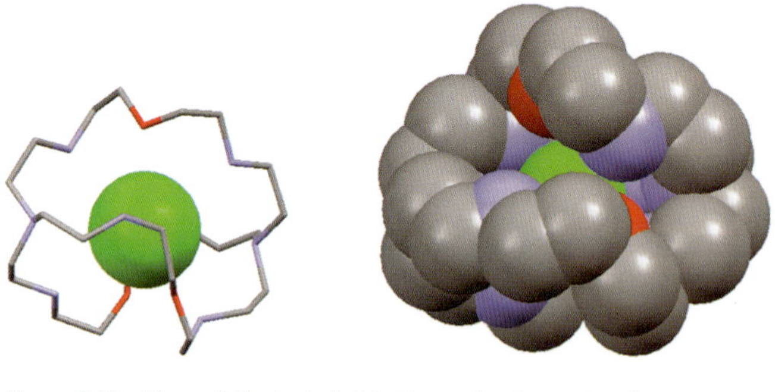

Figure 1.12 View of Cl^- included in the molecular cavity of hexaprotonated **11**. Hydrogen atoms have been omitted.

With respect to fluoride binding, it is worth mentioning that, in 1984, a report by Suet and Handel appeared, describing the ability of different monocyclic tetraazamacrocycles with propylenic and butylenic chains (**12**–**14**) to bind this anion in aqueous solution [121]. The stability constants found for the interaction of fluoride with the tetraprotonated forms of **12**, **13**, and **14** were 1.9, 2.0, and 2.8 logarithmic units, respectively.

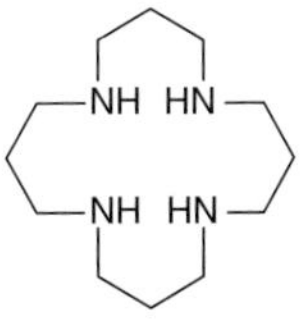

12

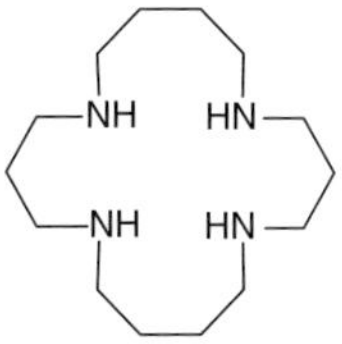

13

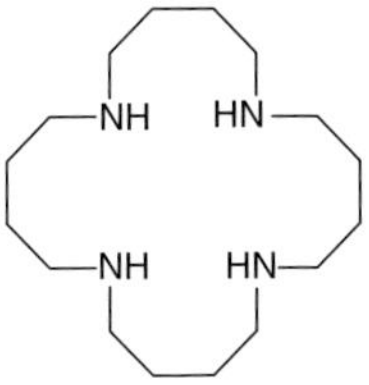

14

In order to obtain a more selective F^- binder, the macrocycle C2-BISTREN was prepared by Lehn and coworkers [122] several years later (**15**) [123].

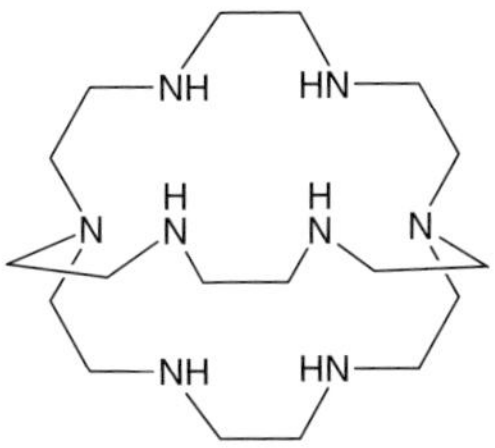

15

The matching in size between fluoride and the cavity was in this case much tighter (Figure 1.13). Unlike in O-BISTREN (**11**), fluoride accepts hydrogen bonds from all six secondary ammonium groups of the cage **15**. Solution studies carried out by Hay, Smith *et al.* in 1995 using potentiometric techniques led to a surprisingly large value of 11.2 logarithmic units for the interaction of the hexaprotonated macrocycle with fluoride anion measured in 0.1 M KNO_3 [124]. The reported F^-/Cl^- selectivity at pH = 5.9 expressed as log K_s(F^- complex)/log K_s (Cl^- complex) was also an exceptionally high 7.5. Lehn and coworkers obtained for this equilibrium a similar value of 10.55 logarithmic units using 0.1 M (Me_4N)TsO as the background electrolyte [123].

The success in obtaining a good fluoride-selective receptor led to the modification of the structure of **15** to obtain receptors that could match the size of the larger halides, Cl^-, Br^-, and I^- (compounds **16** and **17**) [123, 125, 126]. However, the results obtained, although pointing in the desired direction, did not show any particularly relevant selectivity. Receptor **18** (C5-BISTREN) prepared by Lehn's group was studied along with **11** by potentiometry in collaboration with Martell and coworkers. Such studies, and the crystal structure of the azide complex [57], gave the first indications of the possibility of the formation of binuclear or higher nuclear anionic complexes with two encapsulated anions or, even better, with the hydrogen bifluoride anion (HF_2^-) (see below for

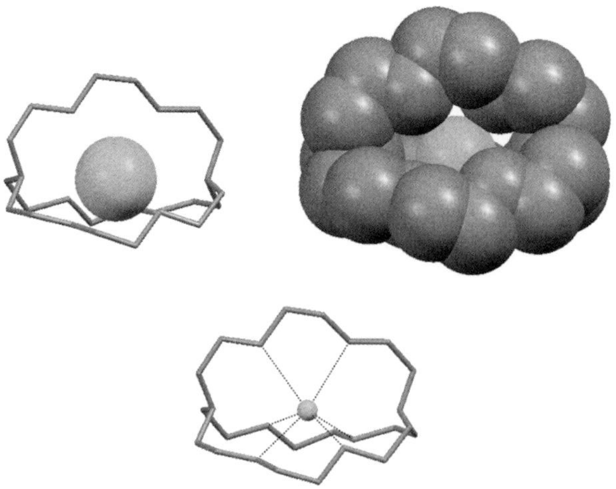

Figure 1.13 Views of the F^- anion included in C2-BISTREN (**15**).

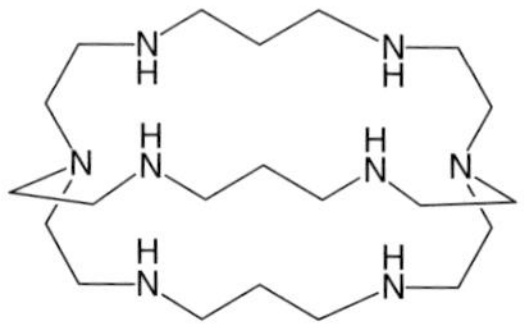

16

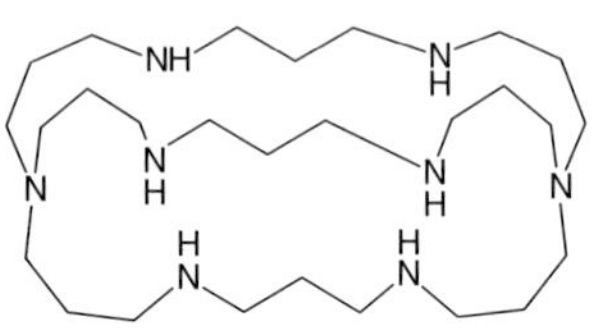

17

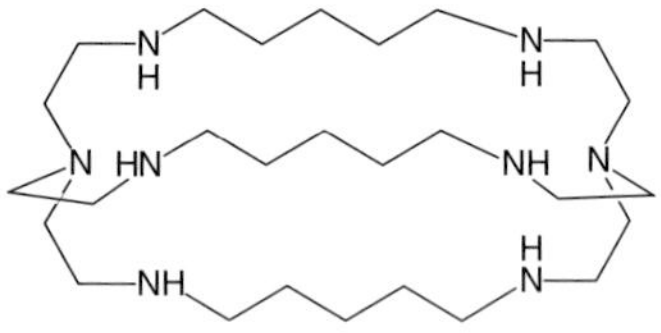

18

further developments) [127, 128]. The constants for the equilibrium between the hexaprotonated receptor and HF_2^- were 6.4 and 5.2 logarithmic units for **11** and **18**, respectively.

O-BISTREN (**11**) was also the first synthetic receptor for which a crystal structure with an included N_3^- was solved by X-ray crystallography. The azide anion fits perfectly along the internal cavity of the receptor, forming each of its terminal nitrogen hydrogen bonds with the three ammonium groups of each of the two *tren* polyamine subunits of the cage (Figure 1.14) [57].

Fortunately and curiously, this structure and the previously discussed one for fluoride, which have proved to be crucial for the development of the field of anion coordination, were accepted for publication in spite of having R factors of 16.2 and 19.8, respectively.

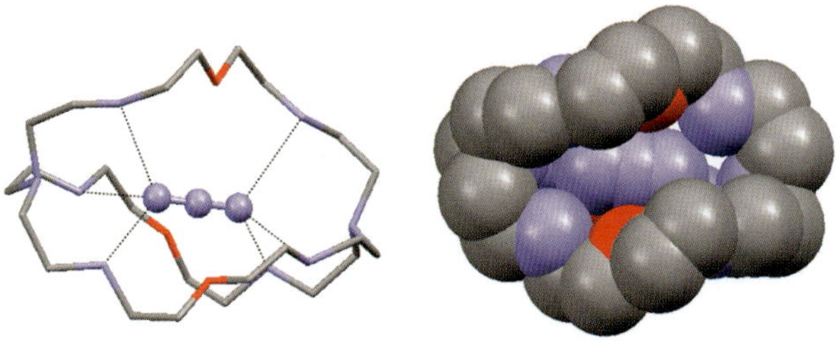

Figure 1.14 Azide anion included in the cavity of receptor **11**.

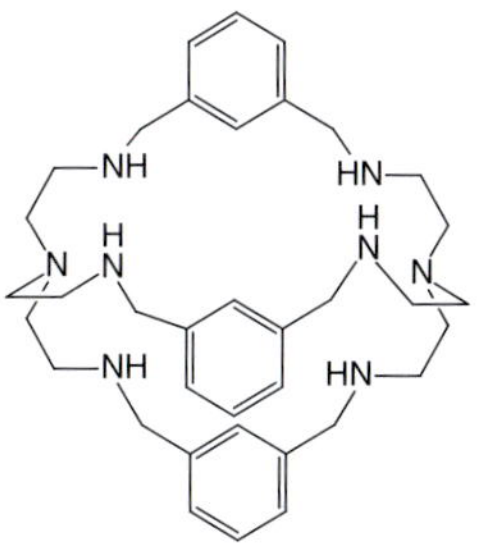

19

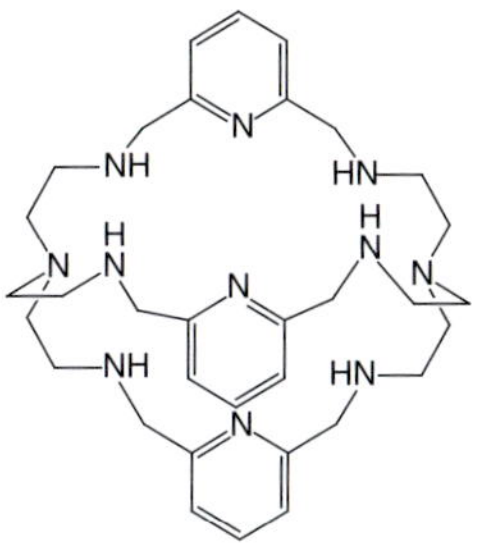

20

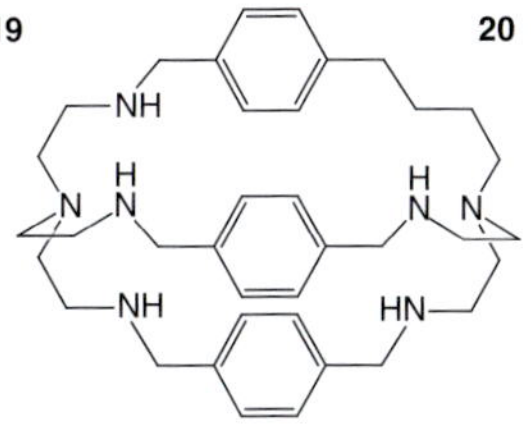

21

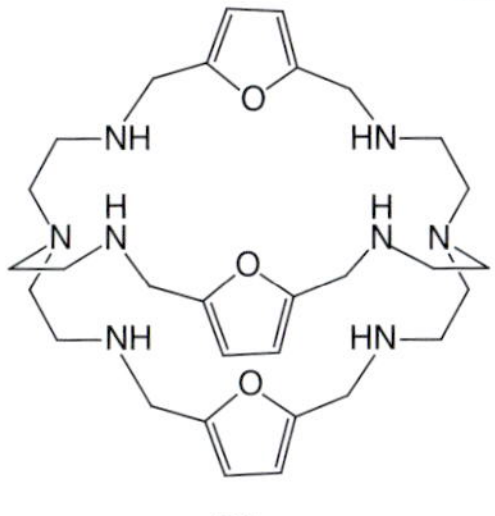

22

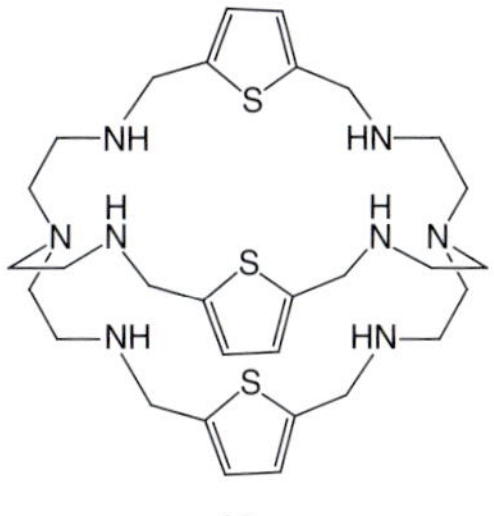

23

Since these initial findings, many efforts have been devoted to halide recognition with different types of receptors. Among polyammonium receptors, probably the most used have been cryptands obtained by 2 + 3 condensation of the tripodal polyamine *tren* and the corresponding aromatic dialdehydes followed by *in situ* reduction with an appropriate reducing species, often $NaBH_4$ (**19–23**). One of the reasons for the large amount of work performed with these receptors is the readiness of its synthesis, which is much more straightforward than those required for preparing cages with aliphatic linkers between the *tren* subunits.

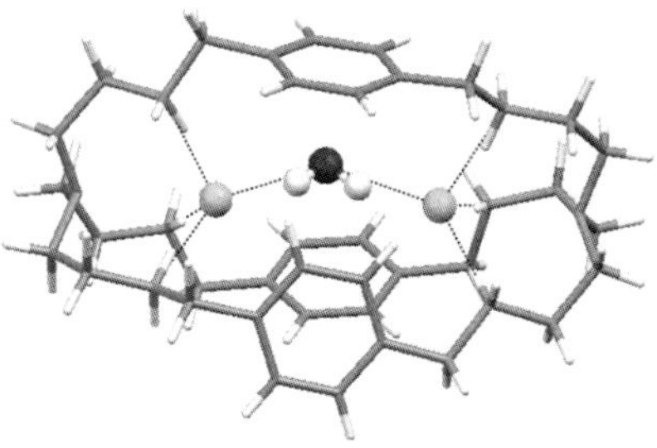

Figure 1.15 View of the fluoride-water cascade complex of $[H_6\mathbf{21}]^{6+}$.

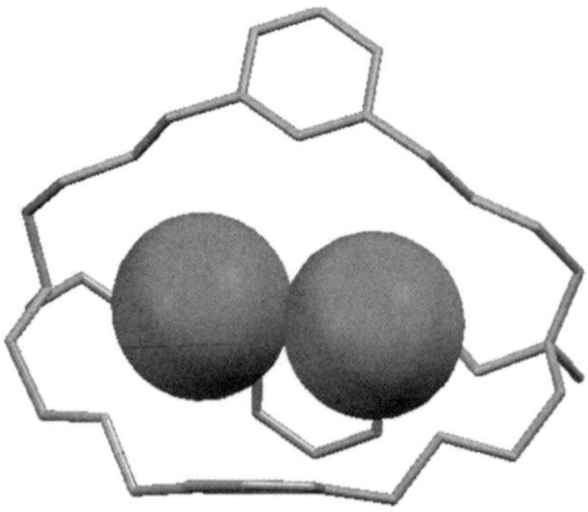

Figure 1.16 Presumed bichloride complex of hexaprotonated **19**, bridging hydrogen not shown.

These latter preps often require tedious protection and deprotection steps in addition to high dilution methods. Examples are receptors **19–23**, which are abbreviated as MEACryp, pyridine azacryptand (PyEACryp), PEACryp, FuEACryp, and ThioEACryp following the short names proposed by the late Robert W. Hay from St. Andrews University [129].

Nelson [21], Bowman-James [20, 129], Fabbrizzi [19], and Ghosh [130], among others, have contributed extensively to this chemistry. Perhaps, one of the most interesting developments in this topic has been the crystallographic evidence that this kind of receptors can lodge two halide anions when they are extensively protonated. Figures 1.15 and 1.16 show the crystal structures of hexaprotonated **21** with two fluorides and a water molecule bridging between them forming an anion "cascade complex" [131], and presumably of a bichloride Cl–H–Cl anion included in hexaprotonated **19** [132].

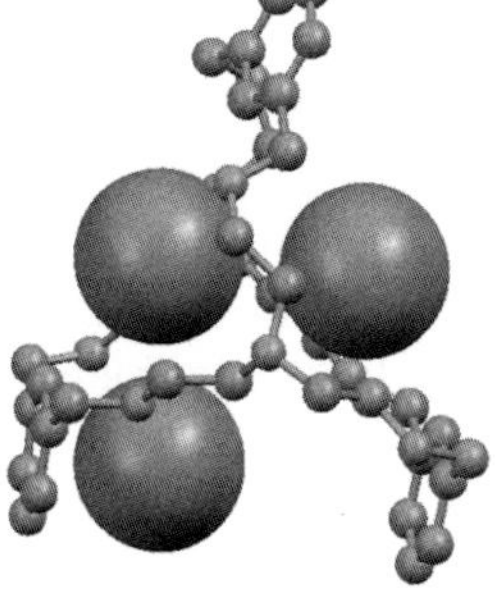

Figure 1.17 View of three bromides partially included in receptor **19**.

Also, in a rather early publication in the field, crystallography was used to prove the almost total inclusion of three bromide anions within the cavity of **19** MEACryp [133] (Figure 1.17).

The polyazacryptands that impose rigidity are a corollary to 1,3,5 trisubstituted benzenes with bridges containing amines (**24**–**26**). This class of compounds, prepared by Lehn's group [134, 135] for the first time, form 1 : 1 halide:receptor complexes with a significant stability in water. Recently, crystal structures of anionic complexes have been obtained for a similar receptor **27** developed by Steed *et al.* [136] (Figure 1.18).

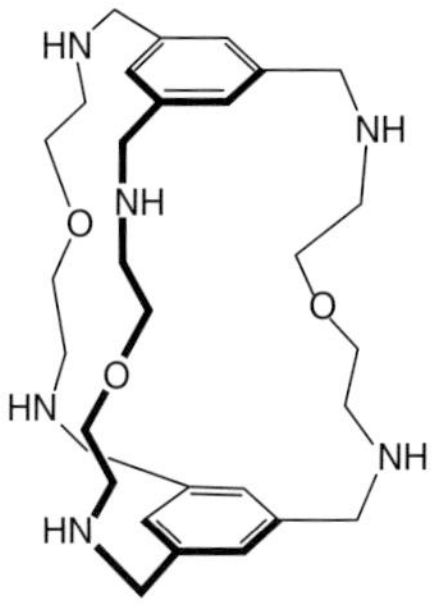

24

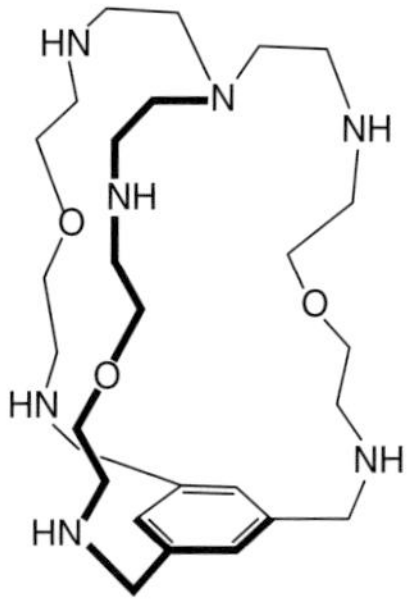

25

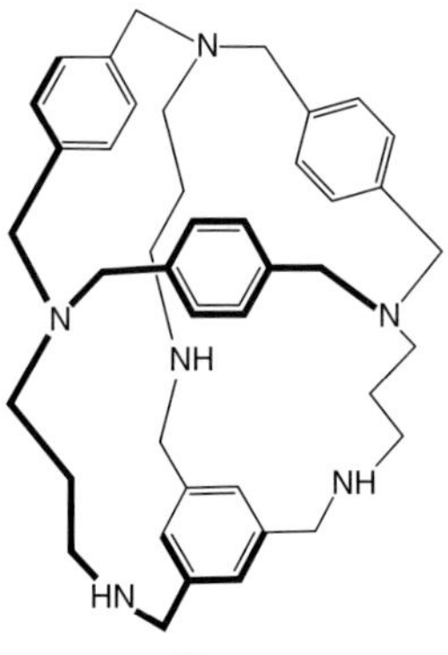

26

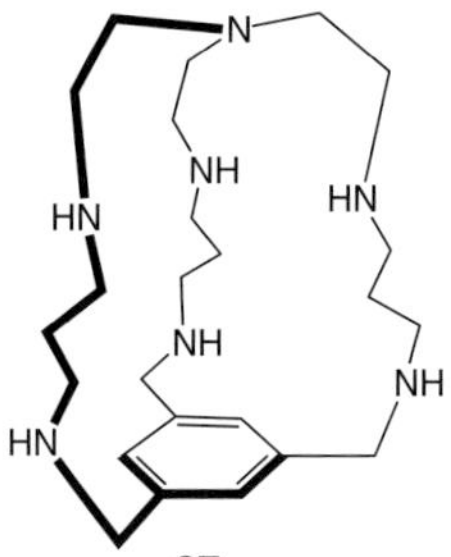

27

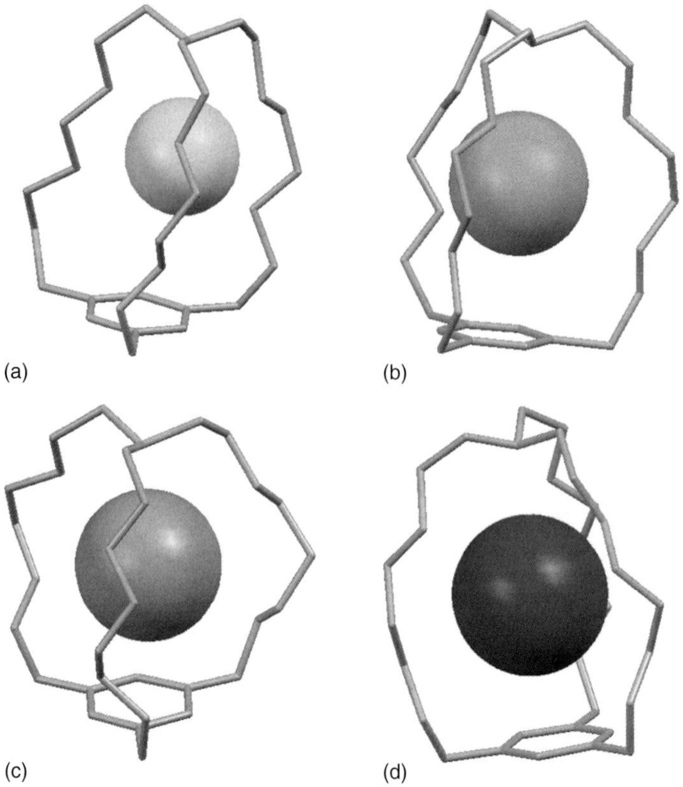

Figure 1.18 Halide anion included in hexaprotonated **27** (a) F^-, (b) Cl^-, (c) Br^-, and (d) I^-.

The initial work in the field was essentially performed with polyammonium receptors to take advantage of their charge. Indeed, the main characteristic distinguishing anions from all other guest species is precisely their negative charge. However, as advanced in Section 1.1, biological receptors, proteins, make use of a combination of binding motifs, which are provided in many instances by the side chains of amino acids and by the amide bonds of their backbones. The environment of protein clefts or pockets, where many binding sites reside, has a pronounced lipophilic character, and therefore, hydrogen bonds become stronger in this ambient condition with reduced water content. On the other hand, extraction strategies of pollutant anions from contaminated aqueous media often require hydrophobic receptors that can be kept soluble in a nonpolar solvent. Moreover, receptors can be grafted in resins or solid supports, making their solubility characteristics less critical.

On the basis of these important considerations, either charged or noncharged receptors containing a variety of hydrogen bonding donor groups came into play.

In this respect, Sessler, Ibers *et al.* [137] were, in 1990, the first to evidence a fluoride anion residing in the central hole of a sapphyrin, a 22-π-electron pentapyrrolic expanded porphyrin (**28–31**).

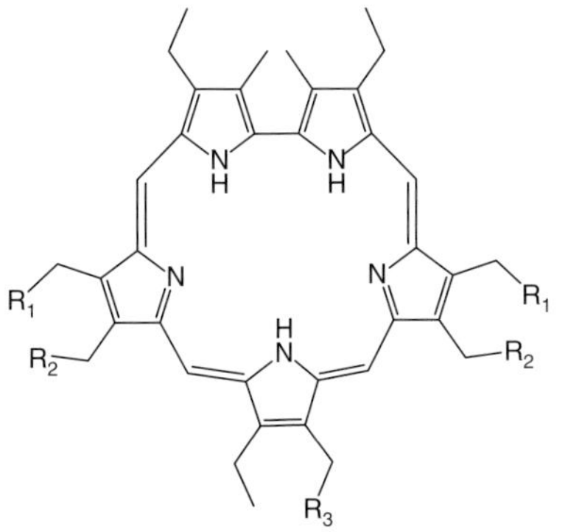

28 $R_1 = CH_3$, $R_2 = H$, $R_3 = H$
29 $R_1 = H$, $R_2 = CH_3$, $R_3 = CH_3$
30 $R_1 = H$, $R_2 = CO_2Et$, $R_3 = CH_3$
31 $R_1 = H$, $R_2 = CO_2Et$, $R_3 = CH_3$

Treatment of sapphyrin (**28**) as its free base with aqueous HCl in dichloromethane followed by adding silver hexafluorophoshate and crystallizing by vapor diffusion led to the isolation of the diprotonated macrocycle with just one hexafluorophosphate counteranion and another anion located at the center of the macrocyclic hole. On the basis of independent synthesis and ^{19}F-NMR studies, it was established that the central anion was fluoride. The anion is hydrogen bonded to all five pyrrolic nitrogens of the macrocycle (Figure 1.19).

This first result on cyclic polypyrrole anion receptors gave rise to the evolution of these ligands and to the understanding of their chemistry and applications in a variety of fields [10, 138]. One of the first applications involved the capacity of these hydrophobic compounds for transporting fluoride anions across lipophilic membranes [139].

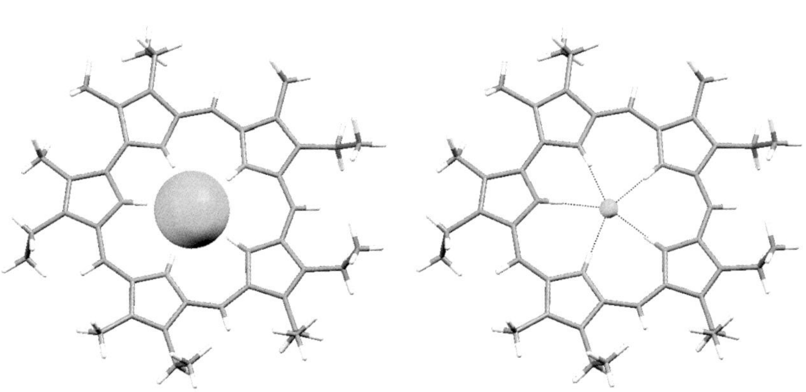

Figure 1.19 Views of the structure of the fluoride complex of diprotonated **28**, sapphyrin.

As commented above, another binding motif relevant to anion coordination in proteins is the amide groups constituting the protein backbone. The first time the amide functionality was introduced in the structure of an abiotic macrocyclic receptor probably dates back to 1986 when Pascal, Spergel, and Van Eggen published the synthesis of compound **32** [140].

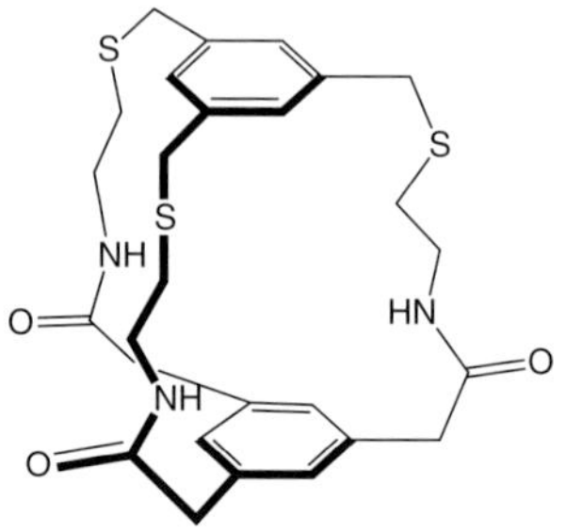

32

The authors stated very enthusiastically that compound **32** "may be prepared from the easily accessible precursors 1,3,5-tris(bromomethyl)benzene and 1,3,5-benzenetriacetic acid in a short convergent synthesis, requiring no chromatographic steps, which may be completed in less than 24 h." In the same paper, the authors indicated that ^{1}H and ^{19}F NMR studies carried out in DMSO-d_6 suggested an association between the macrocycle and fluoride, although there was no certainty at that time about the inclusion of the anion. Since then, many amide-based receptors have been prepared [22, 23, 26, 29, 50, 53].

Kimura, Shiro and coworkers reported in 1989 amino-amide receptors **33**, **34**, along with a crystal structure of receptor **34**, in which two azide anions were trapped between two macrocycles, forming a sort of hydrogen-bonded sandwich complex [141] (Figure 1.20).

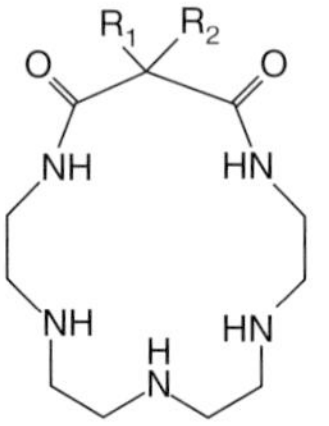

$R_1 = R_2 = H$ **33**
$R_1 = R_2 = CH_3$ **34**

Recently, new cyclic receptors containing mixed amine–amide functions have been developed to take advantage of the charge of the potentially protonated amines and the hydrogen bond formation capabilities of both the amines and the amides [29]. A representative example of this chemistry is provided by the structure of bifluoride or azide anions encapsulated in the cavity of the tricyclic receptor **35** (Figure 1.21).

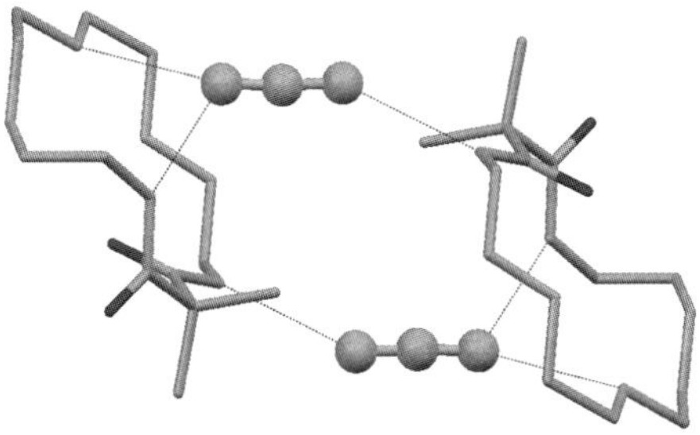

Figure 1.20 Sandwich complex between receptor **34** and azide.

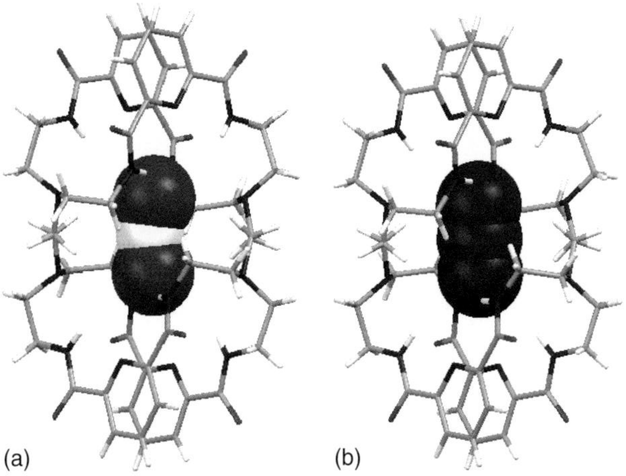

Figure 1.21 HF_2^- and N_3^- anions encapsulated in the tricyclic macrocycle **35**.

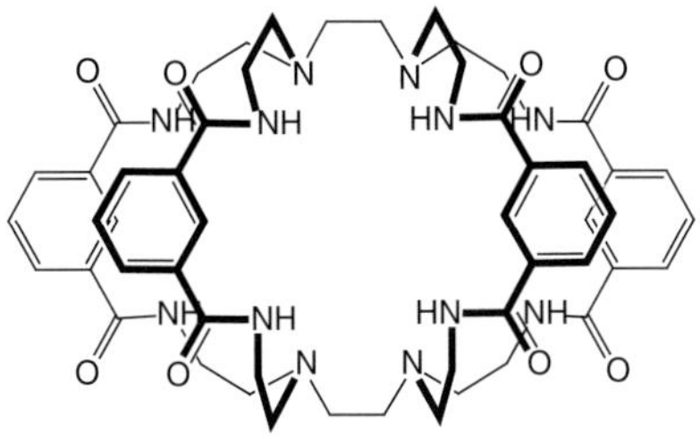

35

Since anions behave as Lewis bases, cyclic or noncyclic receptors containing Lewis acid sites can serve for anion binding. Examples of this chemistry have been advanced in Section 1.1 [78, 81, 83, 85, 87–94], and some other examples are provided in Figure 1.22.

As commented in Section 1.1, the interface between anion coordination chemistry and classical metal coordination chemistry is delimited by the so-called "cascade complexes" [95], a well-known class of multinuclear coordination compounds with bridging ligands. Relevant examples in the field of anion recognition can be found

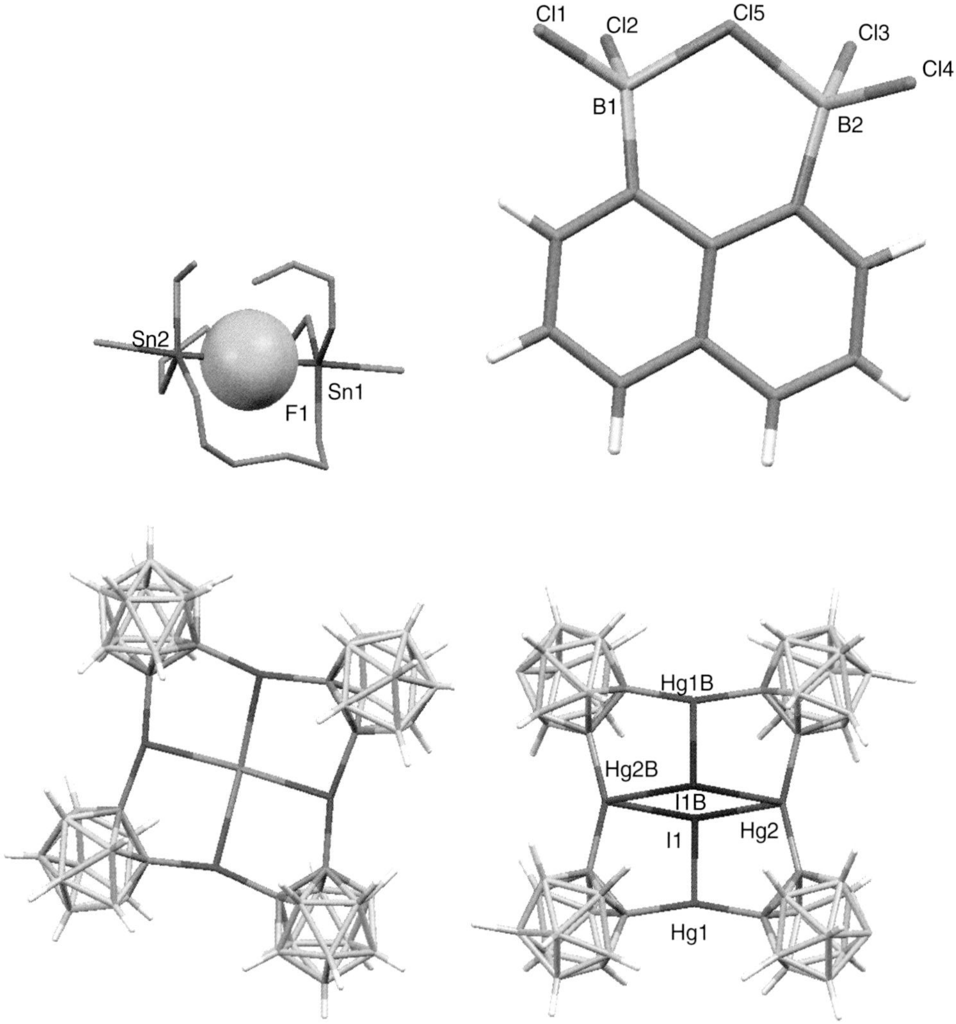

Figure 1.22 Crystal structure of an organotin compound binding fluoride [83] and an organoboron compound binding chloride [86], and of mecuracarborands binding chloride and iodide [78, 79].

in the initial work of Martell and Lehn on O-BISTREN (**11**) and C5-BISTREN (**18**), in which, based on disquisitions about the stability constants, a hydroxide and a fluoride anion were postulated to be included within the metal centers [127, 128]. Fabbrizzi and coworkers, among others, have contributed remarkably to this chemistry with several studies and crystal structures. Figure 1.23 collects three representative structures [142–144]. Reviews dealing with this topic are collected in Refs [19, 31, 47, 145].

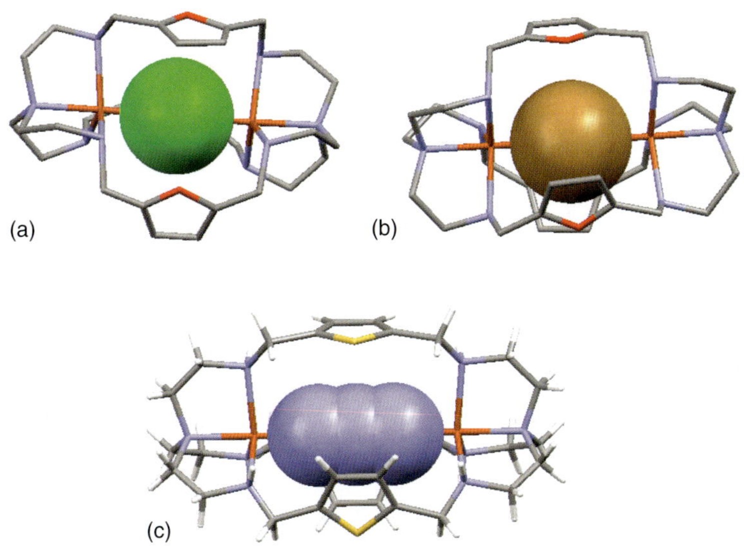

Figure 1.23 View of the "cascade complexes" formed by the binuclear Cu^{2+} complex of **22** with Cl^- and Br^- and between the Cu(II) complex of **23** with Br^-.

1.3 Oxoanions

Oxoanions have triangular, tetrahedral, or more complex shapes resulting from the association of different triangles or tetrahedrons that can be also accompanied by organic residues as in mono- and polynucleotides. On the other hand, if anions are conjugated bases of protic acids, they will undergo protonation processes and their negative charge will depend on their basicity constants. A simple example is provided by phosphate, which displays in water stepwise constants of 11.5, 7.7, and 2.1 logarithmic units for its first, second, and third protonation steps, respectively [146]. Therefore, phosphate exists only as a trivalent anion in a very basic pH range, while at neutral pH it is present in aqueous solution as a mixture of the monovalent and divalent forms. This property can be advantageously used for discriminating between anions of different basicity.

We start this historical description with anions that are conjugated bases of strong acids and that do not change their formal charge with pH.

One of the first studies in this respect was performed by Gelb, Zompa *et al.* on the interaction of nitrate and halide anions with the monocyclic polyamine [18]aneN_6 (**36**) [147]. Apart from deriving stability constants that were relatively low, only slightly above two logarithmic units for the interaction of the tetraprotonated macrocycle with nitrate and below two logarithmic units for its interaction with chloride, the authors described the crystal structure of the compound [(H_4[18]aneN_6)]$(NO_3)_2Cl_2 \cdot H_2O$ (Figure 1.24). The nitrates and chlorides are placed

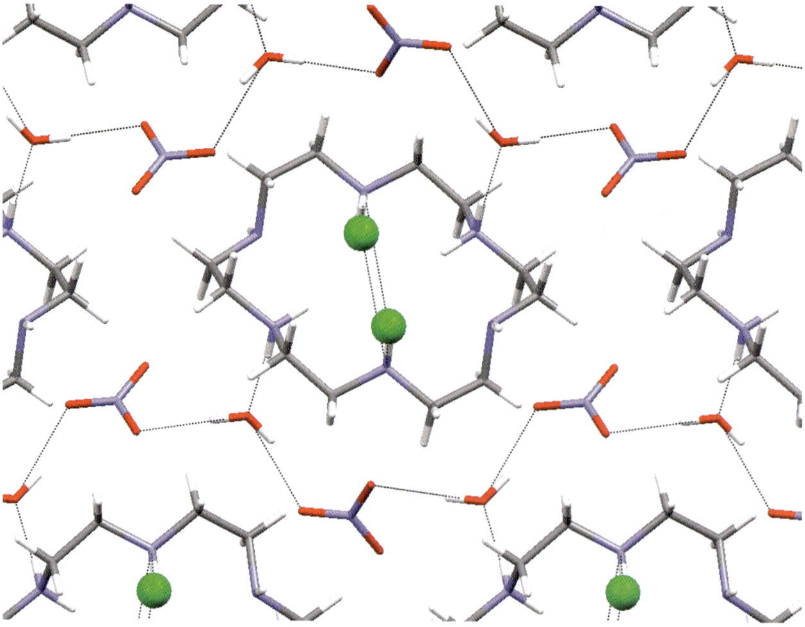

Figure 1.24 View of the hydrogen bonding network in $[H_4(\mathbf{36})](NO_3)_2Cl_2 \cdot H_2O$.

outside the macrocyclic cavity, forming two different hydrogen bonding networks. One of them links the ammonium groups with the nitrate anions through relays of water molecules; in the other, the ammonium groups are directly bound to the chloride anions.

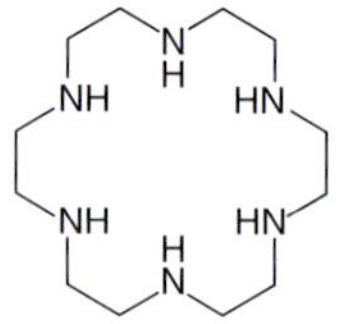

36

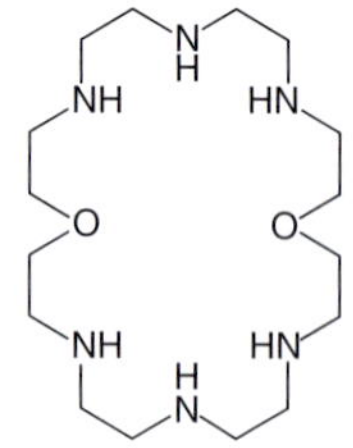

37

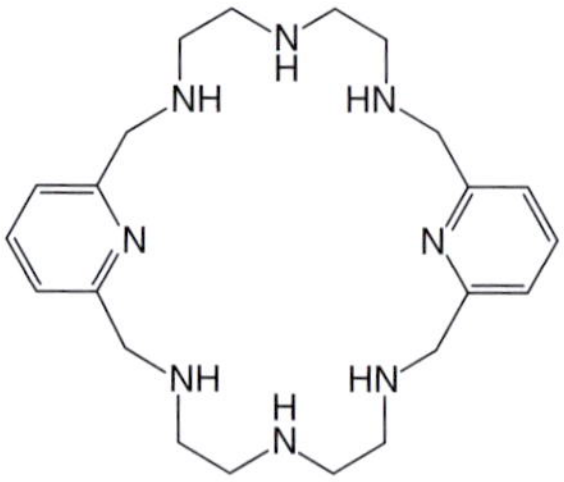

38

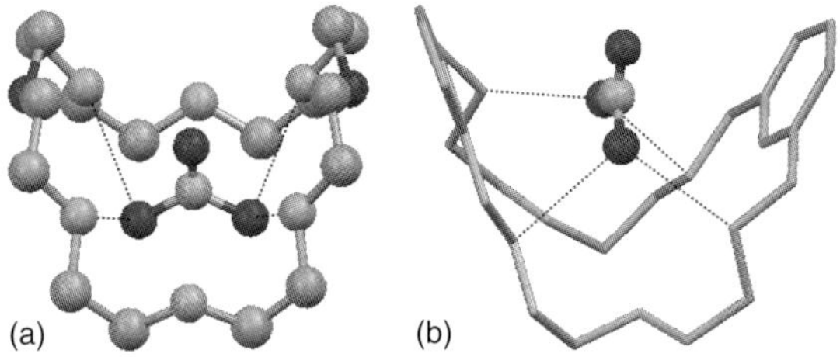

Figure 1.25 Situation of the NO_3^- anion in tetraprotonated macrocycles (a) **36** and (b) **38**.

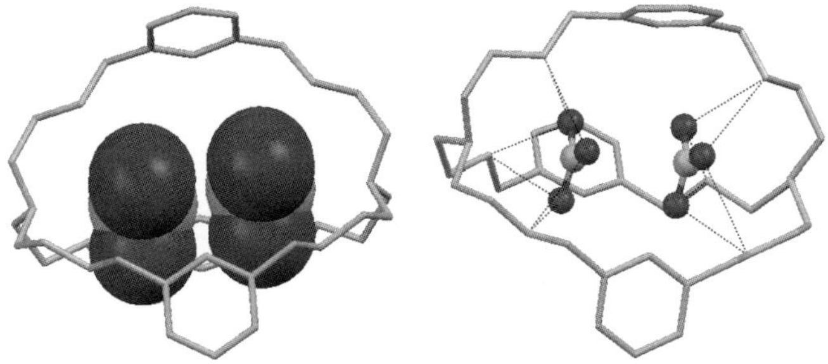

Figure 1.26 Views of the two nitrates included in **19**.

Two related crystal structures that have been more recently reported deserve to be mentioned since they illustrate an inclusive binding of nitrate anion in a monocyclic cavity. The first one is the crystal structure of the 24-membered dioxahexaazamacrocycle (**37**), usually known as *O-BISDIEN*, with nitrate $[(H_4(\mathbf{37})](NO_3)_4$, reported by Bowman-James *et al.* [148], and the second one also corresponds to a 24-membered macrocycle with two meta-substituted pyridine spacers $[(H_4(\mathbf{38})](NO_3)_4$, recently published by Valencia, García-España, and coworkers [149]. In both crystal structures, one of the nitrates is linked through two bifurcated hydrogen bonds to the four protonated amino groups of the macrocycle, which displays a boat-shaped conformation (Figure 1.25). In spite of this similarity, the situation of the nitrate anion with respect to the heteroatoms is different, being symmetrically placed between the two aromatic rings in the pyridine macrocycle.

Azacryptands (**19–23**) can encapsulate nitrate as it was observed crystallographically for **19** (MEACryp) in 1998 [150]. Two nitrate anions were included in the cavity, with a parallel orientation between them (Figure 1.26). Hydrogen bonds were formed between the six secondary amino groups of the cage and all the oxygen atoms of the nitrate anions.

However, the first X-ray crystal structure solved for an oxoanion included in an azacryptand was for a perchlorate. Crystals of perchlorate anion included in the cavity of hexaprotonated **22** (FuEACryp) were obtained serendipitously in the course of an attempt to generate a binuclear manganese complex (Figure 1.27) [151]. Although some disorder obscured the hydrogen bond network of the included perchlorate, the participation of two types of hydrogen bonds, $NH^+\text{–}O_{perchlorate}$ and $NH^+\text{–}O_{water}\text{–}O_{perchlorate}$, seemed clear. Since this first structure, a number of

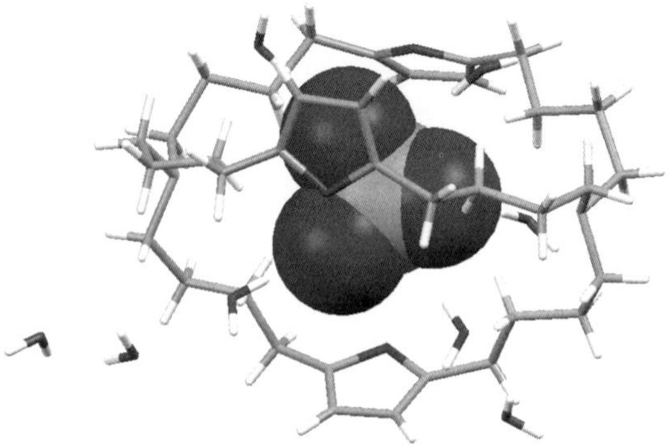

Figure 1.27 View of the perchlorate anion included in the cavity of receptor **22** (FuEACryp) [151].

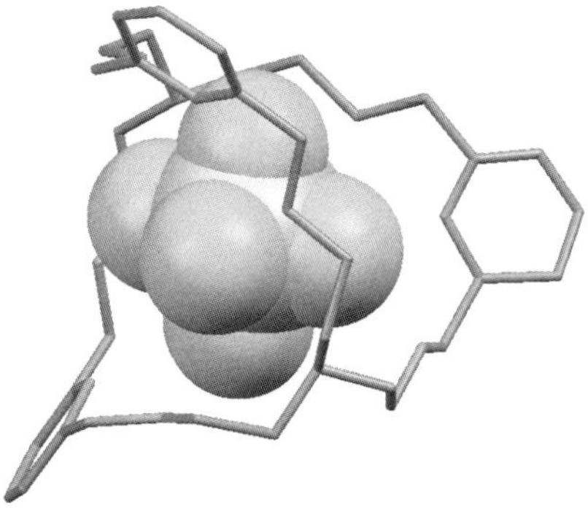

Figure 1.28 View of the anionic complex formed between hexaprotonated **20** and SiF_6^{2-}.

structures of azacryptands have appeared in the literature, in which two main coordination modes are observed that correspond either to inclusive anion binding or to facial binding of three anions similar to that shown in Figure 1.17 for **19**.

In this seminal paper, Nelson's group described another structure in which SiF_6^{2-} was also included in the cavity of **20** (PyEACryp) (Figure 1.28).

ReO_4^- is another anion belonging to this category whose study became relevant because its chemistry parallels that of radioactive $^{99m}TcO_4^-$. At the same time

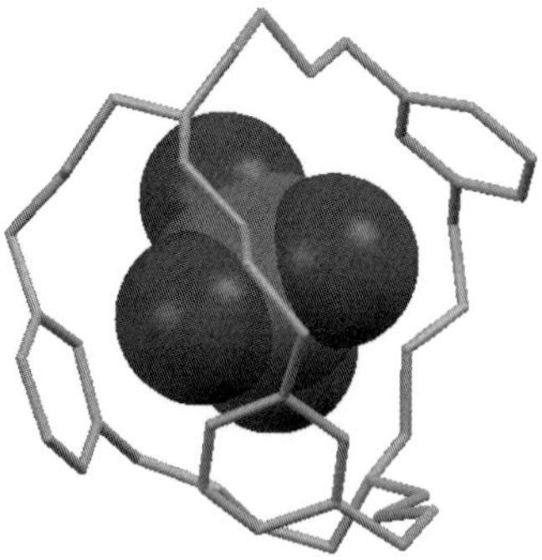

Figure 1.29 ReO_4^- included in MeACryp (**19**).

that $^{189}ReO_4^-$ itself is of medical interest in connection with specific therapeutic and diagnostic applications [49, 152–154]. Cryptands of this series have also been shown to be capable of interacting with ReO_4^-, including the anion within its cavity as shown in Figure 1.29 for **19** (MEACryp) [155]. These studies were devoted to the extraction of pollutant anions from aqueous media, and to do so, Gloe and Nelson also employed a series of hydrophobic polyamines derived from *tren*, as those seen in **39–46**. The extractabilities observed could not be explained solely on the basis of ligand lipophilicity since the level of protonation also played an important role.

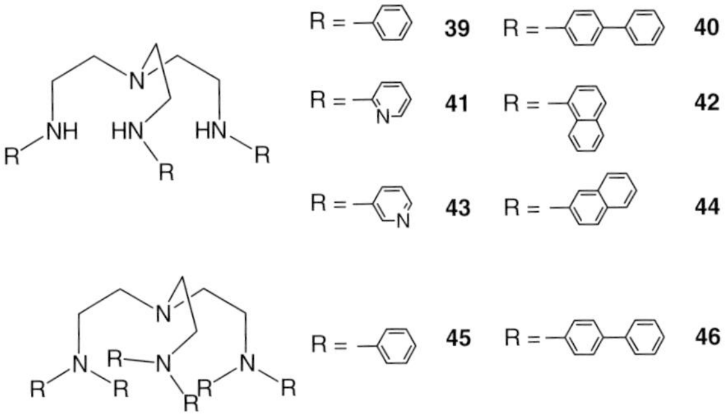

Sulfate anion shows in aqueous solution a protonation step with a pK_a around 1.7 and thus behaves essentially as a divalent anion over a wide pH range, differing from phosphate, which at neutral pH exists as a mixture of mono- and dihydrogenphosphate with formal charges of −2 and −1, respectively. Therefore, while sulfate can only accept hydrogen bonds, phosphate can both donate and accept hydrogen bonds. As mentioned earlier, this property is advantageously used by transport proteins to discriminate between these two anions.

The same kinds of receptors described for halides have also been traditionally used for binding sulfate. For instance, the monocycle [15]aneN_5 (**47**) [156] has been proved to interact in water with several dianions including SO_4^{2-}. Pyridinophane (**48**) interacts with SO_4^{2-} and SeO_4^{2-} among other anions [157] with log K_s values of around 3.5.

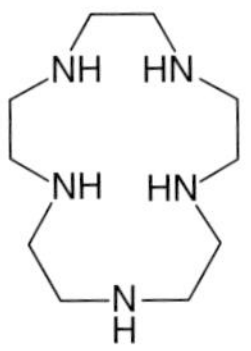

47

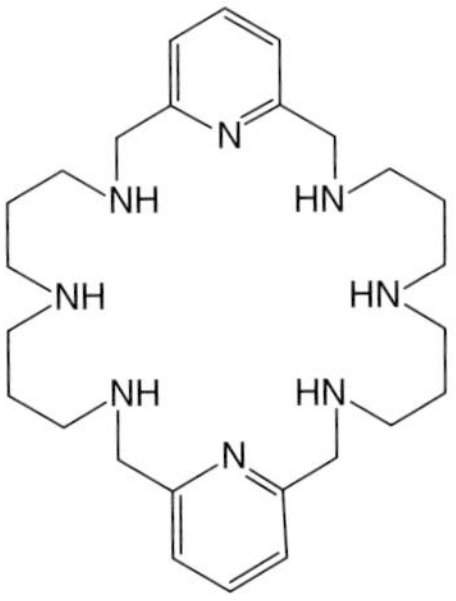

48

However, divalent anions having much larger hydration energies than monovalent anions should be better recognized by cage-type ligands in which the first

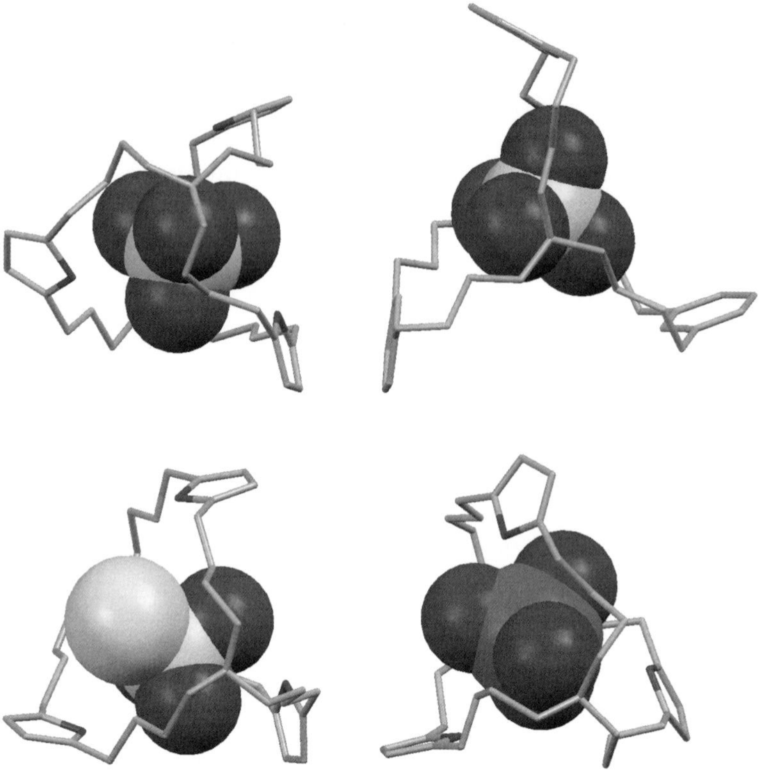

Figure 1.30 Sulfate anion included in the cavities of MEACryp and sulfate, thiosulfate, and chromate anions included in FuEACryp.

hydration sphere of the anion can be completely removed and substituted by the anchoring points of the receptor. In this environment, sulfate reaches a stability constant of 4.9 logarithmic units in its interaction with the hexaprotonated form of O-BISTREN (**11**) [57]. Compound C3-BISPRN (**17**), which can take eight protons, was reported to have log $K_s = 7.45$ for its interaction with sulfate [158].

More recently, the Nelson and Bowman-James groups have published the sulfate crystal structures of FuEACryp (**22**) [159] and MEACryp (**19**) [160], respectively. In the same paper [159], Nelson and coworkers reported the crystal structures of thiosulfate and chromate encapsulated in the furan azacryptand (Figure 1.30). Stability constants for the interaction of the hexaprotonated FuEACryp with sulfate were reported to be over seven logarithmic units.

Receptors with amide functionalities can be appropriate for binding sulfate in less polar solvents than water [22, 29]. An interesting crystal structure of a sandwich sulfate complex was reported by Bowman-James' group in 2001 [161] (**49**, Figure 1.31). The sulfate anion accepts eight hydrogen bonds coming from the amide groups of both macrocycles.

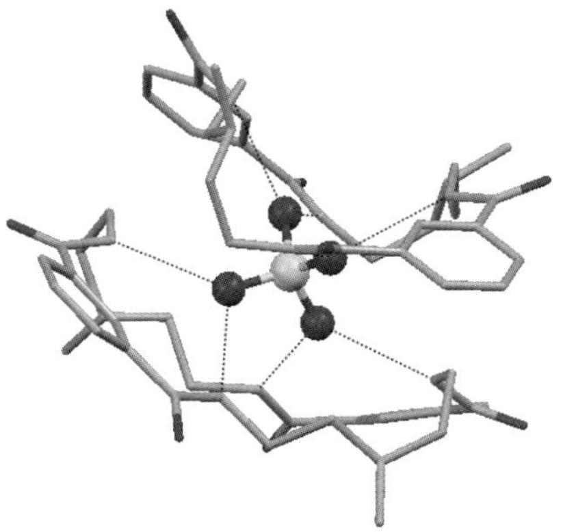

Figure 1.31 Sandwich structure of sulfate anion in macrocycle **49**.

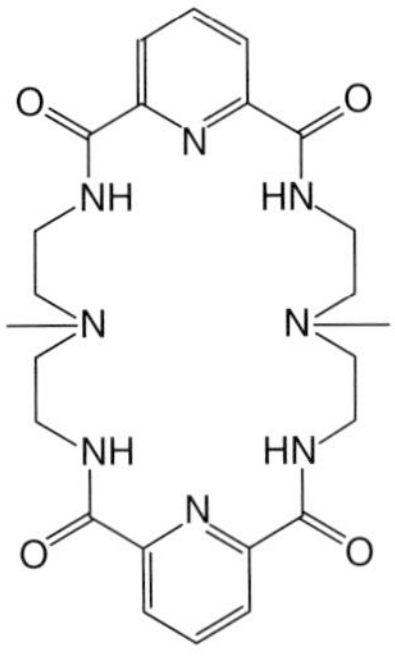

49

1.4 Phosphate and Polyphosphate Anions

As previously commented, at neutral pH phosphate anions coexist as a mixture of the hydrogen and dihydrogenphosphate anionic forms, and therefore, phosphate can either donate or accept hydrogen bonds at this pH.

Some of the earliest research regarding phosphate recognition was carried out by Lehn's group and implied macrocycles **50–52**, which incorporate guanidinium subunits in their framework [162]. Guanidinium groups are present in arginine side chains and are known to have important biological roles related to the maintenance of the tertiary structure of proteins through formation of salt bridges with carboxylate groups and to the binding and recognition of anionic substrates by enzyme receptor sites and antibodies. These roles are based on several important features of this moiety such as its permanent positive charge in aqueous solution at the pH values of biological interest ($pK_a \sim 13.5$) and formation of characteristic pairs of zwitterionic hydrogen bonds (Figure 1.32).

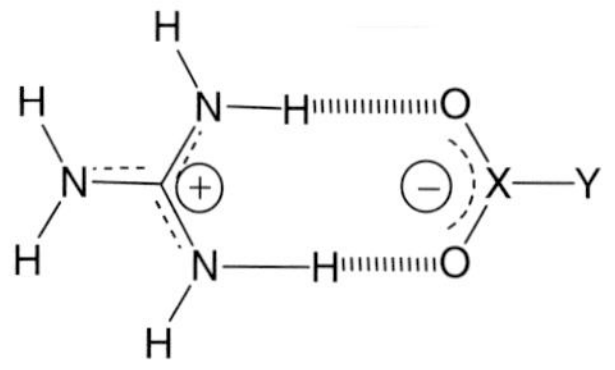

Figure 1.32 Zwitterionic hydrogen bonds formed by guanidinium groups.

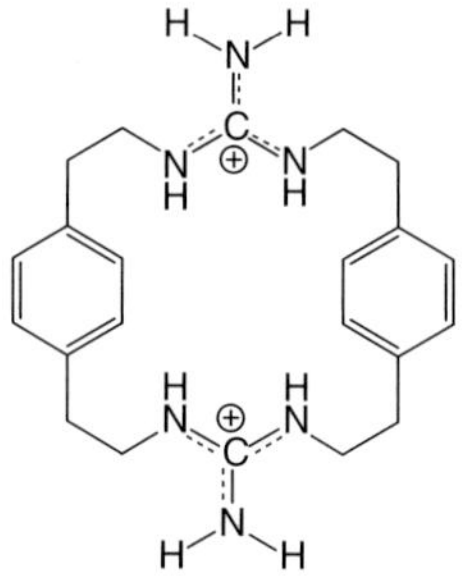

50

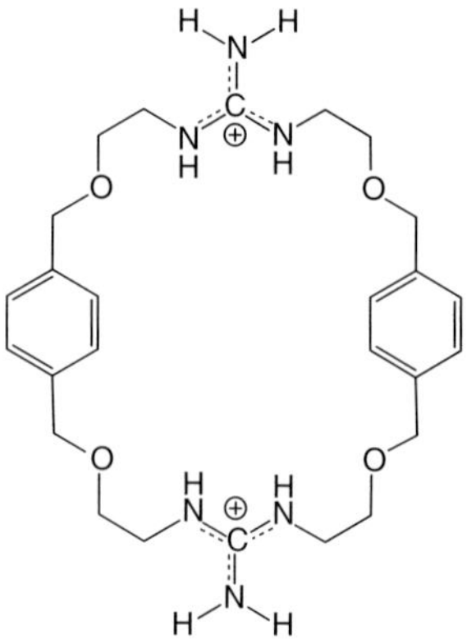

51

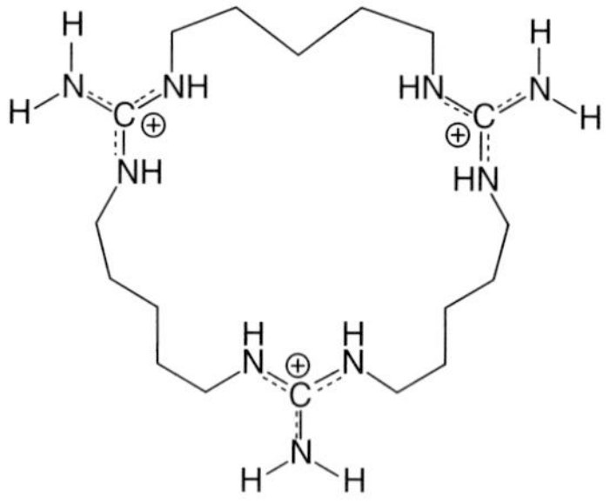

52

In spite of these appealing properties, the constants reported for the formation of 1 : 1 complexes between the macrocycles **50–52** and PO_4^{3-} species in water were not very large; log $K_s = 1.7$, 2.2, and 2.4 for **50**, **51**, and **52**, respectively. The authors stated that, although these constants were still low in comparison with those found for alkali cations, compounds **50–52** represented a step forward in the design of guanidinium-based receptors for phosphate, polyphosphates, and nucleotide anions. One of the reasons for this low affinity might be charge dispersion through the three possible faces of the guanidinium group. Proceeding with this research line, the same research group [163] prepared a series of polyguanidinium salts as anionic complexones. Several of these ligands are shown in **53–61**.

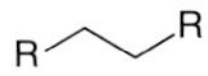

53

54

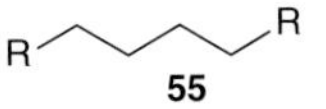

55

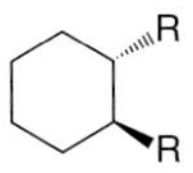

56

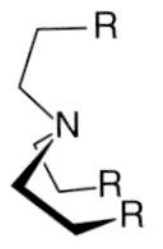

57

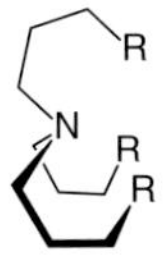

58

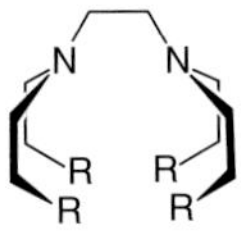

59

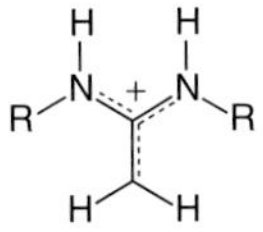

60

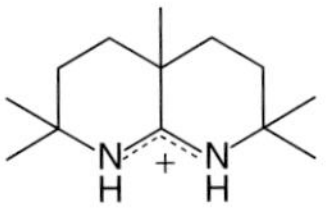

61

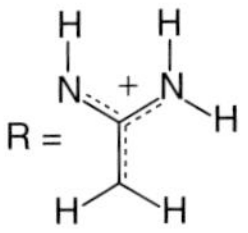

The interaction of these guanidinium-complexone ligands with phosphate, pyrophosphate, and a series of carboxylate and polyacarboxylate anions was studied by pH-metric titration in pure water and in water:methanol mixtures. The largest constants in water for the interaction with PO_4^{3-} were found for receptors **60** and **61**, which displayed values of log K_s of about three logarithmic units, while in the case of $P_2O_7^{4-}$, constants of 4.3 and 4.1 logarithmic units were retrieved from the pH-metric data for the cases of receptors **57** and **59**. In general, protonation of PO_4^{3-} or $P_2O_7^{4-}$ anions to give less negatively charged anionic forms led to stability decreases.

Regarding this kind of receptors, Hamilton *et al.* proposed in 1992 a couple of elegant systems (**62** and **63**) in which internal hydrogen bonding between carbonyl groups of the molecule and the guanidinium moiety induced favorable conformations in the receptors for their interaction with diphenylphosphate (Figure 1.33).

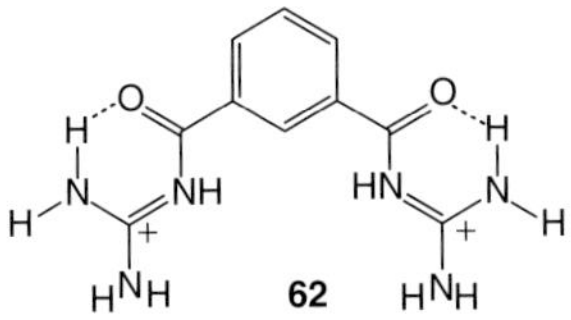

62

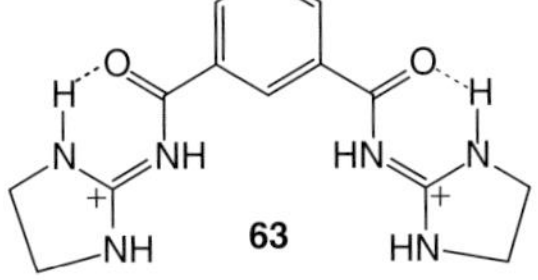

63

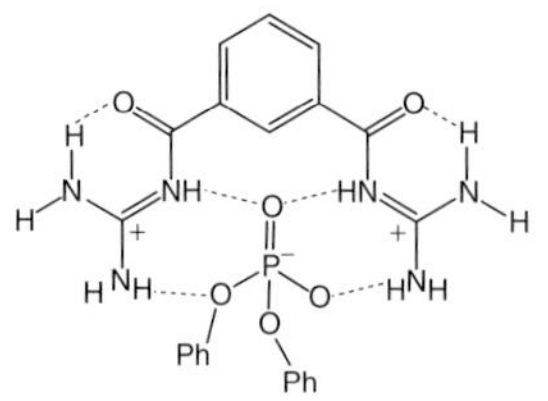

Figure 1.33 Internal C=O–H–N (guanidinium) bonds organizes the receptor for its interaction with phosphate.

Classical azamacrocycles are also appropriate ligands for binding phosphate and polyphosphate anions since the number of protonated and unprotonated amine groups they contain, and thus the overall charge and the number of hydrogen bond donors and acceptors, can be easily controlled by regulating the pH of the solution. Kimura *et al.* described in 1982 the interaction in water of the saturated macrocycles **36** and **64**, in their triprotonated forms, with HPO_4^{2-}, obtaining relatively low values of stability (2.04 and 1.1 logarithmic units, respectively) [99]. The same authors reported years later larger constants for the interaction of the monohydrogenphosphate anion with the tetra- and hexaprotonated forms of the ditopic azamacrocycle **65** (2.9 and 3.8 logarithmic units, respectively) [164].

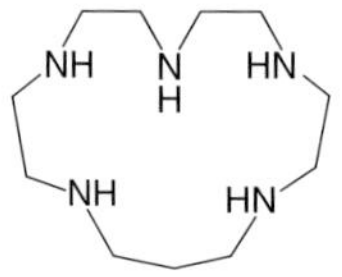

64

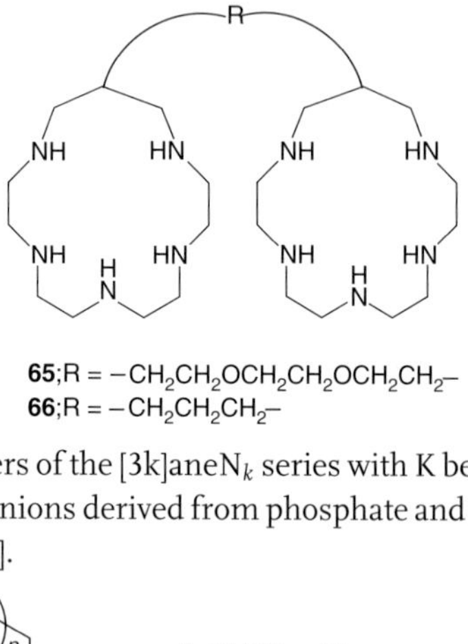

65;R = $-CH_2CH_2OCH_2CH_2OCH_2CH_2-$
66;R = $-CH_2CH_2CH_2-$

Macrocycle **36** and larger congeners of the [3k]aneN$_k$ series with K between 7 and 12 were also shown to interact with anions derived from phosphate and pyrophosphate in aqueous solution (**67**–**72**) [165].

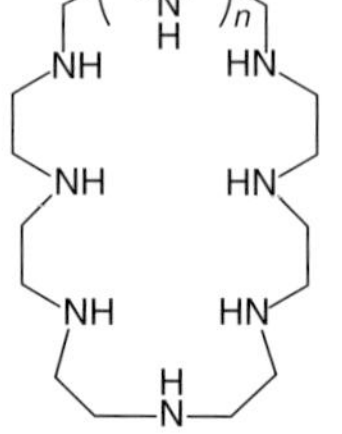

n = 1, **67** [21]aneN$_7$
n = 2, **68** [24]aneN$_8$
n = 3, **69** [27]aneN$_9$
n = 4, **70** [30]aneN$_{10}$
n = 5, **71** [33]aneN$_{11}$
n = 6, **72** [36]aneN$_{12}$

The results obtained for the interaction of phosphate in its HPO_4^{2-} and $H_2PO_4^-$ forms with protonated forms of **70** denoted log K_s limit values of 1.6 for the equilibrium $H_4(\mathbf{70})^{4+} + HPO_4^{2-} = [H_5(\mathbf{70})PO_4]^{2+}$ and 3.5 for $H_9(\mathbf{70})^{9+} + H_2PO_4^- = [H_{11}(\mathbf{70})PO_4]^{8+}$. For pyrophosphate, the constants were clearly higher, varying from 3.7 logarithmic units for $H_4(\mathbf{70})^{4+} + P_2O_7^{4-} = [H_5(\mathbf{70})P_2O_7]$ to 8.3 logarithmic units for $H_8(\mathbf{70})^{8+} + H_2P_2O_7^{2-} = [H_{10}(\mathbf{70})P_2O_7]^{6+}$. For a given equilibrium, the

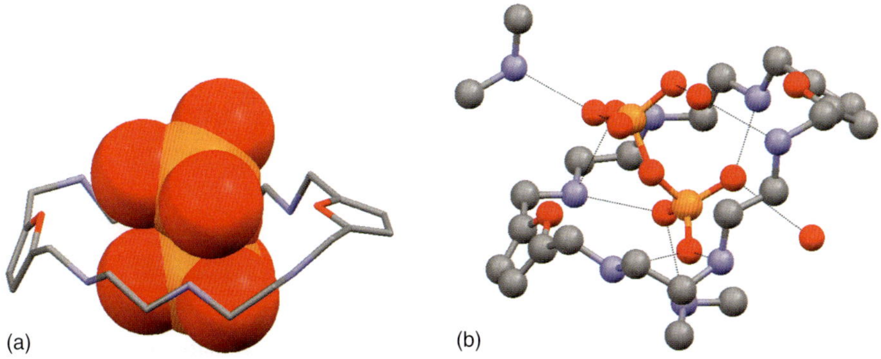

Figure 1.34 (a) View of $H_2P_2O_7^{2-}$ anion included in pentaprotonated $[H_5(\mathbf{73})]^{5+}$. (b) Hydrogen bond contacts of $H_2P_2O_7^{2-}$ in $[H_5(\mathbf{73})\cdot(H_2P_2O_7)]Cl_3\cdot5H_2O$. Parts of the connecting macrocycles are included in the drawing.

progression in the series led to diminished stability constants in agreement with the lower charge density of the receptors as their size is increased. However, in the case of phosphate, this was so until macrocycle [30]aneN_{10} (**70**) was reached; for [33]aneN_{11} (**71**), an increase in stability was observed. This was attributed to a likely inclusion of the anion in the cavity. In Section 1.6, similar trends are discussed for metallocyanides.

Finding crystal structures of phosphate or polyphosphate anions fully or partly included in aza monomacrocycles is not frequent. Early examples of such structures were reported by Martell *et al.* in 1995 and 1996 for azamacrocycles **73** and **74** (Figures 1.34 and 1.35) [166, 167]. The first structure reveals that $H_2P_2O_7^{2-}$ binds to pentaprotonated **73** through multiple hydrogen bonds, with one end of the substrate inserting into the macrocycle and the other one extending outside it. Three oxygen atoms of the inside PO_3 unit and one oxygen atom of the outside PO_3 unit form hydrogen bonds to the macrocycle, while the remaining two oxygen atoms of this outside PO_3 unit are hydrogen bonded to oxygen atoms of a pyrophosphate belonging to another binary complex.

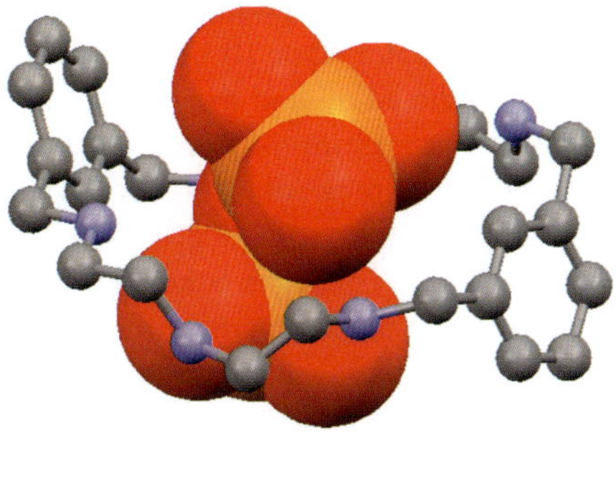

Figure 1.35 View of the $H_2P_2O_7^{2-}$ anion inserted in $[H_4(\mathbf{74})]^{4+}$.

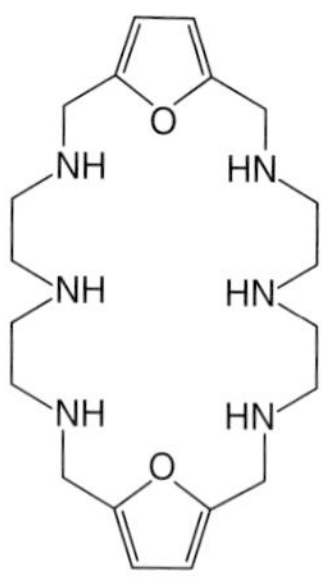

73

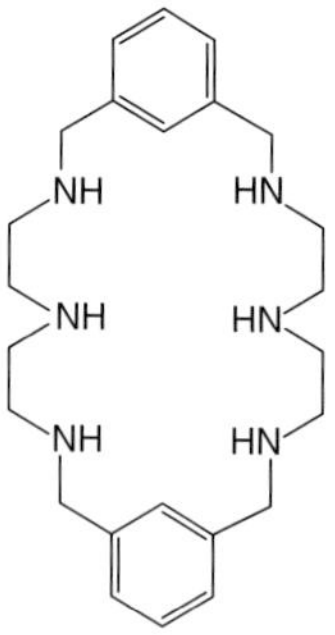

74

Tetraprotonated **74** binds $H_2P_2O_7^{2-}$ anions inside the macrocyclic cavity, with each end of the anion hydrogen bonded to the nitrogen atoms of a *m*-xylyldiamine moiety through two of their oxygen atoms and the protonated third nitrogen atom of each end pointing away from the cavity and hydrogen bonded to nitrogen atoms of adjacent molecules. Other representative structural and/or solution studies regarding 2 + 2 azacyclophanes are included in Refs [168–171].

A study on the thermodynamic terms affecting the interaction of polyammonium receptors either of cyclic or acyclic nature with phosphate and pyrophosphate anions in aqueous solution indicated that there were five modes of hydrogen bond motifs in such systems [170, 171]; four of them involve ammonium or amine groups as donors (types I–IV), and just one involves amine groups as hydrogen bond acceptors (type V):

$$-\text{N–H}\cdots\cdots{}^{-}\text{O–}\quad \Delta H^0 > 0,\quad T\Delta S^0 > 0 \qquad \text{I}$$

$$-\text{N–H}^{+}\cdots\text{OH–}\quad \Delta H^0 > 0,\quad T\Delta S^0 \approx 0 \qquad \text{II}$$

$$-\text{N–H}\cdots{}^{-}\text{O–}\quad \Delta H^0 > 0,\quad T\Delta S^0 \approx 0 \qquad \text{III}$$

$$-\text{NH}\cdots\text{OH–}\quad \Delta H^0 > 0,\quad T\Delta S^0 < 0 \qquad \text{IV}$$

$$-\text{N}\cdots\text{H–O}^{-}\quad \Delta H^0 < 0,\quad T\Delta S^0 < 0 \qquad \text{V}$$

Binding mode I leading to hydrogen-bonded ion pair interactions should be of great importance in association processes occurring in solvents with high dielectric constant, such as water, since they provide synergetic hydrogen bonding and electrostatic attraction.

Type II bonds should be effective only in acidic enough conditions to permit extensive protonation of both the anion and the receptor. Type III bonds will be, however, favored in alkaline media, where both the anion and the receptor are extensively deprotonated. The entropic term associated with this charge transfer process should be favorable. Although types IV and V are possible hydrogen bond modes occurring between amines and compounds possessing –OH groups, mode V is known to be considerably stronger than mode IV. Hydrogen bonding modes I–IV imply a partial deprotonation of the amino group and a partial protonation of a phosphate oxygen. Since deprotonation of an amino group is a strongly endothermic reaction and protonation of HPO_4^{2-} or $H_2PO_4^{-}$ or pyrophosphate anions is weakly endothermic or athermic, formation of hydrogen bonds I–IV should be endothermic, while different contributions depending mostly on the effect the process has on charge separation will be affecting the sign of the entropic term. Conversely, hydrogen bond V, which implies a partial protonation of an amino group and a partial deprotonation of a phosphate oxygen, would be accompanied by a negative enthalpy change and a favorable entropic term.

In the systems studied, I, II, and V are expected to be the principal hydrogen bonding modes with a relative importance directly connected with the extent of proton transfer, which in its turn depends on the N–O separation and the dielectric constant of the medium. Further discussion on this point is included in Chapter 2, devoted to energetics.

The possible use of polyammonium-based receptors as fluorescent chemosensors for phosphate anions was advanced by Czarnik and coworkers in 1989 [172], who proposed that tris(3-aminopropyl)amine derivatives appended with fluorophoric units could be useful for this purpose (**75**–**77**).

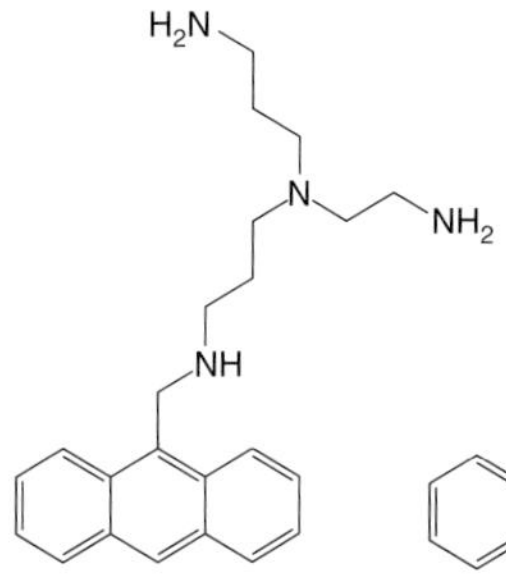

75

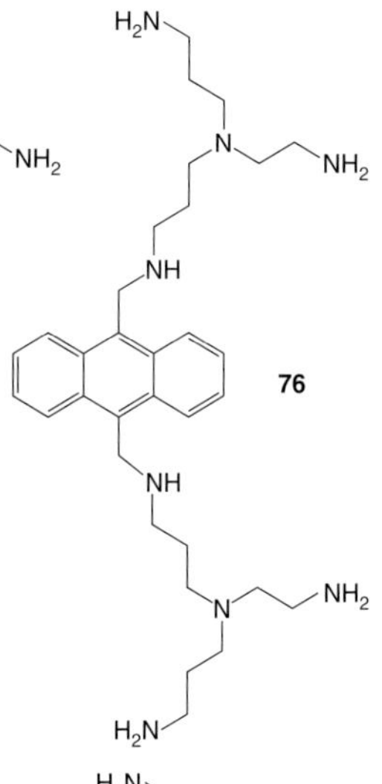

76

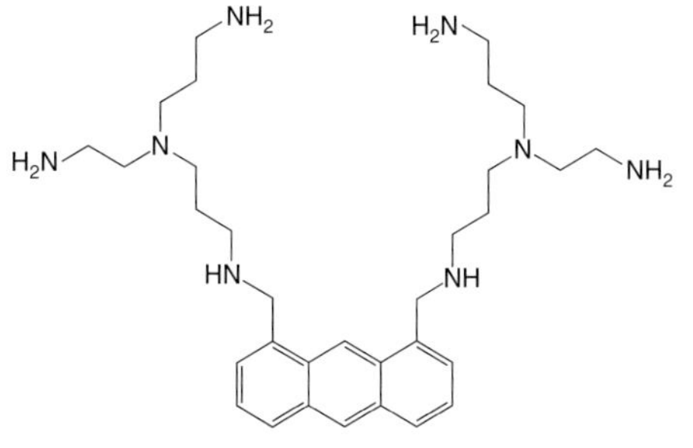

77

These authors proposed that monohydrogenphosphate could deliver a proton to triprotonated **75**, blocking the photoinduced electron transfer from the amine to the excited fluorophore, producing a chelation enhancement of the fluorescence

(CHEF) effect. In 1994, Czarnik reported that receptor **77** containing two tripodal polyamine units could operate as a pyrophosphate chemosensor by a mechanism similar to that just described for phosphate [173] (see Chapter 9).

1.5
Carboxylate Anions and Amino Acids

The study of carbonate and carboxylate anions emerged in the very early years of the supramolecular chemistry of anions. Guanidinium complexones **53–61** were checked for their capability to bind acetate, maleate, and fumarate in methanol:water 9 : 1 mixtures, obtaining relatively high stability values in some instances, as in 5.1 logarithmic units found for the interaction of **59** with maleate dianion [163].

On the basis of the guanidinium platform, Lehn, de Mendoza *et al.* reported receptor **78** for chiral recognition of aromatic carboxylate anions [174]. Sodium *p*-nitrobenzoate was quantitatively extracted from water by a chloroform solution of **78**. Extraction experiments of sodium (*S*)-mandelate and (*S*)-naproxenate [(+)-6-methoxy-a-methyl-2-naphthaleneacetate] with **78**-*SS* and **78**-*RR* afforded the corresponding diastereomeric salts. Since free amino acids in zwitterionic form (valine, phenylvaline, and tryptophan) were not extracted from aqueous solution by **78**, *N*-acetyl and *N*-*tert*-butoxycarbonyl derivatives of tryptophan were examined. It was observed that extraction of an excess of the racemic salts with **78**-*SS* afforded in each case two diastereomeric excesses (de) of ~17% for the L-tryptophan derivative.

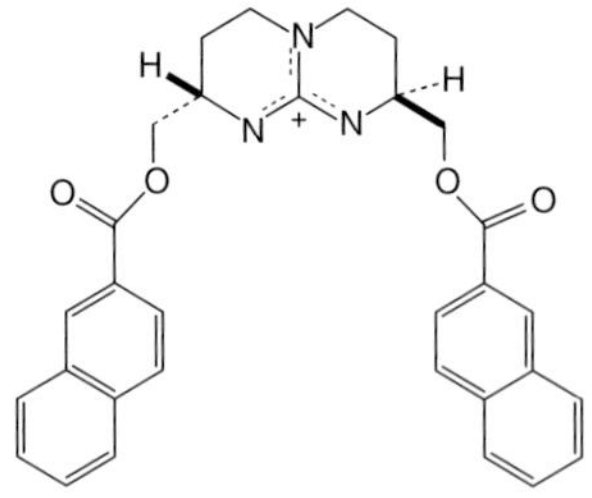

78

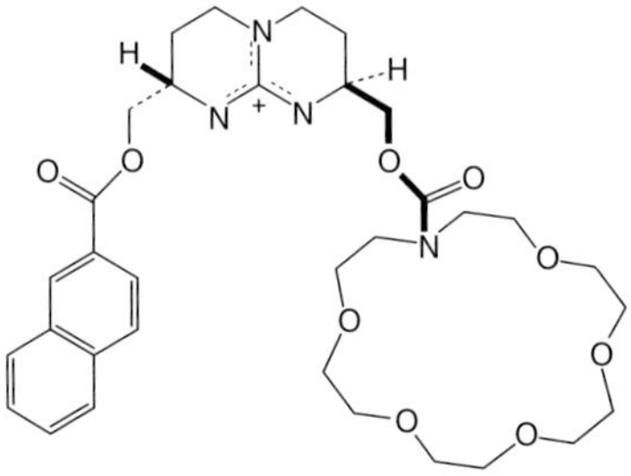

79

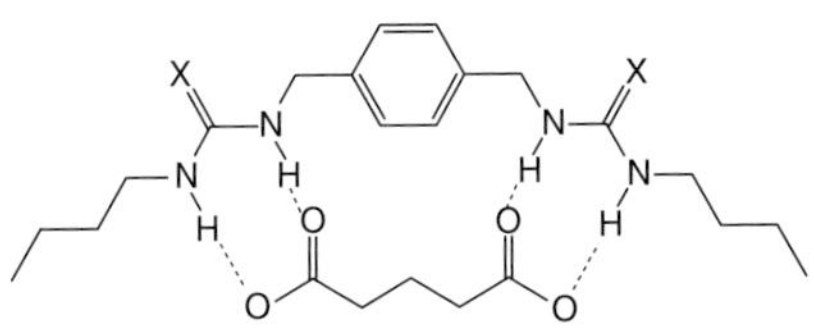

Figure 1.36 Schematic representation of the hydrogen bonding interaction of **81** and **82** with glutamate.

On the basis of this scaffold, a few years later, de Mendoza *et al.* prepared the ditopic receptor **79** for amino acid recognition, including a naphthoyl moiety and a crown ether [175]. Competition liquid–liquid extraction experiments of aqueous solutions containing 13 amino acids showed good selectivity for L-phenylalanine. Chiral recognition was confirmed by NMR since the D-enantiomers were not extracted.

Hamilton and coworkers proved in 1993 that guanidinium receptor **80**, urea, **81**, and thiourea receptors **82** [176] (Figure 1.36) could recognize dicaboxylate anions of matching size in highly competitive solvents such as DMSO. While the association constants of the complex formed between **82** and glutarate in DMSO-d_6 determined by NMR ($\log K_s = (1.0 \pm 0.2) \times 10^4$ M^{-1}) was 15-fold greater than that obtained for **81**, the constant for **80** was too large to be measured by NMR ($\log K_s > 5 \times 10^4$ M^{-1}).

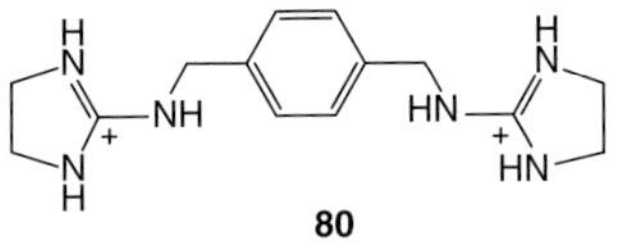

80

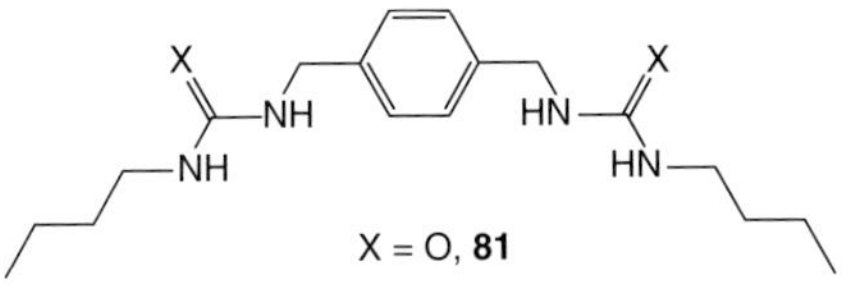

X = O, **81**
X = S, **82**

However, polyazamacrocycles were by far the most studied receptors in the early times of anion coordination chemistry. Lehn and coworkers, in their seminal *JACS* 1981 communication, reported the interaction of receptors **83–85** with several di- and tricarboxylate anions, along with sulfate, cyanometallate, and nucleotide anions [58]. Early cyanometallate and nucleotide anion-binding studies are presented in the next two sections.

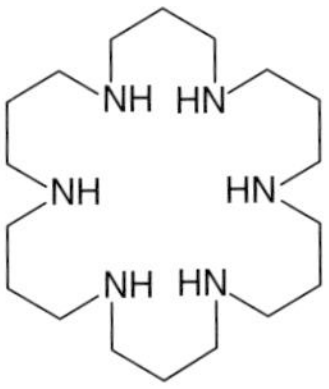

83

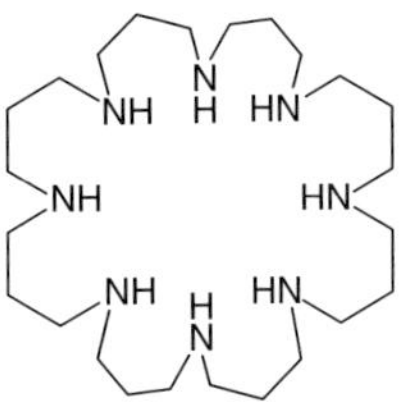

84

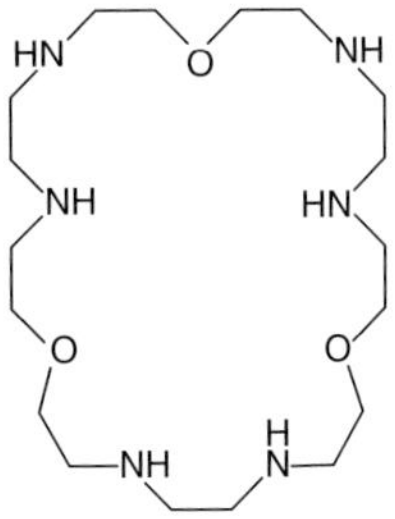

85

The values obtained for the interaction of the hexaprotonated forms of **83** and **85** and the octaprotonated form of **84** with carboxylate anions indicated that electrostatic interactions played a major role in both strength and binding selectivity. In the same year, Kimura *et al.* published a polarographic study concerning the interaction of **36**, the open-chain pentaamine **86**, and the cyclic pentaamines (**47**), **87**, and **88** with several mono- and dicarboxylate anions as well as with the tricarboxylate anion citrate [177].

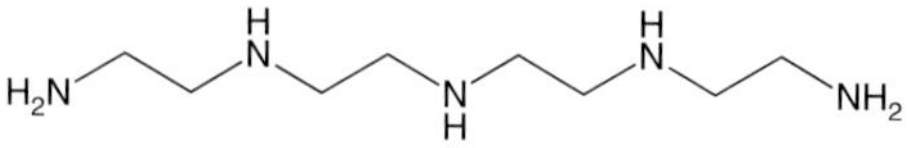

86

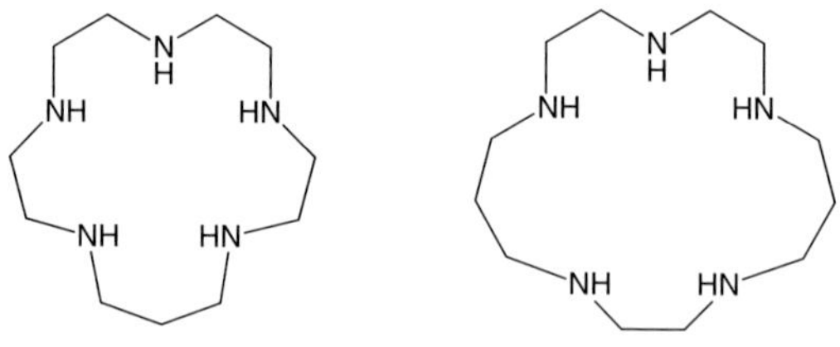

87

88

The authors concluded that macromonocyclic pentaamines and hexaamines specifically interact at neutral pH with polyanions having the carboxylate functions at short distances, such as succinate, malate, citrate, malonate, and maleate, but fail to interact with the other dicarboxylates, fumarate, aspartate, and glutarate, and also with the monocarboxylates, acetate and lactate. In the same year, Kimura and coworkers also established electrophoretic protocols for analyzing polyamines using buffers containing di- or tricarboxylates at pH ≈8. Anomalous electrophoretic behavior of some macrocyclic polyamines migrating in the anode direction in citrate buffer solution at pH ≈6 was discovered [178]. The same group presented a polarographic study about the interaction in aqueous solution of the hexaazamacrocycle **36** and some of the pentaamines, **47**, **48**, **86**–**88**, with carbonate anions. Such interaction persisted for some systems even at slightly acidic pH values [179].

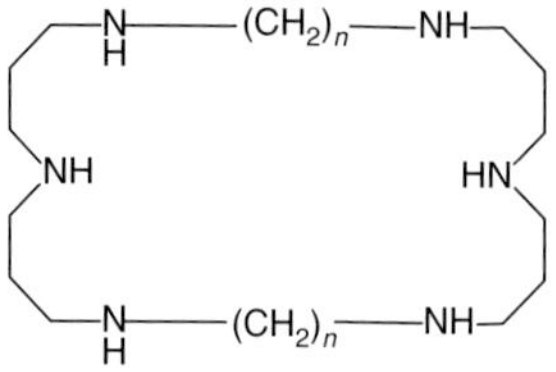

$n = 7$, **89**

$n = 10$, **90**

Regarding selectivity aspects of dicarboxylate recognition, classic contributions were provided by Lehn's group [60, 180]. Hosseini and Lehn studied the interaction of macrocycles constituted by two dipropylenetriamine chains connected by propyl (**17**), heptyl (**89**), and decyl (**90**) hydrocarbon chains in their hexaprotonated forms with the series of dicarboxylate anions progressing from oxalate to sebacate, which

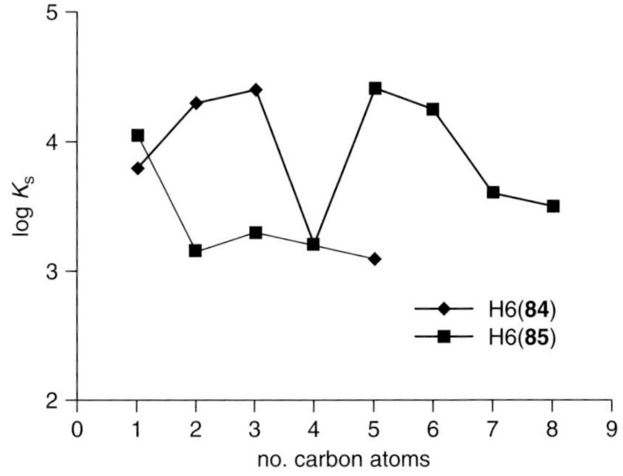

Figure 1.37 Graphical representation of the stability constants (log K_s) of the complexes formed by $[H_6(\mathbf{89})]^{6+}$ and $[H_6(\mathbf{90})]^{6+}$ with dicarboxylates $^{-}O_2C-(CH_2)_n-CO_2^-$ as a function of the number of carbon atoms (n) between the carboxylate functions. Adapted from Ref. [60].

has seven methylene units between the carboxylate groups. The authors concluded that there exists selectivity depending on the respective chain lengths of substrate and receptor; each receptor, $[H_6(\mathbf{89})]^{6+}$ and $[H_6(\mathbf{90})]^{6+}$, shows a marked selectivity peak for a given dicarboxylate (Figure 1.37).

Shape selectivity was found years later by Bianchi, García-España, Luis *et al.* for the interaction of protonated **67** with the anions derived from the isomeric diacids and triacids **91**–**97** [181, 182] using citrate as a reference for a flexible substrate.

Since the basicities of the different anions explored are very different, a criterion balancing this point had to be adopted. The most appropriate way to compare the interaction of two different anions (anion 1 and anion anion 2) with receptor 67 is to calculate the distribution of complexed species as a function of pH for the mixed systems Anion **91** - Anion **92-97** and the overall percentages of formation [182, 183]. This method allows establishing selectivity ratios over the pH range studied and does not require any assumption of the location of protons in the interacting species, which is a frequent source of erroneous interpretation of selectivity. Using this approach, the following general selectivity order was found: 1,2,3-BTC > *cis,cis*-Kemp > 1,3,5-BTC > 1,2-BDC > 1,3-BDC > *cis,trans*-Kemp > citric acid. A more detailed discussion of this approach is included in Chapter 2, devoted to energetics of anion coordination.

Bismacrocycles or cryptand-like compounds of appropriate dimensions have been shown to be well-suited ligands for encapsulating dicarboxylate anions with matching size and functionalities. A classic example of this behavior is represented by receptor **98** [184, 185], which forms in aqueous solution fairly stable complexes with dicarboxylate anions of the $^{-}O_2C-(CH_2)_n-CO_2^-$ series showing selectivity for adipate ($n = 4$). Moreover, hexaprotonated **86** binds terephthalate with a remarkable

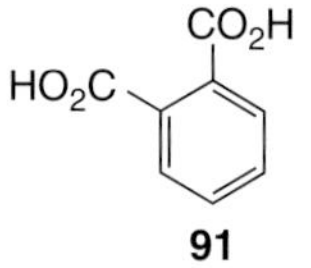

91
1,2-DBC

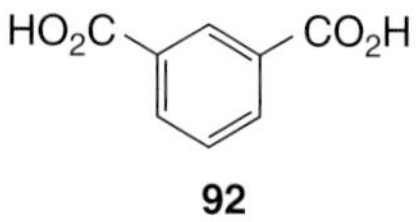

92
1,3-DBC

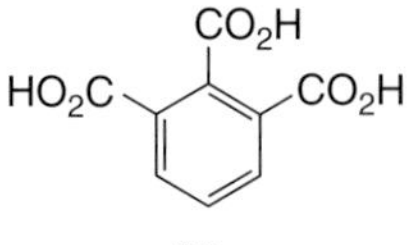

93
12,3-BTC

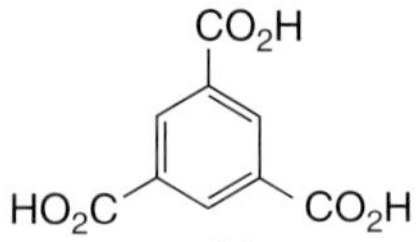

94
1,3,5-BTC

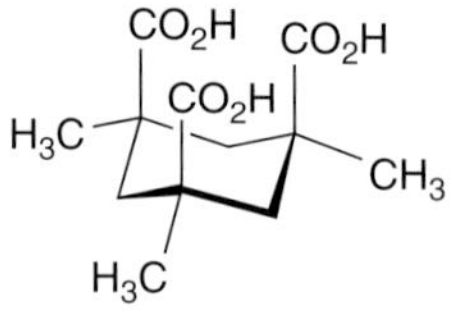

95
cis,cis-KEMP

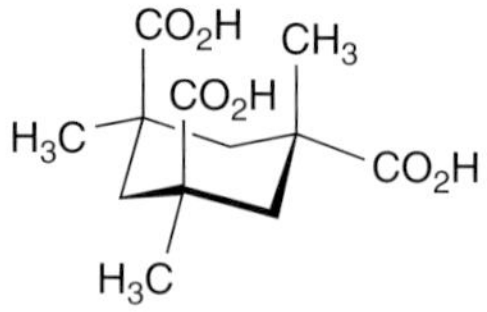

96
cis,trans-KEMP

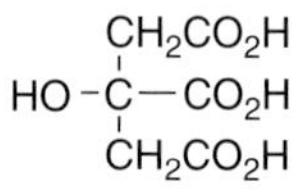

97
cit

strength (log K_s = 60 000 at pH = 5.5, log K_s = 25 000 at pH = 6.0). The higher binding constant obtained at pH 5.5 should be a consequence of the greater percentage of hexaprotonated **98** existing in solution at this pH. However, the most striking finding in this research was the publication of one of the first crystal structures showing a dicarboxylate anion, in this case terephthalate, included in a bicyclic receptor (Figure 1.38).

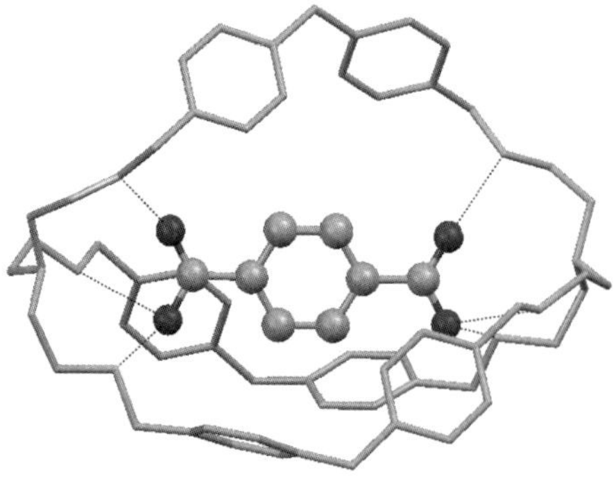

Figure 1.38 Terephthalate dianion included in hexaprotonated **98**.

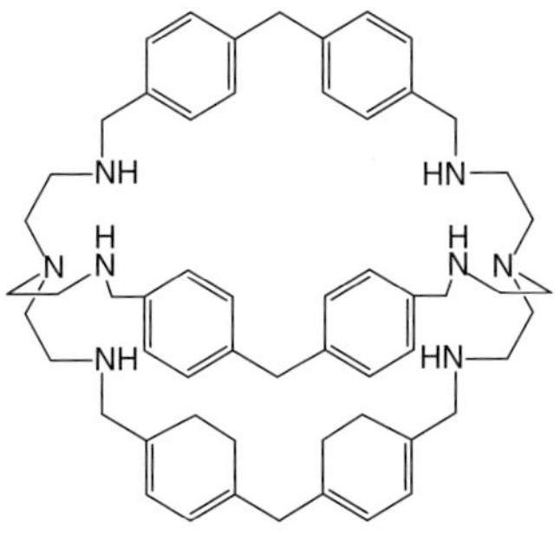

98

A system for the recognition of dicarboxylate anions based on the cooperative interplay of a bis(amidopyridine) hydrogen-bonding donor site and a K^+-binding site constituted by an 18-membered diazacrown ether unit was proposed by Kilburn *et al.* in the early 1990s (**99**). NMR studies performed in chloroform proved the interaction of different dicarboxylate anions and amino acids with this receptor [186].

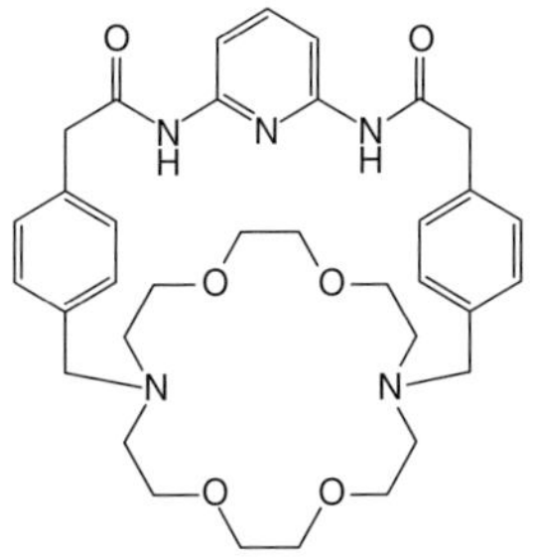

99

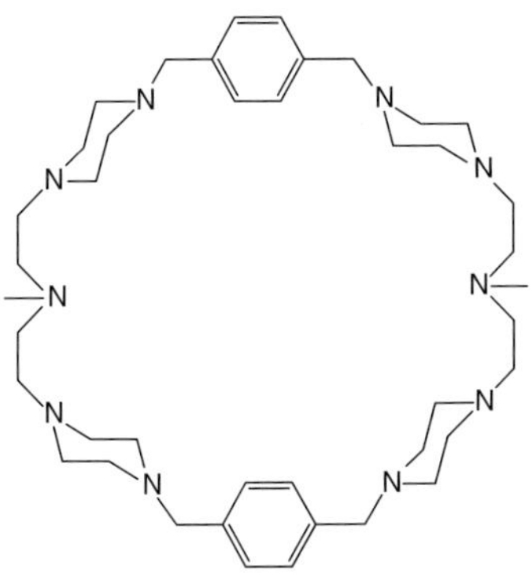

100

Cascade complexes formed between coordinatively unsaturated dinuclear metal complexes and α,ω-dicarboxylate anions can also help in recognizing these kinds of substrates. An example of this chemistry was provided by the Florence and Valencia Supramolecular Chemistry groups with the azacyclophane receptor **100**.

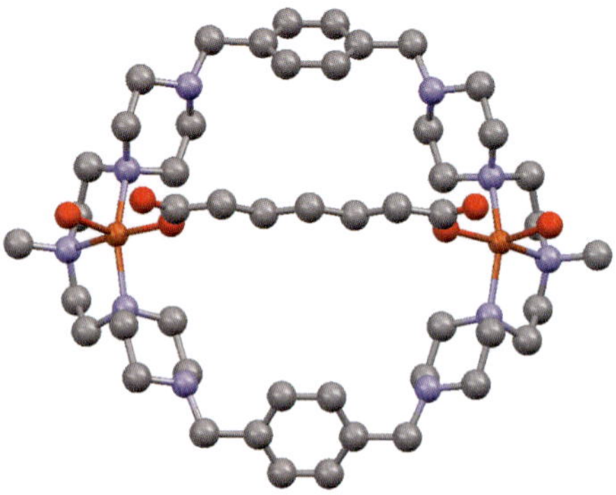

Figure 1.39 Crystal structure of the cascade complex formed by pim^{2-} with the binuclear complex $[Cu_2(\mathbf{100})(H_2O)]^{4+}$ [187]. Hydrogen atoms have been omitted.

The coordinatively unsaturated Cu^{2+} centers can catch a pimelate anion as a bridging bis(monodentate) ligand as shown in the crystal structure presented in Figure 1.39 [187].

1.6
Anionic Complexes: Supercomplex Formation

Cyanometallate anions were used as targets very early in anion coordination chemistry. The resulting supramolecular species formed are called *supercomplexes* or *complexes of the second coordination sphere* in which the polyammonium species is placed in the second coordination sphere, interacting electrostatically and through hydrogen bonding and other weak forces with the complex anion. Such an early appearance of cyanometallates as substrates for the interaction with ammonium receptors is not accidental. These anions offer simple ways of analyzing the influence on host–guest affinity of negative charge increases while other factors such as shape and geometry are kept essentially constant. For instance, apart from solvation effects, the only noticeable change when moving from $[Fe(CN)_6]^{3-}$ to $[Fe(CN)_6]^{4-}$ is the different net charge of both anions. In the same way, exchanging $[Fe(CN)_6]^{3-}$ by $[Co(CN)_6]^{3-}$ should be nonsignificant from a recognition point of view; both anions are octahedral and have a very close size.

On the other hand, the analysis of the changes produced in the electrochemical response of the $[Fe(CN)_6]^{3-}/[Fe(CN)_6]^{4-}$ redox couple on addition of a given polyammonium receptor is very useful. The simple quasi-reversible electrochemical behavior of this couple permits the derivation of stability constants by an alternative method to pH-metric titration or NMR, which are the most widely used techniques in this area. It has to be emphasized that it is always advisable to use more than one independent technique in order to have reliable descriptions of the anion-receptor systems.

Although formation of ion pairs between hexacyanoferrate(III) and quaternary ammonium salts had already been evidenced in 1965 following the shifts in the ^1H NMR [188] spectra of the ammonium salt following the interaction with the anion, Gross, Lehn *et al.* [59, 62] gave impetus to this topic. The group studied the changes produced in the cyclovoltamperograms of $[Fe(CN)_6]^{4-}$ aqueous solutions on addition of increasing amounts of the hexaaza and octaza macrocycles **83** and **84**. Addition of the macrocycles brings about anodic shifts, whose magnitude depends on the stoichiometries of the complexes formed and on their stabilities.

The interaction of the series of large polyazacycloalkanes [21]aneN_7, [24]aneN_8, [27]aneN_9, [30]aneN_{10}, [33]aneN_{11}, and [36]aneN_{10} (**67–72**) and their open-chain counterparts (**101–104**) in their protonated forms with a series of

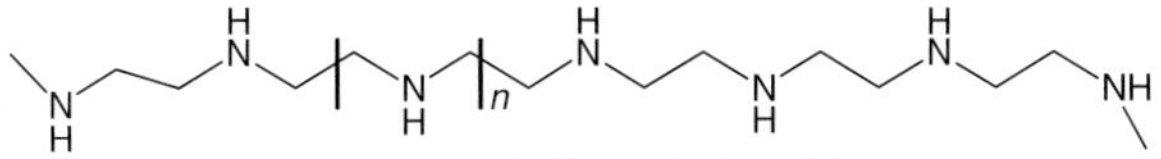

n = 1, Me$_2$hexaen **101**
n = 2, Me$_2$heptaen **102**
n = 3, Me$_2$octaen **103**
n = 4, Me$_2$nonaen **104**

cyanometallates and other anionic metal complexes was an illustrative early example of this chemistry presented by the Supramolecular Chemistry groups of Florence and Valencia during the years 1985–1995 [189–196]. The open-chain counterparts were selected so that they had the same number of carbon and nitrogen atoms and the same class of amine groups as their cyclic counterparts [197]. The analysis of the stepwise stability constants for these systems indicated that the chief driving force in cyanometallate–polyammonium receptor binding is charge–charge interaction, which makes it difficult to modulate the selective discrimination of one guest over another. However, a representation of the constants for the equilibria $[Co(CN)_6]^{3-} + H_p([3\,k]aneN_k)^{p+} = [Co(CN)_6][H_p([3\,k]aneN_k)]^{(3-p)}$ showed that typically the constants for a given protonation degree of the macrocycle (p) steadily decreased when going from one macrocycle to the next until [30]aneN_{10} (**70**) was reached. From there on, an increase in stability was observed (Figure 1.40). This change in the pattern could be attributed to an inclusion of the anion within the macrocyclic cavity, thus favoring shorter charge–charge interactions and the formation of stronger hydrogen bonds.

Inclusion of the anion inside the cavity of the macrocycle was also postulated by early work performed by the groups of Lehn in Strasbourg and Balzani in Bolonia on the interaction of $[Co(CN)_6]^{3-}$ with the polyammonium receptors [24]aneN_6 (**83**) and [32]aneN_8 (**84**) containing all propylenic chains between the nitrogen atoms [198, 199]. These authors observed that the quantum yield of the light-induced aquation reaction of $[Co(CN)_6]^{3-}$ was reduced to one-third of the initial value in the presence of the octaprotonated receptor $H_8([32]aneN_8)^{8+}$, suggesting that inclusion of $[Co(CN)_6]^{3-}$ into the macrocycle through the equatorial plane had occurred. The four CN^- groups in the enclosed equatorial plane would be hydrogen bonded to the ammonium groups of the macrocycle and would not be exchanged by water molecules on light excitation (Figure 1.41) [200].

These analyses allowed for structural conclusions to be drawn for a number of systems [201]. Reduction of the quantum yield by half would suggest an interaction of the polyammonium host and the metallocyanide through one of its triangular

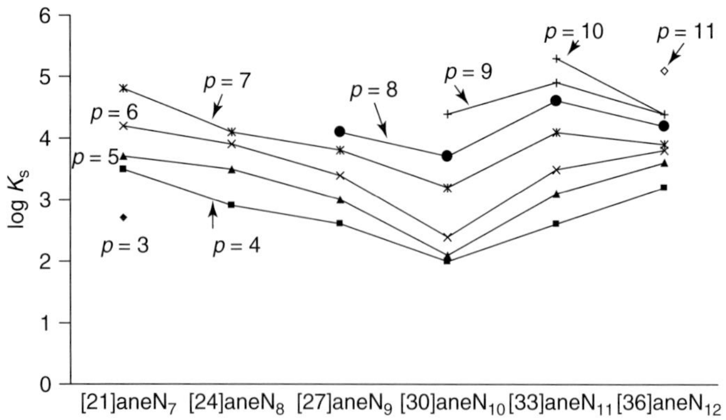

Figure 1.40 Plot of the variation of the constants for the equilibria $[Co(CN)_6]^{3-} + H_p([3k]aneN_k)^{p+} = [Co(CN)_6](H_p([3k]aneN_k))^{(3-p)}$ [191]. Reprinted with permission from Ref. [191]. Copyright 1992 American Chemical Society.

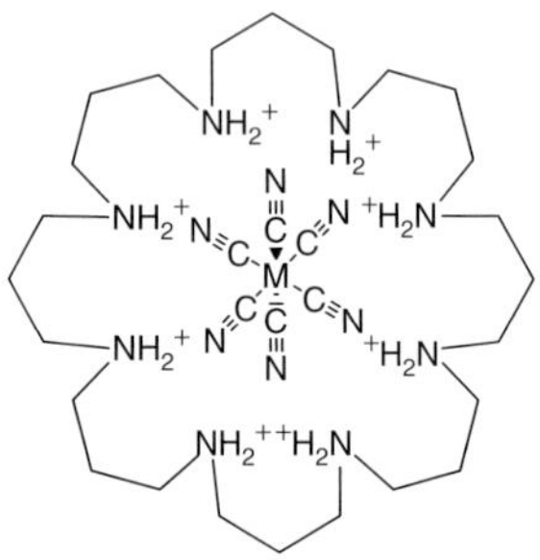

Figure 1.41 Schematic drawing of the inclusion of the equatorial plane of octahedral metallocyanides in the macrocyclic hole of $[H_8([32]aneN_8)]^{8+}$ [198, 199].

faces, while reduction by one-third would imply interaction through one of its edges. Such studies have allowed, for instance, to assume that the interaction between $[Co(CN)_6]^{3-}$ and hexaprotonated $[H_6([24]aneN_6)]^{6+}$ or the tetraprotonated cyclophane 2,6,9,13-tetraza[14]paracyclophane (**105**) [196] involves one face of the octahedron.

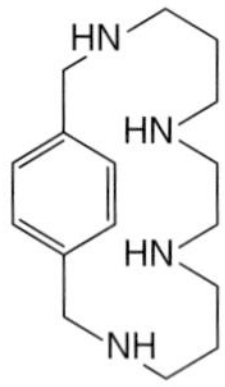

105

Although being relatively small tetraazamcrocycle 14 intercats strongly with metallocyanides in pure water. In fact, Bianchi, Micheloni, Orioli, Paoletti, and Mangani [202] made direct microcalorimetric studies for the interaction of **14** with $[Fe(CN)_6]^{4-}$ and $[Co(CN)_6]^{3-}$, observing that the main contribution

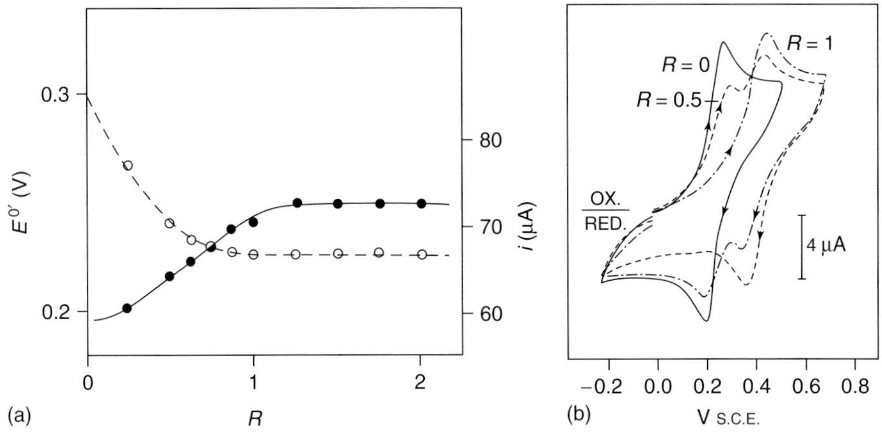

Figure 1.42 (a) Typical effects of the addition of a polyammonium receptor (H_pL^{p+}) on the cyclic voltammogram of $[Fe(CN)_6]^{4-}$. R is the mole ratio receptor:metallocyanide [195]. Reproduced with permission from the Royal Society of Chemistry. (b) Representation of the variation of peak potential (continuous line) and peak current (dashed line) versus R for a system polyammonium receptor H_pL^{p+} – $[Fe(CN)_6]^{4-}$. Reprinted with permission from Ref. [191]. Copyright 1987 American Chemical Society.

to the $-4.94\ \text{kcal mol}^{-1}\ \Delta G^0$ term was coming from the entropic term ($\Delta S^0 = 13\ \text{cal mol}^{-1}$). For both anionic complexes, the enthalpic contribution was slightly exothermic ($\Delta H^0 = -1.1(1)\ \text{kcal mol}^{-1}$ for $[Fe(CN)_6]^{4-}$ and $\Delta H^0 = -2.56$ for $[Co(CN)_6]^{3-}$).

As noted above, the use of $[Fe(CN)_6]^{4-}$ allows for the analysis of the systems by cyclic voltamperometry. Figure 1.42 shows the typical behavior of these systems. Addition of the receptor to an aqueous solution of $[Fe(CN)_6]^{4-}$ yields an anodic shift of the voltammogram until a certain mole ratio R = receptor : $[Fe(CN)_6]^{4-}$, normally 1, is reached (Figure 1.42b). From a plot of the variations of the peak potentials versus R it is possible to deduce the stoichiometry of the formed supercomplex. On the contrary, the effect of supercomplex formation is a decrease in the peak intensity due to the formation of the higher molecular weight adducts species.

Distribution diagrams of the species existing in equilibria calculated from the stability constants determined pH-metrically coupled with plots of the variation of anodic peak currents and formal potentials show plateaus in the zones where single species predominate in solution (Figure 1.43). Application of classic voltammetric and polarographic methods permits calculation of the stability constants for the formation of $[Fe(CN)_6]^{3-}$ adducts. Similar determinations are not possible by means of pH-metric methods because of the very rapid aquation process of the oxidized anion [200]. Table 1.1 shows that, as it could be expected for anions displaying the same charge, geometry, and size, the values of the stability constants for the systems $[Fe(CN)_6]^{3-}$–polyammonium receptors were very close to those obtained by pH-metry for the systems $[Co(CN)_6]^{3-}$–polyammonium receptors.

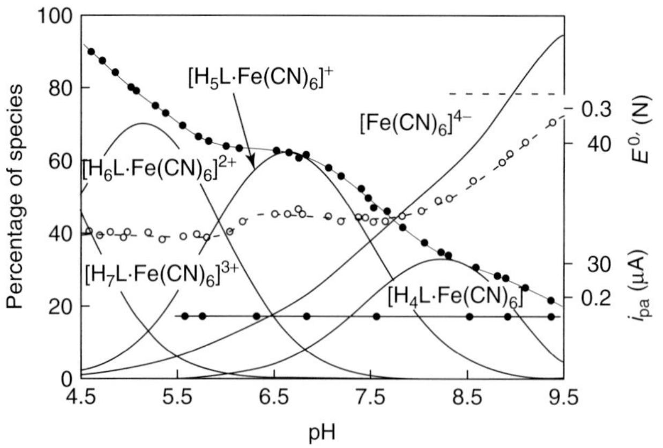

Figure 1.43 Distribution diagram of the species existing in equilibria in the system $[Fe(CN)_6]^{4-}$: Me_2heptaen and anodic peak currents (dashed lines, open circles) and formal potentials (continuous lines, solid circles) of the couple $[Fe(CN)_6]^{3-}$ – $[Fe(CN)_6]^{4-}$ [195]. Reproduced with permission from The Royal Society of Chemistry.

Table 1.1 Logarithms of the equilibrium constants for the supercomplex formation between $[Fe(CN)_6]^{3-}$ and the polyamines [21]aneN_7 (67) and [24]aneN_8 (68) calculated from cyclic voltammetry data. The corresponding values for $[Co(CN)_6]^{3-}$ calculated by pH-metric techniques are included by means of comparison [195].

Reaction[a]	**[21]aneN_7 (67)**		**[24]aneN_8 (68)**	
	$[Fe(CN)_6]^{3-}$	**$[Co(CN)_6]^{3-}$**	**$[Fe(CN)_6]^{3-}$**	**$[Co(CN)_6]^{3-}$**
$A + H_3L = H_3LA$	2.8	2.7	–	–
$A + H_4L = H_4LA$	3.4	3.5	2.9	2.9
$A + H_5L = H_5LA$	3.8	3.7	3.8	3.5
$A + H_6L = H_6LA$	–	–	4.4	3.9

[a]Charges omitted.

One of the first supercomplex structures determined by single-crystal X-ray diffraction corresponded to a solid of chemical formula $[H_8([30]aneN_{10})][Co(CN)_6]_2Cl_2{\cdot}10H_2O$, which evolved from slow evaporation of aqueous solutions of $K_3[Co(CN)_6]$ and [30]aneN_{10}·10HCl (**70**·10HCl) (Figure 1.44) [190]. The crystal structure consisted of octaprotonated macrocycles and two types of hexacyanocobaltate anions placed outside the macrocyclic hole. One of the hexacyanocobaltate anions is hydrogen bonded through four of its six cyanide groups with consecutive macrocycles, forming a chain-like structure. The second

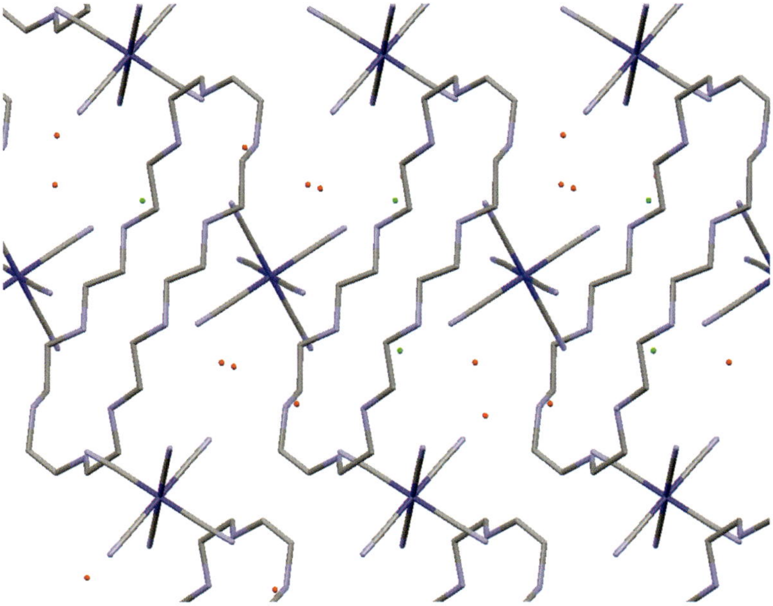

Figure 1.44 Portion of the crystal structure of $[H_8(\mathbf{70})][Co(CN)_6)]_2Cl_2{\cdot}10H_2O$ showing the two types of hexacyanocobaltate(III) anions [190].

type of hexacyanocobaltate anions interconnects macrocycles of different chains by means of two long hydrogen bonds. The macrocycle adopts an elongated elliptical shape, which is mainly due to repulsions between the positively charged ammonium groups. Interestingly, if this structure were kept in solution, the cyanometallates along a single chain would also provide a diminution of two-thirds in the quantum yield of the photoexcited aquation reaction; four of the six cyanide groups would be stabilized by hydrogen bonds.

In order to get further insight into the characteristics of these interactions and into the inclusive nature of the supercomplexes formed, the studies were extended to other types of anionic complexes [192–194]. The interaction of hexachloroplatinate(IV) anions $PtCl_6{}^{2-}$ and the macrocycle [30]aneN_{10} was studied in solution by means of ^{195}Pt NMR spectroscopy. The shift in the ^{195}Pt signal reaches a constant value (Figure 1.45) for molar ratios $R = [H_{10}([30]aneN_{10})]^{10+}/[PtCl_6]^{2-} > 1$ because of the formation of 1 : 1 metal complexes. However, another inflection is observed for $R = 0.5$, suggesting formation of supercomplexes of 2 : 1 anion:receptor stoichiometry.

The crystal structure of $[H_{10}([30]aneN_{10})][PtCl_6]_2Cl_6{\cdot}2H_2O$ (Figure 1.46) consists of a complex hydrogen bond network, which involves the hydrogens of the protonated nitrogen atoms of the receptor, $PtCl_6{}^{2-}$ anions, chloride anions, and water molecules. The complex anions are placed outside the macrocyclic hole.

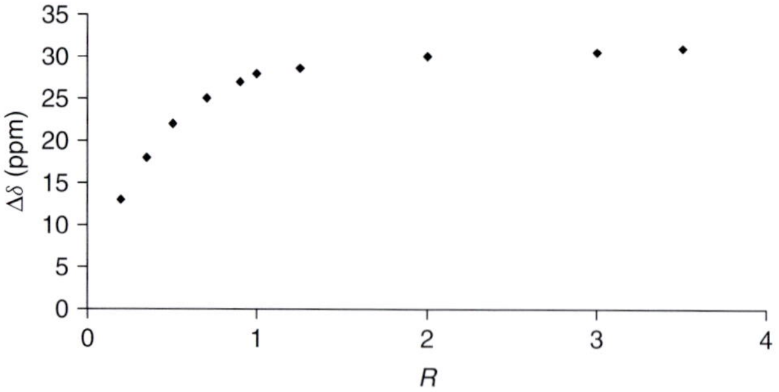

Figure 1.45 Plot of the variation in the ^{195}Pt NMR chemical shifts of $PtCl_6^{2-}$ on addition of increasing amounts of $[H_{10}([30]aneN_{10})]^{10+}$. $R = [H_{10}([30]aneN_{10})]^{10+}/[PtCl_6]^{2-}$. Reprinted with permission from Ref. [194]. Copyright 1987 American Chemical Society.

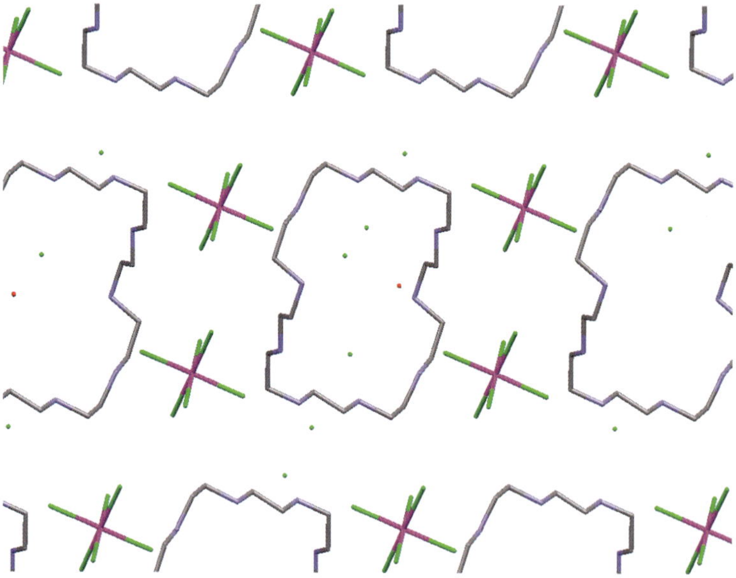

Figure 1.46 Detail of the crystal structure of $[H_{10}(\mathbf{70})][PtCl_6]_2Cl_6{\cdot}2H_2O$ showing the outer location of the $PtCl_6{}^{2-}$ anions. Red dots are water molecules and green are chloride anions [194].

One can imagine that the octahedral shape and the size of $PtCl_6{}^{2-}$ could hinder its inclusion into the macrocycle. However, at least in the solid state, inclusion was also not detected for the square planar anion $[Pt(CN)_4]^{2-}$, even if all calculations and models suggested that there was enough free room for this process to occur. The crystal structure of the solid $[H_{10}([30]aneN_{10})][Pt(CN)_4]_5{\cdot}2H_2O$ showed again an array of hydrogen bonds involving the protonated macrocycle and the

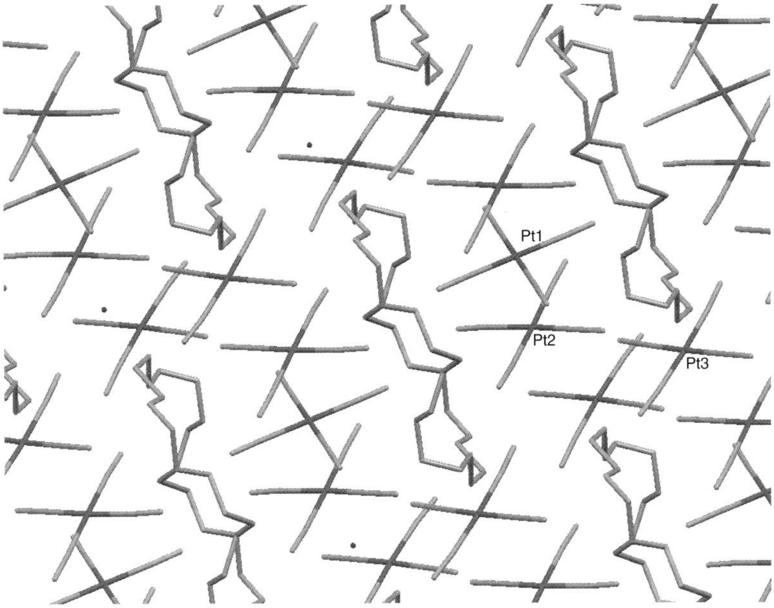

Figure 1.47 Detail of the crystal structure of $[H_{10}([30]aneN_{10})][Pt(CN)_4]_5 \cdot 2H_2O$ showing the three different $[Pt(CN)_4]^{2-}$ anions involved in hydrogen bonding with the decaprotonated receptor [194].

metallocyanide anions whose cyanide groups point directly toward the macrocyclic cavity (Figure 1.47).

Owing to the stronger and more inert character of the Pt-CN bond, it was possible to study these systems by means of pH-metric techniques detecting the formation of 1 : 1 adduct species. A noticeable increase in stability was observed in going from $[30]aneN_{10}$ (**70**) to the larger $[33]aneN_{11}$ (**71**), suggesting possible inclusion of the anion within the macrocyclic hole.

As it can be noticed, there is no agreement between the stoichiometries in the solid state and in solution. ^{195}Pt NMR studies in D_2O failed to give information about the formation of adduct species of higher nuclearity because of the precipitation of the polyammonium salts.

A closer relationship between the events coming up in solution and in the solid state occurs, however, in the system $PdCl_4{}^{2-} - H_{10}([30]aneN_{10})^{10+}$ [192, 193]. Solution studies for the different $H_k([3k]aneN_k)^{k+}$ systems were carried out by batch microcalorimetry in a 0.1 M HCl aqueous medium. The measurements had equilibration times within the timescale of the experiment for $H_6([18]aneN_6)^{6+}$, $H_7([21]aneN_7)^{7+}$, and $H_8([24]aneN_8)^{8+}$ becoming, however, much slower for the next three terms of the series $H_9([27]aneN_9)^{9+}$, $H_{10}([30]aneN_{10})^{10+}$, and $H_{11}([33]aneN_{11})^{11+}$ (Table 1.2). Although all the reactions show slightly exothermic enthalpy terms, the enthalpy changes for $(H_9[27]aneN_9)^{9+}$, $(H_{10}[30]aneN_{10})^{10+}$, and $H_{11}([33]aneN_{11})^{11+}$ are significantly more favorable than for the others. These

Table 1.2 Enthalpy terms and equilibration times for the reactions of fully protonated [3k]aneN$_k$ macrocycles and $PdCl_4^{2-}$ determined in 0.1 M HCl in a BATCH microcalorimeter [192, 193].

Reaction	$-\Delta H^\circ$ (kcal mol^{-1})	Time (min)
$H_6([18]aneN_6)^{6+} + PdCl_4{}^{2-}$	1.5(1)	20
$H_7([21]aneN_7)^{7+} + PdCl_4{}^{2-}$	1.5(1)	20
$H_8([24]aneN_8)^{8+} + PdCl_4{}^{2-}$	1.6(1)	20
$H_9([27]aneN_9)^{9+} + PdCl_4{}^{2-}$	2.9(1)	120
$H_{10}([30]aneN_{10})^{10+} + PdCl_4{}^{2-}$	3.9(1)	110
$H_{11}([33]aneN_{11})^{11+} + PdCl_4{}^{2-}$	3.1(1)	50

Figure 1.48 Drawing of the cation $[(PdCl_4)(H_{10}[30]aneN_{10})]^{8+}$. Hydrogen bonding is indicated with blue dotted lines. Hydrogen atoms are not shown [192, 193].

experimental evidences suggest that $H_{27}([27]aneN_9)^{9+}$ was the first term of the series for which inclusion of the $PdCl_4{}^{2-}$ within the macrocyclic cavity occurred. The slower equilibration time should account for conformational reorganizations following host–guest interaction.

The crystal structure of $[(PdCl_4)(H_{10}[30]aneN_{10})](PdCl_4)_2Cl_4$ (Figure 1.48) shows that one of the $PdCl_4{}^{2-}$ is included into the ellipsoidal cavity of the decaprotonated macrocycle along the shorter axes [192, 193]. The chlorine atoms lie outside of the macrocyclic framework, forming hydrogen bonds with the closest nitrogen atoms of the macrocycle. The outer $PdCl_4{}^{2-}$ anions also participate in hydrogen bonding with several ammonium groups of the receptors. Although, as previously mentioned, the shape of the macrocycle is largely defined by coulombic repulsions between protonated nitrogen atoms. Attractive charge–charge interactions with the anions and hydrogen bonding also play their role in determining the conformation of the protonated receptor. As a matter of fact, the conformations found in each of the structures discussed are somewhat different. Thus, it can be stated that the coarse fitting of the macrocyclic conformation is performed by the charge while hydrogen bonding plays a sort of fine tuning.

Finally, we would like to remark on the case of the inclusion of the very large anions $PW_{12}O_{40}{}^{3-}$, $SiW_{12}O_{40}{}^{4-}$, and $Os_{10}C(CO)_{24}{}^{2-}$ in the cavity of the porphyrin cyclic trimer (**106**) as established by fast atomic bombardment techniques mass spectrometry by Anderson and Sanders in 1992 [203]. The protonated trimer was found to bind giant clusters in a shape-dependent manner, suggesting the application of separating clusters by size using cyclic hosts of varying dimensions.

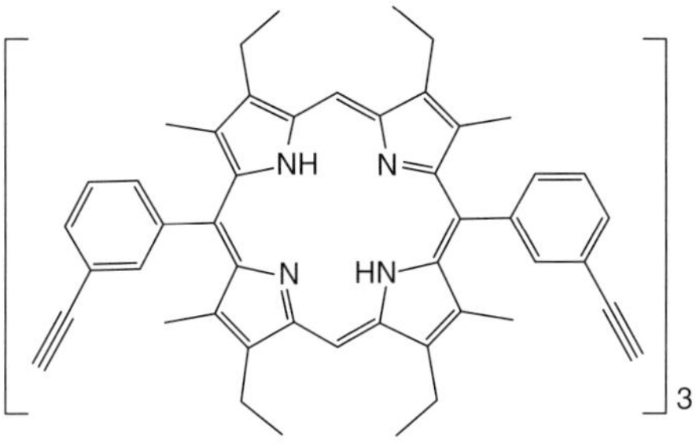

106

1.7 Nucleotides

The study of the interaction between polyammonium receptors and nucleotides meant a real breakthrough in supramolecular chemistry since it provided an early and definitive biological slant to the field. Organized interconversions among AMP, ADP, and ATP are important since the chemical energy supplied from outside is stored through the formation of highly energetic phosphate linkages. In turn, cleavage of such phosphate linkages provides energy sources for biosynthetic reactions that reverse catabolic pathways and drive active transport through cell membranes against electrochemical gradients.

Acyclic and, in particular, cyclic polyammonium salts from very early proved to be very strong nucleotide binders. In a study performed by Tabushi *et al.* [204], it was shown that the quaternized DABCO with stearyl chains (**107**) was able to transfer ADP very efficiently from an aqueous solution at pHs of 3 and 5 into a chloroform phase. The selectivity found for the transfer of ADP over AMP was explained by the formation of a double salt bridge between the phosphate groups and the quaternized receptor.

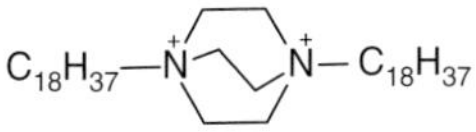

107

The Lehn and Kimura groups provided very initial evidences of the high interaction between polyammonium receptors with the nucleotides ATP, ADP, and AMP [58, 99]. Lehn's group, in a communication to *JACS* [58], gave stability constant data for the interaction of the hexaprotonated [24]aneN_6 (H_6(**83**)$^{6+}$), octaprotonated [32]aneN_8 (H_8(**84**)$^{8+}$), and hexaprotonated [27]aneN_6O_3 (H_6(**85**)$^{6+}$) with AMP^{2-}, ADP^{3-}, and ATP^{4-}. The values obtained by computer analysis of the pH-metric titration curves varied from 3.4 to 9.1 logarithmic units, following the sequence $AMP^{2-} < ADP^{3-} < ATP^{4-}$ for each of the three protonated receptors. Kimura *et al.* reported a full paper in *JACS* [99] in which, by means of polarographic techniques, they had explored the interaction with nucleotides of a series of polyazamacrocycles including [18]aneN_4 (**13**), [15]aneN_5 (**47**), [18]aneN_6 (**36**), the mixed amino-amide receptor (**33**), [16]aneN_5 (**87**), and a tetraazapyridinophane (**108**).

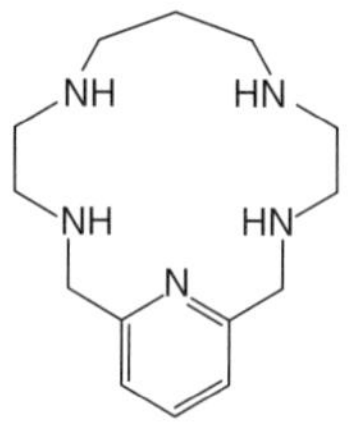

108

The polarographically determined association constants in general followed the order mentioned with higher values for the higher charged ATP anion. However, a reverted sequence seemed to be found for the interaction of diprotonated pyridinophane receptor **108** with the three nucleotides. It was soon realized that the length of the hydrocarbon chains between the nitrogen atoms in a monocyclic receptor is a key factor in determining its basicity and consequently its affinity for the anionic species. In this respect, Bianchi, Micheloni, and Paoletti reported that the tetraaza monocycle [20]aneN_4 (**14**), which has butylenic chains between the amino groups and is fully protonated at the physiological pH of 7.4, has a significant affinity for ATP^{4-} (H_4(**14**)$^{4+}$ + ATP^{4-} = H_4(**14**)·ATP, $\log K_s = 3.81(3)$) [205]. Related to this point, Burrows prepared a chiral [18]aneN_4 derivative including an alcohol C-linked side chain, which was proved to interact with ATP by ^{31}P NMR [206]. Bis(macrocyclic) polyamines **65** and **66** also showed interesting complexing properties toward nucleotide species [164].

Another strategy in nucleotide binding was based on using tetraazacycloalkanes with pendant arms containing additional ammonium groups such as in receptor **109** [207].

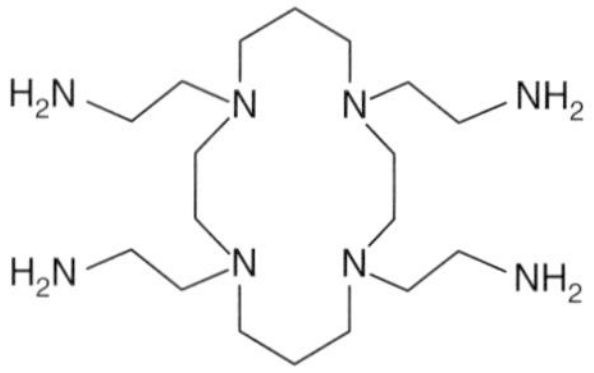

109

However, the aromatic rings in the nucleotide structure provide a very important motif for their binding through stacking interactions. With this purpose, Lehn and coworkers [208] prepared receptor **110**, in which acridine moieties were added to the classical receptor O-BISDIEN to take advantage from electrostatic attractions, hydrogen bonding, and stacking interactions.

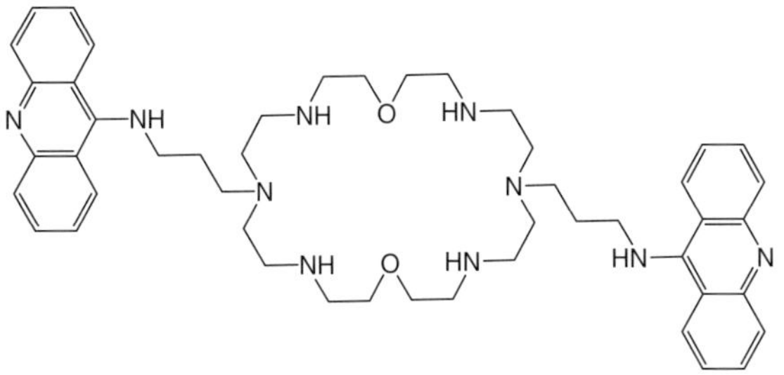

110

Lehn and coworkers also prepared a series of bis-intercaland compounds and their reference mono-intercaland partners to see the affinity enhancement provided by this double binding motif [209] (**111–120**).

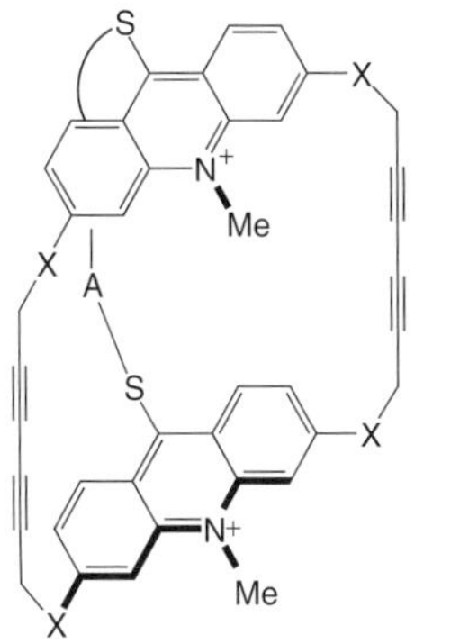

111, X= O, A = $-(CH_2)_6-$
112, X = O, A = $-(CH_2)_2-O-(CH_2)_2-$
113, X = O, A = $-(CH_2)_2-O-(CH_2)_2-O-(CH_2)_2-$
114, X = NH, A = $-(CH_2)_2-O-(CH_2)_2-$

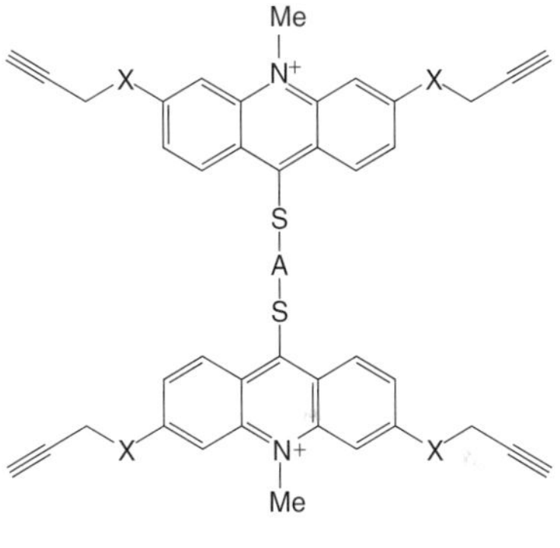

115, X= O, A = $-(CH_2)_6-$
116, X = O, A = $-(CH_2)_2-O-(CH_2)_2-$
117, X = O, A = $-(CH_2)_2-O-(CH_2)_2-O-(CH_2)_2-$
118, X = NH, A = $-(CH_2)_2-O-(CH_2)_2-$

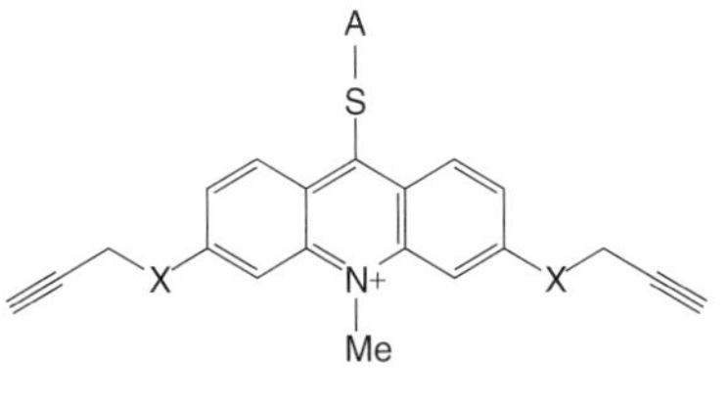

119, X= O, A = $-(CH_2)_2-CH_3$
120, X = NH, A = $-(CH_2)_2-CH_3$

Evidence for intercalation was obtained from the hypochromism observed in the UV–vis spectra of the nucleotide-ligand systems. While differences in stability were obtained for the different nucleic bases, the constants obtained for nucleotides containing the same base pair and different phosphate chains were practically the same. For instance, the stability constants for the interaction of AMP^{2-}, ADP^{3-}, and ATP^{4-} with **112** were 3.79, 3.84, and 3.91 logarithmic units, respectively.

The importance of π-stacking in nucleotide binding is even more evident in the case of receptors **121–124**. For example, the equilibrium constants for the binding of charged ATP^{4-}, ADP^{3-}, and AMP^{2-} are $\log K_s = 5.80$, 5.65, and 5.38, respectively [210] (**121–124**).

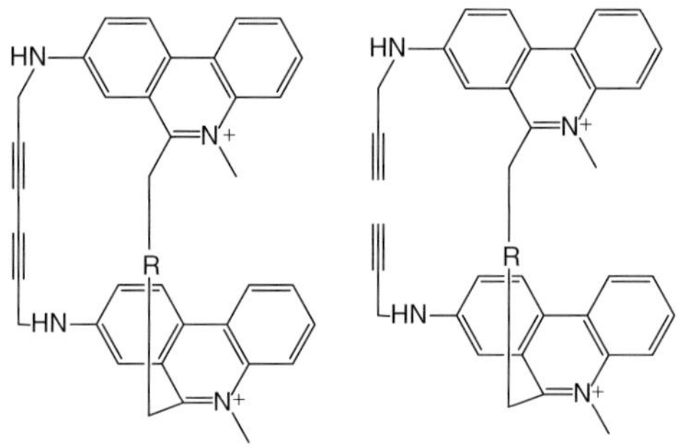

121, R = $(CH_2)_4$
122, R = $(CH_2)_6$
123, R = p-C_6H_4

124, R = $(CH_2)_6$

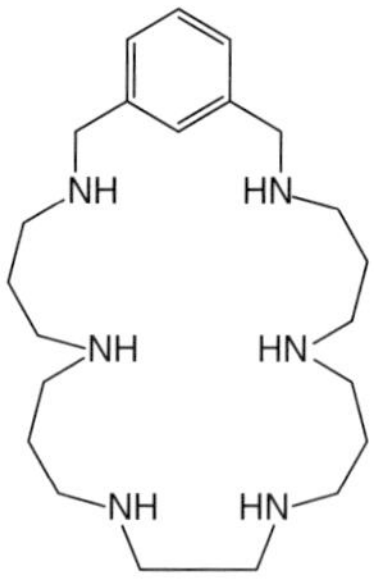

125

As was evidenced for **125** by García-España, Luis, and coworkers using ^1H NMR and potentiometric techniques, monocyclic azacyclophanes with appropriate polyamine chains can behave as multipoint binders of nucleotides. While the phosphate chain is a good electrostatic binding point and a hydrogen bond acceptor, the nucleoside part operates as an adequate site for stacking with the aromatic part of the macrocycle [211, 212]. Recently, Bianchi *et al.* [213] have reported a crystal structure for the interaction of the terpyridinophane macrocycle **126** with thymidine-5′-triphosphate (TTP) that assumes many of the binding mechanisms [214] of nucleotides by polyazamacrocycles containing aromatic functions

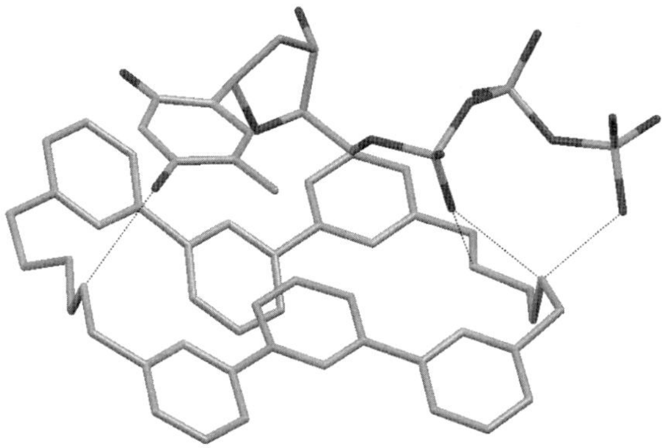

Figure 1.49 View of the crystal structure of complex [(H_4(**126**))HTTP] showing some of the intermolecular hydrogen bonds between the partners.

(Figure 1.49). This structure is discussed in more detail in Chapter 2, devoted to energetic aspects of anion binding.

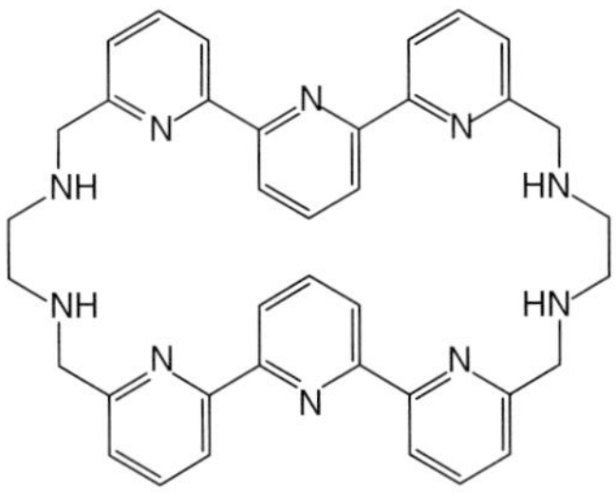

126

ATP and AMP binding to guanidinium monolayer formed by **127** was studied by X-ray photoelectron spectroscopy of Langmuir–Blodget films of the monolayer drawn from substrate-laden aqueous subphases. Binding constants of 1.7×10^7 and $3.2 \times 10^6\ M^{-1}$ were obtained for ATP and AMP, respectively. The respective site occupations were 0.34 and 0.95.

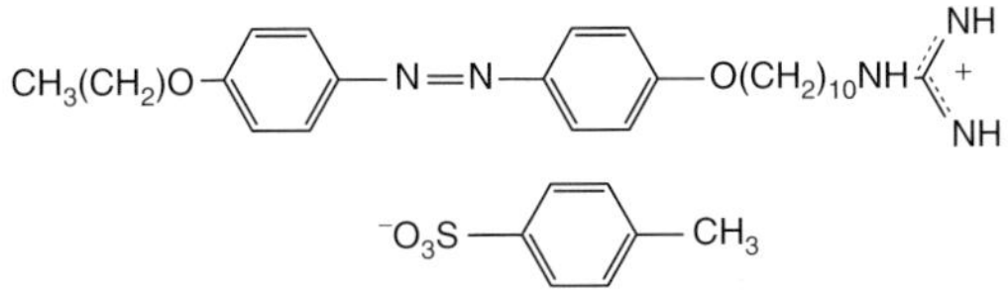

127

Since macrocycle **37** was found to be one of the most powerful complexing agents for ATP [58, 215], it was attached to polysterene beads. The modified polymer was able to take up nucleotides, in particular ATP, at neutral and acidic pHs and release them back into the solution at basic pH values.

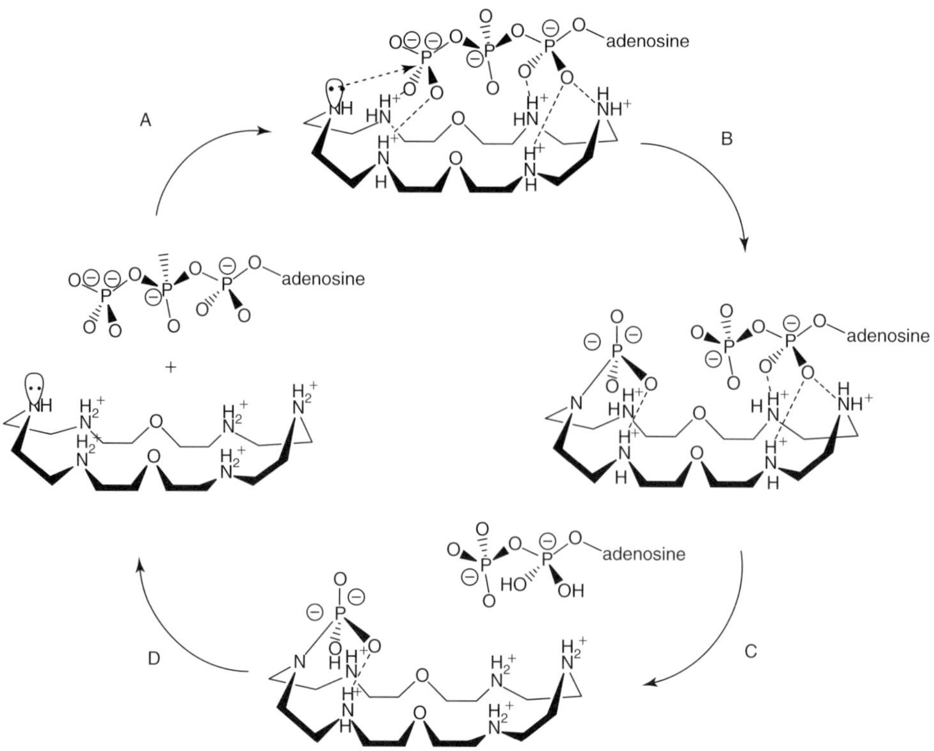

Figure 1.50 Catalytic cycle for ATP in the presence of **37** at neutral pH.

However, the most interesting point of this chemistry concerns the capacity that some polyammonium receptors have for inducing catalytic processes mimicking the behavior of ATPases or kinases. The 24-membered macrocyclic receptor **37** ranks among the abiotic receptors producing highest enhancements of ATP cleavage into ADP and the so-called inorganic phosphate [58, 215–220]. Activation of ATP-cleavage with **37** was observed not only at an acidic pH but also at neutral pH. Under neutral conditions, the hydrolysis of ATP by this receptor was enhanced by a factor of 100. The ^{31}P NMR monitoring of the course of the reaction revealed the formation of a transient phosphorylated species, which appeared in the ^{31}P NMR spectrum as a singlet downfield shift at 10 ppm with respect to an external H_3PO_4 reference. The hydrolysis of ATP in the presence of protonated **36** proved to be a true catalytic process since for a 10-fold excess of ATP with respect to **36**, the change in the ATP concentration was linear with time in the early period. The products of the reaction of ADP and inorganic phosphate did not interfere with the initial course of the reaction because of their reduced affinity for the protonated receptor.

The authors of this work proposed for ATP hydrolysis catalyzed by **37** the classical mechanism depicted in Figure 1.50. One of the central amines of the bridges is not protonated at neutral pH and performs the nucleophilic attack on the terminal phosphorous atom (P_γ) of ATP to form the phosphoramidate transient species (steps A and B in Figure 1.50). The electrostatic and hydrogen bonding interactions of the phosphate chain of the nucleotide with the protonated macrocycle places the γ-phosphorous atom and the nonprotonated amine of the partners in a proper position for facilitating the nucleophilic attack. The final steps of the catalytic cycle are the dissociation of the ADP complex and the hydrolysis of the phosphoramidate, not necessarily in this order (steps C and D).

Studies of ATP cleavage conducted with the series of azacycloalkanes [3k]aneN_k($k = 6-12$) (**36**, **67–72**) revealed the key role played by the macrocycle size [165]. The 21-membered macrocycle [21]aneN_7 (**67**) was shown to be the best catalyst at the examined pHs, with the [24]aneN_8 (**68**) macrocycle providing also very significant rate enhancements. The relevance of macrocyclic size and charge density on rate of catalysis was later checked by introducing structural modifications such as N-methylation of some of the nitrogen atoms in the macrocycles [18]aneN_6 (**36**) and [21]aneN_7 (**67**) [221, 222] to yield macrocycles (**128–130**).

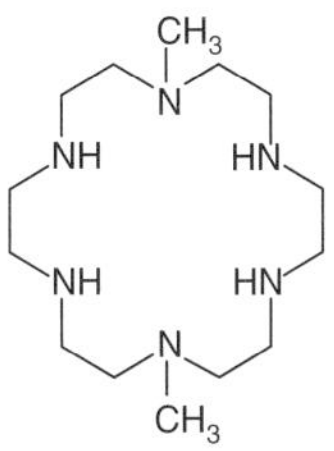
128

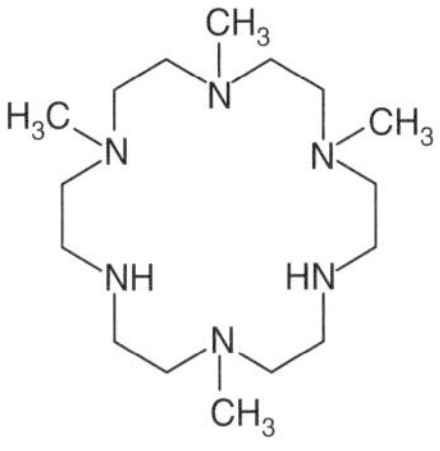
129

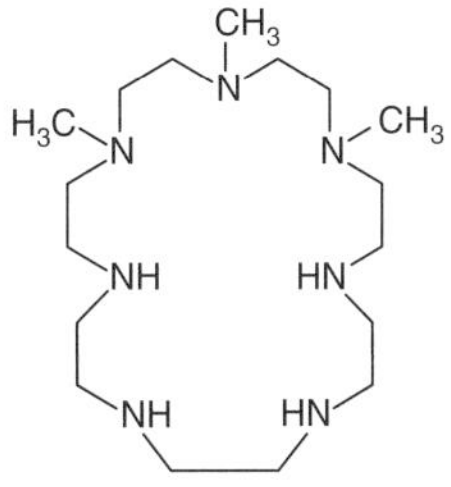
130

While dimethylated **128** leads to a decrease in the rate of the hydrolytic cleavage of ATP with respect to unmethylated **36**, the tetramethylated **129** produces a significant rate enhancement. On the other hand, trimethylation of [21]aneN_7 (**67**) to give Me_3[21]aneN_7 (**130**) produces a diminution in the rate of the process. Taking into account that methyl groups donate electron density to electronegative atoms and should increase the nucleophilicity of the nitrogens, the slower rate of cleavage must be tightly related to the alteration of the optimal cavity size brought about by the functionalization. However, solvation effects need to be always taken into account in these types of considerations.

Additional evidences supporting the key role played by size cavity in ATP hydrolysis were provided by García-España and coworkers for the series of cyclophane receptors **131–133**.

The critical size required is manifested in the fact that both the ortho (**131**) and para isomers (**133**), containing 20- and 22-membered cavities, yield much poorer ATP cleavage rate enhancements than the meta isomer (**132**) with a 21-membered cavity. This lack of activity is particularly noticeable in the case of the ortho derivative.

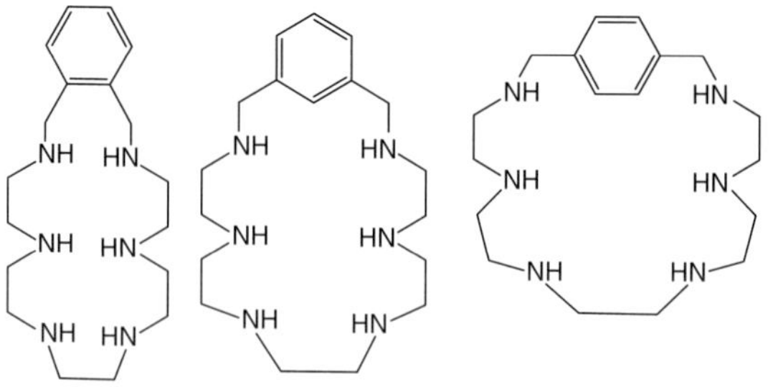

131 132 133

^{31}P NMR studies revealed that **132** produced rate enhancements comparable to those of O-BISDIEN (**37**) and slightly lower than those of [21]aneN_7 (**67**), confirming the critical role of size in the catalytic processes (Figure 1.51) [223]. Interestingly enough, the reaction using **131** as a catalyst is not only efficient but also very specific since it stops in the formation of ADP and does not proceed significantly further. Molecular dynamics studies suggest that the optimal size is one at which just the γ-phosphate of ATP perfectly resides at the macrocyclic cavity. Figure 1.52 shows the perfect fitting between the phosphate group and the macrocyclic cavity of **132**.

O-BISDIEN (**37**) exhibited other characteristics of the naturally occurring enzymes, such as the ability to modify the course of reaction and phosphoryl transfer capabilities to other substrates. Protonated **37** was shown to catalyze acetylphosphate hydrolysis and pyrophosphate formation through formation of a

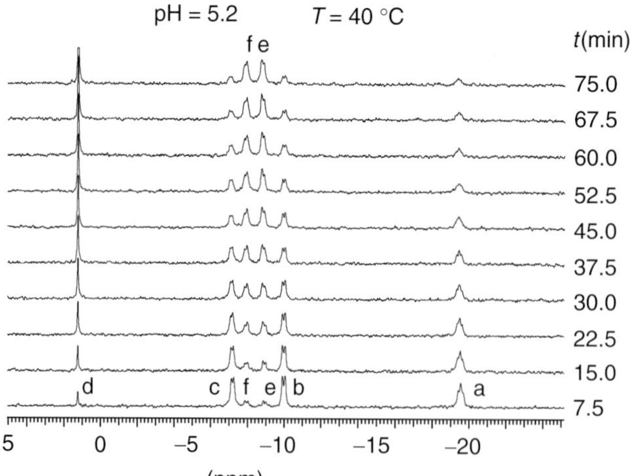

Figure 1.51 Time evolution for ^{31}P NMR spectra of solutions containing ATP and **132** in 10^{-2} M at 40 °C and pH = 5.2. (a) P_β (ATP); (b) P_α(ATP); (c) P_γ(ATP); (d) Pi; (e) P_α(ADP); and (f) P_β(ADP) [222]. Reproduced with permission from the Royal Society of Chemistry.

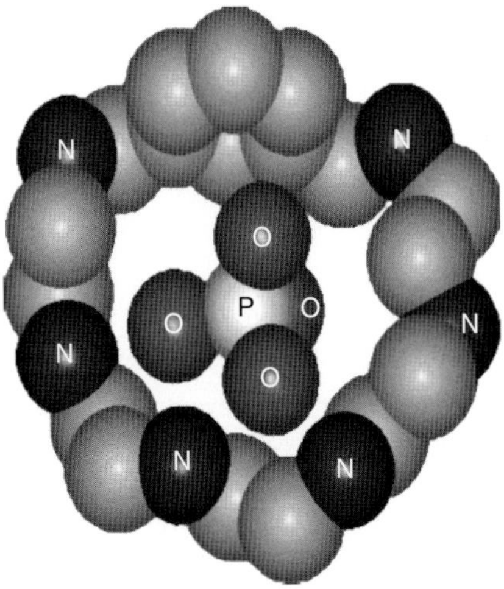

Figure 1.52 Comparison between the size of the macrocyclic cavity of **132** and the γ-phosphate of ATP. The rest of ATP has been obscured for clarity.

phosphorylated macrocycle intermediate, which transfers a phosphoryl group to a phosphate substrate [224]. ATP phosphotransferases, hydrolases, and synthetases, often require mono- or divalent anions in order to carry out their functions. Therefore, to generate systems that resembled even more closely their biological counterpart, protonated **37** was coupled to metal ions [225, 226]. The investigations of the effect of the biologically significant metal ions Ca^{2+}, Mg^{2+}, and Zn^{2+} added to **37** revealed striking influences of the metal ions on ATP hydrolysis and on the formation of the phosphoramidate intermediate and pyrophosphate. While addition of both Ca^{2+} and Mg^{2+} increased the observed percentage of phosphoramidate, only Ca^{2+} provided a significant acceleration in ATP hydrolysis, almost doubling the first order rate constant found for free **37** in the same experimental conditions. Ln(III) also led to a considerable increase in the rate of ATP dephosphorylation. The presence of Mg^{2+} had no apparent effect on the catalytic rate of ATP cleavage, while Zn^{2+} or Cd^{2+} addition to macrocycle-ATP solutions decreased the hydrolytic rate. The most striking finding in this respect was the ready appearance of pyrophosphate in the presence of **37** and either Mg^{2+}, Ca^{2+}, or Ln^{3+} at pH 4.5. No pyrophosphate formation was observed in the absence of metals under these experimental conditions. The pyrophosphate was determined as being formed from nucleophilic attack of inorganic phosphate in solution on the phosphoramidate, aided by metal ion chelation.

Mertes *et al.* indicated that macrocycle **37** was able to activate formate in the presence of ATP and Ca^{2+} or Mg^{2+} ions, yielding as a final product the macrocycle formylated at the central nitrogen of the chain [227]. This finding may represent the first example of a multistep catalytic pathway assisted by a simple macrocyclic host. Furthermore, it resembles what has been proposed to be the pathway followed by N^{10}-formyltetrahydrofolate synthetase in the formylation of the N^{10} nitrogen atom in tetrahydrofolate. The catalytic sequence is presumed to proceed through a formyl phosphate intermediate with the assistance of ATP. The ability of the phosphoramidate intermediate to participate in other phosphorylation reactions was further confirmed when Hosseini and Lehn proved that **37** in the presence of Mg^{2+} as promoter catalyzed the generation of ATP from acetyl phosphate and ADP in dilute aqueous solution at neutral pH [228]. A more detailed description of the evolution of the catalytic capabilities of the simple 24-membered macrocycle can be found in a 2004 review [229].

1.8 Final Notes

In this chapter, we have provided a descriptive overview of the events that encompassed the birth of the field of anion coordination chemistry along with more recent flashes related to such initial research. However, we would like to apologize for any oversight we may have committed. Since these initial findings, and particularly since the appearance of our previous book in 1997, the anion supramolecular chemistry field has experienced a tremendous explosion. Many new research groups have focused their work in this conceptually new area, and many new receptors have been reported in the last two decades. Hence, as noted at the beginning of this chapter, we have covered only a selected group of the seminal findings, primarily limited to ammonium, amide, and guanidinium donor groups. Furthermore, the wide variety of these new hosts is nothing short of astounding, with many containing within their framewords binding motifs that allow for recognizing anions in low-dielectric media trying to mimic the clefts or crevices where anion receptors often occur in nature.

Also, the search for biomedical and environmental applications of anion receptors has gained great impetus. Analytical chemistry has also gained great interest in relation to anion detection and quantification, with novel molecules and devices that permit the identification of ever increasingly low amounts of either pollutants or biomedical relevant anionic species. Aspects related to biomedicine and to the importance that anion regulation have in metabolic roles have also become apparent in these years.

The following chapters address new approaches to anion coordination chemistry in relation to the synthesis of receptors, biological receptors, and metalloreceptors; energetics of anion binding; molecular structures of anionic complexes; sensing devices; and computational studies addressed to understand the different driving forces responsible for anionic complexation. We hope these chapters will provide the reader an actual picture of the state of the art in many aspects of anion coordination. In short, the future is promising for this exciting and constantly evolving field of supramolecular anion coordination chemistry.

References

1. Pedersen, C.J. (1967) Cyclic polyethers and their complexes with metal salts. *J. Am. Chem. Soc.*, **89**, 7017–7036.
2. Dietrich, B., Lehn, J.-M., and Sauvage, J.-P. (1969) Les cryptates. *Tetrahedron Lett.*, **10**, 2889–2992.

3. Dietrich, B., Lehn, J.-M., Sauvage, J.-P., and Blanzat, J. (1973) Cryptates-X: syntheses et proprietes physiques de systemes diaza-polyoxa-macrobicycliques. *Tetrahedron*, **29**, 1629–1645.
4. Cram, D.J. and Cram, J.M. (1974) Host-guest chemistry: complexes between organic compounds simulate the substrate selectivity of enzymes. *Science*, **183**, 803–809.
5. Cram, D.J. and Cram, J.M. (1978) Design of complexes between synthetic hosts and organic guests. *Acc. Chem. Res.*, **11**, 8–14.
6. Lehn, J.-M. (1978) Cryptates: the chemistry of macropolycyclic inclusion complexes. *Acc. Chem. Res.*, **11**, 49–57.
7. Lehn, J.-M. (1988) Supramolecular chemistry. Scope and perspectives. Molecules, supermolecules, and molecular devices (Nobel lecture). *Angew. Chem. Int. Ed. Engl.*, **27**, 89–112.
8. Lehn, J.-M. (1995) *Supramolecular Chemistry. Concepts and Perspectives*, Wiley-VCH Verlag GmbH, Weinheim.
9. Bianchi, A., Bowman-James, K., and García-España, E. (eds) (1997) *Supramolecular Chemistry of Anions*, Wiley-VCH Verlag GmbH, New York.
10. Sessler, J.L., Gale, P.A., and Cho, V.S. (2006) *Anion Receptor Chemistry*, Royal Society of Chemistry, Cambridge.
11. Stibor, I. (ed.) (2005) *Anion Sensing. Topics in Current Chemistry*, Springer, Berlin, Heidelberg, New York.
12. Bianchi, A., Micheloni, M., and Paoletti, P. (1990) Thermodynamic aspects of the polyazacycloalkane complexes with cations and anions. *Coord. Chem. Rev.*, **110**, 17–113.
13. Dietrich, B. (1993) Design of anion receptors: applications. *Pure Appl. Chem.*, **65**, 1457–1464.
14. Izatt, R.M., Pawlak, K., and Bradshaw, J.S. (1995) Thermodynamic and kinetic data for macrocycle interaction with cations, anions and neutral molecules. *Chem. Rev.*, **95**, 2529–2586.
15. Beer, P.D., Gale, P.A., and Chen, G.Z. (1999) Mechanisms of electrochemical recognition of cations, anions and neutral guest species by redox-active receptor molecules. *Coord. Chem, Rev.*, **185–186**, 3–36.
16. Beer, P.D. and Cadman, J. (2000) Electrochemical and optical sensing of anions by transition metal based receptors. *Coord. Chem. Rev.*, **205**, 131–155.
17. Gale, P.A. (2000) Anion coordination and anion-directed assembly: highlights from 1997 and 1998. *Coord. Chem. Rev.*, **199**, 181–233.
18. Beer, P.D. and Gale, P.A. (2001) Anion recognition and sensing: the state of the art and future perspectives. *Angew. Chem. Int. Ed.*, **40**, 486–516.
19. Amendola, V., Fabbrizzi, L., Mangano, C., Pallavicini, P., Poggi, A., and Taglietti, A. (2001) Anion recognition by dimetallic cryptates. *Coord. Chem. Rev.*, **219–221**, 821–837.
20. Llinares, J.M., Powell, D., and Bowman-James, K. (2003) Ammonium based anion receptors. *Coord. Chem. Rev.*, **240**, 57–75.
21. McKee, V., Nelson, J., and Town, R.M. (2003) Caged oxoanions. *Chem. Soc. Rev.*, **32**, 309–325.
22. Bondy, C.R. and Loeb, S.J. (2003) Amide based receptors for anions. *Coord. Chem. Rev.*, **240**, 77–99.
23. Choi, K. and Hamilton, A.D. (2003) Macrocyclic anion receptors based on directed hydrogen bonding interactions. *Coord. Chem. Rev.*, **240**, 101–110.
24. Bowman-James, K. (2005) Alfred Werner revisited: the coordination chemistry of anions. *Acc. Chem. Res.*, **38**, 671–678.
25. Kubik, S., Reyheller, C., and Stüwe, S. (2005) Recognition of anions by synthetic receptors in aqueous solution. *J. Inclusion Phenom. Macrocyclic Chem.*, **52**, 137–187.
26. Hossain, M.A., Kang, S.O., and Bowman-James, K. (2005) in *Current Trends and Future Perspectives* (ed. K. Gloe), Springer, Dordrecht, The Netherlands, pp. 178–183.
27. Gale, P.A. (2006) Structural and molecular recognition studies with acyclic anion receptors. *Acc. Chem. Res.*, **39**, 465–475.
28. García-España, E., Díaz, P., Llinares, J.M., and Bianchi, A. (2006) Anion

coordination chemistry in aqueous solution of polyammonium receptors. *Coord. Chem. Rev.*, **259**, 2952–2986.
29. Kang, S.O., Begum, R.A., and Bowman-James, K. (2006) Amide-based ligands for anion coordination. *Angew. Chem. Int. Ed.*, **45**, 7882–7894.
30. Kang, S.O., Hossain, M.A., and Bowman-James, K. (2006) Influence of dimensionality and charge on anion binding in amide-based macrocyclic receptors. *Coord. Chem. Rev.*, **250**, 3038–3052.
31. Amendola, V., Bonizzoni, M., Esteban-Gómez, D., Fabbrizzi, L., Licchelli, M., Sancenón, F., and Taglietti, A. (2006) Some guidelines for the design of anion receptors. *Coord. Chem. Rev.*, **250**, 1451–1470.
32. Wichmann, K., Antonioli, B., Söhnel, T., Wenzel, M., Gloe, K., Price, J.R., Lindoy, L.F., Blake, A.J., and Schröder, M. (2006) Polyamine-based anion receptors: extraction and structural studies. *Coord. Chem. Rev.*, **250**, 2987–3003.
33. Hirsch, A.K.H., Fischer, F.R., and Diederich, F. (2007) Phosphate recognition in structural biology. *Angew. Chem. Int. Ed.*, **46**, 338–352.
34. Gokel, G.W. and Carasel, I.A. (2007) Biologically active, synthetic ion transporters. *Chem. Soc. Rev.*, **36**, 378–389.
35. Chmielewski, M.J., Zieliski, T., and Jurczak, J. (2007) Synthesis, structure, and complexing properties of macrocyclic receptors for anions. *Pure Appl. Chem.*, **79**, 1087–1096.
36. Custelcean, R. and Moyer, B.A. (2007) Anion separation with metal–organic frameworks. *Eur. J. Inorg. Chem.*, 1321–1340.
37. Blondeau, P., Segura, M., Pérez-Fernández, R., and de Mendoza, J. (2007) Molecular recognition of oxoanions based on guanidinium receptors. *Chem. Soc. Rev.*, **36**, 198–210.
38. Gale, P.A., García-Garrido, S.E., and Garric, J. (2008) Anion receptors based on organic frameworks: highlights from 2005 and 2006. *Chem. Soc. Rev.*, **37**, 151–190.
39. Gale, P.A. (2008) Synthetic indole, carbazole, biindole and indolocarbazole-based receptors: applications in anion complexation and sensing. *Chem. Commun.*, 4525–4540.
40. Alfonso, I. (2008) Recent developments in chiral polynitrogenated synthetic receptors for anions. *Mini-Rev. Org. Chem.*, **5**, 33–46.
41. Lee, C.-H., Miyaji, H., Yoon, D.-W., and Sessler, J.L. (2008) Strapped and other topographically nonplanar calixpyrrole analogues. Improved anion receptors. *Chem. Commun.*, 24–34.
42. Itsikson, N.A., Morzherin, Y.Y., Matern, A.I., and Chupakhin, O.N. (2008) Receptors for anions. *Russ. Chem. Rev.*, **77**, 751–764.
43. Prados, P. and Quesada, R. (2008) Recent advances in macrocyclic and macrocyclic-based anion receptors. *Supramol. Chem.*, **20**, 201–216.
44. Voloshin, Y.Z. and Belov, A.S. (2008) Encapsulation of organic and inorganic anions: synthesis of macropolycyclic ligands and their anion-receptor properties. *Russ.Chem. Rev.*, **77**, 161–175.
45. Nabeshima, T. and Akine, S. (2008) Functional supramolecular systems with highly cooperative and responding properties. *Chem. Record.*, **8**, 240–251.
46. Steed, J.W. (2009) Coordination and organometallic compounds as anion receptors and sensors. *Chem. Soc. Rev.*, **38**, 506–519.
47. Amendola, V.L., and Fabbrizzi, L. (2009) Anion receptors that contain metals as structural units. *Chem. Commun.*, 513–531.
48. Kubik, S. (2009) Amino acid containing anion receptors. *Chem. Soc. Rev.*, **38**, 585–605.
49. Katayev, E.A., Kolesnikov, G.V., and Sessler, J.L. (2009) Molecular recognition of pertechnetate and perrhenate. *Chem. Soc. Rev.*, **38**, 1572–1586.
50. Caltagirone, C. and Gale, P.A. (2009) Anion receptor chemistry: highlights from 2007. *Chem. Soc. Rev.*, **38**, 520–563.
51. Hua, Y. and Flood, A.H. (2010) Click chemistry generates privileged CH hydrogen-bonding triazoles the latest addition to anion supramolecular chemistry. *Chem. Soc. Rev.*, **39**, 1262–1271.

52. Mateus, P., Bernier, N., and Delgado, R. (2010) Recognition of anions by polyammonium macrocyclic and cryptand receptors: influence of the dimensionality on the binding behavior. *Coord. Chem. Rev.*, **254**, 1726–1747.
53. Gale, P.A. and Gunnlaugson, T. (eds) (2010) Themed issue: supramolecular chemistry of anionic species. *Chem. Soc. Rev.*, **39** (10), 3581–4008.
54. Park, C.H. and Simmons, H.E. (1968) Macrobicyclic amines. III. Encapsulation of halide ions by in, in-1, (k + 2)-diazabicyclo[k.l.m]alkane- ammonium ions. *J. Am. Chem. Soc.*, **90**, 2431–2432.
55. Bell, R.A., Christoph, G.G., Fronczek, F.R., and Marsh, R.E. (1975) The cation $H_{13}O_6^+$: a short symmetric hydrogen bond. *Science*, **190**, 151–152.
56. Graf, E. and Lehn, J.-M. (1976) Anion cryptates: highly stable and selective macrotricyclic anion inclusion complexes. *J. Am. Chem. Soc.*, **98**, 6403–6405.
57. Lehn, J.-M., Sonveaux, E., and Willard, A.K. (1978) Molecular recognition. Anion cryptates of a macrobicyclic receptor molecule for linear triatomic species. *J. Am. Chem. Soc.*, **100**, 4914–4916.
58. Dietrich, B., Hosseini, M.W., Lehn, J.-M., and Sessions, R.B. (1981) Anion receptor molecules. Synthesis and anion binding properties of polyammonium macrocycles. *J. Am. Chem. Soc.*, **103**, 1282–1283.
59. Peter, F., Gross, M., Hosseini, M.W., Lehn, J.-M., and Sessions, R.B. (1981) Redox properties and stability constants of anion complexes. An electrochemical study of the complexation of metal hexacyanide anions by polyammonium macrocyclic receptor molecules. *J. Chem. Soc., Chem. Commun.*, 1067–1069.
60. Hosseini, M.W. and Lehn, J.-M. (1982) Anion receptor molecules. Chain length dependent selective binding of organic and biological dicarboxylate anions by ditopic polyammonium macrocycles. *J. Am. Chem. Soc.*, **104**, 3525–3527.
61. Kintzinger, J.-P., Lehn, J.-M., Kauffman, E., Dye, J.L., and Popov, A.I. (1983) Anion coordination chemistry. ^{35}Cl NMR studies of chloride anion cryptates. *J. Am. Chem. Soc.*, **105**, 7549–7553.
62. Peter, F., Gross, M., Hosseini, M.W., and Lehn, J.-M. (1983) Redox properties of metalhexacyanide anions complexed by macrocyclic polyammonium receptors. *J. Electroanal. Chem.*, **144**, 279–292.
63. Bjerrum, N. and Ludwig, E. (1925) Mixtures of strong electrolytes. *K. Danske Videnskab. Selskab. Math. Fys. Medd.*, **6**, 3–20.
64. Amico, P., Daniele, P.G., Rigano, C., and Sammartano, S. (1981) Formation and stability of ammonium-sulphate, phosphate, oxalate and citrate complexes in aqueous solution. *Ann. Chim. (Rome)*, **71**, 659.
65. De Robertis, A., De Stefano, C., Sammartano, S., and Scarcella, R. (1985) Formation and stability of some dicarboxylate-NH_4^+ complexes in aqueous solution. *J. Chem. Res.*, (Suppl.), 322–325.
66. Motekaitis, R.J., Martell, A.E., Lehn, J.-M., and Watanabe, E.I. (1982) Bis(2,2',2''-triaminotriethylamine) cryptates of cobalt(II), nickel(II), copper(II), and zinc(II). Protonation constants, formation constants, and hydroxo bridging. *Inorg. Chem.*, **21**, 4253–4257.
67. Hofmeister, F. (1888) Zur lehre von der wirkung der salze. *Arch. Exp. Pathol. Pharmacol.*, **24**, 247–260.
68. Zhang, Y. and Cremer, P.S. (2006) Interactions between macromolecules and ions: the Hofmeister series. *Curr. Opin. Chem. Biol.*, **10**, 658–653.
69. Kropman, M.F. and Bakker, H.J. (2001) Dynamics of water molecules in aqueous salvation shells. *Science*, **219** (5511), 2118–2120.
70. Batchelor, J.D., Olteanu, A., Triphaty, A., and Pielak, G.J. (2004) Impact of protein denaturants and stabilizers on water structure. *J. Am. Chem. Soc.*, **126**, 1958–1961.
71. Sachs, J.N. and Woolf, T.B. (2003) Understanding the Hofmeister effect in

interactions between chaotropic anions and lipid bilayers: molecular dynamics simulations. *J. Am. Chem. Soc.*, **125**, 8742–8743.
72. Jungwirth, P. and Tobias, D.J. (2006) Specific ion effects in the air/water interface. *Chem. Rev.*, **106**, 1259–1281.
73. Klotz, I.M., Walker, F.M., and Pivan, R.M. (1946) The binding of organic ions by proteins. *J. Am. Chem. Soc.*, **68**, 1486–1490.
74. Colvin, J.R. (1952) The binding of anions by lysozyme, calf thymus histone sulphate and protamine sulphate. *Can. J. Chem.*, **30**, 320–331.
75. Luecke, H. and Quiocho, F.A. (1990) High specificity of a phosphate transport protein determined by hydrogen bonds. *Nature*, **347**, 402–406.
76. Jacobson, B.L. and Quiocho, F.A. (1988) Sulfate-binding protein dislikes protonated oxyacids: a molecular explanation. *J. Mol. Biol.*, **204**, 783–787.
77. Verschueren, K.H.G., Seljée, F., Rozeboom, H.J., Kalk, K.H., and Dijkstra, B.W. (1993) Crystallographic analysis of the catalytic mechanism of haloalkane dehalogenase. *Nature*, **363**, 693–698.
78. Yang, X., Knobler, C.B., and Hawthorne, M.F. (1992) Macrocyclic Lewis acid host-halide ion guest species. Complexes of iodide ion. *J. Am. Chem. Soc.*, **114**, 380–382.
79. Zhenhg, Z., Knobler, C.B., Mortimer, M.D., Kong, G., and Hawthorne, M.F. (1996) Hydrocarbon-soluble mercuracarborands: synthesis, halide complexes and supramolecular chemistry. *Inorg. Chem.*, **35**, 1235–1243.
80. Wedge, T.J. and Hawthorne, M.F. (2003) Multidentate carborane-containing Lewis acids and their chemistry: mercuracarborands. *Coord. Chem. Rev.*, **240**, 111–128.
81. Newcomb, M., Blanda, M.T., Azuma, Y., and Delord, T.J. (1984) Macrocycles containing tin. Synthesis and structure of 1,10-diphenyl-1,10-distannabicyclo[8.8.8] hexacosane. *J. Chem. Soc. Chem. Commun.*, 1159–1160.
82. Newcomb, M., Horner, J.H., and Blanda, M.T. (1987) Macrocycles containing tin. Through space cooperative binding and high size selectivity in the complexation of chloride ion by Lewis acidic macrobicyclic hosts. *J. Am. Chem. Soc.*, **109**, 7878–7879.
83. Newcomb, M., Horner, J.H., Blanda, M.T., and Squatrito, P.J. (1989) Macrocycles containing tin. Solid complexes of anions encrypted in macrobicyclic Lewis acidic hosts. *J. Am. Chem. Soc.*, **111**, 6294–6301.
84. Katz, H.E. (1980) Hydride sponge: complexation of 1,8-naphthalenediylbis(dimethylborane) with hydride, fluoride, and hydroxide. *J. Org. Chem.*, **50**, 5027–5032.
85. Katz, H.E. (1986) Anion complexation and migration in (8-sylyl-1-naphthyl)boranes. Participation of hypervalent silicon. *J. Am. Chem. Soc.*, **108**, 7640–7645.
86. Katz, H.E. (1987) 1,8-Naphthalenediylbis(dichloroborane) chloride: the first bis boron chloride chelate. *Organometallics*, **6**, 1134–1136.
87. Jung, M.E. and Xia, H. (1988) Synthesis and transport properties of 12-silacrown-3, a new type of anion complexing agent. *Tetrahedron Lett.*, **29**, 297–300.
88. Aoyagi, S., Tanaka, K., and Takeuchi, Y. (1994) 1,8-Dimethyl-1,8-dihalo-1,8-digermacyclotetradecanes. The first germamacrocycles with anion transport capability. *J. Chem. Soc. Perkin Trans. 2*, 1549–1543.
89. Jurkschat, K., Kuivila, H.G., Liu, S., and Zubieta, J.A. (1989) 1,5,9-Tristannacyclododecanes as Lewis acids. Novel structure of a chloride complex. *Organometallics*, **8**, 2755–2759.
90. Jurkschat, K., Rühlemann, A., and Tzschach, A. (1990) Synthese und transporteigenschaften von 1,1,6,6,11,11-hexamethyl-1,6,11-tristannacyclopentadecan. *J. Organomet. Chem.*, **381**, C53–C56.
91. Fujita, M., Nagao, S., and Ogura., K. (1995) Guest-induced organization of a three-dimensional palladium gage-like

complex. A prototype for "Induced_Fit" molecular recognition. *J. Am. Chem. Soc.*, **117**, 1649–1650.

92. Leininger, S., Olenyuk, B., and Stang, P.J. (2000) Self-assembly of discrete cyclic nanostructures mediated by transition metals. *Chem. Rev.*, **100**, 853–908.
93. Vilar, R. (2003) Anion-templated synthesis. *Angew. Chem. Int. Ed.*, **42**, 1460–1477.
94. Hasenknopf, B., Lehn, J.-M., Kneisel, B.O., Baum, G., and Fenske, D. (1996) Self-assembly of a circular double helicate. *Angew. Chem. Int. Ed. Engl.*, **35**, 1838–1840.
95. Lehn, J.-M., Simon, J., and Wagner, J. (1973) Mesomolecules. Polyaza-polyoxa macropolycyclic systems. *Angew. Chem. Int. Ed., Engl.*, **85**, 621–622.
96. Bertini, I., Gray, H.B., Lippard, S. J., and Valentine, J. S. (eds) (1994) *Bioinorganic Chemistry*, University Science Books, Mill Valley, CA.
97. Lippard, S.J. and Berg, J.M. (1995) *Principles of Bioinorganic Chemistry*, University Science Books, Mill Valley, CA.
98. Holm, R.H. and Solomon, E.I. (1996) Special issue in bioinorganic enzymology. *Chem. Rev.*, **96**, 2237.
99. Kimura, E., Kodama, M., and Yatsunami, T. (1982) Macromonocyclic polyamines as biological polyanion complexones. 2. Ion-pair association with phosphate and nucleotides. *J. Am. Chem. Soc.*, **104**, 3182–3187.
100. (a) Eriksson, A.E., Jones, T.A., and Liljas, A. (1988) Refined structure of human carbonic anhydrase II at 2.0 Å resolution. *Proteins, Proteins: Structure, Function, and Bioinformatics*, **4**, 278–282. (b) Zalatan, J.G., Fenn, T.D., and Herschlag, D. (2008) Comparative enzymology in the alkaline phosphatase superfamily to determine the catalytic role of an active-site metal ion. *J. Mol. Biol.*, **384**, 1174–1189.
101. Knaff, D.B. (1989) Structure and regulation of ribulose-1, 5-bisphosphate carboxylase/oxygenase. *Trends Biochem Sci.*, **14**, 159–160.
102. Graf, E. and Lehn, J.-M. (1977) Synthesis and cryptate complexes of a spheroidal macrotricyclic ligand with octahedrotetrahedral coordination. *J. Am. Chem. Soc.*, **97**, 5022–5024.
103. Metz, B., Rosalky, J.M., and Weiss, R. (1976) [3] Cryptates: X-ray crystal structures of the chloride and ammonium ion complexes of a spheroidal macrotricyclic ligand. *J. Chem. Soc., Chem. Commun.*, 533–534.
104. Schmidtchen, F.P. (1977) Inclusion of halide anions in macrotricyclic quaternary ammonium salts. *Angew. Chem. Int. Ed. Engl.*, **16**, 720–721.
105. Schmidtchen, F.P. and Müller, G. (1984) Anion inclusion without auxiliary hydrogen bonds: X-ray structure of the iodide cryptate of a macrotricyclic tetra-quaternary ammonium receptor. *J. Chem. Soc., Chem. Commun.*, 1115–1116.
106. Schmidtchen, F.P. (1980) Synthesis of macrotricyclic amines. *Chem. Ber.*, **113**, 864–874.
107. Schmidtchen, F.P. (1981) Macrocyclic quaternary ammonium salts. II. Formation of inclusion complexes with anions in solution. *Chem. Ber.*, **114**, 597–607.
108. Schmidtchen, F.P. (1981) Macrotricyclic quaternary ammonium salts: Enzyme-analogous activity. *Angew. Chem., Int. Ed. Engl.*, **20**, 466–468.
109. Schmidtchen, F.P. (1986) Molecular catalysis by quaternary ammonium salts. Cyclization of 2-(3-halopropyl)-4-nitrophenols. *J. Mol. Catal.*, **37**, 141–149.
110. Schmidtchen, F.P. (1986) Molecular catalysis by polyammonium receptors. *Top. Curr. Chem.*, **132**, 101–133.
111. Schmidtchen, F.P. (1984) Macrocyclic quaternary ammonium salts. IV. Catalysis of nucleophilic aromatic substitutions with azide as a nucleophile. *Chem. Ber.*, **117**, 725–732.
112. Schmidtchen, F.P. (1984) Macrocyclic quaternary ammonium salts, V. Catalysis of aromatic nucleophilic substitutions following enzyme-analogous principles. *Chem. Ber.*, **117**, 1287–1298.
113. Schmidtchen, F.P. (1986) Molecular catalysis by quaternary ammonium salts. Kinetic

analysis of the substitution of 4-chloro-3,5-dinitrobenzenesulfonate by azide. *J. Mol. Catal.*, **38**, 273–278.
114. Schmidtchen, F.P. (1986) Host–guest interactions. The binding mode of 6-nitrobenzisoxazole-3-carboxylate to quaternary ammonium macrocycles. *J. Chem. Soc., Perkin Trans. 2*, 135–142.
115. Schmidtchen, F.P. (1986) Tetazac: a novel artificial receptor for binding ω-amino carboxylates. *J. Org. Chem.*, **51**, 5161–5168.
116. Schmidtchen, F.P. (1986) Probing the design of a novel ditopic anion receptor. *J. Am. Chem. Soc.*, **108**, 8249–8255.
117. Schmidtchen, F.P. (1986) An artificial molecular receptor molecule for ditopic anions. *Tetrahedron Lett.*, **27**, 1987–1990.
118. Worm, K. and Schmidtchen, F.P. (1995) Molecular recognition of anions by zwitterionic host molecules in water. *Angew. Chem. Int. Ed. Engl.*, **34**, 65–66.
119. Lehn, J.-M., Pine, S.H., Watanane, E.I., and Willard, A.K. (1977) Binuclear cryptates. Synthesis and binuclear cation inclusion complexes of bis- *tren* macrobicyclic ligands. *J. Am. Chem. Soc.*, **99**, 6766–6768.
120. Dietrich, B., Guilhem, J., Lehn, J.-M., Pascard, P., and Sonveaux, E. (1984) Molecular recognition in anion coordination chemistry. Structure, binding constants and receptor-substrate complementarity of a series of anion cryptates of a macrobicyclic receptor molecule. *Helv. Chim. Acta*, **67**, 91–104.
121. Suet, E. and Handel, H. (1984) Complexation de l'anion fluoride par des tetramines cycliques protones. *Tetrahedron Lett.*, **25**, 645–648.
122. Dietrich, B., Lehn, J.-M., Guilhem, J., and Pascard, C. (1989) Anion receptor molecules: synthesis of an octaaza-cryptand and structure of its fluoride cryptate. *Tetrahedron Lett.*, **30**, 4125–4128.
123. Dietrich, B., Dilworth, B., Lehn, J.-M., Souchez, J.-P., Cesario, M., Guilhem, J., and Pascard, C. (1996) Synthesis, crystal structures, and complexation constants of fluoride and chloride inclusion complexes of polyammonium macrobicyclic ligands. *Helv. Chim. Acta*, **79**, 569–575.
124. Reilly, S.D., Kalsa, G.R.K., Ford, D.K., Brainard, J.R., Hay, B.P., and Smith, P.H. (1995) Octaazacryptand complexation of the fluoride anion. *Inorg. Chem.*, **34**, 569–575.
125. Hosseini, M.W., Kintzinger, J.-P., Lehn, J.-M., and Zahidi, A. (1989) Chloride binding by polyammonium receptor molecules: ^{35}Cl-NMR studies. *Helv. Chim. Acta*, **72**, 1078–1083.
126. Boudon, S., Decian, A., Fischer, J., Hosseini, M.W., Lehn, J.-M., and Wipff, G. (1991) Structural and anion coordination features of macrocyclic polyammonium cations in the solid, solution and computational phases. *J. Coord. Chem.*, **23**, 113–115.
127. Motekaitis, R.J., Martell, A.E., and Murase, I. (1986) Cascade halide binding by multiprotonated BISTREN and copper(II) BISTREN cryptates. *Inorg. Chem.*, **25**, 938–944.
128. Motekaitis, R.J., Martell, A.E., Murase, I., Lehn, J.-M., and Hosseini, M.-W. (1988) Comparative study of the copper(II) cryptates of C-BISTREN and O-BISTREN. Protonation constants, formation constants, and secondary anion bridging by fluoride and hydroxide. *Inorg. Chem.*, **27**, 3630–3636.
129. Kang, S.O., LLinares, J.M., Day, V.W., and Bowman-James, K. (2010) Cryptand-like anion receptors. *Chem. Soc. Rev.*, **39**, 3980–4003.
130. Lakshminarayanan, P.S., Kumar, D.K., and Ghosh, P. (2005) Counteranion-controlled water cluster recognition in a protonated octaamino cryptand. *Inorg. Chem.*, **44**, 7540–7546.
131. Hossain, A., LLinares, J.M., Mason, S., Morehouse, P., Powel, C., and Bowman-James, K. (2002) Parallels in cation and anion coordination: a new class of cascade complexes. *Angew. Chem. Int. Ed. Engl.*, **41**, 2335–2338.
132. Ravikumar, I., Lakshminarayanan, P.S., Suresh, E., and Ghosh, P. (2008) Spherical versus linear anion

encapsulation in the cavity of a protonated azacryptand. *Inorg. Chem.*, **47**, 7992–7999.

133. Menif, R., Reibenspies, J., and Martell, A.E. (1991) Synthesis, protonation constants, and copper(II) and cobalt(II) binding constants of a new octaaza macrobicyclic cryptand: $(MX)_3(TREN)_2$. Hydroxide and carbonate binding of the dicopper(II) cryptate and crystal structures of the cryptand and of the carbonato-bridged dinuclear copper(II) cryptate. *Inorg. Chem.*, **30**, 3446–3454.

134. Heyer, D. and Lehn, J.-M. (1986) Anion coordination chemistry – synthesis and anion binding features of cyclophane type macrobicyclic anion receptor molecules. *Tetrahedron Lett.*, **35**, 5869–5872.

135. Fujita, T. and Lehn, J.-M. (1988) Synthesis of dome-shaped cyclophane type macrotricyclic anion receptor molecules. *Tetrahedron Lett.*, **29**, 1709–1712.

136. Illoudis, C.A., Tocher, D.A., and Steed, J.W. (2004) A highly efficient, preorganized macrobicyclic receptor for halides based on CH··· and NH··· anion interactions. *J. Am. Chem. Soc.*, **126**, 12395–12402.

137. Sessler, J.L., Cyr, M.J., Lynch, V., Mcghee, E., and Ibers, J.A. (1990) Synthesis and structural studies of sapphyrin, a-22-π-electron pentapyrrolic expanded porphyrin. *J. Am. Chem. Soc.*, **112**, 2810–2813.

138. Sessler, J.L., Camiolo, S., and Gale, P.A. (2003) Pyrrolic and polypyrrolic anion binding agents. *Coord. Chem. Rev.*, **240**, 17–55.

139. Sessler, J.L., Ford, D.A., Cyr, M.J., and Furuta, H. (1991) Enhanced transport of fluoride anion effected using protonated sapphyrin as a carrier. *J. Chem. Soc., Chem. Commun.*, 1733–1735.

140. Pascal, R.A. Jr., Spergel, J., and Van Eggen, D. (1986) Synthesis and X-ray crystallographic characterization of a (1,3,5)cyclophane with three amide N-H groups surrounding a central cavity. A neutral host for anion complexation. *Tetrahedron Lett.*, **27**, 4099–4102.

141. Kimura, E., Anan, H., Koike, I., and Shiro, J. (1989) A convenient synthesis of a macrocyclic dioxo pentaamine and X-ray crystal structure of its monohydrazoic acid salt. *J. Org. Chem.*, **54**, 3998–4000.

142. Amendola, V., Bergmaschi, G., Boiocchi, M., Fabbrizzi, L., Poggi, A., and Zema, M. (2008) Halide ion inclusion into a dicopper(II) bistren cryptate containing 'active' 2,5-dimethylfuran spacers: the origin of the bright yellow colour. *Inorg. Chim. Acta*, **361**, 4038–4046.

143. Amendola, V., Bastianello, E., Fabbrizzi, L., Mangano, C., Pallavicini, P., Perotti, A., Manotti Lanfredi, A., and Uggozzoli, F. (2000) Halide-ion encapsulation by a flexible dicopper(II) bis- *tren* cryptate. *Angew. Chem. Int. Ed.*, **39**, 2917–2920.

144. Amendola, V., Fabbrizzi, C., Mangano, L., Pallavicini, P., and Zema, M. (2002) A di-copper(II) bis- *tren* cage with thiophene spacers as receptor for anions in aqueous solution. *Inorg. Chim. Acta*, **337**, 70–74.

145. O Neil, E.J. and Smith, B.D. (2006) Anion recognition using dimetallic coordination compounds. *Coord. Chem. Rev.*, **250**, 3068–3080.

146. Martell, A.E., Smith, R.M., and Motekaitis, R.J. (1997) NIST Critically Selected Stability Constants of Metal Complexes Database, NIST Standard Reference Database, Version 4.

147. Culliname, J., Gelb, R.I., Margulis, T.N., and Zompa, L.J. (1982) Hexacyclen complexes of inorganic anions: bonding forces, structure, and selectivity. *J. Am. Chem. Soc.*, **104**, 3048–3053.

148. Papoyan, G., Gu, K., Wiórkiewicz-Kuczera, J., Kuczera, K., and Bowman-James, K. (1996) Molecular dynamics simulations of nitrate complexes with polyammonium macrocycles: Insight on phosphoryl transfer catalysis. *J. Am. Chem. Soc.*, **118**, 1354–1364.

149. Valencia, L., Bastida, R., García-España, E., de Julián-Ortiz, V., LLinares, J.M., Macías, A., and Pérez-Lourido, P. (2010) Encapsulation within the cavity

of polyazapyridinophane. Considerations on nitrate-pyridine interaction. *Cryst. Growth Des.*, **10**, 3418–3423.
150. Mason, T., Clifford, T., Seib, L., Kuczera, F., and Bowman-James, K. (1998) Unusual encapsulation of two nitrates in a single bicyclic cage. *J. Am. Chem. Soc.*, **120**, 8999–8900.
151. Morgan, G.G., Nelson, J., and McKee, V. (1995) Caged anions: perchlorate and perfluoroanion cryptates. *J. Chem. Soc., Chem. Commun.*, 1649–1652.
152. Dilwoth, J. and Parrot, S. (1998) The biomedical chemistry of technetium and rhenium. *Chem. Soc. Rev.*, **27**, 43–56.
153. Volkert, W.A. and Hoffman, T.J. (1999) Therapeutic radiopharmaceuticals. *Chem. Rev.*, **99**, 2269–2292.
154. Jurisson, S.S. and Lydon, J.D. (1999) Potential technetium small molecule radiopharmaceuticals. *Chem. Rev.*, **99**, 2205–2218.
155. Farrell, D., Gloe, K., Gloe, K., Goretzki, G., McKee, V., Nelson, J., Nieuwenhuyzen, M., Pál, I., Stephan, H., Town, R.M., and Wichmann, K. (2003) Towards promising oxoanion extractants: azacages and open-chain counterparts. *Dalton Trans.*, 1961–1968.
156. Gelb, R.I., Alper, J.S., and Schwartz, M.H. (1992) Complexes of tri- and tetraprotonated forms of 1,4,7,10,13-pentaazacyclopentadecane with various mono- and divalent anions in aqueous media. *J. Phys. Org. Chem.*, **5**, 443–450.
157. Wu, G., Izatt, R.M., Bruenig, M.L., Jiang, W., Azab, H., Krakowiak, J.S., and Bradshaw, J.S. (1992) NMR and potentiometric determination of the high pK values and protonation sequence of dipyridino-hexaaza-28-crown-8 and its interactions with selenate, sulfate and nitrate ions in aqueous solution. *J. Inclusion Phenom. Mol. Recognit. Chem.*, **13**, 121–127.
158. Hosseini, M.W. and Lehn, J.-M. (1988) Anion receptor molecules: Macrocyclic and macrobicyclic effects on anion binding by polyammonium receptor molecules. *Helv. Chim. Acta*, **71**, 749–756.
159. Nelson, J., Nieuwenhuyzen, M., Pal, I., and Town, R.M. (2004) Dinegative tetrahedral oxoanion complexation: structural and solution phase observations. *Dalton Trans.*, 2303–2308.
160. Kang, S.O., Hossain, M.A., Powell, D., and Bowman-James, K. (2005) Encapsulated sulfates: insight to binding propensities. *Chem. Commun.*, 328–330.
161. Hossain, M.A., LLinares, J.M., Powell, D., and Bowman-James, K. (2001) Multiple hydrogen bond stabilization of a sandwich complex of sulfate between two macrocyclic tetraamides. *Inorg. Chem.*, **40**, 2936–2937.
162. Dietrich, B., Fyles, T.M., Lehn, J.-M., Pease, L.G., and Fyles, D.L. (1978) Anion receptor molecules. Synthesis and some anion binding properties of macrocyclic guanidinium salts. *J. Chem. Soc., Chem. Commun.*, 934–936.
163. Dietrich, B., Fyles, D.L., Fyles, T.M., and Lehn, J.-M. (1979) Anion coordination chemistry; polyguanidinium salts as anion complexones. *Helv. Chim. Acta*, **62**, 2763–2786.
164. Kimura, E., Kuramoto, Y., Koike, T., Fujioka, H., and Kodama, M. (1990) A study of new bis (macrocyclic polyamine) ligands as inorganic and organic anion receptors. *J. Org. Chem.*, **55**, 42–46.
165. Bencini, A., Bianchi, A., García-España, E., Scott, E.C., Morales, L., Wange, B., Deffo, T., Takusagawa, F., Mertes, M.P., Mertes, K.B., and Paoletti, P. (1992) Potential ATPase mimics by polyammonium macrocycles: criteria for catalytic activity. *Bioorg. Chem.*, **20**, 8–29.
166. Lu, Q., Motekaitis, R.J., Reibenspies, J.J., and Martell, A.E. (1995) Molecular recognition by the protonated hexaaza macrocyclic ligand 3,6,9,16,19,22-hexaza-27,28-dioxatricyclo [22.2.1.1^{11,14}]octacosa-1(26),11,13,24-tetraene. *Inorg. Chem.*, **34**, 4958–4964.

167. Nation, D.A., Reibenspies, J., and Martell, A.E. (1996) Anion binding of inorganic phosphates by the hexazamacrocyclic ligand 3,6,9,17,20,23-hexazatricyclo[23.3.111,15] triaconta-1(29),11(30),12,14,25,27-hexaene. *Inorg. Chem.*, **35**, 4597–4603.
168. Lu., Q., Reibenspies, J.H., Carooll, R.I., and Martell, A.E. (1998) Interaction of mono- and triphosphate anions with a protonated poyazamacrocycle and its Cu(II) complexes. *Inorg. Chim. Acta*, **270**, 207–215.
169. Gerasimchuck, O.A., Mason, S., LLinares, J.M., Song, M., Alcock, N.W., and Bowman-James, K. (2000) Binding of phosphate with a simple polyammonium macrocycle. *Inorg. Chem.*, **39**, 1371–1375.
170. Anda, C., Llobet, A., Salvadó, V., Reibenspies, J., Motekaitis, R.J., and Martell, A.E. (2000) A systematic evaluation of molecular recognition phenomena. 1. Interaction between phosphates and nucleotides with hexaazamacrocyclic ligands containing *m*-xylilic spacers. *Inorg. Chem.*, **39**, 2986–2999.
171. Anda, C., Llobet, A., Salvadó, V., Reibenspis, J., Motekaitis, R.J., and Martell, A.E. (2000) A systematic evaluation of molecular recognition phenomena. 2. Interaction between phosphates and nucleotides with hexaazamacrocyclic ligands containing diethyl ether spacers. *Inorg. Chem.*, **39**, 3000–3008.
172. Huston, M.E., Akkaya, E.U., and Czarnik, A.W. (1989) Chelation enhanced fluorescence detection of non-metal ions. *J. Am. Chem. Soc*, **111**, 8735–8737.
173. Vance, D.H. and Czarnik, A.V. (1994) Real-time assay of inorganic pyrophosphate using a high-affinity chelation-enhanced fluorescence chemosensor. *J. Am. Chem. Soc.*, **116**, 9397–9398.
174. Echevarren, A., Galán, A., Lehn, J.-M., and de Mendoza, J. (1989) Chiral recognition of aromatic carboxyltae anions by an optically active abiotic receptor containing a rigid guanidinium binding subunit. *J. Am. Chem. Soc.*, **111**, 4994–4995.
175. Galán, A., Andre, D., Echevarren, A., Prados, P., and de Mendoza, J. (1992) A receptor for the enatioselective recognition of phenylalanine and tryptophan under neutral conditions. *J. Am. Chem. Soc.*, **114**, 1511–1512.
176. Fan, E., Van Arman, S.A., Kincaid, S., and Hamilton, A.D. (1993) Molecular recognition: hydrogen-bonding receptors that function in highly competitive solvents. *J. Am. Chem. Soc.*, **115**, 369–370.
177. Kimura, E., Sakonaka, A., Yatsunami, T., and Kodama, M. (1981) Macromonocyclic polyamines as specific receptors for tricarboxylate-cycle anions. *J. Am. Chem., Soc.*, **103**, 3041–3945.
178. Yatsunami, T., Sakonaka, A., and Kimura, E. (1981) Identification of macrocyclic polyamines and macrocyclic dioxopolyamines by thin-layer chromatography, gas cromatoghraphy, and electrophoresis. *Anal. Chem.*, **53**, 477–480.
179. Kimura, E. and Sakomaka, A. (1981) A carbonate receptor model by macromonocyclic polyamines and its physiological implications. *J. Am. Chem. Soc.*, **104**, 4984–4985.
180. Hosseini, M.W. and Lehn, J.-M. (1986) 61. Anion coreceptor molecules. Linear molecular recognition in the selective binding of dicarboxylate substrates by ditopic polyammonium receptors. *Helv. Chim. Acta*, **69**, 587–603.
181. Bencini, A., Bianchi, A., Burguete, M.I., García-España, E., Luis, S.V., and Ramírez, J.A. (1992) Remarkable shape selectivity in the molecular recognition of carboxylate anions in aqueous solution. *J. Am. Chem. Soc.*, **114**, 1919–1920.
182. Bencini, A., Bianchi, A., Burguete, M.I., Dapporto, P., Doménech, A., García-España, E., Luis, S.V., Paoli, P., and Ramírez, J.A. (1994) Selective recognition of carboxylate anions by polyammonium receptors in aqueous solution. Criteria for selectivity in molecular recognition. *J. Chem. Soc., Perkin Trans. 2*, 569–577.

183. Bianchi, A. and García-España, E. (1999) The use of calculated species distribution diagrams to analyze thermodynamic selectivity. *J. Chem. Educ.*, **76**, 1727–1732.

184. Jawzinski, J., Lehn, J.-M., Lilienbaum, D., Ziessel, R., Guilhem, J., and Pascard, C. (1987) Polyaza macrobicyclic cryptands, crystal structure of a cyclophane type macrobicyclic cryptand and of its dinuclear copper(I) cryptate, and anion binding features. *J. Chem. Soc., Chem. Commun.*, 1691–1694.

185. Lehn, J.-M., Méric, R., Vigneron., J.-P., Bkouche-Waksaman, I., and Pascard, C. (1991) Molecular recognition of anionic binding substrates. Binding of carboxylate by a macrobicyclic coreceptor and crystal structure of its supramolecular cryptate with the terephthlate dianion. *J. Chem. Soc., Chem. Commun.*, 62–64.

186. Flack, S.S., Chaumette, J.-L., Kilburn, J.D., Langley, J., and Webster, M. (1993) A novel receptor for dicarboxylate acid derivatives. *J. Chem. Soc., Chem. Commun.*, 399–401.

187. Bazzicalupi, C., Bencini, A., Bianchi, A., Fusi, V., Gracía-España, E., Giorgi, C., LLinares, J.M., Ramírez, J.A., and Valtancoli, B. (1999) Molecular recognition of long dicarboxylate/dicarboxylic species via supramolecular/coordinative interactions with ditopic receptors. Crystal structure of $\{[Cu_2L(H_2O)_2] \supset$ Pimelate$\}(ClO_4)_2$. *Inorg. Chem.*, **38**, 620–621.

188. Larsen, D.W. and Wahl, A.C. (1965) Nuclear magnetic resonance study of ion association between quaternary ammonium ions and hexacyanoferrate(II) ion. *Inorg. Chem.*, **4**, 1281–1286.

189. García-España, E., Micheloni, M., Paoletti, P., and Bianchi, A. (1985) Anion coordination chemistry. Hexacyanoferrate(II) anion complexed by a large polycharged azacycloalkane. Potentiometric and electrochemical studies. *Inorg. Chim. Acta Lett.*, **102**, L9–L11.

190. Bianchi, A., García-España, E., Mangani, S., Micheloni, M., Orioli, P., and Paoletti, P. (1987) Anion co-ordination chemistry. Crystal structure of the super complex: $[H_8L][Co(CN)_6]_2Cl_2\,10\,H_2O$ (L = 1,4,7,10,13,16,19,22,25,28 – deca – azacyclotriacontane). *J. Chem. Soc., Chem. Commun.*, 729–730.

191. Bencini, A., Bianchi, A., García-España, E., Giusti, M., Mangani, S., Micheloni, M., Orioli, P., and Paoletti, P. (1987) Anion coordination chemistry. 2. Electrochemical, thermodynamic, and structural studies on supercomplex formation between large polyammonium cycloalkanes and the two complex anions hexacyanoferrate(II) and hexacyanocobaltate(III). *Inorg. Chem.*, **26**, 3902–3907.

192. Bencini, A., Bianchi, A., Dapporto, P., García-España, E., Micheloni, M., Paoletti, P., and Paoli, P. (1990) $[PdCl_4]^{2-}$ Inclusion into the decacharged polyammonium receptor $(H_{10}[30]aneN_{10})^{10+}$ ($[30]aneN_{10}$ = 1,4,7,10,13,16,19,22,25,28-deca-azacyclotriaconatne). *J. Chem. Soc., Chem. Commun.*, 753–755.

193. Bencini, A., Bianchi, A., Micheloni, M., Paoletti, P., Dapporto, P., Paoli, P., and García-España, E. (1992) Cation and anion coordination chemistry of palladium(II) with polyazacycloalkanes. Thermodynamic and structural studies. *J. Inclusion Phenom. Mol. Recognit. Chem.*, **12**, 291–304.

194. Bencini, A., Bianchi, A., Dapporto, P., García-España, E., Micheloni, M., Ramírez, J.A., Paoletti, P., and Paoli, P. (1992) Thermodynamic and structural aspects of the interaction between macrocyclic polyammonium cations and complexed anions. *Inorg. Chem.*, **31**, 1902–1908.

195. Aragó, J., Bencini, A., Bianchi, A., Doménech, A., and García-España, E. (1992) Macrocyclic effect on anion binding. A potentiometric and electrochemical study of the interaction of 21- and 24- membered polyazaalkanes with $[Fe(CN)_6]^{4-}$ and $[Co(CN)_6]^{3-}$. *J. Chem. Soc., Dalton Trans. 2*, 319–324.

196. Bernardo, M.A., Parola, A.J., Pina, F., García-España, E., Marcelino, V., Luis, S.V., and Miravet, J.F. (1995) Steady-state fluorescence emission studies on polyazacyclophane macrocyclic receptors and on their adducts with hexacyanocobaltate(III). *J. Chem. Soc., Dalton Trans.*, 993–997.

197. Aragó, J., Bencini, A., Bianchi, A., García-España, E., Micheloni, M., Paoletti, P., Ramírez, J.A., and Paoli, P. (1991) Interaction of "long" open-chain polyazaalkanes with hydrogen and copper(II) ions. *Inorg. Chem.*, **30**, 1843–1849.

198. Manfrin, M.F., Sabbatini, N., Moggi, L., Balzani, V., Hosseini, M.W., and Lehn, J.-M. (1984) Photochemistry of the supercomplex obtained on complexation of the hexacyanocobaltate(III) anion by a polyammonium macrocyclic receptor. *J. Chem. Soc., Chem. Commun.*, **8**, 555–556.

199. Manfrin, M.F., Moggi, L., Castelvetro, V., Balzani, V., Hosseini, M.W., and Lehn, J.-M. (1985) Control of the photochemical reactivity of coordination compounds by formation of supramolecular structures: the case of the hexacyanocobaltate(III) anion associated with polyammonium macrocyclic receptor. *J. Am. Chem. Soc.*, **107**, 6888–6892.

200. Balzani, V., Sabbatini, N., and Scandola, F. (1986) "Second-sphere" photochemistry and photophysics of coordination compounds. *Chem. Rev.*, **86**, 319–337.

201. Pina, F., Moggi, L., Manfrin, M.F., Balzani, V., Hosseini, M.W., and Lehn, J.-M. (1989) Photochemistry of supramolecular systems. Size and charge effects in the photoaquation of adducts of the hexacyanocobaltate(III) anion with polyammonium macrocyclic receptors. *Gazz. Chim. Ital.*, **119**, 65–67.

202. Bianchi, A., Micheloni, M., Orioli, P.L., Paoletti, P., and Mangani, S. (1988) Anion coordination chemistry. 3. Second-sphere interaction between the complex anions hexacyanoferrate(II) and hexacyanocobaltate(III) with polycharged tetraazamacrocycles. Thermodynamic and single crystal X-ray studies. *Inorg. Chim. Acta*, **146**, 153–159.

203. Anderson, J.L. and Sanders, J.K.M. (1992) Recognition of giant cluster anions by a protonated porphyrin trimer: detection by fast-atom bombardment (FAB) mass spectrometry. *J. Chem. Soc., Chem. Commun.*, 946–947.

204. Tabushi, I., Kobuke, Y., and Imuta, J. (1981) Lipophilic diammonium cation having a rigid structure complementary to pyrophosphate dianions of nucleotides. Selective extraction and transport of nucleotides. *J. Am. Chem. Soc.*, **103** (615), 6152–6157.

205. Bianchi, A., Micheloni, M., and Paoletti, P. (1988) Supramolecular interactions between adenosine 5′-triphosphate (ATP) and polycharged tetraazamacrocycles. Thermodynamic and ^{31}P NMR atudies. *Inorg. Chim. Acta*, **151**, 269–272.

206. Marececk, J.F. and Burrows, C.J. (1986) Synthesis of an optically active spermine macrocycle, (*S*)-6-(hydroxyethyl)-1,5,10,14-tetraazacyclotetradecane, and its complexation to ATP. *Tetrahedron Lett.*, **27**, 5943–5946.

207. Bencini, A., Bianchi, A., Burguete, M.I., Doménech, A., García-España, E., Luis, S.V., Niño, M.A., and Ramírez, J.A. (1991) N,N′, N″, N‴-(2-aminoethyl)-1,4,8,11-tetraazacyclotetradecane (TAEC) as a polyammonium receptor for anions. *J. Chem. Soc. Perkin Trans. 2*, 1445–1451.

208. Hosseini, M.W., Blacker, A.J., and Lehn, J.-M. (1990) Multiple molecular recognition and catalysis. A multifunctional anion receptor bearing an anion binding site, an intercalating group, and a catalytic site for nucleotide binding and hydrolysis. *J. Am. Chem. Soc.*, **112**, 3896–3904.

209. Claude, S., Lehn, J.-M., Schmidt, F., and Vigneron, J.-P. (1991) Binding of nucleosides, nucleotides and anionic planar substrates by bis-intercaland receptor molecules. *J. Chem. Soc., Chem. Commun.*, 1182–1185.

210. Cudic, P., Zinic, M., Tomisic, V., Simeon, V., Vigneron, J.-P., and Lehn, J.-M. (1995) Binding of nucleotides in water by phenanthridinium bis(intercaland) receptor molecules.

J. Chem. Soc., Chem. Commun., 1182–1185.

211. Aguilar, J., García-España, E., Guerrero, J.A., Luis, S.V., Llinares, J.M., Miravet, J.F., Ramírez, J.A., and Soriano, C. (1995) Multifunctional molecular recognition of ATP, ADP and AMP nucleotides by the novel receptor 2,6,10,13,17,21,-Hexaaza[22] metacyclophane. *J. Chem. Soc., Chem. Commun.*, 2237–2239.

212. Aguilar, J.A., García-España, E., Guerrero, J.A., Luis, S.V., Llinares, J.M., Ramírez, J.A., and Soriano, C. (1996) Synthesis and protonation behavior of the macrocycle 2,6,10,13,17,21-hexaaza[22] metacyclophane. Thermodynamic and NMR studies on the interaction of 2,6,10,13,17,21,-Hexaaza[22] metacyclophane and of the open-chain polyamine 4,8,11,15-tetraazaoctadecane-1,18- diamine with ATP, ADP and AMP. *Inorg. Chim. Acta*, **246**, 287–294.

213. Bazzicalupi, C., Bencini, A., Bianchi, A., Faggi, E., Giorgi, C., Santarelli, S., and Valtancoli, B. (2008) Polyfunctional binding of thymidine 5′-triphosphate with a synthetic polyammonium receptor containing aromatic groups. Crystal structure of the nucleotide-receptor adduct. *J. Am. Chem. Soc.*, **130**, 2440–2441.

214. Schneider, H.-J. (2009) Binding mechanisms in supramolecular chemistry. *Angew. Chem. Int. Ed.*, **48**, 3924–3977.

215. Hosseini, M.W. and Lehn, J.-M. (1987) Binding of AMP, ADP, and ATP nucleotides by polyammonium macrocycles. *Helv. Chim. Acta*, **70**, 1312–1319.

216. Hosseini, M.W., Lehn., J.-M., and Mertes, M.P. (1983) Efficient molecular catalysis of ATP-hydrolysis by protonated macrocyclic polyamines. *Helv. Chim. Acta*, **66**, 2454–2466.

217. Hosseini, M.W., Lehn., J.-M., and Mertes, M.P. (1985) *Helv. Chim. Acta*, **68**, 818.

218. Hosseini, M.W., Lehn, J.-M., Maggiora, L., Bowman Mertes, K., and Mertes, M.P. (1987) Supramolecular catalysis in the hydrolysis of ATP facilitated by macrocyclic polyamines: mechanistic studies. *J. Am. Chem. Soc.*, **109**, 537–544.

219. Blackburn, G.M., Thatcher, G.R.J., Hosseini, M.W., and Lehn, J.-M. (1987) Evidence for a protophosphatase catalysed cleavage of adenosine triphosphate by a dissociative-type mechanism within a receptor-substrate complex. *Tetrahedron Lett.*, **28**, 2779–2782.

220. Bethell, R.C., Lowe, G., Hosseini, M.W., and Lehn., J.-M. (1988) Mechanisms of the ATPase-like activity of the macrocyclic polyamine receptor molecule [24]N_6O_2. *Bioorg. Chem.*, **16**, 418–428.

221. Andrés, A., Bazzicalupi, C., Bencini, A., Bianchi, A., Fusi, V., García-España, E., Giorgi, C., Nardi, N., Paoletti, P., Ramírez, J.A., and Valtancoli, B. (1994) 1,10-Dimethyl-1,4,7,10,13,16-hexaazacyclooctadecane L and 1,4,7-trimethyl-1,4,7,10,13,16,19-heptaazacycclohenicosane L1: two new macrocyclic receptors for ATP binding. Synthesis, solution equilibria and the crystal structure of $(H_4L)(ClO_4)_4$. *J. Chem. Soc., Perkin Trans.*, **2**, 2367–2373.

222. Bencini, A., Bianchi, A., Giorgi, C., Paoletti, P., Valtanacoli, B., Fusi, V., García-España, E., Llinares, J.M., and Ramírez, J.A. (1996) Effect of nitrogen methylation on cation and anion coordination by hexa- and heptaazamacrocycles. Catalytic properties of these ligands in ATP dephosphorylation. *Inorg. Chem.*, **35**, 1114–1120.

223. Aguilar, J.A., Descalzo, A.B., Díaz, P., Fusi, V., García-España, E., Luis, S.V., Micheloni, M., Ramírez, J.A., Luis, S.V., Micheloni, M., Ramírez, J.A., Romani, P., and Soriano, C. (2000) New molecular catalysts for ATP cleavage. Criteria of size complementarity. *J. Chem. Soc., Perkin Trans. 2*, 1187–1192.

224. Hosseini, M.W. and Lehn, J.-M. (1985) Cocatalysis: pyrophosphate synthesis from acetylphosphate catalysed by a macrocyclic polyamine. *J. Chem. Soc., Chem. Commun.*, 1155–1157.

225. Yohannes, P.G., Mertes, M.P., and Mertes, K.B. (1985) Pyrophosphate formation via a phosphoramidate intermediate in polyammonium macrocycle/metal ion catalyzed hydrolysis of ATP. *J. Am. Chem. Soc.*, **107**, 8288–8289.

226. Yohannes, P.G., Plute, K.E., Mertes M.P., and Mertes, K.B. (1987) Specificity, catalysis, and regulation: effects of metal ions on polyammonium macrocycle catalyzed dephosphorylation of ATP. *Inorg. Chem.*, **26**, 1751–1758.

227. Jahansouz, H., Jiang, Z., Himes, R.H., Mertes, M.P., and Mertes, K.B. (1989) Formate activation in neutral aqueous solution mediated by a polyammonium macrocycle. *J. Am. Chem. Soc.*, **111**, 1409–1412.

228. Hosseini, M.W. and Lehn, J.-M. (1991) Supramolecular catalysis of adenosine triphosphate synthesis in aqueous solution mediated by a macrocyclic polyamine and divalent metal cations. *J. Chem. Soc., Chem. Commun.*, 451–453.

229. Mertes, M.P. and Mertes, K.B. (1990) Polyammonium macrocycles as catalysts for phosphoryl transfer: the evolution of an enzyme mimic. **23**, 413–418.

2
Thermodynamic Aspects of Anion Coordination

Antonio Bianchi and Enrique García-España

2.1
Introduction

By virtue of their negative charge, anions interact with all chemical species carrying net positive charges or permanent polarities or containing groups that can be favorably polarized by neighboring species. This is an inescapable fate dictated by one of the most fundamental laws of Nature, which makes species of opposite charges attract each other. Although all processes involving interactions with anions can be ultimately regarded as anion coordination events, *anion coordination* has evolved as an intentional venture aimed to achieve tight and selective binding and transformation and translocation of anionic species by the use of tailored receptors designed by adjusting tunable parameters such as receptor shapes, sizes and structures, types, and orientation of binding groups. In addition to topological considerations, obviously necessary for efficient matching of anion and receptor binding sites, energetic considerations regarding the different forces involved in the anion–receptor interaction are also of paramount importance for the design of synthetic anion receptors. Noncovalent forces that are relatively strong, such as coulombic attraction and hydrogen bonds, can combine with weaker ones such as dispersive and anion–π interactions to achieve tight anion coordination. The relative strength or weakness of these forces, however, is strictly solvent dependent. Charge–charge attraction and hydrogen bonds, which may furnish a large contribution to complex stability in apolar solvents, become much weaker in a polar protic solvent such as water, while dispersive forces, whose contribution to complex stability in apolar solvents is negligible, become of principal importance in determining the association of apolar species in water and other polar solvents. Hence, solvent effects are other significant parameters to be considered in receptor design. A significant portion of this understanding can be gleaned by probing the thermodynamics of binding interactions. In this respect, the dissection of the Gibbs free energy (ΔG°) of anion binding into its enthalpic (ΔH°) and entropic (ΔS°) components provides useful insight into the nature of the binding interactions, the interplay of their contributions, and contributions due to solvation/desolvation processes occurring on anion coordination.

Anion Coordination Chemistry, First Edition. Edited by Kristin Bowman-James, Antonio Bianchi, and Enrique García-España.

2.2
Parameters Determining the Stability of Anion Complexes

One of the most intriguing aspects of anion coordination involves seeking out receptors that will be able to selectively bind certain anions in preference to others that may be present in solution. This selective process, defining the ability of receptors to recognize anionic substrates, is controlled by various characteristics of both receptors and anions, as well as of reaction media. In the following sections, we shall examine the parameters that influence anion coordination for which thermodynamic data are available.

2.2.1
Type of Binding Group: Noncovalent Forces in Anion Coordination

The formation of anion–receptor complexes commonly occurs through several noncovalent interactions between functionalities of the interacting partners. Depending on anion nature and structure, different functionalities can be introduced into the receptor to target the different groups of the anionic substrates. An example of the different types of interactions that can operate in the binding of a biological substrate by a synthetic receptor is seen in Figure 2.1, which shows the crystal structure of the complex formed by thymidine 5′-triphosphate (TTP) with the tetraprotonated form of the polyamine macrocycle **1** containing two terpyridine units. Three hydrogen bonds between the polyphosphate chain of the

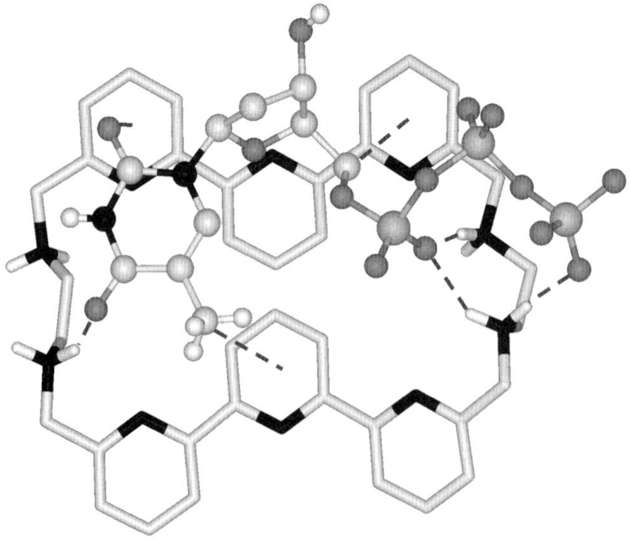

Figure 2.1 Crystal structure of the $[H_4\mathbf{1}(TTP)]$ (TTP = thymidine 5′-triphosphate) complex. (Reproduced with permission from Ref. [1], copyright 2008, American Chemical Society.)

nucleotide and protonated nitrogen atoms of **1**, one hydrogen bond between a carbonyl oxygen of TTP and one ammonium group of **1**, one CH$\cdots\pi$ interaction involving one carbon atom and a ligand pyrimidine unit and one O$\cdots\pi$ interaction between the TTP carbonyl oxygen O(14) and the N(10) pyridine ring of **1** determine the overall solid-state structure of the [H_4**1**(TTP)] complex, whose stability in water ($K = 4.57 \times 10^4$ M^{-1}) is significantly higher than the stability of nucleotide complexes with polyazamacrocycles, bearing the same positive charge and having size comparable with $H_4\mathbf{1}^{4+}$, but not including aromatic groups [1].

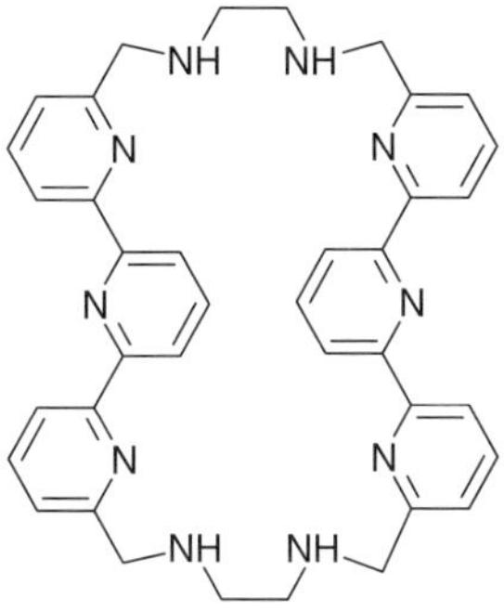

1

Electrostatic attraction, hydrogen bonding, van der Waals and dispersion forces, anion–π, and π-stacking interactions, as well as solvophobic effects, make the principal contribution to the stability of anion complexes in solution. However, even classic coordination to metal centers, involving some covalent character, must be included in the list of binding forces that are used for anion coordination. Thus, quantification of these forces is of major importance for the design of anion receptors and for the understanding of anion–receptor interactions.

Synthetic receptors are useful covalent models for the analysis of noncovalent forces. Indeed, receptors characterized by restricted molecular freedom and differing in the exclusion of specific binding groups can be synthesized and their anion binding ability quantified to correlate the difference in binding energies with the loss of binding interactions determined by the exclusion of those functions. Alternatively, quantification of noncovalent forces can be achieved when a linear correlation is verified between the free energy changes of association and the number of interactions formed in the complex, provided the interactions are of similar type and the binding sites are in the same environment and geometrically match each other. The second method can be extended to the lower limit of interaction between pairs of single binding groups, in those cases for which, despite contemplating weak interaction between small species, it is possible to perform an accurate definition of the interacting model, both qualitatively (stoichiometry and interacting geometry) and quantitatively (binding energies) [2].

For instance, linear correlations between binding free energies and number of salt bridges (hydrogen bonds subtending interacting groups of opposite charge) performed for various anion complexes allowed determining a contribution of -5 ± 1 kJ mol^{-1} per salt bridge in water (ionic strength about 0.02 M), despite binding groups characterized by different size and polarizability being involved with both anions (CO_2^-, SO_3^-, OPO_3H^-, OPO_3^{2-}, phenolate–O^-) and receptors ($R_2NH_2^+$, R_4N^+, pyridinium, R_4P^+) [3–9]. It was also shown that extrapolation to zero ionic strength causes this mean incremental contribution per salt bridge in water to become more favorable (-8 ± 1.7 kJ mol^{-1}) [10]. The unique requirement for the observed role to be followed is that ion-paired complexes without intermolecular strain must be formed.

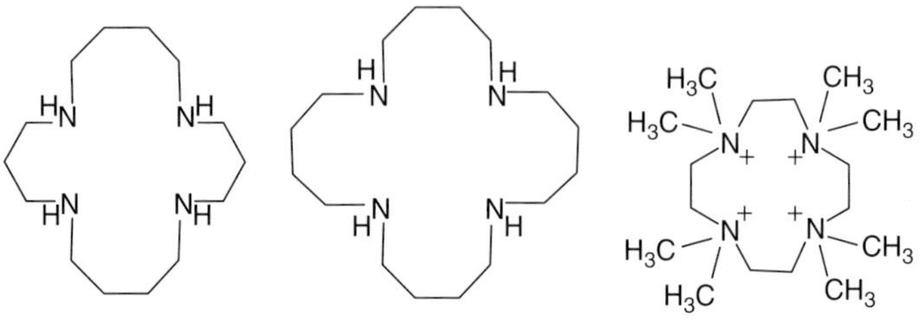

2 3 4

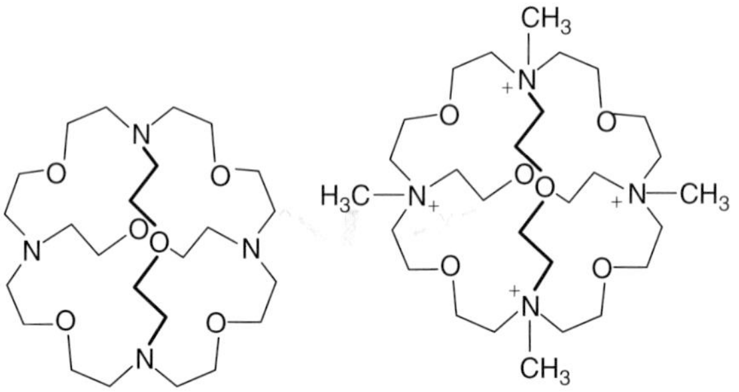

5 6

Formation of salt bridges is the main driving force for the coordination of anions to polyammonium cations, the largest and most studied class of anion receptors, and accordingly, it would be interesting to quantify the individual contributions of hydrogen bonding and charge–charge interactions to their formation. Comparison of anion complexes of protonated amine receptors with those formed by their permethylated analogs, which are unable to form hydrogen bonds, might seem helpful in this respect, but the results obtained are not univocal. For example, unlike the tetraprotonated forms of the azamacrocycles **2** and **3**, the permethylated tetraazamacrocycle **4** is not able to bind ATP^{4-} despite the higher density of positive charge on the quaternized receptor [11, 12]. Full methylation of the macrotricyle **5** yielding **6** gives rise to a dramatic decrease in the stability of the Cl^- complex, while the stability of the Br^- complex is enhanced [13]. Permethylated spermine does not show substantial difference in the binding affinity toward the phosphate residues of DNA with respect to the natural tetraamine [14]. The simplest ammonium group, NH_4^+, does not show any detectable association with halide anions in water, while the ammonium analogs R_4N^+ (R = Me, Et, Pr, Bu) form fairly stable complexes with these anions [15–17]. Such discrepancy between the effects brought about by permethylation of amine groups is a clear evidence of the fact that permethylation of anion receptors does not have the sole outcome of removing hydrogen bonds from salt bridges. Actually, other important structural and electronic modifications of binding sites, such as increasing dimensions, increasing hydrophobicity, increasing contribution to van der Waals interactions, and displacement of charge from hydrogen atoms to the central nitrogen, affect the overall binding properties of anion receptors. Nonetheless, with the exclusion of specific cases requiring individual evaluations, differentiation between the energetics of salt bridges and simple ion pairs does not seem to be of great importance since a mean incremental contribution of about $-5\ \mathrm{kJ\ mol^{-1}}$ was found for a large number of single charge–charge interactions, including ion pairs and salt bridges formed by a variety of charged partners [4, 5, 8]. It is to be noted that many authors do not differentiate salt bridges from strong hydrogen bonds [18–20]. As shown later on, however, hydrogen bonding can play fundamental roles in anion coordination.

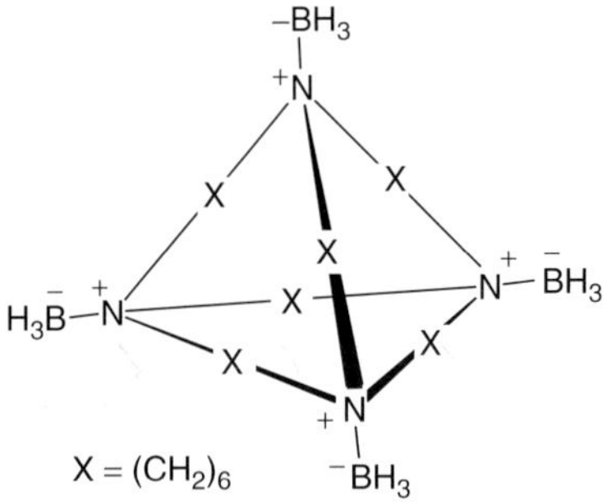

7

Electrostatic attraction between anions and permanent dipoles is weaker than the anion–cation interaction regarded above and is more easily hindered by solvent effects, in particular, in the presence of high-polarity solvents that reduce, or may even extinguish, the anion-binding ability of dipolar groups. Furthermore, the effectiveness of the anion–dipole interaction is also dependent on orientation, and thus receptors based on dipolar binding groups require stricter conformational preorganization to be successful in binding processes. Such considerations were fully accounted for in the design of the macrotricyclic uncharged receptor **7** characterized by the presence of four tetrahedrally oriented dipolar $^{+}N{-}BH_3^{-}$ binding sites [21]. ^1H NMR measurements performed in 20% $CD_2Cl_2/CDCl_3$ (v/v) solution showed that inclusive coordination of Cl^-, Br^-, I^-, and CN^- occurs according to the stability sequence $Br^- > Cl^- > I^- > CN^-$. In the case of the best match occurring with Br^-, a free energy change of complexation of $-6.3\ \mathrm{kJ\ mol^{-1}}$ (297 K) was determined, corresponding to about $-1.6\ \mathrm{kJ\ mol^{-1}}$ contribution for a single anion–dipole interaction. As expected, this value is smaller than those obtained for the formation of single ion pairs and salt bridges (about $-5\ \mathrm{kJ\ mol^{-1}}$). Relatively greater free energy contributions for single anion–dipole interactions of about $-3\ \mathrm{kJ\ mol^{-1}}$ were calculated from experimental data obtained in chloroform or chloroform containing small amounts of methanol [3–5, 7].

Owing to their negative charge, anions can modify the local charge distribution of neighboring species. These species assume a dipole moment that is favorably orientated for attractive interaction with the inducing anions. The resulting anion-induced dipole interactions, however, are even weaker than the interactions of anions with permanent dipoles, are more affected by solvent effects, and require a rather tight contact to be generated. Interesting polarization effects can occur when an anion is close to π-electron clouds. Nevertheless, a permanent polar character can be associated with π-electron clouds, and then anion–π interactions may arise from both anion-induced dipole and anion–dipole attractive forces. Calculated interaction potentials for Cl–π interactions demonstrate that a significant attractive force is generated even when the interacting species are relatively distant, indicating that the major source of the attraction are electrostatic and dispersive forces [22]. Accordingly, free energy contributions to anion–π association in solution have been reported [4, 5, 22–26] in the range from about -2.5 kJ mol^{-1} for a single anion–arene interaction in water/methanol (80 : 20), such as those shown by **8** with diphenylamine [4], to -9.9 kJ mol^{-1} for the interaction of Cl^- with receptor **9** [23, 24]. A value of -8.9 kJ mol^{-1} was determined for the interaction of sulfate with the pyrimidine ring of **10** in water as the residual free energy contribution to complex formation at zero receptor charge obtained by regression of the linear relation between the formation free energy of $[SO_4(H_n\mathbf{10})]^{n-2}$ ($n = 1$–3) complexes and the receptor charge n ($n = 1$–3) (Figure 2.2) [26]. A similar value (about -8 kJ mol^{-1}) was obtained by means of *ab initio* calculations using benzene–chlorohydrocarbon model systems [22]. Chapter 6 is dedicated to anion–π interactions, where a detailed description of structural and energetic properties of these weak forces can be found.

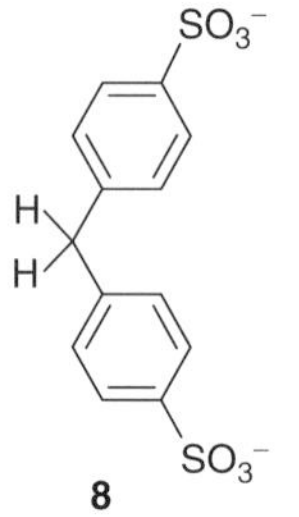

8

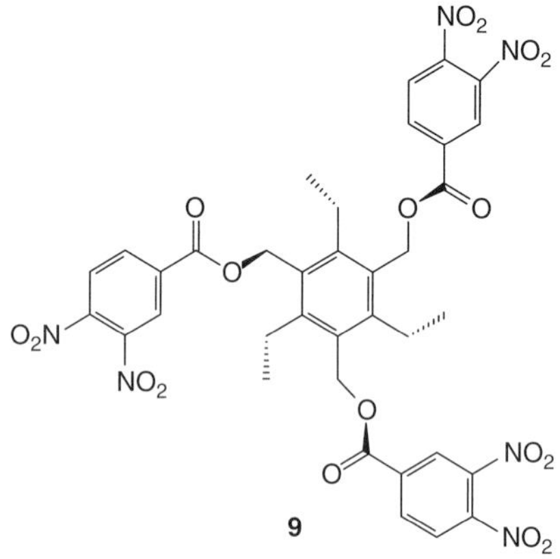

9

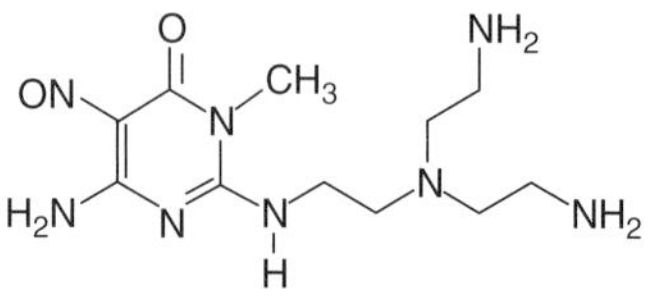

10

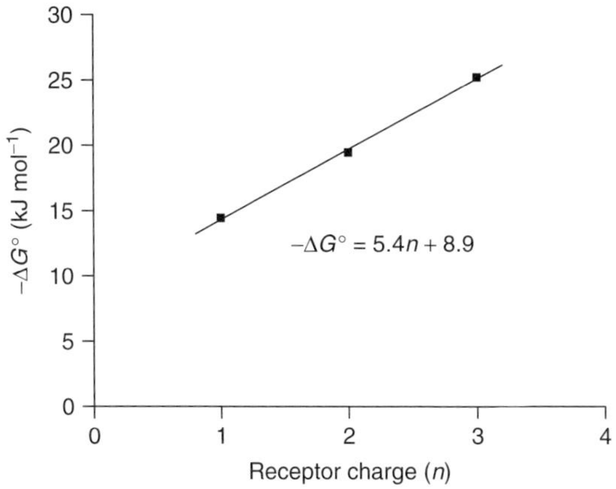

Figure 2.2 Linear relation between the formation free energy of $[SO_4(H_n\mathbf{10})]^{n-2}$ ($n = 1$–3) complexes and the receptor charge n [26].

Hydrogen bonding [18–20] is likely the most striking force involved in anion–dipole interactions. Despite its fundamental electrostatic nature, other factors, such as charge transfer and dispersive and covalent interactions, may furnish important contributions to hydrogen bonding and make it worthy of special consideration among other weak forces. Hydrogen bonds are formed when the electronegativity of D in a D–H covalently bonded donor is such as to withdraw electrons and leave H partially unshielded. To interact with D–H, the acceptor A must have lone-pair electrons or polarizable π electrons (Figure 2.3). Despite its simplicity, such definition clearly refers to a wide range of interactions spanning from very strong hydrogen bonds resembling covalent bonds, as in the case of the difluoride anion (HF_2^-), to very weak hydrogen bonds, which are very similar to van der Waals interactions. Hydrogen bond energies, as derived from thermodynamic studies of binding equilibria in the gas phase or apolar solvents, range from 5–15 kJ mol^{-1} for weak bonds to 15–60 kJ mol^{-1} for moderate bonds and 60–170 kJ mol^{-1} for strong bonds [18–20]. Weak and moderate hydrogen bonds are very soft interactions displaying a wide range of bond length and angles. These types of hydrogen bonds are generally formed by neutral donor and acceptor species. Weak bonds are observed when H is covalently linked to slightly more electronegative D atoms, as in C–H or Si–H or when the acceptor uses π electrons (arenes, C≡C), while moderate bonds are found for D atoms much than H and A acceptors with lone pairs available (i.e., >N–H, –O–H, –N(H) –H, O=C, O<, aromatic N); also, NH_4^+ and aromatic NH^+ form moderate hydrogen bonds. On the other hand, strong hydrogen bonds involve charged donor (i.e., R_3NH^+, –O–H^+) and/or acceptor (i.e., F^-, H–O^-, C–O^-, P–O^-, > N^-) groups. Salt bridges, as well as other hydrogen bonds involving anions, pertain to this category. It is to be noted that hydrogen bonds have group-pair properties, the nature of a hydrogen bond depending on the nature of both donor and acceptor groups. For instance, H_2O is moderate in both its donor and acceptor hydrogen bond properties, but the hydrogen bond in $F^- \cdots HOH$ is strong, with a bond energy amounting to 96 kJ mol^{-1}, as determined by means of ion cyclotron resonance spectroscopy [27].

Three structural parameters define the geometry of a hydrogen bond: the D–H covalent bond length, the H···A hydrogen bond length, and the D···A hydrogen bond distance, which define the D–H···A hydrogen bond angle. Only in the case of strong hydrogen bonds this angle is close to 180°, while for weaker bonds deviation from linearity is very common. Accordingly, only for strong hydrogen bonds the D···A distance is a reliable measure of the H···A hydrogen bond length. It has been accepted for many years that when the D···A distance is shorter than the sum of D and A van der Waals radii, D and A are hydrogen bonded. This condition, however, is sufficient, but not necessary, and while it generally holds for strong and moderate bonds, weak hydrogen bonds can occur with significantly longer D···A distances. There is a qualitative correlation between hydrogen bond lengths and angles, lengths becoming shorter as the angles increase [20]. Hence, shortest lengths are observed for linear or almost linear hydrogen bonds. There are also evidences that, at least for certain types of donor–acceptor pairs, hydrogen bonds tend to align with the lone pairs of the acceptor. For instance, statistical analysis of 1500 NH···O=C hydrogen bonds showed a significant tendency of these bonds to occur along the directions of the conventionally viewed sp^2 lone pair [28]. A successive computational study performed by an *ab initio* method substantially agreed with this observation, although it pointed out that there is poor dependence of the strength of hydrogen bonds on hydrogen bond angles, which cannot be explained by steric hindrance [29]. Figure 2.3 is a schematic representation of the principal hydrogen bond types.

There is also a rough correlation between hydrogen bond energies and distances, bonds tending to become stronger as the distance decreases. *Ab initio* molecular

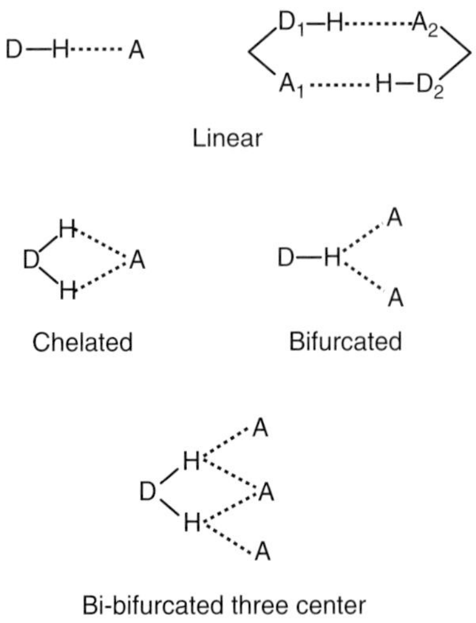

Figure 2.3 Principal hydrogen bond types.

orbital calculations performed for hydrogen-bonded dimers with a variety of possible configurations and conformations showed that, for weak and moderate hydrogen bonds, electrostatic forces may account for more than 80% of the interaction energy, while for very strong hydrogen bonds, the charge transfer term becomes more important. In the case of the difluoride $(F–H–F)^-$ anion, for instance, the charge transfer term, accounting for 44% of the attractive energy, is greater than the electrostatic contribution [30–32].

Hydrogen bonds are responsible for strong association events in the gas phase, but the strength of these interactions is attenuated by medium effects and can be enormously reduced depending on the medium nature. For instance, a reliable energy value for the $F–H\cdots F^-$ bond, one of the strongest hydrogen bonds, was determined to be 163 kJ mol^{-1} by means of ion cyclotron resonance spectroscopy [27], while the free energy change for the formation of HF_2^- from F^- and HF in water does not exceed −3.5 kJ mol^{-1} at 298 K [15–17]. Accordingly, it is widely accepted that strong hydrogen bonds are quite rare in host–guest solution chemistry.

The transition between the hydrogen-bonded $HCOOH\cdots NHCH_2$ (formic acid–methyleneimine) and the salt-bridged $HCOO^-\cdots{}^+HNHCH_2$ pairs was studied by *ab initio* calculations as a function of the hydrogen bond distance and the medium dielectric constant (ε) [33]. In the absence of external influence ($\varepsilon = 1$), the formation of the hydrogen-bonded pairs from the separated partners produces a stabilization of some 40 kJ mol^{-1}, while the formation of the salt-bridged pair would be disfavored by about 200 kJ mol^{-1} because of the difficulty in achieving the relevant charge separation in the absence of dielectric effects. Calculations showed that, on increasing ε, the salt-bridged adduct is stabilized much more than the neutral hydrogen-bonded one, the first becoming more stable for $\varepsilon > 4$, and the energy barrier for proton transfer decreases, vanishing for very short hydrogen bond distances (2.5 Å).

Total or partial proton transfer occurring in hydrogen bond formation is related to the acidity or basicity of donors and acceptors. An interesting study correlating the energetics of hydrogen bonds with the difference in acidity constants ΔpK_a for homologous series of compounds, under nonaqueous conditions leading to the formation of low-barrier hydrogen bonds (LBHBs), showed a linear correlation between the increase in hydrogen bond energy and the decrease in ΔpK_a, the equalization of acidic properties ($\Delta pK_a = 0$) being the best condition for strong hydrogen bonding [34]. Extension of these results to anion coordination points out that anions that are expected to take more advantage from hydrogen bonding with receptors are those resulting from deprotonation of weak acids in agreement with generally observed trends ($F^- > Cl^-$, $PO_4^{3-} > RCO_2^- > NO_3^-$).

Taking into account the available data for hydrogen bond association it is possible to say that, under the best matching and medium conditions, the energetic contribution of a single hydrogen bond to the formation of anion complexes in solution does not exceed 30 kJ mol^{-1}, being much smaller in water, where a value of about −5 kJ mol^{-1}, as for salt bridges, seems to be a good contribution.

In the case of anions containing organic moieties, other weak forces, such as van der Waal and π-stacking interactions, in addition to those discussed above, as well as hydrophobic effects, can be operative through the organic moieties in determining the overall stability of anion complexes. The analysis of van der Waals and π-stacking interactions, as well as of the hydrophobic effect, is not central to anion coordination, although it is so for supramolecular chemistry, and can be found in specific literature [2, 35–38].

2.2.2
Charge of Anions and Receptors

According to Coulomb's law, it is reasonable to expect that the strength of the interaction between anions and positively charged receptors increases with the charge, or the charge density, on the interacting partners. While this expectation is completely satisfied when point charge particles in the gas phase are considered, deviations from this rule can be observed with real anions and receptors for which the individual structure and localization of charge, as well as the involvement of forces other than electrostatic attractions, and solvation effects may characterize the pairing process. Nevertheless, Coulomb's law is frequently followed in anion coordination. This is a simple and valuable rule that can be successfully used when the binding of anions is addressed, but charge–charge interactions can be rarely exploited to modulate the selective discrimination of a certain anion over another, unless they bear largely different charges. Some examples of anion coordination processes that follow, or not follow, the Coulomb's law are presented in other sections of this chapter. We describe here the illustrative case of metallocyanide anion binding by protonated forms of polyamine receptors, which is mostly based on charge–charge attraction.

Metallocyanide anions, such as $Co(CN)_6^{3-}$ and $Fe(CN)_6^{4-}$, offer a very simple way of analyzing the influence on anion complex stability of negative charge increase, while other factors such as size, shape, and geometry are kept essentially constant. On the other hand, polyammonium species formed by protonation of polyamine ligands offer the possibility of considering anion receptors spanning a large variation of positive charge. Complexes deriving from the interaction of such anions and receptors are called *supercomplexes* or *complexes of second coordination sphere*, since the polyammonium species is placed in the second coordination sphere of the anionic metal complex.

The metallocyanide anions $Co(CN)_6^{3-}$ and $Fe(CN)_6^{4-}$ were shown to form second-sphere complexes with protonated forms of the polyazacycloalkanes **11** and **12** bearing three to seven positive charges [39]. The stability constants determined for these complexes showed that (i) for a given receptor, the stability increases with increasing ligand protonation (increasing positive charge) and (ii) for a given protonation state (positive charge) of the receptors, the highest stability is displayed by the receptor with the highest charge density, namely, the smaller macrocyclic ligand **11**; for a given protonation state of the receptors, the more charged $Fe(CN)_6^{4-}$ forms more stable complexes than $Co(CN)_6^{3-}$. All these data confirm that the formation of similar complexes is mainly driven by electrostatic attraction. It was also shown that the acyclic analogs of **11** and **12** follow the same trends, while forming less stable complexes than **11** and **12**, which are able to gather a greater density of positive charge on protonation, because of their cyclic nature. It was also shown that these trends are followed by larger macrocycles containing greater numbers (up to 12) of amine groups, although increased stability was observed from the complexes formed by the receptor containing 10 amine groups to those of the receptors having 11 such groups, a phenomenon that was ascribed to a change from external to inclusive coordination of the metallocyanide anions [40].

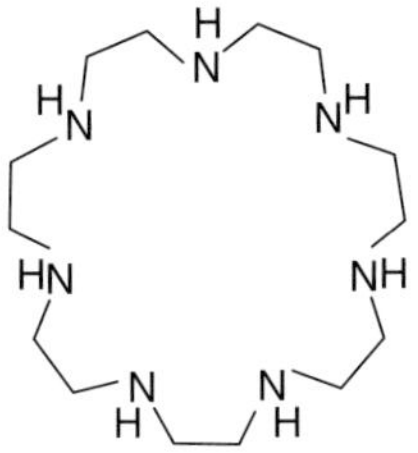

11

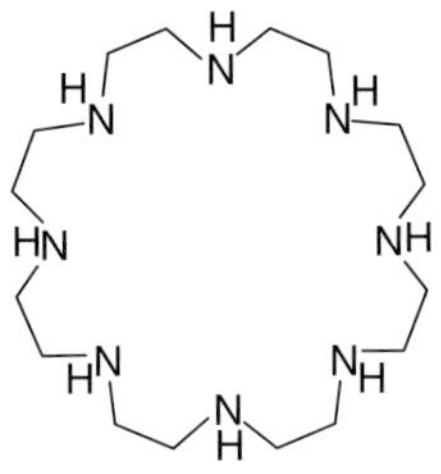

12

2.2.3
Number of Binding Groups

Although only weak forces are involved in anion coordination, the participation of a large number of binding groups makes it possible to achieve a considerable stability of the anion complexes. An outstanding example of multiple interactions in anion binding is offered by the HPO_4^{2-} complex of the periplasmatic phosphate-binding protein (PBP) responsible for phosphate transport in bacteria once the anion has diffused across the outer cell membrane. An extraordinarily well-resolved crystal structure of this complex allowed the direct observation of 12 hydrogen bonds holding the completely desolvated anion into the specific protein-binding site, buried 8 Å from the protein surface [41] (see Figure 1.3, Chapter 1). Eleven hydrogen bonds involve phosphate oxygen atoms and NH groups of the main chain or from arginine side chains or OH groups from serine and threonine side chains. An additional hydrogen bond, with the anion acting as a donor, is found between the proton of HPO_4^{2-} and a carboxylate group of the side chain of Asp56. A high stability constant ($K = 6.51 \times 10^6\ M^{-1}$) is associated with the formation of this complex [42]. Likewise, synthetic anion receptors can be constructed in such a way to introduce a number of appropriate functions matching as many functions as possible of the anionic counterparts to achieve strong binding. This implies the effectiveness in anion coordination of a paradigm similar to that valid for metal ion coordination, according to which, within a limited range of coordination possibilities, complex stability increases with the number of donor atoms.

2.2.3.1 Additivity of Noncovalent Forces

As for other supramolecular assemblies, the idea that each one of the several functionalities included in a given receptor–anion system can make a separate and additive contribution to the stability of anion complexes has given impulse to the development of anion coordination and to the design of anion receptors. An analysis of the additivity of weak forces [2], developed by extending to supramolecular assemblies a former interpretation of the classic chelate effect [43], gave a clear definition of additivity, showing that even very weak interactions may lead to strong host–guest association provided the total number of binding sites is sufficiently large. According to this analysis, the simultaneous formation of n weak interactions between the anion A and the receptor R could be described, to a first approximation, by assuming that the free energy change, ΔG_t, for the equilibrium A + R = AR is the sum of the n free energy changes associated to the n interactions interpreted as binding equilibria between pairs of hypothetical individual species bearing the interacting functionalities.

$$\Delta G_t = \Delta G_1 + \Delta G_2 + \Delta G_3 + \cdots + \Delta G_n \tag{2.1}$$

Accordingly, the total binding constants, K_t, would be the product of the n hypothetical single equilibrium constants.

$$K_t = K_1 \times K_2 \times K_3 \times \cdots \times K_n \tag{2.2}$$

It must be stressed, however, that Eqs. (2.1) and (2.2) are not correct. In fact, in Eq. (2.2), there is incongruence between the dimension of K_t (M^{-1}) and the dimension of the product $K_1 \times K_2 \times K_3 \times \cdots \times K_n(M^{-n})$. On the other hand, each term of the sum $\Delta G_1 + \Delta G_2 + \Delta G_3 + \cdots + \Delta G_n$ in Eq. (2.1) contains an entropy term referring to the conversion of two species into one complex, while ΔG_t involves only one such term. One way to overcome this problem consists in transforming all equilibrium constants into dimensionless quantities by expressing concentrations in molar fractions instead of molarities. This means that, for sufficiently diluted solutions, molar concentration must be divided by 55.6 (moles of H_2O per dm^3 of H_2O) to obtain molar fractions. The total binding constant expressed in molar fraction, K_t^X, will then take the form

$$K_t^X = K_1^X \times K_2^X \times K_3^X \times \cdots \times K_n^X = (55.6)^n K_1 \times K_2 \times K_3 \times \cdots \times K_n \tag{2.3}$$

where

$$K_t^X = 55.6 K_t \tag{2.4}$$

and consequently,

$$K_t = (55.6)^{n-1} K_1 \times K_2 \times K_3 \times \cdots \times K_n \tag{2.5}$$

This means that, if ideal additivity is assumed, the n weak interactions between the anion A and the receptor R give rise to a complex whose stability constant is equal to the product of the n hypothetical single equilibrium constants magnified by a factor of $(55.6)^{n-1}$.

Interestingly, this approach, which is condensed in Eq. (2.5), furnishes a quantitative explication of the classic chelate effect, which can be extended to all interaction processes involving species containing various binding sites. Unfortunately, Eq. (2.5) is not amenable to verification since stability constants such as $K_1 - K_n$ are not generally available, given that the corresponding binding equilibria involving single isolated functionalities are very weak and generally difficult to be studied.

We have seen in Section 2.2.1 that the individual contribution of a single bond to the overall free energy of complexation can be quantified when a linear correlation is verified between the free energy changes of association and the number of interactions formed in the complex. Although some requirements (interactions of similar type and binding sites in the same environment and geometrically matching each other) must be verified and the number of interactions correctly evaluated for observing similar correlations, several successful cases have been reported in which the complexation free energy responds to the additivity of single incremental contributions. The linear correlations affording the mean contribution of -5 ± 1 kJ mol^{-1} for the formation of a salt bridge in water presented in Section 2.2.1 are examples of these cases.

2.2.4 Preorganization

Molecular preorganization is a relative property, since it is related to the interaction between specific species. It is a property of large consideration for all aspects of interaction processes. A preorganized receptor does not need to modify its structure to achieve optimal interaction with a certain substrate, and consequently, it does not need to spend energy and time to get the most appropriate conformation, thus leading to more favorable interaction processes than a poorly preorganized one, both from kinetic and thermodynamic viewpoints. An example of the effect of ligand preorganization on anion coordination is shown by the sulfate complexes of the two parent bisguanidinium receptors **13** and **14** (Figure 2.4). A key feature of the guanidinium group is its ability to provide two protons that point in almost

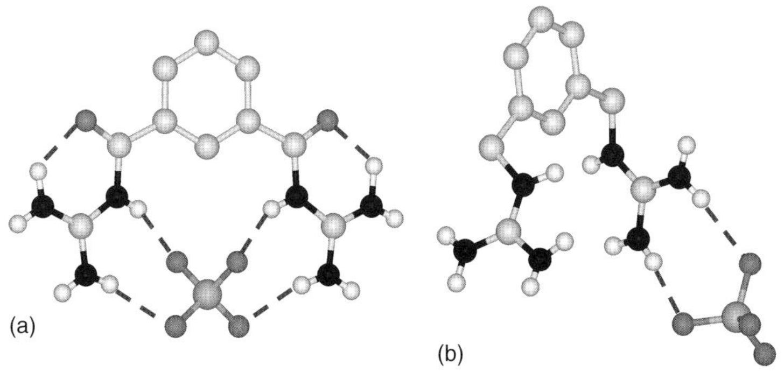

Figure 2.4 Structures of the sulfate complexes with **13** (a) [45] and **14** (b) [46].

the same direction, allowing the stabilization of two parallel hydrogen bonds. It was reported that intramolecular hydrogen bonding between the carbonyl group and guanidine N–H preorganizes the receptor, forcing a planar geometry that promotes coordination of tetrahedral oxoanion into the receptor cavity [44]. Indeed, in the crystal structure of the sulfate complex with **13**, the anion is hosted into the rigid receptor cavity, held by four hydrogen bonds with the two convergent guanidinium groups acting as chelating functions (Figure 2.4a) [45]. By contrast, the parent receptor **14**, endowed with more conformational freedom, does not act as a bis-chelating agent in the crystal structure of [SO_4(**14**)] displaying the anion coordinated by a single guanidinium group pointing outside the receptor cavity (Figure 2.4b) [46]. Although **13** displays successful preorganization for sulfate binding, some flexibility exists at the phenyl–carbonyl bonds, which prevents a greater preorganization of this receptor.

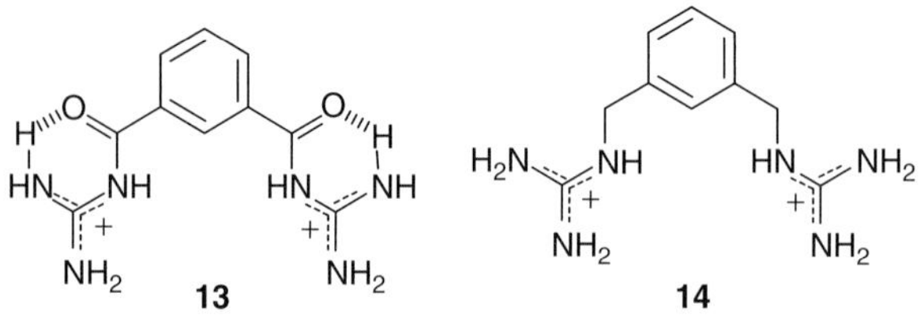

13 **14**

The effect of flexible single bonds in determining the preorganization of linear systems was analyzed by considering the variation of the free energy changes determined for the formation of several anion complexes involving the association of α,ω-dianions and α,ω-dications (Figure 2.5) with the total number of flexible (freely rotatable) single bonds connecting the binding sites. Although the number of such single bonds differs from 6 to 13, the difference between the free energy change variation between the strongest and the weakest complexes was small. The fairly linear correlation observed between the free energy of association and the number of single bonds yielded an energetic disadvantage of only 0.5 kJ mol^{-1} for one single bond [10].

One system to reduce the conformational freedom of ligand molecules, in search of enhanced preorganization, consists in constructing macrocyclic or even more structured three-dimensional receptors. Actually, there is not a survey of data large enough to convincingly establish if macrocyclic receptors give rise to more stable anion complexes than their acyclic analogs, which could be regarded as a *macrocyclic effect* by analogy with metal ion coordination (see next section), and even fewer data are available for polycyclic ones. Nonetheless, macrocyclic and macropolycyclic receptors have been the basis for the evolution of anion coordination; probably they have made possible the observation of many anion coordination events, and surely, they have established the largest basis for selectivity in anion coordination.

Preorganization in macrocycles is principally determined by convergence of binding groups inside the receptor cavity. A good example is the fluoride complex with the diprotonated form of the sapphyrin **15**. The X-ray structure of this

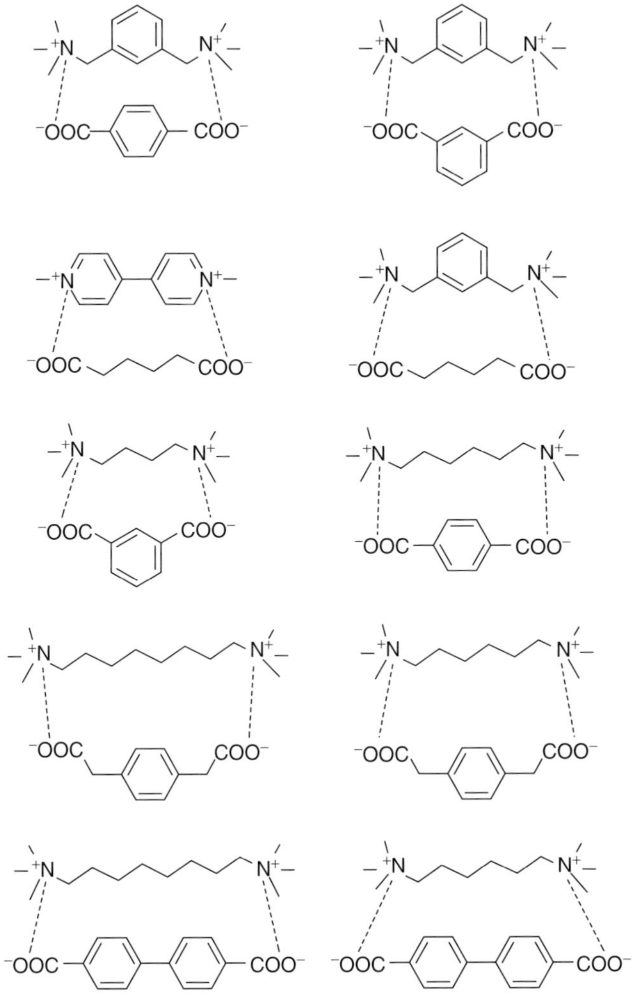

Figure 2.5 Anion complexes between α,ω-diammonium cations and α,ω-dicarboxylate anions [10].

complex showed an almost perfect complementarity between the ligand and the fluoride anion, which is lodged inside the center of the sapphyrin to form five nearly linear hydrogen bonds of very similar length (2.7–2.8 Å), the ligand nitrogen atoms and the coordinated fluoride anion being essentially coplanar [47]. Successive studies evidenced that, in agreement with the stable in-plane hydrogen bond arrangement observed in this crystal structure, $H_2\mathbf{15}^{2+}$ is capable of forming highly stable encapsulated complexes with fluoride in CH_2Cl_2 and MeOH solutions, and displayed a binding selectivity toward this anion of at least 10^3 relative to either chloride or bromide even in the competitive MeOH solvent [48].

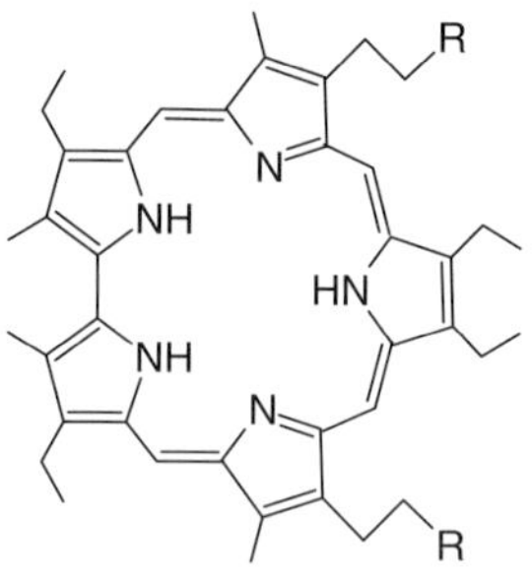

15

An important category of anion receptors is that of ammonium-based macrocycles and macropolycycles. The anion-binding ability of these ligands is due to the positive charge accumulated by protonation or quaternization of amine groups around the ligand cavity. Protonated cyclic and polycyclic amines take further advantage by hydrogen bonding with the anions. The high preorganization of such cyclic receptors is hampered, however, by the *out* conformation that ammonium groups tend to assume as the positive charge on the receptor increases. While an increase in the number of protonated amine groups substantially increases their affinity for anions, an opposite effect is to be expected in consequence of the *in–out* interconversion that ammonium groups can undergo to reduce the electrostatic repulsion exerting between them. This effect was already known since the historical work by Park and Simmons on halogenide anion binding with diprotonated forms of the macrobicyclic diaza ligands named *katapinands* (Figure 2.6) [49]. Although the *out* conformation of protonated amine groups generally reduces ligand preorganization toward anion binding, the *out* conformation of quaternary ammonium groups can be used in the design of anion receptors as an element of their preorganization. This is the case, for instance, of the macrotricycle **6** obtained by permethylation of **5**. The *out* conformation of all ammonium groups in **6** gives rise to an expanded ligand cavity and eliminates the possibility of hydrogen bonding with anions coordinated into the cavity. Accordingly, **6** displays a greater preorganization to bind Br^- among other halogenide anions, while fully protonated **5** forms a more stable complex with Cl^- [13].

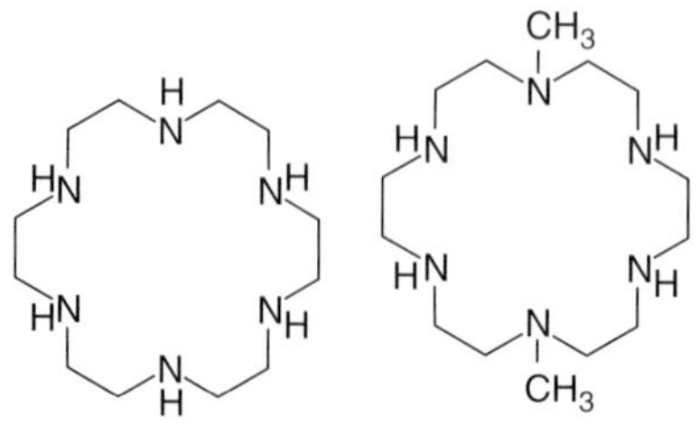

16 17

In addition to conformation effects, protonation of polyazamacrocycles may affect ligand preorganization through localization of charge. It has been shown

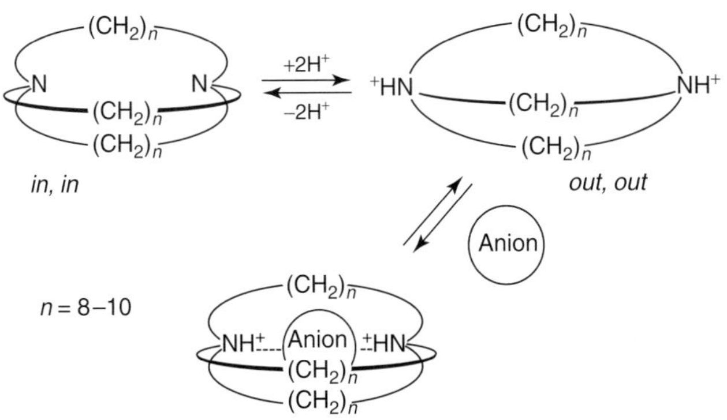

Figure 2.6 Conformational changes in the katapinands on protonation and anion coordination.

that for symmetrical ligands such as **16**, protonation occurs in such a way that the ammonium groups are located at the greatest distance between them in order to disperse the accumulated positive charge as much as possible and minimize the electrostatic repulsion. Different localization of charge can be achieved by removing the equivalence of the amino groups. Methylation of secondary amines, for instance, reduces their basicity in water [50]. Accordingly, in the tetraprotonated forms of **17**, $H_4\mathbf{17}^{4+}$, the lower basicity of tertiary amino groups orientates protonation on secondary ones, thus determining a greater localization and gathering of charge with respect to $H_4\mathbf{16}^{4+}$. Such an arrangement of binding groups in $H_4\mathbf{17}^{4+}$ was shown to be better preorganized for the binding of certain anions such as ATP^{4-} and $P_2O_7{}^{4-}$ forming more stable complexes with $H_4\mathbf{17}^{4+}$ than with $H_4\mathbf{16}^{4+}$ [51, 52].

In order to avoid conformational modifications and the ambiguity in establishing protonation site location that may occur on protonation of polyamine ligands, preorganized distributions of charge have been obtained by inserting quaternary ammonium groups into the receptor structures. Several example of quaternary ammonium-based anion receptors can be found in the literature for linear [4, 5, 10, 53–57], macrocyclic [12, 58–64], and macropolycyclic [13, 65–71] systems.

2.2.4.1 Macrocyclic Effect

The greater stability of macrocyclic complexes versus the analogous species formed by the corresponding acyclic ligands, otherwise known as the *macrocyclic effect*, is not as largely observed for anion coordination as for metal ion coordination. Anions exist in a wide range of shapes, and optimal matching with receptors may require the definition of more structural features than the simple cyclic/acyclic one. Nevertheless, several examples of anion coordination responding to the macrocyclic effect can be outlined, provided that appropriate cyclic and acyclic counterparts are considered. The last assumption also applies to the effect observed for metal ions, but because of the lower stability commonly displayed by anion complexes, the connected macrocyclic effect is less evident and necessitates a strict observance of this rule. Accordingly, the cyclic and the acyclic receptors should differ from each other for a single connection determining the opening of the ring without modifying the nature of the binding groups.

It was reported, for instance, that the macrocyclic effect is observed when the binding of $Fe(CN)_6^{4-}$ and $Co(CN)_6^{3-}$ anions by the polyprotonated forms of the macrocyclic polyamine ligands **11** and **12** is compared with the formation of analogous complexes with the acyclic polyamines **18** and **19** in water [39]. In this case, **18** and **19** can be ideally generated by the cleavage of one C–C bond in **16** and **17**, respectively, followed by saturation of the separated carbon atoms. Accordingly, all nitrogen atoms in the four ligands are secondary and are spaced by ethylenic connectors. As depicted in Figure 2.7, a small to moderate macrocyclic effect is observed for all protonation states of the ligands, the difference between the stability constants of the macrocyclic and the corresponding acyclic complexes ($\Delta \log K = \log K_{mac} - \log K_{acy}$) varying in the range $\Delta \log K = 0.3{-}1.9$. A more evident effect ($\Delta \log K = 2.1{-}4.6$) is found when comparing the binding ability of **11** and **18** toward $H_2PO_4^-$ in water [72]. In this case, large $\Delta \log K$ values, up to 4.6, are reached because **18** does not show any detectable tendency to bind the anion. A similar case was reported for the complexes of **11** and **12** with SO_4^{2-} [73]. Since the acyclic polyamines **18** and **19** do not give rise to appreciable interaction with this anion in water, in any ligand protonation state, $\Delta \log K$ values up to 5.4 were found. The last cases are a clear demonstration that the macrocyclic effect can be concretely operative also in anion coordination, as the cyclic nature of the ligands is decisive in determining their ability to bind or not bind the target anions. Dissection of the free energy changes for $H_2PO_4^-$ and SO_4^{2-} binding by **11** and **12** into the

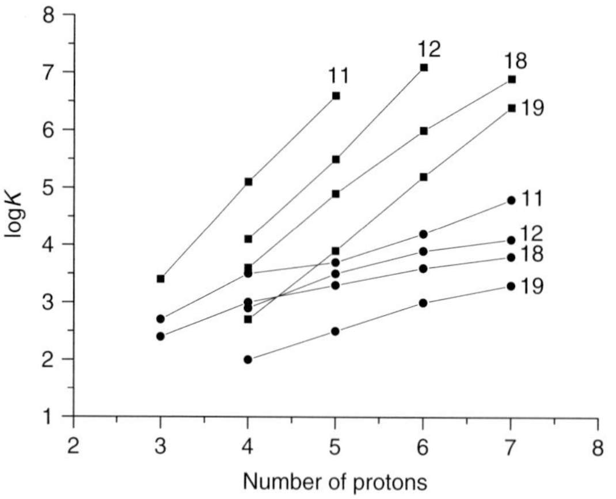

Figure 2.7 Plot of the logarithms of the stability constants for the equilibria $M(CN)_6^{(n-6)} + H_pL^{p+} = [M(CN)_6(H_pL)]^{(n+p-6)}$ (n = metal ion charge) versus the number of positive charges in the receptor for the interaction of $Fe(CN)_6^{4-}$ (squares) and $Co(CN)_6^{3-}$ (circles) with receptors **11**, **12**, **18**, and **19** [39].

enthalpic and entropic contributions, performed by means of microcalorimetric measurements, showed that the binding of these anion is promoted either by favorable enthalpic ($H_2PO_4^-$) or entropic (SO_4^{2-}) contributions [72, 73], extending to anion coordination the unresolved question about the enthalpic/entropic nature of the macrocyclic effect for metal ion complexation.

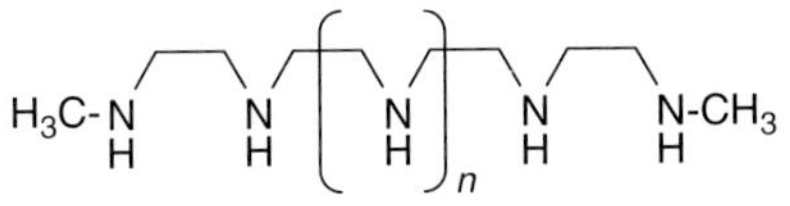

$n = 3$, **18**
$n = 5$, **19**

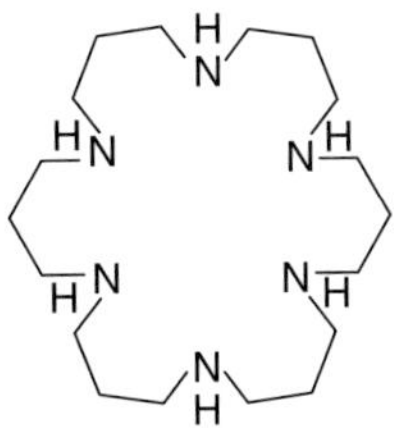

20

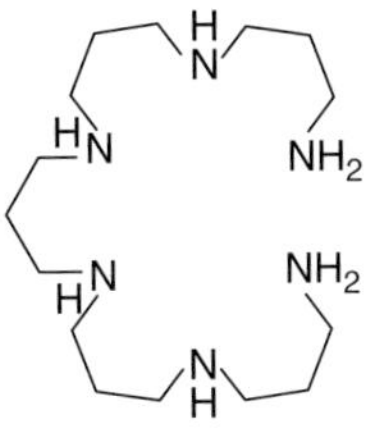

21

A small to moderate macrocyclic effect was also found for SO_4^{2-} ($\Delta \log K = 1\text{–}1.5$), oxalate^{2-} ($\Delta \log K = 1.2\text{–}1.4$), maleate^{2-} ($\Delta \log K = 0.6\text{–}1.3$), and fumarate^{2-} ($\Delta \log K = 0.2\text{–}0.4$) binding by protonated forms of the azamacrocycle **20** and the polyamine **21**, which can be considered a close acyclic counterpart of **20** because of the presence of terminal secondary amine groups in **21**, while a larger and decisive effect for malonate^{2-} binding ($\Delta \log K = 2.4\text{–}3.3$) is determined by the inability of **21** to bind the anion [74].

More recently, a moderate macrocyclic effect was reported for the neutral macrocyclic amide receptor **22** versus its acyclic counterpart **23** for the binding of Cl^- ($\Delta \log K = 1.7$), acetate$^-$ ($\Delta \log K = 1.3$), and $H_2PO_4^-$ ($\Delta \log K = 1.4$) in DMSO. Crystal structures of these ligands and their Cl^- complexes suggested that the observed macrocyclic effect is determined by an ill-preorganization of the acyclic receptor [75].

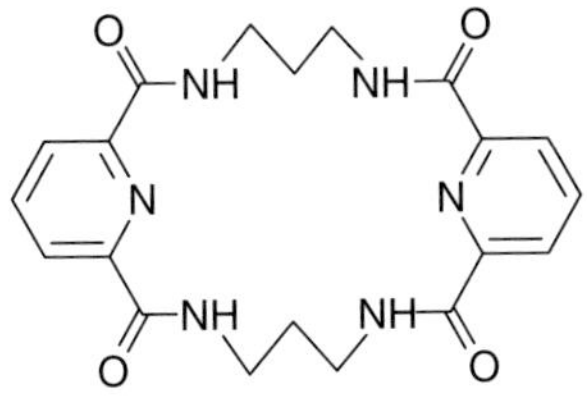

22

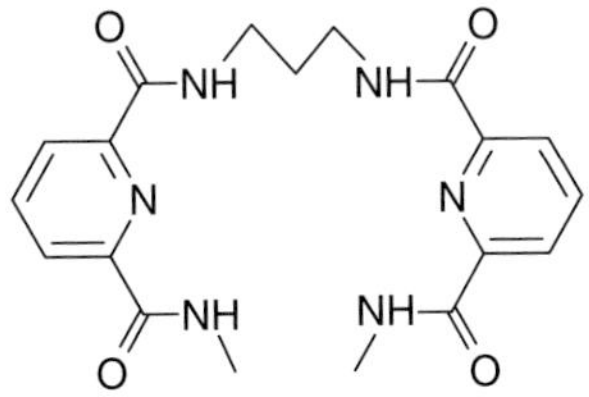

23

2.2.5 Solvent Effects

Almost all anion coordination processes are experimentally studied in solution. Although it might seem trivial, the prime action of solvents in equilibrium studies is to ensure the solubility of reagents and products. This is accomplished by solvent–solute interactions that determine an organized gathering of solvent molecules around the solvated species, which is called *solvation*.

Solvation of a given species (S) in a given solvent (solv) is defined as the process of transfer of this species from the gas phase to the given solvent,

$$S_{(g)} + solv_{(l)} = S_{(solv)} \tag{2.6}$$

which is associated to a free energy change of solvation ΔG_{solv}.

According to a simple electrostatic model, the formation of ion pairs between hard anions and cations (hard spheres with embedded point charges) in the gas phase is expected to be mainly determined by electrostatic attraction, and the process is largely exothermic and accompanied by a small entropic loss. In the presence of a structureless homogeneous solvent, the electrostatic energy is strongly reduced by the dielectric shielding of charges, and the ion-pairing process is much less exothermic or even endothermic, depending on the value of the solvent dielectric constant and mainly promoted by the increase in translational entropy due to desolvation of the interacting ions [76]. Nevertheless, when the actual structures of the interacting ions, as well as of the solvent, in the association process are considered, a number of different short-range interactions must be taken into account. Although computational methods can be used with a certain success to model these interaction processes, the use of experimentally measured thermodynamic parameters is preferable for energetic considerations.

For a general 1 : 1 anion–receptor complexation reaction, A + R = AR, the relevant thermodynamic equilibrium constants K_0 can be expressed by Eq. (2.7)

$$K_0 = \frac{a_{AR}}{a_A a_R} \tag{2.7}$$

Assuming that the activity (a) of each component can be expressed as its concentration multiplied by its activity coefficient (γ), Eq. (2.7) becomes

$$K_0 = \frac{[AR]\gamma_{AR}}{([A]\gamma_A[R]\gamma_R)} = K\left(\frac{\gamma_{AR}}{\gamma_A\gamma_R}\right) \tag{2.8}$$

which can be rewritten in the form

$$K = K_0\left(\frac{\gamma_A\gamma_R}{\gamma_{AR}}\right) \tag{2.9}$$

where K is the stoichiometry equilibrium constant.

For a given species in a given medium, the activity coefficient can be expressed as a function of the free energy change of transfer (ΔG_{tr}) of this species from a reference medium to the given medium:

$$\Delta G_{tr} = RT\ln\gamma \tag{2.10}$$

Combining Eqs. (2.9) and (2.10), one obtains

$$\ln K = \ln K_0 + [\Delta G_{tr}\,(A) + \Delta G_{tr}\,(R) - \Delta G_{tr}\,(AR)]\,/RT \tag{2.11}$$

where K_0 is the equilibrium constant in the reference medium, or the equivalent expression

$$\Delta G = \Delta G_0 - \Delta G_{tr}(A) - \Delta G_{tr}(R) + \Delta G_{tr}(AR) \tag{2.12}$$

Table 2.1 Anion hydration free energies (ΔG°_{h}, kJ mol^{-1}) and transfer free energies (ΔG°_{tr}, kJ mol^{-1}) from water to organic solvents at 298 K [77–80].

	ΔG°_{h}	ΔG°_{tr}							
		H_2O to MeOH	H_2O to EtOH	H_2O to MeCN	H_2O to $MeNO_2$	H_2O to DMSO	H_2O to DMF	H_2O to Me_2CO	H_2O to 1,2-DCE
F^-	−472	16	25.8	71	–	73	51	85	–
Cl^-	−347	13.2	20.2	42.1	37	40.3	48.3	57	52
Br^-	−321	11.1	18.2	31.3	30	27.4	36.2	42	38
I^-	−283	7.3	12.9	16.8	17	10.4	20.4	25	25
N_3^-	−287	9.1	17.0	37	28	25.8	36	43	–
CN^-	−305	8.6	7	35	–	35	40	48	–
SCN^-	−287	5.6	–	14.4	15	9.7	18.4	–	–
NO_3^-	−306	–	14	21	–	–	–	–	7
ClO_4^-	−214	6.1	10	2	(−5)	–	4	–	16
Ac^-	–	16.0	–	61	–	(50)	66	–	–
BPh_4^-	–	−24.1	−21.2	−32.8	–	−37.4	−38.5	−32	−33

where ΔG_0 is the free energy change of complexation in the reference medium. When the reference medium is the gas phase, the free energies of transfer in Eq. (2.12) are free energies of solvation, and Eq. (2.12) can be written in the form

$$\Delta G = \Delta G_0 - \Delta G_{solv}(A) - \Delta G_{solv}(R) + \Delta G_{solv}(AR) \qquad (2.13)$$

According to the last equation, binding of anions in condensed phase is successful only if the host can provide sufficient favorable interactions to outmatch the difference between the solvation free energies of the complex and of the interacting partners ($\Delta G_{solv}(AR) - \Delta G_{solv}(A) - \Delta G_{solv}(R)$).

Table 2.1 lists the free energy changes of solvation in water (hydration free energy changes, ΔG°_{h}) along with the free energy changes of transfer (ΔG°_{tr}) from water to other solvents for several anions. Transfer free energies are generally preferred to solvation free energy in the analysis of solute–solvent interactions. The changes that a solute undergoes when it is transferred from the gas phase to solution are profound, while those occurring on transfer between solvents are largely smaller. The thermodynamic quantities of transfer reflect more nearly the changes in the solute–solvent interactions in different solvents. Accordingly, if these interactions with a solvent are well known, as in the case of water, the interaction with other solvents can be understood better if transfer occurs from this specific solvent than directly from the gas phase [77].

Some physical properties (dielectric constants, dipole moments, polarizability) and polarity indices (donor number (DN), acceptor number (AN), $E_T(30)$) of common solvents, which are of interest in the present context, are reported in Table 2.2 [77, 81–88]. The Gutmann's donor number (DN) is the negative of the molar enthalpy of reaction, expressed in kcal mol^{-1}, of the solvent in question

Table 2.2 Physical parameters and polarity indices of common solvents [77, 80–87]. Values determined at 298 K unless otherwise noted.

Solvent	Dielectric constant	Dipole moment (D)	Molecular polarizability (Å^3)	Donor number (kcal mol^{-1})	Acceptor number	$E_T(30)$
H_2O	78.54	1.85	1.45	18	54.8	63.1
MeOH	32.63	1.70	3.32	19	41.3	55.5
EtOH	24.30	1.69	5.11	20	37.1	51.9
1-PrOH	20.1	1.68	6.74	–	34	50.7
NH_3 (liquid)	16.9	1.47	2.26	59	–	–
Pyridine	12.3	2.19	9.5	33.1	14.2	40.2
CCl_4	2.24	0.0	10.49	0	8.6	32.5
$CHCl_3$	4.81[a]	1.01	9.5	4	23.1	39.1
CH_2Cl_2	8.93	1.14	6.49	1	20.4	41.1
1,2-Dichloroethane	10.36	1.86	8.33	0.0	–	41.9
Me_2SO	48.9	3.96	7.99	29.8	19.3	45.1
Me_2CO	20.7	2.88	6.39	17.0	12.5	42.2
THF	6.4	1.63	7.93	20	8.0	37.4
Et_2O	4.33[a]	1.15	10.2	19.2	3.9	34.6
Dioxane	2.21	0	8.60	14.8	10.8	36.0
DMF	36.7	3.82	7.81	26.6	16.0	43.9
MeCN	37.5	3.92	4.48	14.1	19.3	46.0
$MeNO_2$	35.8	3.46	7.37	2.7	20.5	46.2
$C_6H_5NO_2$	34.8	4.21	14.7	4.4	14.8	41.9
C_6H_6	2.28[a]	0	10.32	0.1	8.2	34.5
Hexane	1.89[a]	0	11.9	0.0	0.0	30.9

[a] At 293 K.

with $SbCl_5$ in 1,2-dichloroethane [84]. The acceptor number AN is defined as a dimensionless number related to the relative chemical shift of ^{31}P in Et_3PO in the solvent in question, with hexane as a reference solvent on one hand, and $Et_3PO–SbCl_5$ in 1,2-dichloroethane on the other, to which AN values of 0 and 100 have been assigned, respectively [87, 89]. The electron pair acceptance polarity index E_T is the lowest energy transition, expressed in kcal mol^{-1}, of the pyridinium phenol betaine indicator dissolved in the solvent in question [88]. The acceptor number, by definition, measures the ability of a solvent to interact with donor solutes such as anions. Accordingly, AN and the polarity index E_T, which is linearly correlated with AN [88], are largely used to analyze solvent effects in anion coordination, while bulk solvent properties, such as dielectric constants and dipole moments, are not of much help.

If the solvation free energies of ligands and their anion complexes are not very dissimilar, as can be expected to occur for complexation of monocharged anions with large and poorly charged receptors, by analogy with metal cation complexation with cryptands in aprotic solvents [90–93], $\Delta G_{solv}(R)$ and $\Delta G_{solv}(AR)$ contributions in Eq. (2.13) cancel at significant extent, thus rendering of major importance the contribution of anion solvation, $\Delta G_{solv}(A)$. Actually, it was shown that binding of Cl^- by the aryl-1,2,3-triazole receptor **24** in acetone, CH_2Cl_2, $CHCl_3$, CH_3CN, and 1 : 1 mixtures of these solvents in acetone gives rise to complexes whose stability, spanning from $K = 1260$ in acetone to $K = 18$ in $CHCl_3$, correlates with Cl^- solvation, decreasing with increasing AN of the solvents [94]. In DMSO and DMSO/acetone 1 : 1 mixture, however, the stability of the chloride complex is considerably smaller than expected on the basis of solvent ANs. DMSO is an excellent acceptor of localized positive charge, such as that of a hydrogen bond donor, and thus a strong competition of this solvent with chloride for the triazole CH donors was hypothesized to explain this particular behavior. In the same study, it was also noted that, in contrast with the correlation found with solvent AN, there is no correlation between binding constants and bulk solvent properties such as dielectric constant and dipole moment. A similar correlation between stability constants and AN was also found for the complexes of the ferrocenyl-based receptors **25** and **26** with Cl^-, $PhCO_2^-$, and $MeCO_2^-$ in acetone, CH_2Cl_2, and CH_3CN, while no obvious correlation was observed with bulk solvent properties [95].

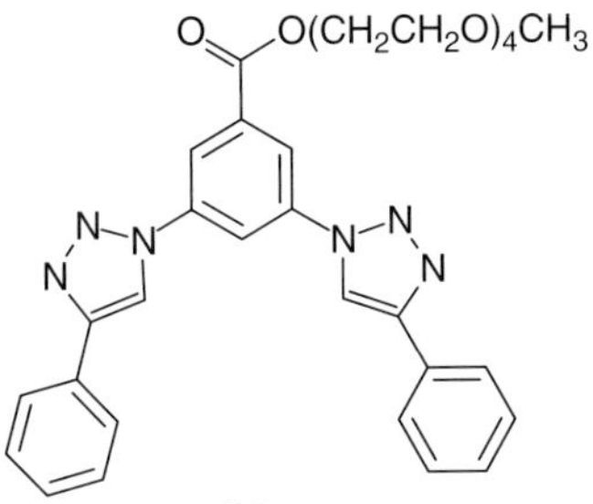

24

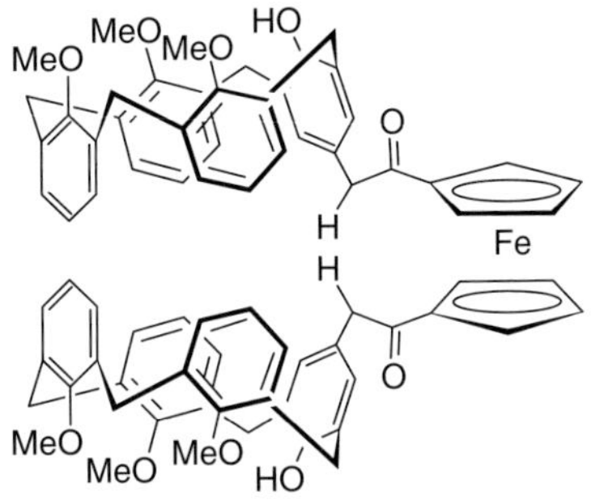

25

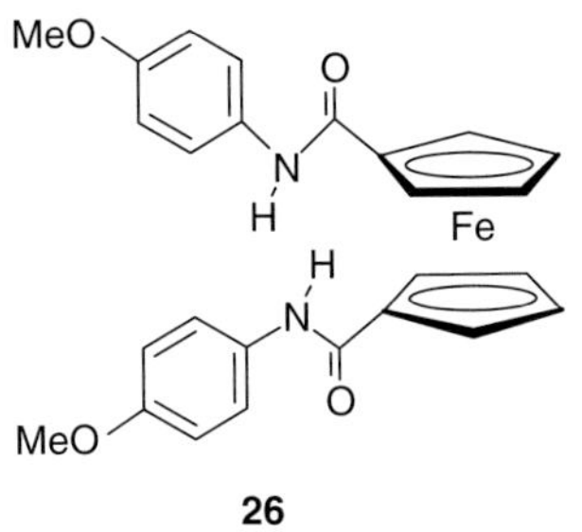

26

Also ligand solvation is, of course, of great importance, in particular, when positively charged receptors are considered and anion binding takes place in highly polar protic solvents such as water. As discussed later in more detail, the association of charged partners in similar solvents produces a large release of solvent molecules, due to charge neutralization, and this phenomenon can be the principal driving force of the association process. As reported just above for the binding of chloride by **24** in the presence of DMSO, neutral ligands may suffer from strong interaction with certain solvents that reduce their anion-binding ability. Another interesting example is given by the octamethylcalix[4]pyrrole receptor **27**. The interaction of **27** with chloride, in the presence of various counterions, was studied in different solvents (CH_3CN, CH_3NO_2, DMSO, CH_2Cl_2, 1,2-dichloroethane) by means of NMR and isothermal titration calorimetry, which gave concordant stability constants characterized by large solvent dependence but with no apparent correlation with any bulk property or polarity index of the solvent [96]. For example, the highest and the lowest complex stabilities were observed in CH_3CN ($K \approx 10^5$) and in DMSO ($K \approx 10^3$), respectively, despite the identical AN of these solvents. Strong association of **27** with DMSO constitutes a strong competition with chloride binding, thus leading to a weaker chloride complex. It was previously shown that **27** is able to strongly interact with solvent molecules, such as DMSO, and the interaction can cause structural modifications of this receptor [97]. Accordingly, it can also be expected that different preorganizations of the receptor might occur in different solvents. Furthermore, the effect of different counterions (tetraethyl-, tetrapropyl-, and tetrabutylammonium; tetraethyl-, tetrabutyl-, and tetraphenylphosphonium), while generally small, was found to be large in CH_2Cl_2, likely due to ion pairing in this nonpolar solvent. From these results, the recommendation arose that "anion binding studies involving new receptors has to be carried out in several different solvents and with several different countercations before a detailed understanding of the anion binding properties or receptor-based selectivity is claimed" [96], a recommendation that should be largely acknowledged.

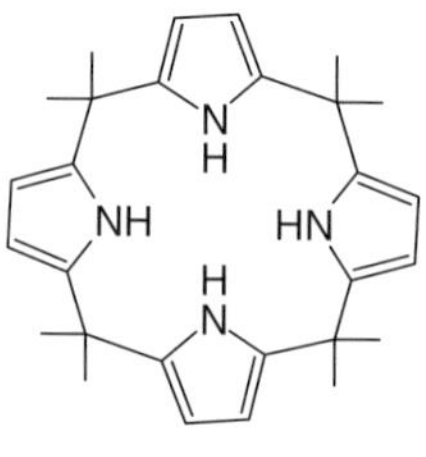

27

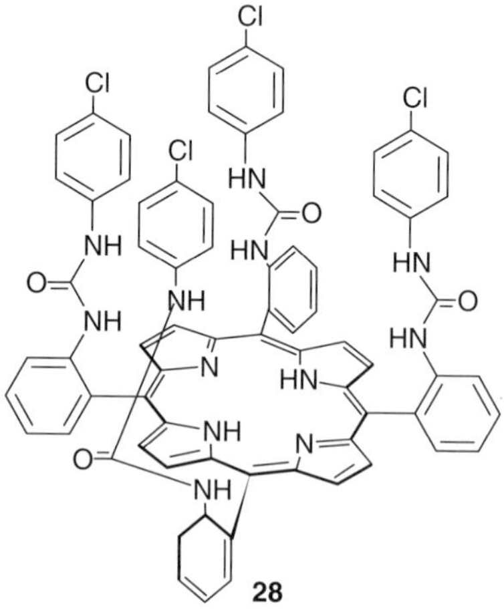

28

Competitive solvents generally have a negative effect on anion complexation. Nevertheless, when such solvent molecules are incorporated into the binding motif, they can contribute to strengthen anion binding rather than weaken it. In other words, a receptor's binding cavity, which is not properly complementary for a certain anion, can be made complementary with the binding and appropriate positioning of a ubiquitous solvent molecule. This is the case, for instance, of chloride and bromide binding with the tetraurea picket porphyrin **28**. Crystal structures of chloride and bromide complexes with **28** showed the anions to be situated between two adjacent ureas and hydrogen bonded via four NH protons [98]. The receptor also includes a DMSO molecule and utilizes it as an active participant of its anion recognition unit. Actually, the DMSO molecule is positioned in the center of the receptor pocket, bound to a urea group via two hydrogen bonds, and oriented such that the electron-deficient sulfur atom is in near van der Waals contact with the included anions. Solution studies demonstrated that the bound solvent molecule determines the anion binding affinity, selectivity, and binding stoichiometry. On binding of a DMSO molecule, **28** is a highly selective receptor for chloride over $H_2PO_4^-$ and forms anion complexes with 1 : 1 stoichiometry. The DMSO molecule precludes $H_2PO_4^-$ from the formation of strong interactions with the receptor without the additional expense for complete desolvation of the receptor pocket. Conversely, in the absence of the bound DMSO molecule, it is selective for $H_2PO_4^-$ and forms complexes with 1 : 1 and 1 : 2 receptor–anion stoichiometry with $H_2PO_4^-$ and chloride, respectively [98, 99].

The solvation power of neat solvents may fall short of the mark for certain applications, but in many such cases, solvent mixtures can be used instead. The different components of a solvent mixture may interact with different portions of the solute, thus having a synergistic effect on solvation. In some cases, even small amounts of a cosolvent are enough to render the solvation power of the mixture sufficiently large for the particular application. We have seen that the polarity indices of solvents are good measures of their ability to solvate ions. The donor and acceptor properties of a solvent mixture are expected to be intermediate to those of the neat solvents, but they are not necessarily linearly dependent on the mixture composition. This is due to the fact that the environment of the solute may have a different composition relative to the bulk solvent, that is, called *preferential solvation*, in a general sense, or *selective solvation*, when the preference for one solvent in the mixture is so large that the other components are practically excluded from the solvation sphere of the solute. Preferential and selective solvation is not limited to ions but also occurs with neutral molecules, such as anion receptors. The above-described anion binding features displayed by **24**–**28** in the presence of DMSO are examples of how selective solvation of the receptor may determine its anion binding properties, including binding affinity, selectivity, and stoichiometry.

The effect of solvent mixture composition on anion solvation is shown in Figure 2.8 for the case of halides in aqueous acetone mixtures. The free energy of transfer, ΔG°_{tr}, from water to aqueous acetone shows a monotonic increase (solvation becomes less favorable) with decreasing water concentration, the increase becoming steeper at low water concentrations as expected for anions preferentially solvated by water [77, 100, 101]. Solvation becomes less favorable in the order $F^- < Cl^- < Br^- < I^-$, with differences becoming more pronounced as the water content decreases [80]. This pattern, which is found for other solvents [100–103], is consistent with the loss of hydrogen bonding ability of halide ions as their density of charge decreases from F^- to I^-. Also, the trend of increasing free energy of transfer with decreasing water contents is observed for other aqueous organic

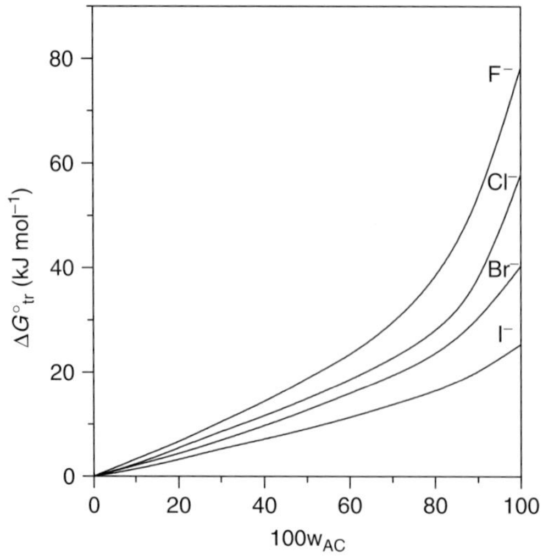

Figure 2.8 Free energy of transfer of halides in acetone/water mixture at 298 K as a function of solvent composition. (Reproduced with permission from Ref. [80].)

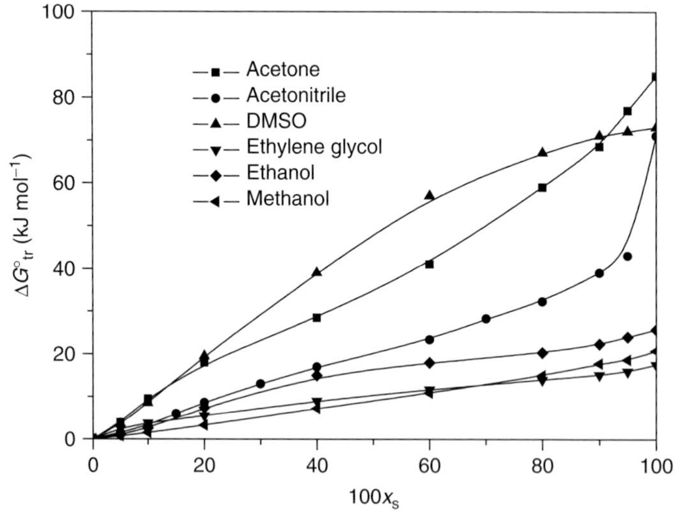

Figure 2.9 Free energy of transfer of fluoride in aqueous mixed solvents at 298 K as a function of solvent composition. (Reproduced with permission from Ref. [100].)

mixtures, although the observed profiles are not necessarily parallel [100–103]. As an example, the variations of the free energy of transfer of fluoride from water to aqueous acetone, CH_3CN, DMSO, MeOH, EtOH, and ethylene glycol mixtures are represented in Figure 2.9. The preferential solvation of fluoride by water is particularly apparent in the very sharp variation occurring at very low water concentrations in the weak acceptor dipolar aprotic acetone and CH_3CN solvents. Noteworthy is the exceptional variation displayed by CH_3CN mixtures, the ΔG°_{tr} decreasing by about 30 kJ mol^{-1} on addition of only 0.05 mole fraction of water to the neat solvent. Accordingly, even a tiny amount of water in aqueous mixtures of similar solvents, especially CH_3CN, in the very low range of water concentration may be responsible for very large variations of fluoride properties in these solvents. For example, the stability constant of the fluoride complex with the calix[4]pyrrole **27** determined by means of isothermal titration calorimetry in absolute CH_3CN (H_2O < 10 ppm) with carefully dried reagents, and taking the precaution of performing calorimetric titrations both by adding the guest into host solutions ($\log K = 1.5 \times 10^5$) and vice versa ($\log K = 1.3 \times 10^5$), is higher by at least a power of 10 than that obtained in CH_3CN containing only 0.5% of water [104]. Molecular dynamics coupled to thermodynamic integration calculations showed that even the presence of water traces introduced by the use of the commercially available trihydrated tetrabutylammonium fluoride can spoil the interaction of fluoride with **27**, and involvement of the counterion seems to be of great importance [105].

On the contrary, less evident effects are found for the mixture of water with the reasonably strong acceptor alcohols. Similarly, because of the particular behavior of aqueous DMSO mixtures, in the low water concentration range (Figure 2.9), fluoride solvation is not much affected by the small addition of water to neat DMSO. Analogous behaviors are found for other anions characterized by a marked ability to interact with protic solvents, in particular, with water.

The monotonous variation of ΔG°_{tr} from water to aqueous mixtures with solvent composition results from the cancellation of complex changes in its free enthalpy (ΔH°_{tr}) and entropy (ΔS°_{tr}) components deriving from a subtle interplay of anion–solvent and solvent–solvent interactions [101]. The variations of enthalpic and entropic contributions to the transfer of fluoride from water to aqueous CH_3CN mixtures are shown in Figure 2.10. Such behavior has been explained [80] by considering that the presence of small amounts of CH_3CN reinforces the water structure about F^-, leading to favorable ΔH°_{tr} and unfavorable ΔS°_{tr} contributions. Since a limited amount of water molecules can be accommodated in the solvation sphere of fluoride, further increase in the concentration of CH_3CN reinforces the water structure within the water domains, rendering fluoride transfer enthalpically unfavorable but entropically favorable. At much higher CH_3CN concentrations, the water structure breaks down, increasing the number of water molecules available for fluoride solvation (favorable ΔH°_{tr} but unfavorable ΔS°_{tr} contributions) until, at very high CH_3CN concentrations, acetonitrile molecules replace water in the coordination sphere of fluoride, likely determining a decrease in the coordination number of the anion, leading to the sharp rise (unfavorable effect) in ΔH°_{tr}. Unlike ΔH°_{tr}, ΔS°_{tr} does not display a sharp reversal at high CH_3CN concentrations, but entropy values in such concentration range must be treated with great caution because of accumulation of error [80].

According to these thermodynamic data, it is evident that the use of organic solvent–water mixtures as solvents for anion binding studies can be a potent tool for regulating both binding strength and selectivity patterns. It must be

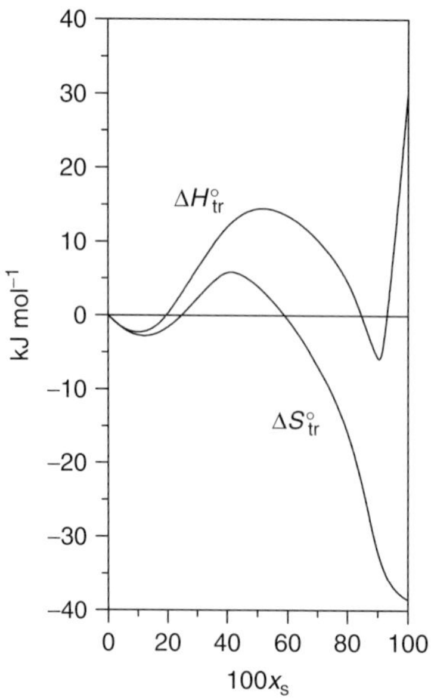

Figure 2.10 Enthalpy and entropy of transfer of fluoride in acetonitrile/water mixtures at 298 K as a function of solvent composition. (Reproduced with permission from Ref. [80].)

stressed, however, that the strong dependence of ΔG°_{tr} and ΔH°_{tr} observed for many solvent mixtures with the mixture composition, at least in particular ranges of composition, can be the source of large errors in the determination of such equilibrium parameters.

2.3 Molecular Recognition and Selectivity

Molecular recognition refers to the specific interaction between two or more species through noncovalent forces [106]. This implies that a receptor is able to perform the recognition of a particular substrate when is able to selectively bind it in the presence of other potential substrates. Under thermodynamic control, such selective binding, which is the expression of a recognition process, is determined by the number and the strength of the receptor–substrate interactions and by the steric and electronic complementarity of the interacting partners, which lead to tighter association with the recognized substrate. Nevertheless, also the binding interactions that are lost on formation of the receptor–substrate complex, including interactions with the solvent and eventual intramolecular interactions, as well as possible competitive reactions, contribute to the success of the recognition process. Accordingly, binding selectivity can be understood as the preferential binding of a receptor to a substrate over one or more other substrates under equivalent experimental conditions (analytical concentration of substrates and receptors, pH, temperature, etc.). Considering the intimate relationship between molecular recognition and binding selectivity, these aspects are treated within the same section.

First of all, let us consider a crucial point in the analysis of binding selectivity. Usually, quotients of stability constants are used to express selectivity ratios. This method is correct when processes developing with the same stoichiometry are considered, the formed complexes coexist in the same range of experimental conditions, and no competing reactions occur. These conditions are often met, in particular, when anion coordination is studied in aprotic solvents, and both anions and receptors do not suffer protonation. Nevertheless, there are a large number of anion binding processes that are the result of competitive reactions and modification of the environmental conditions (pH, etc.). In such cases, the analysis of selectivity may be not very straight, and the use of selectivity ratios calculated as quotients of stepwise stability constants is to be considered generally incorrect. This issue was addressed for the binding of phosphate-type anions with polyammonium receptors by using for the first time a method based on calculating the distribution diagram for a system containing one receptor and two substrates, all in the same concentration, and representing the overall percentages of free and complexed substrates as a function of pH [107]. Another method [108], successively suggested, implies the comparison of the effective (conditional) stability constants calculated as a function of pH for the individual interactions of the receptor (L) with each substrate (A) under analysis. If the total amount of free receptor ($\sum[H_iL]$), free substrate ($\sum[H_jA]$), and complex formed ($\sum[H_{i+j}AL]$) are known, one can define an effective stability constant as

$$K_{\mathrm{eff}} = \sum \frac{[H_{i+j}AL]}{(\Sigma[H_iL] \times \Sigma[H_iA])} \tag{2.14}$$

Both methods, which can be extended to systems containing more substrates and more receptors, have the advantage of not requiring any assumption on the location of protons within the complex.

Application of the two methods, performed by using literature stability data [72, 73], is represented in Figure 2.11 for the binding of phosphate and sulfate by protonated forms of **16**. The information provided by both methods is quite similar, pointing out that preferential binding of phosphate over sulfate is possible below pH 9, while a reversed selectivity occurs in more alkaline solutions. Plots of the ratios between the percentages of coordinated phosphate and sulfate (Figure 2.11a, inset) as well as of the ratios between the effective stability constants of phosphate and sulfate complexes (Figure 2.11b, inset) versus pH show a selectivity peak centered at about pH 7, indicating the optimal pH conditions for selective binding of phosphate. The reciprocal of these ratios would afford the best conditions for selective binding of sulfate that occur above pH 10.

The macrocyclic effect described in the previous section highlights the role played by receptor structure and, in particular, its preorganization in determining binding selectivity. One of the most basic criterion for selective binding consists in the dimensional discrimination of anionic guests, which is also related to

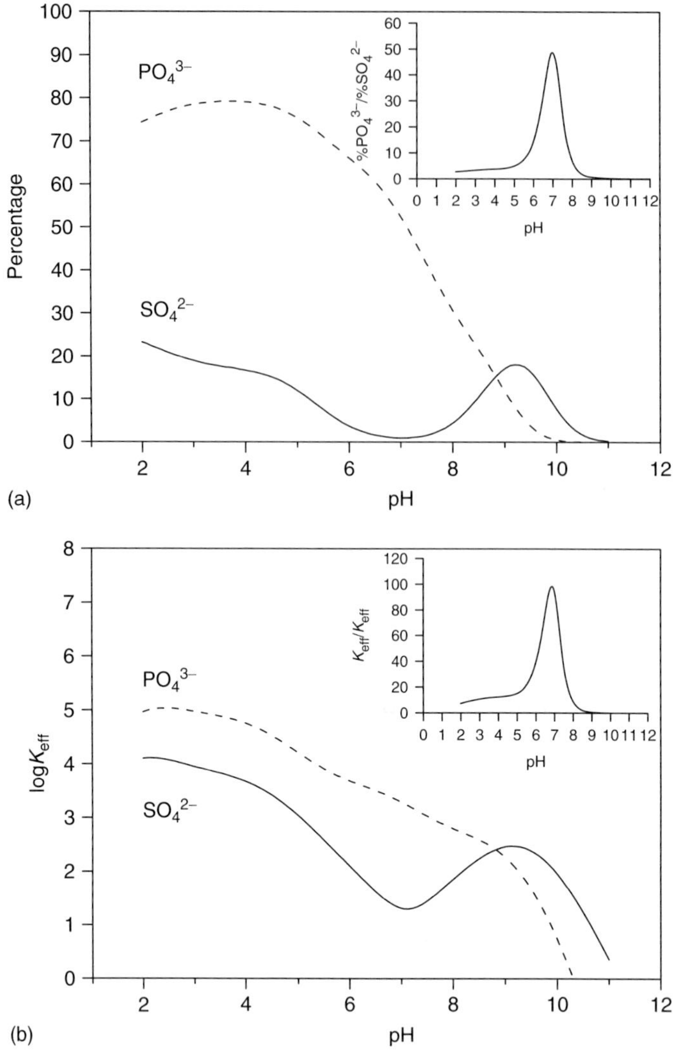

Figure 2.11 (a) Overall percentages of sulfate and phosphate, respectively, bound to protonated forms of **16** calculated for equimolar concentrations of anions and receptor. Inset: ratio between the percentages of coordinated phosphate and sulfate. (b) Logarithms of the conditional stability constants of sulfate and phosphate complexes with **16**. Inset: ratio between the conditional stability of phosphate and sulfate complexes.

receptor structure. An early example in this line was the spherical recognition of Cl^-, among halogenide anions, performed by the tetraprotonated form, $H_4\mathbf{5}^{4+}$, of the macrotricyclic receptor **5** [109]. A crystal structure of the chloride complex $[Cl(H_4\mathbf{5})]^{3+}$ showed the anion held inside the receptors cavity by a tetrahedral arrangement of four $NH^+\cdots Cl^-$ salt bridges [110]. The ligand displays a cavity size (3.2 Å from N to cavity center) appropriately matching the Cl^- dimensions, the other halogenides being too small (F^-) or too large (Br^-, I^-) for a good cavity fitting. Accordingly, ^{13}C NMR measurements in aqueous solution showed that $H_4\mathbf{5}^{4+}$ forms inclusion complexes with F^-, Cl^-, and Br^-, while it is not able to bind I^- or polyatomic anions such as NO_3^-, $CF_3CO_2^-$, and ClO_4^-. The stability of the Cl^- complex ($\log K > 4$) was found to be at least 3 orders of magnitude greater than that of the Br^- complex ($\log K < 1$), showing a selectivity ratio of at least 10^3 for chloride over the other halogenide anions [109].

An extremely larger selectivity in F^- over Cl^- binding was achieved by the smaller macrobicycle **29** containing two tripodal subunits, which defines a spheroidal cavity [111]. Increasing stability of the F^- complex was observed in water when the receptor positive charge (protonation) was progressively increased from $H_3\mathbf{29}^{3+}$ to $H_6\mathbf{29}^{6+}$ ($\log K = 3.30$ for $[F(H_3\mathbf{29})]^{2+}$, $\log K = 10.55$ for $[F(H_6\mathbf{29})]^{5+}$). Such enormous stability for complexes of monocharged anions can be ascribed to the ligand ability to form up to six strong hydrogen bonds with F^-, as shown by a crystal structure [112] containing the $[F(H_6\mathbf{29})]^{5+}$ complex (Figure 2.12). Conversely, complexation of Cl^- was appreciable only with the ligand in its highest charged $H_6\mathbf{29}^{6+}$ form ($\log K < 2$), defining, for this protonation state, a selectivity ratio greater than 10^8.

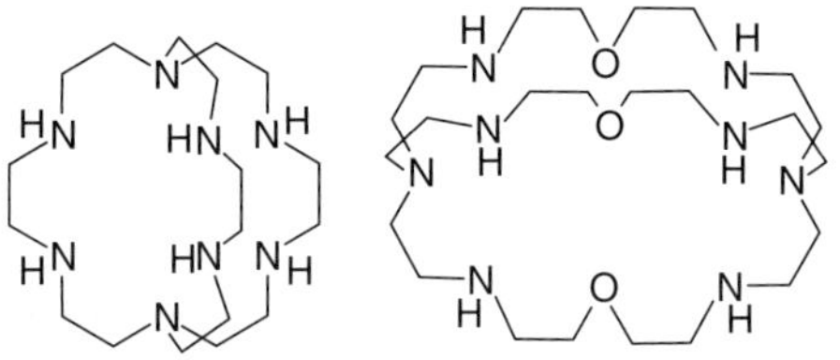

29 **30**

Insertion of CH_2OCH_2 chains between the two tripodal subunits of **29** makes the ligand (**30**) to assume an ellipsoidal shape and a relatively larger size. As a result of these structural modifications, the stability of the F^- complex with the hexaprotonated ligand drops down to $\log K = 4.1$, while that of the Cl^- complex is enhanced to $\log K = 3.0$. The most interesting consequence of such modifications, however, is the change of selectivity from spherical to linear-shaped anions, such as N_3^-. Indeed, high stability was found for the azide complex with $H_6\mathbf{30}^{6+}$ ($\log K = 4.3$) in water, reaching the same selectivity observed for F^- over Cl^-, Br^-

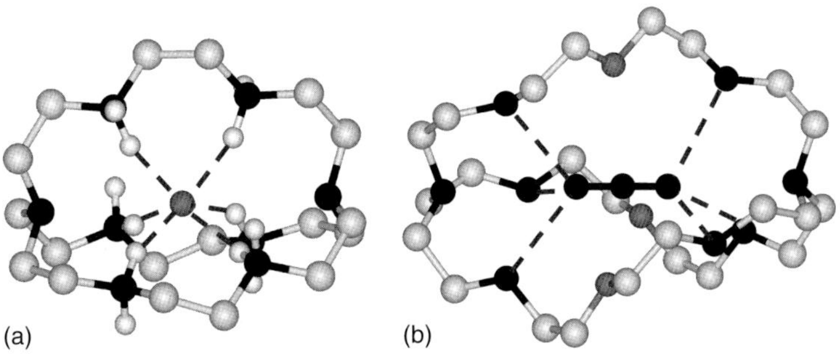

Figure 2.12 Structures of the $[F(H_6\mathbf{29})]^{5+}$ and $[N_3(H_6\mathbf{30})]^{5+}$ complexes [112–114].

(log$K$ = 2.6), and I^- (log$K$ = 2.15) despite the azide anion displaying lower charge density and hydrogen bonding ability than F^- [113, 114]. As shown by the crystal structure of the azide complex, recognition of the triatomic anion occurs via the formation of three salt bridges between the protonated amine groups of each tripodal moiety of H_6**30**$^{6+}$ and each terminal nitrogen atom of N_3^-, which lays almost exactly on the bridgehead N–N ligand axis (Figure 2.12).

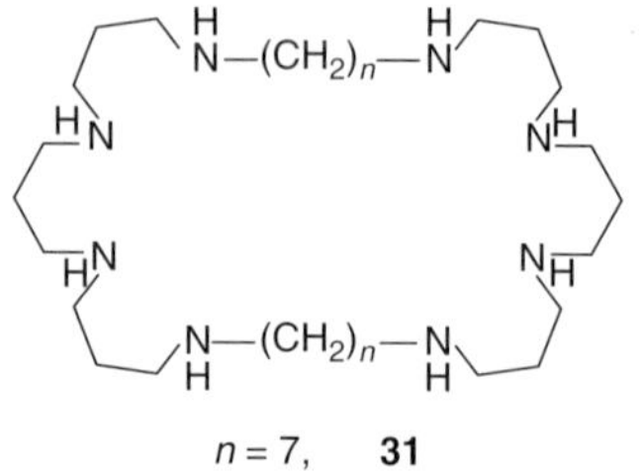

n = 7, **31**
n = 10, **32**

Linear recognition of ditopic anions in aqueous solution was achieved by spatial matching of complementary functionalities between the hexaprotonated ditopic ligands **31** and **32**, containing two 1,5,9-triazanonane subunits linked by hydrocarbon bridges and α,ω-dicarboxylate anions $^-O_2C{-}(CH_2)_n{-}CO_2^-$ of the series oxalate^{2-}-sebacate^{2-} (n = 0–8) [115]. The fully protonated species H_6**31**$^{6+}$ and H_6**32**$^{6+}$ form strong complexes with such dicarboxylate anions showing structural dependence of binding selectivity. H_6**31**$^{6+}$ associates more strongly with succinate^{2-} and glutarate^{2-} (n = 2, 3) than with shorter and longer dicarboxylates, while the selectivity peak shifts to pimelate^{2-} and suberate^{2-} (n = 5, 6) for H_6**32**$^{6+}$ corresponding to the same increase in length, by three CH_2 groups, both for the most strongly bound dicarboxylates and for the hydrocarbon chains linking the triammonium subunits of the protonated receptors. A similar linear recognition of α,ω-dicarboxylate anions was reported for the protonated forms of the ditopic macrocyclic (**33**) and macrobicyclic (**34**) polyamines. Selective binding of glutarate^{2-} (n = 3), within the series of homologous dianions, was achieved by both H_4**33**$^{4+}$ and H_5**33**$^{5+}$ [116], while terephthalate^{2-} was the preferred dianionic substrate forming a cryptate complex with H_6**34**$^{6+}$ [117].

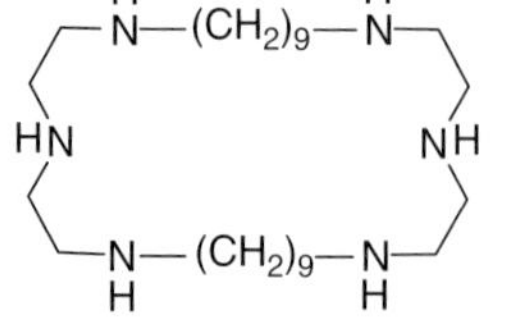

33

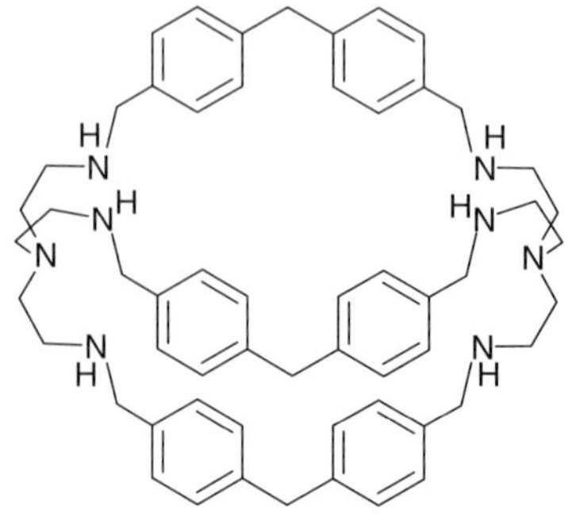

34

A more structured recognition scheme for linear dianions was elaborated by adding metal ions, as additional discriminating parameters, to the simpler H^+/polyamine systems. Indeed, striking recognition of pimelate^{2-}, hydrogenpimelate$^-$, and pimelic acid was accomplished in water, by the different species formed by the ditopic polyazacyclophane **35** in the presence of 2 equiv. of Cu(II) [118]. Selective binding of pimelate^{2-}, hydrogenpimelate$^-$, and pimelic acid takes place by means of binuclear, mononuclear polyprotonated complexes and metal-free polyammonium species, respectively, in successive pH ranges (pimelate^{2-}, pH > 6; hydrogenpimelate$^-$, pH 4–6; pimelic acid, pH < 4) as sketched in Figure 2.13. The crystal structure of the $\{[Cu_2\mathbf{35}(H_2O)_2]\supset$pimelate$\}(ClO_4)_2$ complex showed the pimelate dianions fitting the long distance between the two metal centers located 11 Å apart (Figure 2.13) [118].

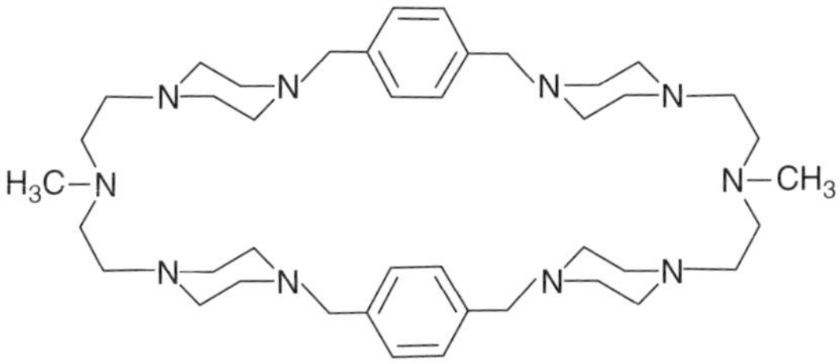

35

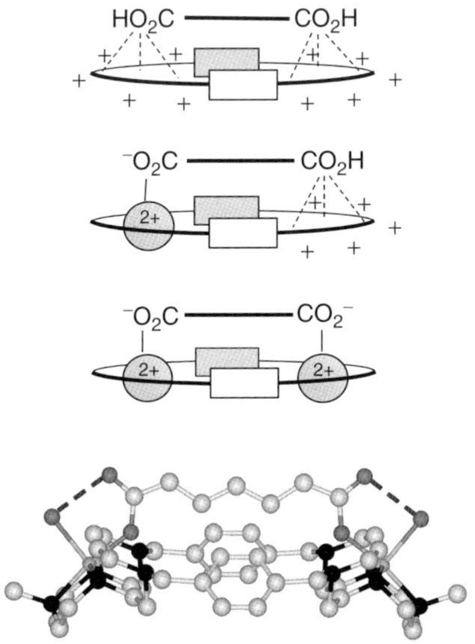

Figure 2.13 Binding modes of pimelate in the system **35**/Cu(II)/pimelate and crystal structure of the $\{[Cu_2\mathbf{35}(H_2O)_2]\supset$pimelate$\}^{2+}$ complex [118].

Positively charged (protonated) polyamine cryptands are efficient receptors for oxoanions recognition because of their preorganized molecular arrangement, which allows many binding groups to converge inside the receptor cavity [119]. Good examples of tetrahedral oxoanion complexes were obtained with protonated forms of **36** [120]. Solution studies performed in water showed that tri- to hexa-protonated **36** form very stable complexes with dicharged anions such as SO_4^{2-}, SeO_4^{2-}, and $S_2O_3^{2-}$. Increasing stability was observed along this anion series for all ligand protonation states. For instance, the fully protonated $H_6\mathbf{36}^{6+}$ generates the stability constants $\log K = 6.57$ (SO_4^{2-}), $\log K = 7.24$ (SeO_4^{2-}), and $\log K = 8.51$ ($S_2O_3^{2-}$). Although this stability trend follows the expected order of decreasing anion hydration energies, a comparison with analogous complexes of similar receptors evidenced that anion salvation is not determinant. A successful anion–receptor fitting, instead, is thought to be responsible for this selectivity trend across the anion series. Indeed, the crystal structure of the $S_2O_3^{2-}$ cryptate (Figure 2.14a) shows the anion held in the cavity of $H_6\mathbf{36}^{6+}$ and firmly coordinated to the receptor through nine salt bridges. The NH_2^+ functions of a tripodal ligand subunit chelate each oxygen atom of the anion and in turn are individually chelated (via bifurcated H–bonds) by a pair of adjacent anion oxygen atoms. At the other side, the anion sulfur atom is hydrogen bonded to the three NH_2^+ functions of the second ligand subunit. The anion is thus bound by six direct NH···O and three NH···S interactions, which are evidence of good host–guest complementarity and fitting.

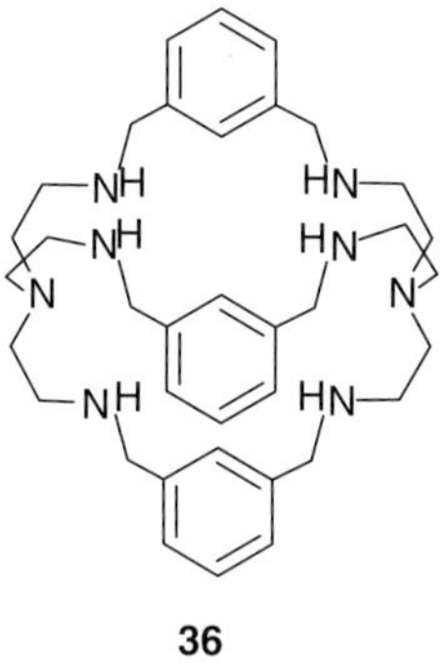

36

The same ligand was shown to be also well suited for binding of trigonal oxoanions such as NO_3^-. Two independent crystal structures showed that one nitrate anion can be encapsulated into each tripodal moiety of $H_6\mathbf{36}^{6+}$, forming a binuclear anion complex (Figure 2.14b) [121, 122]. As in the case of thiosulfate, each nitrate oxygen atom is chelated by a pair of NH_2^+ functions, so each nitrate anion is tethered by six direct hydrogen bonds and each NH_2^+ makes a pair of bifurcated hydrogen bonds with adjacent nitrate oxygen atoms. Equilibrium constants for the association of the first and the second nitrate anions to $H_6\mathbf{36}^{6+}$ were determined to be in the range $\log K = 3.0$–3.7 and $\log K = 1.8$–2.4, respectively [121, 122]. Although these stability constants are relatively small, they are large enough to allow selective complexation by $H_6\mathbf{36}^{6+}$ of NO_3^- over other monocharged oxoanions, such as ClO_4^-, which form with the hexaprotonated receptor only 1 : 1 complexes [119].

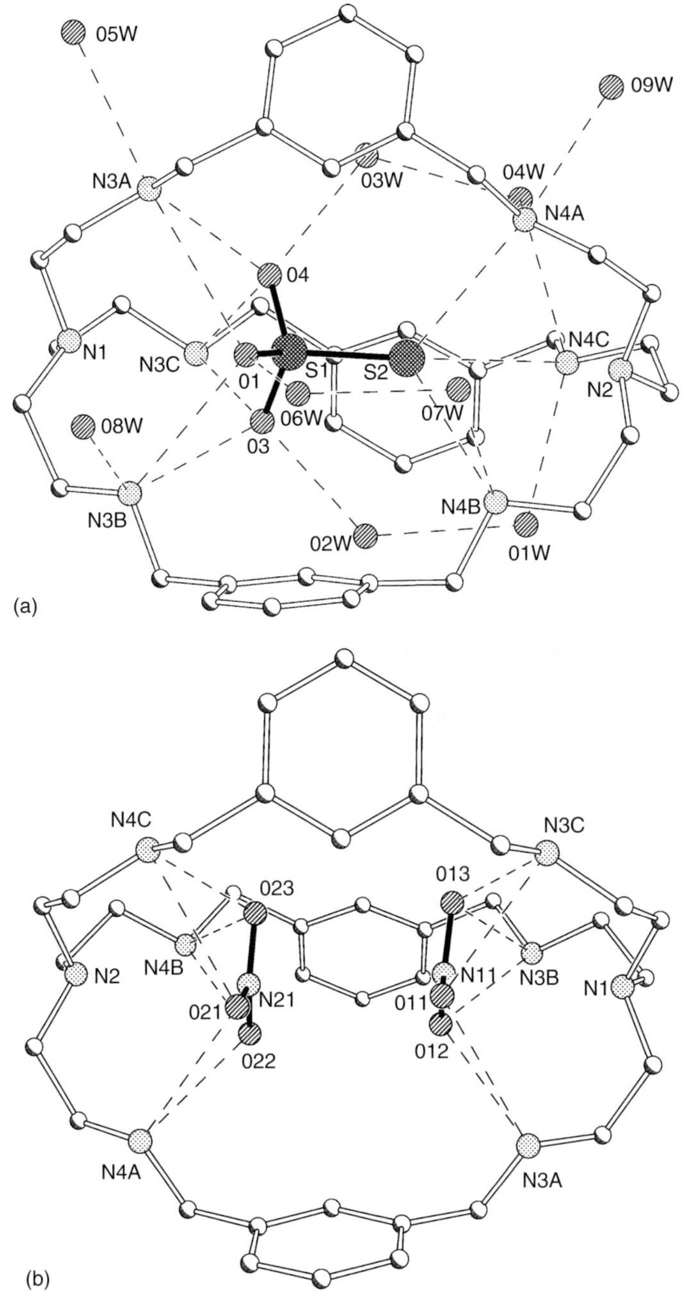

Figure 2.14 Crystal structures of the cryptate complexes formed by $H_6\mathbf{36}^{6+}$ with $S_2O_3^{2-}$ (a) and NO_3^- (b) [119]. (Reproduced with permission from the Royal Society of Chemistry.)

2.4
Enthalpic and Entropic Contributions in Anion Coordination

The knowledge of the enthalpic (ΔH°) and the entropic (ΔS°) components of the complexation free energy (ΔG°) are of great help for proper evaluation of anion complex formation in solution where changes in solvent structure and solvation spheres of anions and receptors contribute to the overall enthalpy and entropy changes, in addition to the effects due to the specific interaction occurring between anions and receptors.

The enthalpy change can be determined either by direct calorimetry or by temperature dependence of the complexation constant according to the van't Hoff isochore. Once the ΔH° has been determined, the entropic component ΔS° can be calculated by using the relationship $\Delta G^\circ = \Delta H^\circ - T\Delta S^\circ$. The standard free energy change is related to the complexation constants by the expression $\Delta G^\circ = -RT\ln K$.

From the van't Hoff isochore

$$R\ln K = \frac{-\Delta H^\circ}{T} + \Delta S^\circ$$

ΔH° is calculated as the slope of a plot of $R\ln K$ versus $1/T$, under the assumption that in the studied temperature interval, ΔH° can be reasonably assumed as temperature independent. Since there is a logarithmic correlation between the complexation constant, K, and ΔH°, experimental errors on K has an exponential propagation on ΔH°. Accordingly, to obtain accurate enthalpy changes by this method, the complexation constants have to be of great accuracy and the studied range of temperature should be as wide as possible. Unfortunately, the temperature range that can be explored is limited by experimental conditions, such as the range of liquidity of the solvent and the solubility of reactants, and the assumed invariance of ΔH° with temperature is less likely to occur the wider the temperature range employed. For the last reason, the temperature dependence of ΔC_p should be known to correct ΔH° for its temperature dependence $\Delta(\delta H^\circ)/\delta \mathrm{T} = \Delta C_p$.

The calorimetric method, consisting of measuring the amount of heat involved in the complexation reaction, is largely more accurate, provided appropriate calorimeters are used. The present availability of very sensitive microcalorimeters, which can operate with very small samples and under high-dilution conditions, makes calorimetry rather popular, not only for the determination of enthalpy changes but also for the speciation of simple complex systems and the determination of the relevant stability constants.

Early determinations of the enthalpy and entropy changes of anion complex formation were derived from the temperature dependence of the stability constants, according to the van't Hoff isochore, for complexes formed in water by the tetraprotonated form of the hexaazacycloalkane **16** with the monocharged anions Cl^-, NO_3^-, ClO_4^-, IO_3^-, $CF_3CO_2^-$, $C_6H_5SO_3^-$ [123, 124] and the dicharged SO_4^{2-} [125]. These complexation processes occurring between charged species were interpreted to be mostly determined by modification of the solvation spheres of the interacting partners rather than by specific interactions between receptor and anions. When two ions of opposite charge interact in solution to form an

Table 2.3 Enthalpic and entropic changes for the complexation of anions by $H_4\mathbf{16}^{4+}$ [123–125].

	ΔH (kcal mol^{-1})	ΔS (cal mol^{-1}K^{-1})
ClO_4^-	−2.5	−3.8
Cl^-	4.9	24
$CF_3CO_2^-$	6.5	25.9
$C_6H_5SO_3^-$	6.6	24.6
NO_3^-	−0.4	12
IO_3^-	1.3	17.1
SO_4^{2-}	5.6	37.8

ion pair, the occurring neutralization of charge produces a mobilization of solvent molecules that are more weakly attracted by the ion pair than by the separated ions. The resulting desolvation effect is accompanied by a positive (unfavorable) enthalpy change (bonds are broken) and a positive (favorable) entropy change (solvent molecules are released). The values obtained (Table 2.3) were classified according to three complexation behaviors. The first, shown by Cl^-, $CF_3CO_2^-$, and $C_6H_5SO_3^-$, give rise to marked entropy-driven complexation reactions with positive enthalpy changes, which were attributed to the disruption of a highly ordered solvent network near $H_4\mathbf{16}^{4+}$ and to the decreased long-range ion-solvent interaction of the complex, compared with the reagents, due to charge neutralization. The second behavior, typical of NO_3^- and IO_3^-, is characterized by entropy-driven reactions with enthalpy changes near zero, ascribed to the formation of solvent-mediated complexes, in agreement with the interaction mode shown by NO_3^- in the crystal structure of $(H_4\mathbf{16})(NO_3)_2Cl_2{\cdot}2H_2O$ [123] (Figure 1.24, Chapter 1), taking place with lower desolvation effect compared to the previous case. Finally, a third behavior, displayed by ClO_4^-, consists in an enthalpy-driven reaction having only a small negative entropy change, which is indicative of a net ordering of complexation products, compared to reactants, and was rationalized by assuming the formation of "outer-sphere" complexes, not requiring significant solvent reorganization. SO_4^{2-} substantially showed a behavior similar to Cl^-, $CF_3CO_2^-$, and $C_6H_5SO_3^-$, with a larger entropy change due to the greater charge neutralization of the complex produced by the greater charge of the anion.

After these early enthalpic and entropic parameters of anion complexation were derived from the temperature dependence of the stability constant, the first calorimetrically measured enthalpy changes for the formation of anion complexes with polyammonium azamacrocycles as second-sphere ligands were determined for the interaction of the tetraprotonated form of **3** with $Fe(CN)_6^{4-}$ ($\Delta G^\circ = -4.94$ kcal mol^{-1}, $\Delta H^\circ = -1.1$ kcal mol^{-1}, $T\Delta S^\circ = 3.9$ kcal mol^{-1}) and $Co(CN)_6^{3-}$ ($\Delta G^\circ = -3.25$ kcal mol^{-1}, $\Delta H^\circ = -2.56$ kcal mol^{-1}, $T\Delta S^\circ =$ 0.7 kcal mol^{-1}) in water [126]. The complexation reactions were found to be promoted by both the enthalpy- and the entropy-favorable changes, the thermodynamic parameters being consistent with the main contributions to complex stability deriving from anion charge and solvent effects. Indeed, the more charged $Fe(CN)_6^{4-}$ anion forms a more stable complex with $H_4\mathbf{3}^{4+}$ than $Co(CN)_6^{3-}$, as expected on the basis of electrostatic considerations, its greater stability being determined by a more favorable entropic contribution accompanied by a less favorable enthalpic contribution. Since complete charge neutralization occurs in $Fe(CN)_6(H_4\mathbf{3})$, the desolvation effect occurring on the formation of this complex is larger, and accordingly, it occurs with greater enthalpy loss but with greater entropy gain compared to the formation of $[Co(CN)_6(H_4\mathbf{3})]^+$.

Calorimetric measurements for the determination of reaction enthalpies for anion complex formation in water were successively extended to a large number of polyammonium receptors, including acyclic, macrocyclic and tris-macrocyclic ligands, and macrobicyclic cleft systems [72, 73, 127–129].

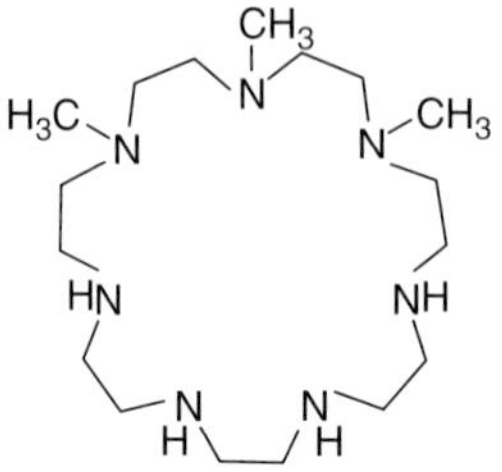

37

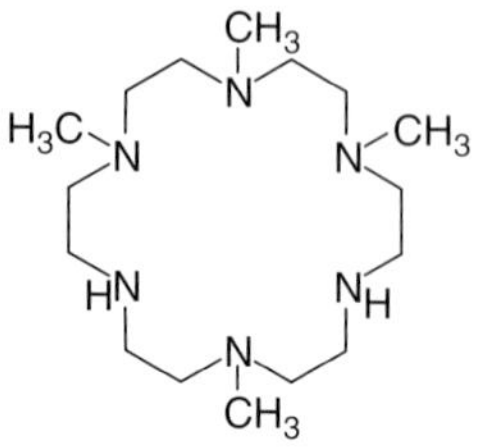

38

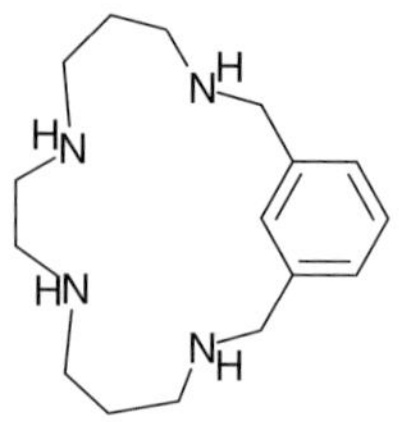

39

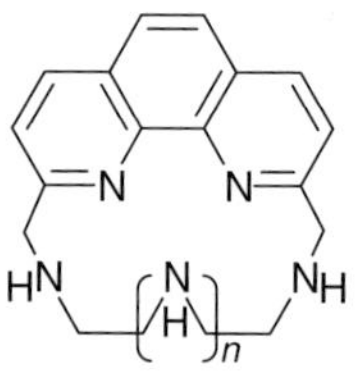

$n = 2$, **40**
$n = 2$, **41**

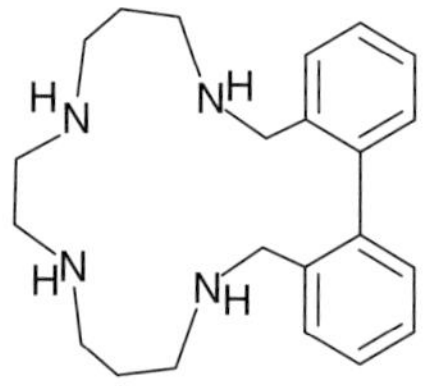

42

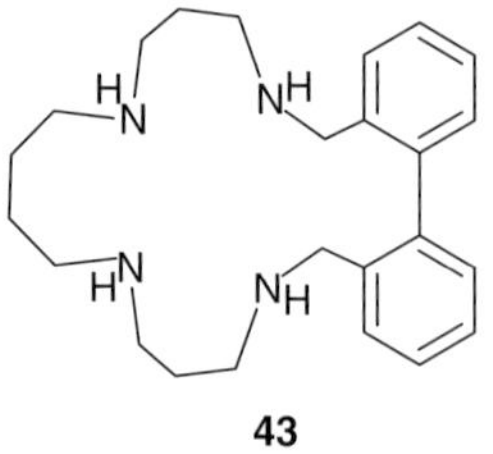

43

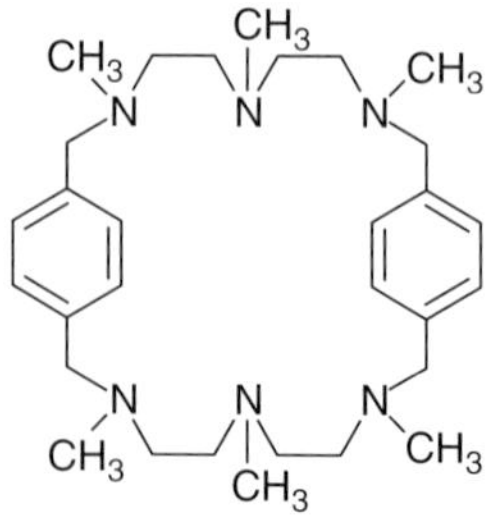

44

The binding of SO_4^{2-} by protonated forms of ligands **11**, **16–19**, **37–44** in water was studied by means of calorimetric measurements showing that complexation reactions are driven by invariably favorable entropic contributions, the enthalpic terms being endothermic or almost athermic, in agreement with association

Table 2.4 Selected thermodynamic parameters for the binding of phosphate and pyrophosphate by **43** and **44**, respectively, in water at 298 K [72].

	ΔG° (kcal mol^{-1})	ΔH° (kcal mol^{-1})	$T\Delta S^\circ$ (kcal mol^{-1})
$HPO_4^{2-} + H\mathbf{43}^{+}$	−4.67	−0.66	4.01
$HPO_4^{2-} + H_2\mathbf{43}^{2+}$	−4.12	−0.43	3.69
$HPO_4^{2-} + H_3\mathbf{43}^{3+}$	−3.40	0.82	4.22
$HP_2O_7^{3-} + H_4\mathbf{44}^{4+}$	−3.49	2.25	5.74
$H_2P_2O_7^{2-} + H_4\mathbf{44}^{4+}$	−3.78	0.93	4.71
$H_3P_2O_7^{-} + H_4\mathbf{44}^{4+}$	−4.33	−6.20	−1.87

processes mostly controlled by desolvation of the interacting species [73]. The thermodynamic parameters for the interaction of PO_4^{3-} and $P_2O_7^{4-}$, and their protonated forms, with most of these receptors (**11**, **16–19**, **37–39**, **42–44**), in addition to spermine, 1,5,8,12-tetraazadodecane, and 1,5,9,13-tetraazatridecane, were also obtained by determining the complexation constants by means of potentiometric titrations and the complexation enthalpies by means of isothermal titration calorimetry [72]. This study evidenced that, although electrostatic attraction was found to be the principal driving force of anion complexation, the more charged species interacting more strongly, and the enthalpic and entropic contributions are consistent with desolvation effects similar to those found for the binding of SO_4^{2-}; some complexation reactions are in conflict with electrostatic expectations. For instance, the stability of the complexes formed by HPO_4^{2-} with mono-, di-, and triprotonated forms of **43** decreases with increasing charge on the ligand, while the stability of the complexes formed by $H_4\mathbf{44}^{4+}$ with $HP_2O_7^{3-}$, $H_2P_2O_7^{2-}$, and $H_3P_2O_7^{-}$ increases with decreasing charge on the anion (Table 2.4). Similar behaviors were ascribed to the particular ability of phosphate species to behave as acceptors and donors of hydrogen bonds.

When hydrogen bonds are formed between chemical species characterized by marked acid/base properties such as amine, ammonium, phosphate, and protonated phosphate compounds, contributions due to proton transfer from the donor to the acceptor groups may be important. There are four possible modes of hydrogen bonding (Type 1) involving amine or ammonium groups as donors, and only one (Type 2) involving the amine groups as acceptors in the formation of the anion complexes of the above receptors with these phosphate species.

$$\text{N–H}^{+}\cdots{}^{-}\text{O} \quad \Delta H^\circ > 0, T\Delta S^\circ > 0 \tag{1.1}$$

$$\text{N–H}^{+}\cdots\text{OH} \quad \Delta H^\circ > 0, T\Delta S^\circ \approx 0 \tag{1.2}$$

$$\text{N–H}\cdots{}^{-}\text{O} \quad \Delta H^\circ > 0, T\Delta S^\circ \approx 0 \tag{1.3}$$

$$\text{N–H}\cdots\text{OH} \quad \Delta H^\circ > 0, T\Delta S^\circ < 0 \tag{1.4}$$

$$\text{NI}\cdots\text{H–O} \quad \Delta H^\circ < 0, T\Delta S^\circ < 0 \tag{2}$$

Type (1) hydrogen bonds determine a partial deprotonation of the amino group and a partial protonation of a phosphate oxygen. According to general behaviors, deprotonation of ligand amino groups was shown to be strongly endothermic while protonation of HPO_4^{2-}, $H_2PO_4^-$, and pyrophosphate anions are athermic or weakly endothermic [72]. Hence, charge transfer in Type (1) hydrogen bonds is accompanied by an unfavorable enthalpic contribution ($\Delta H^\circ > 0$), while from the entropic point of view, different contributions are expected, mostly depending on the effect the process has on the separation of charge: (i) an entropy gain is expected for Type (1.1) bonds, because of the release of solvent molecules determined by charge neutralization; (ii) an entropy loss should accompany charge transfer in Type (1.4) bonds; and (iii) no evident entropic effects should be determined by Types (1.2) and (1.3), which are expected not to alter significantly the charge separation. On the other hand, the formation of Type (2) hydrogen bonds, which consists of partial protonation of an amino group and partial deprotonation of a phosphate oxygen, is promoted by negative enthalpy changes, the entropic terms being unfavorable.

In agreement with these observations, the binding reactions involving completely deprotonated anions, which can form only Type (1.1) hydrogen bonds, were found to be accompanied by unfavorable enthalpic and favorable entropic contributions. Also, the stability decrease previously noted for the complexes formed by HPO_4^{2-} with the mono-, di-, and tri-protonated forms of **43** was interpreted in terms of increasing hydrogen bond donor properties (Type (1) bonds) of the receptors, determining unfavorable enthalpic contributions (Table 2.4), while the stability increase of the complexes formed by pyrophosphate and H_4**44**$^{4+}$, as the charge on the anion decreases from $HP_2O_7^{3-}$ to $H_3P_2O_7^-$, was attributed to the greater donor ability of the more protonated anions (Type (2) bonds) determining more favorable enthalpic and less favorable entropic contributions (Table 2.4) [72].

Considering the complete set of thermodynamic data obtained in this work for the interaction of phosphate and pyrophosphate with nine macrocyclic and five acyclic ligands, good linear ΔH° versus $T\Delta S^\circ$ correlations were obtained for both cyclic (92 data points) and acyclic (29 data points) receptors, according to the relation $T\Delta S^\circ = \alpha\Delta H^\circ + I$, showing that enthalpy–entropy compensatory relationships hold for all such complexation reactions [72]. The obtained α parameters, which are interpreted as a measure of receptor rigidity, pointed out that, despite the different structures of these molecules (polyazamacrocycles, polyazacyclophanes, linear polyamines), and despite the increased stiffening they experience on increasing their protonation state, all these receptors show very good and almost identical adaptability to these anionic substrates. Evidently, their adaptability is determined not by their molecular rigidity but by their ability in organizing hydrogen bonds and salt bridges in the complexes. The determined I parameters furnished further interesting information. This parameter represents the common contribution to the stability of all complexes of the considered class of ligands ($\Delta G^\circ = (1 - \alpha)\Delta H^\circ - I$). While α is identical, within experimental errors, for all ligand protonation states, indicating that the adaptability of receptors is independent of their charge, the I

parameter increases considerably with receptor charge, accounting for a greater intrinsic contribution to the stability of complexes with more charged receptors.

Further examples of increasing complex stability with decreasing charge (increasing protonation) of phosphate-type anions were found for the binding of HPO_4^{2-}, $H_2PO_4^-$, $P_2O_7^{4-}$, $HP_2O_7^{3-}$, $H_2P_2O_7^{2-}$, $H_3P_2O_7^-$, $P_3O_{10}^{5-}$, $HP_3O_{10}^{4-}$, $H_2P_3O_{10}^{3-}$ and nucleotidic anions H_2ATP^{2-} and H_3ATP^- by protonated forms of the macrobicyclic cleft system **45** [127]. In these cases, however, the enthalpic and entropic contributions to anion coordination were seen to be mostly controlled by dominating desolvation effects, the reactions being promoted by invariably favorable entropic changes that are only partially cancelled by unfavorable enthalpy changes, because of the inclusion of the phosphate ions into the ligand cleft [127].

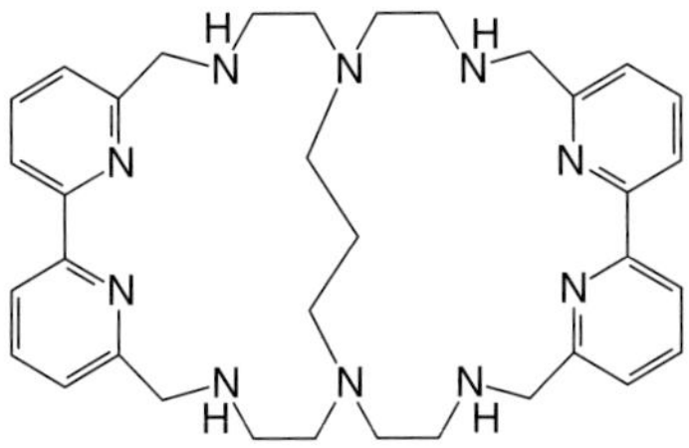

45

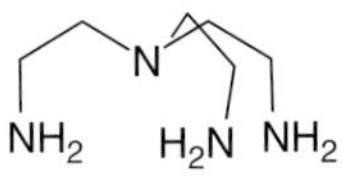

46

Also the coordination of HPO_4^{2-} to di- and triprotonated species of the tripodal ligand tren (**46**) seems to be characterized by the formation of different hydrogen bond interactions [129]. Indeed, while the interaction of HPO_4^{2-} with $H_2\mathbf{46}^{2+}$ is driven by a largely favorable enthalpy change ($\Delta H^\circ = -20.1$ kJ mol^{-1}) partially cancelled by an unfavorable entropy term ($T\Delta S^\circ = -9.3$ kJ mol^{-1}) in agreement with the formation of one hydrogen bond contact of Type (2) (Figure 2.15), the interaction of the same anion with $H_3\mathbf{46}^{3+}$, not allowing this type of hydrogen bonds, is accompanied by small favorable contributions, both enthalpic and entropic ($\Delta H^\circ = -9.0$ kJ mol^{-1}, $T\Delta S^\circ = 6.2$ kJ mol^{-1}).

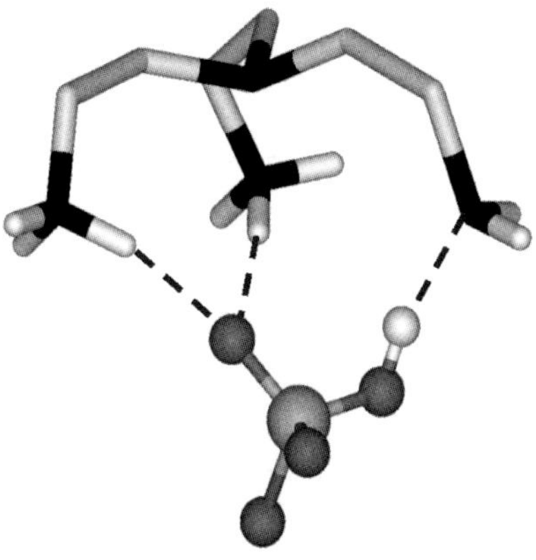

Figure 2.15 Schematic representation of the interaction mode in [$H_2\mathbf{46}(HPO_4)$] [129].

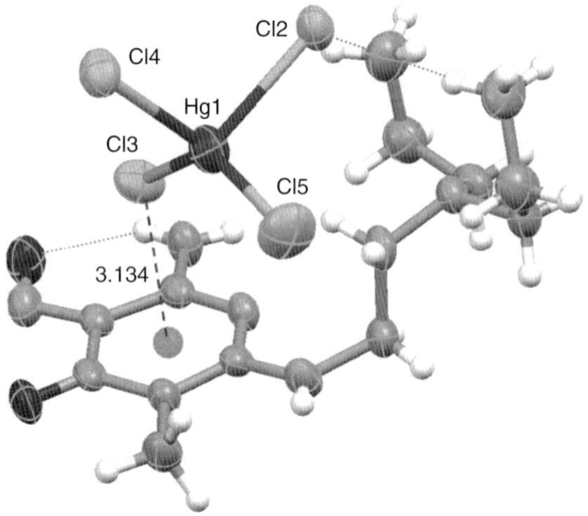

Figure 2.16 Structure of the $[HgCl_4(H_2\mathbf{10})]$ complex showing salt bridge and anion–π interactions. (Reproduced with permission from Ref. [26], copyright 2008, American Chemical Society.)

Similar features are also manifested by the ligand **10**, which is obtained by functionalization of **46** with a pyrimidine residue [26]. Compound **10** displays enhanced anion binding ability relative to **46**. Indeed, for equally charged forms of the two receptors, more favorable free energy changes by 6.5–12.5 kJ mol^{-1} were found for the complexes of **10** with sulfate and phosphate anion, the greater stability being due to more favorable entropic terms. It was shown that anion–π interactions afford significant contributions to the stability of anion complexes with protonated forms of **10**; for instance, it was estimated that similar interactions contribute −8.9 kJ mol^{-1} to the stability of the $[SO_4(H_n\mathbf{10})]^{n-2}$ ($n = 1$–3) species [26]. An anion–π interaction, similar to that shown in Figure 2.16 for the $[HgCl_4(H_2\mathbf{10})]$ complex, brings the anion in the hydrophobic domain of the pyrimidine ligand moiety, causing a strong anion desolvation, which is expected to be at the origin of the favorable entropic contributions.

The formation of anion complexes in water between the tripodal tris-macrocycles **47** and **48** and the benzenetricarboxylate (BTC) anions 1,2,3-BTC, 1,2,4-BTC, and 1,3,5-BTC offered other examples of the interplay of enthalpic and entropic contributions in determining the stability of anion complexes [128]. Owing to the high number of complex species formed by these systems (10 with **47** and up to 13 with **48**), the relevant thermodynamic parameters were analyzed by considering the pH dependence of enthalpic and entropic conditional contributions. Such conditional parameters are defined as the sum ($\sum \Delta H_i^\circ \alpha_i, \sum \Delta S_i^\circ \alpha_i$) of the individual contributions ($\Delta H_i^\circ, \Delta S_i^\circ$) of each complex species multiplied by the corresponding molar fractions α_i at a given pH. The results obtained for both receptors are similar; for this reason, only those obtained for **47** are visualized in Figure 2.17. These results clearly showed that, depending on pH, different

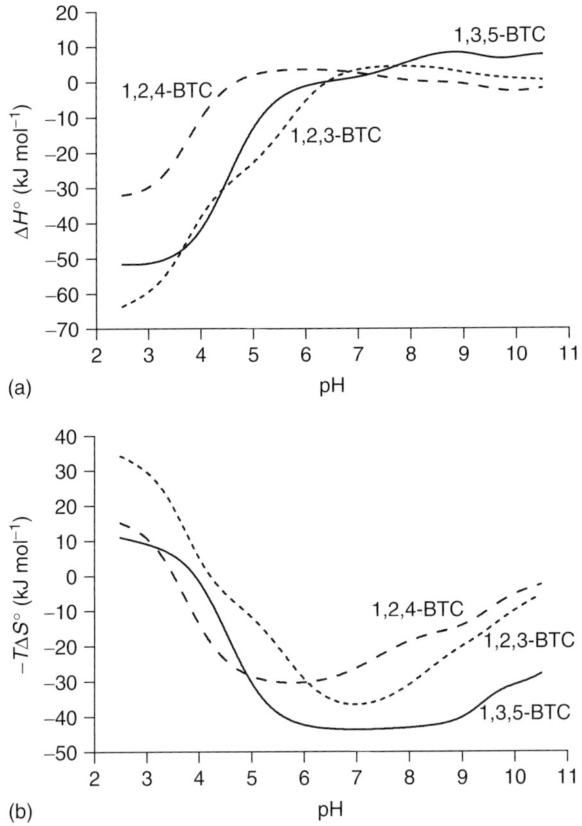

Figure 2.17 pH dependence of the enthalpic (a) and entropic (b) conditional contributions for the complexation of 1,3,5-BTC (solid lines), 1,2,4-BTC (dashed lines), and 1,2,3-BTC (dotted lines) with **47**. (Reproduced with permission from Ref. [128], copyright 2005, American Chemical Society.)

driving forces are responsible for the complexation processes. From alkaline to slightly acidic pH values, the conditional enthalpy changes are almost negligible and complex formation is entropy driven, while moving toward more acidic solutions, complex formation becomes enthalpy driven, the entropy changes being increasingly unfavorable. In the first pH region, complexation takes place between differently protonated forms of the ligands (from H_2L^{2+} to H_9L^{9+}, $\mathbf{L} = \mathbf{47}, \mathbf{48}$) and the completely deprotonated BTC trianions that form $NH^+\cdots{}^-O$ contacts, while in more acidic solutions, where the trianions become increasingly protonated, interactions of the type $NH^+\cdots OH$ or $N\cdots HO$ become the principal binding modes. Again, the enthalpy and entropy changes accompanying these binding modes are concordant with the overall thermodynamic properties of the observed binding processes.

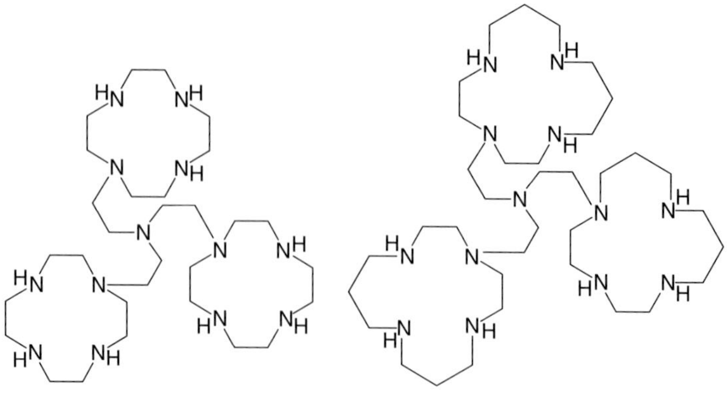

47 48

Also the binding of citrate by the tris-guanidinium receptor **49** to form a 1 : 1 complex in water was shown to be mainly driven by a large entropy term, the enthalpic one being much less favorable. However, the formation of a 2 : 1 **49**–citrate complex, most likely containing the anion sandwiched between two receptor cations, is entirely entropy driven in agreement with the large desolvation processes expected for a similar assembly [130].

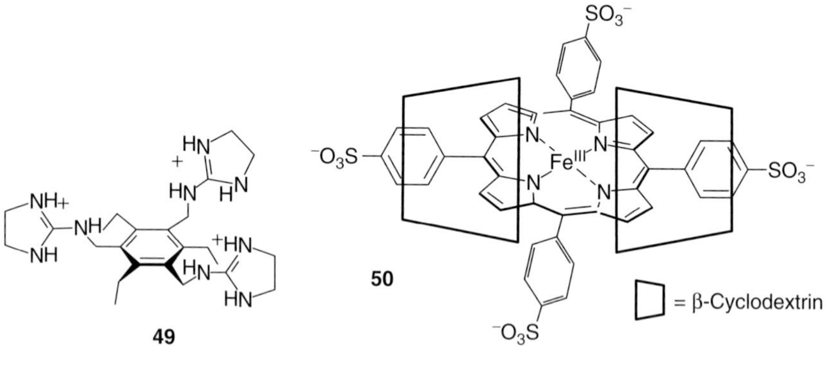

49 50

The importance of desolvation effects in determining the enthalpic and entropic components of the free energy of association of anions with positively charged receptors was confirmed by the use of a metal-based hydrophobic anion binding site. 5,10,15,20-Tetrakis(4-sulfonatophenyl)porphinato iron(III) forms the very stable 1 : 2 complex (**50**) with heptakis(2,3,6-tri-*O*-methyl)-*β*-cyclodextrin, whose iron(III) center is located into the hydrophobic cleft formed by the face-to-face pair of cyclodextrins [131]. The presence of the two cyclodextrins is of paramount importance in determining the anion-binding ability of the iron(III) center of **50**, as only F^- binds the porphinato iron(III) complex in water in their absence. Conversely, several anions of suitable size, including Cl^-, Br^-, N_3^-, and SCN^- bind to the metal center of **50**. The thermodynamic parameters determined by isothermal titration calorimetry for complex formation of **50** with Cl^- ($\Delta G^\circ = -13.5$ kJ mol^{-1},

$\Delta H^\circ = -2.2$ kJ mol^{-1}, $\Delta S^\circ = 37.9$ J mol^{-1}K^{-1}), Br^- ($\Delta G^\circ = -9.9$ kJ mol^{-1}, $\Delta H^\circ = -1.2$ kJ mol^{-1}, $\Delta S^\circ = 29.2$ J mol^{-1}K^{-1}), N_3^- ($\Delta G^\circ = -23.4$ kJ mol^{-1}, $\Delta H^\circ = -13.9$ kJ mol^{-1}, $\Delta S^\circ = 31.9$ J mol^{-1}K^{-1}), and SCN^- ($\Delta G^\circ = -12.1$ kJ mol^{-1}, $\Delta H^\circ = -15.1$ kJ mol^{-1}, $\Delta S^\circ = -10.1$ J mol^{-1}K^{-1}) show that the entropic contributions increases in the order $SCN^- < Br^- < N_3^- < Cl^-$ corresponding to the order of hydrophobicity in the Hofmeister series. In the case of Cl^-, Br^-, and N_3^-, the highly favorable entropic contributions were ascribed to an extensive desolvation of these more hydrophilic anions during penetration into the cleft between the faced cyclodextrin molecules, while for the most hydrophobic SCN^- anion, the lower desolvation effect is not sufficient to compensate for the entropy loss occurring on complexation, and the overall entropy change becomes unfavorable. The negative values of the heat capacity changes (ΔC_P) determined for the coordination of Cl^- ($\Delta C_P = -320$ J mol^{-1}K^{-1}) and Br^- ($\Delta C_P = -345$ J mol^{-1}K^{-1}) from the linear variation of ΔH° with temperature supported the effect of anion desolvation on the complexation equilibria [131].

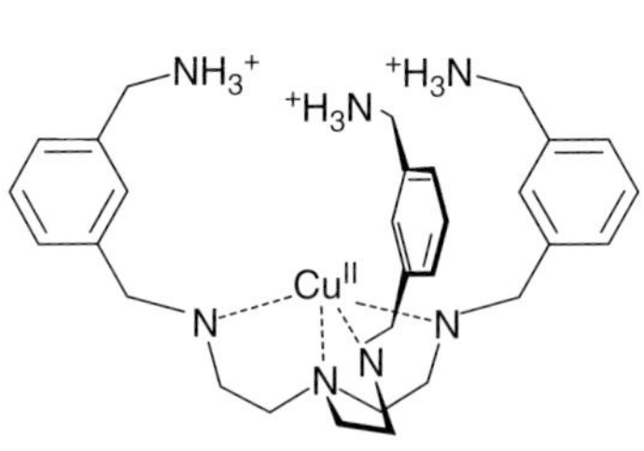

51

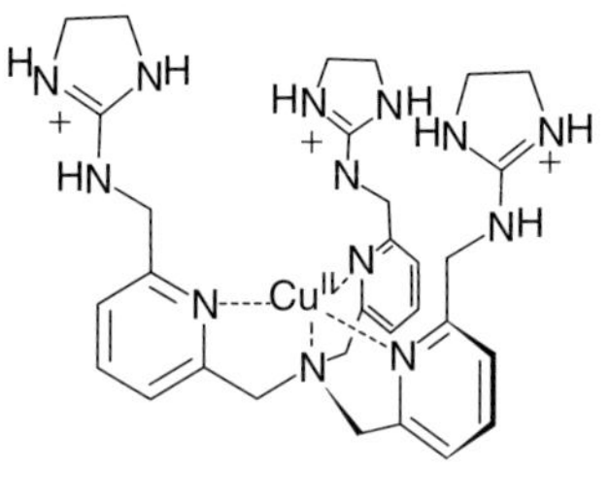

52

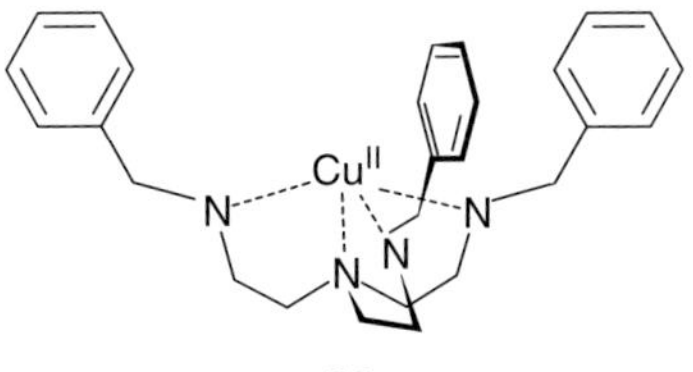

53

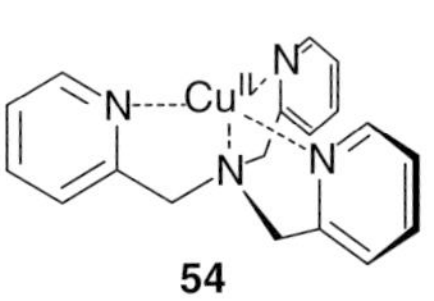

54

The thermodynamic parameters for anion binding to other metal-based receptors (**51–54**) in water were determined by means of ITC measurement or by the van't Hoff method employing stability constants determined from UV–vis data at different temperatures [132]. The data, listed in Table 2.5, showed that solvation effects are of major importance also for these systems. Coordination of HPO_4^{2-} to **51**, taking place at the metal center with further participation of ammonium groups, displays a thermodynamic behavior typical of anion-binding processes involving charged receptors in water, being promoted by the favorable entropy change of an almost athermic reaction principally driven by the release of solvent molecules determined by the neutralization of charge occurring with complex formation. This behavior, which is more evident for highly solvated anions, such as HPO_4^{2-}, is attenuated (Table 2.5) in the case of ReO_4^-, which is characterized by a larger molar volume and has a more loosely held solvation shell because of its lower charge

Table 2.5 Thermodynamic parameters for the complexation of anions by **51–54** in water at 298 K [132].

Receptor	Anion	$\Delta G°$ (kcal mol^{-1})	$\Delta H°$ (kcal mol^{-1})	$T\Delta S°$ (kcal mol^{-1})
51[a]	HPO_4^{2-}	−6.5	−0.6	5.9
51[b]	ReO_4^-	−3.7	−2.2	1.5
51[b]	AcO^-	−3.4	0.7	4.1
52[a]	HPO_4^{2-}	−5.3	−3.8	1.5
53[b]	HPO_4^{2-}	−3.8	−0.9	2.9
54[b]	HPO_4^{2-}	−4.1	−0.8	3.3

[a]Values determined using the van't Hoff method.
[b]Values determined by ITC.

density relative to HPO_4^{2-}. As a matter of fact, binding of ReO_4^- by **51** is favored by similar enthalpic and entropic contributions. On the other hand, enthalpic and entropic effects for the binding of the well-solvated acetate anion are similar to those observed for HPO_4^{2-}.

Enthalpic and entropic contributions to anion binding are also affected by the solvation of receptor binding groups, as shown by coordination of HPO_4^{2-} to **52**, a guanidinium-based analog of **51** [132]. Unlike **51**, binding of HPO_4^{2-} by **52** is promoted by a predominant enthalpy change, the entropic change being less favorable (Table 2.5). Neutron diffraction experiments on aqueous solutions showed that guanidinium has no recognizable hydration shell and is one of the most weakly hydrated cations yet characterized [133]. Thus, release of the poorly organized water molecules from the hydration spheres of guanidium groups on binding of HPO_4^{2-} by **52** does not represent a large entropy advantage and takes place at a very low enthalpic cost, since these water molecules recover their well-organized hydrogen bond arrangement in the bulk solvent. Accordingly, it was postulated that the more exothermic HPO_4^{2-} binding to **52** relative to **51** arises from a smaller enthalpic contribution to solvation of guanidinium groups in **52**, relative to ammonium groups in **51**, allowing the enthalpic effects of phosphate interactions to manifest themselves to a greater extent than with **51** [132].

The binding of HPO_4^{2-} to the control compound **53** showed that, in the absence of ammonium groups, the primary mode of binding to **53** is through a slightly exothermic coordination of HPO_4^{2-} to the metal center with a predominant entropic driving force that was ascribed to the release of solvent, or counterions, from the interacting species. It is also interesting to note that the advantage of appending the guanidinium groups to **54**, creating **52**, is only a free energy stabilization of about 1 kcal mol^{-1}, but the driving force has switched from primarily entropic with **54** to primarily enthalpic with **52**.

Enthalpic and entropic components of complexation free energy have also been determined for anion coordination in nonaqueous solvents. In organic solvents, the receptors used are generally neutral (not charged) and both anions and receptors are involved neither in the protonation/deprotonation equilibria nor in the multiple complexation reactions between the resulting protonated species, such as those observed in water for polyammonium receptors. From this point of view, anion binding in organic solvents are characterized by simpler systems, but other complications, such as incomplete dissociation of the salts used to introduce the anions into the solution, which are less frequent in a highly polar solvent such as water, can bias the results. Furthermore, the properties of many organic solvents are strictly dependent on the presence of cosolvents, in particular of water, which can be easily absorbed from the atmosphere. We have shown before that even traces of water in CH_3CN can significantly modify its solvation properties. Other solvents, such as DMSO, give rise to strong thermal effects upon dilution with water, and accordingly, small differences between the water contents in titrand and titrant solutions during a calorimetric titration may afford large heats of dilutions. All these potential sources of error need to be taken into account when determining the thermodynamic parameters for binding equilibria in organic solvents.

All these aspects were held in high regard in a study of chloride complexation by calix[4]pyrrole **27** in different solvents (CH_3CN, CH_3NO_2, DMSO, CH_2Cl_2, 1,2-dichloroethane) and adopting different countercations [96]. The thermodynamic parameters obtained by isothermal titration calorimetry are reported in Table 2.6. As already observed for the complex stability constants, the enthalpy and entropy changes of complexation are also characterized by large solvent dependence but with no apparent correlation with any bulk property and any polarity index of the solvent. The enthalpic and entropic contributions were found to be invariably favorable and unfavorable, respectively, in all solvents except DMSO. Anion complexation by neutral receptors in organic solvents does not give rise to large desolvation processes such as those occurring in water with charged receptors. Hence, the considerable entropic gains and enthalpic costs for desolvation in water do not exist in organic solvents, and the observed thermodynamic parameters are more

Table 2.6 Thermodynamic parameters for the complexation of chloride by **27** in different solvents at 298 K [96].

Solvent	Counteraction	$\Delta G°$ (kcal mol^{-1})	$\Delta H°$ (kcal mol^{-1})	$T\Delta S°$ (kcal mol^{-1})
CH_3CN	TEA[a]	−7.19	−10.10	−3.07
	TBA[b]	−7.29	−10.16	−2.91
CH_3NO_2	TEA	−5.84	−7.54	−1.83
	TBA	−5.68	−7.49	−1.80
CH_2Cl_2[c]	TEM	−6.33	−10.96	−4.63
1,2-Dichloroethane	TEA	−6.59	−9.87	−3.28
	TBA	−6.06	−10.39	−4.32
DMSO	TEA	−4.19	−1.93	2.26
	TBA	−4.17	−1.87	2.30

[a]TEA = tetraethylammonium.
[b]TBA = tetrabutylammonium.
[c]At 293 K.

representative of the anion–receptor interactions. Accordingly, favorable enthalpy changes arise from hydrogen bonds between **27** and chloride, while the association process manifests its entropy loss. In DMSO, the complexation reaction displays different features, being favored by both enthalpic and entropic contributions. These contributions are small, in agreement with a lower stability of the complex in DMSO relative to the other solvents. It was reported that **27** is able to strongly interact with DMSO molecules [97]. For this reason, more energy is consumed to remove DMSO from the receptor cavity than for the other solvents, but more entropy is gained when the DMSO molecules, which are highly organized by the receptor, are released into the bulk solvent. As shown in Table 2.6, this entropic effect prevails over the enthalpic one. The effect of countercations was found to be small and only appreciable in 1,2-dichloroethane [96].

Enthalpic and entropic contributions to the formation of complexes of **27** with other anions in dry CH_3CN (F^-, Br^-, HPO_4^{2-}) [104] and 1,2-dichloroethane (CN^-, NO_2^-, acetate) [134] were also determined, showing that these anions also hold the same trend of favorable complexation enthalpies partially cancelled by unfavorable entropy changes. Interestingly, it was found that, from an enthalpic point of view, HPO_4^{2-} is a much better guest to **27** than fluoride, but a much more unfavorable entropic contribution leads to a greater stability of the fluoride complex [104]. Similarly, chloride and acetate complexes have the same stability, within experimental errors, in 1,2-dichloroethane, since the enthalpic gain for the formation of the chloride complex is offset by a much greater entropy loss [134].

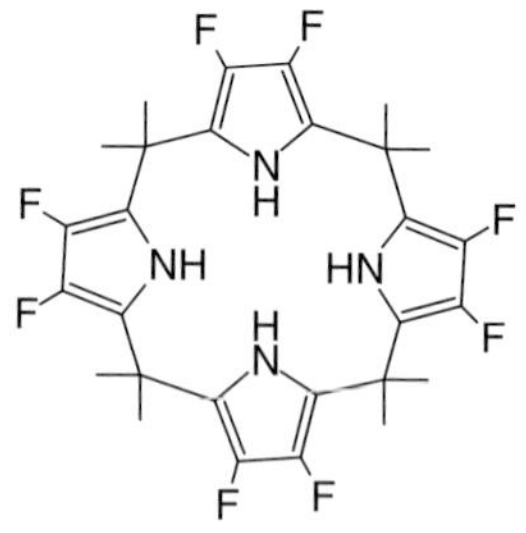

55

Anion binding by various derivatives of **27** has been analyzed from the thermodynamic point of view, and the results obtained offer the basis for interesting considerations [135–138]. Various modifications of **27** have been made to enhance its binding ability and anion selectivity [135]. Fluorination of **27** in the β-pyrrolic positions afforded **55**, which is a better anion receptor, relative to **27**, owing to the electron withdrawing effect of the fluorine substituents. The free energy and enthalpy changes of chloride complexation by **55** were determined by isothermal titration calorimetry in the presence of different countercations; those obtained with tetrabutylammonium at 303 K are $\Delta G^\circ = -7.94$ kcal mol^{-1}, $\Delta H^\circ = -7.78$ kcal mol^{-1}, $T\Delta S^\circ = 0.16$ kcal mol^{-1}. The enhanced binding ability of the fluorinated receptor, relative to **27** (Table 2.6), is due to an entropic gain, the enthalpic contribution being much less favorable [135].

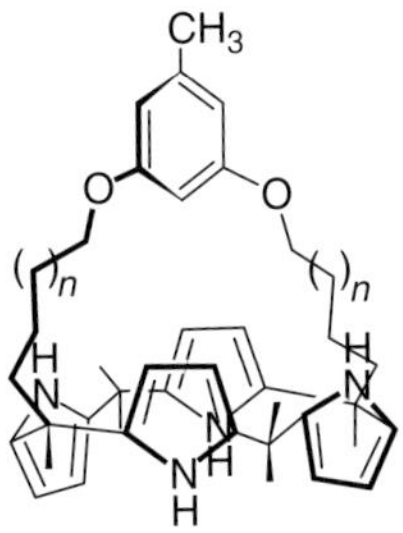

n = 1–3, **56–58**

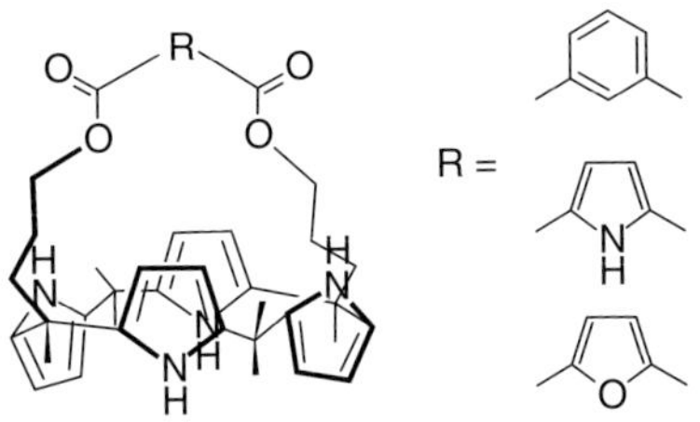

59

60

61

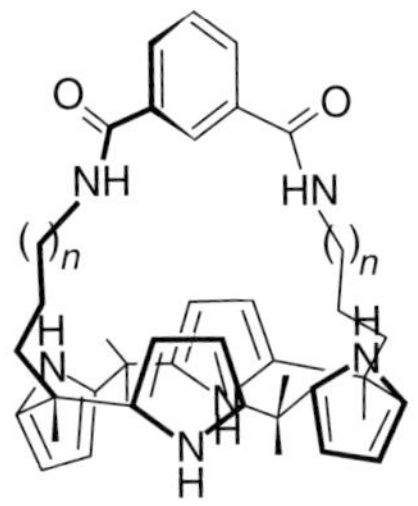

n = 1–3, **62–64**

Improvements in anion binding ability of **27** were also obtained by construction of bridging straps across the macrocyclic ring of **27** to enhance the receptor preorganization, insert additional binding sites, and limit the solvent accessibility to the binding domain. Thermodynamic parameters for anion binding with strapped calix[4]pyrroles **56–64** were obtained by isothermal titration calorimetry in dry CH_3CN [135–137]. Chloride complexation was studied with all these receptors, showing that the association reactions are largely driven by favorable enthalpy changes, while entropic contributions can be either positive or negative but are almost ineffective in determining the complex stabilities. A similar behavior was also observed for the coordination of bromide and iodide by **62–64** [136]. Such entropic contributions, which are favorable, or less unfavorable, with respect to complexation with **27**, likely reflect the greater preorganization of the strapped receptors relative to the parent **27**. It was also observed that appropriately sized straps may allow bridging functionalities to favorably interact with the anion, enhancing complex stability. Indeed, the high enthalpy changes for chloride binding to **56** ($\Delta H^\circ = -12.65$ kcal mol^{-1}) [135] and **60** ($\Delta H^\circ = -11.34$ kcal mol^{-1}) [137], corresponding to high complex stabilities, were ascribed to the additional interaction of chloride with the phenyl CH proton and the pyrrole NH proton in the straps of **56** and **60**, respectively.

Interesting modifications of the thermodynamic properties of chloride complexation was found for **65**, a sulfonium derivative of calix[4]pyrrole [138]. For instance, binding of chloride to the tetracationic **62** in DMSO ($\Delta G^\circ = -5.7$ kcal mol^{-1},

$\Delta H^\circ = 0.1$ kcal mol^{-1}, $T\Delta S^\circ = 5.8$ kcal mol^{-1}) exhibits a free energy increase of 1.5 kcal mol^{-1} relative to **27** (Table 2.6). The association enthalpy is raised to become almost nil, and the complex owes its stability entirely to the much enhanced entropy of association. Since the binding sites in **27** and **65** are similar, it seems reasonable that also similar desolvation changes occur on chloride binding. Under this assumption, the much increased entropic contributions observed for **65** would be determined by a considerable modification of the receptor structure on complexation. The presence of four positive charges in **65** is expected to stiffen its structure because of electrostatic repulsion, and relaxation would occur on coordination of the negatively charged guest, determining a favorable entropy change [138].

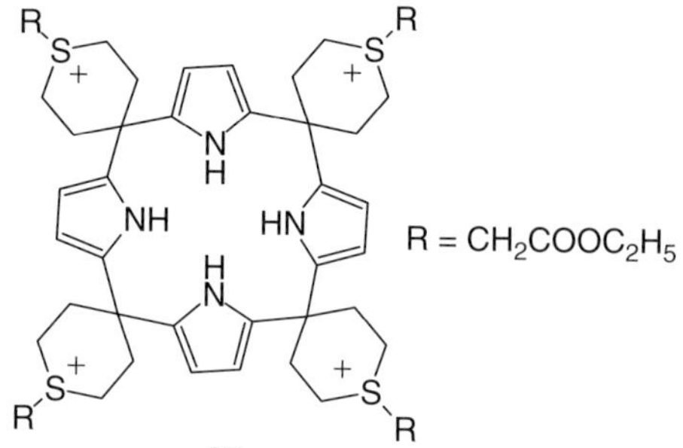

65

Pyrrole units have also been used for the construction of several pyrrole-based linear receptors, which were principally studied as synthetic analogs of prodigiosines, a family of naturally occurring tripyrroles pigments showing promising immunosuppressive and anticancer activity [139]. Analysis by isothermal titration calorimetry of the interaction between chloride and monoprotonated HI salts of **66–70** in CH_3CN showed high association free energy changes ($-\Delta G$ in the range 6.94–8.23 kcal mol^{-1}), despite the possible competition with iodide. The reactions were found to be driven by both favorable enthalpy ($-\Delta H$ in the range 1.59–4.2 kcal mol^{-1}) and entropy changes (ΔS in the range 3.8–5.35 kcal mol^{-1}). The enthalpic contribution prevails over the entropic one only in the case of **66**, while larger entropic terms are preponderant in determining the stability of the chloride complexes with **67–70** [140].

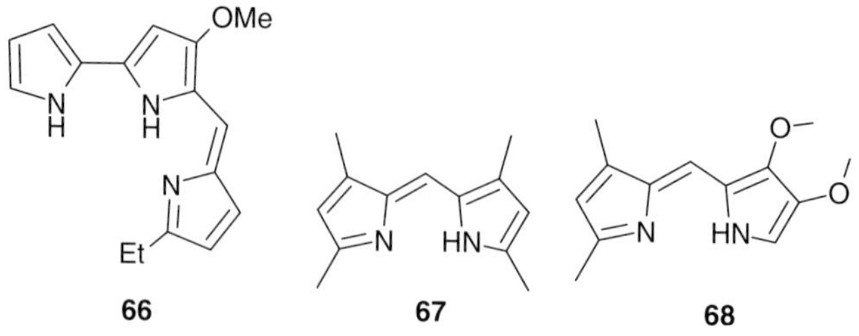

66 **67** **68**

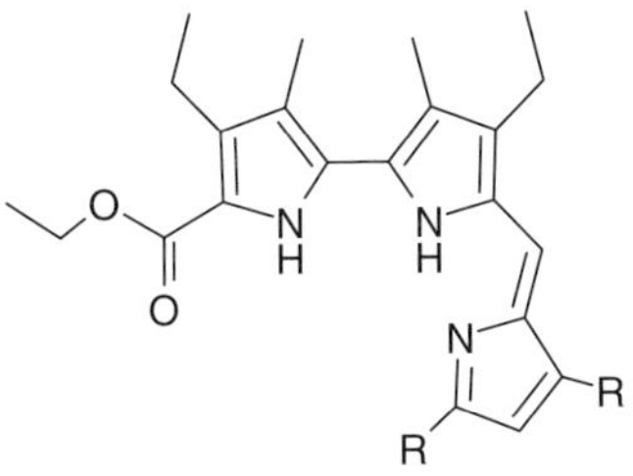

R = H, **69**
R = Me, **70**

The cationic receptors **71** and **72** (Figure 2.18) constituted by a $M(bpy)_3$ (M = Ru(II), Rh(III); bpy = bipyridine) core functionalized with two pyrrole units arranged in a chelate manner, were analyzed for chloride and benzoate binding in DMSO by isothermal titration calorimetry [141]. As previously observed for **27**, the formation of anion complexes with **71** and **72** in DMSO is also largely driven by entropy. In the case of benzoate anion, about 65–80% of the free energy of complexation is ascribable to entropic factors, while entropic contributions are completely responsible for the free energy associated with the binding of chloride,

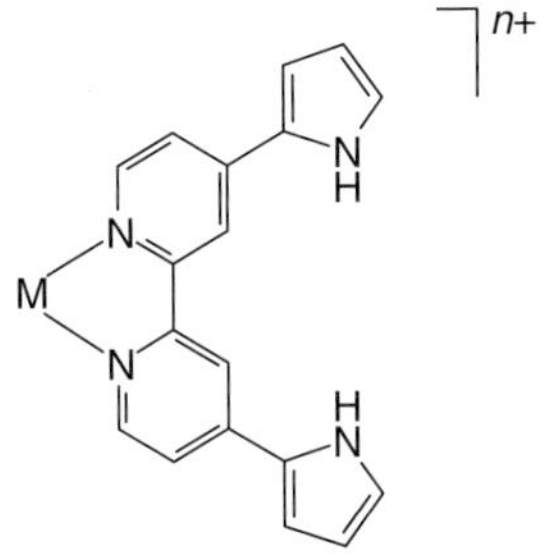

M = $bpy_2Ru(II)$, $n = 2$ **71**
M = $bpy_2Rh(III)$, $n = 3$ **72**

	ΔG° (kcal mol^{-1})	ΔH° (kcal mol^{-1})	$T\Delta S^\circ$ (kcal mol^{-1})
71 + Cl^-	−3.56	0.41	3.96
71 + $PhCO_2^-$	−4.35	−1.60	2.75
72 + Cl^-	−4.13	0.49	4.63
72 + $PhCO_2^-$	−4.97	−1.03	3.93

Figure 2.18 Thermodynamic parameters for the interaction of **71** and **72** with chloride and benzoate in DMSO at 298 K [141].

the enthalpic ones being slightly unfavorable (Figure 2.18). It was suggested that DMSO solvent molecules are associated strongly to free **71** and **72**, as already seen for **27**, and accordingly, their desolvation occurring upon anion binding requires a considerable energetic cost ($\Delta H^\circ > 0$) but produces a large entropy gain ($T\Delta S^\circ > 0$). This would explain the large positive entropic contributions and the slightly unfavorable or moderately favorable enthalpic factors observed for these anion binding processes [141].

A different effect of DMSO had been suggested to explain the different thermodynamic parameters determined for acetate anion complexation with the bicyclic guanidinium receptor **73** in this solvent ($\Delta G^\circ = -22.1$ kJ mol^{-1}, $\Delta H^\circ = -14.2$ kJ mol^{-1}, $\Delta S^\circ = 50.3$ J mol^{-1}K^{-1}), relative to CH_3CN ($\Delta G^\circ = -30.7$ kJ mol^{-1}, $\Delta H^\circ = -15.5$ kJ mol^{-1}, $\Delta S^\circ = 26.1$ J mol^{-1} K^{-1}) [142]. The lower stability of the acetate complex in DMSO is primarily due to a less positive entropy change, which might reflect a smaller number of bulkier DMSO molecules released on anion binding.

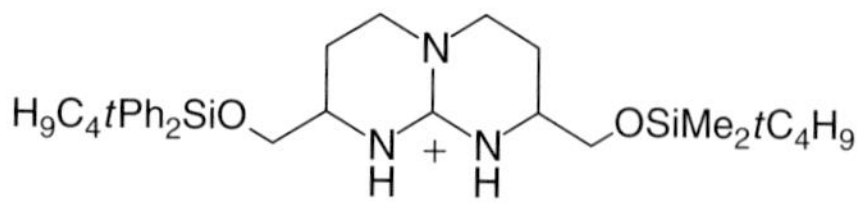

73

Interesting results were obtained for the coordination of SO_4^{2-} to the analogs **74** and **75** receptors and to the bisguanidinium derivatives **76–80** (Figure 2.19) in methanol (DMSO for **76**) [142, 143]. Isothermal titration calorimetry showed that these anion complexation reactions are endothermic and driven by favorable entropy changes, such characteristics being largely more marked in methanol, which is a solvent characterized by high AN (41.3, Table 2.2), hence by good acceptor ability in forming hydrogen bonds with anions. Owing to solubility problems, DMSO was used in the case of **76**, which is present in solution as a dimer under these conditions and, in fact, calorimetric measurements revealed the formation of a complex with 1 : 2 anion to receptor stoichiometry. Accordingly, this species cannot be used for a congruent comparison with the other complexes, although it follows this general enthalpic/entropic trend. For receptors **77–80**, containing two guanidinium groups, the sulfate binding is strongly entropy driven, with the entropic contribution by far outmatching the unfavorable enthalpic contribution, leading to consequently larger complex stability relative to receptors **74** and **75** having only one guanidinium binding unit. The removal of the silyl ether functionalities together with the introduction of a fluorine substituent in the aromatic group of **79** reduces the stability of the sulfate complex by a factor of about 10 with respect to **77** and **78**. This loss of stability is due to a less favorable entropy term, the enthalpic contribution being almost insensitive to such structural modifications. The different entropic contributions were ascribed to some solvophobic interaction occurring between silyl moieties of **77** and **78** when the two guanidinium groups are brought in proximity by sulfate coordination. With **80**, the monocyclic guanidinium analogous of **79**, the stability of the sulfate

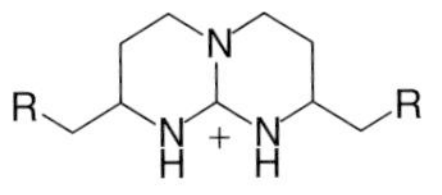

R = OH **74**

R = N_3 **75**

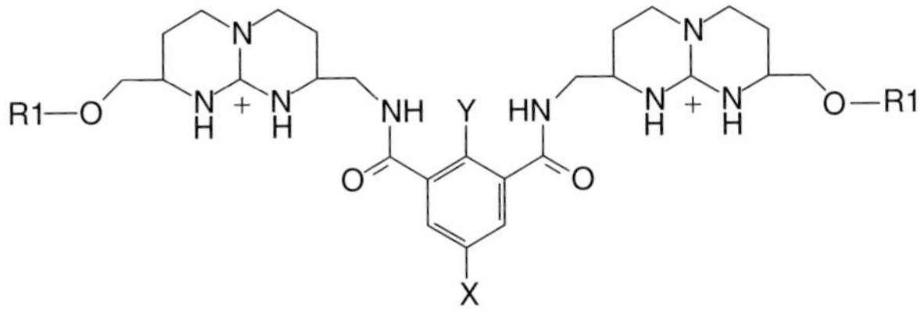

X	Y	R^1	
$CH_2SB_{12}H_{11}^{2-}$	H	Si*t*BuPh$_2$	**76**
$OCH_2C_6H_5$	H	Si*t*BuPh$_2$	**77**
OH	H	Si*t*BuPh$_2$	**78**
Br	Y	H	**79**

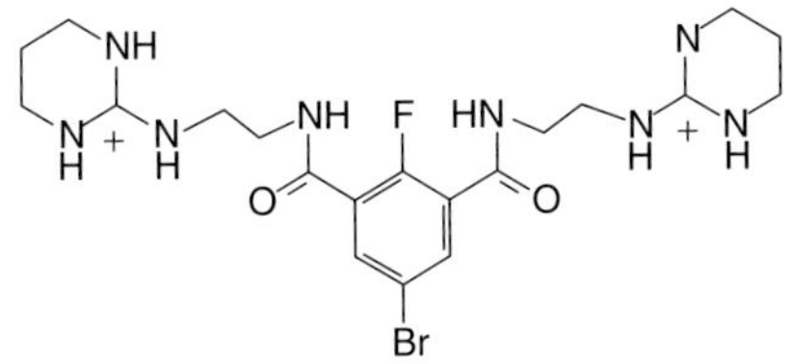

80

	ΔG° (kcal mol^{-1})	ΔH° (kcal mol^{-1})	$T\Delta S^\circ$ (kcal mol^{-1})
74	−3.43	3.76	7.22
75	−3.80	2.76	6.56
76	−3.40	1.50	4.90
77	−9.47	7.71	17.18
78	−9.27	7.07	16.34
79	−7.82	7.28	15.10
80	−8.91	5.46	14.37

Figure 2.19 Thermodynamic parameters for the interaction of **74–80** with sulfate in methanol (DMSO for **74**) at 303 K [142, 143].

complex increases, returning to be similar to those found for **77** and **78**. The greater flexibility of **79** was thought to facilitate a lower accumulation of intramolecular strain on anion binding, leading to a better anion–receptor fitting expressed by a more favorable (less unfavorable) enthalpy change, which outweighs the expected entropy loss [143].

A similar pattern of entropy-driven reactions accompanied by enthalpically unfavorable contributions was also observed for the association in methanol between the bicyclic guanidinium receptors **81–84** and several oxoanions (benzoate, naphtalene dicarboxylate, phthalate, terephthalate, oxalate, fumarate, isophthalate, squarate, $H_2PO_4^-$) [144]. Despite the good ability of methanol to stabilize anions through the formation of hydrogen bonds, the complexation reactions showed high stability constants for the association of monocharged species ($K = 10^3–10^4\ M^{-1}$).

Complex stability increases enormously (to above $K = 10^6\ M^{-1}$) in CH_3CN, exceeding the stability commonly found for anion pairing between monocharged partners, and the energetic profile changes substantially; the reactions become enthalpically favorable, but the entropic terms remain the leading contribution in determining the spontaneity of these association processes [144].

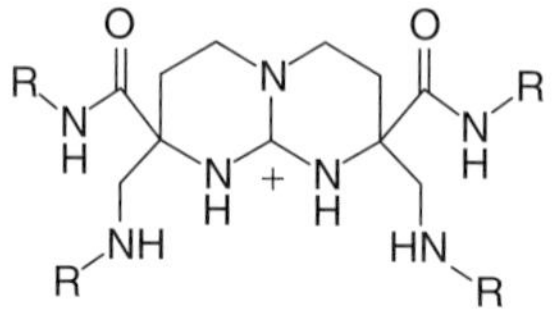

R = C_3H_7 **81**
R = $C_2H_4{-}OCH_3$ **82**
R = C_6H_5 **83**

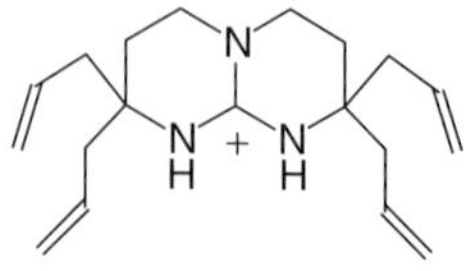

84

A considerable participation of entropic contributions to compose the free energy changes of anion coordination in CH_3CN were also revealed by means of isothermal titration calorimetry for the binding of benzoate anion by the diamide pyrrole derivatives **85**, **86** and their thioamides analogs **87** and **88** (Figure 2.20), although, only for the binding process involving **85** the enthalpy changes are prevalent over the enthalpic ones [145]. Despite complex stability being apparently invariant with respect to receptor structures, the thioamide ligands bind benzoate anion with more exothermic (more favorable) reactions relative to the amide ones, in agreement with their greater acidity and their stronger hydrogen bond donor ability, and compensation to an almost constant free energy of complexation is given by the entropic component.

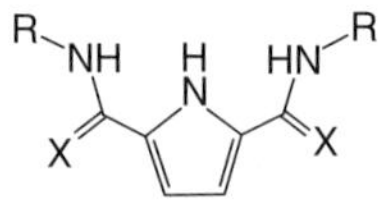

X = O, R = *n*-Bu **85**
X = O, R = Ph **86**
X = S, R = *n*-Bu **87**
X = S, R = Ph **88**

	$\Delta G°$ (kJ mol^{-1})	$\Delta H°$ (kJ mol^{-1})	$T\Delta S°$ (kJ mol^{-1})
85	−24.1	−11.43	12.66
86	−24.6	−14.71	9.90
87	−24.6	−13.67	10.98
88	−24.9	−16.40	8.54

Figure 2.20 Thermodynamic parameters for the interaction of **85**–**88** with benzoate in acetonitrile (DMSO for **74**) at 303 K [145].

Biscyclopeptides, such as **89–94**, were developed by linkage of two hexapeptidic rings through different bridging units. These molecules, which are able to bind anions in a sandwich-like manner, proved to be among the most efficient neutral anion receptors in aqueous organic solvents [146–149]. The thermodynamic parameters for the coordination of SO_4^{2-} and I^- with **89–91** (Table 2.7) were obtained by means of isothermal titration calorimetry in CH_3CN/H_2O 2 : 1 (v/v) solution [147]. The binding of sulfate by the three receptors is endothermic and strongly entropy driven, suggesting that desolvation of the interacting partners, principally of sulfate, is the main driving force. This interpretation found supporting elements in the crystal structure of the sulfate complex with **89**, showing that the anion completely desolvated and sandwiched between the two peptide rings [148]. Conversely, for the less solvated iodide anion, the desolvation effect is much reduced and the complexation reactions become exothermic and accompanied by largely less favorable entropic contributions. Interesting results were obtained by analyzing the solvent dependence of the thermodynamic parameters of sulfate and iodide complexes with **89** and **90** [148]. The free energy of sulfate complexation becomes less negative with increasing amount of water present in the CH_3CN/H_2O solvent mixture in the range of about 50–90 mol% of water. While this variation is not very large, the enthalpy and entropy changes with solvent composition are much more pronounced. Enthalpy and entropy changes become less unfavorable and less favorable, respectively, with increasing water content. A similar result was found for iodide, although, in this case the observed variation of the thermodynamic parameters of complexation are much less solvent dependent. Analysis of these thermodynamic parameters provided indication of the involvement of hydrophobic intrareceptor interactions on complex formation.

The influence of the linking unit between the two cyclopeptide rings on the stability of the anion complexes was studied in 1 : 1 (v/v) water/methanol mixture using the receptors **91–94** and sulfate, chloride, bromide, and iodide anions as guests [146, 149]. The thermodynamic parameters for the association processes, determined by means of isothermal titration calorimetry, showed that, although the differences in the affinity and selectivity of these receptors toward a given anion are not very large, there are important differences in the thermodynamics of

Table 2.7 Thermodynamic parameters for the complexation of sulfate and iodide by receptors **89–91** determined in 2 : 1 (v/v) acetonitrile/water at 298 K [147].

	$\Delta G°$ (kJ mol^{-1})	$\Delta H°$ (kJ mol^{-1})	$T\Delta S°$ (kJ mol^{-1})
89 + SO_4^{2-}	−38.4	1.8	40.1
90 + SO_4^{2-}	−39.0	3.7	42.7
91 + SO_4^{2-}	−30.2	10.7	41.0
89 + I^-	−25.5	−20.7	4.8
90 + I^-	−27.1	−13.4	13.7
91 + I^-	−20.0	−4.3	15.7

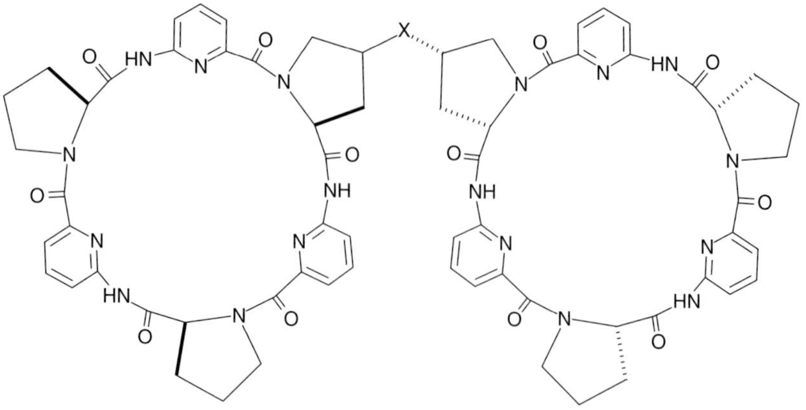

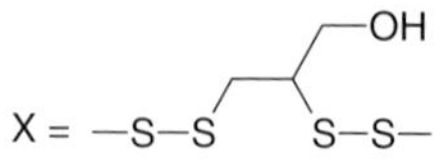
89

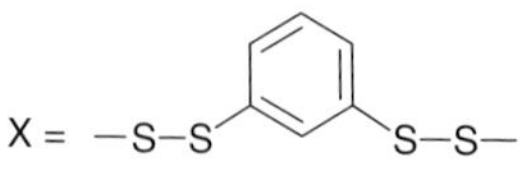
90

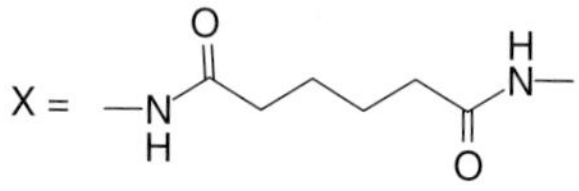
91

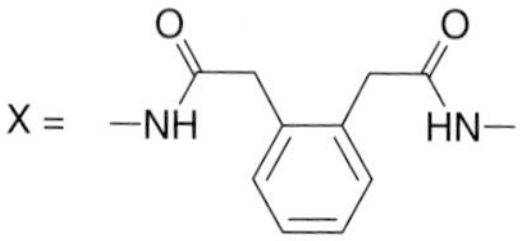
92

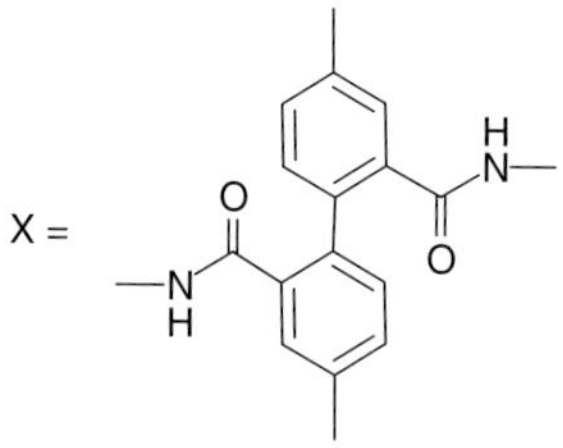
93

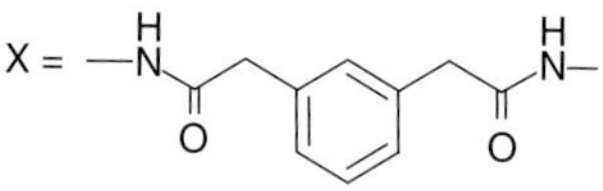
94

anion complexation. With the exception of chloride complexation with **93**, which resulted to be almost athermic preventing the determination of the thermodynamic parameters, all complexation reactions are driven by favorable enthalpic and entropic contributions. All four receptors bind these anions according to the affinity trend $SO_4^{2-} > I^- > Br^- > Cl^-$, in contrast to expectations based on the hydrogen binding ability of the anions, $SO_4^{2-} > Cl^- > Br^- > I^-$, and better interpreted in terms of size fitting of the halogenide anions. Compound **92** is the weakest sulfate receptor, while halides are bound the least strongly by **93**. The complexation enthalpy changes become less favorable in the order **91** > **92** > **93** to increase for **94**, which shows similar or even larger enthalpic contributions than **92**, while an opposite trend was found for the entropy changes. This behavior was related, at least for **91–93**, to the rigidity of the linkers bridging the two cyclopeptide rings, increasing in the order **91** < **92** < **93**. The more rigid linkers prevent the binding subunits from achieving optimal contacts with the coordinated anions, thus generating less favorable enthalpy changes of complexation. On the other hand, more flexible linkers allow the receptors to better match the coordinated anions, but this takes place at the expense of a greater loss of conformational mobility, which corresponds to a greater entropy loss. The behavior of **94** was justified considering that, despite its rigidity, the structure of the linker allows the biscyclopeptide to efficiently bind the anions. The interplay of these enthalpic and entropic contributions determines the observed binding selectivity [149].

The enthalpic and entropic components of free energy of anion complexation reactions described in this section stimulate some reflections on the nature of similar binding events, on the role played by the receptors, and how the design of receptors could be modeled on the basis of the different information given by the enthalpy and entropy terms. Anion coordination in solution is the successful result of the anion–receptor interaction emerging from a number of parallel and competitive reactions involving solvent, background electrolytes, counterions, and any other species present in the reaction medium. The complexation stability constants, and the connected free energy changes, quantify these successful results in a massive and rather anonymous way, while dissection of the free energy change into its enthalpic and entropic components moves the viewpoint more deeply, allowing consideration at the molecular level.

We have seen that anion complexation reactions are promoted by a variable interplay of enthalpy and entropy changes. Anion binding by charged receptors in water, or other strongly competing solvents, is commonly athermic and favored by large entropic contributions, while in apolar solvents anion complexation reactions are more typically enthalpy driven, with minor entropic contributions. In some cases, both enthalpy and entropy are favorable, and even tight associations can be achieved by summation of moderately favorable contributions. In a general sense, a larger variability is observed for the enthalpies of anion coordination, the entropic terms being frequently favorable or moderately unfavorable, and limitation to the efficiency of anion coordination due to enthalpy–entropy compensation effects is commonly encountered.

Even though solvation/desolvation phenomena have been recognized as primarily responsible of these behaviors, the effects of ligand structure, preorganization, rigidity, or flexibility on the enthalpic and entropic changes of complexation have also been noted. Since solvent effects are in a first approximation inaccessible to molecular design, the portions of enthalpic and entropic contributions to anion coordination that are beyond the solvent reign are the ground for the optimization of anion–receptor binding. Nevertheless, optimization may be asked to satisfy different functions. For instance, for sequestration or extraction purposes, the anion complexes should be as stable as possible, irrespective of their structures, while for regioselective/stereoselective binding and catalysis or in self-assembly, the desired function may depend on a unique structure. These issues were deeply analyzed, giving strong emphasis to the influence of entropic contributions [150]. The design of anion receptors for binding optimization has mostly been based on the search of more preorganized and complementary ligands, addressing the enthalpy contribution connected with the formation of anion–receptor bonds as a fundamental tool to enhance complex stability. On the other hand, binding on entropic grounds may provide the appropriate instruments to satisfy functional requirements. The observed entropy change ΔS_{obs} is composed of a solvent-related term ΔS_{solv} and an intrinsic contribution due to all modifications of the interacting partners $\Delta S_{intrinsic}$. As already noted, ΔS_{solv} is essentially inaccessible to molecular design. A detailed interpretation of the second term ($\Delta S_{intrinsic}$) was performed by considering appropriate additive components:

$$\Delta S_{intrinsic} = \Delta S_{trans + rot} + \Delta S_{vibration} + \Delta S_{conformation} + \Delta S_{configuration}$$

$\Delta S_{trans+rot}$ is the translational and rotational entropy change experienced by anion and receptor on binding. $\Delta S_{vibration}$ refers to the generation of vibrational entropy in the complex. $\Delta S_{conformation}$ is the entropy change related to the loss of internal rotations on complexation, and $\Delta S_{configuration}$ refers to the different geometric arrangements of anion and receptor.

With the exception of $\Delta S_{trans+rot}$, which can be excluded from design considerations since it is almost invariant for the kind of synthetic receptors we are dealing with, all the other terms are important to molecular design. $\Delta S_{vibration}$ and $\Delta S_{conformation}$ are closely related to the tightness of the anion–receptor association. Loose binding of two flexible partners takes place with small entropic cost, since a moderate freezing of internal rotations occurs and the vibrations of the anion–receptor bonds have large amplitude. The generation of low-frequency (large-amplitude) motion increases entropy, enhancing the binding affinity. On the contrary, tight association on enthalpic basis severely impedes internal rotations and the bonding vibrations increase their frequency, leading to an entropy loss, compared to the uncomplexed state, that is, unfavorable to association. Furthermore, the $\Delta S_{configuration}$ component also affords its contribution, which is related to the distribution of geometrical arrangements found by the interacting partners in the complex. Several, if not many, arrangements are possible, even in the case of tight association, which reduce the structural definition of the complex but increase the entropic advantage. On this basis, strong anion binding can be achieved at different levels of structural specificity, for functional purposes, by receptor design on entropic grounds, the structural specificity of the association process increasing with decreasing association entropy [150]. On the other hand, appropriate design on the enthalpic ground may ensure a consistent energetic support to association affinity and selectivity. Receptor design on both enthalpic and entropic grounds could be a possible way to escape the occurrence of enthalpy–entropy compensation effects, largely observed in anion coordination, and move toward more efficient anion receptors.

References

1. Bazzicalupi, C., Bencini, A., Bianchi, A., Faggi, E., Giorgi, C., Santarelli, S., and Valtancoli, B. (2008) Polyfunctional binding of thymidine 5'-triphosphate with a synthetic polyammonium receptor containing aromatic groups. Crystal structure of the nucleotide-receptor

adduct. *J. Am. Chem. Soc.*, **130**, 2440–2441.
2. Schneider, H.-J. and Yatsimirsky, A. (2000) *Principles and Methods in Supramolecular Chemistry*, John Wiley & Sons, Ltd, Chichester.
3. Schneider, H.-J. and Theis, I. (1989) Additivities of electrostatic and hydrophobic interactions in host-guest complexes. *Angew. Chem. Int. Ed. Engl.*, **28**, 753–754.
4. Schneider, H.-J. (1991) Mechanisms of molecular recognition: investigation of organic host-guest complexes. *Angew. Chem. Int. Ed. Engl.*, **30**, 1417–1436.
5. Schneider, H.-J. (2009) Binding mechanisms in supramolecular complexes. *Angew. Chem.*, **48**, 3924–3977.
6. Schneider, H.-J., Blatter, T., Palm, B., Pfingstag, U., Rüdiger, V., and Theis, I. (1992) Complexation of nucleosides, nucleotides, and analogs in an azoniacyclophane. Van der Waals and electrostatic binding increments and NMR shielding effects. *J. Am. Chem. Soc.*, **114**, 7704–7708.
7. Schneider, H.-J., Schiestel, T., and Zimmermann, P. (1992) The incremental approach to noncovalent interactions: Coulomb and van der Waals effects in organic ion pairs. *J. Am. Chem. Soc.*, **114**, 7698–7703.
8. Schneider, H.-J. (1994) Linear free energy relationships and pairwise interactions in supramolecular chemistry. *Chem. Soc. Rev.*, **22**, 227–234.
9. Schneider, H.-J., Eblinger, F., Sartorius, J., and Rammo, J. (1996) Experimental results from host-guest complexes for the design of effectors in biological systems and of enzyme analogous catalysts. *J. Mol. Rec.*, **9**, 123–132.
10. Hossain, M.A. and Schneider, H.J. (1999) Flexibility, association constants, and salt effects in organic ion pairs: how single bonds affect molecular recognition. *Chem. Eur. J.*, **5**, 1284–1290.
11. Kimura, E., Kodama, M., and Yatsunami, T. (1982) Macromonocyclic polyamines as biological polyanion complexons. 2. Ion-pair association with phosphate and nucleotides. *J. Am. Chem. Soc.*, **104**, 3182–3187.
12. Bianchi, A., Micheloni, M., and Paoletti, P. (1988) Supramolecular interaction between adenosine 5'-triphosphate (ATP) and polycharged tetraazamacrocycles. Thermodynamic and ^{31}P NMR studies. *Inorg. Chim. Acta*, **151**, 269–272.
13. Schmidtchen, F.P. (1977) Inclusion of anions in macrotricyclic quaternary ammonium salts. *Angew. Chem. Int. Ed. Engl.*, **16**, 720–721.
14. Schneider, H.-J. and Blatter, T. (1992) Interactions between acyclic and cyclic peralkylammonium compounds and DNA. *Angew. Chem. Int. Ed. Engl.*, **31**, 1207–1208.
15. Högfeldt, E. (1983) *Stability Constants of Metal-Ion Complexes. Part A: Inorganic Ligands*, Pergamon, Oxford.
16. Pettit, L.D. and Powell, K.J. (1997) *Stability Constant Database*, Academic Software, New York.
17. Martell, A.E. and Smith, R.M. (1976, 1982) *Critical Stability Constants*, Vol. 4 (1976), Vol. 5 (1982), Plenum Press, New York.
18. Gilli, G. and Gilli, P. (2009) *The Nature of the Hydrogen Bond: Outline of a Comprehensive Hydrogen Bond Theory*, Oxford University Press, New York.
19. Gautam, D. (2001) *The Weak Hydrogen Bond*, Oxford University Press, New York.
20. Jeffrey, G.A. (1997) *An Introduction to Hydrogen Bonding*, Oxford University Press, New York.
21. Worm, K., Schmidtchen, F.P., Schier, A., Schäfer, A., and Hesse, M. (1994) Macrotricyclic borane-amine adducts: the first uncharged synthetic host compounds without Lewis acid character, for anionic guests. *Angew. Chem. Int. Ed. Engl.*, **33**, 327–329.
22. Imai, Y.N., Inoue, Y., Nakanishi, I., and Kitaura, K. (2008) Cl-π interactions in protein-ligand complexes. *Protein Sci.*, **17**, 1129–1137.
23. Berryman, O.B. and Johnson, D.W. (2009) Experimental evidence for interactions between anions and electron-deficient aromatic rings. *Chem. Commun.*, 3143–3153.

24. Berryman, O.B., Sather, A.C., Hay, B.P., Meisner, J.S., and Johnson, D.W. (2008) Solution phase measurement of both weak σ and C-H·X- hydrogen bonding interactions in synthetic anion receptors. *J. Am. Chem. Soc.*, **130**, 10895–10897.
25. Rosokha, Y.S., Lindeman, S.V., Rosokha, S.V., and Kochi, J.K. (2004) Halide recognition through diagnostic "anion-π" interactions: molecular complexes of Cl^-, Br^-, and I^- with olefinic and aromatic π receptors. *Angew. Chem. Int. Ed.*, **43**, 4650–4652.
26. Arranz, P., Bianchi, A., Cuesta, R., Giorgi, C., Godino, M.L., Gutiérrez, M.D., López, R., and Santiago, A. (2010) Binding and removal of sulphate, phosphate, arseniate, tetrachloromercuriate and chromate in aqueous solution by means of an activated carbon functionalized with a pyrimidine-based anion receptor (HL). Crystal structures of $[H_3L(HgCl_4)]\cdot H_2O$ and $[H_3L(HgBr_4)\cdot]H_2O$ showing anion-π interactions. *Inorg. Chem.*, **49**, 9321–9332.
27. Hibbert, F. and Emsley, J. (1990) Hydrogen bonding and chemical reactivity. *Adv. Phys. Org. Chem.*, **26**, 255–379.
28. Taylor, R., Kennard, O., and Versichel, W. (1983) Geometry of the NH···O=C hydrogen bond. 1. Lone-pair directionality. *J. Am. Chem. Soc.*, **105**, 5761–5766.
29. Adalsteinsson, H., Maulitz, A.H., and Bruice, T.C. (1996) Calculation of the potential energy surface for intermolecular amide hydrogen bonds using semiempirical and *ab initio* methods. *J. Am. Chem. Soc.*, **118**, 7689–7693.
30. Morokuna, K. (1971) Molecular orbital studies of hydrogen bonds, III. C=O–H···O hydrogen bond in H_2CO–H_2O and H_2CO–$2H_2O$. *J. Chem. Phys.*, **55**, 1236–1244.
31. Kitaura, K. and Morokuna, K. (1976) A new energy decomposition scheme for molecular interactions within the Hartree-Fock approximation. *Int. J. Quant. Chem.*, **10**, 325–340.
32. Umeyana, H. and Morokuma, K. (1977) The origin of hydrogen bonding: an energy decomposition study. *J. Am. Chem. Soc.*, **99**, 1316–1332.
33. Scheiner, S. and Kar, T. (1995) The nonexistence of specially stabilized hydrogen bonds in enzymes. *J. Am. Chem. Soc.*, **117**, 6970–6975.
34. Shan, S., Loh, S., and Herschlag, D. (1996) The energetics of hydrogen bonds in model systems: implications for enzymatic catalysis. *Science*, **272**, 97–101.
35. Atwood, J.L. and Steed, J.W. (2004) *Encyclopedia of Supramolecular Chemistry*, Marcel Dekker, New York.
36. Atwood, J.L. and Steed, J.W. (2009) *Supramolecular Chemistry*, 2nd edn, Wiley-VCH Verlag GmbH, New York.
37. Tanford, C. (1980) *The Hydrophobic Effect*, 2nd edn, Wiley-VCH Verlag GmbH, New York.
38. Ben-Naim, A. (1980) *Hydrophobic Interactions*, Plenum Press, New York.
39. Aragó, J., Bencini, A., Bianchi, A., Domenech, A., and García-España, E. (1992) Macrocyclic effect on anion binding. A potentiometric and electrochemical study of the interaction of 21- and 24-membered polyazaalkanes with $[Fe(CN)_6]^{4-}$ and $[Co(CN)_6]^{3-}$. *J. Chem. Soc. Dalton Trans.*, 319–324.
40. García-España, E., Diaz, P., Llinares, J.M., and Bianchi, A. (2006) Anion coordination chemistry in aqueous solution of polyammonium receptors. *Coord. Chem. Rev.*, **250**, 2952–2986.
41. Luecke, H. and Quiocho, F.A. (1990) High specificity of a phosphate transport protein determined by hydrogen bonds. *Nature*, **347**, 402–406.
42. Yao, N., Ledvina, P.S., Choudhary, A., and Quiocho, F.A. (1996) Modulation of a salt link does not affect binding of phosphate to its specific active transport receptor. *Biochemistry*, **35**, 2079–2085.
43. Adamson, A.W. (1954) A proposed approach to the chelate effect. *J. Am. Chem. Soc.*, **76**, 1578–1579.
44. Dixon, R.P., Geib, S.J., and Hamilton, A.D. (1992) Molecular recognition: bis-acylguanidiniums provide a simple family of receptors for phosphodiesters. *J. Am. Chem. Soc.*, **114**, 365–366.

45. Grossel, M.C., Merckel, D.A.S., and Hutchings, M.G. (2003) The effect of preorganisation on the solid state behaviour of simple 'aromatic-cored' bis(guanidinium) sulfates. *Cryst. Eng. Commun.*, **5**, 77–81.
46. Hutchings, M.G., Grossel, M.C., Merckel, D.A.S., Chippendale, A.M., Kenworthy, M., and McGeorge, G. (2001) The structure of *m*-xylylenediguanidinium sulfate: a putative molecular tweezer ligand for anion chelation. *Cryst. Grouth Des.*, **1**, 339–342.
47. Sessler, J.L., Cyr, M.J., Lynch, V., McGhee, E., and Ibers, J.A. (1990) Synthetic and structural studies of sapphyrin, a 22-π-electron pentapyrrolic "expanded porphyrin". *J. Am. Chem. Soc.*, **112**, 2810–2813.
48. Shionoya, M., Furuta, H., Lynch, V., Harriman, A., and Sessler, J.L. (1992) Diprotonated sapphyrin: a fluoride selective halide anion receptor. *J. Am. Chem. Soc.*, **114**, 5714–5722.
49. Park, C.H. and Simmons, H.E. (1968) Macrobicyclic amines. III. Encapsulation of halide ions by *in, in*-1, (k + 2)-diazabicyclo[*k.l.m.*]alkane ammonium ions. *J. Am. Chem. Soc.*, **90**, 2431–2432.
50. Bencini, A., Bianchi, A., García-España, E., Micheloni, M., and Ramirez, J.A. (1998) Proton coordination by polyamine compounds in aqueous solution. *Coord. Chem. Rev.*, **188**, 97–156.
51. Bencini, A., Bianchi, A., Giorgi, C., Paoletti, P., Valtancoli, B., Fusi, V., García-España, E., Llinares, J.M., and Ramirez, J.A. (1996) Effect of nitrogen methylation on cation and anion coordination by hexa- and heptaazamacrocycles. Catalytic properties of these ligands in ATP dephosphorylations. *Inorg. Chem.*, **35**, 1114–1120.
52. Andrés, A., Bazzicalupi, C., Bencini, A., Bianchi, A., Fusi, V., García-España, E., Giorgi, C., Nardi, N., Paoletti, P., Ramirez, J.A., and Valtancoli, B. (1994) 1,10-Dimethyl-1,4,7,10,13,16-hexaazacyclooctadecane L and 1,4,7-trimethyl-1,4,7,10,13,16,19-heptaazacycloheneicosane L1: two new macrocyclic receptors for ATP binding. Synthesis, solution equilibria and the crystal structure of $(H_4L)(ClO_4)_4$. *J. Chem. Soc. Perkin Trans. 2*, 2367–2373.
53. Schneider, H.-J., Schiestel, T., and Zimmermann, P. (1992) Host-guest supramolecular chemistry. 34. The incremental approach to noncovalent interactions: coulomb and van der Waals effects in organic ion pairs. *J. Am. Chem. Soc.*, **114**, 7698–7703.
54. Beer, P.D., Fletcher, N.C., Grieve, A., Wheeler, J.W., Moore, C.P., and Wear, T. (1996) Halide anion recognition by new acyclic quaternary polybipyridinium and polypyridinium receptors. *J. Chem. Soc. Perkin Trans.*, 1545–1551.
55. Jeong, K.-S. and Cho, Y.L. (1997) Highly strong complexation of carboxylates with 1-alkylpyridinium receptors in polar solvents. *Tetrahedron Lett.*, **38**, 3279–3282.
56. Prohens, R., Rotger, M.C., Piña, M.N., Deyà, P.M., Morey, J., Ballester, P., and Costa, A. (2001) Thermodynamic characterization of the squaramide–carboxylate interaction in squaramide receptors. *Tetrahedron Lett.*, **42**, 4933–4936.
57. Piña, M.N., Rotger, M.C., Costa, A., Ballester, P., and Deyà, P.M. (2004) Evaluation of anion selectivity in protic media by squaramide-Cresol Red ensembles. *Tetrahedron Lett.*, **45**, 3749–3752.
58. Schneider, H.-J., Kramer, R., Simova, S., and Schneider, U. (1988) Host-guest chemistry. 14. Solvent and salt effects on binding constants of organic substrates in macrocyclic host compounds. A general equation measuring hydrophobic binding contributions. *J. Am. Chem. Soc.*, **110**, 6442–6448.
59. Shinoda, S., Tadokoro, M., Tsukube, H., and Arakawa, R. (1998) One-step synthesis of a quaternary tetrapyridinium macrocycle as a new specific receptor of tricarboxylate anions. *Chem. Commun.*, 181–182.
60. Schneider, H.-J., Blatter, T., Simova, S., and Theis, I. (1989) Large binding constant differences between aromatic

and aliphatic substrates in positively charged cavities indicative of higher order electric effects. *J. Chem. Soc. Chem. Commun.*, 580–581.

61. Menger, F.M. and Catlin, K.K. (1995) Octacationic cyclophanes: binding of ATP and other anionic guests in water. *Angew. Chem. Int. Ed. Engl.*, **34**, 2147–2150.
62. Hinzen, B., Seiler, P., and Diederich, F. (1996) Mimicking the vancomycin carboxylate binding site: synthetic receptors for sulfonates, carboxylates, and *N*-protected α-amino acids in water. *Helv. Chim. Acta*, **79**, 942–960.
63. Prohens, R., Martorell, G., Ballester, P., and Costa, A. (2001) A squaramide fluorescent ensemble for monitoring sulfate in water. *Chem. Commun.*, 1456–1457.
64. Hossain, M.A., Kang, S.O., Powell, D., and Bowman-James, K. (2003) Anion receptors: a new class of amide/quaternized amine macrocycles and the chelate effect. *Inorg. Chem.*, **42**, 1397–1399.
65. Schmidtchen, F.P. (1986) Probing the design of a novel ditopic anion receptor. *J. Am. Chem. Soc.*, **108**, 8249–8255.
66. Claude, S., Lehn, J.-M., Schmidt, F., and Vigneron, J.-P. (1991) Binding of nucleosides, nucleotides and anionic planar substrates by bis-intercaland receptor molecules. *J. Chem. Soc. Chem. Commun.*, 1182–1185.
67. Hossain, M.A. and Ichikawa, K. (1994) A novel macrotricyclic receptor for the inclusion of fluoride ion. *Tetrahedron Lett.*, **35**, 8393–8396.
68. Murakami, Y., Hayashida, O., Ito, T., and Hisaeda, Y. (1993) Molecular recognition by novel cage-type azaparacyclophanes bearing chiral binding sites in aqueous media. *Pure Appl. Chem.*, **65**, 551–556.
69. Murakami, Y., Hayashida, O., and Nagai, Y. (1994) Enantioselective discrimination by cage-type cyclophanes bearing chiral binding sites in aqueous media. *J. Am. Chem. Soc.*, **116**, 2611–2612.
70. Eudić, P., Žinić, M., Tomišić, V., Simeon, V., Vigneron, J.-P., and Lehn, J.-M. (1995) Binding of nucleotides in water by phenanthridinium bis(intercaland) receptor molecules. *J. Chem. Soc. Chem. Commun.*, 1073–1075.
71. Hayashida, O., Ono, K., Hisaeda, Y., and Murakami, Y. (1995) Specific molecular recognition by chiral cage-type cyclophanes having leucine, valine, and alanine residue. *Tetrahedron*, **51**, 8423–8436.
72. Bazzicalupi, C., Bencini, A., Bianchi, A., Cecchi, M., Escuder, B., Fusi, V., García-España, E., Giorgi, C., Luis, S.V., Maccagni, G., Marcelino, V., Paoletti, P., and Valtancoli, B. (1999) Thermodynamics of phosphate and pyrophosphate anions binding by polyammonium receptors. *J. Am. Chem. Soc.*, **121**, 6807–6815.
73. Arranz, P., Bencini, A., Bianchi, A., Diaz, P., García-España, E., Giorgi, C., Luis, S.V., Querol, M., and Valtancoli, B. (2001) Thermodynamics of sulfate binding by macrocyclic polyammonium receptors. *J. Chem. Soc. Perkin Trans. 2*, 1765–1770.
74. Hosseini, M.W. and Lehn, J.-M. (1988) Anion-receptor molecules: macrocyclic and macrobicyclic effects on anion binding by polyammonium receptors molecules. *Helv. Chim. Acta*, **71**, 749–756.
75. Chmielewski, M.J. and Jurczak, J. (2005) Anion recognition by neutral macrocyclic amides. *Chem. Eur. J.*, **11**, 6080–6094.
76. Bianchi, A. and García-España, E. (1990) in *Supramolecular Chemistry of Anions* (eds A. Bianchi, K. Bowman-James, and E. García-España), Wiley-VCH Verlag GmbH, New York, pp. 217–275.
77. Marcus, Y. (1985) *Ion Solvation*, John Wiley & Sons, Ltd, Chichester.
78. Johnsson, M. and Persson, I. (1987) Determination of Gibbs free energy of transfer for some univalent ions from water to methanol, acetonitrile, dimethylsulfoxide, pyridine, tetrahydrothiophene and liquid ammonia; standard electrode potentials of some

couples in these solvents. *Inorg. Chim. Acta*, **127**, 15–24.

79. Cox, B.G. and Waghorn, W.E. (1980) Thermodynamics of ion–solvent interactions. *Chem. Soc. Rev.*, **9**, 381–411.

80. Hefter, G.T. (1991) Fluoride solvation–the case of the missing ion. *Pure Appl. Chem.*, **63**, 1749–1758.

81. Weast, R.C. (ed.) (1987) *Handbook of Chemistry and Physics*, 67th edn, CRC Press, Boca Raton, FL.

82. Landolt-Börnstein (1950) *Atoms and Ions*, vol. **1**, Part 1, Springer-Verlag, Berlin.

83. Applequist, J., Carl, J.R., and Fung, K.-K. (1972) Atom dipole interaction model for molecular polarizability. Application to polyatomic molecules and determination of atom polarizabilities. *J. Am. Chem. Soc.*, **94**, 2953–2960.

84. Gutmann, V. and Wychera, E. (1966) Coordination reactions in non aqueous solutions. The role of the donor strength. *Inorg. Nucl. Chem. Lett.*, **2**, 257–260.

85. Gutmann, V. (1978) *The Donor-Acceptor Approach to Molecular Interactions*, Plenum, New York.

86. Mayer, U., Gutmann, V., and Gerger, W. (1975) The acceptor number. A quantitative empirical parameter for the electrophilic properties of solvents. *Monatsh. Chem.*, **106**, 1235–1257.

87. Gutmann, V. (1976) Empirical parameters for donor and acceptor properties of solvents. *Electrochim. Acta*, **21**, 661–670.

88. Reichardt, C. (1988) *Solvent and Solvent Effects in Organic Chemistry*, Verlag Chemie, Weinheim.

89. Mayer, U. (1975) Ionic equilibria in donor solvents. *Pure Appl. Chem.*, **41**, 291–326.

90. Cox, B.G. and Schneider, H. (1989) Influence of macrocyclic ligands on electrolyte solvation and solubility. *Pure Appl. Chem.*, **61**, 171–178.

91. Cox, B.G., Firman, P., García-Rosas, J., and Schneider, H. (1982) Free energies of transfer from water to non-aqueous solvents of univalent (2,2,2) cryptand and 18-crown-6 complexes. *Tetrahedron Lett.*, **23**, 3777–3780.

92. Danil de Namor, A.F., Ghousseini, L., and Lee, W.H. (1985) Stability constants and free energies of complexation of metal-ion cryptates in nitromethane. Derived parameters for the extraction of cations by cryptand 222 from water to pure nitromethane. *J. Chem. Soc. Faraday Trans. 1*, **81**, 2495–2502.

93. Danil de Namor, A.F., Ghousseini, L., and Hill, T. (1986) Alkali-metal cryptates in nitromethane. Thermodynamic parameters for complex formation and for transfer among different solvents. *J. Chem. Soc. Faraday Trans. 1*, **82**, 349–357.

94. Juwarker, H., Lenhardt, J.M., Castillo, J.C., Zhao, E., Krishamurthy, S., Jamiolkowski, R.M., Kim, K.-H., and Craig, S.L. (2009) Anion binding of short, flexible aryl triazole oligomers. *J. Org. Chem.*, **74**, 8924–8934.

95. Beer, P.D. and Shade, M. (1997) Solvent dependent selectivity exhibited by neutral ferrocenyl receptors. *Chem. Commun.*, 2377–2378.

96. Sessler, J.L., Gross, D.E., Cho, W.-S., Lynch, V.M., Schmidtchen, F.P., Bates, G.W., Light, M.E., and Gale, P.A. (2006) Cailx[4]pyrrole as a chloride anion receptor: solvent and countercation effects. *J. Am. Chem. Soc.*, **128**, 12281–12288.

97. Allen, W.E., Gale, P.A., Brown, C.T., Lynch, V.M., and Sessler, J.L. (1996) Binding of neutral substrates by calix[4]pyrroles. *J. Am. Chem. Soc.*, **118**, 12471–12472.

98. Jagessar, R.C., Shang, M., Scheidt, W.R., and Burns, D.H. (1998) Neutral ligands for selective chloride anion complexation: ($\alpha,\alpha,\alpha,\alpha$)-5,10,15,20-tetrakis(2-(arylurea) phenyl)porphyrins. *J. Am. Chem. Soc.*, **120**, 11684–11692.

99. Burns, D.H., Calderon-Kawasaki, K., and Kularatne, S. (2005) Buried solvent determines both anion-binding selectivity and binding stoichiometry with hydrogen-bonding receptors. *J. Org. Chem.*, **70**, 2803–2807.

100. Hefter, G.H. (2005) Ion solvation in aqueous-organic mixtures. *Pure Appl. Chem.*, **77**, 605–617.

101. Hefter, G., Marcus, Y., and Waghorne, W.E. (2002) Enthalpies and entropies of transfer of electrolytes and ions from water to mixed aqueous organic solvents. *Chem. Rev.*, **102**, 2773–2836.
102. Hefter, G.T. (1989) Solvation of fluoride ions. 3. A review of fluoride solvation thermodynamics in nonaqueous and mixed solvents. *Rev. Inorg. Chem.*, **10**, 185–223.
103. Hefter, G.T. and McLay, P.J. (1988) Solvation of fluoride ions. II. Enthalpies and entropies of transfer from water to aqueous methanol. *Aust. J. Chem.*, **41**, 1971–1975.
104. Schmidtchen, F.P. (2002) Surprises in the energetics of host-guest anion binding to calix[4]pyrrole. *Org. Lett.*, **4** (3), 431–434.
105. Blas, J.R., Márquez, M., Sessler, J.L., Luque, F.J., and Orozco, M. (2002) Theoretical study of anion binding to calix[4]pyrrole: the effects of solvent, fluorine substitution, cosolute, and water traces. *J. Am. Chem. Soc.*, **124**, 12796–12805.
106. Lehn, J.-M. (1995) *Supramolecular Chemistry. Concepts and Perspectives*, Wiley-VCH Verlag GmbH, Weinheim.
107. Andrés, A., Aragó, J., Bencini, A., Bianchi, A., Domenech, A., Fusi, V., García-España, E., Paoletti, P., and Ramírez, J.A. (1993) Interaction of hexaazaalkanes with phosphate type anions. Thermodynamic, kinetic, and electrochemical considerations. *Inorg. Chem.*, **32**, 3418–3424.
108. Albelda, M.T., Bernardo, M.A., García-España, E., Godino-Salido, M.L., Luis, S.V., Melo, M.J., Pina, F., and Soriano, C. (1999) Thermodynamic and fluorescence emission studies on potential molecular chemosensors for ATP recognition in aqueous solution. *J. Chem. Soc., Perkin Trans. 2*, 2545–2549.
109. Graf, E. and Lehn, J.-M. (1976) Anion cryptates: highly stable and selective macrotricyclic anion inclusion complexes. *J. Am. Chem. Soc.*, **98**, 6403–6405.
110. Metz, B., Rosalky, J.M., and Weiss, R. (1976) [3] Cryptates: X-ray crystal structures of the chloride and ammonium ion complexes of a spheroidal macrotricyclic ligand. *J. Chem. Soc., Chem. Commun.*, 533b–5534.
111. Dietrich, B., Dilworth, B., Lehn, J.-M., Souchez, J.-P., Cesario, M., Guilhem, J., and Pascard, C. (1996) Synthesis, crystal structures, and complexation constants of fluoride and chloride inclusion complexes of polyammonium macrobicyclic ligands. *Helv. Chim. Acta*, **79**, 569–587.
112. Dietrich, B., Lehn, J.-M., Guilhem, J., and Pascard, C. (1989) Anion receptor molecules: synthesis of an octaaza-cryptand and structure of its fluoride cryptate. *Tetrahedron Lett.*, **30**, 4125–4128.
113. Lehn, J.-M., Soveaux, E., and Willard, A.K. (1978) Molecular recognition. Anion cryptates of a macrobicyclic receptor molecule for linear triatomic species. *J. Am. Chem. Soc.*, **100**, 4914–4916.
114. Dietrich, B., Guilhelm, J., Lehn, J.-M., Pascard, C., and Sonveaux, E. (1984) Molecular recognition in anion coordination chemistry. Structure, binding constants and receptor-substrate complementarity of a series of anion cryptates of a macrobicyclic receptor molecules. *Helv. Chim. Acta*, **67**, 91–104.
115. Hosseini, M.W. and Lehn, J.-M. (1982) Anion receptor molecules. Chain length dependent selective binding of organic and biological dicarboxylate anions by ditopic polyammonium macrocycles. *J. Am. Chem. Soc.*, **104**, 3525–3527.
116. Hosseini, M.W. and Lehn, J.-M. (1986) Anion coreceptor molecules. Linear molecular recognition in the selective binding of dicarboxylate substrates by ditopic polyammonium macrocycles. *Helv. Chim. Acta*, **69**, 587–603.
117. Lehn, J.-M., Méric, R., Vigneron, J.-P., Bkouche-Waksman, I., and Pascard, C. (1991) Molecular recognition of anionic substrates. Binding of carboxylates by a macrobicyclic coreceptor and crystal structure of its supramolecular cryptate with the terephthalate dianion. *J. Chem. Soc., Chem. Commun.*, 62–64.

118. Bazzicalupi, C., Bencini, A., Bianchi, A., Fusi, V., García-España, E., Giorgi, C., Llinares, J.M., Ramirez, J.A., and Valtancoli, B. (1999) Molecular recognition of long dicarboxylate/dicarboxylic species via supramolecular/coordinative interactions with ditopic receptors. Crystal structure of $\{[Cu_2L(H_2O)_2] \supset \text{pimelate}\}(ClO_4)_2$. *Inorg. Chem.*, **38**, 620–621.

119. McKee, V., Nelson, J., and Town, R.M. (2003) Caged oxoanions. *Chem. Soc. Rev.*, **32**, 309–325.

120. Nelson, J., Nieuwenhuyzen, M., Pál, I., and Town, R.M. (2004) Dinegative tetrahedral oxoanion complexation; structural and solution phase observations. *J. Chem. Soc. Dalton Trans.*, 2303–2308.

121. Mason, S., Clifford, T., Seib, L., Kuczera, K., and Bowman-James, K. (1998) Unusual encapsulation of two nitrates in a single bicyclic cage. *J. Am. Chem Soc.*, **120**, 8899–8900.

122. Hynes, M.J., Maubert, B., McKee, V., Town, R.M., and Nelson, J. (2000) Protonated azacryptate hosts for nitrate and perchlorate. *J. Chem. Soc., Dalton Trans.*, 2853–2859.

123. Cullinane, J., Gelb, R.J., Margulis, T.N., and Zompa, L.J. (1982) Hexacyclen complexes of inorganic anions: bonding forces, structure, and selectivity. *J. Am. Chem. Soc.*, **104**, 3048–3053.

124. Gelb, R.J., Lee, B.T., and Zompa, L.J. (1985) Hexacyclen complexes of anions. 2. Bonding forces, structures, and selectivity. *J. Am. Chem. Soc.*, **107**, 909–916.

125. Gelb, R.J., Schwartz, L.M., and Zompa, L.J. (1986) Hexacyclen complexes of sulphate anion. *Inorg. Chem.*, **25**, 1527–1535.

126. Bianchi, A., Micheloni, M., Orioli, P., Paoletti, P., and Mangani, S. (1988) Anion coordination chemistry. 3. Second-sphere interaction between the complex anions hexacyanoferrate(II) and hexacyanocobaltate(III) with polycharged tetraazamacrocycles. Thermodynamic and single crystal X-ray studies. *Inorg. Chim. Acta*, **146**, 153–159.

127. Anda, C., Bazzicalupi, C., Bencini, A., Berni, E., Bianchi, A., Fornasari, P., Llobet, A., Giorgi, C., Paoletti, P., and Valtancoli, B. (2003) A thermodynamic and spectrophotometric study of anion binding with a multifunctional dipyridine-based macrobicyclic receptor. *Inorg. Chim. Acta*, **356**, 167–178.

128. Bazzicalupi, C., Bencini, A., Bianchi, A., Borsari, L., Giorgi, C., and Valtancoli, B. (2005) Tren-based tris-macrocycles as anion hosts. Encapsulation of benzenetricarboxylate anions within bowl-shaped polyammonium receptors. *J. Org. Chem.*, **70**, 4257–4266.

129. Bazzicalupi, C., Bencini, A., Bianchi, A., Danesi, A., Giorgi, C., and Valtancoli, B. (2009) Anion binding by protonated forms of the tripodal ligand tren. *Inorg. Chem.*, **48**, 2391–2398.

130. Rekharsky, M., Inoue, Y., Tobey, S., Metzger, A., and Anslyn, E. (2002) Ion-pairing molecular recognition in water: aggregation at low concentrations that is entropy-driven. *J. Am. Chem. Soc.*, **124**, 14959–14967.

131. Kano, K., Kitagishi, H., Tamura, S., and Yamada, A. (2004) Anion binding to a ferric porphyrin complexed with per-*o*-methylated β-cyclodextrin in aqueous solution. *J. Am. Chem. Soc.*, **126**, 15202–15210.

132. Tobey, S.L. and Anslyn, E.V. (2003) Energetics of phosphate binding to ammonium and guanidinium containing metallo-receptors in water. *J. Am. Chem. Soc.*, **125**, 14807–14815.

133. Mason, P.E., Neilson, G.W., Dempsey, C.E., Barnes, A.C., and Cruickshank, J.M. (2003) The hydration structure of guanidinium and thiocyanate ions: Implications for protein stability in aqueous solution. *Proc. Nat. Acad. Sci. U.S.A.*, **100**, 4557–4561.

134. Nielsen, K.A., Cho, W.-S., Lyskawa, J., Levillain, E., Lynch, V.M., Sessler, J.L., and Jeppesen, J.O. (2006) Tetrathiafulvalene-calix[4]pyrroles: synthesis, anion binding, and electrochemical properties. *J. Am. Chem. Soc.*, **128**, 2444–2451.

135. Lee, C.-H., Na, H.-K., Yoon, D.-W., Won, D.-H., Cho, W.-S., Lynch, V.M.,

Shevchuk, S.V., and Sessler, J.L. (2003) Single side strapping: a new approach to fine tuning the anion recognition properties of calix[4]pyrroles. *J. Am. Chem. Soc.*, **125**, 7301–7306.

136. Lee, C.-H., Lee, J.-S., Na, H.-K., Yoon, D.-W., Miyaji, H., Cho, W.-S., and Sessler, J.L. (2005) Cis- and trans-strapped calix[4]pyrroles bearing phthalamide linkers: synthesis and anion-binding properties. *J. Org. Chem.*, **70**, 2067–2074.

137. Yoon, D.-W., Gross, D.E., Lynch, V.M., Sessler, J.L., Hay, B.P., and Lee, C.-H. (2008) Benzene-, pyrrole-, and furan-containing diametrically strapped calix[4]pyrroles – an experimental and theoretical study of hydrogen-bonding effects in chloride anion recognition. *Angew. Chem. Int. Ed.*, **47**, 5038–5042.

138. Valik, M., Král, V., Herdtweck, E., and Schmidtchen, F.P. (2007) Sulfoniumcalixpyrroles: the decoration of a calix[4]pyrrole host with positive charges boosts affinity and selectivity of anion binding in DMSO solvent. *New J. Chem.*, **31**, 703–710.

139. Sessler, J.L., Gale, P.A., and Cho, W.-S. (2006) *Anion Receptor Chemistry*, RSC Publishing, Cambridge.

140. Sessler, J.L., Eller, L.R., Cho, W.-S., Nicolaou, S., Aguilar, A., Lee, J.T., Lynch, V.M., and Magda, D.J. (2005) Synthesis, anion-binding properties, and in vitro anticancer activity of prodigiosin analogues. *Angew. Chem. Int. Ed.*, **44**, 5989–5992.

141. Plitt, P., Gross, D.E., Lynch, V.M., and Sessler, J.L. (2007) Dipyrrolyl-functionalized bipyridine-based anion receptors for emission-based selective detection of dihydrogen phosphate. *Chem. Eur. J.*, **13**, 1374–1381.

142. Berger, M. and Schmidtchen, F.P. (1999) Zwitterionic guanidinium compounds serve as electroneutral anion hosts. *J. Am. Chem. Soc.*, **121**, 9986–9993.

143. Berger, M. and Schmidtchen, F.P. (1998) The binding of sulphate anions by guanidinium receptors is entropy-driven. *Angew. Chem. Int. Ed. Engl.*, **37**, 2694–2696.

144. Jadhav, V.D., Herdtweck, E., and Schmidtchen, F.P. (2008) Addressing association entropy by reconstructing guanidinium anchor groups for anion binding: design, synthesis, and host-guest binding studies in polar and protic solutions. *Chem. Eur. J.*, **14**, 6098–6107.

145. Zieliński, T. and Jurczak, J. (2005) Thioamides versus amides in anion binding. *Tetrahedron*, **61**, 4081–4089.

146. Kubik, S., Kirchner, R., Nolting, D., and Seidel, J. (2002) A molecular oyster: a neutral anion receptor containing two cyclopeptide subunits with a remarkable sulfate affinity in aqueous solution. *J. Am. Chem. Soc.*, **124**, 12752–12760.

147. Otto, S. and Kubik, S. (2003) Dynamic combinatorial optimization of a neutral receptor that binds inorganic anions in aqueous solution. *J. Am. Chem. Soc.*, **125**, 7804–7805.

148. Rodriguez-Docampo, Z., Pascu, S.I., Kubik, S., and Otto, S. (2006) Noncovalent interactions within a synthetic receptor can reinforce guest binding. *J. Am. Chem. Soc.*, **128**, 11206–11210.

149. Reyheller, C., Hay, B.P., and Kubik, S. (2007) Influence of linker structure on the anion binding affinity of biscyclopeptides. *New J. Chem.*, **31**, 2095–2102.

150. Schmidtchen, F.P. (2006) Reflections on the construction of anion receptors. Is there a sign to resign from design? *Coord. Chem. Rev.*, **250**, 2918–2928.

3
Structural Aspects of Anion Coordination Chemistry

Rowshan Ara Begum, Sung Ok Kang, Victor W. Day, and Kristin Bowman-James

3.1
Introduction

As structural data for supramolecular molecules have accumulated in the past 50 years, analogies with transition metal coordinate covalent (or dative) interactions have accumulated. Our early understanding of transition metal coordination chemistry comes from the nineteenth century scientist Alfred Werner, who had incredible structural insight at a time when X-rays had just been discovered, and thus with no input from crystallographic data. At that time, transition metal or "molecular" compounds were considered to be "double salts," since, in addition to the appropriate number of counter ions, they often had other molecular or ionic species that appeared in chemical formulas. This gave rise to more complex species, such as $CoCl_2{\cdot}6NH_3$. Werner received the Nobel Prize in 1913 for his brilliant observation that transition metal complexes had not just one but two valencies. He assigned the primary valence to the appropriate number of counterion(s) that served to neutralize the charge. He then defined the secondary valence: "Even when they are saturated in the sense of the older theory of valence, the elementary atoms still possess sufficient chemical affinity to bind other seemingly also saturated atoms and groups of atoms, under generation of clearly defined atomic bonds" [1]. This concept revolutionized the understanding of transition metal chemistry.

About two-thirds of a century later, the first entry of anions into supramolecular chemistry began with a little noticed report by Park and Simmons [2]. They observed that bicyclic organic compounds they termed *katapinands* containing two nitrogen bridgehead atoms could protonate. Subsequently these bicycles could form various isomers depending on the orientation of the NH hydrogen atoms *exo* or *endo* with respect to the molecular cavity. Furthermore, they found from NMR data that in the *endo–endo* conformer, katapinands could bind halides internally. This was later confirmed by a partial crystal structure report [3].

The chemistry of anion recognition slowly gained momentum starting in the mid-1970s with the contributions of Lehn and coworkers, who reported anion encapsulation in macrotricyclic hosts [4]. By the 1980s, a number of researchers were beginning to pursue the tricky chemistry of designing hosts for anions of

Anion Coordination Chemistry, First Edition. Edited by Kristin Bowman-James, Antonio Bianchi, and Enrique García-España.

various shapes and sizes [5, 6]. Now, more than 40 years later, supramolecular chemistry of anions has blossomed into its own field, with many structural analogies to the simple coordination geometries observed in transition metal chemistry. This idea is not limited to anions, but to the coordination of other host–guest species. However, as anions are the focus of this treatise, only those ions (with just a smattering of cations) are dealt with herein.

3.2
Basic Concepts of Anion Coordination Chemistry

As is commonly understood, supramolecular chemistry refers to the interactions between molecular and/or ionic species that do not involve covalent bonds, that is, from Lehn "It is chemistry beyond the molecule. Not just the chemistry of the covalent bond but how molecules that are already covalently saturated will interact with other species and bind them" [7]. It is to be noted that even here Lehn's wording is similar to that of Werner. Perhaps the most important interactions in this chemical class derive from the appropriate placement of hydrogen bonding functional groups that leads to selective or complementary (in terms of guest and host shape) binding.

It should be noted, however, that electrostatic interactions and simple physical properties such as hydrophilicity and size also play roles in anion binding propensities, creating a series of affinities based on *bias* as proposed by Moyer and Bonnesen [8]. Dimensionality (acyclic vs monocyclic vs bicyclic and higher order hosts) and the chelate/macrocyclic/cryptate effects can also influence binding.

Another class of noncovalent interactions in anion binding chemistry involves "anion–π" interactions. Interest in this area has gained momentum in recent years [9–12]. Indeed, an entire chapter (Chapter 6) in this book is focused on anion–π interactions. It should be noted that C–H$\cdots$A$^-$ interactions of varying strengths in π systems, rather than distinct interactions with π clouds, are sometimes included in this category [9, 10]. Anion-π complexes are not included here, and the reader is encouraged to refer to Chapter 6 for structural information on this intriguing and growing field of anion coordination chemistry.

As described in the introduction, Alfred Werner was the first to introduce the concept of "double valence" in coordination complexes. Here, we consider that a similar concept exists for supramolecular *coordination* chemistry, especially for host interactions with ionic guests. There is a charge requirement – the "primary valence" according to Werner – and a coordination number – Werner's "secondary valence," – usually provided by associated hydrogen bonds as opposed to the coordinate covalent bonds observed in transition metal complexes. One of the key differences between the two types of coordination when anions are the guests is in the direction of electron flow. In the case of positively charged metal ions in transition metal coordination, the flow is from lone pairs on the "ligand" to the metal ion (Figure 3.1a). This type of interaction is also operable in the

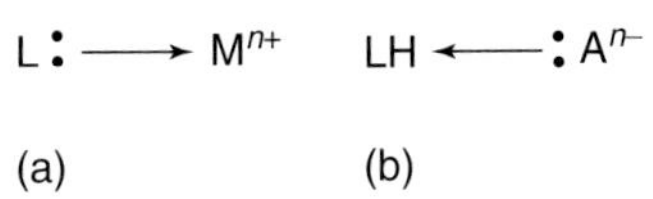

Figure 3.1 (a) Transition metal coordinate covalent (dative) bond and (b) anion supramolecular H-bond.

supramolecular chemistry of other positively charged guests such as alkali and alkaline earth metal ions (electrostatic interactions) and polyatomic cations such as ammonium ion (H-bonds). In anions, the electron flow is in the opposite direction, from the anion to the ligand (Figure 3.1b). The term anion coordination chemistry was first introduced by Lehn in 1978 [13, 14], but the analogy to transition metal coordination was made later [15].

3.3
Classes of Anion Hosts

There are a number of different donor groups for anions, with the list expanding greatly in the past five years or so. This chapter mainly focuses on H-bonding

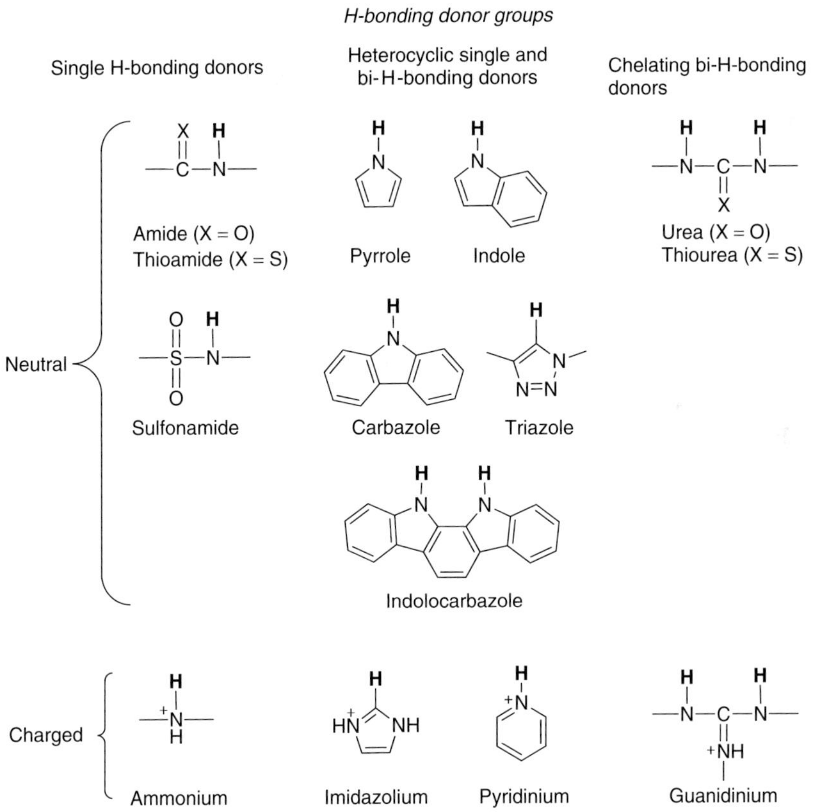

Figure 3.2 Sampling of H-bonding donor groups.

ligands for anions. Figure 3.2 shows the H-bonding groups that are included in structures described here. In another analogy to transition metal ligands, ligands for anions can also be described in terms of the denticity of the donor groups. There are single H-bond donors (mono-H donors), such as amines, amides, thioamides, sulfonamides, and the heterocyclic pyrroles, pyridiniums, imidazoles, indoles, and carbazoles. There are also chelating H-bond donors (bi-H donors or mini-chelates) such as ureas, thioureas, guanidiniums, and the heterocyclic imidazoliums, triazines, and indolocarbazoles. Another way to classify ligands is by dimensionality, that is, acyclic, monocyclic, bicyclic, and polycyclic, and this is the outline that is followed here. There is also a class of H-bonding ligands that are assembled with transition metal assistance. Other Lewis acid donor groups embedded in the ligands also have been used in direct linkages with anions. These two groups are also described, although the main focus of this chapter is on H-bonding ligands.

3.4 Acycles

Acyclic ligands can be found in a variety of different structural patterns. There are acycles that are built from a grouping of mono-H-bond donor functional groups, that is, amides and thioamides, pyrroles and imidazoliums, and so on, while others are designed incorporating the "preorganized" bi-H-bond donor group or groups, that is, urea, thiourea, and guanidinium. Additional organization can be provided by heterocyclic groups with substituents at various sites on the heterocycle that hold the functional groups in proximity for a chelate effect in binding. Indeed, while early anion coordination chemistry often involved macrocyclic or polycyclic hosts, the introduction of a wide variety of functional groups as well as an increased focus on clever host design has led to a huge number of successful acyclic hosts.

3.4.1 Bidentate

Two main types of bidentate binding can be achieved via placement of H-bonding functional groups. These include two mono-H donors or a single bi-H donor group. There are a number of very simple acyclic hosts containing a single bi-H donor, which works because it is already essentially preorganized for binding.

A seminal paper in anion coordination involved a surprisingly simplistic isophthalamide-based ligand that binds bromide ion, **1** (Figure 3.3). This communication by Kavallieratos, Hwang, and Crabtree in 1997 heralded the onslaught of amide-based anion hosts [16]. In the structure, the bromide is held by two H-bonds with the amide NH groups at N · · · Br distances of 3.634 and 3.437 Å. The distance to the phenyl hydrogen atom is even shorter at 3.576 Å, but undoubtedly is a result of imposed proximity. At the time, this was one of just a few H-bonds reported shorter than 2.8 Å for NH· · ·Br. Solution evidence for retention of the

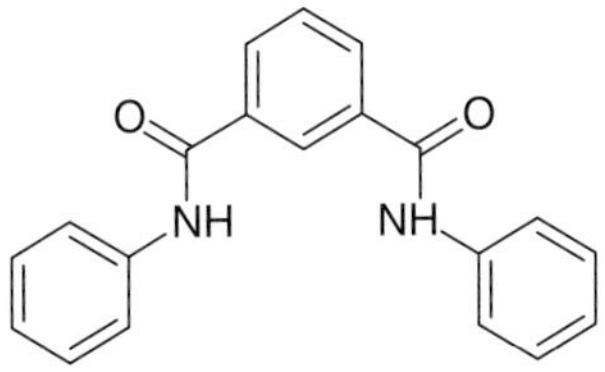

1

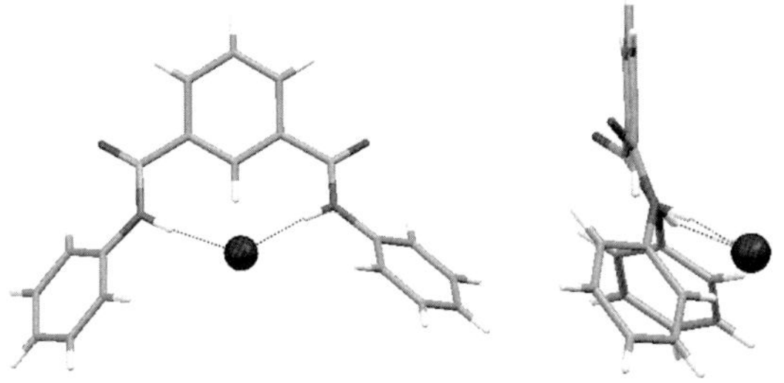

Figure 3.3 (a) Overhead and (b) side views of the bromide complex with the simple isophthalamide-derived ligand **1**. The $[Ph_4P]^+$ counterion is not shown for clarity.

H-bond was obtained from FT-IR data. Crabtree noted that the isophthalamide spacer would even be superior to pyridine [16, 17], citing that the electron pair on the pyridine nitrogen atom tends to disrupt H-bonding with the anion because of its interaction with the amide NH groups.

A number of authors have noted, however, that pyridine spacers can facilitate preorganization and can result in enhanced anion binding [18, 19]. Indeed, Gale and coworkers reported that a 2,6-dicarboxyamidopyridine-containing host, **2**, shows even better affinity for chloride compared with isophthalamide, to the extent that it facilitates HCl transport [20]. The ligand, **2**, was synthesized to mimic the naturally occurring prodigiosins. The chloride complex crystallizes as a dimer, with the chloride forming H-bonds with the two amide NH groups at 3.11 and 3.48 Å. Instead of forming a third H-bond with the chloride ion, however, the imidazolium NH group links to a neighboring molecule to form a dimer pair (Figure 3.4).

While one might argue that these isophthalamide frameworks are too simple and flexible to provide any real advantage to binding anions, some clever functionalizations can enhance preorganization in a manner similar to pyridine-containing systems. For example, Davis, Gale, Quesada, and coworkers examined the structural differences afforded by ligands with appended methoxy (**3**) or hydroxy (**4**) groups [21]. The difference in H-bonding capabilities influences the conformation in the solid state, so that a substituted methoxy group promotes formation of the *anti–anti* and a substituted hydroxy group promotes the formation of the *syn–syn* isomer (Figure 3.5a,b, respectively). In the crystal structure of **4**, two of the three

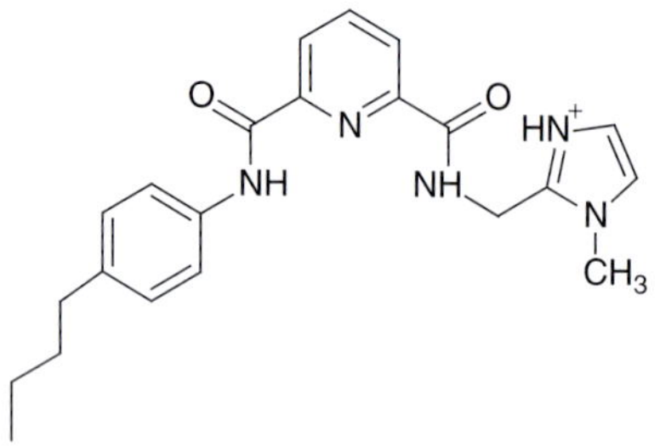

2

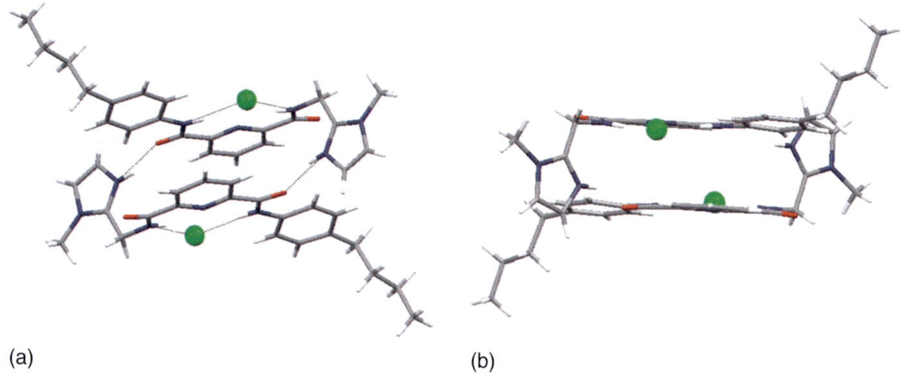

Figure 3.4 Views of the pyridine-based host **2** showing (a) the dimer formation and (b) the side view with the nearly coplanar pyridine and phenyl groups for the two ligands.

independent units are in the *syn–syn* form but are intertwined with the third *anti–anti* form. The dihydroxy-substituted isophthalamide was found to be an excellent transmembrane transporter of chloride ion.

As noted earlier, there are a number of bi-H-bond donors in acycles, including a very simple functionalized urea (**5**) that serves to illustrate the binding capabilities of the adjacent H-bond donors [22]. On the left, the urea perfectly fits the acetate guest, with N···O distances of 2.69 and 2.77 Å (Figure 3.6a). Similar binding is observed crystallographically with bicarbonate. However, unlike acetate, bicarbonate has a third site that can participate in H-bonding – the OH group. This host–guest complex forms dimers because of a crystallographic inversion center between two bicarbonates, with the OH groups linking the two carbonates (Figure 3.6b). Distances are slightly longer in the case of bicarbonate, with N···O = 2.790 and 2.811 Å for the urea interactions. The OH groups link the two bicarbonate ions at an O···O distance of 2.617 Å.

A heteroditopic (ureido) crown ether receptor **6**, which can trap both cations and anions, was reported by Barboiu [23]. The NaCl and $NaNO_3$ structures show different modes of binding. In the nitrate structure, two of the three oxygen atoms are associated with the sodium ion at 2.666 and 2.516 Å. The third oxygen atom

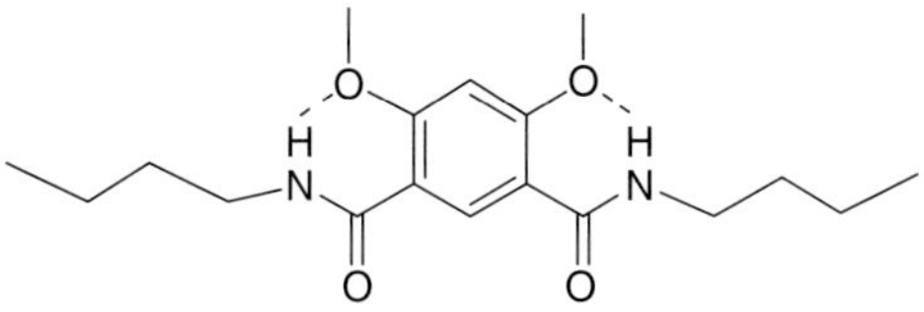

3

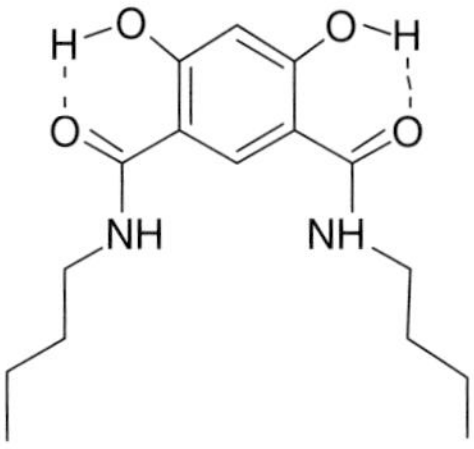

4

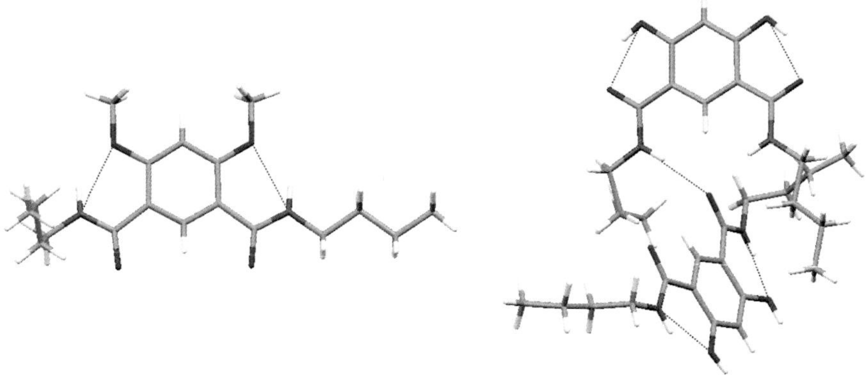

(a) (b)

Figure 3.5 (a) Methoxy-substituted *anti–anti* form of **3** and (b) hydroxy-substituted mixture of the *syn–syn* and *anti–anti* forms of **4**.

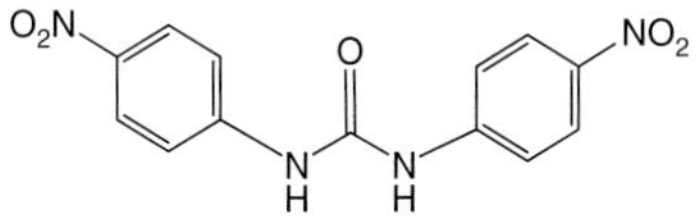

5

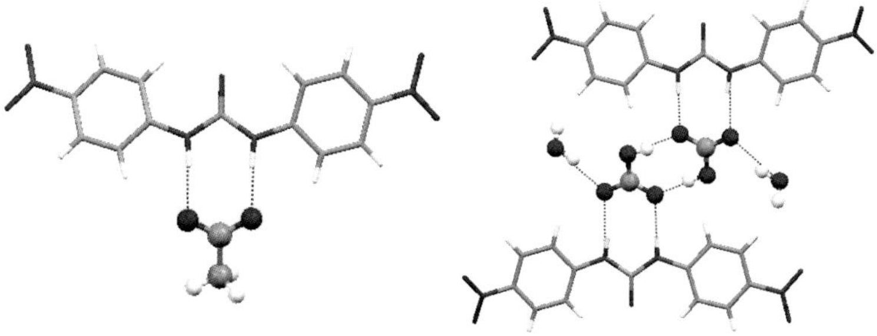

(a) (b)

Figure 3.6 (a) Acetate and (b) bicarbonate structures with **5**. The $[n\text{Bu}_4\text{N}]^+$ counterions are not shown for clarity.

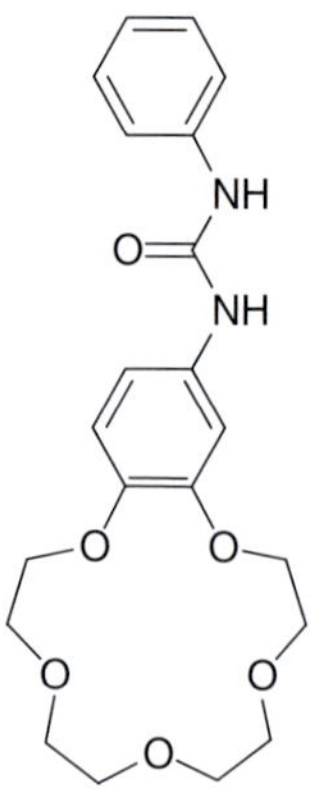

6

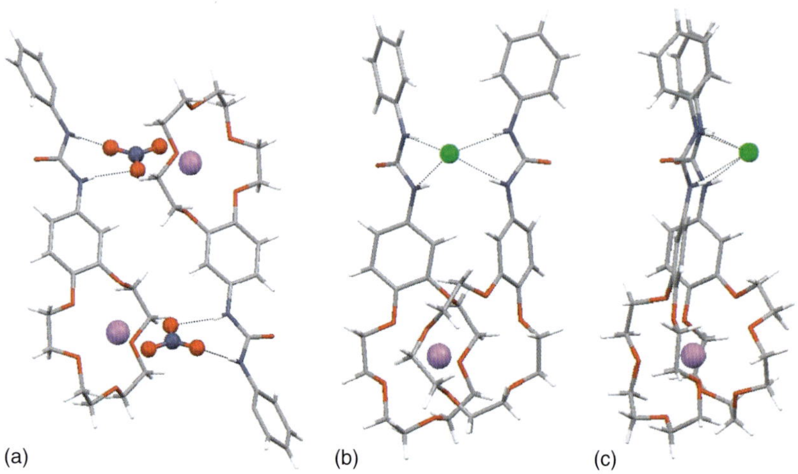

Figure 3.7 Crystal structures of the dual-host crown ether dimer **6**, showing (a) the bidentate coordination in the sodium nitrate complex, (b) a clear view of the four co-ordination in the chloride complex, and (c) the pyramidal geometry of the chloride.

is H-bonded to the urea of a neighboring molecule via one rather long H-bond at 3.063 Å. One of the oxygen atoms associated with the sodium also interacts with the neighbor molecule, binding to the other NH group via a relatively long H-bond of 3.072 Å, making this a two-coordinate linkage (Figure 3.7a). In the NaCl structure with **6**, two of the crown ethers form a sandwich complex with the sodium ion, and the chloride is held in a fourfold square pyramidal association with the ureas at distances ranging from 3.333 to 3.430 Å (Figure 3.7b,c). While the anion is four coordinate in the chloride complex, each of the ligands only binds in a bidentate manner.

3.4.2
Tridentate

A number of acyclic pyrrole-containing hosts are dual donors that often have two incorporated amides as a second donor group, logically resulting in three H-bond donors, as shown in **7** with three juxtapositioned donors (Figure 3.8a) [24]. The chloride is associated with the two amide NH groups at 3.334 and 3.391 Å and with the central pyrrole NH group at a quite short distance of 3.076 Å. The chloride ion has three additional close approaches to CH_2 groups of the $[n\mathrm{Bu}_4\mathrm{N}]^+$ ion at 3.609, 3.671, and 3.674 Å.

Gale and coworkers have an expanded form of this ligand, **8**, that sports chloride substituents [25, 26]. The ligand binds two chloride ions in an *anti* manner (Figure 3.8b). The H-bonds from the chloride to the NH groups again fall into two categories, a very short linkage to the pyrrole NH group at 3.068 Å and the two longer amide H-bonds at 3.269 and 3.276 Å. Again, the chloride is surrounded by the $[n\mathrm{Bu}_4\mathrm{N}]^+$ counterions with approaches to three additional CH_2 groups (3.76–3.81 Å).

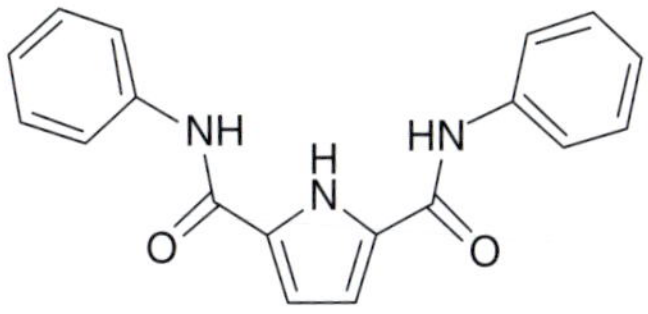

7

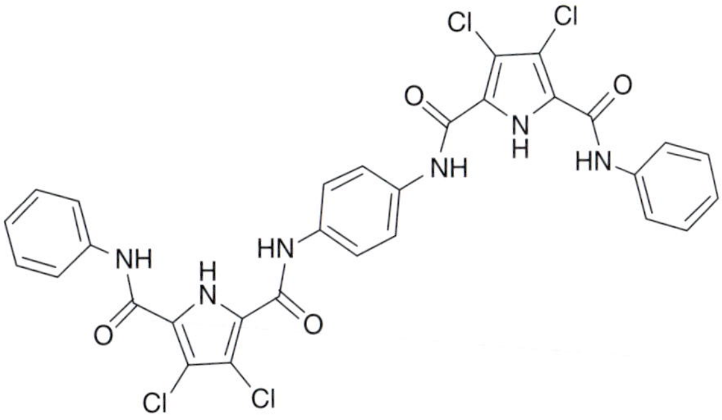

8

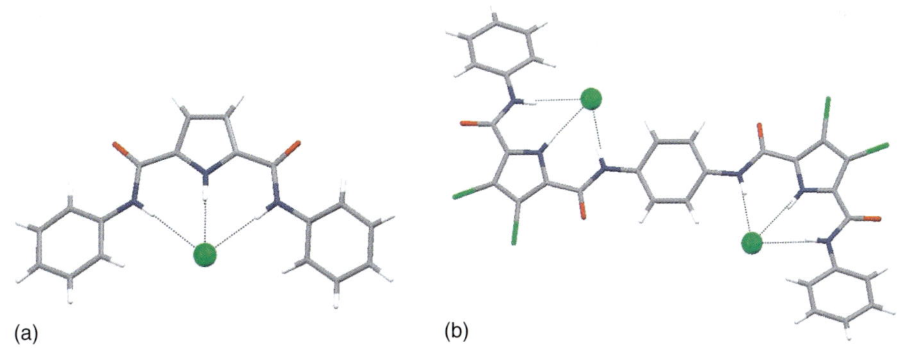

Figure 3.8 Crystal structures of the chloride complexes of (a) **7** and (b) **8**. The surrounding $[n\mathrm{Bu}_4\mathrm{N}]^+$ counterions are not shown for clarity.

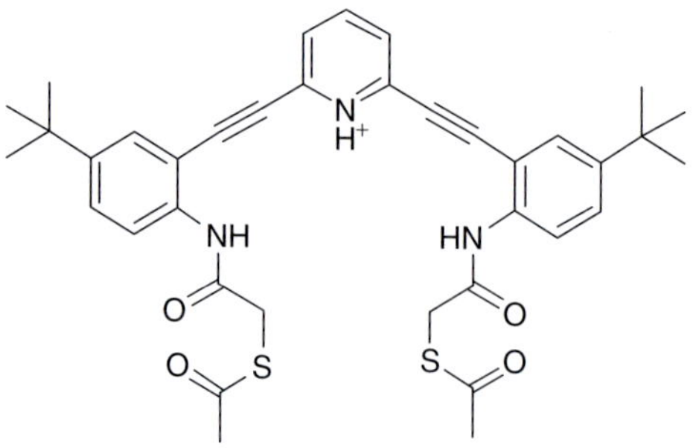

9

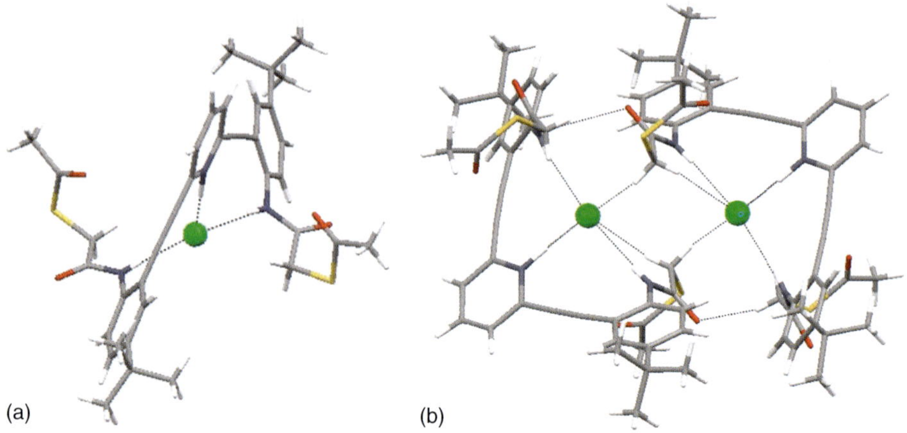

Figure 3.9 The structure of **9** showing (a) the tridentate coordination mode of a single complex of chloride and (b) the dimer pair showing the "bridging" CH_2 groups.

Acetylene spacers have become a player in ligand design, providing additional rigidification to a host framework. For example, Johnson and coworkers have added acetylene groups in the 2,5-positions of a pyridine (which actually protonates to become a pyridinium), providing a more rigid core framework, **9**. This tactic also adds fluorescence capabilities to the host [27]. In the chloride complex, the anion is H-bonded with the pyridinium NH group at 3.002 Å and the two amide groups at 3.258 and 3.277 Å (Figure 3.9a). The short and long distances are reminiscent of the pyrrole/amide ligands described above (Figure 3.8) [24–26]. There are two other longer associations with CH_2 groups in the same ligand, at 3.634 and 3.643 Å, for which H-bonding interactions could definitely be argued and the ligand might well be considered as providing pentadentate coordination. However, owing to the rather large differences in the H-bonding distances with the CH_2 groups, **9** remains in the tridentate section. The complex folds in a helical manner, possibly promoted by a rather short $Cl\cdots CH_2$ interaction (noted above) with a neighboring ligand at 3.638 Å. Hence, one might say that the two chlorides are linked by two bridging

CH_2 groups in a cascade-like structure (Figure 3.9b). There is also a relatively short carbonyl–CH_2 interaction between the two ligands at 3.104 Å.

A tridentate complex also forms with the triazole **10** [28]. Triazoles are relatively new in anion binding studies, and the simplified synthetic route reported in 2002 of 1,3-dipolar cycloadditions between azides and acetylene catalyzed by Cu(I) have made their fabrication readily accessible [29, 30]. The triazine resulting from these reactions has a remaining C–H donor capable of forming H bonds with anions. While the use of CH groups for H-bonding is still somewhat rare, reports are rapidly developing as the area progresses. Here, the triazole takes the place of the acetylene linkers in **9** (Figure 3.9) [27] while additionally providing an H-bonding site for interacting with an anionic guest. Two water molecules serve to link the ligands by H-bonding to the para-substituted carboxylate (Figure 3.10). No longer is the two long and one short H-bonding pattern observed as in the pyrrole hosts (Figure 3.8); rather the two triazole C···Cl distances are 3.620 and 3.476 Å and the C_{ph}···Cl separation is 3.644 Å. The

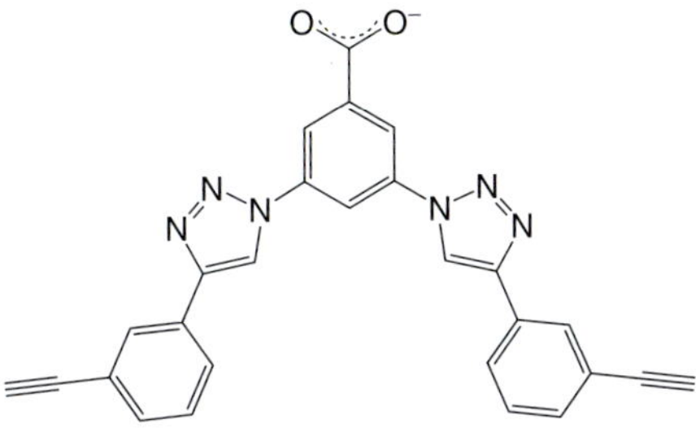

10

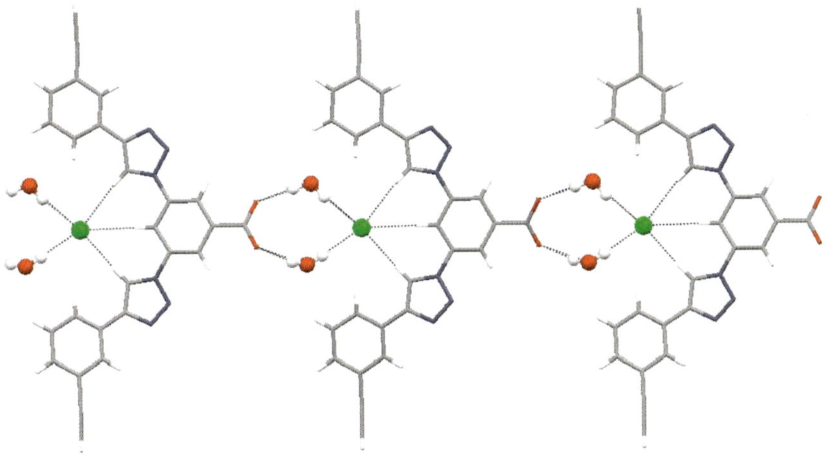

Figure 3.10 Water-linked chloride complexes of the triazole-containing ligand **10**. The $[nBu_4N]^+$ counterions are not shown for clarity.

distances from the chloride ion to the water molecules are shorter at 3.244 and 3.277 Å, and the geometry around the chloride is almost pentagonal planar.

In the pyridinium host **11**, the ligand utilizes all three of its H-bonding sites by binding to two halide ions [31]. Two H-bonds are formed to one halide, with a third H-bond to a second halide that is bound to the neighboring ligand, resulting in a dimer pair (Figure 3.11). Johnson and coworkers found this to be a preferred mode of binding for this ligand for both methyl and nitro substituents on the dangling phenyl group. In the chloride structure, each halide binds via three H-bonds, two with amide NH groups at 3.157 and 3.181 Å, and one with the pyridinium NH group at 2.926 Å. Since one of the amide H-bonds is with a neighboring ligand, the result is a dimer pair that has a short Cl···Cl distance of 3.099 Å (Figure 3.11a). The same coordination mode occurs also for the bromide complex (Figure 3.11b). The H-bonds are a bit longer at 3.127 Å for the pyridinium NH group and 3.337 and 3.440 for the amide NH groups. The Br···Br distance is 4.079 Å. The free base (with a dimer of water molecules) as well as a mixed chloride and water complex also exhibit similar dimerization.

A popular acyclic building block has been tren (*N*,*N*′,*N*″-tris(2-benzylaminoethyl)amine) and substituted trens. The popularity of this framework derives from the fact that it can hold an anion in a threefold (or with bi-H donors, sixfold) H-bonding vice. The simplest of this series is unsubstituted tren, **12**, which is shown, complexed with molybdate (Figure 3.12a) [32]. The three ammonium units form H-bonds with the apical oxygen atom of the molybdate ion, with distances ranging from 2.76 to 2.82 Å. The macrocycles are linked via a H-bonding contact between one of the ammonium groups and a neighboring molybdate at 2.978 Å, resulting in a zigzag array of cationic hosts and anionic guests, another motif that is often seen (Figure 3.12b).

Ghosh and coworkers have examined simple tren amide hosts with nitro substituents (**13**) (Figure 3.13) [33]. They isolated a beautifully intertwined complex of two ligands with two nitrates, each nitrate associated with one of the tren units. A view down the threefold axis shows the two nitrates to be oriented in a *gauche* conformation. Actually, only one shows significant H-bonding with the amide NH groups, with N···O distances of 2.984, 3.040, and 3.132 Å. The other nitrate is much more distant from the amides, with two N···O separations of 3.108 and 3.181 Å. Other contacts are 3.3 Å and longer. The nitrates lie almost parallel to the bridgehead phenyl groups, with distances from the two nitrogen atoms to the phenyl centroid of phenyl rings of 3.225 and 3.229 Å. The nitrate N···N distance is 3.562 Å. It would appear almost as though the second nitrate was trapped when the cage formed. Note that in Figure 3.13b, the nitrates taken together almost align with the hexagonal carbon pattern of the two bridgeheads.

An interesting class of complexes is developing from a need for extracting chlorometallates. Because these ligands bind outside the first coordination sphere of transition metal complexes, they are considered by transition metal coordination chemists to be outer sphere complexes. An example is the $CoCl_4^{2-}$ complex with the amidopyridyl extractant **14** (Figure 3.14) [34]. The two ligands are fixed in a more or less linear conformation by the two amide oxygen atoms, the latter of

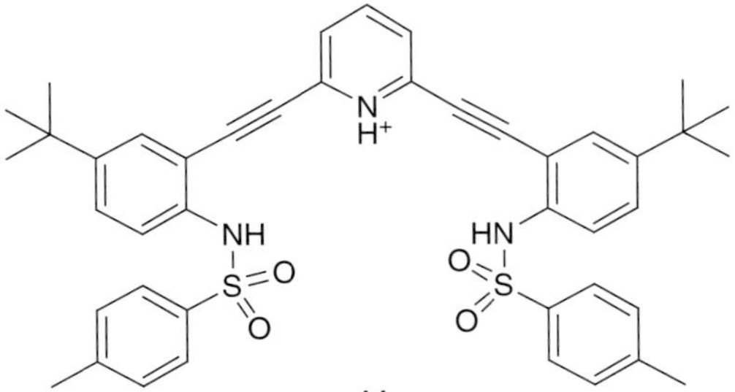

11

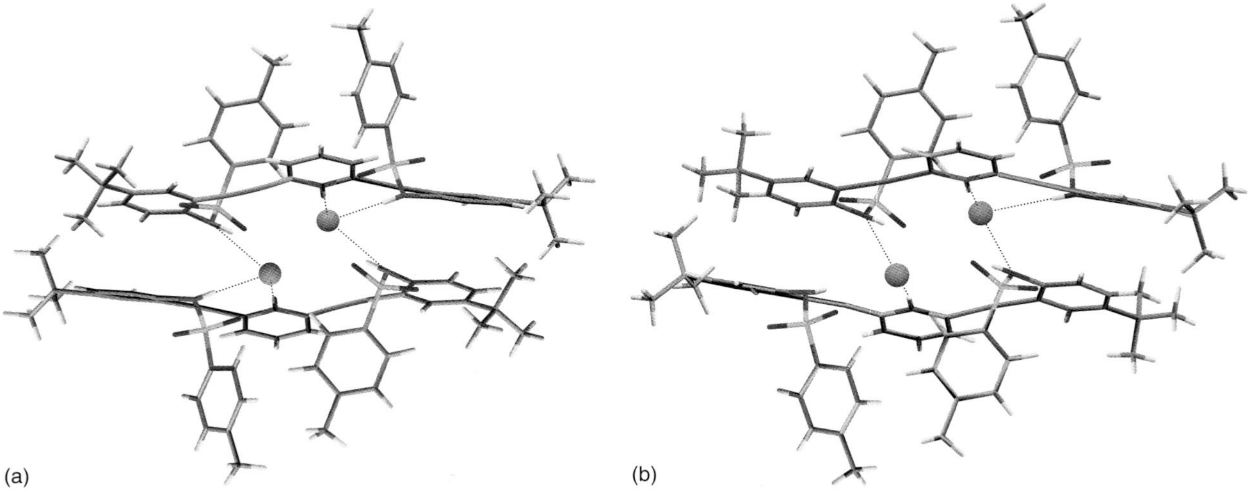

Figure 3.11 Structures of **11** with two different halides: (a) the dichloride complex and (b) the dibromide complex.

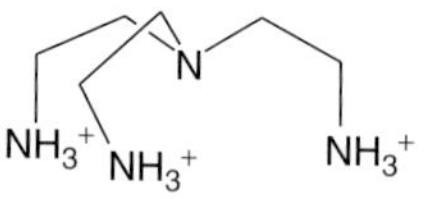

12

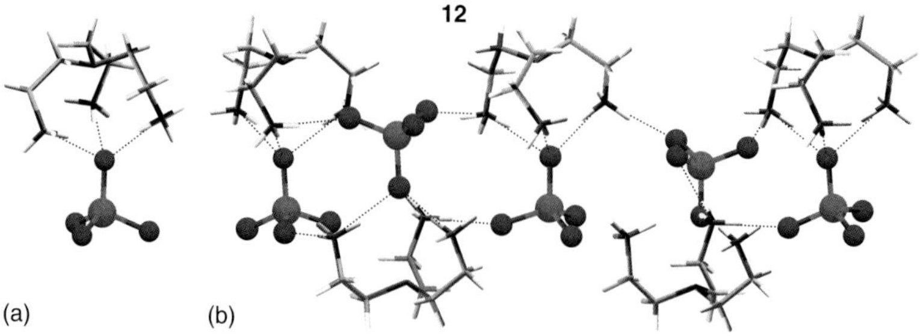

Figure 3.12 Perspective views of the molybdate complex with triprotonated tren (**12**), showing (a) a single complex and (b) the zigzag packing motif.

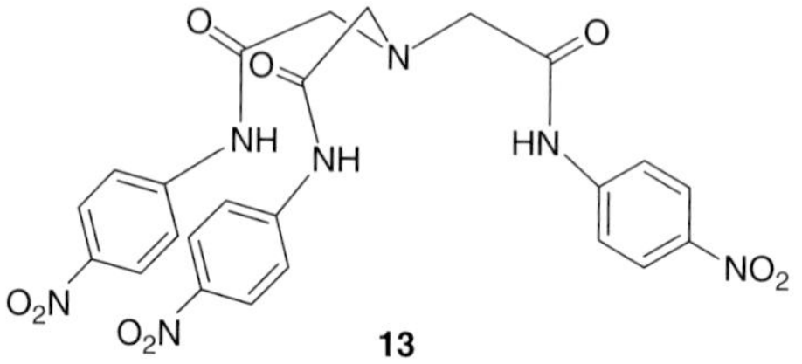

13

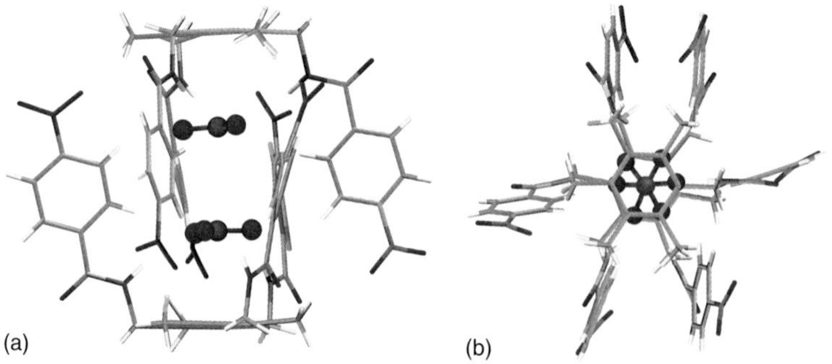

Figure 3.13 Two views of the intertwined dinitrate dimer complex of **13**: (a) side view showing the almost parallel orientation of the two nitrates and (b) overhead view showing the staggered orientation of the nitrate oxygen atoms. The $[n\mathrm{Bu_4N}]^+$ counterions are not shown for clarity.

which are intramolecularly H-bonded with the pyridinium NH group at 2.612 and 3.052 Å. This leaves the ligand in a position for nicely packing around the tetrachlorocobaltate dianion and satisfies the overall charge requirement. Two chlorides are associated via weak interactions with the two amides of the two surrounding ligands at 3.428 and 3.326 Å. The other two chlorides are weakly

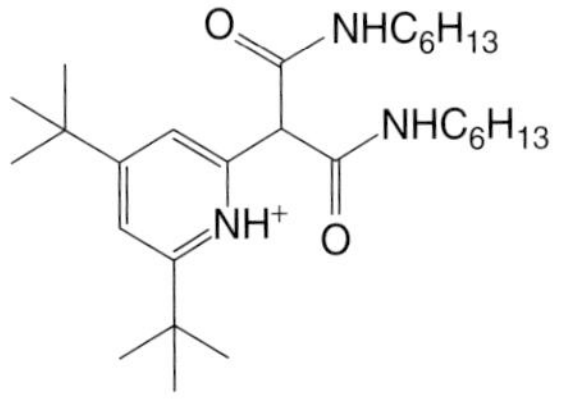

14

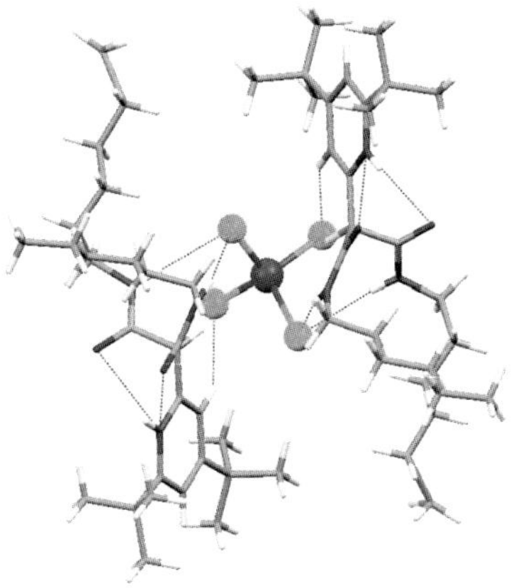

Figure 3.14 "Outer sphere complex" of the tetrachlorocobaltate(2-) complex with the amidopyridinium ligand **14**.

associated with a pyridine CH group at 3.634 Å. The authors note that these are very weak interactions, but these types of preorganization may well lead to stable "assemblies" in the nonpolar solvents used in the extraction process.

3.4.3
Tetradentate

Tetradentate chelates are common for simple acycles. Within this series, the opportunity for diverse binding groups increases. For example, one could place four mono-H donors, one bi-H donor and two mono-H donors, or two bi-H donors within a ligand. The following discussion includes examples of all three types of binding sites.

A popular class of chelating ligands derives from meta-difunctionalized phenyl or other heterocyclic groups. In these cases, there is the possibility of some preorganization of the binding site in the absence of preorganization that can be introduced by having bi-H donor groups. An expanded *meta*-substituted heterocycle with a pyridine spacer (**15**) was reported by Chmielewski and Jurczak (Figure 3.15) [35]. The chloride is held by four H-bonds with the surrounding amide hydrogen atoms ranging from 3.246 to 3.307 Å and a fifth H-bond to an axial water molecule at 3.180 Å. The presence of an axial water molecule is a common occurrence in these complexes and reflects yet another analogy to transition metal complexes, in which an axial water molecule is often observed [15].

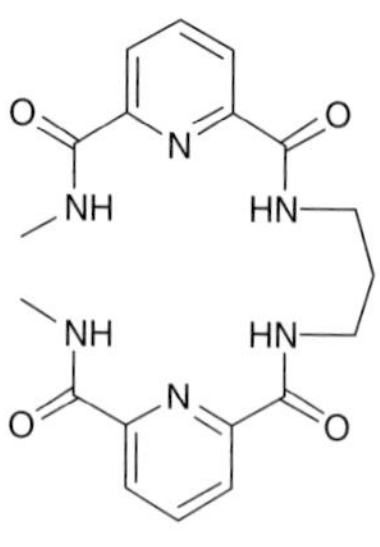

15

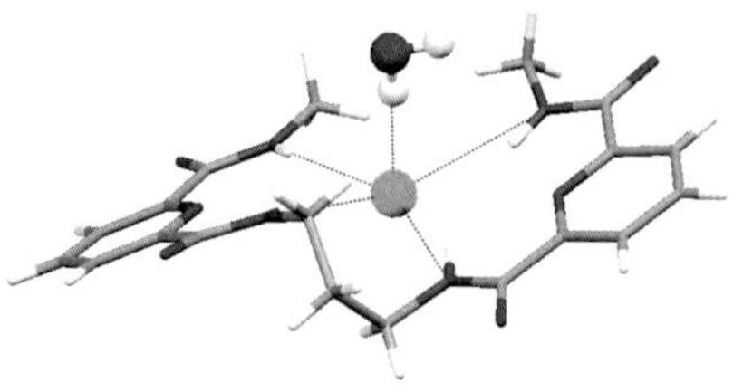

Figure 3.15 Structure of the chloride complex with **15**. The $[nBu_4N]^+$ counterion is not shown for clarity.

Next in order of complexity is the addition of mixed mono- and bi-H donor groups. One can vary the number of binding sites by mixing together two mono-H donor groups (e.g., amides, pyrroles, indoles) with a bi-donor (e.g., urea, guanidinium). Below are examples of such a combination: diindolylureas that consist of two indoles bridged by urea or thiourea as backbone, **16** and **17**, respectively.

In the structure of **16** with benzoate, isolated as the $[nBu_4N]^+$ salt, the ligand is slightly twisted and the benzoate is canted with N···O H-bonds ranging from 2.846 to 2.905 Å (Figure 3.16a) [36]. Instead of a 1 : 1 anion:ligand complex, **16** coordinates as a tris-chelate around phosphate, resulting in the formation of a 12-coordinate complex (Figure 3.16b) [37], also isolated as the $[nBu_4N]^+$ salt. A crystallographically imposed threefold axis runs through one of the phosphate oxygen atoms and the phosphorus. The result is that three of the phosphate oxygen atoms are bound identically to two NH groups of a urea on one ligand (2.756 and 2.850 Å) and an indole NH on a neighboring ligand (2.722 Å). The unique oxygen atom is held by three indole NH groups at 2.763 Å (lower left, Figure 3.16b).

It is worthwhile to compare two complexes of the urea (**16**) and thiourea (**17**) ligands [37]. In the structure of **16** with carbonate, isolated as the $[Et_4N]^+$ salt, two of the ligands surround the carbonate ion and N···O distances range from 2.739 to 2.938 Å (Figure 3.17a). One of the ligands is H-bonded to only two of the three oxygen atoms, while the other is H-bonded to all three. Nonetheless, each ligand still behaves in a tetradentate manner, although the actual coordination number of the carbonate ion is eight. In the thiourea analog **17** with bicarbonate, also isolated as the $[Et_4N]^+$ salt, chains of bicarbonate ions line a channel between chains of the tetradentate hosts (Figure 3.17b). The H-bonds range from 2.798 to 2.887 Å. The bicarbonate hydrogen atoms were not located because of disorder along the bicarbonate chain and hence are not depicted. However, it is intriguing to see that two closely related ligands can exhibit different modes of coordination with

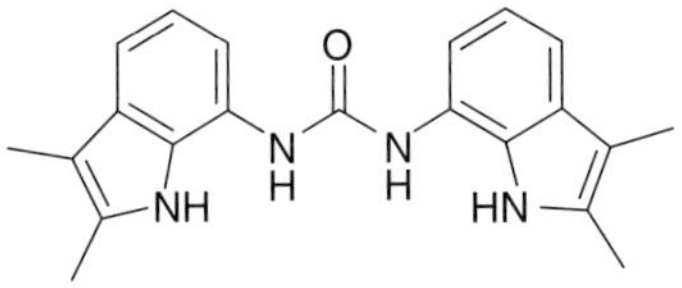

16

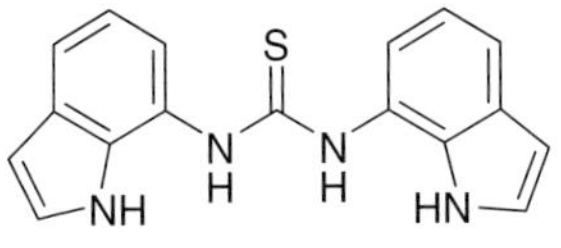

17

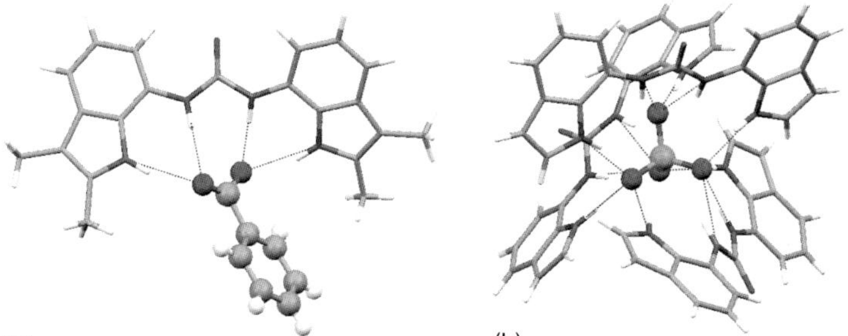

(a) (b)

Figure 3.16 Crystal structures of (a) the benzoate complex with **16**, and (b) the 12-coordinate complex of **17** with phosphate, $[\mathbf{17}_3{\cdot}PO_4]^{3-}$. The $[nBu_4N]^+$ counterions are not shown for clarity.

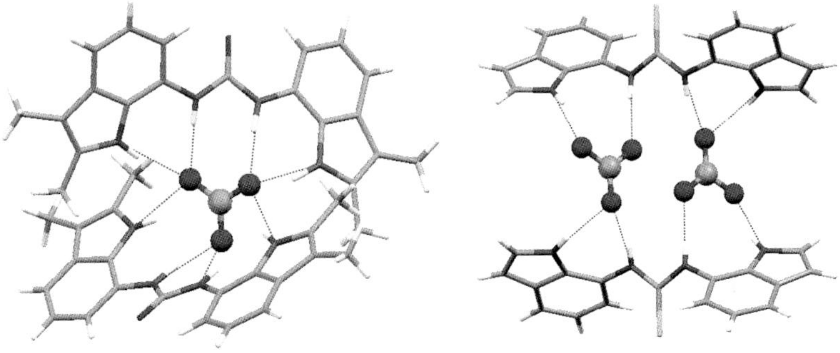

(a) (b)

Figure 3.17 Crystal structures of (a) the carbonate complex with **16**, $[\mathbf{16}_2{\cdot}CO_3]^{2-}$ and (b) the bicarbonate complex with **17**, $[\mathbf{17}_2{\cdot}2HCO_3]^{2-}$. The $[Et_4N]^+$ cations are not shown for clarity.

that differ in charge, again a similarity with the different coordination preferences of transition metal elements in different oxidation states. (It should be noted, however, that even anions of the same charge can exhibit different binding modes depending on crystallization conditions, including the identity of the counterion.)

A carbazole-based host containing the rarely used hydroxyl group for H-bonding (**18**) was reported by Jeong and coworkers [38]. The ligand framework also contains

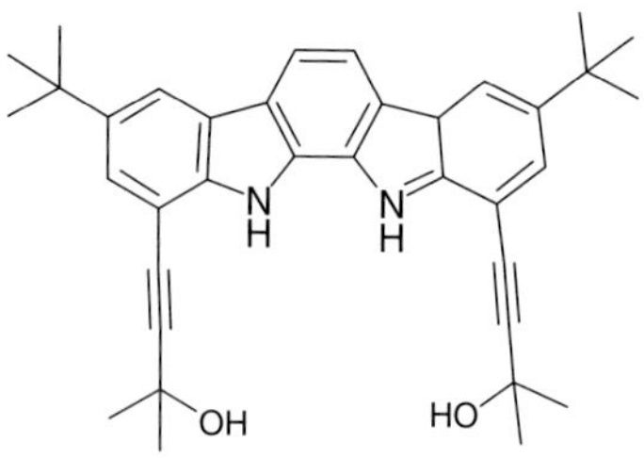

18

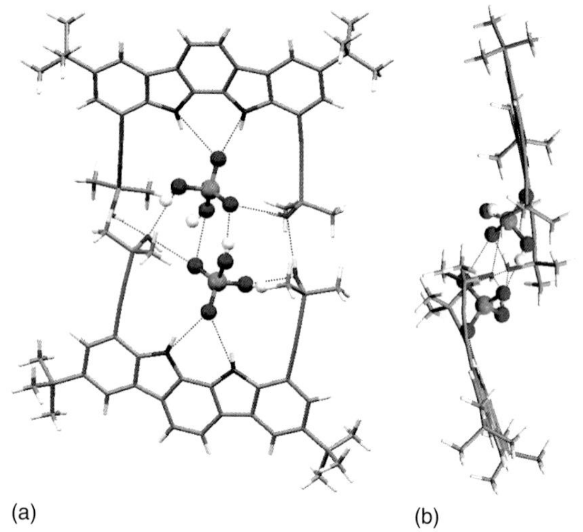

Figure 3.18 (a) Overhead and (b) side views of the $H_2PO_4^-$ structure of **18**, $[\mathbf{18}_2{\cdot}2H_2PO_4]^{2-}$. The $[nBu_4N]^+$ counterions are not shown for clarity.

two acetylene arms. This is an unusual case of OH bonding with an alcohol side chain. In the $H_2PO_4^-$ complex (Figure 3.18), each deprotonated anion is H-bonded via two bonds to the carbazoles, two H-bonds to the adjacent $H_2PO_4^-$ ion, and two H-bonds to hydroxyl groups, one from the "home" ligand and one from the neighbor. One of the hydroxyl groups binds to an unprotonated oxygen atom of one $H_2PO_4^-$ with an O···O separation of 2.627 Å and also to a neighboring $H_2PO_4^-$ with O···O = 2.688 Å. The two ligands are linked via H-bonding between the two hydroxyl groups (O···O = 2.683 and 2.895 Å). The carbazole $H_2PO_4^-$ distances range from N···O = 2.690 to 2.967 Å. While the ligand itself is tetradentate, each of the $H_2PO_4^-$ anions has six H-bonds that include two between the two anions. Hence, the anions also play a role in holding the complex together. The ligand shows high affinities for acetate, chloride, and $H_2PO_4^-$ in that order (K_a = 29 000 for $H_2PO_4^-$ in 1% H_2O–CD_3CN by NMR titration).

The ligand **19** was designed to explore the efficacy of *ortho*-phenylenediamine frameworks (rather than the more commonly used *meta* substitutions), in this

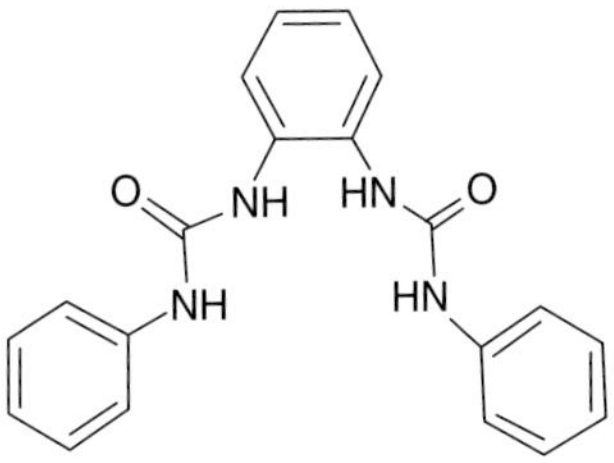

19

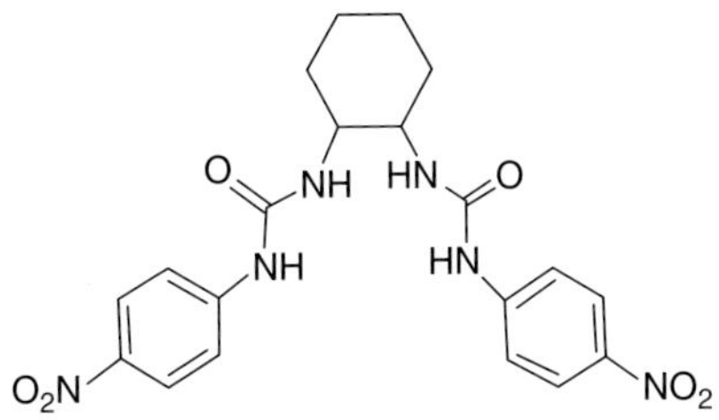

20

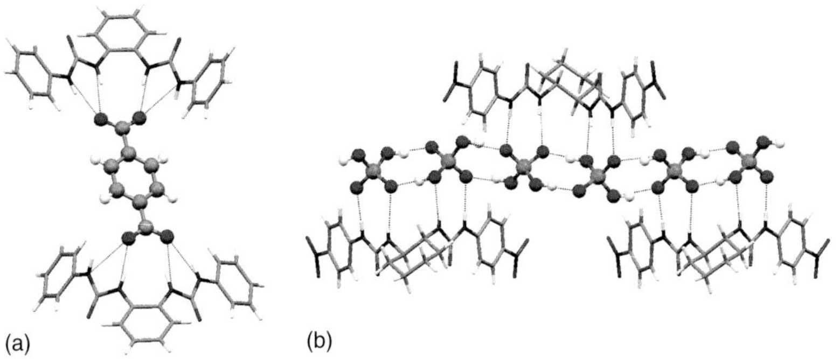

Figure 3.19 (a) Crystal structure of the terephthalate complex of **19** and (b) the chain-like array of the 2 : 1 complex of **20** with $H_2PO_4^-$. The $[nBu_4N]^+$ counterions are not shown for clarity.

instance by appending a bis-urea complex [39]. This is an example of a potentially tetradentate ligand design incorporating two bi-H-bond donors. The target guests were to be oxoanions, particularly carboxylates. The resulting terephthalate complex shows a complementary bis-coordination of two ligands at the *para* carboxylate sites. Each urea is bound to just a single carboxylate oxygen, with bond lengths ranging from 2.759 to 3.082 Å. The phenylurea arms are rotated slightly from the phenyl group, undoubtedly in part due to the proximity of the two adjacent phenyl hydrogen atoms (Figure 3.19a). Compared to **5**, the simple monourea host [22], this structure adds an additional chelating moiety and should exhibit an even greater chelate effect, although the actual thermodynamic gain has not been determined by comparing **19** with **5**.

Since *ortho*-substituted frameworks are rather rare, it is appropriate to compare the *o*-phenylene-containing ligand (**19**) with the saturated *ortho*-substituted cyclohexyl-containing ligand (**20**) [40]. The $H_2PO_4^-$ structure of host **20** fabricated in the Fabbrizzi group also binds in a tetradentate manner (Figure 3.19b). However, in this complex, each urea group binds to a single anion. It is surmised that the noncoplanar arrangement of the ureas may predispose the ligand to a propensity for oxoanions. The 2 : 1 complex is obtained when an additional equivalent of $H_2PO_4^-$ is added. From spectrophotometric studies in CH_3CN, log K_2 is even higher than log K_1, (3.46 and 2.96, respectively), which is thought to be due to the

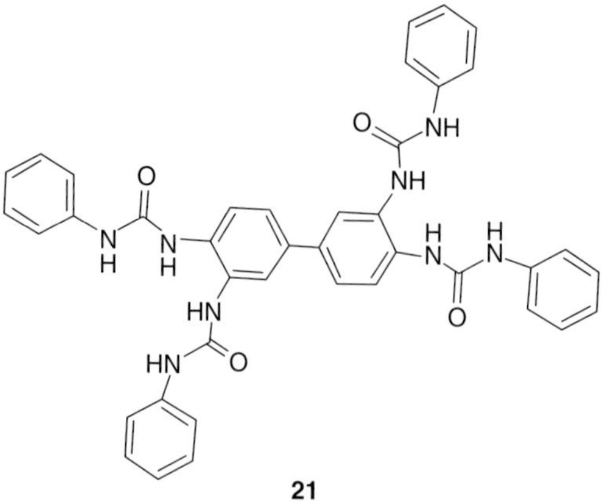

21

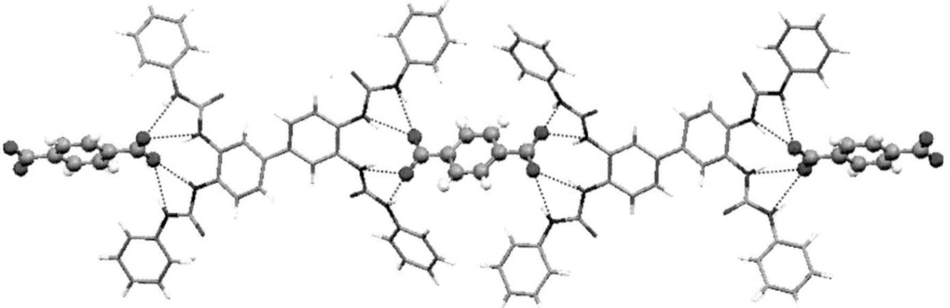

Figure 3.20 View of the alternating "series-linked" chain of terephthalate with **21**. The $[n\mathrm{Bu_4N}]^+$ counterions are not shown for clarity.

mode of dual $H_2PO_4^-$ binding with a single receptor and the chain-like packing interactions. The hydrogen atoms on the phosphate ions serve to link the anions in a chain throughout the crystal lattice, with ligands lining the corridor of anions above and below in similar chains.

Brooks *et al.* also synthesized a chain-like array of anions and cations, although in this case, in an interesting example of an expanded form of **19** using a biphenyl linker, **21** [41]. The resulting structure with terephthalate provides an engineered crystalline chain of host and guest (Figure 3.20). As in the structure of **19** with terephthalate, each oxygen atom is doubly coordinated to its own urea moiety, with distances ranging from 2.805 to 2.919 Å. As seen from the packing diagram, the phenylurea arms are again rotated slightly.

Simple guanidinium hosts, **22** and **23**, were reported by Hutchings and coworkers [42]. In **22** the ligand contains just two guanidinium groups linked via a *m*-xylyl spacer. The guanidinium groups form intramolecular H-bonds with the carbonyl oxygen atom, resulting in a well-rigidified and preorganized ligand primed for complexing a guest anion. The structure is somewhat reminiscent of the intentionally

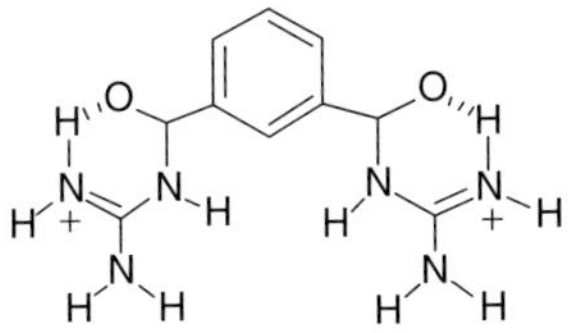

22

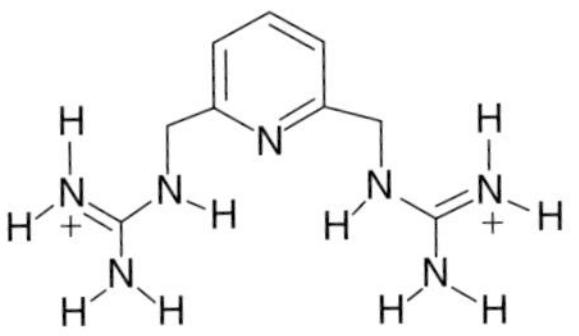

23

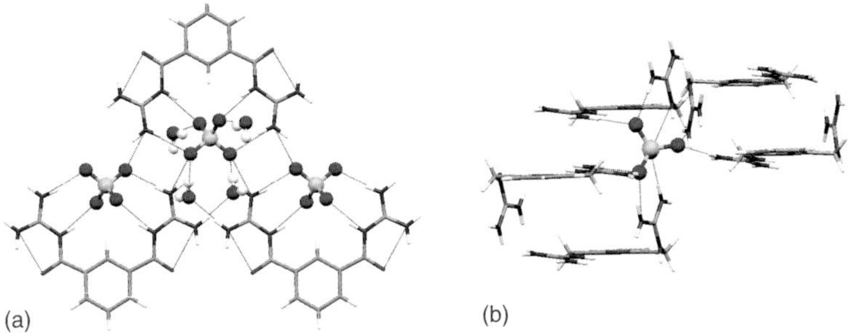

Figure 3.21 Crystal structures showing (a) the trimer of sulfates in **22** and (b) the packing of five of the surrounding ligands of **23** with sulfate (water molecules are shown only for the central sulfate for clarity).

preorganized hosts **3** and **4** (Figure 3.5) [21]. In the sulfate structure, the dianion is held by four H-bonds of the two guanidinium groups (Figure 3.21a) at distances of 2.843 and 2.857 Å as a result of two symmetry-related NH groups per ligand. In addition to four N···O bonds, each to adjacent oxygen atoms, the central sulfate forms two additional H-bonds with two neighboring ligands. The coordination of the sulfate is completed by four additional O···O H-bonds, resulting in a 10-coordinate sulfate ion.

The group also examined the pyridine analog **23** without the carbonyl being present [42], which was previously reported by Hamilton and coworkers [43, 44]. In the absence of the preorganization afforded by the carbonyl–guanidinium O···HN interaction, the guanidinium groups no longer line up on a single plane and the sulfate does not lie snugly in the semicircle provided by the neatly preorganized **22**. Instead, the sulfate is surrounded by five of the ligands with a total of nine H-bonds holding it in place (Figure 3.21b). Of these, one ligand uses three of its binding sites, one ligand uses two, and three of the surrounding ligands bind only via one of their NH groups. N···O H-bonds range from 2.798 to 2.916 Å.

3.4.4 Pentadentate

Haley, Johnson, and coworkers designed a urea-based pyridinium analog of the amide-containing host **9** (Figure 3.9) [27], **24** [45]. This ligand also binds halides in a 1 : 1 stoichiometry but has two bi-H donors available for H-bonding, making it

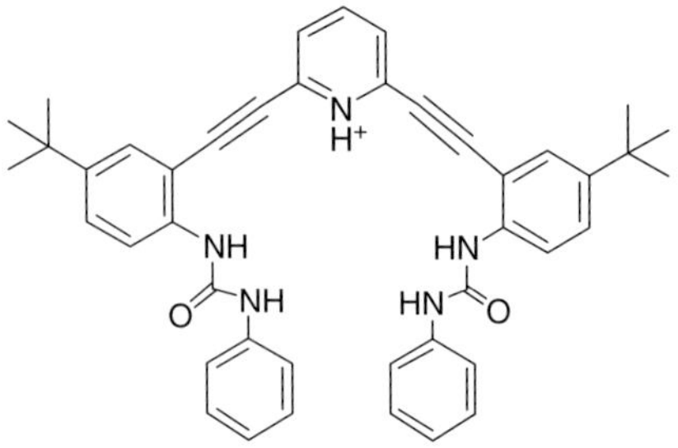

24

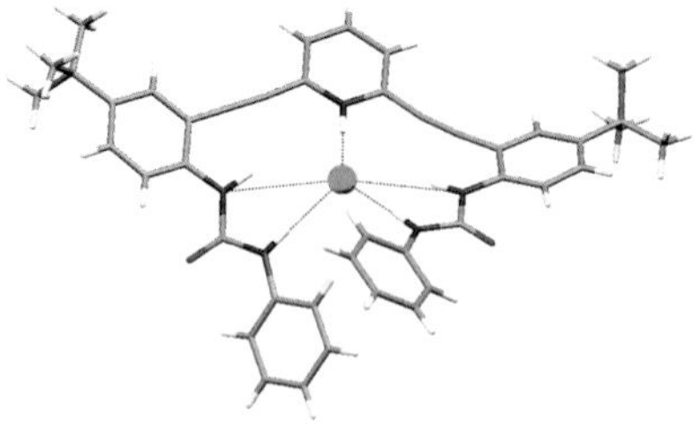

Figure 3.22 Perspective view of the pentadentate pyridinium complex of **24** with chloride ion.

a potentially pentadentate ligand if the pyridine is protonated. The authors found enhanced affinity on protonating the pyridine, and, as anticipated, a five-coordinate pseudopentagonal planar coordination site for chloride (Figure 3.22). The N···Cl distances range from an unusually short 3.03 Å for the contact with the pyridinium NH to 3.64 Å. The two longest Cl···N interactions are with the more strained urea NHs adjacent to the phenyl coming off the acetylene (3.54 and 3.63 Å), undoubtedly a result of the enforced rigidity caused by the acetylene linkage.

3.4.5 Hexadentate

With three bi-H donor groups in a 1,3,5-phenyl-substituted guanidinium host, six-coordination is anticipated. This is confirmed for the potentially hexadentate tricarballate complex with the tricationic citrate analog, **25** [46]. The ligand is selective for citrate in highly dielectric solutions. Tricarballate (1,2,3-propanetricarboxylate) is an analog of citrate that lacks an OH group. Binding with tricarballate was found to be extremely high in aqueous conditions, $K_a \sim 7 \times 10^3$, as determined by NMR titrations in D_2O. The crystal structure has two independent complexes, and only one is shown (Figure 3.23a). The two NH groups of one of the guanidiniums binds to the two oxygen atoms of one of the carboxylate groups of the guest (2.758 and 2.793 Å); a second guanidinium forms H-bonds to oxygen atoms on two different carboxylate groups separated by two CH_2 groups (2.824 and 2.756 Å); and the third forms two H-bonds to a single oxygen atom, albeit one weak bond (2.674 and 3.078 Å). What is unanticipated about this structure is that one of the ethyl groups is pointed in

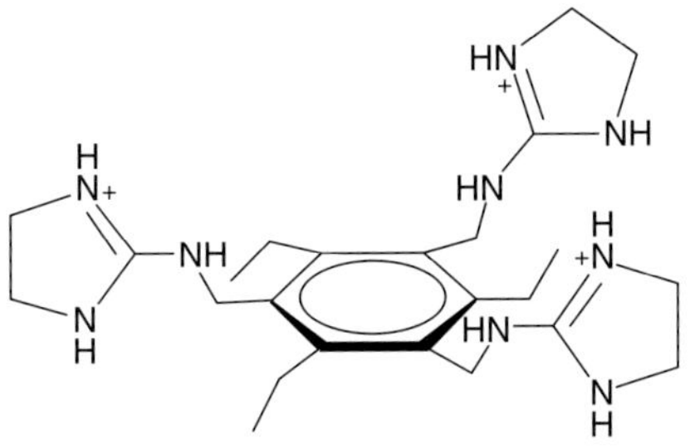

25

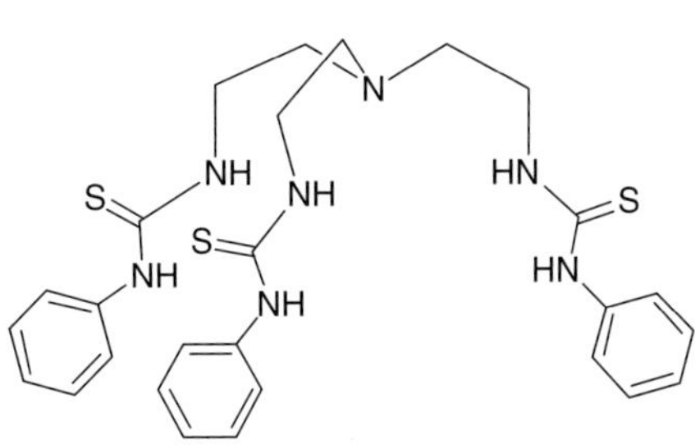

26

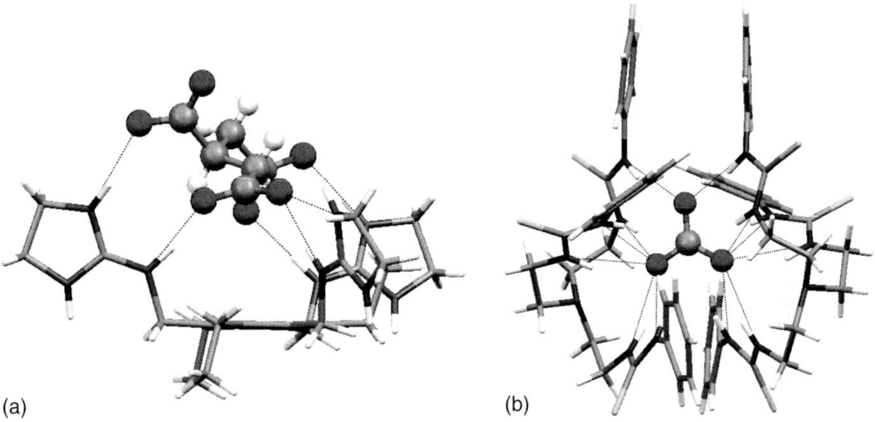

Figure 3.23 Crystal structure of (a) the tricarballate complex with **25** and (b) the carbonate complex of **26**.

the same direction as the adjacent guanidinium groups (seen on the right, inbetween the two guanidinium groups and in the ChemDrawing of **25**). Indeed, Anslyn's use of 1,3,5-ethyl-substituted ligands was based on the premise that the ethyl groups will force the alternating H-bonding groups so that an up–down–up–down orientation around the benzene ring would be achieved. So, as the authors note, the observed structure is not expected to be the thermodynamically preferred conformation but probably is a result of crystal packing influences.

The tren framework provides an excellent source for a hexadentate ligand by using bi-H donors or mini-chelates in the framework, as seen with a thiourea-appended ligand, **26** [47]. The ligand was synthesized in a quest for chloride/bicarbonate transport ligands, and it was found that **26** functioned well in this regard. The ligand forms a bis-complex with the carbonate, with each ligand donating six H-bonds (Figure 3.23b). Of particular interest is that five of the six H-bonds are to the same oxygen atom, ranging from 2.872 to 3.069 Å. The "apical" oxygen atom is only held by two H-bonds, one from each of the ligands at 2.824 and 2.831 Å. The authors do note several longer NH···O interactions, which are not included here.

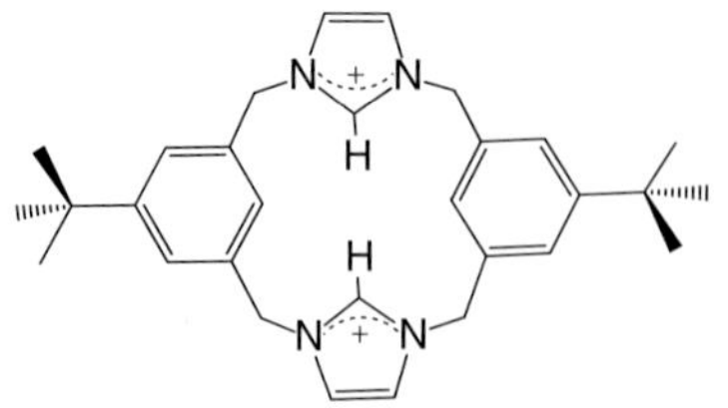

27

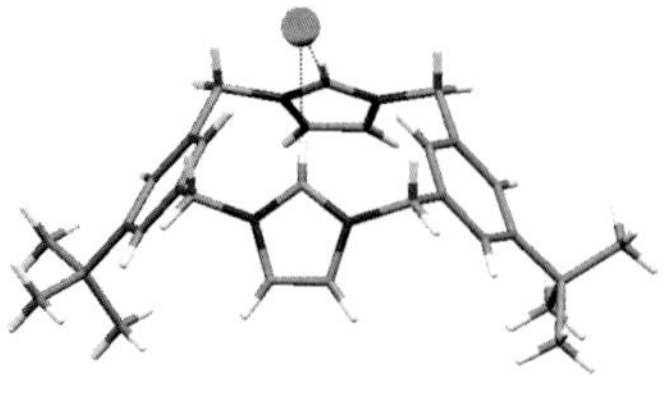

Figure 3.24 Side view of the complex of **27** with chloride showing the pyramid-like structure.

3.5 Monocycles

3.5.1 Bidentate

Macrocyclic hosts with just two H-bond donors are somewhat rare, but two examples provide some interesting structures. Alcalde and coworkers published the structure of an imidazolium macrocycle, **27** [48]. The chloride complex shows a chloride ion poised above the macrocycle, with $C_{im}\cdots Cl$ distances of 3.549 and 3.553 Å (Figure 3.24). The phenyl CH hydrogen atoms are also pointed in the direction of the chloride but at significantly longer distances, 3.716 and 3.968 Å. A second chloride resides outside the macrocycle in the crystal lattice, H-bonded with neighboring water molecules.

Unsymmetrical ring systems can provide some unique chemistry, and in the case below, another two-coordinate complex. Here, the ligand (**28**) contains only two (or potentially three, considering the tertiary amine) anion binding sites [49]. This mixed amide/amine/ether macrocycle of Smith displays interesting solution chemistry. The amine lone pair can mount a nucleophilic attack on the CH_2Cl_2 solvent, covalently binding the $-CH_2Cl$ group and releasing a chloride to be bound internally in the macrocyclic cavity. The resulting two-coordinate complex is neutral because of the resulting quaternization of the amine (Figure 3.25).

A third and final example of bidentate coordination was obtained by the Bowman-James group as an artifact of a decomposition product, **29**, of a tricyclic anion host (described in Section 3.6.7) [50]. Its guest is an isophthalate ion, which is one of two isophthalates cleaved off of the original tricycle (Figure 3.26). Distances from the bridgehead amine to the carboxylate oxygen atoms are 2.664 and 2.714 Å. The two fused ring caps cant away from the anionic guest, as seen in the side view (Figure 3.26b), and are 42° away from being parallel in order to allow the entrance of the isophthalate into the cavity.

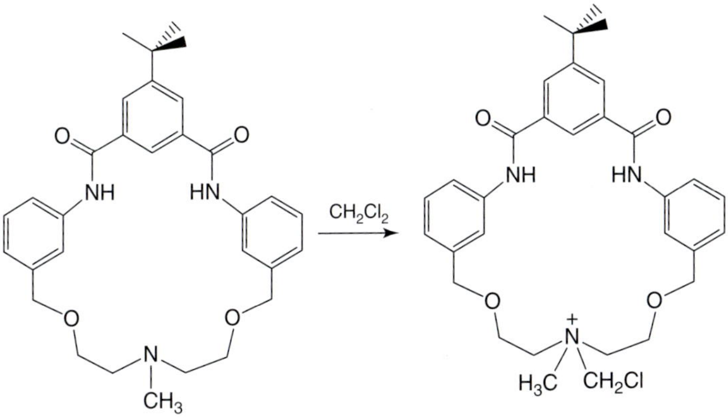

28 **28**·CH_2Cl^+

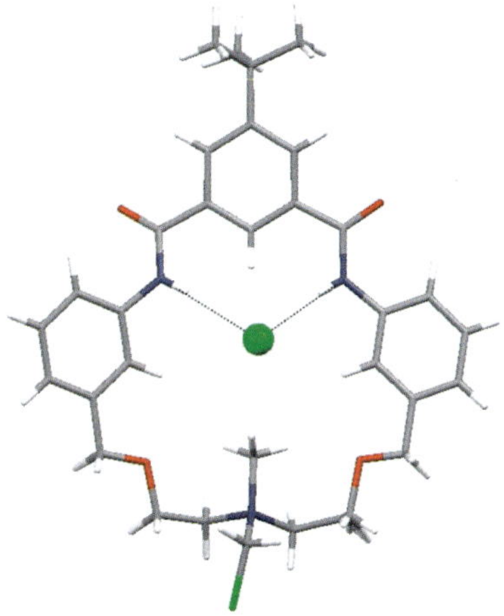

Figure 3.25 Chloride complex resulting from nucleophilic attack of the tertiary amine of **28** on CH_2Cl_2.

3.5.2 Tridentate

Three is not necessarily a common coordination denticity for a macrocyclic host, given that donor patterns often increase in increments of two. However, the beautiful cyclic hexapeptide of Kubik and coworkers is an excellent model for a tridentate ligand, **30** [51]. The ligand was found to interact (based on NMR spectroscopy) with iodide, bromide, chloride, carbonate, sulfate, and benzenesulfonate. The structure of the iodide complex with a 2 : 1 sandwich encapsulation of iodide ion is shown in Figure 3.27. The crystals are hydrated with 17 water molecules for every four of the sandwich complexes. The authors describe the sandwich formation as being gear-like in the way that the two macrocycles interlock. Each of the ligands is tridentate, resulting in six-coordination for the iodide. The coordination geometry

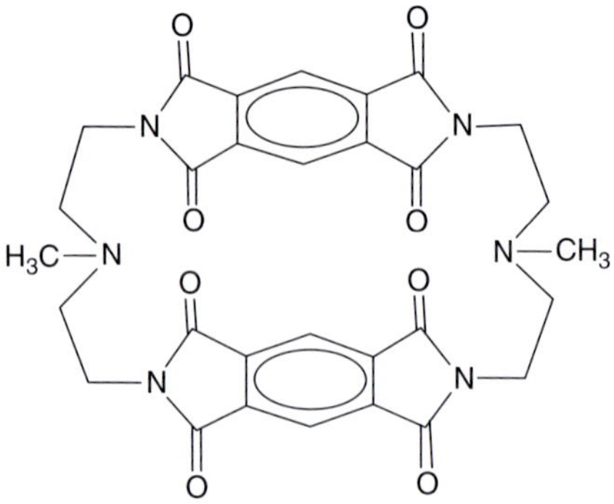

29

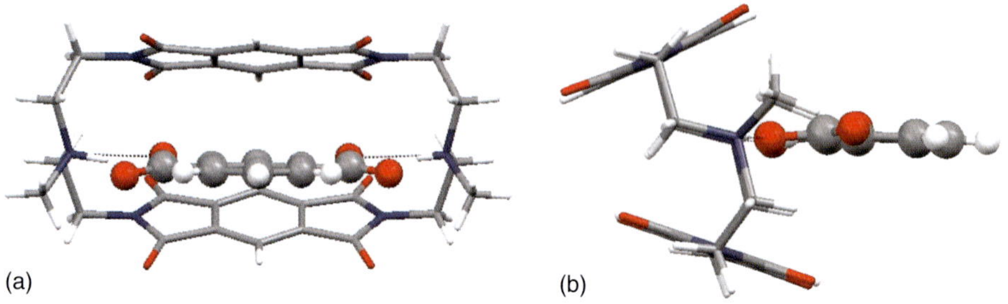

Figure 3.26 Perspective drawings of the crystal structure of isophthalate with **29**: (a) view perpendicular to the amine axis and (b) view down the bridgehead amine axis.

is very close to trigonal prismatic (Figure 3.27b). The iodide is coordinated to the three amide NH groups, which point into the cavity, at distances that range from 3.723 to 3.925 Å.

3.5.3
Tetradentate

A calix[4]imidazolium[2]pyridine macrocycle was synthesized by Kim, Hwang and coworkers, which includes two pyridine spacers at each end of the ligand, **31** [52]. The macrocycle shows a selectivity for fluoride ion with a binding constant in DMSO-d_6 of $K_1 = 28\,900\ M^{-1}$. Affinities are 1–2 orders of magnitude less for chloride, bromide, and iodide. The structure of the fluoride complex shows an almost perfect fit of the tiny anion within the calix cavity (Figure 3.28a). The complex crystallizes with three PF_6^- counterions to satisfy the charge. One of the counterions lies below the macrocycle and H-bonds with the imidazolium CH groups. The two independent N···F distances are 2.826 and 2.911 Å. The fluoride is also H-bonded to an axial water molecule at 2.569 Å as seen for the tetradentate acyclic ligand with chloride (**15**, Figure 3.15). The presence of an axial water molecule reinforces the observation that an axial water molecule is a common coordination motif as in transition metal complexes. However, K_2 for

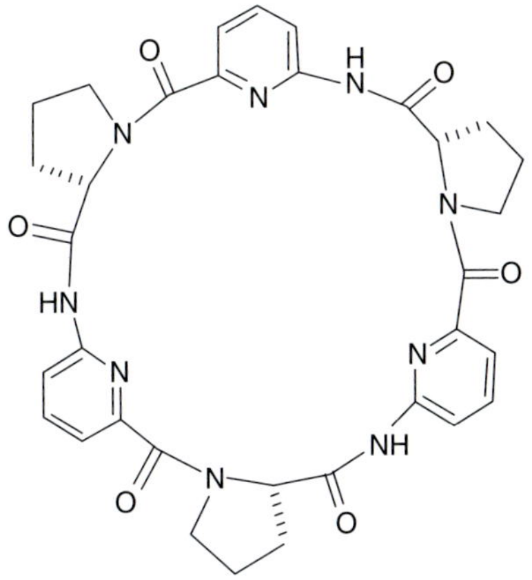

30

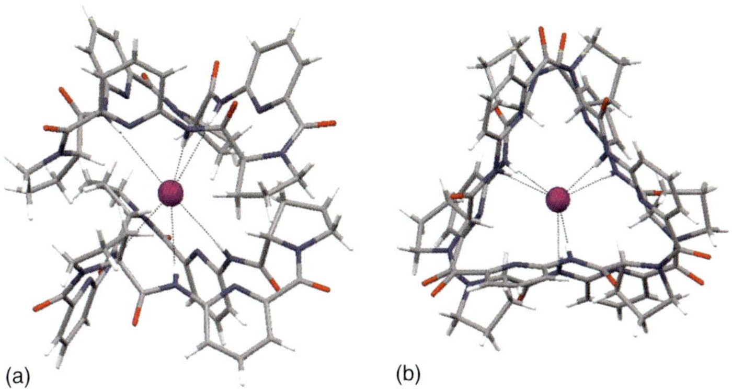

Figure 3.27 Two views of the crystal structure of the iodide complex with **30**: (a) side view perpendicular to the two rings and (b) overhead view showing the nearly trigonal prismatic coordination.

both bromide and iodide exceeds K_1, indicating 1 : 2 L:anion complex formation ($K_2 = 10\,700\ M^{-1}$ for bromide in DMSO-d_6). Indeed, the ditopic coordination is confirmed crystallographically. The macrocycle folds in a chair-like manner and binds to two bromide ions above and below the "seat." Distances from the bromide to the imidazolium carbon atoms range from 3.516 to 3.806 Å (Figure 3.28b,c). Two other bromide ions lie within the unit cell to balance charge.

Some of the simple early amide-based macrocycles (**32**) were from the Jurczak group [35]. By virtue of their design, with four amide groups, these macrocycles naturally form tetradentate complexes. In the chloride complex with **32**, the pyridine-based macrocycle folds around the anion (Figure 3.29a). As opposed to the chloride structure with the acyclic analog **15** (Figure 3.15), this structure is more open. The complex is isolated as the $[n\mathrm{Bu_4N}]^+$ salt. The authors also reported several structures with acetate, with the stoichiometry and structure varying with

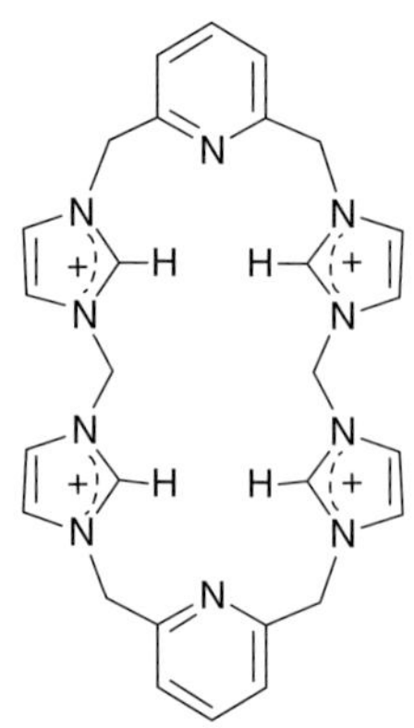

31

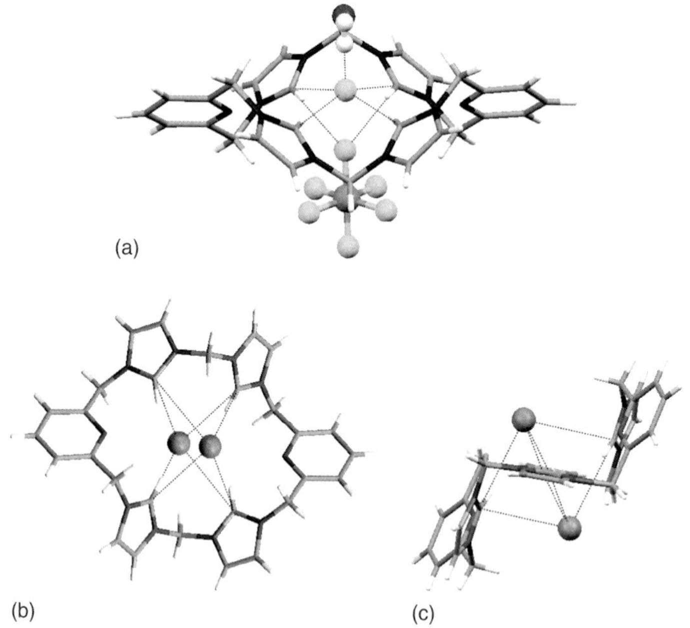

Figure 3.28 Views of the crystal structures of halide complexes with **31**: (a) view of the fluoride complex showing the axial water molecule and PF_6^- counterion below; (b) overhead view of the dibromide complex; and (c) side view of the dibromide complex showing the chair conformation.

the counterion [53]. For example, when a smaller counterion such as $[Me_4N]^+$ is used, a simple 1 : 1 complex is formed. The oxygen atom that is not bonded with the ligand is coordinated to a water molecule, somewhat like the axial water molecules in the halide complexes (Figure 3.29b). When the more bulky $[n\text{Bu}_4\text{N}]^+$ is the counterion, a 2 : 1 sandwich complex is observed (Figure 3.29c), and the

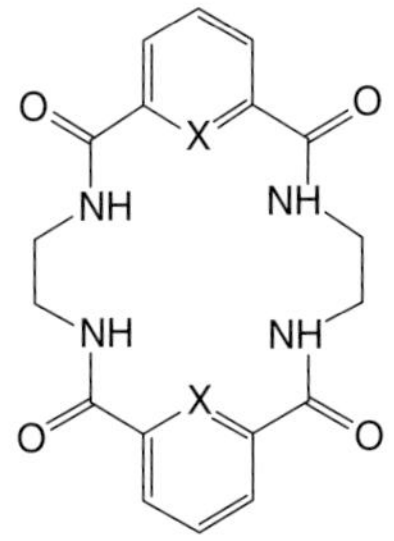

32

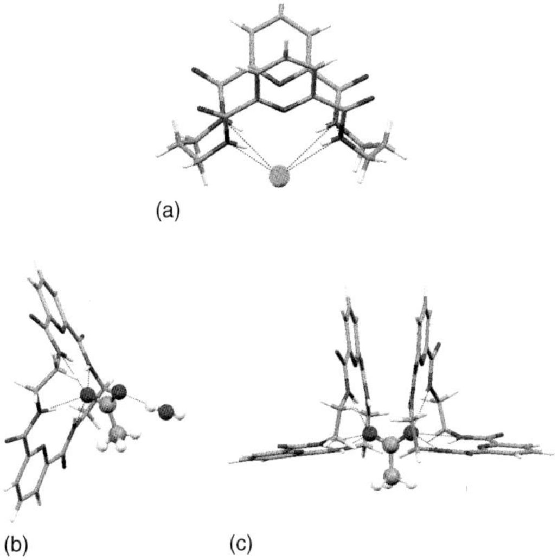

Figure 3.29 Crystal structures of anion complexes with **32**: (a) the structure of the chloride complex showing the folded macrocycle; (b) the 1 : 1 complex of acetate showing an "axial" water molecule; and (c) the acetate sandwich complex. The counterions are not shown for clarity.

two tetradentate macrocycles each bind one of the acetate oxygen atoms via four H-bonds, resulting in an overall eight-coordinate complex.

An eight-coordinate sandwich structure was observed for sulfate binding to a similar but expanded ring system, **33** (Figure 3.30) [54]. The macrocycles are neutral and form a sandwich structure much like the acetate sandwich of Jurczak just described, however, here with staggered conformations of the two macrocycles that match the S_4 symmetry of the sulfate guest. N···O H-bond distances range from 2.855 to 2.980 Å. Each sulfate oxygen atom is coordinated via H-bonds to two amide NH groups associated with the same *m*-xylyl spacer. The resulting structure is somewhat like a square prism. Each ligand, however, displays tetradentate coordination. Interestingly, the counterions are $[n\text{Bu}_4\text{N}]^+$, as in the acetate sandwich above (Figure 3.29c) [53].

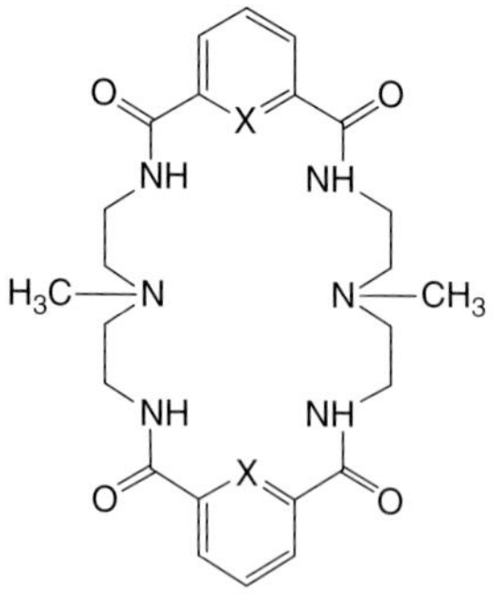

33:X = CH
34:X = N

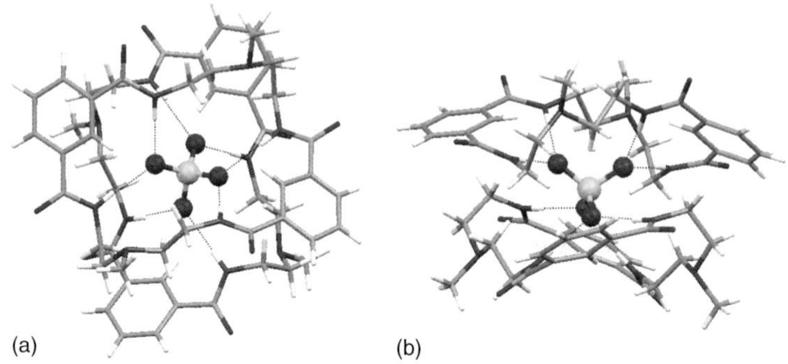

Figure 3.30 (a) Overhead and (b) side views of the sulfate structure with **33** $[\mathbf{33}_2{\cdot}SO_4]^{2-}$. The $[n\mathrm{Bu}_4\mathrm{N}]^+$ counterions are not shown for clarity.

Folding appears to be a common trait for pyridine-containing macrocycles and is seen for the pyridine analog of the mixed amide/amine macrocycle, **34** [55]. In this case, however, the ligand seems to prefer protonation rather than to remain neutral, allowing it to become potentially a hexadentate donor (as seen later in Section 3.5.5). However, although the ligand is protonated, obviating the need for a counterion with a dianionic guest, it only H-bonds with its amide NH groups, as seen for the dichromate complex (Figure 3.31). In the case of these larger oxometallates, the anion appears to be too large to be swept into the cavity and lies below it. The spacing between the two chromium atoms fits well with the spacing between the amide nitrogen atoms joined to opposing pyridine rings: 3.761 and 3.666 Å for the amide nitrogen atoms and 3.764 for the two oxygen atoms H-bonded to the amides. The fate of the amine proton is to be pulled away from the anion to intramolecularly H-bond with one of the amide oxygen atoms. The perrhenate structure (not shown) is very similar and consists of two of the monometallic perrhenates hanging below the folded pocket. While the two structures are virtually superimposable, the rhenium atoms are farther apart compared to the dichromate

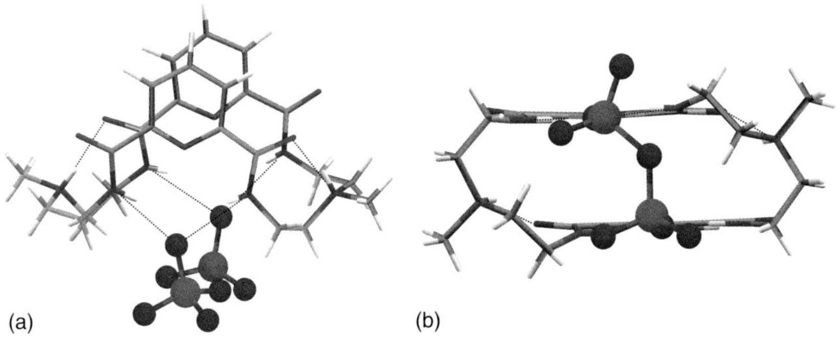

Figure 3.31 Two views of the dichromate complex with **34**: (a) side view showing the H-bond network and (b) view from underneath showing how the span of the chromium atoms matches that of the macrocycle.

chromium atoms, at about 4.8 Å compared to 3.222 Å for the dichromate Cr–Cr distance.

In an attempt to add a definitive charge to the host to achieve charge neutrality on anion binding (as opposed to relying on possible amine protonation), hosts **33** and **34** were quaternized with MeI to yield ligands **35** and **36** [56]. Again, the aromatic or heterocyclic groups appear to govern the conformation. For the isophthalamide (**35**), the macrocycle elongates with the two charged ends at the farthest distances of the elliptical host. Disappointingly, in the chloride structure, the carbonyl groups were all pointed inward and the anions tend to lie outside of the cavity (Figure 3.32a). As a result of the inversion center, there are only two independent N···Cl distances, which are 3.227 and 3.258 Å. Thus, although the ligand itself is tetradentate by donating four H-bonds, each is donated to a different anion. This is just one possible orientation of the amide carbonyls for this splayed-out dicationic ligand **35**.

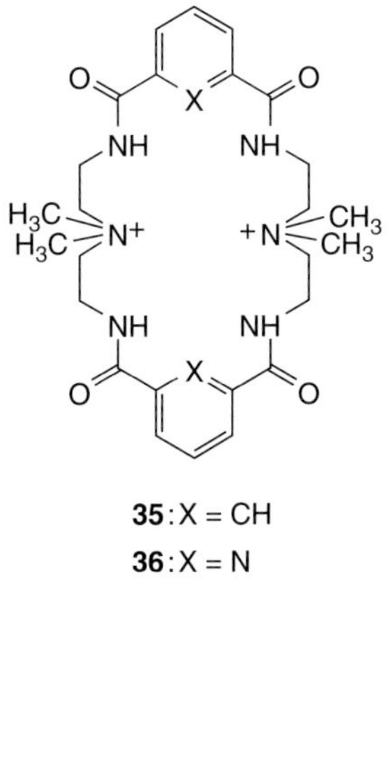

35: X = CH
36: X = N

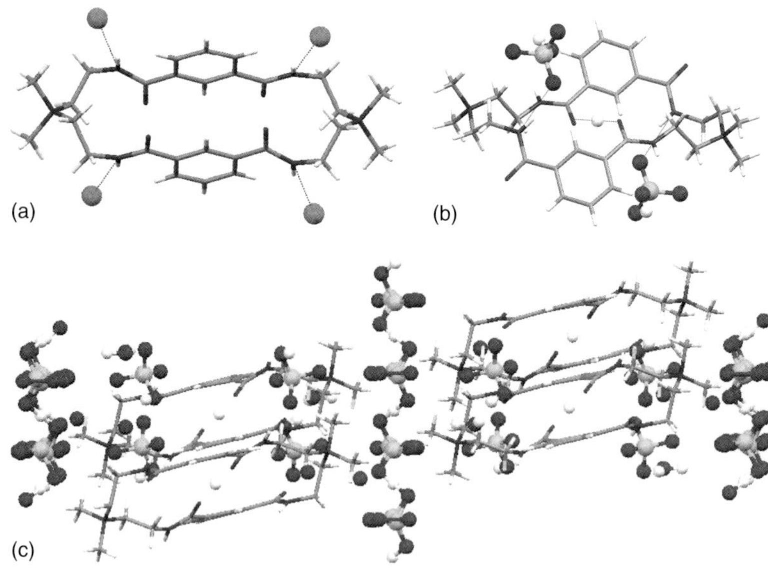

Figure 3.32 Crystal structures of **35**: (a) tetranuclear chloride complex; (b) encircled proton complex; and (c) the packing diagram of the proton/bisulfate structure, showing the channel of the third bisulfate chain as viewed looking down the *b* axis.

In another conformation of **35**, the ligand carbonyl groups alternate *in–out–in–out*, and a highly unanticipated structure was obtained. Rather than encapsulating an anion, the ligand trapped a lone proton between the two catty-corner *in–in* protons [57]. Adding to the unusual occurrence is that the ligand itself is already dipositively charged by virtue of the quaternization. The extremely short O···O separation makes this bond an example of a low barrier hydrogen bond (LBHB) (Figure 3.32b) [58], with an O···O distance of 2.45 Å. Two bisulfate ions lie just outside the cavity as was seen for the chloride. The third required negative charge is provided by a third disordered bisulfate, which lies in a channel between the quaternized macrocycles (Figure 3.32c).

The ligand folds in the quaternized pyridine macrocycle, **36**, as seen in the iodide structure (Figure 3.33a) [56]. In this case, the iodides lie outside the folded macrocycle. These two iodides are symmetry related, with distances to the amide nitrogen atoms of 3.667 and 3.694 Å. Each of the iodide ions is associated with a neighboring ligand resulting in four-coordinate iodides with a pyramidal coordination geometry.

The cleverly designed indolocarbazole-containing macrocycle, **37**, can be expanded by adding a second two-carbon acetylene bridge to the spacer between the two H-bonding units to give **38** [59]. The smaller macrocycle binds azide, but the distance between the two indolocarbazoles is too short to accommodate the triatomic anion within the macrocyclic cavity (Figure 3.34a). In this case,

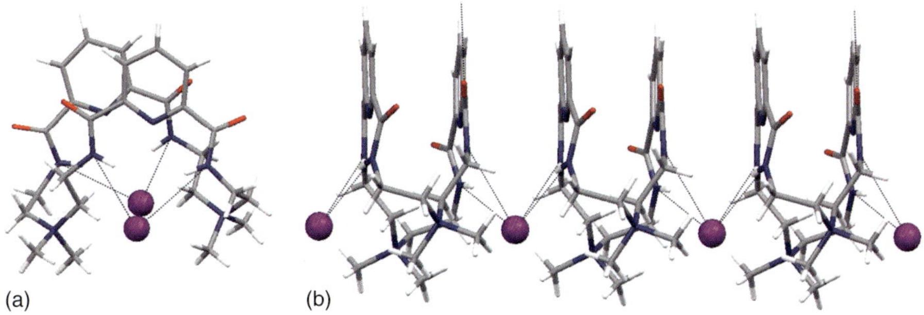

Figure 3.33 Two views of the iodide complex of **36** showing (a) the four H-bonds, two each to an iodide ion; and (b) the side view showing the positioning of the iodides between macrocycles.

the terminal nitrogen atom of the azide ion is H-bonded to the four ligand NH groups at distances that range from 2.898 to 2.935 Å, and the azide points upward from the ligand (Figure 3.34a). The N–N azide distances are 1.194 and 1.175 Å, with the slightly longer bond associated to the H-bonded nitrogen atom. In the larger macrocycle, the azide seems to be a perfect fit, snuggly situated as the authors note in a "μ_4-1,1,3,3 (end-to-end) coordination mode" (Figure 3.34b). The H-bond distances to the macrocyclic NH groups are 2.856 and 2.848 Å, and the N–N azide distances are both 1.187 Å because of the crystallographic inversion center. The smaller macrocycle forms an almost 2 : 1 rectangle, considering the distances directly across the macrocycle between the two acetylene carbon atoms (8.24 Å), compared to the analogous distance measured between opposing amide nitrogen atoms (4.40 Å). The larger macrocycle is closer to being square with corresponding distances of 7.0 and 8.13 Å.

Calixpyrroles and other expanded porphyrin-like molecules have provided a large variety of ligands for anions. The simple calix[4]pyrrole, **39**, acts as a dual host or ion pair receptor for alkali metal halides and for carbonate as well [60]. The CsF complex is the most symmetrical of the group of halides, and the pattern is a pleasing vertical stacking of Cs^+–**39**···F^-–Cs^+–**39**···F^-, and so on, as shown in Figure 3.35a. The distance between the Cs^+ ion and the center of the pyrrole rings is 3.39 Å. On the other side of the macrocycle, the fluoride ion is bound to all four of the calixpyrrole NH groups, with an N···F distance of 2.79 Å. The Cs–F contact distance is 2.765 Å to its nearest neighbor (above) and 3.69 Å to the cesium ion below the macrocyclic ring. As the authors state, the result is "like a one-dimensional coordination polymer."

The Cs_2CO_3 structure (Figure 3.35b) was a serendipitous isolation as a result of trying to obtain a hydroxide complex and was obtained most probably from CO_2 fixation in air. The Cs_2CO_3 complex crystallizes as a dimer with four H-bonds to two carbonate oxygen atoms on one side (2.776–2.836 Å) and four H-bonds to a single oxygen atom to the other ligand, 2.819 and 2.887 Å. The two cesium ions balance the carbonate charge of −2. Furthermore, the authors note that this may

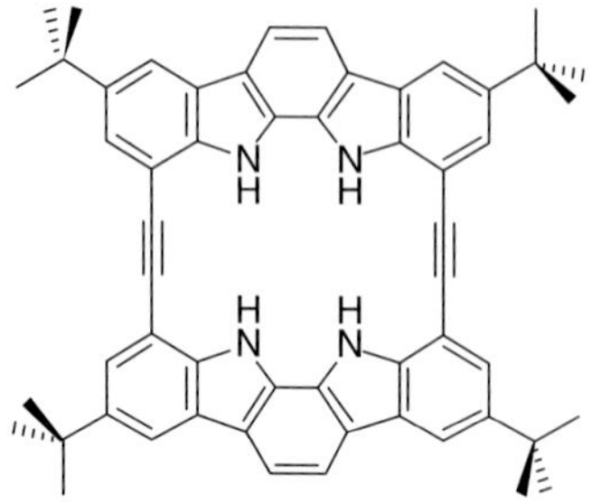

37

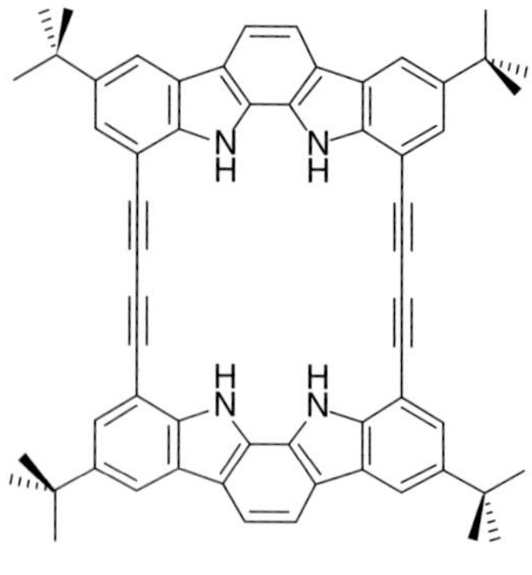

38

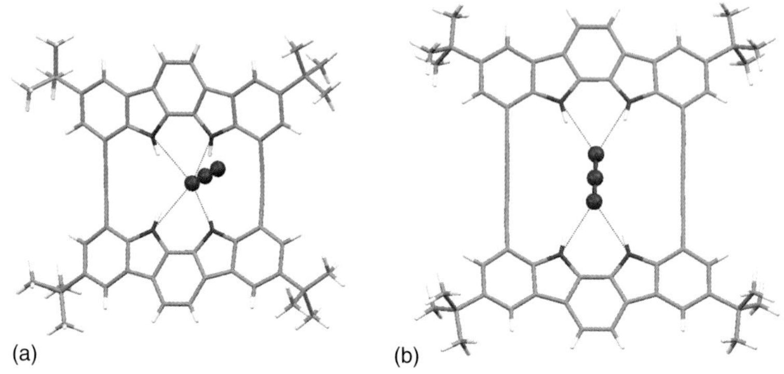

Figure 3.34 Indolocarbazole rectangles and complexes with azide ion: (a) $[\mathbf{37}\cdot N_3]^-$ and (b) $[\mathbf{38}\cdot N_3]^-$.

be the first anion sandwich structure of a neutral pyrrole-based ligand, where a single anion is sandwiched between two of the calixpyrrole units.

3.5.4 Pentadentate

The guanidinium/urea-based macrocycle **40** [61] was designed based on theoretical studies that indicated that such a combination would be favorable for binding nitrate [62, 63]. Here, the guanidinium group provides the scaffold lynchpin for the two urea binding sites as well as the charge complementarity for binding mononegative anions (Figure 3.36a). Of the three macrocycles the authors studied (with different carbon chain lengths from $-(CH_2)_4-$ to $-(CH_2)_6-$), the longest, **40**, had the highest affinity for nitrate ion, 7.27×10^4, as determined by isothermal calorimetry in CH_3CN. In the X-ray structure, the ligand binds in a pentadentate manner with H-bonds ranging from 2.842 to 3.048 Å (Figure 3.36a). The sixth N···O distance is a bit longer, and it could be argued that this is a H-bond, even at 3.114 Å.

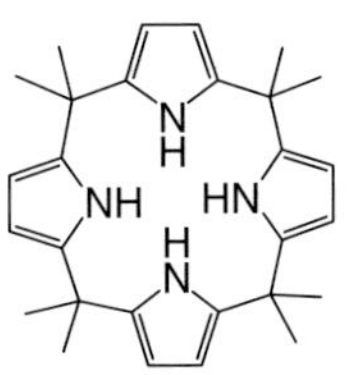

39

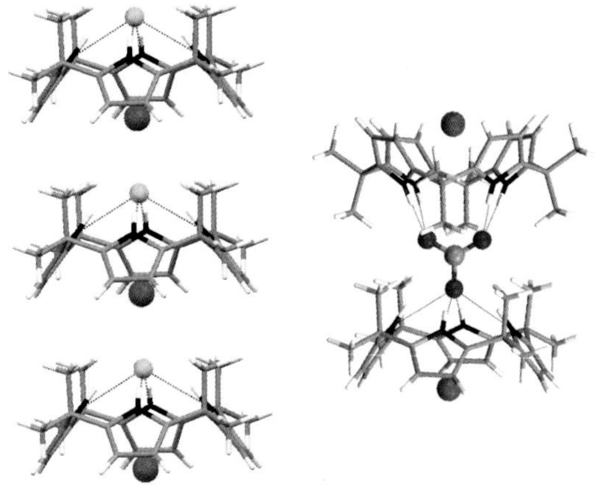

(a) (b)

Figure 3.35 Structures of **39** with (a) CsF showing the linear polymer chain and (b) Cs_2CO_3.

Although known for their calixpyrrole and related host work, Sessler's group has also incorporated other "modules" into the calixpyrrole frameworks, including the 2,6-diamidopyridine group (**41**) that has been such a favorite among many anion researchers [64]. The ligand crystallizes as the dipositive anion, thus again providing a handy complementary charge to dinegative anions. This molecule binds sulfate in a fivefold H-bonding motif with H-bond distances ranging from 2.809 to 2.889 Å and an additional "axial" contact with a neighboring water molecule at 2.712 Å (Figure 3.36b). The sixth NH group makes contact with one of the sulfate oxygen atoms at 3.127 Å, which, as above, is arguably a very weak H-bond. However, since the other five O···N contacts were so much shorter, the authors of this chapter have chosen it as an example of pentadentate binding. So, as in the nitrate structure with **40** (Figure 3.36a), the ligand is potentially hexadentate, and one could argue that both ligands are indeed hexadentate. The reader is encouraged to make his or her own decision based on the evidence.

3.5.5
Hexadentate

In pyridine-containing macrocycles, many of the complexes tended to exhibit a folded conformation as seen for compounds **32** (Figure 3.29), **34** (Figure 3.31),

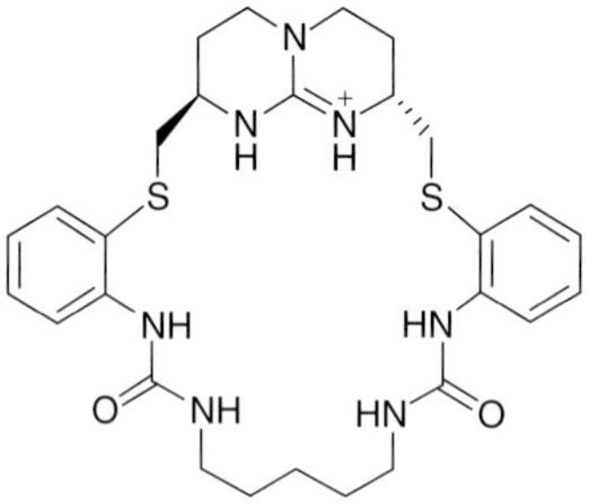

40

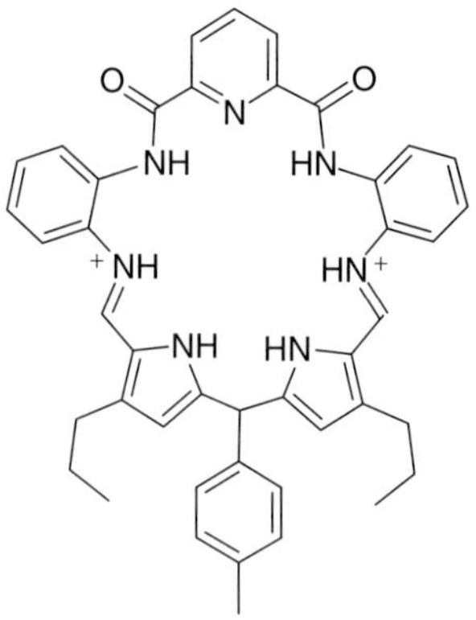

41

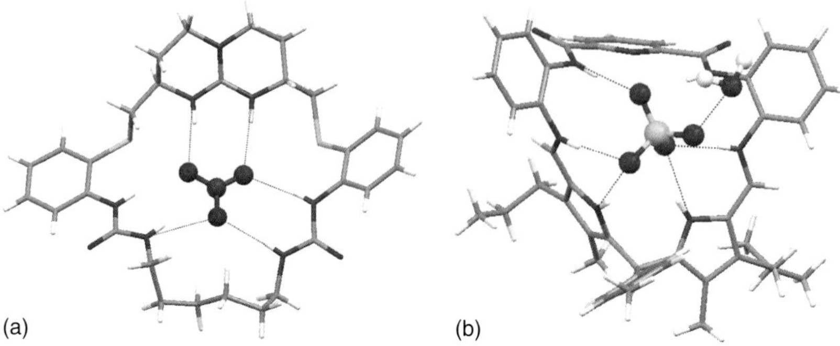

Figure 3.36 Two ligands displaying pentadentate coordination: (a) nitrate structure with **40** and (b) sulfate structure with **41**.

and **36** (Figure 3.33). In all these cases, however, the ligands exhibited tetradentate behavior toward the anion. Also, in general, the pyridine-containing macrocycles tend to show higher binding affinities for anions compared to the *m*-xylyl systems [17, 18] possibly because of preorganization of the pyridine NH groups inward as a result of H-bonding with the pyridine nitrogen lone pair. In the sulfate complex with **34**, the ligand binds in a hexadentate manner to the sulfate dianion held in the ligand pocket (Figure 3.37) (unpublished results). It should be noted that the amide nitrogen atoms are 2.961 Å from the upper sulfate oxygen atoms; however, the N–H–O angle is slightly less than 100°, which is thus probably not indicative of an actual H-bond.

An early example of a macrocyclic urea was **42**, which incorporates both urea and amide linkages [65]. The goal of the design was to add a 2,6-dicarboxamidopyridine binding site to a urea framework and to use an *o*-phenylene link from the urea to an adjacent amide to provide a convergent set of H-bonds. The result is a potentially hexadentate ligand with four amide and one urea binding sites

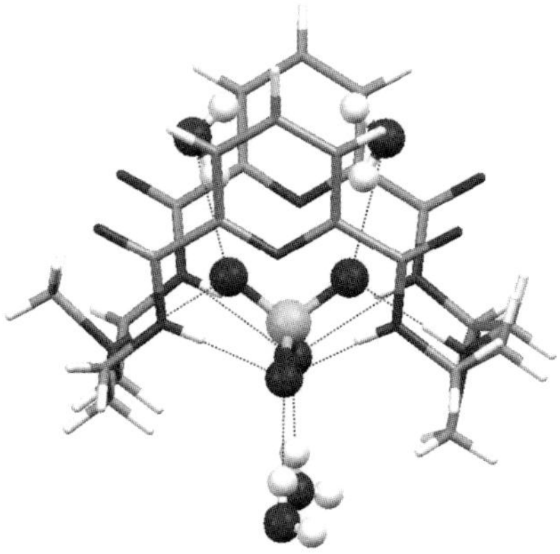

Figure 3.37 Structure of the folded pyridine-containing macrocycle, **34**, with sulfate dianion.

(Figure 3.38a). The authors note that it was not easy to crystallize the macrocycle using tetrabutylammonium salts; however, they were able to obtain a mixed fluoride/carbonate complex, in which the carbonate ions were bound within the macrocycle. The carbonate evidently was incorporated as a result of abstracting carbon dioxide from the air during crystallization, that is, "CO_2 fixation" by the fluoride complex of **42**. In the complex, each carbonate oxygen atom is held via two H-bonds (Figure 3.38a). In an earlier paper, Hay also predicted that the ideal binding for nitrate would be twofold H-bonding for each oxygen atom [62]. In this case, two of the oxygen atoms of the carbonate are held by one of the urea NH bonds and one from the adjacent amide NH, with distances ranging from 2.765 to 2.842 Å. The bottom two amide N· · ·O distances are 2.814 and 2.844 Å.

Another macrocyclic urea-based ligand that contains only urea H-bonding sites was designed based on molecular modeling studies, during which focus was on utilizing a rigid 4,5-substituted xanthene skeleton. The authors also decided to examine diphenyl ether units as comparatively more flexible than xanthene and obtained the cyclic tris-urea-containing **43** (Figure 3.38b) [66]. The expanded xanthene ligand is seen in the section on dodecahedron coordination (Section 3.5.7, Figure 3.40, ligand **45**). The chloride complex of **43** crystallizes with no solvent in the cavity, and the macrocycle forms a "three-bladed-propeller"-like environment around the anion, with N· · ·Cl distances ranging from 2.47 to 2.75 Å.

3.5.6
Octadentate

In the sulfate complex with the cyclo[8]pyrrole (**44**), the macrocycle has no meso bridges. The design was intentional to provide a more rigid framework to a large (8-pyrrole) ring so that it would not twist or fold in figure-eight-like conformations [67]. The resulting ring system is dicationic and contains 30 π-electrons and hence is aromatic according to the $4n+2$ Hückel rule, not unlike the famous porphyrin, which is aromatic with 18 π-electrons. There are two crystallographically independent complexes, one with sulfate linked axially to a neighboring methanol

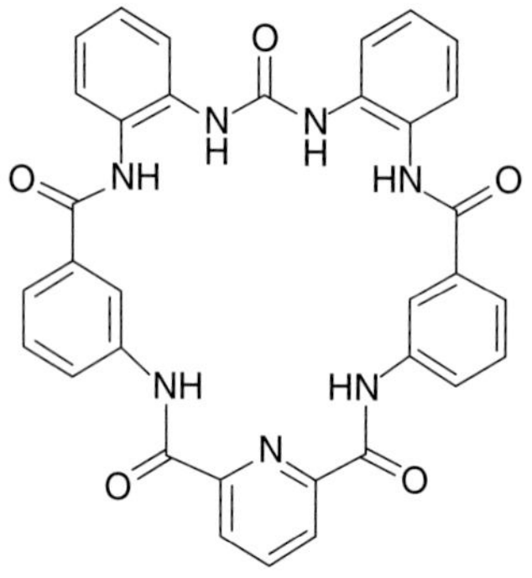

42

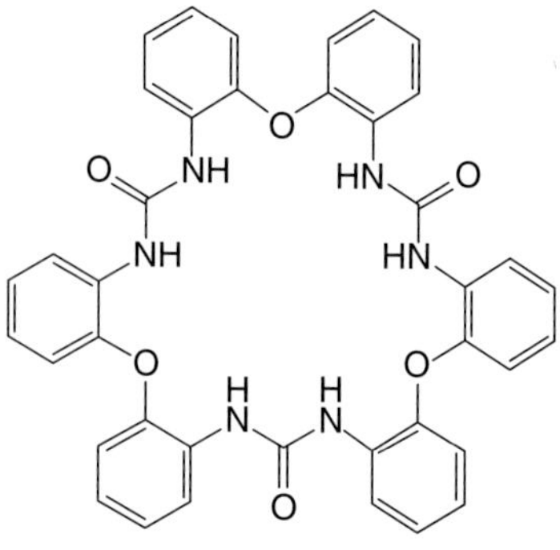

43

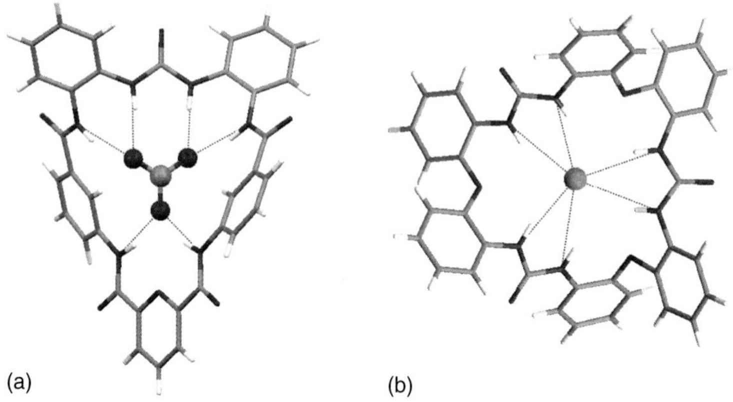

Figure 3.38 Hexadentate structures with urea-containing ligands: (a) carbonate structure with **42** and (b) chloride structure with **43**. The $[n\text{Bu}_4\text{N}]^+$ counterions are not shown for clarity.

of crystallization and one with disordered sulfate and no appended methanol. The former complex is shown (Figure 3.39). The ligand provides seven H-bonds to the sulfate oxygen atoms and one to the neighboring solvent, which bridges from the pyrrole NH to the sulfate oxygen atom. N···O distances range from 2.751 to 3.059 Å. The shortest 2.751 Å H-bond is flanked by two longer H-bonds (3.013 and 3.059 Å), while the other N···O bonds range from 2.802 to 2.879 Å. The result is octa-coordination for the sulfate, a common coordination number since each oxygen atom can easily accommodate two H-bonds. Because the ligand also binds a methanol molecule, it is considered to be an octadentate provider. The sulfate sits slightly out of the cavity, as can be seen in the side view (Figure 3.39b), but is anchored by methanol on the opposite side. The methanol is coordinated via its OH group (hydrogen atom not shown) at 2.646 Å. This pyrrole NH group is much farther from the nearest sulfate oxygen atom (3.327 Å).

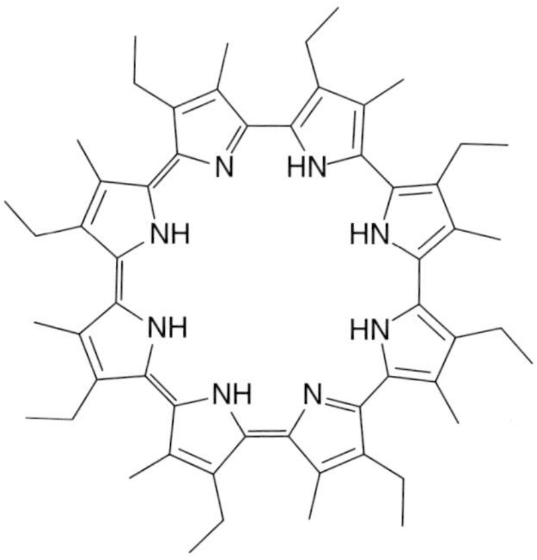

44

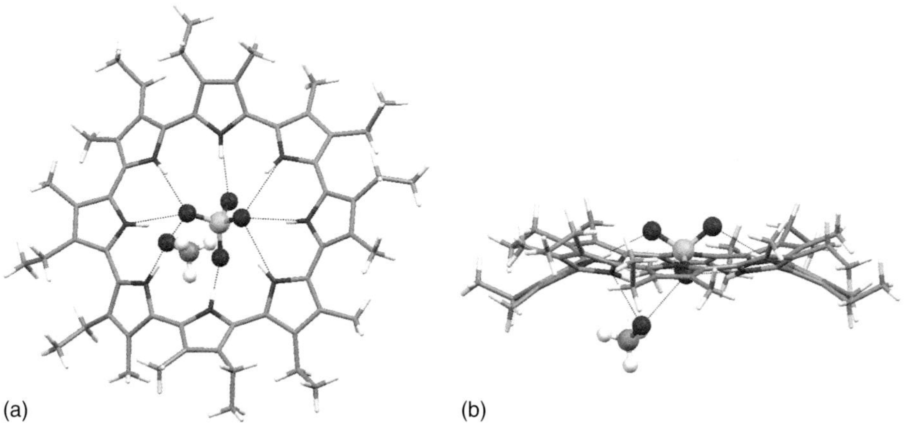

Figure 3.39 Overhead (a) and side (b) views of the cyclo[8]pyrrole **44** with sulfate.

3.5.7
Dodecadentate

Coordination numbers higher than eight are rather unusual for macrocyclic ligands. However, an interesting example is a hexaurea xanthene-derived macrocycle, **45** [68]. The ligand was isolated in decent yield (20%) as a 2 × 2 side-product from a condensation designed to provide a tris-urea host. Two chloride ions served as a template for isolation of the figure-eight-like complex, which subsequently was obtained in crystalline form. Each of the chloride ions is located in one of the loops of the figure eight (Figure 3.40). The N···Cl distances range from 3.18 to 3.34 Å, indicating a relatively strongly held anion. The Cl···Cl separation is 6.029 Å. There is a pseudo C_2 axis that lies between the two loops of the "eight" (vertical to the perspective view in Figure 3.40).

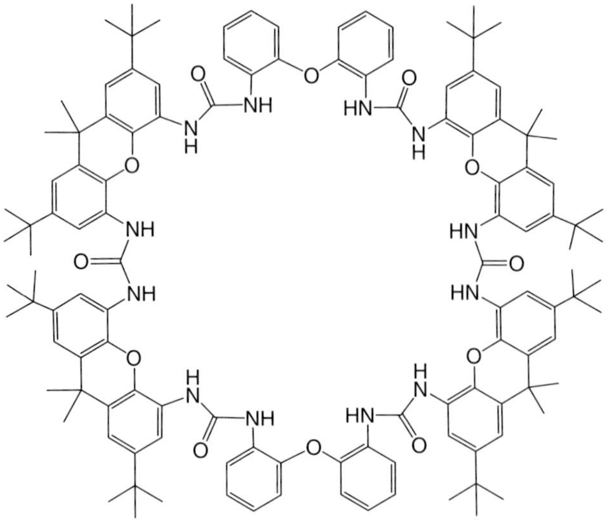

45

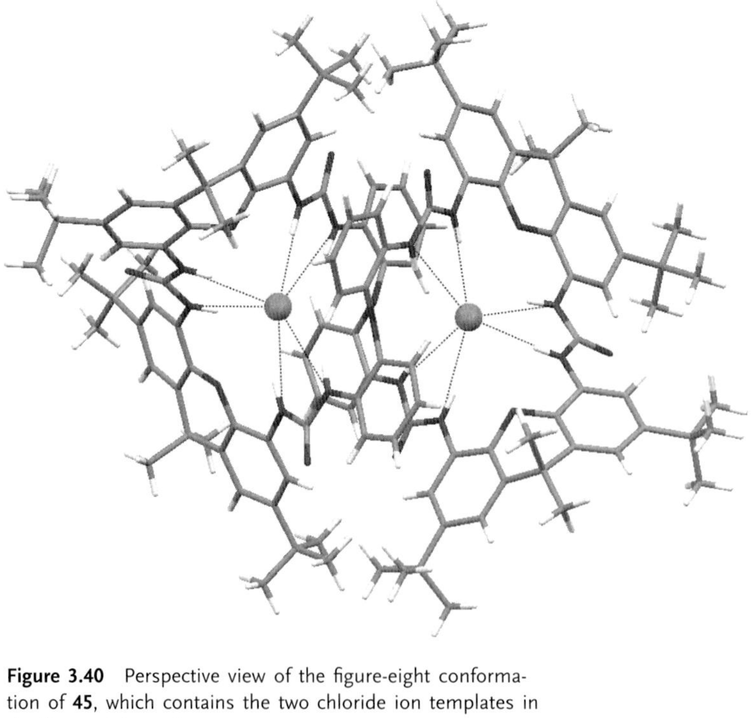

Figure 3.40 Perspective view of the figure-eight conformation of **45**, which contains the two chloride ion templates in the figure-eight pockets.

3.6 Cryptands

3.6.1 Bidentate

One would surmise that in the case of bicyclic or cryptand cavities, bidentate coordination would be rare. However, this is not necessarily the case. Of course there is the first anion complex of the diaza *katapinand* of Park and Simmons [2] and Bell *et al.* [3]; however, there are also other examples. In the thiophene complex with chloride ion of Hossain and coworkers, the anion lies midcavity in the octaprotonated cryptand **46**, bound only to the two apical NH donors (Figure 3.41) [69]. The bridgehead separation in this complex is 6.096 Å, and the N· · ·Cl distance is 3.048 Å. Binding (from ^{1}H NMR titrations at pH 2 in water) gives a log K of 3.6. By binding only to the two apical NH groups, this complex provides a polyammonium analogy to the historical *katapinand*.

The Bowman-James group fashioned a bicyclic cryptand with 1,3,5-substituted phenyl bridgeheads and three flexible aliphatic chains with both amide and amine H-bond donors, **47** [50]. Rather than sequestering ions between the two phenyl groups, this host binds anions between two of the three cis-oriented arms. It

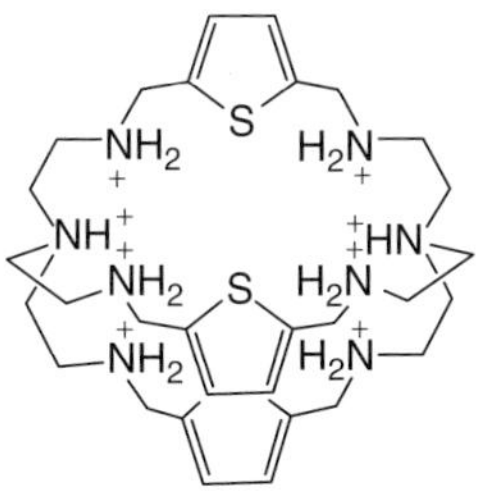

46

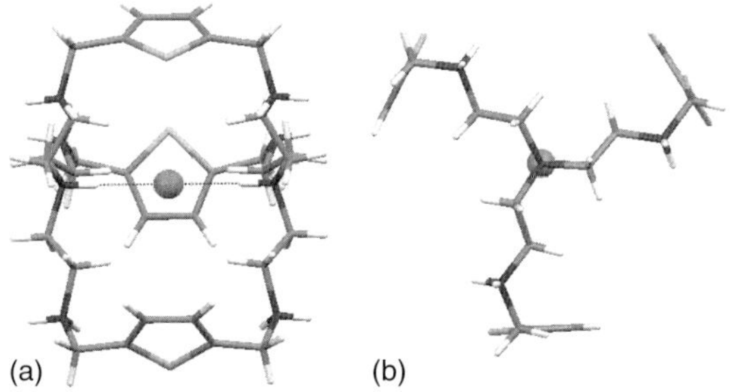

Figure 3.41 Crystal structure of the two-coordinate linear chloride complex with the octaprotonated ligand **46** showing (a) the side view with the linear H-bond and (b) the view down the bridgehead amines. The other chloride ions in the unit cell are not shown for clarity.

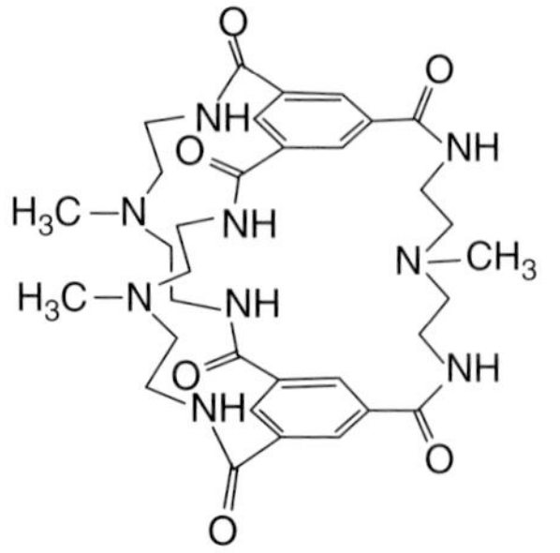

47

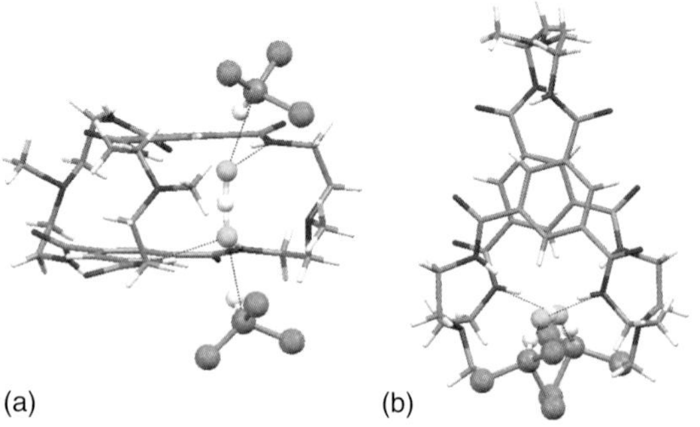

(a) (b)

Figure 3.42 The bifluoride complex with the neutral ligand **47** showing (a) side view and (b) view down the phenyl bridgeheads. The $[Me_4N]^+$ counterions are not shown for clarity.

appears to be suited for binding linear ions such as bifluoride, which fits well into the clam-shell-like opening (Figure 3.42). Two $CHCl_3$ "corks" appear to hold the bifluoride in place, with a C···F distance 3.001 Å. There are only two bonds from the ligand NH groups to the bifluoride ion, with an N···F distance of 2.83 Å. The F–F distance is a short 2.18 Å, a reflection of the weak interactions with the ligand donor groups, which does little to weaken the strong FHF interaction.

Smith and coworkers are noted for their work with "dual-host" cryptands such as **48**. In **48**, the base macrocycle is an aza-oxo crown ether, designed to hold the cation, while the top chain contains H-bond donors to capture the anion (Figure 3.43) [70]. The authors were the first to utilize this approach and reported an initial host in 2000, which contained both a sodium and a chloride in the cavity, separated by solvent [71]. They refined this prototype and in 2001 reported the seminal sodium chloride structure of a ditopic host that could simultaneously bind both the anion and cation of an ion pair in proximity (Figure 3.43a) [70]. Potassium seems best suited in terms of cation enhancement of binding in this dual host. In the larger earlier example, neither potasium nor sodium showed much effect on enhancement of chloride binding. The distances from the chloride ion to the amide hydrogen atoms are 3.189 and 3.292 Å, and the distance to the phenyl hydrogen atom is 3.314 Å. The Na–Cl distance is 2.989 Å.

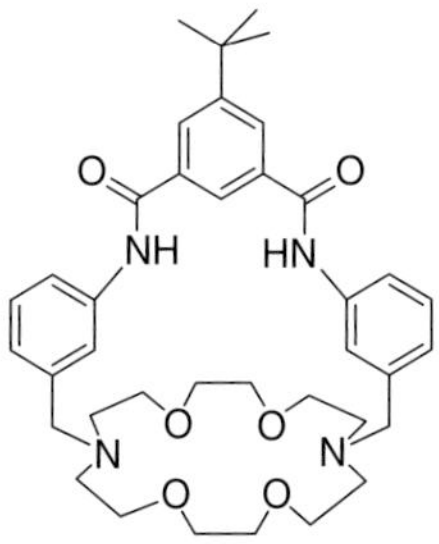

48

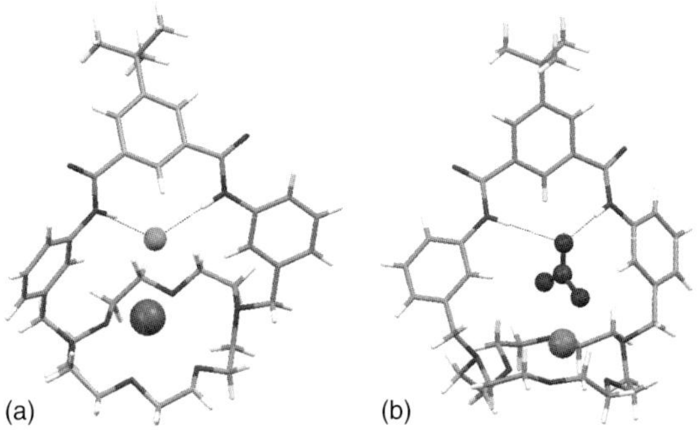

Figure 3.43 Two structures of the dual-host ligand **48**: (a) the NaCl complex and (b) the $NaNO_3$ complex.

The dual host, basket crown ether with a H-bonding handle, approach can be applied to other nonhalide examples such as the sodium nitrate complex shown (Figure 3.43), again with ligand **48** [72]. While H-bonds are drawn to the apical oxygen atom, the associations are really quite long at 3.145 and 3.256 Å, indicating only very weak interactions. The shortest distance is from the nitrate oxygen atom to the phenyl hydrogen atom at 3.059 Å; however, since the angle is so small, $\sim 120^\circ$, it is not included as an interaction. The stronger interactions are most probably the electrostatic interactions between the sodium and nitrate ions. Distances from the lower oxygen atoms of the nitrate to the potassium ion are 2.460 and 2.490 Å. However, this example is included here to show the versatility of these dual hosts.

3.6.2
Tridentate

By adding a H-bond donor to the dual-host crown-ether-containing ligands, a tridentate corollary to the bidentate prototype ligands described in Section 3.6.1 can be obtained. This is the case with the sodium chloride complex of the pyrrole-based dual host of Smith, Gale *et al.*, **49** (Figure 3.44) [73]. The chloride distances to the two amides are 3.380 and 3.320 Å, and the pyrrole sits more closely to the guest at 3.168 Å. This H-bonding motif (long–short–long) is reminiscent of the acyclic

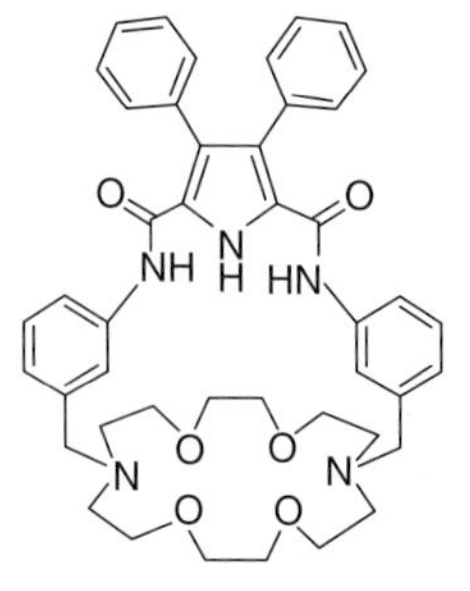

49

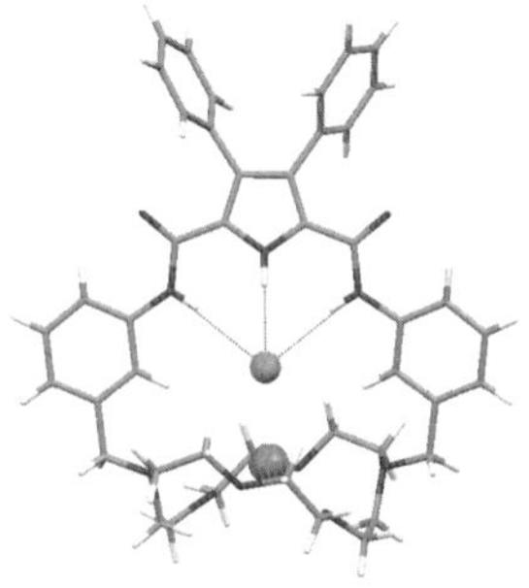

Figure 3.44 Structure of the tridentate pyrrole-containing dual-host (**49**) complex with sodium and chloride guests.

mixed amide/pyrrole complexes in Sections 3.4.2 and 3.4.3, ligands **7–9** [23–25]. The distance from the chloride ion to the sodium ion is 2.654 Å.

Bharadwadj and coworkers synthesized a potentially dual-host mixed oxa-aza macrocycle in which a nitrate lies in one side of the cavity, **50** (Figure 3.45) [74]. With the possibility of the bridgehead amines becoming protonated, this ligand has an optimal denticity of five. However, in the nitrate complex, the actual denticity is three, although the nitrate is held by six bifurcated H-bonds, two each from the three trigonally oriented amines. This is a common coordination mode for nitrate; however, it leaves a question as to whether bifurcated bonds count as monodentate or bidentate. In this chapter, the authors classify each H-bond donor as a single denticity. As is also common with these bifurcated bonds, there is one shorter distance averaging around 2.8 Å and a second longer bond averaging around 3.0 Å. The protonated bridgehead amine is quite close to the central nitrate nitrogen atom at 3.011 Å, and the distance between the bridgehead amines is 8.609 Å.

3.6.3
Tetradentate

Sessler and coworkers expanded their calixpyrrole systems into a cryptand by adding a strap to a calix[4]pyrrole (**51**) (Figure 3.46) [75]. This is the reverse of the earlier basket with a different strap concept: in this case, all of the H-bonding groups reside in the basket portion of the ligand. Chloride binds well in the cavity, with a secondary role of attracting all four amide hydrogen atoms into the *cone*

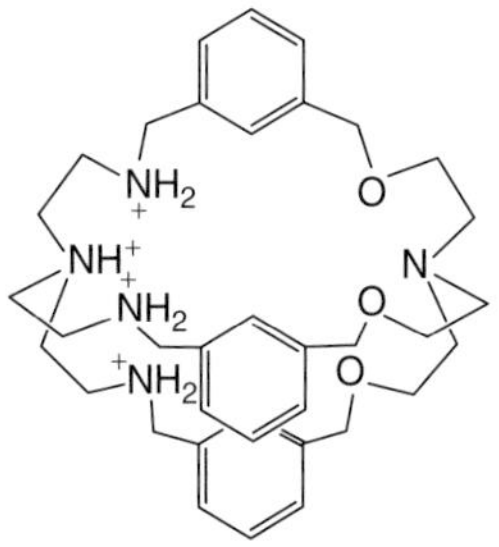

50

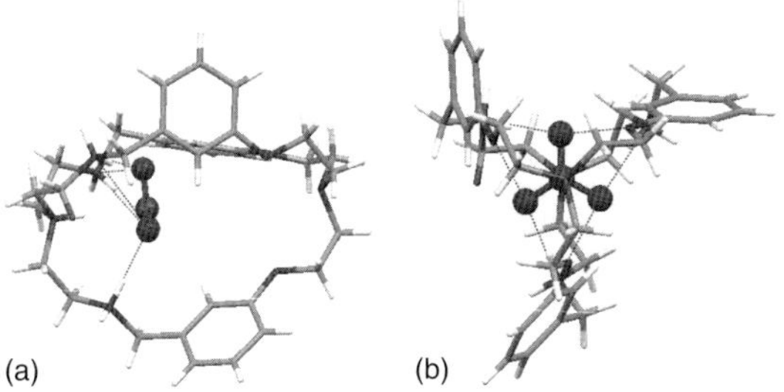

Figure 3.45 Crystal structure of the nitrate complex of **50** showing (a) the side view and (b) the view down the bridgehead axis.

configuration. The phenyl group folds over to accommodate possibly either the chloride ion resulting in a $C_{ph}\cdots Cl$ of 3.793 Å, or the $[Et_4N]^+$ counterion, which sits just above the face of the macrocycle (Figure 3.46b). Considering only H-bonds to the chloride from the pyrrole nitrogen atoms, the range is from 3.239 to 3.318 Å, and the chloride is held in a square pyramidal geometry. However, the authors point out that the two methylene CH groups connected to the bridgehead carbon atom are also within H-bonding range, 3.546 and 3.541 Å. Counting all the close interactions, this complex could potentially be considered as six coordinate or, as the authors conclude, seven coordinate including the *m*-xylyl proton as well. Only the four-coordination is shown here, but readers should be aware that as H-bonding understanding is advancing, more credence is given to $CH\cdots A^-$ interactions.

A tricyclic host, which technically has multiple binding sites, binds linear triatomic anions quite effectively. This is seen for both the azide and bifluoride complexes of **52** (Figure 3.47) [76, 77]. While the host could use a total of 14 H-binding sites for coordination to a variety of guests, in these two cases just four are involved, and the two structures are essentially superimposable. In the bifluoride structure, the $N\cdots F$ distances are shorter than in the earlier bifluoride structure (Section 3.6.1, Figure 3.42, **47**) [50], averaging 2.74 Å compared to 2.83 in **47**. However, the result is a lengthening of the bifluoride F–F distance to a long (for bifluoride) 2.475 Å. This was the first example of bifluoride encapsulated in an organic (as opposed to transition metal) host. In the azide structure, the $N_{amide}\cdots N_{azide}$ distances average 2.90 Å and the terminal $N_{azide}-N_{azide}$ distance is

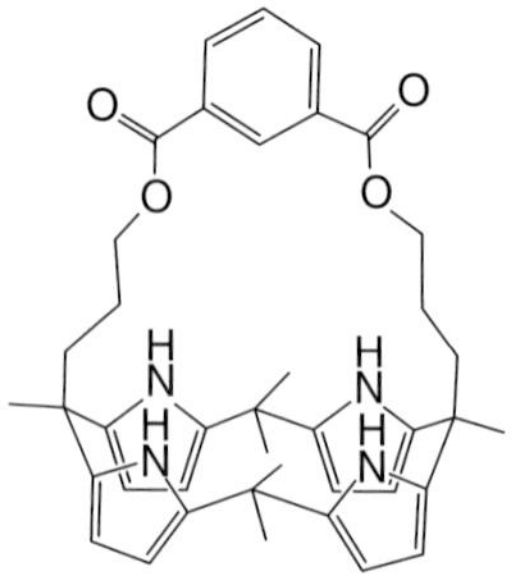

51

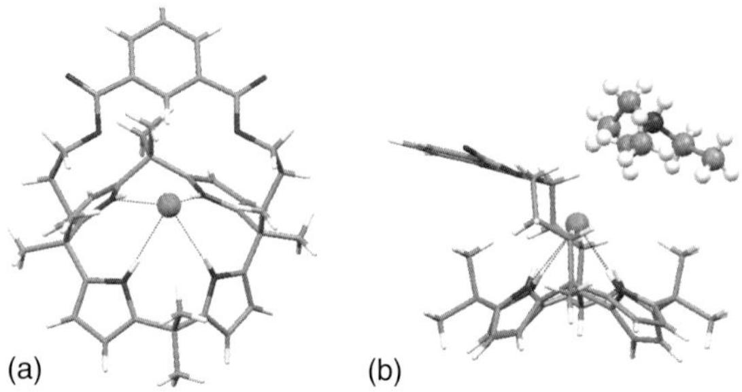

Figure 3.46 Views of the crystal structure of the chloride complex with the calix[4]pyrrole **51** showing (a) the fourfold H-bonding network and (b) the side view of the tilted phenyl group. The counterion $[Et_4N]^+$ is sitting above the face of the macrocycle.

a very short 2.355 Å, which may have been an artifact of the twinned crystal. An extensive H-bonding network that runs from the amine lone pairs through the amide hydrogen atoms and across to the pyridine nitrogen lone pairs (not shown) is what holds this ligand in the same conformation in both the azide and bifluoride structures. Essentially the same network is also present in the free base, which is also virtually superimposable with only minor shifting of two of the amide C=O groups.

3.6.4
Pentadentate

In the strapped calix[4]pyrrole (**53**), a pentadentate coordination is anticipated [78]. This study was done to explore H-bonding effects in calixpyrroles with benzene, pyrrole, and furan. The benzene analog was shown in Section 3.6.3, **51**, Figure 3.46. The interactions with the calixpyrrole NH groups average 3.261 to 3.313 Å, while the distance to the apical pyrrole is 3.093 Å (Figure 3.48). Here too, there are close interactions with the methylene CH_2 groups at 3.689 and 3.673 Å, so this could also be classified as a seven-coordinate complex if the CH_2 interactions are

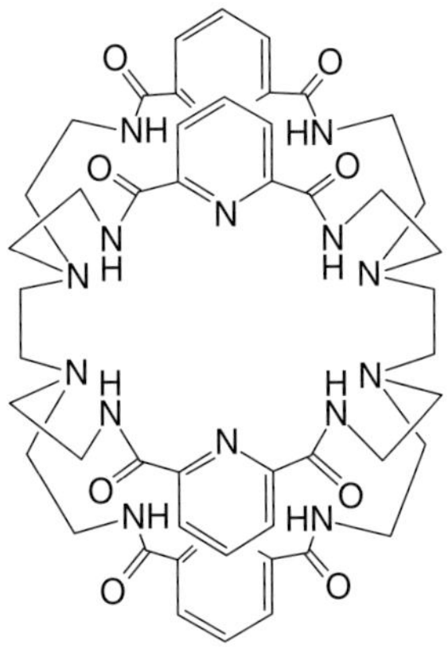

52

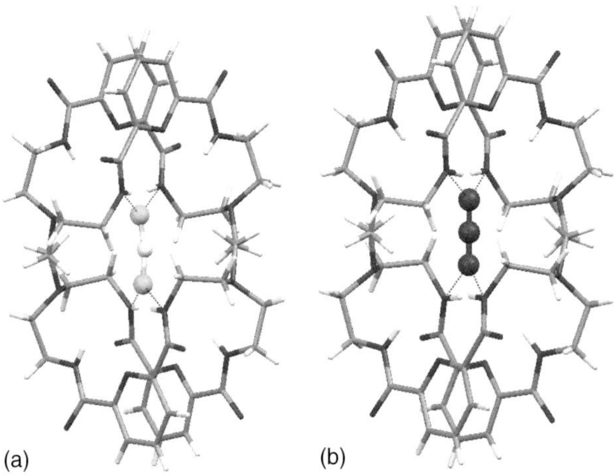

Figure 3.47 Crystal structures of the ligand **52** with (a) bifluoride and (b) azide.

included. The authors also performed electronic structure calculations on chloride interactions with pyrrole NH, benzene CH, and methane CH. The resulting optimized distances were N···Cl = 3.070 and 3.104 Å for pyrrole, C···Cl = 3.491 and 3.497 Å for benzene, and C···Cl = 3.752 and 3.754 Å for methane, the former distances from MP2/aug-cc-pVDZ and the latter from B3LYP/DZVP2 calculations. The authors concluded that in the two crystal structures **51** and **53**, "the CH···Cl contacts were attractive in nature and make a significant contribution to the overall binding." In this case, the $[n\mathrm{Bu_4N}]^+$ counterions do not impinge on the complex.

The imidazolium cryptand **54** provides an opportunity for pentadentate coordination via its three imidazolium groups and two protonated bridgehead amines [79]. The ligand was found to form an unusually strong complex with fluoride (log $K = 12.5$ from ^{1}H NMR competition experiments in water). The +5-charged

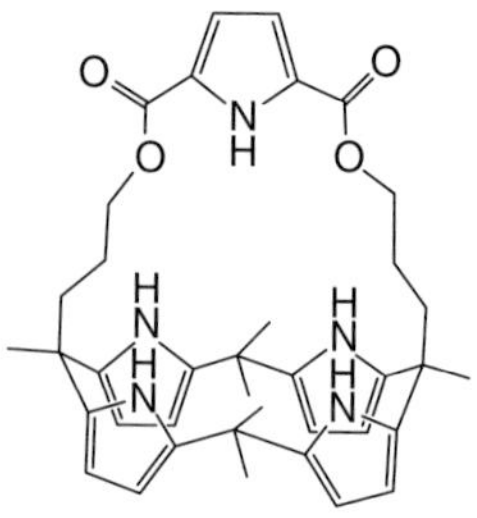

53

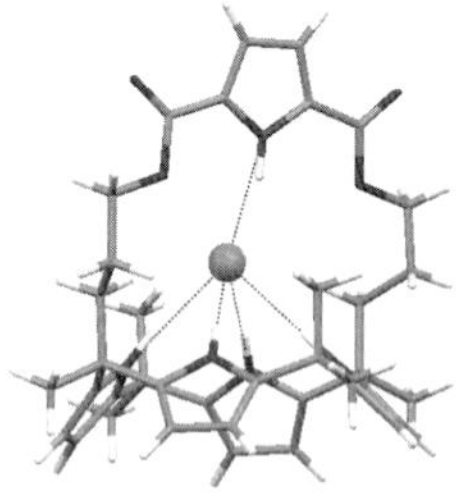

Figure 3.48 Crystal structure of the five-coordinate chloride complex with **53**. The $[nBu_4N]^+$ counterion is not shown for clarity.

ligand forms an inclusion complex with fluoride, with N···F distances from 2.590 to 2.608 Å and C···F distances from 2.582 to 2.607 Å. The $C-H_{im}-F$ angles are all less than 90°: 80.83, 81.06, and 81.34°. The bridgehead N–N distance is very short, 5.198 Å (Figure 3.49).

3.6.5 Hexadentate

Probably one of the most famous of the hexadentate halide complexes is that of the tiny octaaza cryptand, **55**. Because of its small size, only fluoride ion was expected to bind inside the cavity [80, 81]. An extremely high log K value of 10.55 in aqueous solutions was initially reported. However, a later study showed that chloride ion could also be encapsulated [82, 83] and that binding to chloride commenced below pH 2.5. As seen earlier, an even smaller cavity was capable of encapsulating fluoride, the five-coordinate imidazolium ligand, **54** (Figure 3.49) [79]. The N···F distances in **55** range from 2.762 to 2.875 Å and the N···Cl distances range from 2.988 to 3.179 Å, with bridgehead distances of 6.644 for the fluoride and 6.597 for the chloride complexes (Figure 3.50). Both these distances are considerably longer than the 5.198 Å for the imidazolium complex **54**.

An asymmetric cryptand with one phenyl and one tren-based bridgehead, **56**, also binds halides effectively, encapsulating even iodide ion [84] (Figure 3.51). In addition to an internally bound iodide, the charge on the complex is balanced by four triiodide ions and an additional iodide that surround the cryptand. Three of the triiodides are bound within the clefts of the cryptand arms (Figure 3.51b). Only three N···I H-bonds are observed within the cryptand cavity on the tren side of the ligand, at distances ranging from 3.429 to 3.473 Å. The next closest interactions are with the CH_2 groups between the secondary amines, ranging from

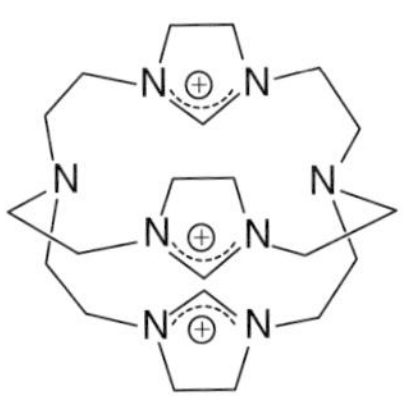

54

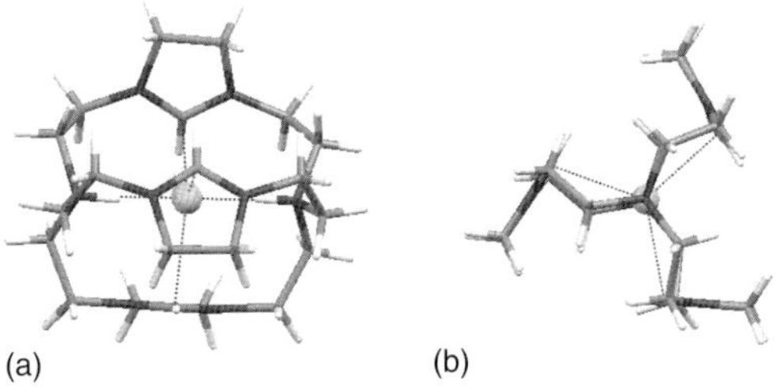

Figure 3.49 Two views of the crystal structure of the fluoride complex with the imidazolium-based ligand **54** showing (a) the side view and (b) the view down the bridgehead axis. The ClO_4^- counterions are not shown for clarity.

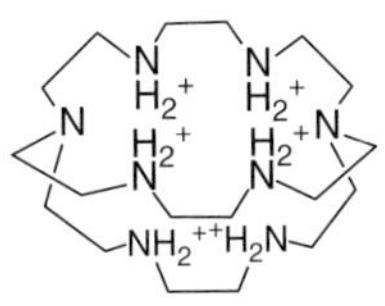

55

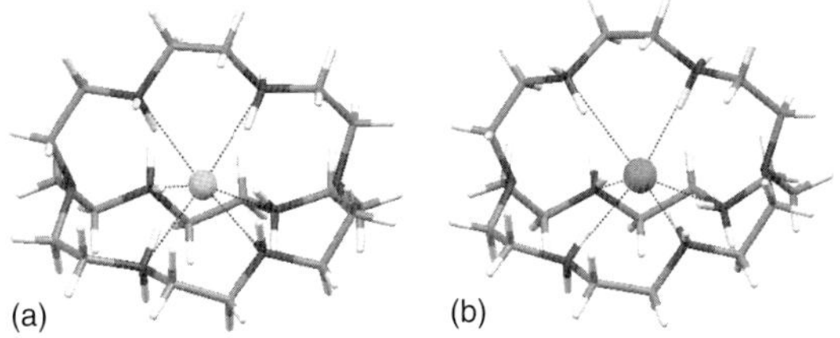

Figure 3.50 Crystal structures of the ligand **55** with (a) fluoride and (b) chloride. The other external anions are not shown for clarity.

around 3.6 to 3.8 Å. The other amines are about 4.2 Å away from the iodide. So in this case, all of the anticipated amine donors are not utilized for binding with the internally held iodide, but rather the CH_2 groups serve in that capacity. The bridgehead $N{-}C_{ph}$ distances fall in a relative narrow range, 7.583–7.617 Å. Several of the amine NH groups are occupied with H-bonding to the external triiodide ions (Figure 3.51b). If the additional external H-bonds are included, this ligand could also be said to be decadentate.

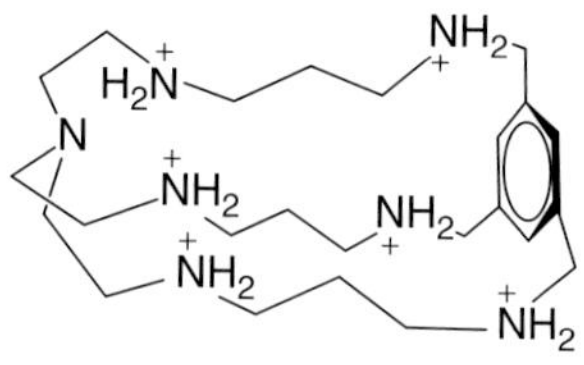

56

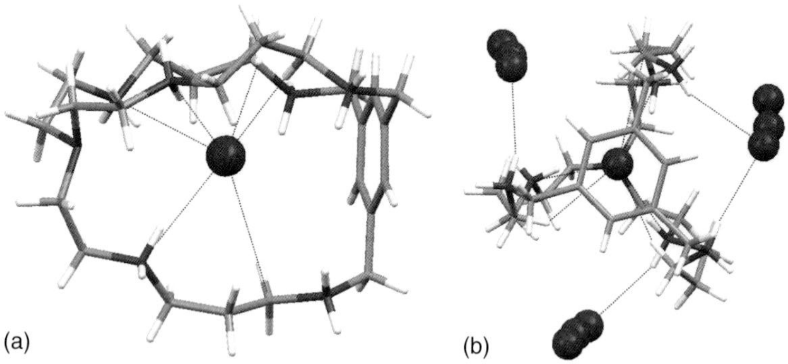

Figure 3.51 (a) Asymmetric aza cryptand **56** showing hexadentate coordination of iodide in the cryptand cavity. (b) View down the bridgehead phenyl ring showing the triiodide contacts with the cryptand.

The hexaprotonated cryptand **57** is an ideal host to bind two nitrate ions within its cavity (Figure 3.52) [85, 86]. Each of the ammonium groups displays bifurcated H-bonds to two oxygen atoms of the nitrates, very similar to the tridentate coordination observed in the unsymmetrical aza cryptand of Bharadwadj, Section 3.6.2, **50** (Figure 3.45) [74]. A similar long/short pattern of H-bonds occurs, with distances ranging from 2.851 to 2.923 Å for the shorter H-bonds and 2.980 to 3.033 Å for the longer H-bonds. The distances from the unprotonated bridgehead amines to the central nitrogen atoms of the respective nitrates are 3.066 and 3.094 Å, and the nitrate N–N distance is 3.339 Å, possibly indicative of some π-interaction. The bridgehead N–N distance is 9.485 Å.

A slightly expanded version of the hexaprotonated polyammonium cryptand **57** with *m*-xylyl spacers is the *p*-xylyl version, **58**. In the fluoride complex, **58** becomes a tritopic host, binding two fluoride ions separated by a water molecule (Figure 3.53a,b) [87, 88]. The distances from the fluoride ions to the NH groups range from 2.604 to 2.724 Å; the distances from the bridgehead amines to the fluorides are 2.994 and 3.002 Å; and the bridgehead N–N distance is 10.717 Å. Thus, the cavity is slightly larger than that of **57** (Figure 3.52). The distances from the two fluoride ions to the water oxygen atom are 2.709 and 2.717 Å, and the F–O–F angle is 121.6°. This structural motif is another link to that seen in transition metal coordination chemistry, where a number of "cascade" complexes have been structurally characterized with aza cryptands in their neutral form.

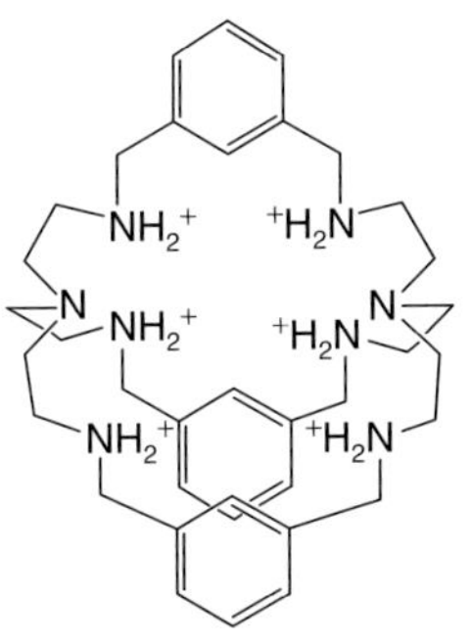

57

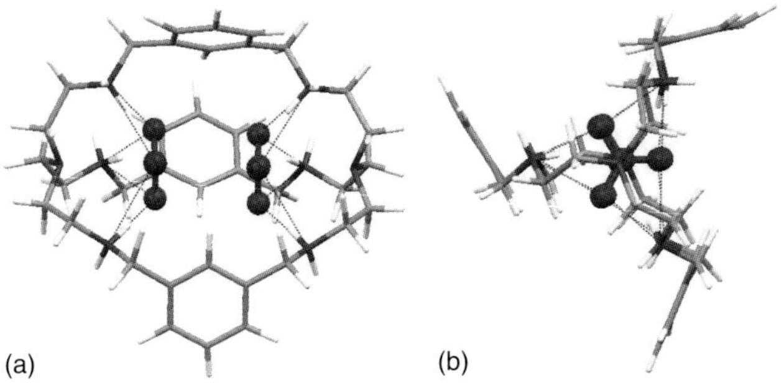

Figure 3.52 Two views of the crystal structure of the nitrate complex with the polyammonium cryptand **57** showing (a) the side view and (b) the view down the bridgehead axis. The external nitrates are not shown for clarity.

In transition metal cascade complexes the ligands tend to bind two metal ions, which are often bridged by a "cascading" anion [89, 90].

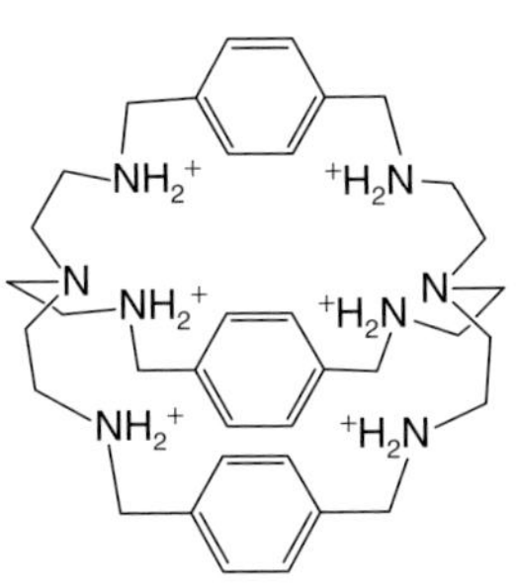

58

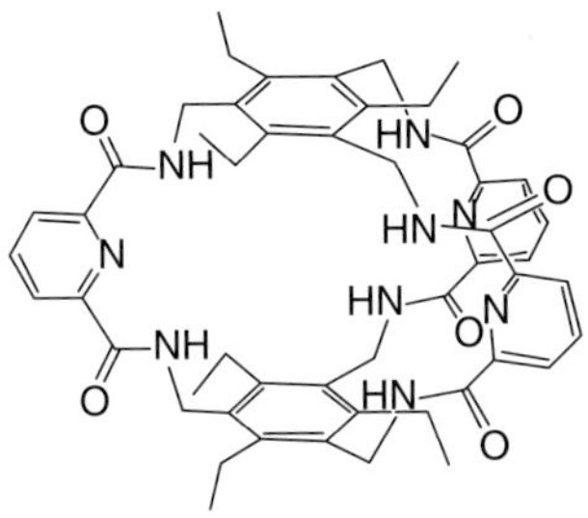

59

While most of the cryptands examined so far have had amine or carbon bridgeheads, phenyl bridgeheads are another possibility that has been explored by the Anslyn group. The strategy was to use 1,3,5-ethyl-substituted cyclophane as the two caps and attach rigid straps at the 2, 4, and 6 positions to form the cylindrical host, **59** (Figure 3.53c,d) [91]. The idea around the positioning of the

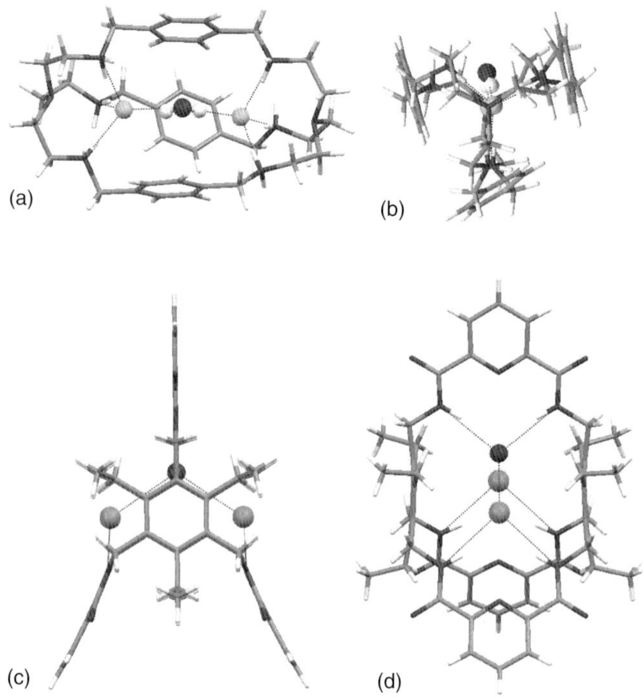

Figure 3.53 Views of the crystal structures of the two cascade complexes, the fluoride complex of **58** and the chloride complex with **59**, showing (a,c) the "cascade" side view and (b,d) a perpendicular view. The external ions are not shown for clarity.

ethyl groups was to force the longer chains all in the same direction (by avoiding steric interactions). The plan worked, and ligand **59** was obtained. Selectivity for acetate and nitrate was observed, and the chloride complex was isolated and crystallographically characterized. Its structure was an additional (and indeed earlier) example of the cascade complex described above (Figure 3.53a,b). Here again, the ligand behaves in a hexadentate manner, but all three of the inhabitants are coordinated directly to the ligand, each to two of the amide groups. The molecule is very symmetric, having crystallized in the orthorhombic space group *Cmcm*; hence, all four N···Cl distances are 3.202 Å, N···O distances are 3.064 Å, and Cl···O distances are 3.239 Å. The Cl–Cl distance is 5.419 Å, and the Cl–O–Cl angle is 113.8°.

3.6.6 Septadentate

Another interesting structure of the cryptand with mixed phenyl and amine bridgeheads, **56**, is the same ligand with the bridgehead amine protonated, **60**.

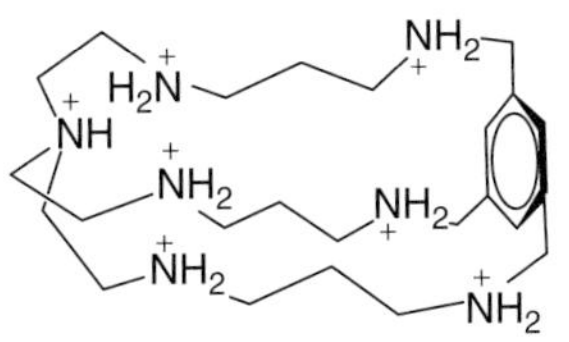

60

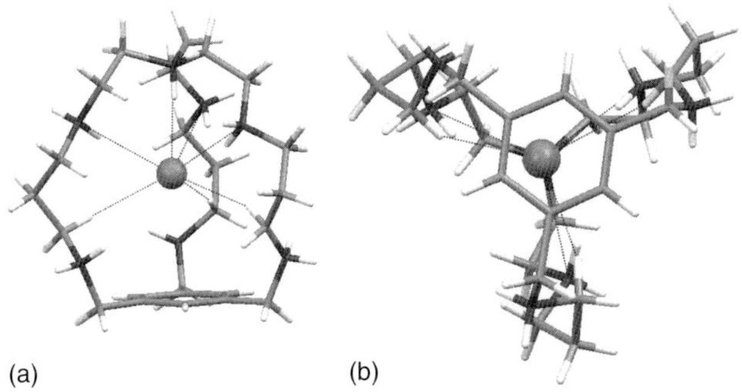

Figure 3.54 Two views of the crystal structure of the bromide complex with the polyammonium cryptand **60** showing (a) the side view and (b) the view down the phenyl group. The external bromides are not shown for clarity.

In this case, a seven-coordinate complex results with bromide ion (Figure 3.54) [84]. In the crystal structure, there are two independent encapsulated bromide complexes. In one, the bromide ion is positioned toward the apical protonated amine and is bound to all three of the secondary amines as well as the CH_2 groups on the other side of the cryptand, similar to that seen with the iodide structure. However, here, due to the additional H-bond, rather than six (Figure 3.51) in the earlier structure, seven-coordination is obtained. Distances from the bromide ion to the secondary amine nitrogen atoms range from 3.30 to 3.34 Å, and the distance to the apical amine nitrogen is 3.19 Å. The Br···C distances range from 3.68 to 3.84 Å. The molecule is disordered around one of the secondary nitrogen atoms at the phenyl side of the ligand. The second independent complex is also disordered but at the amine bridgehead side of the ligand. Since this latter disorder interferes with determining the Br···N interactions, it is not shown. However, this bromide appears to be shifted more toward the phenyl end of the cryptand and is not H-bonded to all of the amine nitrogen atoms (also complicated by the disorder).

3.6.7 Octadentate

The protonated mixed amine/amide cryptand **61**, when utilizing all its NH sites is an octadentate ligand. As such it was found to be a selective receptor for oxoanions, especially sulfate ions (Figure 3.55) [92]. Owing to the protonated bridgehead amines, the ligand provides charge complementarity with sulfate, so the complex

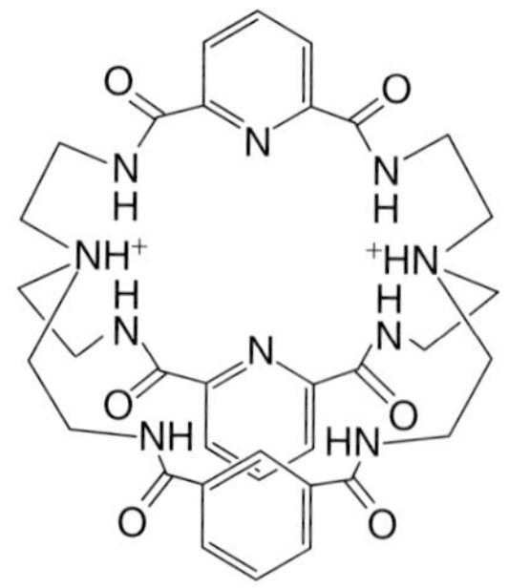

61

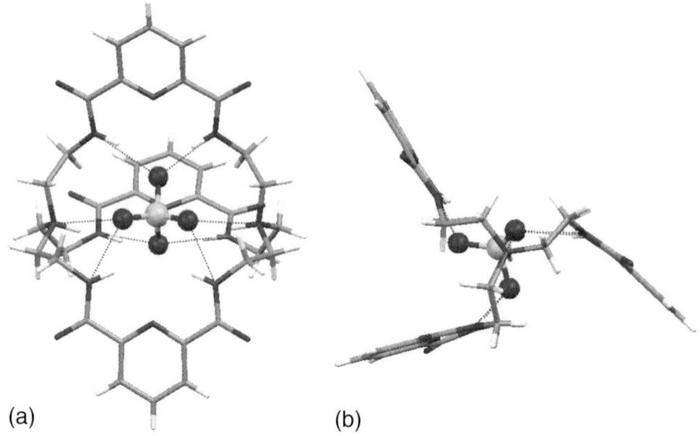

Figure 3.55 Two views of the crystal structure of the sulfate complex with **61** showing (a) the side view and (b) the view down the bridgehead axis.

is neutral. The sulfate binds within the cryptand in a pseudo-bicapped trigonal prism geometry, with the two amine protons forming the two caps. The $N_{amine}\cdots O$ distances are 2.643 and 2.714 Å, and the $N_{amide}\cdots N$ distances range from 2.785 to 3.047 Å. The two apical oxygen atoms of the sulfate are doubly H-bonded to the amine NH and a neighboring amide NH.

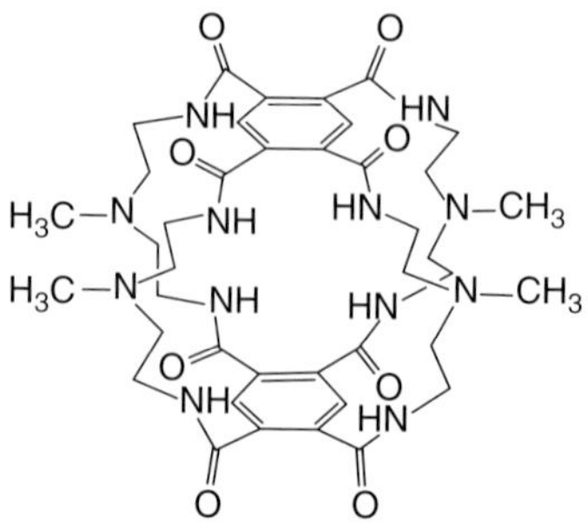

62

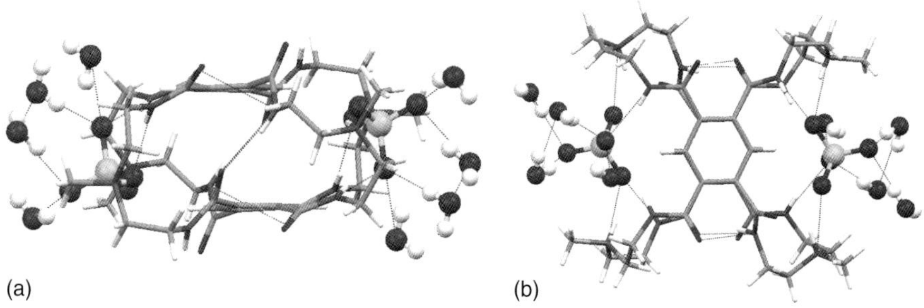

Figure 3.56 (a) Side view and (b) overhead view of the sulfate complex with **62**.

In an attempt to maximize the binding efficiency of the cyclophane-based cryptand (**47**), a tricyclic host with four arms bridging between two phenyl units (**62**) was designed and synthesized [50]. As in **47**, the anion guests on **62** were found to bind within the clefts afforded by the *ortho*-oriented bridging arms. This binding is exemplified for the sulfate complex (Figure 3.56). In this case, two sulfate dianions lie on either side of the ligand. Each sulfate is H-bonded to three of the four available amide hydrogen atoms at distances ranging from 2.758 to 2.882 Å and to the protonated amine as well at an N···O distance of 2.711 Å. On each side of the tricycle, the fourth amide is turned away from the sulfate because of an intramolecular H-bond with an amide oxygen atom on the other side of the host. Additionally, the sulfate is coordinated with four molecules of water that form part of the water channels between neighboring complexes at distances ranging from 2.791 to 2.853 Å. Hence, each sulfate ion is eight coordinate, a common coordination number for sulfate, while the ligand is also octadentate–tetradentate on each side. The host is unstable, losing two of its chains over time to give a simple monocycle (Section 3.5.1, **29**, Figure 3.26).

A slightly larger analog of **61**, **63**, with propylene linkers, also forms a halide cascade complex (Figure 3.57) [93]. The structure is similar to those seen from the Bowman-James and Anslyn groups. (Section 3.6.5, **58** and **59**, and Figure 3.53 for the other halide cascade complexes.) As in Anslyn's complex, all three of the guests form bonds with the ligands. The water molecules, in addition to interactions with the chloride, have longer associations with the upper amides (3.048 Å and a very distant 3.173 Å). The chlorides each form two H-bonds with the amide NH groups ranging from 3.227 to 3.559 Å and with the apical amines at shorter distances of 3.192 and 3.165 Å. The shortened distances to apical NH groups appear to be a general pattern in these ligands (in some ways mirroring the types of distortion in octahedral complexes with axial and equatorial bond distances being different). Here, the Cl···O distances are 3.249 and 3.151 Å, the Cl–Cl distance is 4.949 Å, and the Cl–O–Cl angle is 101.3°. Note again that because of the protonated amines, the ligand provides charge complementarity to the chloride guests.

Another example of a higher order polycycle, **64**, is an expanded version of **52** (Figure 3.47), with longer spacers between the two macrocyclic units. This

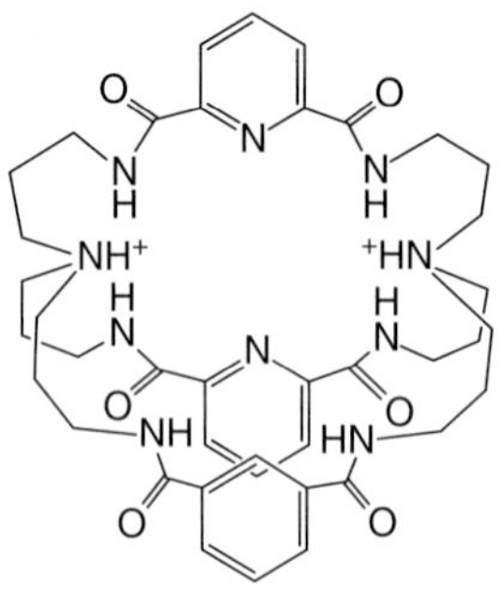

63

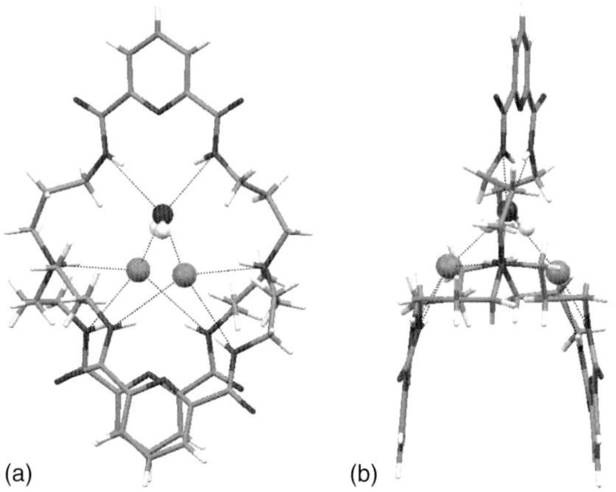

Figure 3.57 Two views of the crystal structure of the chloride cascade complex with **63** showing (a) the side view and (b) the view down the bridgehead axis.

larger ligand binds pseudolinear dicarboxylates internally as evidenced by the crystal structures of terephthalate and succinate complexes. Figure 3.58a,b shows the octadentate-coordinated succinate ion with **64**. For the terephthalate complex, the ligand behaves in a higher order decadentate manner and is described in Section 3.6.8. The approximate dimensions of the cylindrical ligand are 21.7 Å across the long side (measuring from the *p*-pyridine carbon atoms) and 5.49 and 3.69 Å across the short (capping) side, measuring from the tertiary amines and the *p*-pyridine carbon atoms, respectively. (The free base form of **64** is elongated in the opposite direction but is not shown here.) In the succinate complex, each oxygen atom of the two carboxylate ends of the acid are H-bonded to two amide NH groups at distances ranging from 2.807 to 3.086 Å. The distance from the carboxylate oxygen atoms to the symmetry-related chloroform corks is 3.167 Å, making for an approximate square pyramidal symmetry around the carboxylates.

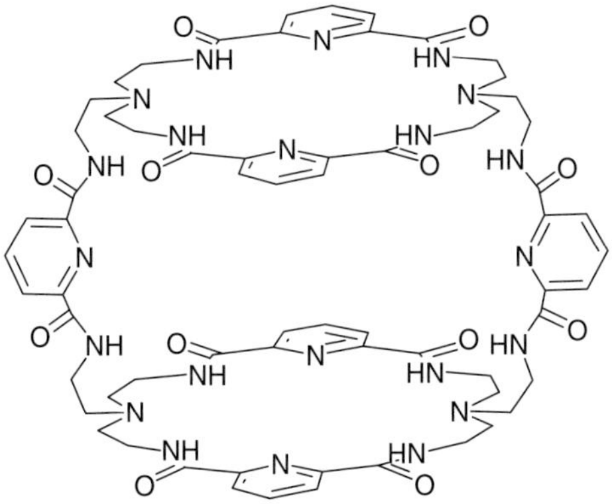

64

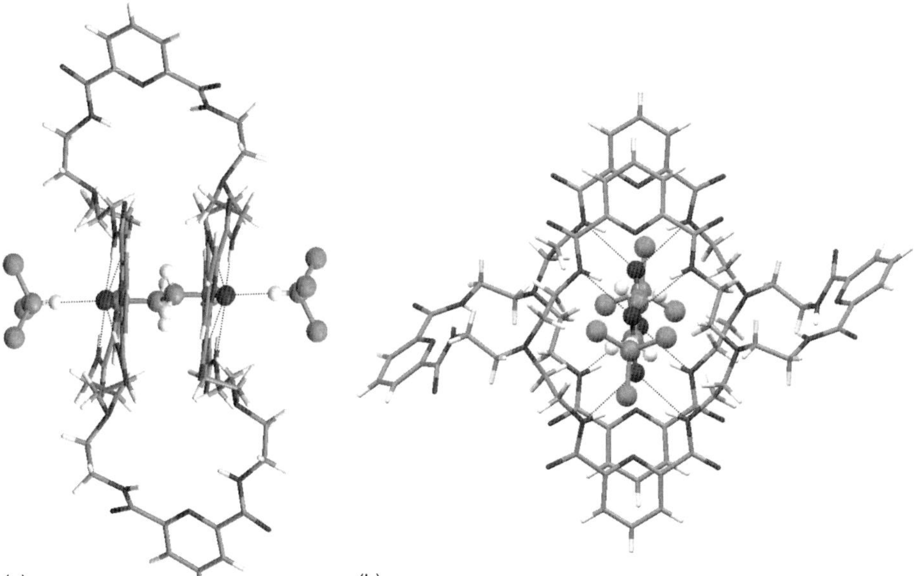

(a) (b)

Figure 3.58 Two views of the crystal structure of the succinate complex with **64** showing (a) the side view across the "cylinder" and (b) the view down the cylinder through the succinate. The $[Me_4N]^+$ counterions are not shown for clarity.

3.6.8 Nonadentate

While the mixed amine/amide cryptands frequently crystallize in the diprotonated form, such is not always the case as with the neutral *m*-xylyl-containing cryptand **65**. In the encapsulated fluoride complex, the ligand is not protonated, rather an

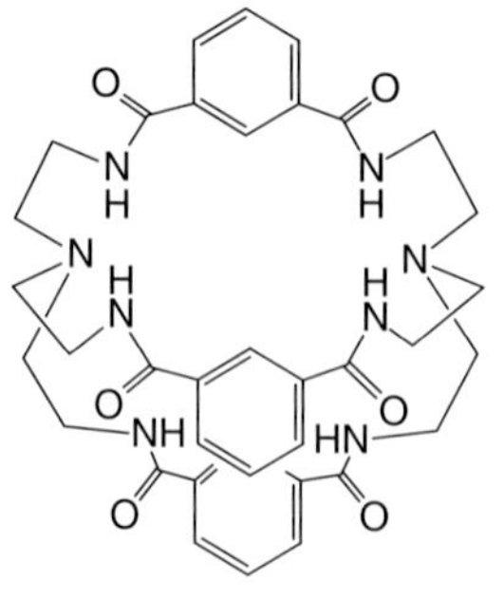

65

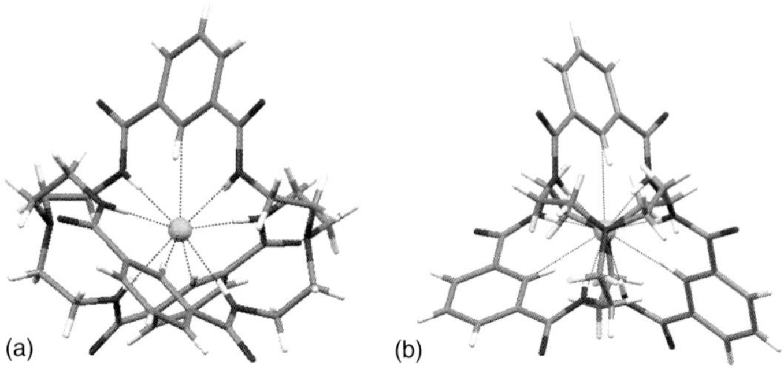

Figure 3.59 Two views of the crystal structure of the fluoride complex with **65** showing (a) the side view and (b) the view down the bridgehead axis.

$[nBu_4N]^+$ cation satisfies the charge requirement. The sequestered fluoride shows H-bonding (albeit not so strong) to the six amide NH groups as well as to the *m*-xylyl hydrogen atoms dipping into the cavity (Figure 3.59) [94, 95]. N···F distances range from 2.946 to 3.113 Å, and the N···C distances range from 3.031 to 3.076 Å. Of particular interest is that the ^{19}F NMR supports the encapsulation of the fluoride ion even in solution (DMSO), displaying a multiplet pattern reflective of coupling of the internally held fluoride ion with all nine protons. A similar situation was observed for the fluoride complex in the pyridine analog; however, in this case, it was a septet pattern indicative of association with the six amide hydrogen atoms. In both cases, deuterium exchange occurs between the amide hydrogen atoms and the solution, resulting in a series of multiplets reflecting the varying degrees of deuteration of the NH groups on the cryptand [94–96].

3.6.9 **Decadentate**

The terephthalate complex of the cylindrical expanded ligand **64** displays decadentate coordination via encapsulation of two additional water "guests" in the side

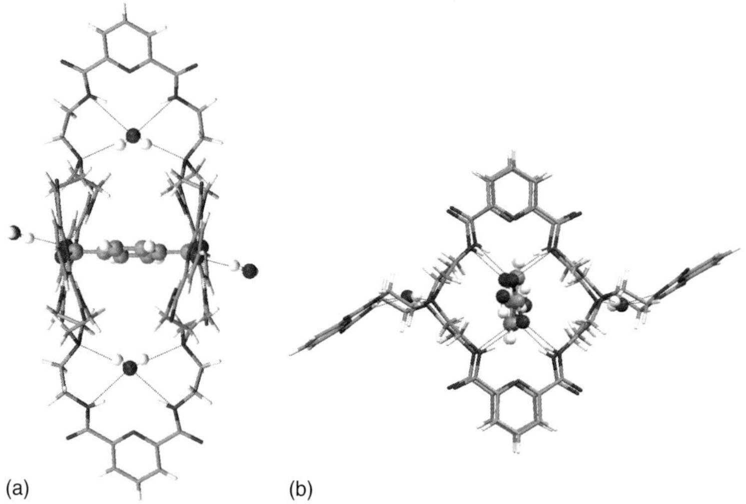

Figure 3.60 Two views of the crystal structure of the terephthalate complex with **64** showing (a) the side view across the "cylinder" and (b) the view down the cylinder through the succinate. The $[Me_4N]^+$ counterions are not shown for clarity.

arms, thus making it a tritopic ligand (Figure 3.60). Here, the dimensions of the ligand are approximately the same: 23.1 Å across the long side (side measuring from the *p*-pyridine carbon atoms) and 5.42 and a longer 7.82 Å across the short (capping) side, measuring from the tertiary amines and the *p*-pyridine carbon atoms, respectively. The latter expanded opening may be what allows for the water molecules to enter the cavity. The amide N···O distances to the carboxylate oxygen atoms range from 2.836 to 2.889 Å, and the amide N···O distances to the encapsulate water molecules are 3.001 and 2.996 Å.

3.6.10
Dodecadentate

There has been considerable interest in outer sphere coordination as exemplified in Tasker's metallate anion complexes (Section 3.4.2, **14** in Figure 3.14), and the following is a case where, rather than it being an outer sphere complex of a transition metallate complex, the Bowman-James group has observed it to be an outsphere complex of a fluoride hydrate complex. In this rather unique host–guest complex, two different spheres of coordination are involved with an expanded tetrahedron-like host, **66** (Figure 3.61) (Wang *et al.* unpublished results). The complex consists of a hydrated fluoride ion surrounded by a tetrahedral array of water molecules, which in turn is encapsulated within an expanded amine/amide-containing **66**. The fluoride forms four H-bonds with surrounding water molecules, ranging from 2.585 to 2.615 Å, in a distorted tetrahedral array, with

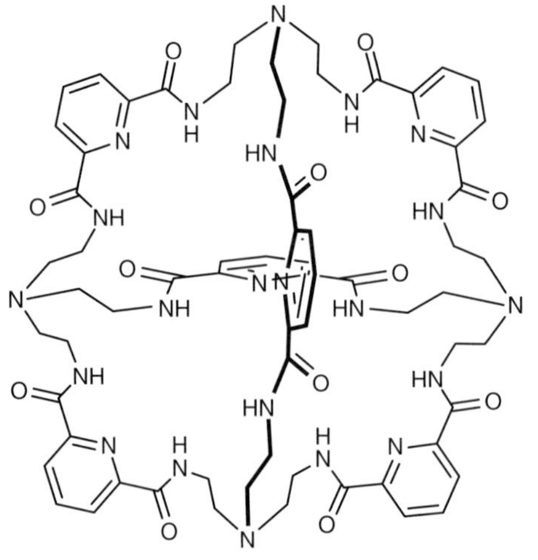

66

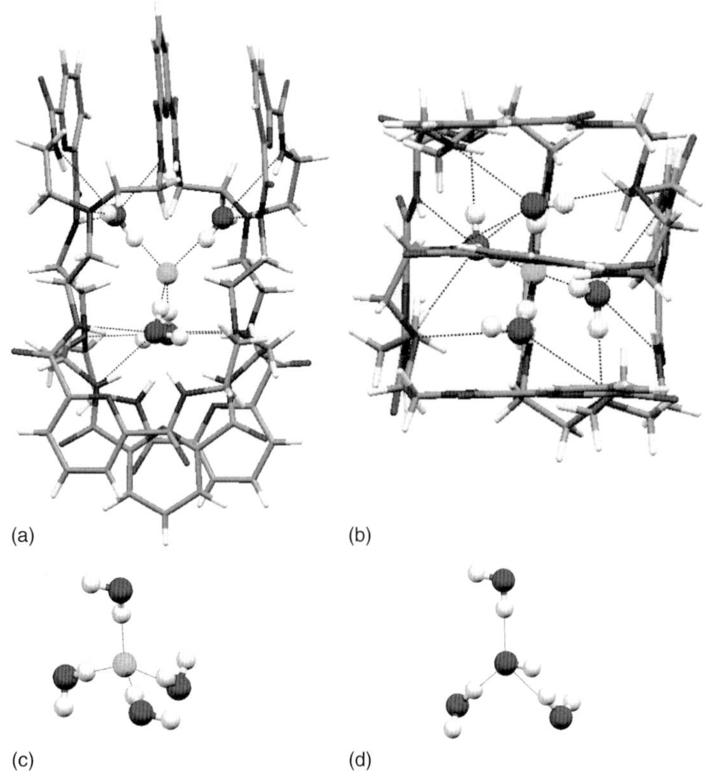

Figure 3.61 Crystal structure views of the tetrahedron water cage complex with fluoride and **66** showing (a) a view of the cavity with the tetrahedral arrangement of water molecules; (b) a view down the pyridine rings showing the almost central placement of the fluoride ion; (c) a view of just the tetrahedral water complex showing the tetrahedral array; and (d) a view down an O–F axis.

two small (83 and 89°), two intermediate (110 and 119°), and two large (130 and 131°) angles. All the water molecules have potential sites for three additional H-bonds that include relatively significant N···H–O interactions with the pyridine nitrogen lone pairs (O···N ranging from 2.928 to 2.968 Å). None of the water molecules are H-bonded to each other, and O–O distances range from 3.440 to 4.741 Å. While this complex could potentially be described as an example of a decadentate, in this case, because of the beautiful symmetry aspect, it will be considered as dodecadentate.

3.7 Transition-Metal-Assisted Ligands

Another facet to anion coordination chemistry involves hosts that contain metal ions. The metal ions can serve different purposes as outlined by Steed in a recent review [97]. Function depends on both the chemical properties of the metal and the resulting role that the metal plays in anion complexation. In three of the classes, the metal ions serve in a structural–organizational mode. These would be (i) an inert transition metal complex that serves as a lynchpin holding the organic ligand in place for binding; (ii) labile transition metal associations in which an anion serves to act as a template for organizing the host; and (iii) a variation of the first two classes, where the metal and organic parts of the host are part of an extended framework. The last classification is also known as the popular and rising field of metal-organic frameworks (MOFs). Two other classes of transition-metal-assisted hosts are (iv) where the metals behave in a Lewis acid capacity, binding directly to the anion and (v) where the metal plays the role of "reporter," that is, is involved in a redox, fluorescent, or other colorimetric or detection capacity. In this section, several key examples of the structure-organizing hosts are described, with the focus again on the general anion coordination concept where applicable. Lewis acid hosts are described in Section 3.8. These include both metal and nonmetal Lewis acids. In some cases, the Lewis acid hosts are probably more appropriately considered as transition metal coordination complexes rather than the organic anion coordination complexes described previously.

3.7.1 Bidentate

An interesting dimetallic host with a dual metal scaffold self-assembles in the presence of nitrate ion (Figure 3.62) [98]. The multidentate ligand, **67**, wraps into a triple helicate that encompasses two cobalt(II) ions. This complex provides an example of a host scaffold with labile metal ions (Co(II) = d^7). The helicate is chiral and crystallizes with both ΔS and ΛS diastereoisomers in the complex. Only two of the three available amide NH donors are fixed in position to bind the nitrate via two H-bonds (N···O = 2.942 and 2.848 Å). The third NH is slightly farther away at 3.169 Å and so is only weakly associated. Since the entire metal complex is

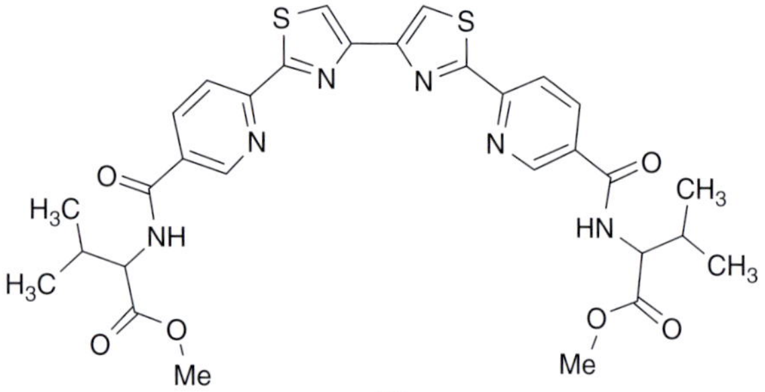

67

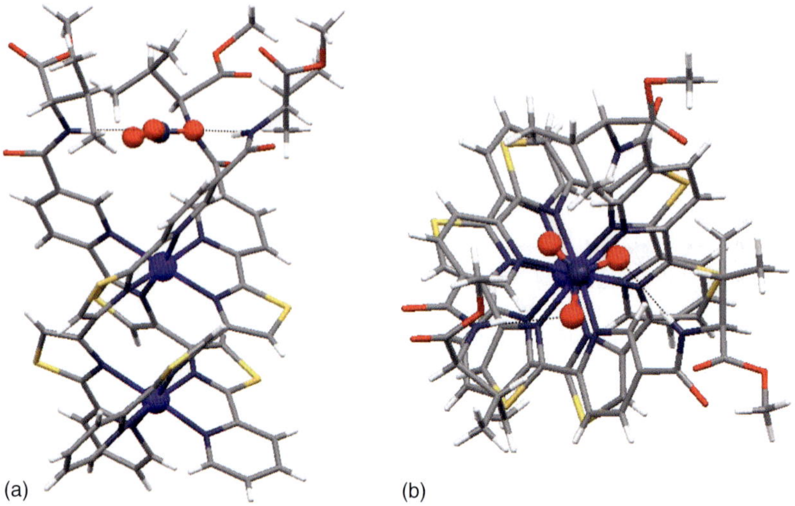

Figure 3.62 Crystal structure of the triple helicate complex $[Co_2(\mathbf{67})_3(NO_3)]^{3+}$ as viewed (a) along the side of the helix and (b) down the helix.

considered in this case to be the ligand, the actual individual ligand coordination is considered to be bidentate.

Loeb, Gale, and Lighta introduced an interesting platinum(II) complex with four simple pyridine/pyrrole-based scaffolding ligands, **68** [99]. This complex provides an example of a transition metal serving as a scaffold involving an inert transition metal ion. The pyridine binds to the platinum in a *cis*-conformation, and the pyrroles bind in a bidentate manner to an axial methyl sulfonate anion (Figure 3.63) [99]. What appears to be a straightforward logical binding of the two pyrrole NH groups with sulfonate oxygen atoms is more complex than at first glance. In solution, there is ambiguity as to what forms the major H-bonding contribution to the complex. In $MeNO_2$, both the pyrrole and α-pyridine CH groups undergo downfield shifts (1.77 and 1.24 ppm, respectively), indicating H-bonding with the anion. However, in DMSO, the pyrrole NH group does not shift, while the pyrrole CH group does – albeit to a much lesser extent (0.17 ppm). The solid-state structure consists

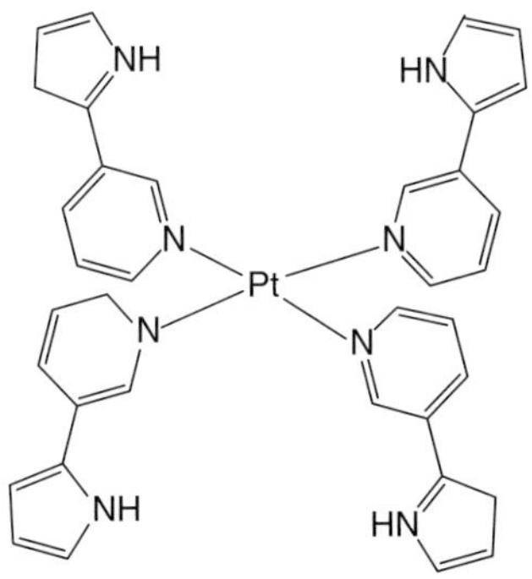

68

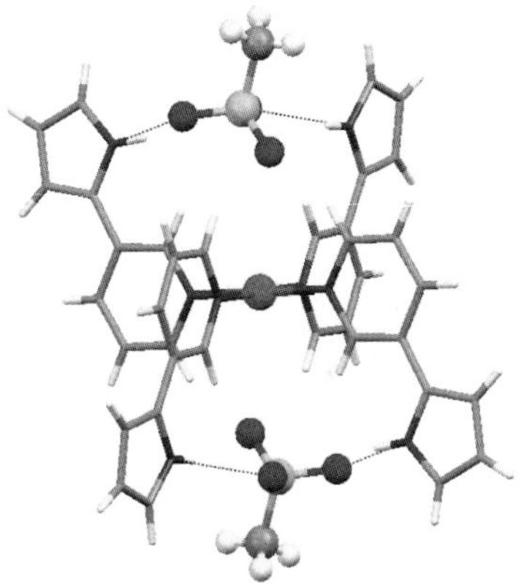

Figure 3.63 Crystal structure of the scaffold complex **68** with methanesulfonate ion.

of two orientations of the sulfonate oxygen atoms, which could be an artifact of the flexible conformation of the pyrrole groups in the scaffold as indicated by the NMR spectra. The pyrrole N···O distances are 2.850 and 2.927 Å. Some of the pyridine α carbon atoms are about 3.4–3.5 Å from the oxygen atoms, and for one (on an adjacent pyridine), this distance is actually 3.100 Å. The Pt–O distance is 3.404 Å.

3.7.2 **Tridentate**

The tris-2-(3-methylindolyl)phosphine host of Browning and coworkers can bind with or without transition metal assistance (**69**) [100]. As for the copper(I) complex, the phosphorus binds to the copper ion, while the tripodal indolyl groups provide a pocket for the BF_4^- counterion (Figure 3.64a). N···F distances range from 2.879 to 2.994 Å. Without the copper ion, **69** also acts as an anion ligand by binding fluoride ion in a 1 : 1 manner. N···F distances range from 2.632 to 2.690 Å (Figure 3.64b). In this form, it functions somewhat like the previously discussed dual-host ligands and could have just as appropriately been discussed in the section on acycles but is included here for comparative purposes with the transition metal analog.

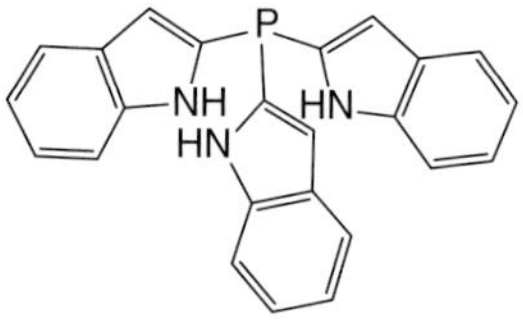

69

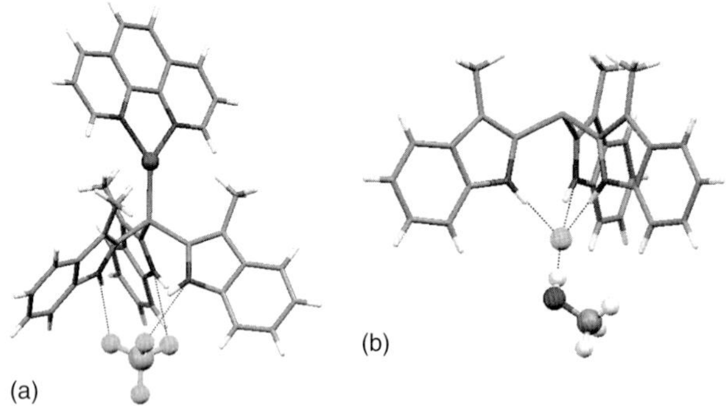

Figure 3.64 Two versions of anion hosting for **69**: (a) the $CuBF_4$ complex and (b) the fluoride complex.

3.7.3
Tetradentate

There have been several reports of the use of tris(bipyridyl) ligands to serve as link to transition-metal-templated scaffolds that are ultimately oriented for anion binding sites. Fabbrizzi reported an interesting azide complex using a metal tris(bipyridyl) scaffold combined with a triethylbenzene cap, to yield "pinwheel"-based organic receptors, **70** (Figure 3.65a) [101]. The low-spin Fe(II) complex was formed on reaction of the imidazolium- and bipyridine-containing ligand system. The use of $Fe^{II}(CF_3SO_3)_2$ is an example of an inert (low-spin d^6) transition metal acting as template. The charge on the complex without anions is +5. Azide ion is bound in the cavity but is disordered, with one of the disordered ions (not shown) sticking partly outside the cavity. The part of the disordered azide ion that is ensconced in the cavity is held by three relatively loose H-bonds to the three imidazolium CH groups (3.178, 3.225, and 3.234 Å) with an additional shorter interaction with one of the bipyridine CH groups (3.029 Å). The distance from the azide to the iron ion is greater than 4.5 Å. The other counterions in the unit cell are PF_6^- ions.

3.7.4
Hexadentate

The iron(II) complex of the tris-bipyridine ligand shown above (**70**) also binds bromide ion (Figure 3.65b) [101]. In this case, the coordination geometry is hexacoordinate, utilizing the α-carbon atoms of the bipyridine in addition to the

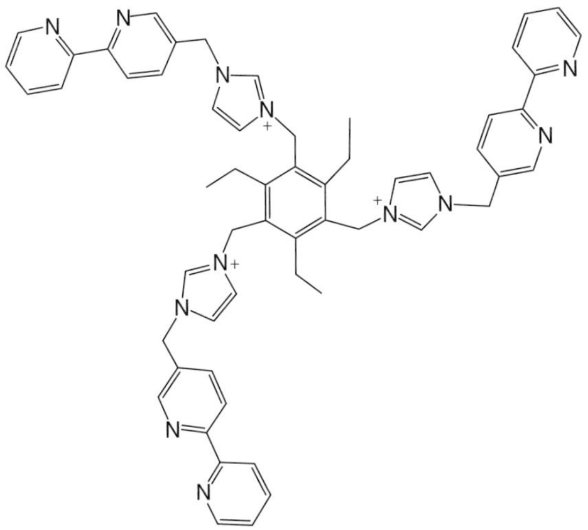

70

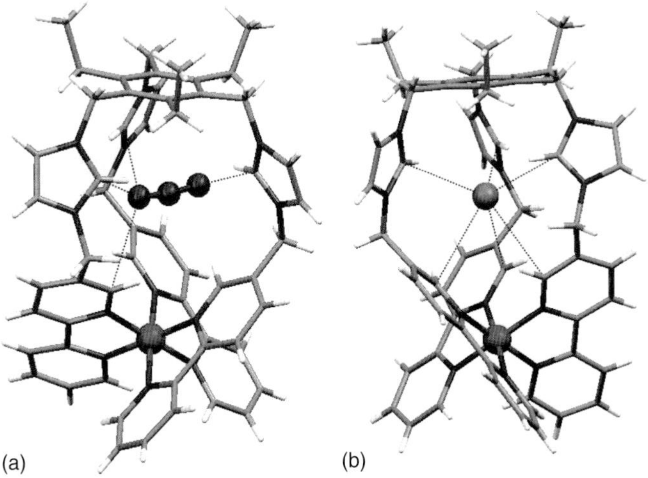

Figure 3.65 Views of (a) the tetracoordinate azide complex and (b) the six-coordinate bromide complex with **70**. The four PF_6^- counterions are not shown for clarity.

imidazolium CH groups. The complex has a fairly high affinity for bromide ion, $\log K = 3.26$ in CD_3CN/D_2O, although the preference is for chloride over bromide. The imidazolium C···Br distances range from 3.344 to 3.400 Å, while the α-carbon atoms from the bipyridines are also fairly close, C···Br ranging from 3.347 to 3.497 Å. The bromide ion is 4.759 Å away from the iron, and the geometry around it is close to trigonal pyramidal.

An added benefit of the use of scaffolds provided by transition metal assistance is the expanded degree of potential H-bonding that can be achieved [102], for example,

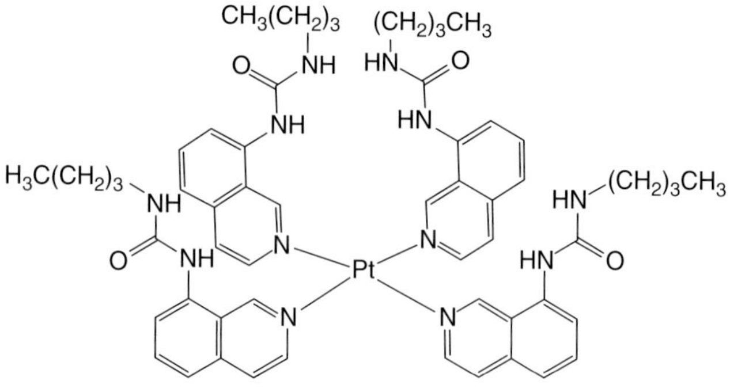

71

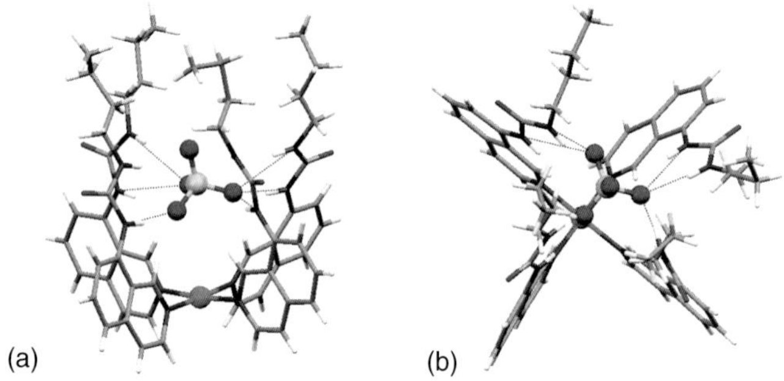

Figure 3.66 (a) Side view and (b) overhead view of the structure of the platinum-based scaffold **71** with sulfate.

by using a bi-H donor within four walls to provide eight donor groups. This is the case in another example of a platinum(II) inert transition metal-based scaffold. Instead of up–up–down–down *cis* orientations (of slightly different groups) (as seen in Section 3.7.1, **68**, Figure 3.63), all four of the walls orient in the same direction to provide maximized H-bonding interactions with the sulfate guest, **71** (Figure 3.66). A rigid foundation is provided by 8-(*n*-butylurea)isoquinoline groups. As seen in Figure 3.61, the sulfate sits in the cavity with an apical oxygen atom pointing upward. The H-bonds from the urea hydrogen atoms to the lower oxygen atoms of the sulfate range from 2.890 to 3.061 Å. The upper oxygen atom has three very weak associations with the upper urea NH groups, with distances of 3.136, 3.258, and 3.301 Å. As can be seen from the overhead view (Figure 3.66), one of the oxygen atoms is poised above the platinum. The Pt–O distance is 3.719 Å.

3.7.5 Septadentate

A 3 : 2 host:anion complex **72** was isolated by Loeb, Gale, and coworkers, in which the scaffolds coordinate to platinum ions in a trans configuration [103], as opposed to the *cis* coordination seen earlier (Section 3.7.1; Figure 3.63, **68**; and Figure 3.66,

71). The entire ensemble is a bit more complex than a single platinum complex, and a trimeric array of platinum complexes serve to fill out the coordination environment of the sulfate ions [103]. The trimer preferentially binds (or enfolds) sulfate rather than the other counterion in the cell, BF_4^-, and two sulfate ions are bound on opposite sides of the central platinum ion (Figure 3.67). The BF_4^- ions satisfy the remaining charge, with one above and one below the dicationic complex (not shown). Each of the sulfate ions is surrounded by a sevenfold H-bonding array of urea NH groups, ranging from 2.771 to 3.108 Å. The longer H-bond is bifurcated from the urea NH to two different sulfate oxygen atoms at 3.102 and 3.108 Å. There are two short C· · ·N contacts slightly over 3 Å, but these are from CH groups that are forced into proximity by the rigid scaffold, so they will not, in this case, be counted. Seven-coordinate geometries for sulfate ion are rather rare, and this finding makes the structure somewhat reminiscent of the sulfate binding protein, which also exhibits seven H-bonding interactions with its sulfate guest [104]. The authors note, however, that if one includes all of the aromatic interactions from the α-carbon atoms in the pyridine and isoquinoline of 72, there are 14 interactions around each of the sulfate ions.

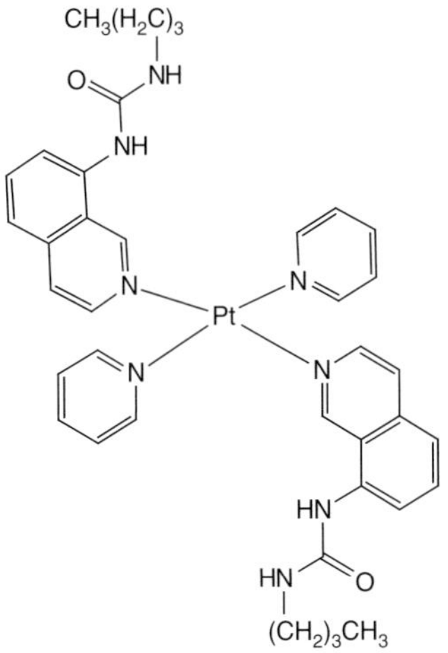

72

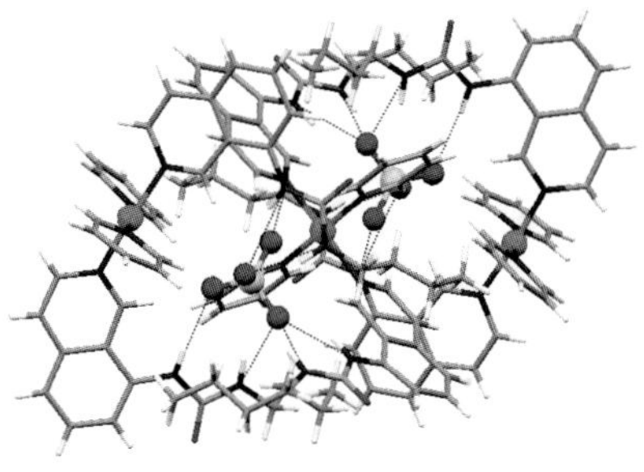

Figure 3.67 Structure of the 3 : 2 host:guest complex $[\mathbf{72}_3(SO_4)_2]^{2+}$.

3.7.6
Dodecadentate

Anion chemistry has also now expanded into the MOF world. Custelcean *et al.* utilized silver ion as a template to self-assemble two tren-based urea hosts, **73**, resulting in an MOF with an encapsulated sulfate [105]. The key in the design strategy was to functionalize the urea groups with phenyl-group-containing cyano substituents in para positions in relation to the ureas. The purpose of the cyano groups was to bind silver(I) ions. The result is a cryptand-like unit that extends throughout the crystal lattice, $[(Ag_2\mathbf{73}_2)\cdot(SO_4)]_n$ (Figure 3.68). This structure was a prototype illustrating Hay's prediction that an optimal coordination number for sulfate ion should be 12, with three preferred H-bonding sites on each sulfate

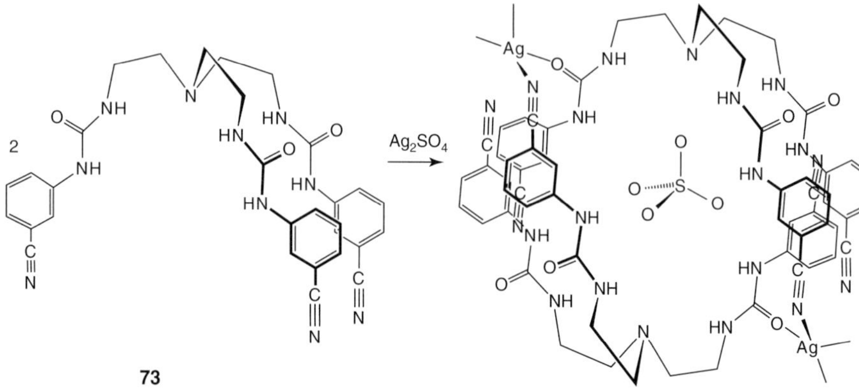

73

$[\mathbf{73}_2\cdot Ag_2(SO_4)]$

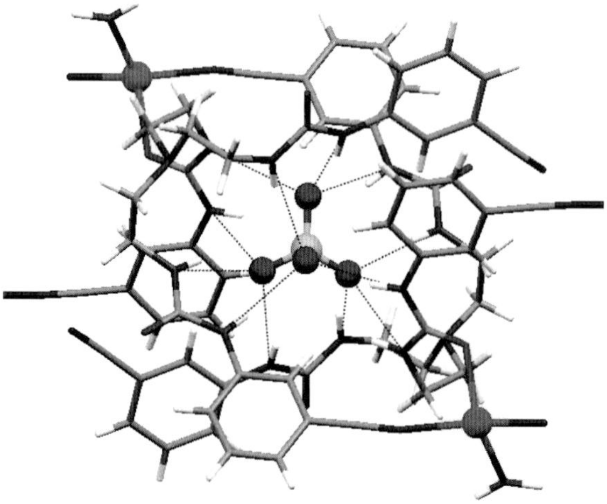

Figure 3.68 Partial view of one "unit" of the extended 12-coordinate silver(I) MOF $[\mathbf{73}_2\cdot Ag_2(SO_4)]$ complexed with sulfate ion.

oxygen atom [63]. The N···O distances range from 2.852 to 3.174 Å. Other silver salts did not produce the MOFs, including the tetrahedral BF_4^-, possibly indicating influences such as size, H-bonding tendencies, and perhaps charge in forming the MOF.

Another interesting case of a metal ion template host, in this case not an extended MOF, is again urea-based, with bipyridine groups as the metal-binding units (**74**) [106]. The inclusion complex was templated by $NiSO_4$ to form a tetrahedral binding cage with six ligands (Figure 3.69). In this case, the N···O distances are even shorter than in the 12-coordinate **73** described above (Figure 3.68) and range from 2.904 to 2.942 Å.

It would appear that dodecadentate coordination is becoming more common. Another example is the elegant Borromean ring of Steed and coworkers, which represents another example of an MOF. A Borromean ring is an interwoven group of three rings linked together without actually interpenetrating each other. A characteristic of such a ring is that if any of the rings are cut, the entire association falls apart. In this case, the rings consist of a mixture of pyridine and urea functional groups separated by an ethylene chain, **75**, and the Borromean framework consists of six of the fragment ligands, **75**, held together by silver(I) ions (Figure 3.70) [107]. The nitrate ion is surrounded by portions of macrocycles from three different rings. The H-bond distances from the ureas to the nitrates range from N···O = 2.944 to

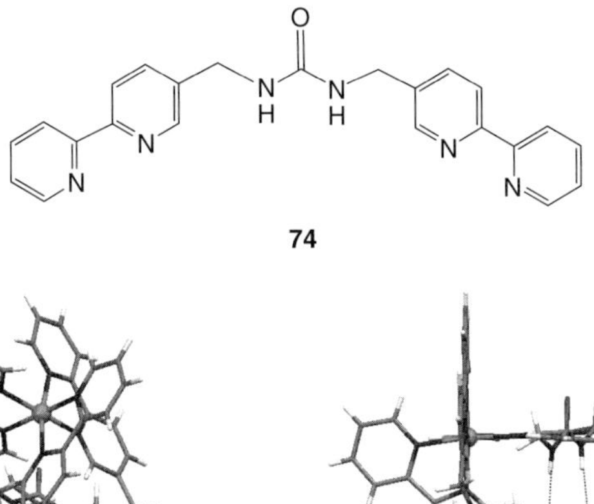

74

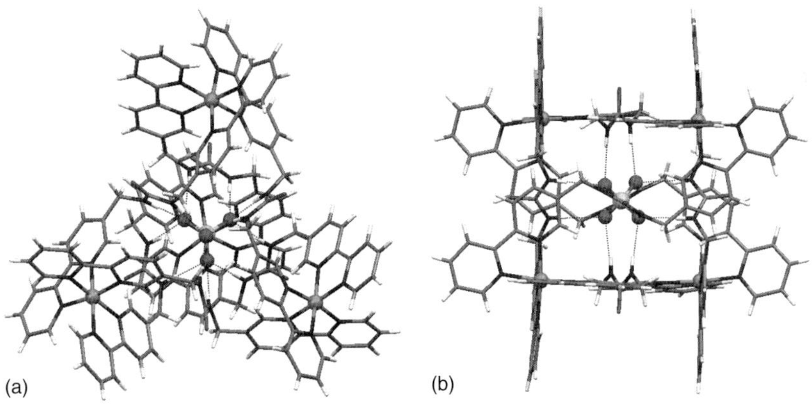

Figure 3.69 Two views of the Ni-templated complex [$\mathbf{74}_6 \cdot Ni_3(SO_4)$]: (a) down the sulfate C_3 axis and (b) down the sulfate S_4 axis. The three other sulfate ions (outside the cage) are omitted for clarity.

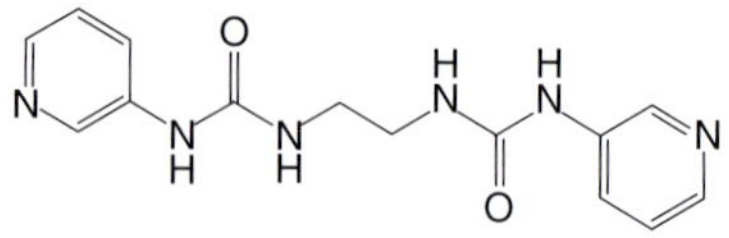

75

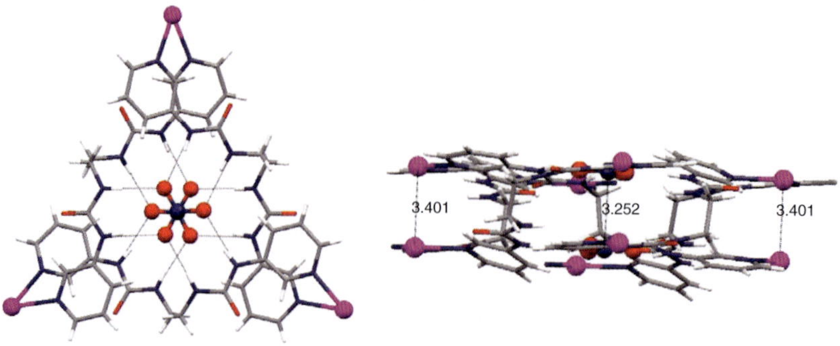

(a) (b)

Figure 3.70 Borromean weave consisting of portions of three of the large macrocycles held together by silver(I) ions, and surrounding two staggered nitrate ions: (a) overhead view showing the staggered nitrate ions and (b) side view showing the interlayer separation. Only an individual unit is shown for clarity.

2.960 Å. The end result is a 12-coordinate array of H-bonds for the two nitrates, with each nitrate oxygen atom held by two H-bonds. The distance between the nitrate nitrogen atoms is 3.252 Å, and the Ag(I)–Ag(I) separation between layers is 3.401 Å.

3.8
Lewis Acid Ligands

Much of the field involving the supramolecular chemistry of anions has resulted in the design and fabrication of organic frameworks that range from simple to elegant and that function primarily as H-bonding donors to snare elusive anions. However, other Lewis acid hosts have also been employed, including some from early on in anion coordination, before it became a field of its own, such as transition metal cascades. In this section, the structural aspects of a few select models from other Lewis acid hosts are described.

3.8.1
Transition Metal Cascade Complexes

Transition metal cascade complexes are a totally different class of metal-assisted anion complexes compared to those described in Section 3.7. Here, the metal interacts directly with the anion, rather than residing in the ligand host or assisting ligand formation. The name for this class of complexes was coined many years ago [108], with the observation that polyaza macrocycles and cryptands of a certain size could incorporate two metal ions and that anions often bridged (cascaded) between the two metals. The anions included halides, pseudohalides and oxoanions, among others. Copper(II) is one of the key ions that readily participates in this type of complex; however, examples with other metal ions such as nickel(II), cobalt(II), and others can be found [109, 110]. Below are presented a few classic examples of cascade complexes.

A cascade complex with a simple monocycle (**76**) was reported in the mid-1990s by Martell and coworkers [111]. The dicopper(II) complex readily incorporates a bridging oxalate to complete the copper coordination sphere (Figure 3.71). The copper(II) ions are five coordinate and coordinate to the three ethylenediamine nitrogen atoms and two of the oxalate oxygen atoms. Each of the copper ions is five coordinate, with Cu–N distances ranging from 1.94 to 2.09 Å. Both copper ions exhibit one short and one long bond to the oxalate, 1.82 and 2.15 Å for one and 1.95 and 2.21 Å for the other. The Cu–Cu distance is 5.34 Å, which, according to the authors, is the same distance as in the complex without oxalate cascade [111]. The coordination around the copper ion is quite distorted and intermediate between a square pyramid and trigonal bipyramid.

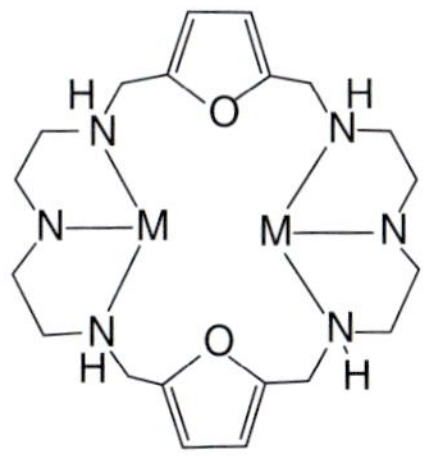

76

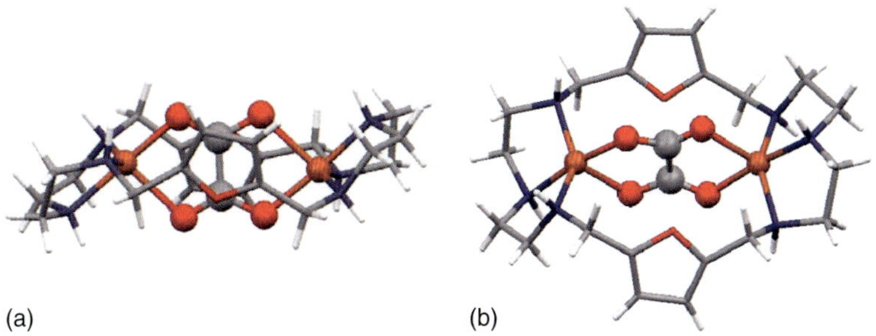

(a) (b)

Figure 3.71 Two views of the copper(II) complex **76** with a cascading oxalate ion: (a) view perpendicular to the oxalate plane and through the macrocycle and (b) view perpendicular to the macrocyclic "plane." The other two BF_4^- counterions are not shown for clarity. Chem. 2005, **44**, 2143–2149.

In addition to macrocycles, cryptands also form cascade complexes when dimetallated. Two examples are shown (Figure 3.72) to illustrate the versatility of these ligands. In the bromide complex (Figure 3.72a), Fabbrizzi and coworkers observed an unusual affinity of the furan-containing cryptand (**77A**) toward halide ions, which can be evidenced by the appearance of a bright yellow color. The bromide complex is illustrative of this, with the bromide forming an almost linear conduit between the two metal ions (Cu–Br–Cu = 179.4°) [112]. The Cu–Cu distance is 4.866 Å, and the copper is coordinated in a trigonal bipyramidal manner, with Cu–N distances ranging from 2.046 to 2.132 Å. The Cu–Br distances are nearly identical at 2.432 and 2.435 Å. The bridgehead distance is 8.96 Å.

In the azide complex with the thiophene analog **77B** (Figure 3.72b), the anion also forms a linear bridge between the two copper ions [113]. These cryptands (and macrocycles) are quite flexible in terms of stretching and shrinking to hold their guests, since here the Cu–Cu distance has stretched to 6.150 Å to incorporate the linear (constrained by crystallographic symmetry) azide. The bridgehead distance is 10.19 Å. The copper ion is once again in a trigonal bipyramidal coordination mode, with Cu–N distances in the trigonal plane of 2.16 Å and axial distances of 2.02 Å. The Cu–N_{azide} distance is 1.94 Å.

Many of the cascade complexes have, in the past, involved polyaza macrocycles and cryptands. However, **78** is an example of a metallocycle, with silver(I) ion both serving as a template to link the two organic halves into a cyclic host as well as binding a cascading nitrate ion [114]. The macrocycle self-assembles around the two silver ions, forming a cavity that can bind a variety of anions, including nitrate,

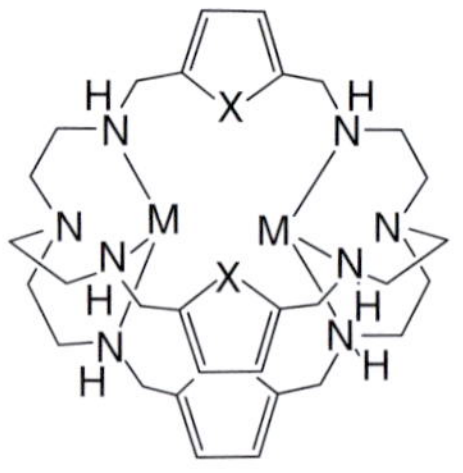

77A:X = O
77B:X = S

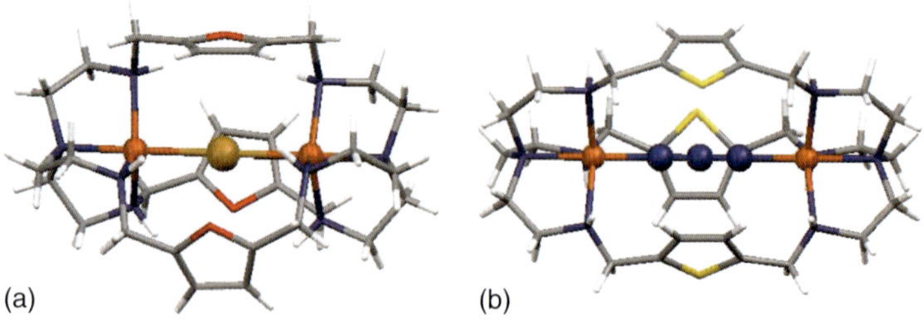

Figure 3.72 Copper(II) cascade complexes: (a) **77A** with bromide ion and (b) **77B** with azide. The perchlorate counterions are not shown for clarity.

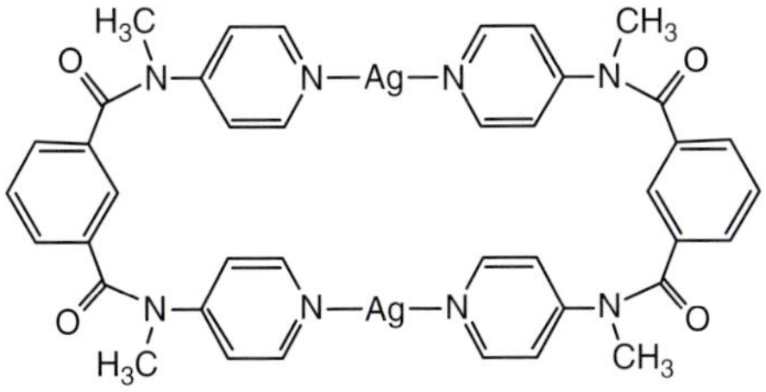

78

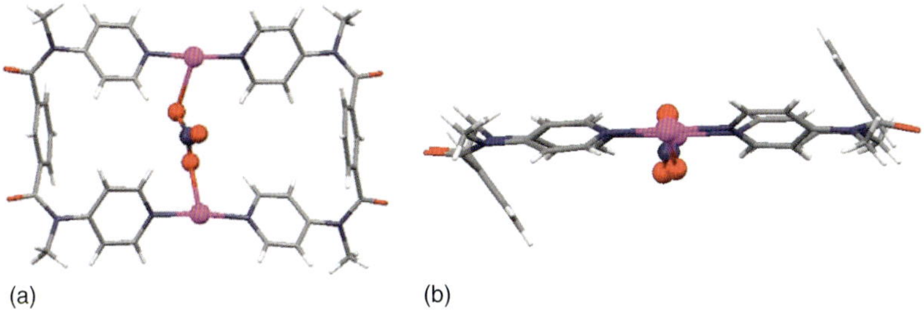

Figure 3.73 (a) Overhead view of **78** looking down into the macrocyclic cavity and (b) side view showing the orientation of the nitrate in the macrocyclic plane.

within the circle (Figure 3.73). The Ag–N distances are all very similar, ranging from 2.154 to 2.167 Å, while the Ag–O distances are longer at 2.558 and 2.528 Å. The Ag–Ag distance is 6.370 Å, which is quite comparable to some of the traditional cascade complexes.

3.8.2
Other Lewis Acid Donor Ligands

Examples of other Lewis base complexes with anions have been steadily increasing, and a variety of different elements, both metal and nonmetal, have been reported. Since the focus of this chapter has been primarily on structural aspects of supramolecular anion coordination chemistry, only some key representative examples of other Lewis acid complexes are described in this section.

3.8.2.1 **Boron-Based Ligands**

Of the other examples of non-hydrogen bonding Lewis acid complexes, perhaps boron is the most frequently studied. Some of the early prototypes of this area were the 1,8-naphthalendiylbis-(dimethylboranes), **79** and **80**. Compound **79** heralded the entrance of boron into the anion realm and was the first example of a rigid and uncharged multidentate Lewis acid [115]. The simple diborane was observed to extract hydrides from a variety of potential hydride donors. This handy feat led to its being named the "*proton sponge*." Although interaction with chloride was not very strong for the original borane (**79**), two years later, the related dichloroborane (**80**),

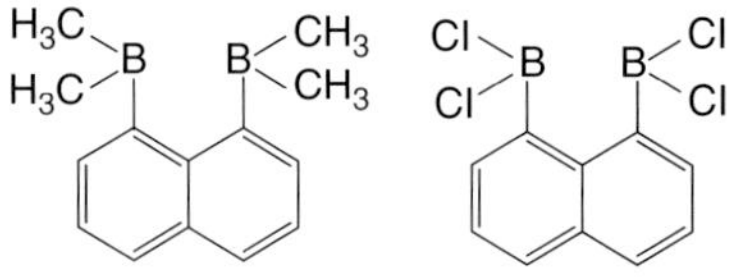

79 80

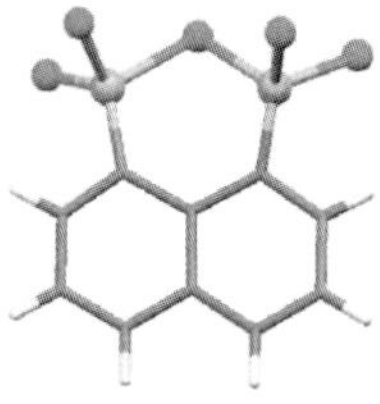

Figure 3.74 Crystal structure of the boron complex [**80**·Cl]$^-$. The $[Ph_3PNPPh_3]^+$ counterion is omitted for clarity.

a stronger Lewis acid, was found to bind chloride ion. The resulting anion provided the "first definitive example of chloride bridging between two otherwise trigonal borons" (Figure 3.74) [116]. The geometries around the two boron atoms are a distorted tetrahedron and the chloride is displaced somewhat from the naphthalene plane. The B–Cl distances are 1.920 and 2.012 Å, while the other B–Cl distances range from 1.770 to 1.935 Å.

As this area developed, so did the options for varying the heteroatom. Examples of heteronuclear Lewis acid bidentate "ligands" are **81** and **82** [117, 118]. The heteronuclear Lewis acid **81** is a phosphorescent sensor for fluoride ion (Figure 3.75a,b) [117]. Here, the B–F distance is 1.483 Å, while the Hg–F distance is 2.589 Å. The authors note that this interaction is more probably electrostatic in nature rather than covalent-like in the B–F case, since the empty 6p orbital on the mercury that accepts the fluoride electron pair is probably too high to mix with the fluoride atomic orbitals.

The Gabbai group has also expanded the Lewis acid host arsenal to the heavier chalcogenides, with the report of the mixed boron-telluronium host **82** (Figure 3.75c,d) [118]. The utility of the heteronuclear Lewis acids is their greater stability in protic solvents and potential use in extracting toxic anions. For the telluronium and other onium Lewis acids, another appeal is the positive charge that adds charge complementarity when singly charged anions are bound to the ligand, resulting in a neutral complex. In the telluronium complex, the B–F bond distance is 1.514 Å, while the Te–F distance is 2.506 Å. (This is actually slightly less than in the sulfur analog that is not shown.) The authors note that the affinity of the ligand for fluoride is possibly traced to the formation of a B–F→Te chelate, which is enhanced by interaction of the electron pair on fluoride with a σ^* Te–C orbital.

3.8.2.2 **Tin-Based Ligands**

Tin complexes have also been examined as Lewis acid ligands. For example, the simple ligand **83**, with the two tin atoms separated by a two-carbon phenylene spacer, forms 1 : 1 adducts with Lewis bases such as chloride and fluoride. The stability of the resulting complexes is attributed to the stability of the five-member

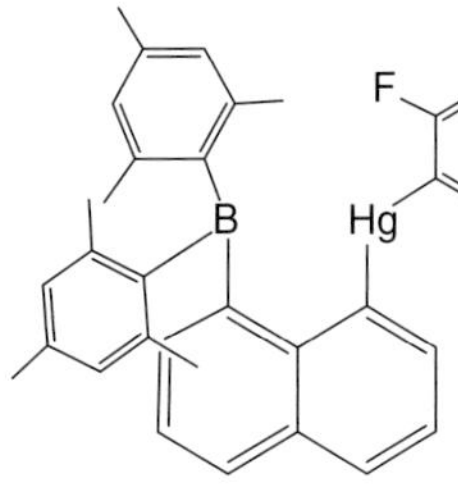

81

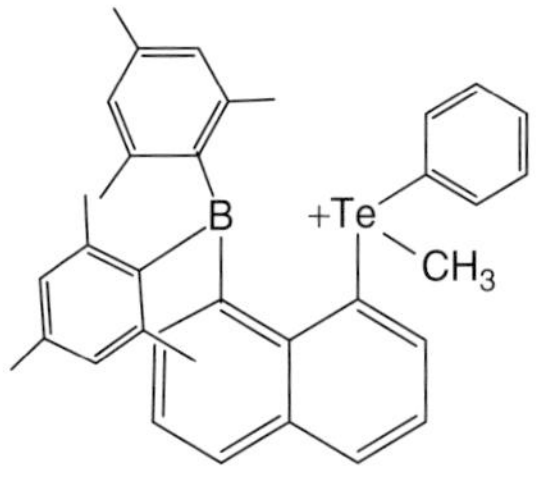

82

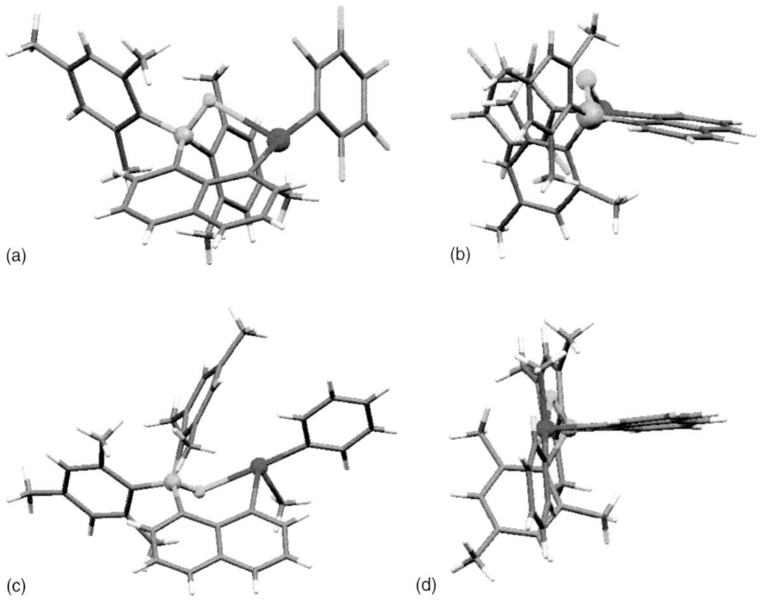

Figure 3.75 Views of the crystal structures of the heteronuclear Lewis acid ligands: (a,b) perpendicular to the B–F–Hg bridge and down the bridge in [**81**·F]$^-$, respectively, and (c,d) perpendicular to the B–F–Te bridge and down the bridge in [**82**·F], respectively. The $[S(NMe_2)_3]^+$ counterion is not shown for [**81**·F]$^-$.

C–Sn–X–Sn–C ring [119]. In the complex with **83**, the Sn–Cl distances are 2.742 and 2.803 Å for the bridging chloride, and once again, the centralized chloride is tilted up from the phenylene plane (Figure 3.76a,b). The other terminal Sn–Cl distances are significantly shorter at 2.475 and 2.525 Å. The tin has a trigonal bipyramidal geometry with the three carbon atoms (two methyl groups and a phenylene carbon atom) lying on the equatorial plane and the two bound chlorides at the axial sites. The counterion is $[(Ph_3P)_2N]^+$.

In a related fluoride complex, the cation is the [K·dibenzo-18-crown-6]$^+$, and it interacts with the complex (Figure 3.76c,d) [119]. In this complex, the Sn–F distances are 2.213 and 2.140 Å. The Sn–Cl distances are 2.532 and 2.608 Å, the latter

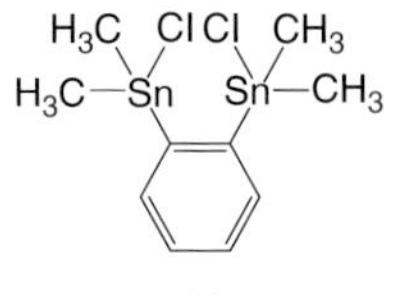

83

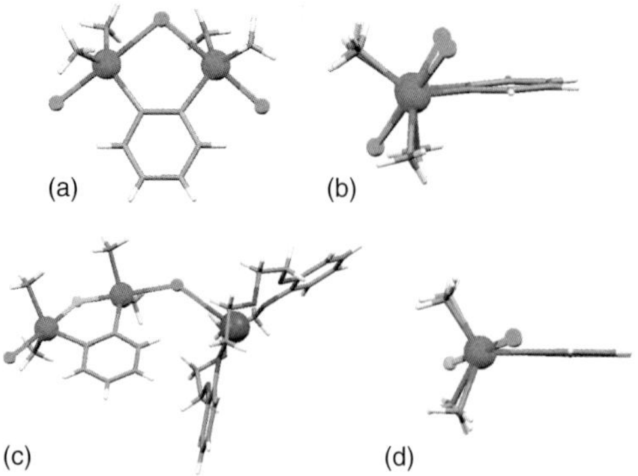

Figure 3.76 Views of the crystal structures of the ditin Lewis acid ligands with **83**: (a,b) views of the chloride complex perpendicular to the Sn–Cl–Sn bridge and down the bridge, respectively (counterion not shown for clarity), and (c,d) views of the fluoride complex perpendicular to the Sn–F–Sn bridge showing the interaction with the potassium·crown, and down the bridge (without the counterion), respectively.

chloride also being associated with the potassium ion at 3.131 Å. Compared to the chloride complex, the fluoride is closer to being coplanar with the phenylene ring.

The ditin katapinand complex (**84**) is a reminiscent of the original aza katapinands [120]. In the chloride structure, the crystallographic results clearly show that the chloride is bound to a single tin atom (Sn–Cl = 2.611 and 3.387 Å) (Figure 3.77), making the complex a stannane/stannate. The two external chloride ion distances are 2.415 (stannane) and 2.745 Å (stannate). While the angles around the three-coordinate stannane side of the cryptand are distorted somewhat from 109.5° (ranging from 114 to 120°), these are not much different from those in the uncomplexed free host. Solid-state MAS ^{119}Sn NMR also indicates two tin environments as well, indicative of stannane and stannate tin. The results for the fluoride complex with the tin katapinand with a six-carbon chain are more clearly indicative of a true distannate complex, although there is still about 0.15 Å difference in the Sn–F bond lengths (SnF = 2.276 and 2.128 Å). However, owing to the poor data available for this structure, it was decided not to include it here.

3.8.2.3 **Hg-Based Ligands**

A mercury analog of the crown ethers, **85**, provides an interesting cyclic Lewis acid sandwich host for anions [121]. The perfluorinated *o*-phenylene-derived host displays a high affinity toward a number of anions, forming sandwich complexes

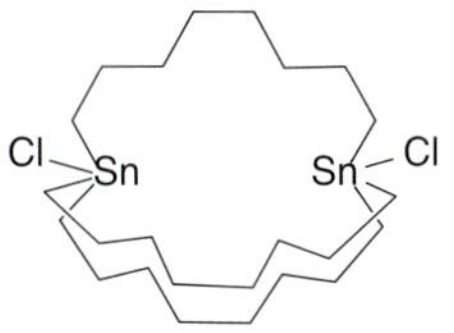

84

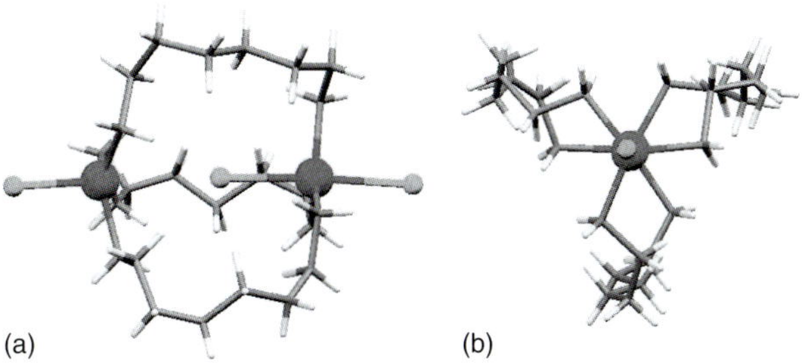

Figure 3.77 (a) Side view and (b) view down the bridgehead axis for the mixed stannane/stannate complex of **84** with chloride ion. The $[Ph_4P]^+$ counterion is not shown for clarity.

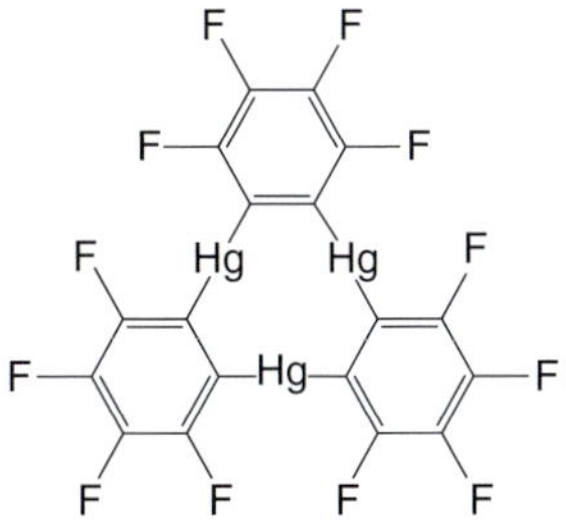

85

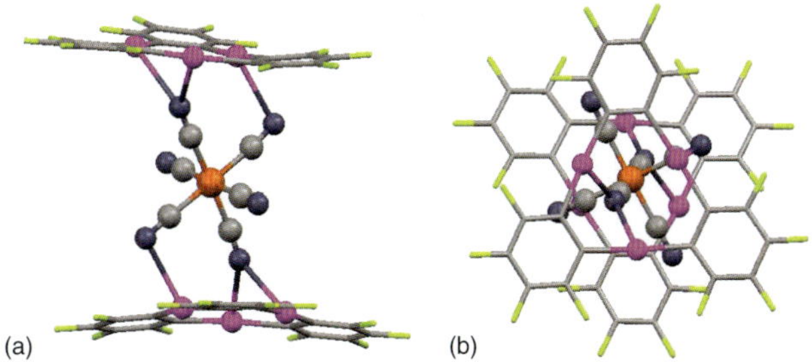

Figure 3.78 Views of the hexacyanaferrate(III) complex with the mercury-based ligand **85**: (a) side view perpendicular to the macrocyclic planes and (b) view down through the macrocycles. The $[(Ph_3P)_2N]^+$ counterions are not shown for clarity.

as seen in hexacyanoferrate(III) (Figure 3.78). The Hg–N distances range from 2.719 to 2.820 Å. These distances include the two "axial" nitrogen atoms, which are closer to pointing directly toward the macrocycles, each interacting with two mercury atoms on opposing rings, and the "equatorial" nitrogen atoms, slightly off to the side, interacting with a third mercury on each ring. Because of this "equatorial interaction," the $Fe(CN)_6{}^{3-}$ is tilted rather than sitting straight up and down between the two metallamacrocycles. The third singly "coordinated" mercury on each ring also has a fairly close interaction with the axial nitrogen atom, 2.903 and 2.915 Å. The Fe–C distances range from 1.908 to 1.953.

3.9 Conclusion

The analogies between anion coordination chemistry and transition metal coordination chemistry continue to increase. In this chapter, we have seen examples of the chelate, macrocyclic, and cryptand effects; the presence of axial water molecules in a number of anion complexes; anion corollaries to transition metal cascade complexes; the use of anions as templates for ligand formation; sandwich complexes; and compartmental ligands that hold anions at two different and separated sites. Foremost to the analogy, however, is the basic premise that transition metal ions have two valencies, a primary valence that refers to the charge on a metal ion and a secondary valence of a certain coordination number, which may be dictated by the metal ion's preferences [1]. Now it is clearly seen that anions have two valencies, a primary valence that is the charge on the anion and a secondary valence that is the supramolecular coordination number, which, as in metal ions, varies according to anion preferences. Indeed, the structural understanding of anion complexes has come a long way since the first text devoted specifically to the area, *Supramolecular Chemistry of Anions* (1997) [5]. The future of anion coordination chemistry, an area that began with a hardly noticed publication about halide inclusion in 1968, now seems without bounds.

Acknowledgments

The work described in this chapter that was done in the authors' laboratory was supported by the NSF through two awards: CHE0316623 and CHE0809736, and the DOE through two awards DE-FG 96ER62307 and DE-FG 02ER63741. This support is gratefully acknowledged.

References

1. See the Nobel Prize in Chemistry 1913 at Nobelprize.org: *http://nobelprize.org/nobel_prizes/chemistry/laureates/*.
2. Park, C.H. and Simmons, H.E. (1968) Macrobicyclic amines. III. Encapsulation of halide ions by *in, in*-1, (*k* + 2)-diazabicyclo[*k.l.m.*]alkane-ammonium ions. *J. Am. Chem. Soc.*, **90**, 2431–2433.

3. Bell, R.A., Christoph, G.G., Fronczek, F.R., and Marsh, R.E. (1975) The cation $H_{13}O_6{}^+$: a short, symmetric hydrogen bond. *Science*, **190**, 151–152.
4. Graf, E. and Lehn, J.-M. (1976) Anion cryptates: highly stable and selective macro-tricyclic anion inclusion complexes. *J. Am. Chem. Soc.*, **98**, 6403–6405.
5. Bianchi, A., Bowman-James, K., and García-España, E. (eds) (1997) *Supramolecular Chemistry of Anions*, Wiley-VCH Verlag GmbH, New York.
6. Gale, P.A. (ed.) (2003) Special issue: 35 years of synthetic anion receptor chemistry. *Coord. Chem. Rev.* **240**, 1–2.
7. Nelson, S. (1988) Chemistry beyond the molecule. *Chem. Br.*, 752–753.
8. Moyer, B.A. and Bonnesen, P.V. (1997) *Supramolecular Chemistry of Anions*, Wiley-VCH Verlag GmbH, New York, pp. 1–44.
9. Hay, B.P. and Custelcean, R. (2009) Anion-π interactions in crystal structures: commonplace or extraordinary? *Cryst. Growth. Des.*, **9**, 2539–2545.
10. Hay, B.P. and Bryantsev, V.S. (2008) Anion–arene adducts: C–H hydrogen bonding, anion–π interaction, and carbon bonding motifs. *Chem. Commun.*, 2417–2428.
11. Gamez, P., Mooibroek, T.J., Teat, S.J., and Reedijk, J. (2007) Anion binding involving π-acidic heteroaromatic rings. *Acc. Chem. Res.*, **40**, 435–444.
12. Schottel, B.L., Chifotides, H.T., and Dunbar, K.R. (2008) Anion-π interactions. *Chem. Soc. Rev.*, **37**, 68–83.
13. Lehn, J.-M. (1978) Cryptates: the chemistry of macropolycyclic inclusion complexes. *Acc. Chem. Res*, **11**, 49–57.
14. Lehn, J.-M. (1978) Cryptates: inclusion complexes of macropolycyclic receptor molecules. *Pure Appl. Chem.*, **50**, 871–892.
15. Bowman-James, K. (2005) Alfred Werner revisited: the coordination chemistry of anions. *Acc. Chem. Res.*, **38**, 671–678.
16. Kavallieratos, K., Gala, S.R., Austin, D.J., and Crabtree, R.H. (1997) A readily available non-preorganized neutral acyclic halide receptor with an unusual nonplanar binding conformation. *J. Am. Chem. Soc.*, **119**, 2325–2326.
17. Kavallieratos, K., Bertao, C.M., and Crabtree, R.H. (1999) Hydrogen bonding in anion recognition: a family of versatile, nonpreorganized neutral and acyclic receptors. *J. Org. Chem.*, **64**, 1675–1683.
18. Chmielewski, M.J., Zieliński, T., and Jurczak, J. (2007) Synthesis, structure, and complexing properties of macrocyclic receptors for anions. *Pure Appl. Chem.*, **79**, 1087–1096.
19. Hossain, M.A., Kang, S.O., Powell, D., and Bowman-James, K. (2003) Anion receptors: A new class of mixed amide/quaternized amine macrocycles and the chelate effect. *Inorg. Chem.*, **42**, 1397–1399.
20. Gale, P.A., Garric, J., Light, M.E., McNally, B.A., and Smith, B.D. (2007) Conformational control of HCl co-transporter: imidazole functionalised isophthalamide vs. 2,6-dicarboxamidopyridine. *Chem. Commun.*, 1736–1738.
21. Santacroce, P.V., Davis, J.T., Light, M.E., Gale, P.A., Iglesias-Sánchez, J.C., Prados, P., and Quesada, R. (2007) Conformational control of transmembrane Cl^- transport. *J. Am. Chem. Soc.*, **129**, 1886–1887.
22. Boiocchi, M., Del Boca, L., Esteban-Gómez, D., Fabbrizzi, L., Licchelli, M., and Monzani, E. (2004) Nature of urea-fluoride interaction: incipient and definitive proton transfer. *J. Am. Chem. Soc.*, **126**, 16507–16514.
23. Barboiu, M., Vaughan, G., and van der Lee, A. (2003) Self-organized heteroditopic macrocyclic superstructures. *Org. Lett.*, **5**, 3073–3076.
24. Zieliński, T. and Jurczak, J. (2005) Thioamides versus amides in anion binding. *Tetrahedron*, **61**, 4081–4089.
25. Light, M.E., Gale, P.A., Navakhun, K., and Maynard-Smith, M. (2005) 4-Aminomethyl-phenylamino-bis-(3,4-dichloro-5-phenylcarbamoyl-1*H*-pyrrole-2-carboxamide)tetrabutyl-ammonium salt. *Acta Crystallogr., Sect. E*, **E61**, o1300–o1301.
26. Gale, P.A., Navakhun, K., Camiolo, S., Light, M.E., and Hursthouse,

M.B. (2002) Anion-anion assembly: A new class of anionic supramolecular polymer containing 3,4-dichloro-2,5-diamido-substituted pyrrole anion dimers. *J. Am. Chem. Soc.*, **124**, 11228–11229.
27. Johnson, C.A. II, Berryman, O.B., Sather, A.C., Zakharov, L.N., Haley, M.M., and Johnson, D.W. (2009) Anion binding induces helicity in a hydrogen-bonding receptor: crystal structure of a 2,6-bis(anilinoethynyl)pyridinium chloride. *Cryst. Growth Des.*, **9**, 4247–4249.
28. Juwarker, H., Lenhardt, J.M., Pham, D.M., and Craig, S.L. (2008) 1,2,3-Triazole CH···Cl^- contacts guide anion binding and concomitant folding in 1,4-diaryl triazole oligomers. *Angew. Chem. Int. Ed.*, **47**, 3740–3743.
29. Rostovtsev, V.V., Green, L.G., Fokin, V.V., and Sharpless, K.B. (2002) A stepwise Huisgen cycloaddition process: copper(I)-catalyzed regioselective ligation of azides and terminal alkynes. *Angew. Chem. Int. Ed.*, **41**, 2596–2599.
30. Tornøe, C.W., Christensen, C., and Meldal, M. (2002) Peptidotriazoles on solid phase: [1,2,3]-triazoles by regiospecific copper(I)-catalyzed 1,3-dipolar cycloadditions of terminal alkynes to azides. *J. Org. Chem.*, **67**, 3057–3062.
31. Berryman, O.B., Johnson, C.A., Zakharov, L.N., Haley, M.M., and Johnson, D.W. II (2008) Water and hydrogen halides serve the same structural role in a series of 2+2 hydrogen-bonded dimers based on 2,6-bis(2-anilinoethynyl)pyridine sulfonamide receptors. *Angew. Chem. Int. Ed.*, **47**, 117–120.
32. Alyea, E.C., Ferguson, G., and Xu, Z. (1995) Hydrogen-bonded self-assembly of tris(2-ammonioethyl)amine molybdate hydrate, $4[N(CH_2CH_2NH_3)_3]^{3+}\cdot 6[MoO_4]^{2-}\cdot 6H_2O$. *Acta Crystallogr.*, **C51**, 353–356.
33. Arunachalam, M. and Ghosh, P. (2010) A versatile tripodal amide receptor for the encapsulation of anions or hydrated anions via formation of dimeric capsules. *Inorg. Chem.*, **49**, 943–951.
34. Ellis, R.J., Chartres, J., Sole, K.C., Simmance, T.G., Tong, C.C., White, F.J., Schröder, M., and Tasker, P.A. (2009) Outer-sphere amidopyridyl extractants for zinc(II) and cobalt(II) chlorometallates. *Chem. Commun.*, 583–585.
35. Chmielewski, M.J. and Jurczak, J. (2005) Anion recognition by neutral macrocyclic amides. *Chem. Eur. J.*, **11**, 6080–6094.
36. Caltagirone, C., Gale, P.A., Hiscock, J.R., Brooks, S.J., Hursthouse, M.B., and Light, M.E. (2008) 1,3-Diindolylureas: high affinity dihydrogen phosphate receptors. *Chem. Commun.*, 3007–3009.
37. Caltagirone, C., Hiscock, J.R., Hursthouse, M.B., Light, M.E., and Gale, P.A. (2008) 1,3-Diindolylureas and 1,3-diindolylthioureas: anion complexation studies in solution and the solid state. *Chem. Eur. J.*, **14**, 10236–10243.
38. Ju, J., Park, M., Suk, J.-M., Lah, M.S., and Jeong, K.-S. (2008) An anion receptor with NH and OH groups for hydrogen bonds. *Chem. Commun.*, 3546–3548.
39. Brooks, S.J., Edwards, P.R., Gale, P.A., and Light, M.E. (2006) Carboxylate complexation by a family of easy-to-make *ortho*-phenylenediamine based bis-ureas: studies in solution and the solid state. *New J. Chem.*, **30**, 65–70.
40. Amendola, V., Boiocchi, M., Esteban-Gómez, D., Fabbrizzi, L., and Monzani, E. (2005) Chiral receptors for phosphate ions. *Org. Biomol. Chem.*, **3**, 2632–2639.
41. Brooks, S.J., Gale, P.A., and Light, M.E. (2005) *Ortho*-phenylenediamine bis-urea-carboxylate: a new reliable supramolecular synthon. *CrystEngComm*, **7**, 586–591.
42. Grossel, M.C., Merckel, D.A.S., and Hutchings, M.G. (2003) The effect of preorganisation on the solid state behaviour of simple 'aromatic-cored' bis(guanidinium) sulfates. *CrystEngComm*, **5**, 77–81.
43. Dixon, R.P., Geib, S.J., and Hamilton, A.D. (1992) Molecular recognition:

bis-acylguanidiniums provide a simple family of receptors for phosphodiesters. *J. Am. Chem. Soc.*, **114**, 365–366.
44. Jubian, V., Dixon, R.P., and Hamilton, A.D. (1992) Molecular recognition and catalysis. Acceleration of phosphodiester cleavage by a simple hydrogen-bonding receptor. *J. Am. Chem. Soc*, **114**, 1120–1121.
45. Carroll, C.N., Berryman, O.B., Johnson, C.A. II, Zakharov, L.N., Haley, M.M., and Johnson, D.W. (2009) Protonation activates anion binding and alters binding selectivity in new inherently fluorescent 2,6-bis(2-anilinoethynyl)pyridine bisureas. *Chem. Commun.*, 2520–2522.
46. Metzger, A., Lynch, V.M., and Anslyn, E.V. (1997) A synthetic receptor selective for citrate. *Angew. Chem. Int. Ed. Engl.*, **36**, 862–865.
47. Busschaert, N., Gale, P.A., Haynes, C.J.E., Light, M.E., Moore, S.J., Tong, C.C., Davis, J.T., and Harrell, W.A. Jr. (2010) Tripodal transmembrane transporters for bicarbonate. *Chem. Commun.*, **46**, 6252–6254.
48. Alcalde, E., Mesquida, N., Vilaseca, M., Alvarez-Rua, C., and García-Granda, S. (2007) Imidazolium-based dicationic cyclophanes. Solid-state aggregates with unconventional $(C-H)^{+}\cdots Cl^{-}$ hydrogen bonding revealed by X-ray diffraction. *Supramol. Chem.*, **19**, 501–509.
49. Lee, J.-J., Stanger, K.J., Noll, B.C., Gonzalez, C., Marquez, M., and Smith, B.D. (2005) Rapid fixation of methylene chloride by a macrocyclic amine. *J. Am. Chem. Soc.*, **127**, 4184–4185.
50. Kang, S.O., Day, V.W., and Bowman-James, K. (2008) Cyclophane capsule motifs with side pockets. *Org. Lett.*, **10**, 2677–2680.
51. Kubik, S., Goddard, R., Kirchner, R., Nolting, D., and Seidel, J. (2001) A cyclic hexapeptide containing L-proline and 6-aminopicolinic acid subunits binds anions in water. *Angew. Chem. Int. Ed.*, **40**, 2648–2651.
52. Chellappan, K., Singh, N.J., Hwang, I.-C., Lee, J.W., and Kim, K.S. (2005) A calix[4]imidazolium[2]pyridine as an anion receptor. *Angew. Chem. Int. Ed.*, **44**, 2899–2903.
53. Szumna, A. and Jurczak, J. (2001) A new macrocyclic polylactam-type neutral receptor for anions 2 structural aspects of anion recognition. *Eur. J. Org. Chem.*, 4031–4039.
54. Hossain, M.A., Llinares, J.M., Powell, D., and Bowman-James, K. (2001) Multiple hydrogen bond stabilization of a sandwich complex of sulfate between two macrocyclic tetraamides. *Inorg. Chem.*, **40**, 2936–2937.
55. Ghosh, S., Roehm, B., Begum, R.A., Kut, J., Hossain, M.A., Day, V.W., and Bowman-James, K. (2007) Versatile host for metallo anions and cations. *Inorg. Chem.*, **46**, 9519–9521.
56. Hossain, M.A., Kang, S.O., Powell, D., and Bowman-James, K. (2003) Anion receptors: a new class of amide/quaternized amine macrocycles and the chelate effect. *Inorg. Chem.*, **42**, 1397–1399.
57. Day, V.W., Hossain, M.A., Kang, S.O., Powell, D., Lushington, J., and Bowman-James, K. (2007) Encircled proton. *J. Am. Chem. Soc.*, **129**, 8692–8693.
58. Hibbert, F. and Emsley, J. (1990) Hydrogen bonding and chemical reactivity. *Adv. Phys. Org. Chem.*, **26**, 255–279.
59. Kim, N.-K., Chang, K.-J., Moon, D., Lah, M.S., and Jeong, K.-S. (2007) Two distinct anion-binding modes and their relative stabilities. *Chem. Commun.*, 3401–3403.
60. Custelcean, R., Delmau, L.H., Moyer, B.A., Sessler, J.L., Cho, W.S., Gross, D., Bates, G.W., Brooks, S.J., Light, M.E., and Gale, P.A. (2005) Calix[4]pyrrole: an old yet new ion-pair receptor. *Angew. Chem. Int. Ed.*, **44**, 2537–2542.
61. Blondeau, P., Benet-Buchholz, J., and de Mendoza, J. (2007) Enthalpy driven nitrate complexation by guanidinium-based macrocycles. *New J. Chem.*, **31**, 736–740.
62. Hay, B.P., Gutowski, M., Dixon, D.A., Garza, J., Vargas, R., and Moyer, B.A. (2004) Structural criteria for the rational design of selective ligands: convergent hydrogen bonding sites for

the nitrate anion. *J. Am. Chem. Soc.*, **126**, 7925–7934.
63. Hay, B.P., Firman, T.K., and Moyer, B.A. (2005) Structural design criteria for anion hosts: strategies for achieving anion shape recognition through the complementary placement of urea donor groups. *J. Am. Chem. Soc.*, **127**, 1810–1819.
64. Sessler, J.L., Katayev, E., Pantos, G.D., Scherbakov, P., Reshetova, M.D., Khrustalev, V.N., Lynch, V.M., and Ustynyuk, Y.A. (2005) Fine tuning the anion binding properties of 2,6-diamidopyridine dipyrromethane hybrid macrocycles. *J. Am. Chem. Soc.*, **127**, 11442–11446.
65. Brooks, S.J., Gale, P.A., and Light, M.E. (2006) Anion-binding modes in a macrocyclic amidourea. *Chem. Commun.*, 4344–4346.
66. Meshcheryakov, D., Arnaud-Neu, F., Böhmer, V., Bolte, M., Hubscher-Bruder, V., Jobin, E., Thondorf, I., and Werner, S. (2008) Cyclic triureas – synthesis, crystal structures and properties. *Org. Biomol. Chem.*, **6**, 1004–1014.
67. Seidel, D., Lynch, V., and Sessler, J.L. (2002) Cyclo[8]pyrrole: a simple-to-make expanded porphyrin with no meso bridges. *Angew. Chem., Int. Ed.*, **41**, 1422–1425.
68. Meshcheryakov, D., Böhmer, V., Bolte, M., Hubscher-Bruder, V., Arnaud-Neu, F., Herschbach, H., Dorsselaer, A.V., Thondorf, I., and Mögelin, W. (2006) Two chloride ions as a template in the formation of a cyclic hexaurea. *Angew. Chem. Int. Ed.*, **45**, 1648–1652.
69. Saeed, M.A., Fronczek, F.R., and Hossain, M.A. (2009) Encapsulated chloride coordinating with two in-in protons of bridgehead amines in an octaprotonated azacryptand. *Chem. Commun.*, 6409–6411.
70. Mahoney, J.M., Beatty, A.M., and Smith, B.D. (2001) Selective recognition of an alkali halide contact ion-pair. *J. Am. Chem. Soc.*, **123**, 5847–5848.
71. Deetz, M.J., Shang, M., and Smith, B.D. (2000) A macrobicyclic receptor with versatile recognition properties: simultaneous binding of an ion pair and selective complexation of dimethylsulfoxide. *J. Am. Chem. Soc.*, **122**, 6201–6207.
72. Mahoney, J.M., Stucker, K.A., Jiang, J., Carmichael, I., Brinkmann, N.R., Beatty, A.M., Noll, B.C., and Smith, B.D. (2005) Molecular recognition of trigonal oxyanions using a ditopic salt receptor: evidence for anisotropic shielding surface around nitrate anion. *J. Am. Chem. Soc.*, **127**, 2922–2928.
73. Mahoney, J.M., Marshal, R.A., Beatty, A.M., Smith, B.D., Camiolo, S., and Gale, P.A. (2001) Complexation of alkali chloride contact ion-pairs using a 2,5-diamidopyrrole crown macrobicycle. *J. Supramol. Chem.*, **1**, 289–292.
74. Das, M.C., Ghosh, S.K., and Bharadwaj, P.K. (2010) Binding of various anions in laterally non-symmetric aza-oxa cryptands through H-bonds: characterization of water clusters of different nuclearity. *CrystEngComm*, **12**, 413–419.
75. Lee, C.H., Na, H.-K., Yoon, D.-W., Won, D.-H., Cho, W.-S., Lynch, V.M., Shevchuk, S.V., and Sessler, J.L. (2003) Single side strapping: a new approach to fine tuning the anion recognition properties of calix[4]pyrroles. *J. Am. Chem. Soc.*, **125**, 7301–7306.
76. Kang, S.O., Powell, D., Day, V.W., and Bowman-James, K. (2006) Trapped bifluoride. *Angew. Chem. Int. Ed.*, **45**, 1921–1925.
77. Kang, S.O., Day, V.W., and Bowman-James, K. (2010) Tricyclic host for anions. *Inorg. Chem.*, **49**, 8629–8636.
78. Yoon, D.-W., Gross, D.E., Lynch, V.M., Sessler, J.L., Hay, B.P., and Lee, C.H. (2008) Benzene-, pyrrole-, and furan-containing diametrically strapped calix[4]pyrroles – an experimental and theoretical study of hydrogen-bonding effects in chloride anion recognition. *Angew. Chem. Int. Ed.*, **47**, 5038–5042.
79. Zhang, B., Cai, P., Duan, C., Miao, R., Zhu, L., Niitsu, T., and Inoue, H. (2004) Imidazolidinium-based robust crypt with unique selectivity for fluoride anion. *Chem. Commun.*, 2206–2207.

80. Dietrich, B., Lehn, J.-M., Guilhem, J., and Pascard, C. (1989) Anion receptor molecules: synthesis of an octaaza-cryptand and structure of its fluoride cryptate. *Tetrahedron Lett.*, **30**, 4125–4128.
81. Reilly, S.D., Khalsa, G.R.K., Ford, D.K., Brainard, J.R., Hay, B.P., and Smith, P.H. (1995) Octaazacryptand complexation of the fluoride ion. *Inorg. Chem.*, **34**, 569–575.
82. Hossain, M.A., Llinares, J.M., Miller, C.A., Seib, L., and Bowman-James, K. (2000) Further insight to selectivity issues in halide binding in a tiny octaazacryptand. *Chem. Commun.*, 2269–2270.
83. Hossain, M.A., Llinares, J.M., Alcock, N.W., Powell, D., and Bowman-James, K. (2003) Structural and anion binding aspects of the tiny octaazacryptand. *J. Supramol. Chem.*, **2**, 143–149.
84. Iloudis, C.A., Tocher, D.A., and Steed, J.W. (2004) A highly efficient, preorganized macrobicyclic receptor for halides based on CH··· and NH··· anion interactions. *J. Am. Chem. Soc.*, **126**, 12395–12402.
85. Mason, S., Clifford, L., Seib, K., Kuczera, K., and Bowman-James, K. (1998) Unusual encapsulation of two nitrates in a single bicyclic cage. *J. Am. Chem. Soc.*, **120**, 8899–8900.
86. Clifford, T., Danby, A., Llinares, J.M., Mason, S., Alcock, N.W., Powell, D., Aguilar, J.A., García-España, E., and Bowman-James, K. (2001) Anion binding with two polyammonium macrocycles of different dimensionality. *Inorg. Chem.*, **40**, 4710–4720.
87. Hossain, M.A., Llinares, J.M., Mason, S., Morehouse, P., Powell, D., and Bowman-James, K. (2002) Parallels in cation and anion coordination: a new class of cascade complexes. *Angew. Chem. Int. Ed.*, **41**, 2335–2338.
88. Hossain, M.A., Morehouse, P., Powell, D., and Bowman-James, K. (2005) Tritopic (cascade) and ditopic complexes of halides with an azacryptand. *Inorg. Chem.*, **44**, 2143–2149.
89. Amendola, V., Fabbrizzi, L., Mangano, C., Pallavicini, P., Poggi, A., and Taglietti, A. (2001) Anion recognition by dimetallic cryptates. *Coord. Chem. Rev.*, **219–221**, 821–837.
90. Amendola, V. and Fabbrizzi, L. (2009) Anion receptors that contain metals as structural units. *Chem. Commun.*, 513–531.
91. Bisson, A.P., Lynch, V.M., Monahan, M.K.C., and Anslyn, E.V. (1997) Recognition of anions through NH–π hydrogen bonds in a bicyclic cyclophane – selectivity for nitrate. *Angew. Chem. Int. Ed.*, **36**, 2340–2342.
92. Kang, S.O., Hossain, M.A., Powell, D., and Bowman-James, K. (2005) Encapsulated sulfates: insight to binding propensities. *Chem. Commun.*, 328–330.
93. Kang, S.O., Powell, D., and Bowman-James, K. (2005) Anion binding motifs: topicity and charge in amidocryptands. *J. Am. Chem. Soc.*, **127**, 13478–13479.
94. Kang, S.O., Day, V.W., and Bowman-James, K. (2010) Fluoride: solution- and solid-state structural binding probe. *J. Org. Chem.*, **75**, 277–283.
95. Kang, S.O., VanderVelde, D., Powell, D., and Bowman-James, K. (2004) Fluoride-facilitated deuterium exchange from DMSO-d_6 to polyamide-based cryptands. *J. Am. Chem. Soc.*, **126**, 12272–12273.
96. Kang, S.O., Llinares, J.M., Powell, D., VanderVelde, D., and Bowman-James, K. (2003) New polyamide cryptand for anion binding. *J. Am. Chem. Soc.*, **125**, 10152–10153.
97. Steed, J.W. (2009) Coordination and organometallic compounds as anion receptors and sensors. *Chem. Commun.*, **38**, 506–519.
98. Harding, L.P., Jeffery, J.C., Riis-Johannessen, T., Rice, R.R., and Ze, Z. (2004) Anion control of the formation of geometric isomers in a triple helical array. *Dalton Trans.*, 2396–2397.
99. Vega, I.E.D., Gale, P.A., Lighta, M.E., and Loeb, S.J. (2005) NH *vs.* CH hydrogen bond formation in

metal-organic anion receptors containing pyrrolylpyridine ligands. *Chem. Commun.*, 4913–4915.
100. Yu, J.O., Browning, C.S., and Farrar, D.H. (2008) Tris-2-(3-methylindolyl)phosphine as an anion receptor. *Chem. Commun.*, 1020–1022.
101. Amendola, V., Boiocchi, M., Colasson, B., Fabbrizzi, L., Douton, M.-J.R., and Ugozzoli, F. (2006) A metal-based trisimidazolium cage that provides six C-H hydrogen-bond-donor fragments and includes anions. *Angew. Chem. Int. Ed.*, **45**, 6920–6924.
102. Bondy, C.R., Gale, P.A., and Loeb, S.J. (2004) Metal-organic anion receptors: arranging urea hydrogen-bond donors to encapsulate sulfate ions. *J. Am. Chem. Soc.*, **126**, 5030–5031.
103. Fisher, M.J., Gale, P.A., Light, M.E., and Loeb, S.J. (2008) Metal-organic anion receptors: trans-functionalised platinum complexes. *Chem. Commun.*, 5695–5697.
104. Pflugrath, J.W. and Quiocho, F.A. (1985) Sulphate sequestered in the sulphate-binding protein of *Salmonella typhimurium* is bound solely by hydrogen bonds. *Nature*, **314**, 257–260.
105. Custelcean, R., Moyer, B.A., and Hay, B.P. (2005) A coordinatively saturated sulfate encapsulated in a metal–organic framework functionalized with urea hydrogen-bonding groups. *Chem. Commun.*, 5971–5973.
106. Custelcean, R., Bosano, J., Bonnesen, P.V., Kertesz, V., and Hay, B.P. (2009) Computer-aided design of a sulfate-encapsulating receptor. *Angew. Chem. Int. Ed.*, **48**, 4025–4029.
107. Byrne, P., Lloyd, G.O., Clarke, N., and Steed, J.W. (2008) A "compartmental" Borromean weave coordination polymer exhibiting saturated hydrogen bonding to anions and water cluster inclusion. *Angew. Chem. Int. Ed.*, **47**, 5761–5764.
108. Lehn, J.-M., Pine, S.H., Watanabe, E., and Willard, A.K. (1977) Binuclear cryptates. Synthesis and binuclear cation inclusion complexes of bis-tren macrobicyclic ligands. *J. Am. Chem. Soc.*, **99**, 6766–6768.
109. Izatt, R.M., Pawlak, K., Bradshaw, J.S., and Bruening, R.L. (1991) Thermodynamic and kinetic data for macrocycle interaction with cations and anions. *Chem. Rev.*, **91**, 1721–2085.
110. Bianchi, A., Micheloni, M., and Paoletti, P. (1991) Thermodynamic aspects of the polyazacycloalkane complexes with cations and anions. *Coord. Chem. Rev.*, **110**, 17–113.
111. Lu, Q., Reibenspies, J.J., Martell, A.E., and Motekaitis, R.J. (1996) Copper(II) complexes of the hexaaza macrocyclic ligand 3,6,9,16,19,22-hexaaza-27,28-dioxatricycle [22.2.1.1^{11,14}]octacosa-1(26),11,13,24-tetraene and their interaction with oxalate, malonate, and pyrophosphate anions. *Inorg. Chem.*, **35**, 2630–2636.
112. Amendola, V., Bastianello, E., Fabbrizzi, L., Mangano, C., Pallavicini, P., Perotti, A., Lanfredi, A.M., and Ugozzoli, F. (2000) Halide-ion encapsulation by a flexible dicopper(II) bis-tren cryptate. *Angew. Chem. Int. Ed.*, **39**, 2917–2920.
113. Amendola, V., Fabbrizzi, L., Mangano, C., Pallavicini, P., and Zema, M. (2002) A di-copper(II) bis-tren cage with thiophene spacers as receptor for anions in aqueous solution. *Inorg. Chim. Acta*, **337**, 70–74.
114. Yue, N.L.S., Jennings, M.C., and Puddephatt, R.J. (2005) Disilver(I) macrocycles: variation of cavity size with anion binding. *Inorg. Chem.*, **44**, 1125–1131.
115. Katz, H.E. (1985) Hydride sponge: 1,8-naphthalenediylbis(dimethylborane). *J. Am. Chem. Soc.*, **107**, 1420–1421.
116. Katz, H.E. (1987) 1,8-Naphthalenediylbis(dichloroborane) chloride: the first bis boron chloride chelate. *Organometallics*, **6**, 1134–1136.
117. Melaimi, M. and Gabbaï, F.P. (2005) A heteronuclear bidentate Lewis acid as a phosphorescent fluoride sensor. *J. Am. Chem. Soc.*, **127**, 9680–9681.
118. Zhao, H. and Gabbaï, F.P. (2010) A bidentate Lewis acid with a telluronium ion as an anion-binding site. *Nat. Chem.*, **2**, 984–990.
119. Altmann, R., Jurkschat, K., Schürmann, M., Dakternieks, D., and

Duthie, A. (1998) *o*-Bis (haloorganostannyl)benzenes as powerful bidentate Lewis acids toward halide ions. *Organometallics*, **17**, 5858–5866.

120. Newcomb, M., Horner, J.H., Blanda, M.T., and Squattrito, P.J. (1989) Macrocycles containing tin. Solid complexes of anions encrypted in macrobicyclic Lewis acidic hosts? *J. Am. Chem. Soc.*, **111**, 6294–6301.

121. Tikhonova, I.A., Dolgushin, F.M., Tugashov, K.I., Ellert, O.G., Novotortsev, V.M., Furin, G.G., Antipin, M.Y., and Shur, V.B. (2004) Crown compounds for anions. Sandwich complexes of cyclic trimeric perfluoro-*o*-phenylenemercury with hexacyanoferrate(III) and nitroprusside anions. *J. Organomet. Chem.*, **689**, 82–87.

4
Synthetic Strategies

Andrea Bencini and José M. Llinares

4.1
Introduction

In the past decades, many new kinds of receptors for anions incorporating a variety of different functionalities have been prepared. Therefore, it is difficult to discuss all the synthetic procedures developed for such purpose in this chapter. Other chapters of this book comment on the synthesis of specific receptors, and opportune literature references have been included. Here, we focus just on two classes of synthetic receptors for anions: polyammonium and polyamide receptors. The first class is representative of charged receptors, which usually operate in water and use their positive net charge as the major driving force in anion binding, while other motifs such as hydrogen bonding and hydrophobic interactions can cooperate in the binding event. Indeed, as commented in Chapter 1 devoted to historical aspects, polyamines were the first type of compounds exploited as anion receptors following the pioneering work of Park and Simmons and Lehn. On the other hand, amide bonds constitute the building block of the backbone of proteins and are involved in anion coordination through their acidic NH groups. Amide functions participate in the intimate mechanism of anion transport through membranes and therefore in the regulation of charge and osmotic balance in cells. Therefore, it is not surprising that this type of bonds were incorporated since very early in the structure of synthetic receptors to achieve anion recognition, particularly in nonaqueous media. In these receptors, hydrogen bonding constitutes the main driving force in anion binding, while, analogous to the previous class of compounds, other forces such as π-stacking and anion–π may provide additional contribution to the binding (see Chapter 2).

4.2
Design and Synthesis of Polyamine-Based Receptors for Anions

Anion recognition by artificial receptors relies on the cooperative work of weak intermolecular forces such as hydrogen bonding, π-stacking interactions, electrostatic

Anion Coordination Chemistry, First Edition. Edited by Kristin Bowman-James, Antonio Bianchi, and Enrique García-España.

interactions, and hydrophobic effects. Therefore, the different binding sites of the receptor need to be appropriately organized in order to achieve a topological complementarity with the binding moieties of the anionic substrates. Stereochemical complementarity is of particular relevance for anion binding in water, a solvent which can strongly compete with the receptor in the coordination process.

A further level of complexity to the problem of receptor design is added by the fact that many anions exist in water, at neutral pH, in different protonated states bearing different negative charges. Therefore, it is reasonable to consider as a starting point in the design process easily available positively charged receptors featuring in their structures appropriately positioned hydrogen bonding donor and acceptor groups. In consequence, aliphatic polyamines have been extensively studied for the binding of anions in aqueous media [1–8]. As a matter of fact, the amine groups can easily protonate in water at neutral pH, thus serving both as a bearer of a positive charge to interact electrostatically with anions and to increase water solubility and as hydrogen bonding donors. Furthermore, the amine groups can also act as hydrogen bonding acceptors if they are not protonated, which is useful for binding protonated anions. Therefore, amine groups have been implemented into a variety of different structural architectures, from linear structures to monocyclic- and polycyclic-geometry-based organizations of varying flexibility, in an attempt to achieve selectivity toward target substrates.

In the case of inorganic anion binding, the most important parameters to be considered for both linear and cyclic aliphatic polyamines is their degree of protonation in water at around neutral pH and the spatial disposition of the protonated basic centers in order to optimize electrostatic and hydrogen bonding interactions with inorganic anions. Conversely, organic anions can also contain other functional groups, which can participate in the binding process. For instance, nucleotides, which are among the most studied organic substrates in anion coordination chemistry, contain nucleobases, capable not only of giving π-stacking and hydrophobic interactions with aromatic sections of the receptor but also interacting via hydrogen bonding with the receptor, because of their H-bonding donor or acceptor functional groups. Similarly, the hydroxyl groups of the sugar moiety can be used to interact with the receptor via hydrogen bonding. While receptors for inorganic anions generally need to contain only appropriately organized binding sites capable of giving charge–charge and/or hydrogen bonding interactions, the design of host molecules for organic anions implies the introduction in the receptor structure of binding sites to target not only the inorganic moiety but also the binding sites of the "organic" sections of these guests.

Here, starting from the simple considerations outlined above, we intend to review the development in design and synthesis of polyamine receptors for anions from a topological point of view, through the lens of the structural and geometrical modifications brought to these compounds, in order to optimize the directional controlling noncovalent interactions involved in the binding process.

4.2.1
Acyclic Polyamine Receptors

Among alkyl polyamines, the acyclic category features the simplest anion receptors in terms of design. The discovery that spermidine and spermine, two naturally occurring linear tetraamines widely distributed in living cells, are involved in different important biological processes based on anion recognition prompted the use of other linear polyamines as anion binders. It is known that an increasing number of amine functions can lead to the formation of polyammonium cations with increased charge at neutral pH values and, therefore, with increased binding ability for anionic substrates. With this in mind, Bianchi, Paoletti, and García-España, in the course of their analysis on the anion-binding ability of cyclic and acyclic aliphatic polyamines, synthesized a series of linear polyamines containing from 6 to 10 amine groups (**10–13** in Scheme 4.1) in order to compare their binding abilities toward anions with that of their cyclic counterparts (*vide infra*) [9]. These terminally methylated aliphatic polyamines, containing only secondary nitrogen atoms connected by ethylenic chains, were prepared by following the synthetic procedure depicted in Scheme 4.1, which uses the protected precursors **1–4**, easily accessible from commercially available polyamines, as starting materials. Reaction of **1–4** with the sodium salt of *N*, *N*′-ditosyl-*N*-methyl-ethylendiamine (**5**) affords the tosylated polyamines (**6–9**), which were then deprotected in H_2SO_4 to give the desired polyamine (**10–13**). According to this procedure, linear polyamines containing *n* nitrogen atoms can be converted into longer receptors containing $n + 4$ amine groups.

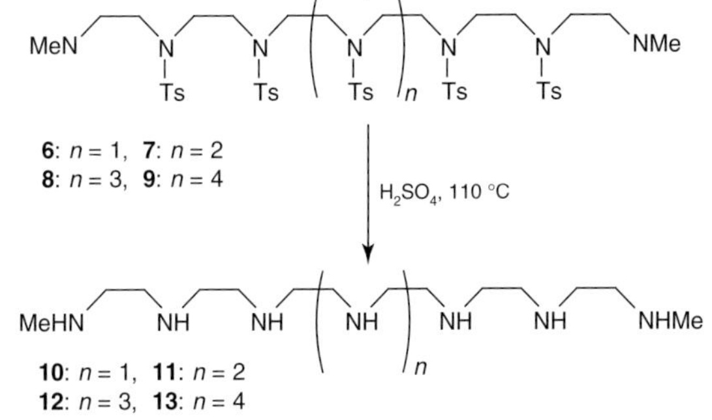

Scheme 4.1

Other procedures to increase the number of amine donors in linear polyamines can be used to achieve both receptors able to bind anionic species in aqueous solutions and synthetic precursors to obtain macrocyclic systems suitable for anion coordination studies. Symmetric elongation can be also achieved either by Michael condensation of a tosylated polyamine with acrylonitrile followed by reduction with B_2H_6 and removal of the tosyl groups [10] in HBr/CH_3COOH mixture or by reaction of the tosylated polyamine with *N*-(3-bromopropyl)phthalimide in the presence of a base (K_2CO_3), followed by deprotection of the terminal nitrogen atoms with hydrazine and subsequent removal of the tosyl groups [11–13]. Examples of these two alternative methods, widely used to symmetrically elongate polyamine chains with *N*-propylamine moieties, are sketched in Scheme 4.2 [11] and Scheme 4.3 [12], respectively.

Linear polyamines have also been appropriately functionalized in order either to improve their binding ability toward anionic species or to obtain chemosensors capable of signaling the binding process because of a change in the physical properties of the functional groups on substrate coordination. From this point of view, García-España and coworkers have designed a number of linear polyamines of different lengths and a number of nitrogen donors featuring one or two terminal benzyl, naphthalen-1-ylmethyl, or anthracen-9-ylmethyl groups [14–18]. The naphthalene and anthracene moieties can behave as additional binding sites for the ATP anion via π-stacking with the nucleobase and, at the same time, can be used as signaling units because of the modification of their fluorescence-emission properties generated on nucleotide binding [14, 17]. These receptors were synthesized in high yields by the reaction of the linear polyamines with aldehyde derivatives

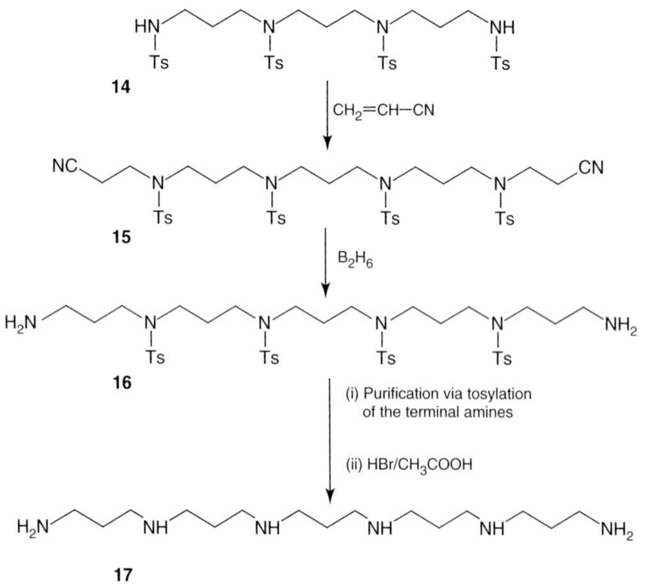

Scheme 4.2

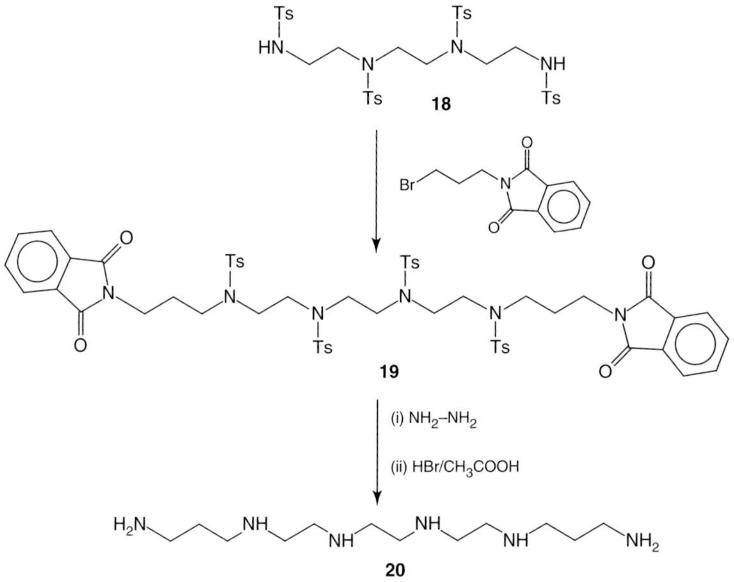

Scheme 4.3

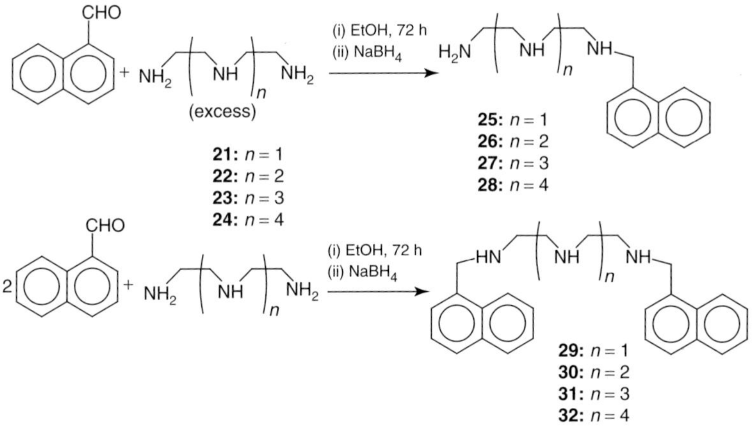

Scheme 4.4

of the aromatic compounds, followed by reduction *in situ* with $NaBH_4$ of the resulting Schiff bases (see Scheme 4.4 for the naphthalene-containing polyamines) [15]. While the difunctionalized receptors **29–32** were obtained by using a 2 : 1 molar ratio between the aldehyde derivatives and the polyamine, synthesis of the monofunctionalized compounds **25–28** was achieved only in the presence of large excesses of the starting linear polyamines.

The introduction of rigid or conformationally constrained elements within an open-chain polyamine can also induce geometrical preorganization in linear acyclic polyamines for chelate anion binding in order to favor a convergent projection of binding sites, which would enable the systems to bind anions with size and shape selectivity. A simple example is represented by the linear octaamine **43** featuring a *p*-xylenyl rigid unit [19]. This receptor, in its protonated forms, presents a surprising binding ability for triphosphate in aqueous medium. In fact, in the adduct with triphosphate, the two tetraprotonated pendant arms are rolled up to bind simultaneously both sides of the $H_2P_3O_{10}^{3-}$ anion. This receptor was obtained by using a general procedure developed by Handel and coworkers (Scheme 4.5) for N-monoalkylation of linear tetraamines 1,4,7,10-tetraazadecane (**22**) and 1,4,8,11-tetraazaundecane (**33**) via bis-aminals intermediates [20]. Basically, reaction of **22** and **33** with glyoxal or pyruvic aldehyde affords the bis-aminals **34–38**. In the case of **22**, these reactions gave mixtures of the *gem* and *vic* structural isomers in both *cis* and *trans* conformations. Interestingly, reaction of **33** with glyoxal, carried out in boiling ethanol, afforded only the *gem-trans* isomer **36**. Conversely, the same reaction, carried out at 0 °C led to the formation of the *gem-cis* isomer **37** as a unique product.

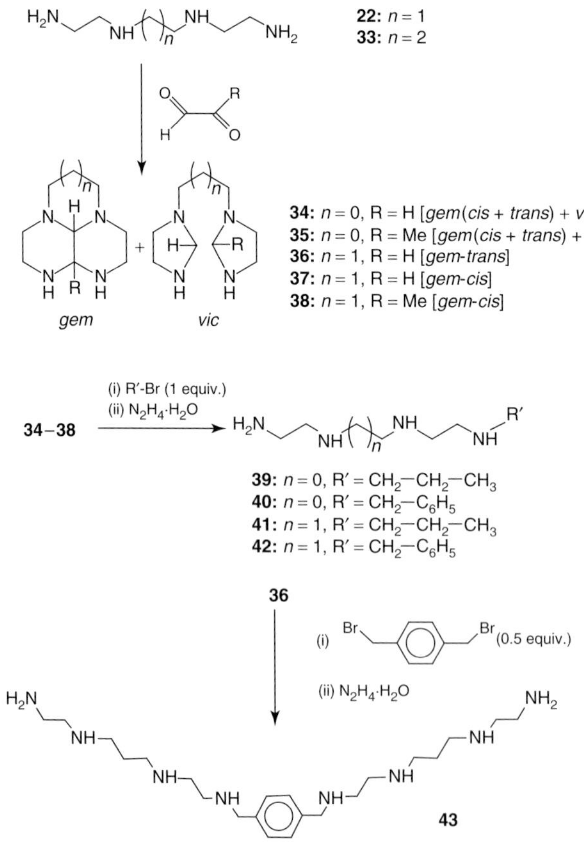

Scheme 4.5

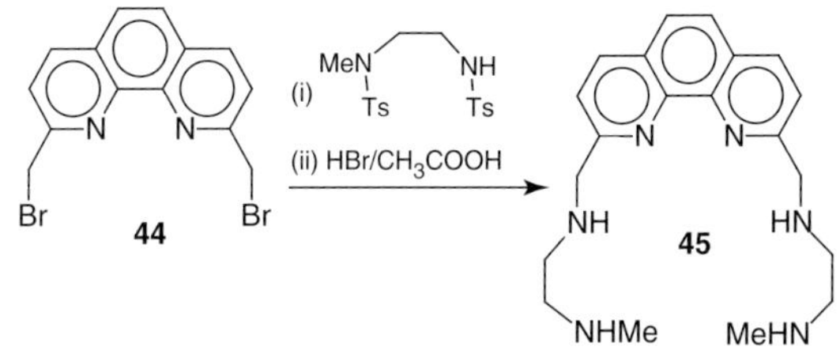

Scheme 4.6

The reaction of 1 equiv. of an electrophilic agent, such as 1-bromopropane or benzyl bromide, with the different bis-aminals led, after deprotection with hydrazine hydrate, to the N-monoalkylated compounds **39–42**. Monoalkylation was achieved with good yields irrespective of the electrophile and starting bis-aminal (single compound or a mixture of isomers). The authors also used this synthetic strategy to synthesize linear octaamines by using bis-electrophilic reagents. In particular, reaction of compound **36**, which gave the best yield in the monoalkylation reactions, with the bis-electrophile *p*-dibromo-xylene affords octaamine (**43**). This synthetic strategy appears particularly promising to construct linear polyamines constituted by two tetraamine chains linked by aliphatic or aromatic bridges of different nature.

Larger aromatic units, inserted between two polyamine chains, can represent not only a preorganizing building block for the receptor but also a binding site for anions containing an aromatic section, such as nucleotides. This is the case of **45**, featuring a phenanthroline unit separating two monomethylated ethylendiamine moieties. Compound **45** was obtained by reaction of 2,9-bis(bromomethyl)-1,10-phenanthroline (**44**) with *N, N′*-ditosyl-*N*-methylethylenediamine in the presence of K_2CO_3 as a base (Scheme 4.6) [21]. The resulting tosylated amine was subsequently deprotected in a HBr/CH_3COOH mixture. This receptor in its diprotonated form shows a remarkable selectivity for ATP over the triphosphate anion, because of the beneficial effect on complex stability of π-stacking interactions between phenanthroline and the adenine unit of the nucleotide [22].

Other functional groups have been used to organize polyamine chains. In **53**, the structural role of preorganizing the receptor in a near-macrocyclic arrangement is played by a conformationally constrained backbone including two amide units hinged on a *cis* carbon–carbon double bond. Compound **53** displays a high binding affinity for orthophosphate. Conversely, the *trans*-analog **52** shows only a very moderate affinity for phosphate, which is typical of open-chain receptors [23]. The synthesis of the open-chain receptors **52** and **53** (Scheme 4.7) started with mono-Boc protection of *N*-methyl-*N,N*-bis(3-aminopropyl)amine (**46**) to generate **47**. Two equivalents of compound **47** were then coupled with the diacid chloride of either fumaric (**48**) or maleic (**49**) acid to yield the Boc-protected intermediates (**50** and **51**), respectively, which were subsequently deprotected using gaseous HCl to provide receptors **52** and **53**.

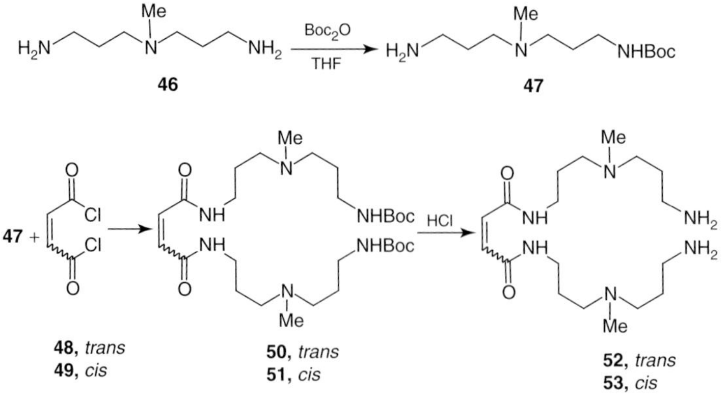

Scheme 4.7

4.2.2 Tripodal Polyamine Receptors

Tripodal acyclic polyamines represent the natural synthetic structural evolution of the very flexible linear acyclic analogs toward preorganized architectures for anion chelate binding. The simplest member of this class of polyamines is tren (**54**, tren = tris(2-aminoethyl)amine), which, since the beginning of anion coordination chemistry, has largely been used for the construction of three-dimensional receptors featuring cavities conformed in such a way as to promote anion encryption (see below).

In spite of the central stage occupied by tren in the chemistry of anion receptors, anion-binding properties of **54** in aqueous solution have only recently been studied systematically toward the oxoanions NO_3^-, SO_4^{2-}, TsO^-, PO_4^{3-}, $P_2O_7^{4-}$, and $P_3O_{10}^{5-}$ [24].

Tren-based receptors generally offer a tripodal conformation of convergent binding groups to the anionic substrates, leading to beneficial effects on the stability of the anion complexes with respect to simple linear polyamines. This property can be exploited to design ligands capable of extracting toxic oxoanions from aqueous medium to organic phases. In this context, the attachment of lipophilic aromatic side arms to the three primary amine groups of tren was used by Nelson and coworkers to construct anion receptors (**55**–**61** in Scheme 4.8) for perrhenate and pertechnetate extraction from aqueous solutions to chloroform [25]. Furthermore, Bowman-James and coworkers found that **55** displays selective coordination of tetrahedral anions, such as dihydrogen phosphate and hydrogen sulfate, over nitrate and halides [26].

Receptor **55** and **57**–**59** were synthesized via Schiff base condensation of the amine tren with the appropriate aldehyde followed by borohydride reduction of the resulting imines to amines, as sketched in Scheme 4.8 for **55** [25, 26]. In the case of **56**, the Schiff base was reduced via hydrogenation in the presence of palladium on activated carbon [27]. Compounds **60** and **61** were prepared by reaction of tren

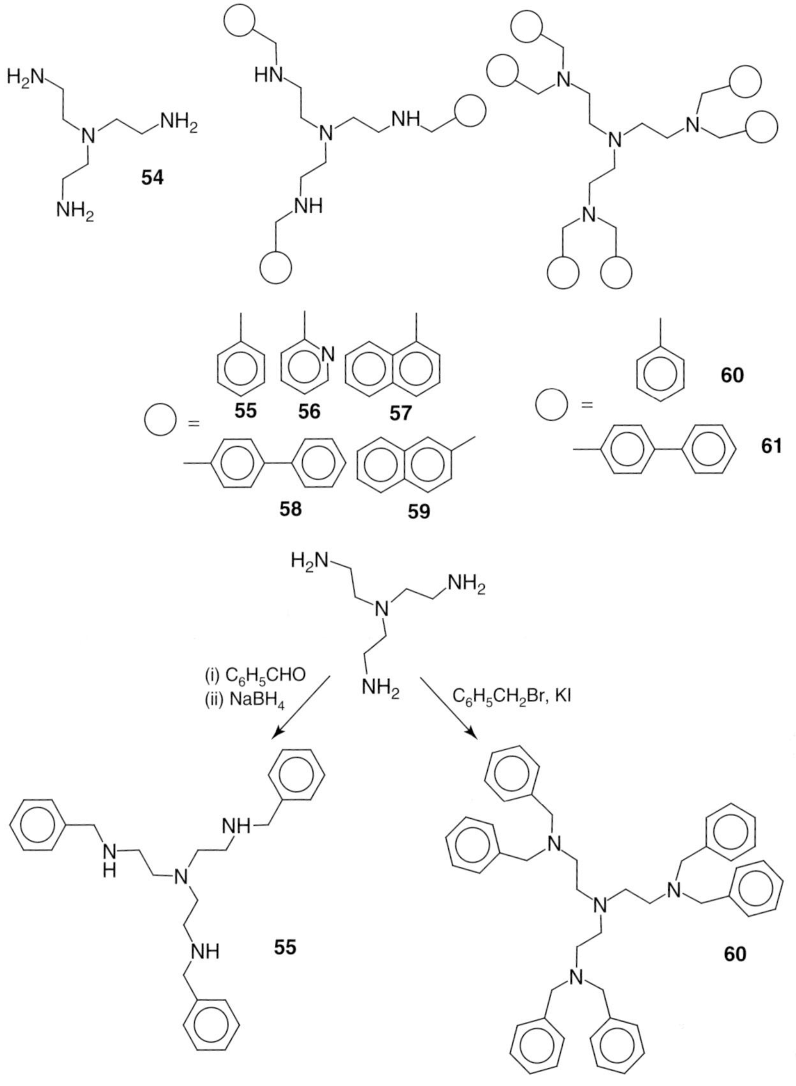

Scheme 4.8

with bromomethylbenzene or 4-bromomethylbiphenyl in the presence of KI (see Scheme 4.8 for **60**).

The ability of tripodal polyamines to bind oxoanions has also been used to develop tripodal receptors able not only to bind but also to signal anionic species because of quantifiable changes in the fluorescence emission of a fluorogenic unit linked to a tren unit. The most famous examples are the anthracene-based receptors **62** and **63** developed by Czarnik [28, 29]. In fact, receptor **62** (Scheme 4.9), composed of a trpn unit (**64**, trpn = tris(3-aminopropyl)amine) linked to an anthracene moiety

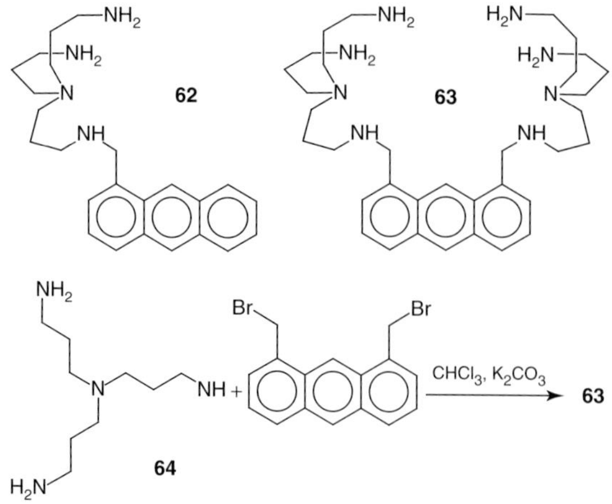

Scheme 4.9

through a methylenic spacer, displays an increase in its fluorescent emission intensity by more than 145% in the presence of HPO_4^{2-} in aqueous solution [28]. Receptor **63**, featuring two trpn units linked by an anthracene moiety, is even able to discriminate between PO_4^{3-} and $P_2O_7^{4-}$. In fact, the tripodal receptor groups are geometrically disposed at the right distance to fit the stereochemical requirements and size of the pyrophosphate anion, thus encapsulating it within the cavity formed by their facing protonated branched arms [29]. Receptors **62** and **63** were synthesized by reaction of 9-(chloromethyl)-anthracene or 1,8-bis(bromomethyl)anthracene, respectively, with trpn (in the case of **63**, a 10-fold excess of trpn was used) in the presence of K_2CO_3 as base, as sketched in Scheme 4.9 for **63**.

4.2.3
Macrocyclic Polyamine Receptors with Aliphatic Skeletons

As compared to tripodal acyclic polyamines or rigidified open-chain systems, cyclic aliphatic polyamines conceptually represent a step further in the design of even more preorganized receptors for anions [30–32]. In fact, their cyclic structure allows the formation of polyammonium receptors in aqueous solution with high positive charge density with respect to open-chain polyamines containing the same number of amine groups. In turn, they are still flexible enough to be able to wrap around bulky anions according to their ring size and to self-adapt to the stereochemical requirements of the anionic guests, thus forming adducts with optimized H-bond ion pairing. These characteristics prompted the synthesis of an enormous number of aliphatic polyamine macrocycles (see **65–76** for a few examples), either containing only amine groups separated by ethylenic [30] or propylenic chains [31] (see **65–71** and **72** or **73**) or both nitrogens and oxygen atoms [32], such as **74**, or featuring two polyamine moieties separated by longer aliphatic chains (**75** and **76**) [33].

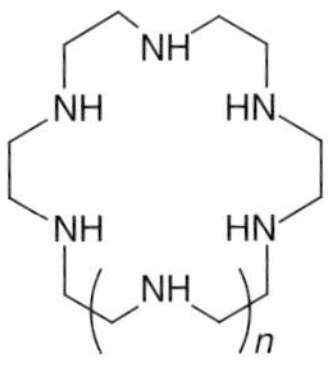

[3k]aneN$_k$
k = $n + 6$

65: $n = 1$
66: $n = 2$
67: $n = 3$
68: $n = 4$
69: $n = 5$
70: $n = 6$
71: $n = 7$

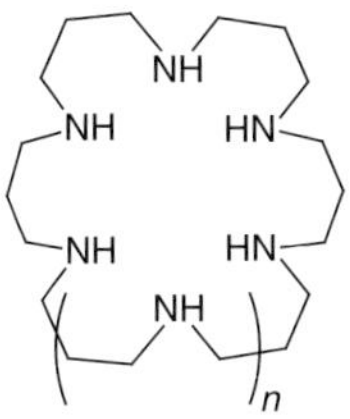

[4k]aneN$_k$
k = $n + 5$

72: $n = 1$
73: $n = 3$

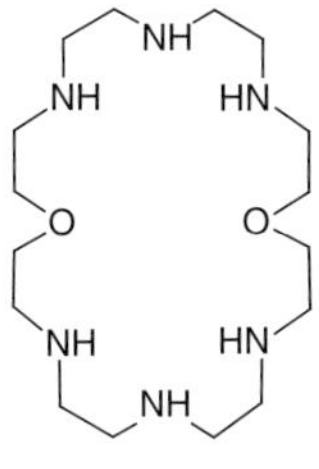

74

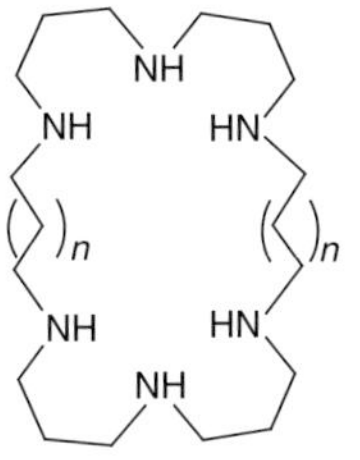

75: $n = 5$
76: $n = 8$

The pioneer work of Lehn pointed out that the protonated forms of these macrocycles behave as efficient receptors for a variety of anionic species, from inorganic anions, including $[Fe(CN)_6]^{4-}$ or $[Co(CN)_6]^{3-}$ [2, 3], to organic species, such as ATP or carboxylates [2, 3, 8]. Macrocycle **74** has probably been the most studied binder for nucleotides and inorganic phosphates, because of its ability not only to form stable complexes with these anions but also to efficiently catalyze a variety of chemical transformations on the bound substrate, including ATP hydrolysis to ADP, acetyl phosphate hydrolysis to orthophosphate, and pyrophosphate synthesis [3].

The synthesis of these receptors was generally achieved by using the procedure developed by Richman and Atkins for **65** [34] or its modifications (Scheme 4.10). In this procedure, cyclization is achieved by reaction in DMF of the sodium salt of a tosylated polyamine, most often obtained by treatment with NaH, with a second tosylated polyamine fragment appropriately functionalized with terminal $-(CH_2)_n-$OMs or $-(CH_2)_n-$OTs groups. The insertion of these groups is generally obtained via reaction of a tosylated polyamine with ethylene carbonate (Scheme 4.10) or 2-chloroethanol to introduce two terminal $-(CH_2)_2-OH$ moieties or with 3-chloro-1-propanol to append two terminal $-(CH_2)_3-OH$ units. The alcoholic functions are then mesylated with MsCl. The tosylated macrocycles, resulting from the 1 : 1 cyclization, are subsequently deprotected with concentrated H_2SO_4 or HBr/CH_3COOH in the presence of phenol as an antioxidant to afford the final polyamine macrocycles.

Functional groups have also been appended to these macrocycles either to increase their ability to bind anionic species or to confer specific functions to the resulting receptors. The most famous example is receptor **90**, featuring an acridine moiety appended to **74**. This large heteroaromatic unit was inserted as an additional binding site for nucleotides, such as ATP, because of its ability to interact via π-stacking with nucleobases. Its synthesis implies specific functionalization of a single amine group and requires the multistep procedure depicted

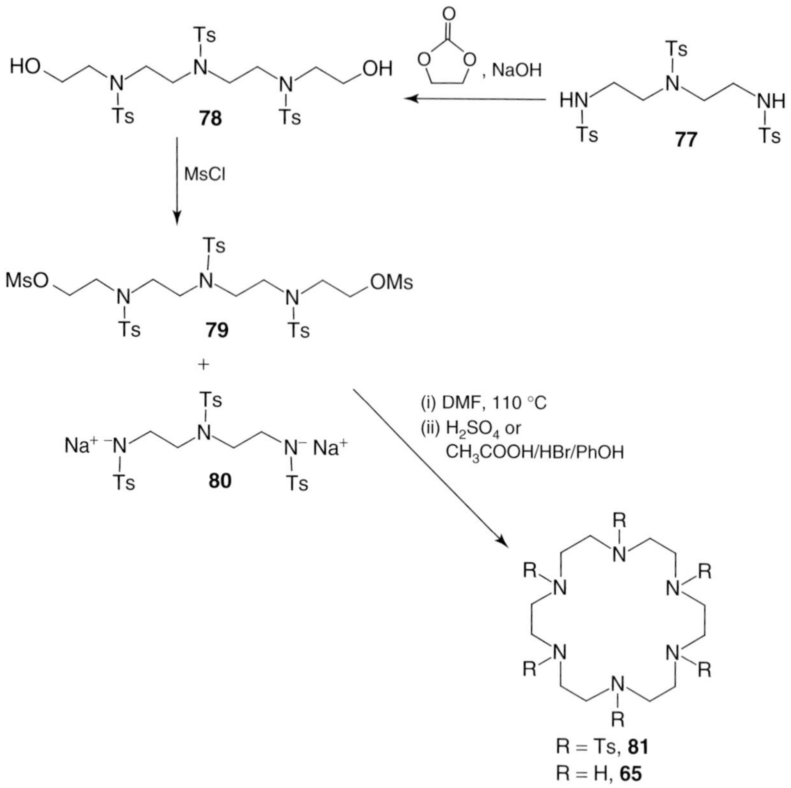

Scheme 4.10

in Scheme 4.11 to obtain the pentatosylated derivative of **74** (Scheme 4.11) [35]. Reaction of tosyl aziridine (**82**) with monotosylated ethylendiamine (**83**) affords ditosylated triamine (**84**), which is subsequently protected at the central nitrogen via reaction with benzoyl chloride, affording compound **85**. Reaction of the latter with 2-(2-chloroethoxy)ethanol is followed by conversion of the resulting diol in the corresponding dimesylated derivative **86**, which is used as the starting material in the cyclization with tosylated amine (**77**) to give macrocycle **87** and, after removing the benzoyl group, **88**. The latter is a versatile precursor to obtain macrocyclic receptors containing a single lateral side arm [35].

Actually, reaction with acrylonitrile, followed by reduction of the nitrile group with diborane, affords the propylamine derivative **89**. Finally, condensation with 9-chloro-acridine produces **90** [36]. Of note, replacement in the cyclization reaction of **77** with **85** leads to the achievement of a tetratosylated macrocycle containing two benzoyl-protected amine groups, which can be used to obtain macrocyclic receptors bearing two acridine moieties, following the same synthetic procedure described for **90** [35, 36].

The binding features of a polyamine receptor can be altered not only via insertion of pendant arms containing additional binding sites for anionic species but also by

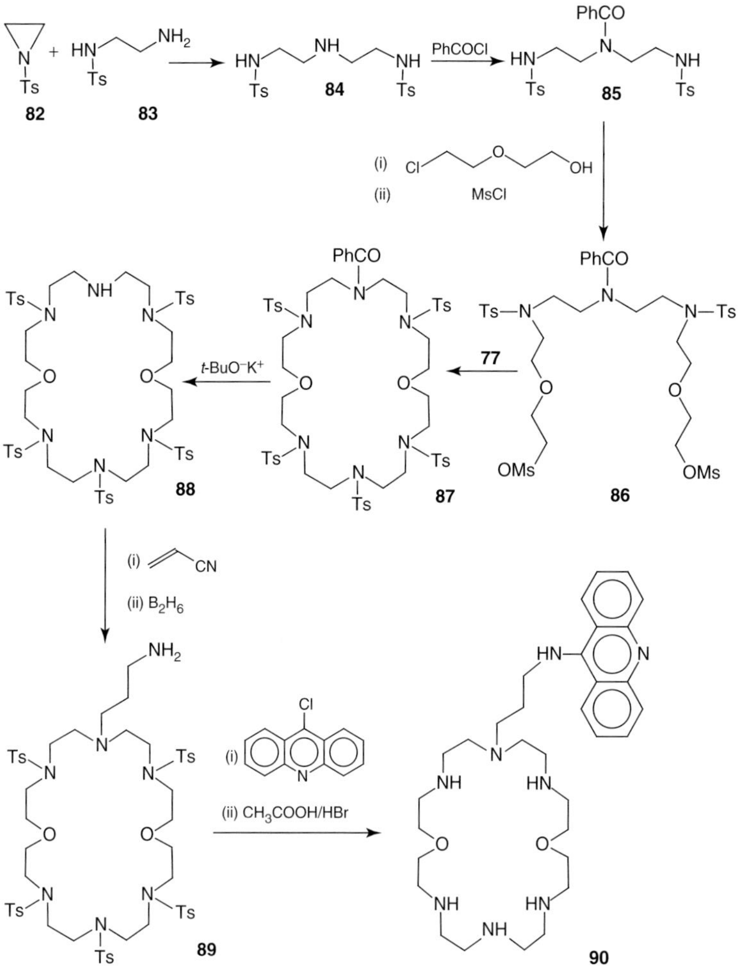

Scheme 4.11

simply changing the characteristics of the amine groups, that is, by replacing the secondary nitrogens with tertiary ones. In fact, N-methylation modifies the proton binding affinity of the receptor because of the lower basicity of tertiary nitrogens and increases its rigidity and hydrophobic characteristics.

From this point of view, the effect of N-methylation on the binding ability of polyazacycloalkanes toward inorganic phosphate anions and nucleotides have been analyzed in detail by using macrocycles (**91–93**) (Scheme 4.12) [37, 38], which are formally partially methylated derivatives of **65** and **66**. Synthesis of partially *N*-methylated polyazacycloalkanes generally requires a multistep procedure, outlined in Scheme 4.12 for **92** [37]. Reaction of bis(2-chloroethyl)methylamine (**94**) with the sodium salt of tosylated *N*-methylamine (**95**) affords the ditosyl derivative

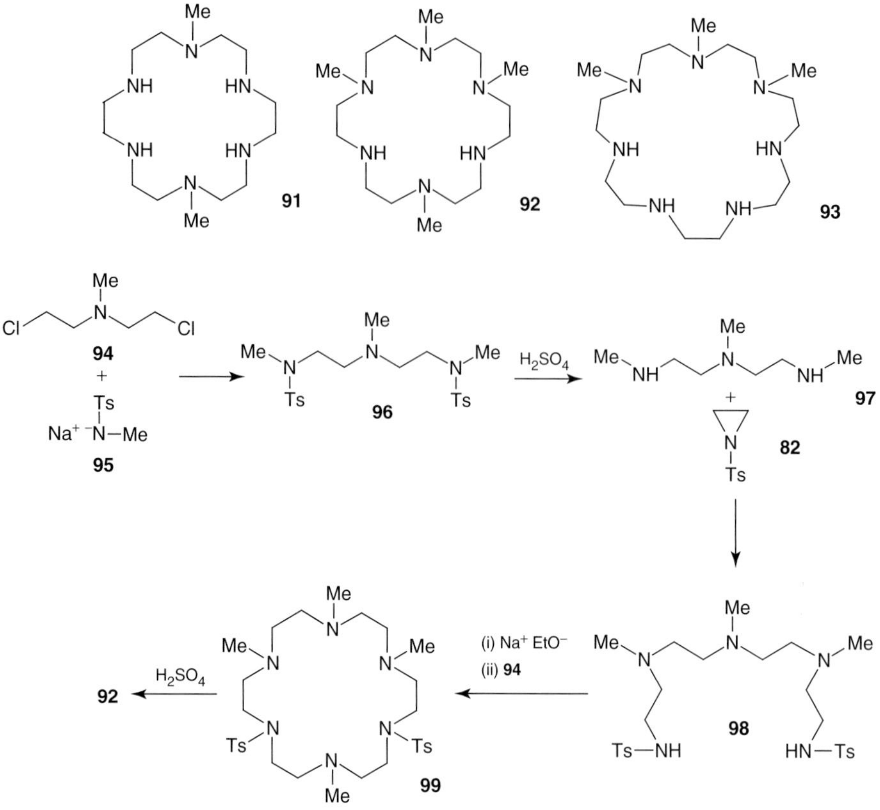

Scheme 4.12

of the *N*, *N′*, *N″*-trimethyltriamine (**96**). The tosyl groups are subsequently removed in concentrated H_2SO_4 to give **97**, which is used to obtain the partially methylated pentaamine (**98**) via reaction with tosyl aziridine (**82**). Cyclization between **94** and the disodium salt of **98** affords the ditosylated macrocycle (**99**). Subsequent removal of the tosyl groups gives the desired tetramethylated hexaamine macrocycle (**92**).

Of note, **92** contains two secondary amine groups, and therefore, it represents a versatile precursor to develop other anion receptors. In fact, the two secondary nitrogens can be used to append functional groups to the macrocyclic framework as potential anion-binding sites. Actually, reaction of **92** with 2-chloromethylpyridine leads to the insertion of two methylpyridine side arms. The resulting receptor displays an increased binding ability for ATP because of hydrophobic and π-stacking interactions between the pyridine pendants and adenine [39]. Alternatively, the two secondary amine groups can be connected by an appropriate bridging moiety, producing a cryptand-like architecture, which can be used for anion encapsulation (Scheme 4.28).

4.2.4
Macrocyclic Receptors Incorporating a Single Aromatic Unit

The acridine-bearing macrocycle (**90**) represents one of the most known examples of anion binding through multiple forces, that is, electrostatic, hydrogen bonding, and π-stacking interactions. In **90**, the aromatic moiety is appended to the macrocyclic backbone, and therefore, it is unlikely to affect the binding characteristics of the polyamine aliphatic moiety. Conversely, the insertion of aromatic or heteroaromatic subunits as an integral part of the cyclic structure can also be used to modulate the overall rigidity of the macrocyclic receptor, to increase the hydrophobic characteristics of the cavity, and above all, to organize the binding moieties of the receptor in order to obtain an appropriate disposition of the recognition sites for coordination of the selected anion.

From this point of view, the simplest example is the insertion of a benzene ring within an aliphatic polyamine backbone. The effect of this structural modification in anion binding with polyazacycloalkanes has been extensively analyzed by García-España and coworkers [3] by the synthesis of a variety of polyazacyclophanes differing in number and type of amine groups and in the dimension of the cavity, by using *ortho*, *meta*, or *para*-benzene spacers or inserting oxygen atoms within the aliphatic chain [40–42]. Examples of polyazacyclophanes used in anion coordination studies are shown in **100–104**.

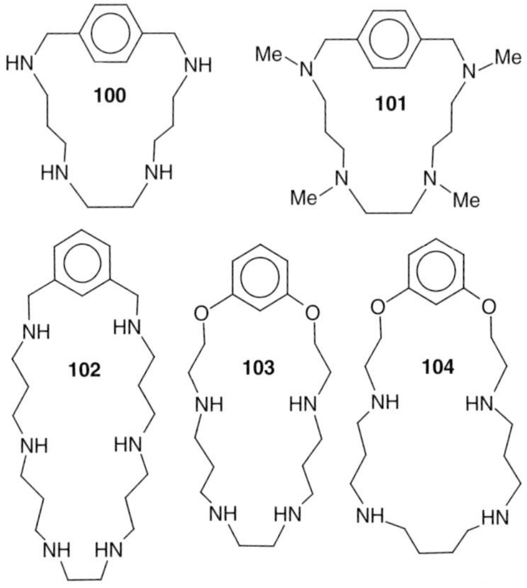

As exemplified in Scheme 4.13, synthesis of receptors such as **100** and **102** was generally achieved via cyclization of a tosylated polyamine with 1,4-, 1,3-, or 1,2-bis(bromomethyl)benzene in CH_3CN in the presence of K_2CO_3 or Cs_2CO_3 [40]. In fact, in most cases, the classic Richman and Atkins method [34], which uses the sodium salts of the tosylated polyamines as starting materials, failed. Removal of the tosyl groups was generally accomplished by using $CH_3COOH/HBr/PhOH$. In some cases, however, the strong acidic conditions of this detosylation method led to

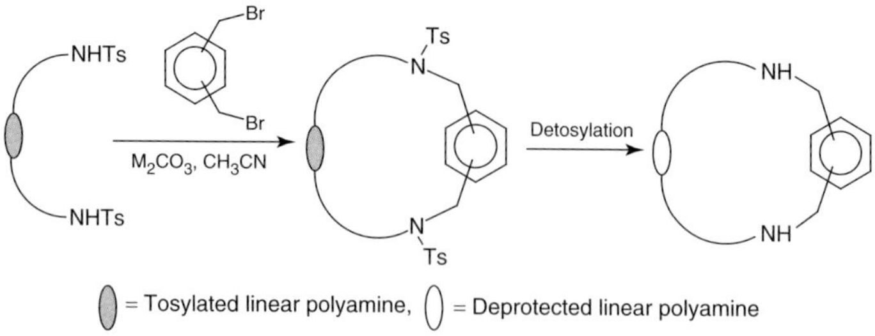

Scheme 4.13

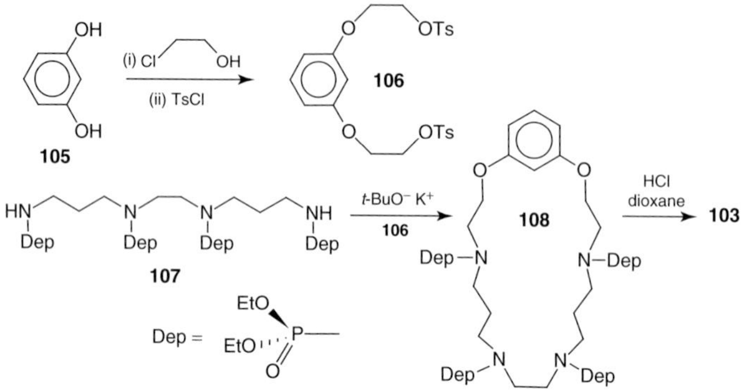

Scheme 4.14

cleavage of the benzylic N–CH_2 bonds, and alternative procedures, such as lithium in liquid ammonia [12, 40] or sodium amalgam in buffered methanol solution [40], had to be used. Methylation of nitrogen groups to produce **101** was achieved by treatment of the corresponding nonmethylated analog with HCHO/HCOOH [41].

Conversely, a different synthetic procedure was used to obtain resorcinol-based macrocycles, such as **103** and **104**. In fact, these macrocycles were obtained using linear polyamines protected with diethoxyphosphoryl (Dep) groups as starting materials, instead of the most common tosyl groups [42].

As exemplified in Scheme 4.14 for **103**, cyclization of Dep-protected tetraamine (**107**) in the presence of potassium *tert*-butoxide with the resorcinol derivative **106** affords the protected macrocycle **108**. Removal of the Dep groups, performed in 3 M HCl in dioxane, gives the final product **103**. Compound **106** was obtained by reaction of resorcinol (**105**) with 2-chloro-ethanol followed by tosylation under standard conditions. Although this synthetic procedure is not of general applicability, the use of Dep instead of tosyl groups leads to cyclization products more soluble in common organic solvents, facilitating their purification. Furthermore, the Dep groups can be removed in remarkably milder conditions than tosyl groups.

A common characteristic of cyclophanes (**100–104**) is their ability to bind nucleotides through multiple forces, including π-stacking pairing between the aromatic ring and the nucleobases. In principle, a larger aromatic or heteroaromatic unit can give stronger interactions with aromatic substrates and may represent a better recognition site for the bases of nucleotides than a simple benzene unit. Therefore, the insertion of polycyclic heteroaromatic subunits, such as phenanthroline, within a cyclic polyamine backbone can exert a beneficial effect on the stability of the resulting complexes with nucleotides. Actually, receptors **109–111**, featuring a polyamine chain connecting the 2,9-positions of a 1,10-phenanthroline, form highly charged polyammonium cations at neutral pH value, capable of strongly interacting with nucleotides [43–45]. Even if charge–charge and hydrogen bonding contacts remain the major interaction mode, the ability of phenanthroline to behave as a binding site for nucleobases strongly reinforces the receptor–substrate interaction and, above all, determines selectivity in nucleotide triphosphate binding. In fact, **111** is able to selectively bind ATP over CTP, TTP, and GTP, because of slightly different interaction modes between the nucleobases and phenanthroline. The synthesis of these receptors was accomplished by using a procedure analogous to that reported for **100** and **102**, for example, by cyclization of 2,9-bis(bromomethyl)phenanthroline with the appropriate tosylated polyamine. Removal of the tosyl groups was performed in $CH_3COOH/HBr/PhOH$ [43, 44].

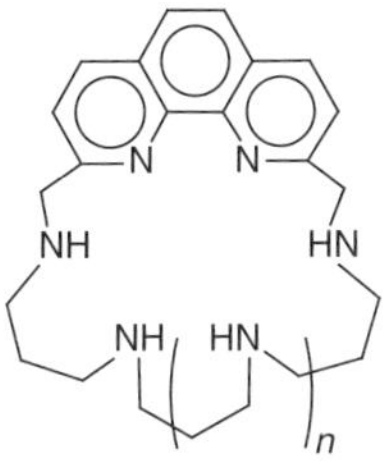

109: $n = 1$
110: $n = 2$
111: $n = 3$

4.2.5
Macrocyclic Receptors Incorporating Two Aromatic Units

The insertion of two aromatic or heteroaromatic units within the cyclic structure can be used to separate two equal or different polyamine moieties, increasing the organization of the receptor for anion binding. The simplest examples are constituted by polyamine macrocycles featuring two triamine units separated by two aromatic rings, in most cases benzene, pyridine, or furan (**112–118**), whose ability in the coordination of ATP, ADP, inorganic phosphate anions, and carboxylic acids has been extensively analyzed by Martell and coworkers [45].

These ligands were synthesized by 2 + 2 condensation of diethylenetriamine (**119**) or dipropylenetriamine with the appropriate dialdehyde, followed by reduction with borohydride of the resulting Schiff bases, as schematically shown in Scheme 4.15 for the synthesis of diethylenetriamine-based macrocycles [46, 47].

Unlike cyclization occurring via double nucleophilic substitution (Scheme 4.13), cyclization via Schiff base condensation of aliphatic linear triamines with aromatic dialdehydes does not afford the 1 + 1 condensation products. The different reaction pathway is likely to be ascribed to the higher rigidity of the dialdehyde precursors with respect to the corresponding bis(bromomethyl) aromatic precursors used in the cyclizations via nucleophilic substitution. The use of short triamine chains

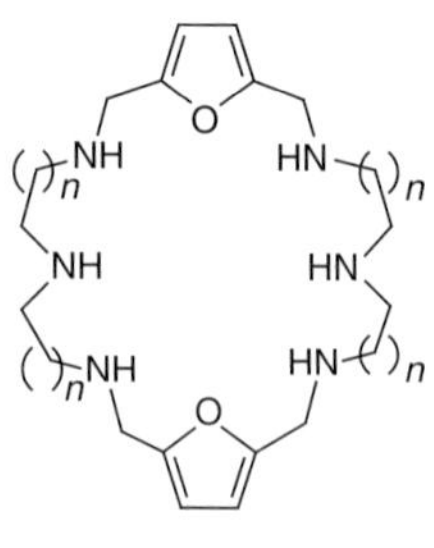

112: $n = 1$
113: $n = 2$

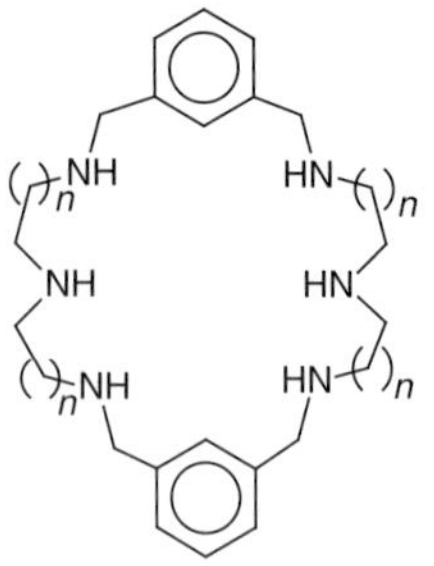

114: $n = 1$
115: $n = 2$

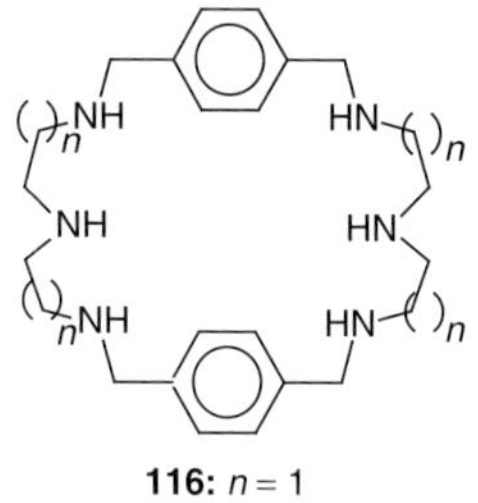

116: $n = 1$
117: $n = 2$

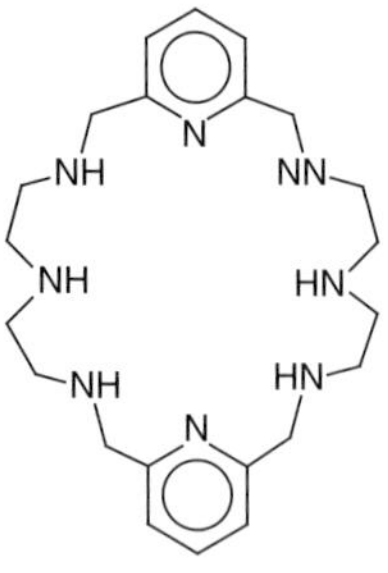

118

appears to play a minor role in determining the reaction pathway. In fact, cyclization of *p*-bis(bromomethyl)benzene with tosylated diethylenetriamine or dipropylenetriamine invariably leads to the formation of the 1 + 1 macrocycles [40]. However, other factors can influence the preferential formation of 1 + 1

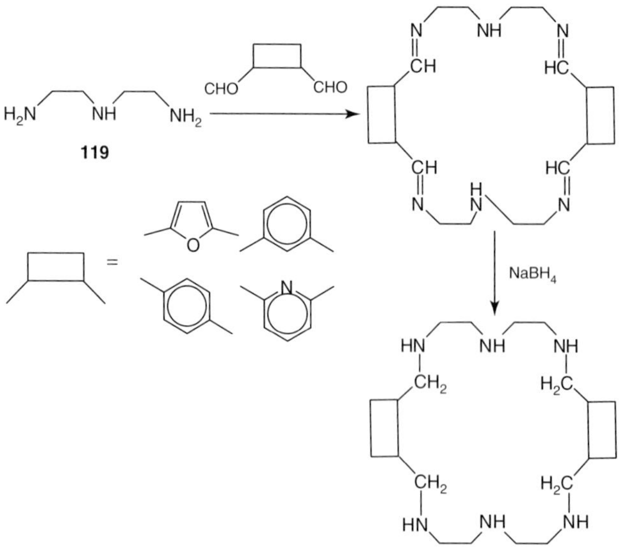

Scheme 4.15

macrocycles over oligomerization products, including concentration of reactants, reaction rate, solvent used, and functionalization of the amine groups. From this point of view, the presence of bulky tosyl groups reduces the number of conformational degrees of freedom, such as bond rotation, in the reactants and/or intermediates. This reduction is thought to facilitate 1 + 1 cyclization with respect to the formation of oligo- and polymerization products [48].

The Schiff base procedure has subsequently been used to assemble a variety of ditopic macrocycles featuring two polyamine units, including 1,5-diamino-3-oxa-pentane (see ligand **120**) [49] and *N,N*-bis(aminoethyl)methylamine (**121**) [49, 50] or *N,N*-bis(aminopropyl)methylamine (**122**) [51], and aliphatic diamines of different lengths (**123–125**) [52], separated by benzene or pyridine spacers. Owing to their ditopic molecular architecture, these macrocycles have mainly been used as receptors for substrates containing two separated anionic functions, such as aromatic or aliphatic dicarboxylates [4, 49, 51, 52], as well as for the simultaneous binding of two halogenide anions [49].

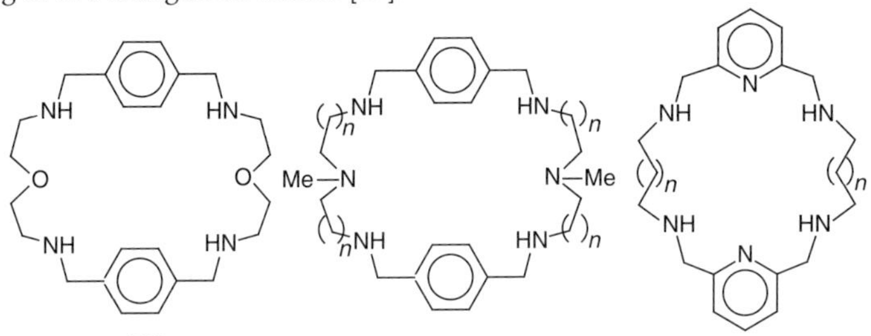

120 **121:** $n = 1$ **122:** $n = 2$ **123:** $n = 2$ **124:** $n = 3$ **125:** $n = 4$

Larger aromatic or heteroaromatic bridges, including naphthalene [53], phenanthroline [54, 55], terpyridine [56], acridine [57], or quinacridine [58, 59], connecting two polyamine moieties, have also been synthesized to enhance the binding ability toward anions containing aromatic portions because of the stabilizing effect of hydrophobic and π-stacking interactions (see, for example, receptors **126–132**). Furthermore, these receptors may define a large hydrophobic cavity, where aromatic sections of substrates can be conveniently lodged [54, 55, 57, 59]. This is the case, for instance, of receptors **128–132**, which were designed to target nucleotides or aromatic carboxylate anions [55, 57]. Of note, encapsulation of the aromatic subunits of the substrates may give rise to peculiar binding properties, such as selective recognition of guanosine mono-, di-, and triphosphates in the case of **128** [57], or simultaneous binding of two guanosine monophosphate anions within the hydrophobic cavity of **131** [59]. These receptors were generally synthesized starting from the appropriate dialdehyde derivative of the aromatic moieties via the procedure depicted in Scheme 4.15, with the only exception of **126**. The latter was synthesized by 2 + 2 cyclization between ditosylated ethylenediamine and 6,6″-bis(bromoethyl)-[2,2′ : 6′,2″]terpyridine, followed by removal of the tosyl groups in CH_3COOH/HBr in the presence of phenol [56].

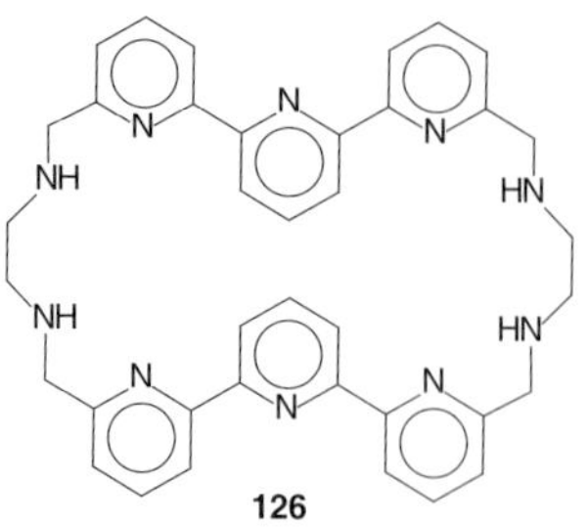

126

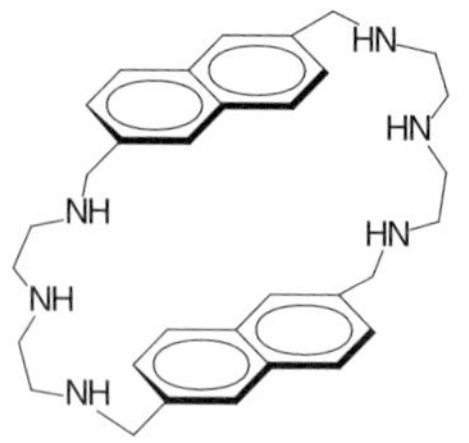

127

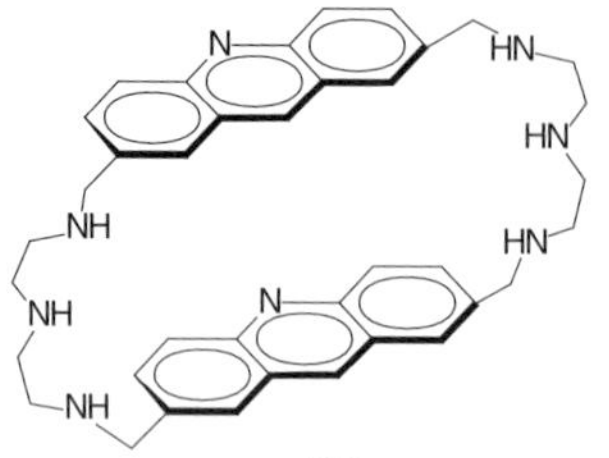

128

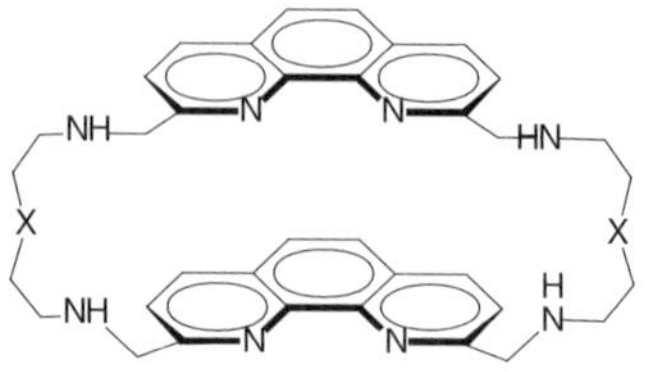

129: X = NH
130: X = O

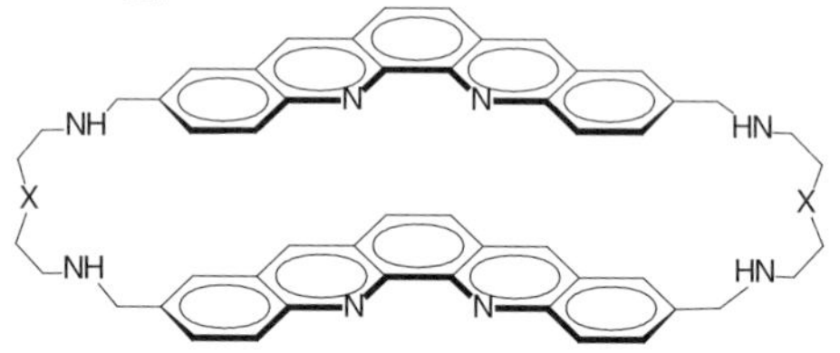

131: X = NH
132: X = O

The 2 + 2 cyclizations described above represent an elegant and relatively simple two-step procedure to assemble macrocyclic structures. On the other hand, this method can be used only to synthesize macrocyclic receptors containing two equal polyamine-binding subunits, mostly triamine moieties, connected by two equal spacers. To overcome this limit, several multistep procedures have been developed to build up macrocyclic architectures featuring two different polyamine units linked by aromatic or heteroaromatic spacers.

Receptors **138–140**, designed in the course of a study on ATP and ADP binding by polyammonium macrocycles [60, 61], represent the first example of ditopic macrocycles containing two different polyamine moieties. As sketched in Scheme 4.16, the tris-methylated triamine (**97**) (Scheme 4.12) is used to achieve the bis(*p*-(α-chloromethyl)benzyl derivative (**135**). The latter is obtained by reaction of **97** with the methyl ester of *p*-(α-chloromethyl)benzoic acid (**133**). The resulting diester **134** is reduced to the corresponding dialcohol, which is subsequently converted to **135**. Reaction of **135** with the reinforced pentaamine (**136**), the triamine (**97**), or the oxa-azamacrocycle (**137**) afforded receptors **138–140**, respectively. Compound **135** represents a potentially powerful precursor to assemble ditopic macrocyclic structures containing two different polyamine units because of the simultaneous presence of two chloromethyl groups available to react with bis-nucleophilic agents, such as **136**, **97**, or **137** in the present case.

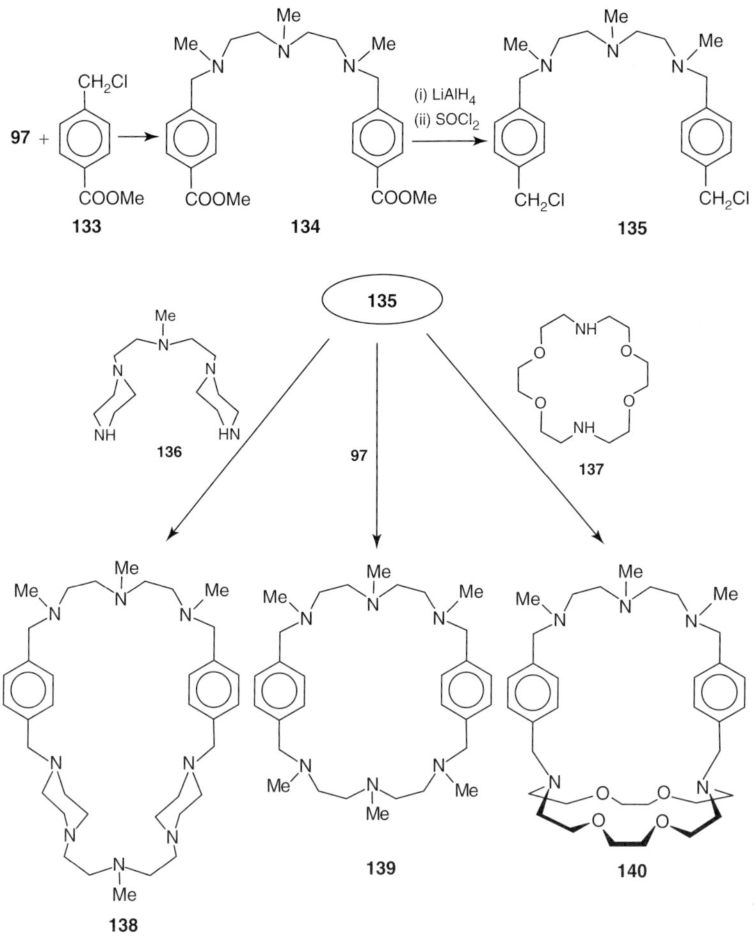

Scheme 4.16

More recently, Teulade-Fichou and coworkers have reported an example of [1 + 1] cyclocondensation to produce ditopic macrocycles containing two polyamine units, one of which contains a lateral aminopropyl side arm (**141** in Scheme 4.17). The primary amine group of the monobrachial macrocycle (**141**) may be employed for introduction of various functional units, including a pyrene unit for fluorescence sensing of aromatic dicarboxylate anions [62].

The procedure to obtain **141** uses dialdehyde **142** and the monoprotected tripodal amine **143** as starting materials. Their 1 + 1 condensation leads to the Boc-protected macrocyclic Schiff base (**144**), which is reduced *in situ* with $NaBH_4$ to **145**. Purification of the cyclization product was achieved via Boc protection of the two amine groups of **145** to give **146**, followed by simple column chromatography. Removal of the Boc groups with HCl affords the final product **141**. The pyrene unit was then introduced by Schiff base condensation of the primary amine group

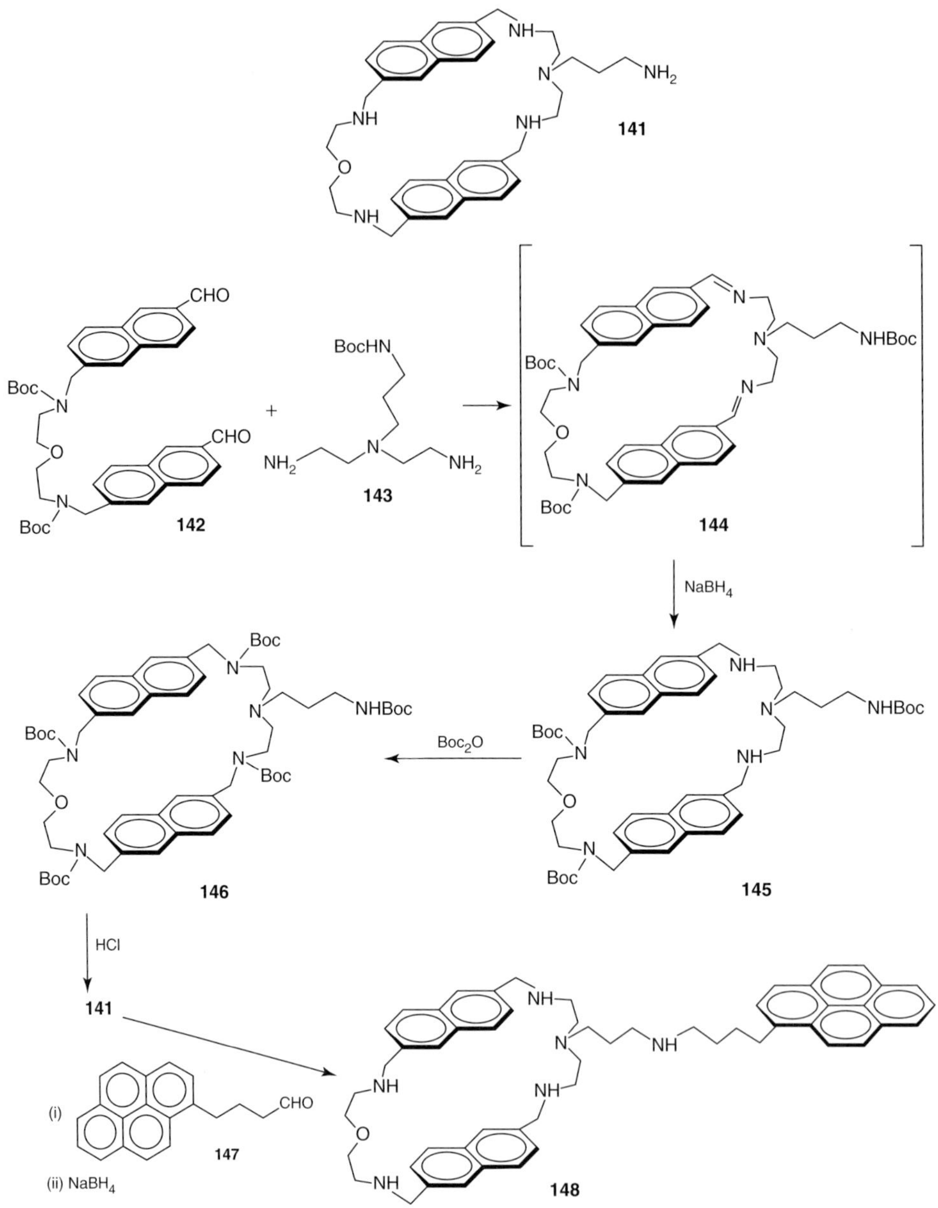

Scheme 4.17

of the side arm with the aldehyde derivative of pyrene (**147**). Of note, the starting reactants **142** and **143** are also obtained via multistep pathways. Therefore, the synthetic procedure described above points out that the achievement of receptors with appropriately tailored molecular architectures, in this case a versatile precursor to assemble monobrachial ditopic macrocycles bearing specific functional groups, may require an accurate design and an appropriate tuning of many sequential chemical reactions.

4.2.6
Anion Receptors Containing Separated Macrocyclic Binding Units

In this reconstruction of the structural evolution of polyamines toward more and more preorganized receptors for anion binding, ligands **149–152** (Scheme 4.18) represent a sort of "connection link" between foldable and rigid hosts, the latter having the necessary geometrical complementarity to selectively recognize the anionic partner via encapsulation without undergoing excessive conformational changes on host–guest interaction (see below).

These receptors, developed by Handel and coworkers to bind inorganic phosphate anions and AMP, ADP, or ATP [63, 64], feature two cyclen units facing each other at a distance that depends on the aromatic spacer and on their degree of protonation, thus creating pseudocavities of varying dimensions for phosphate anion hosting. Their synthesis was achieved via a synthetic strategy involving tetraazamacrocyclic bis-aminals, as sketched in Scheme 4.18 [65].

This synthetic pathway implies reaction of the cyclen bis-aminal (**153**), easily attainable via condensation of cyclen with glyoxal, with the appropriate bis(bromomethyl) derivatives of benzene or pyridine. The bis-quaternary ammonium cations **154–157** are isolated as bromide salts and treated with hydrazine to give the final bis-cyclen derivatives (**149–152**). Bis-aminals of tetraazamacrocycles,

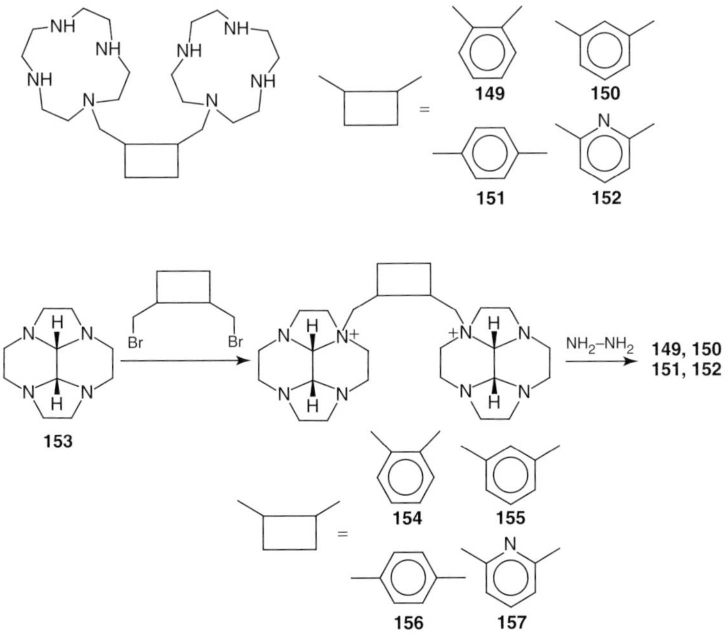

Scheme 4.18

such as cyclen or cyclam, present a *cis* configuration of the central two-carbon bridge and possess a folded geometry that induces discrimination between the two pairs of opposite nitrogen atoms. In fact, the lone pairs of two opposite nitrogen atoms of cyclen or cyclam are directed toward the convex side of the molecule and are more nucleophilic, while the lone pairs of the remaining two nitrogen donors point out from the concave fold and are more suitable to react with eletrophilic agents. Therefore, double functionalization of bis-aminals occurs on two nonadjacent nitrogen atoms, for example, on nitrogens in 1 and 7 positions of the cyclen ring (or 1 and 8 of cyclam).

Bis-aminals are versatile tools to assemble a variety of molecular architectures of potential interest for anion binding [65]. For instance, bis-macrocycles containing two cyclam or homocyclam (homocyclam = 1,4,8,12-tetraazacyclotetradecane) rings bridged by aromatic or aliphatic spacers can be produced by using the bis-aminals of these macrocycles as starting reagents. Furthermore, depending on the solvent used, reaction of bis-aminals with bis(bromoalkyl) or bis(bromoaryl) derivatives may afford monofunctionalized monocharged cations, which can be versatile precursors to assemble either bis-macrocycles containing two different tetraazamacrocyclic moieties (**165–167** in Scheme 4.19, for example) or bis-macrocycles containing two functionalized pendant arms (**170** in Scheme 4.20) [65].

A way to constrain two cyclen rings to stay face to face at a fixed distance is to link them with two aromatic spacers in rigid macrotricyclic systems such as **175–178** (Scheme 4.21) [66, 67]. On protonation, these receptors exhibit a higher

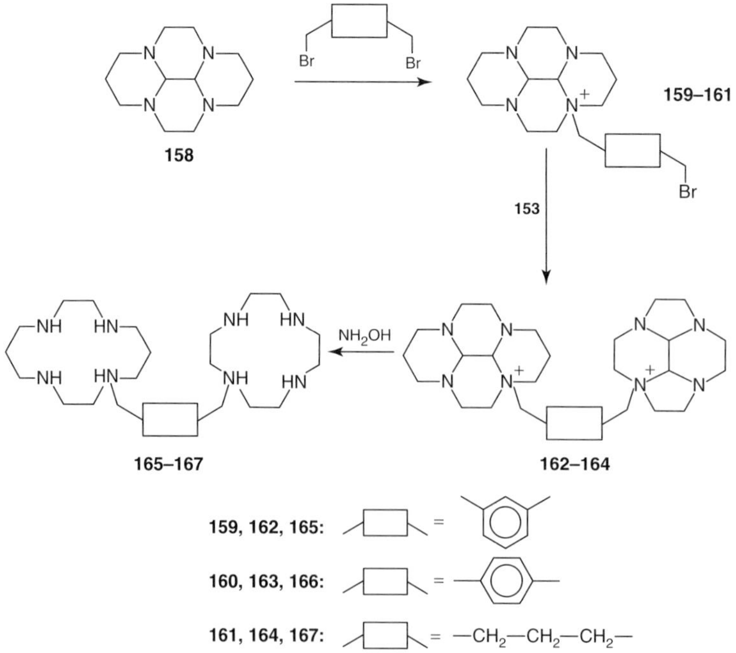

Scheme 4.19

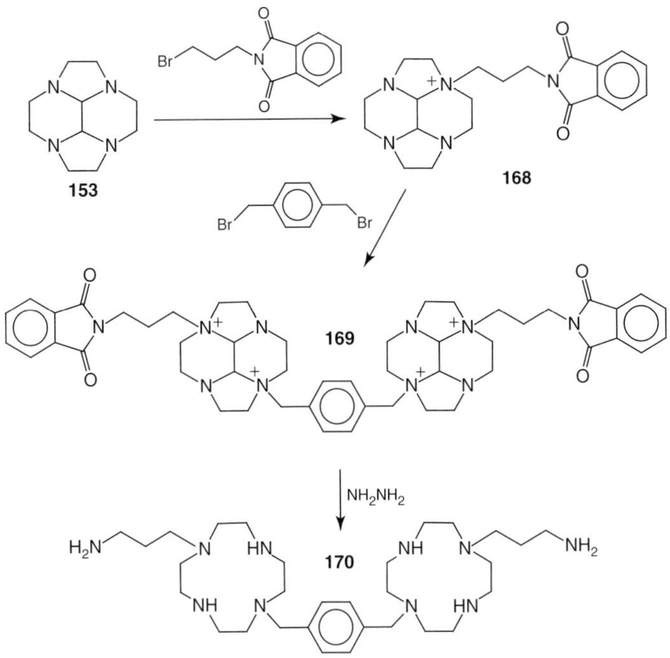

Scheme 4.20

binding affinity for inorganic phosphate anions than their monobridged homologs **149–152** (Scheme 4.18), due to the rigid cavities delimited by the two facing cyclen units and by the aromatic linkers, which are preorganized for a size-dependent encapsulation of phosphate anions [66]. These receptors were obtained either by a one-pot reaction of the bis-aminal of cyclen (**153**) with *para*-, *meta*-, or *ortho*-bis(bromomethyl)benzene or 2,6-bis(bromomethyl)pyridine or by using the bis-quaternary ammonium cations **154–157** (Scheme 4.18) as starting materials, as shown in Scheme 4.21 [67]. In both cases, the resulting tetraammonium cations are subsequently transformed in the bis-macrocyclic receptors **175–178** with hydrazine. Of note, the use of **171–174** as reactants may enable achievement of not only bis-macrocycles containing two equal spacers separating the two cyclen moieties, such as **175–178**, but also dissymmetric receptors featuring two different aromatic linkers between the two tetraazamacrocyclic rings.

However, alternative strategies to bis-aminal-based syntheses to obtain bis- or tris-macrocyclic structure are known. As a representative example, a three-step procedure to obtain tripodal tris-macrocycles is sketched in Scheme 4.22 [68, 69]. The C_3 symmetry displayed by these receptors allows them to selectively bind the 1,3,5-benzene-tricarboxylate anion over the other benzene-tricarboxylate anions [69]. The procedure developed for the synthesis of these tris-macrocycles utilizes the simple starting material **179**, which can be obtained by reaction of tren with *N*-tosylated aziridine in high yields [68]. Reaction of its hexasodium salt, obtained with standard methods (treatment with sodium ethanolate), with tosylated

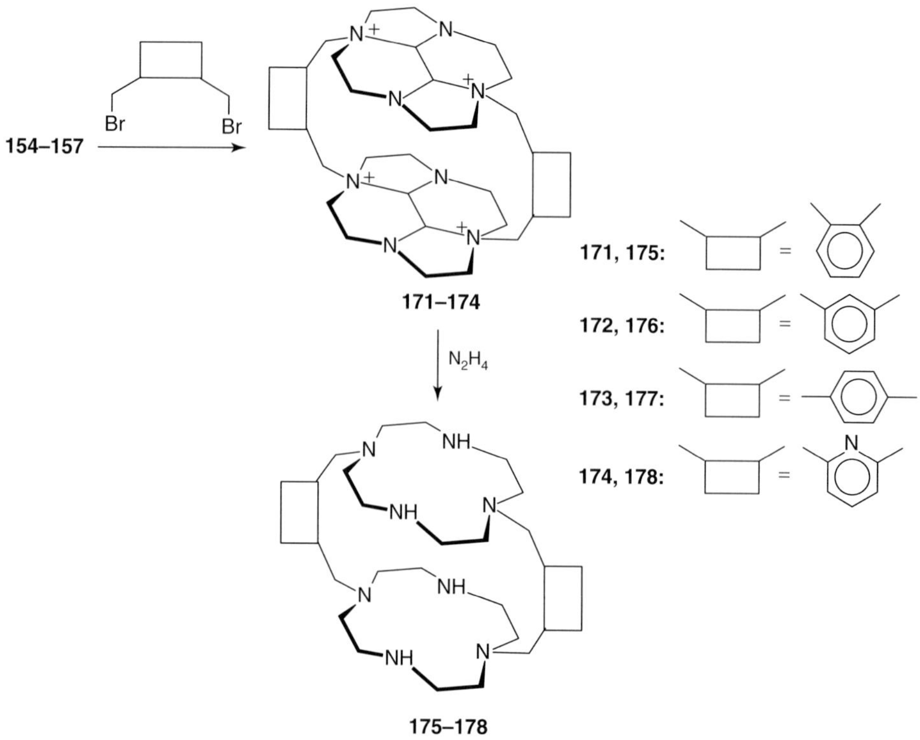

Scheme 4.21

diethanolamine (**180**), diethylene glycol (**181**), or dipropanolamine (**182**) in 1 : 3 molar ratio was carried out by using the Richman and Atkins procedure and afforded, after purification by column chromatography, the tosylated macrocycles **183**, **184**, and **185**, respectively. Finally, removal of the tosyl groups to give **186–188** was performed in concentrated H_2SO_4 (**186**) or, in the case of **187** and **188**, in HBr/CH_3COOH mixture in the presence of phenol. In fact, H_2SO_4 induces oxidative cleavage of both the ether linkages of **187** and the propylenic chains of **188** [68].

4.2.7 Cryptands

The idea that a three-dimensional inclusion of an anionic guest within the rigid intramolecular cavity of a hollow receptor molecule would increase the selectivity and strength of the host–guest interaction, following the principle of a geometrical size/cavity and binding complementarity, has guided and driven the design and synthesis of numerous cagelike frameworks featuring coordination centers well preorganized for complexation. In particular, bicyclic tren-based cryptands, featuring different kinds of spacers (aliphatic, aromatic, heteroaromatic) between the two

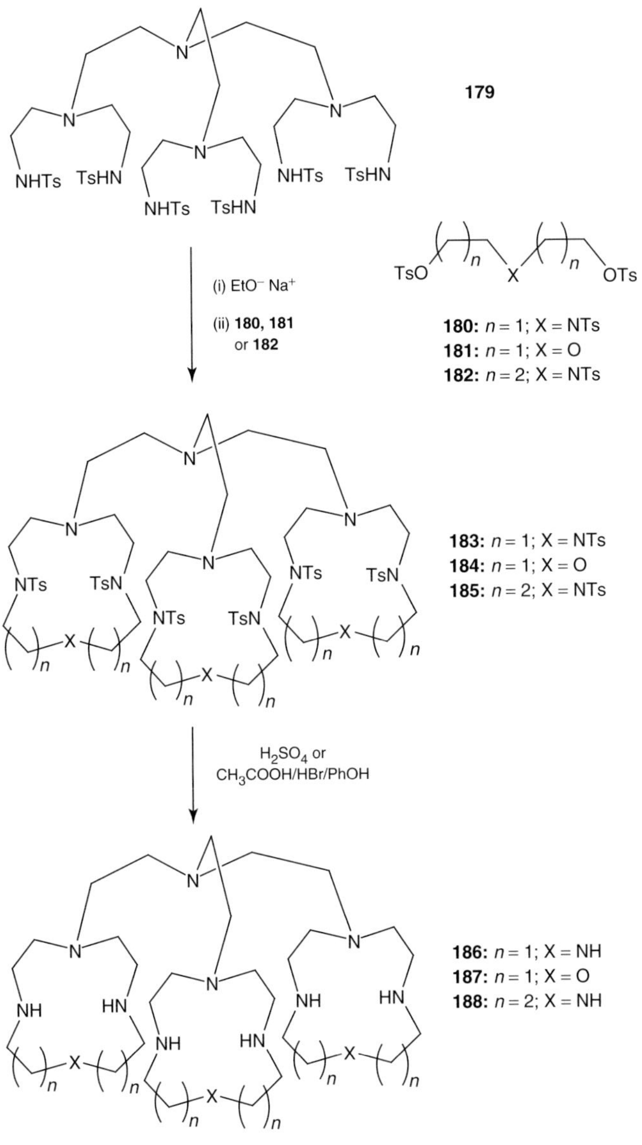

Scheme 4.22

tripodal moieties for providing specific hollow sizes and binding sites within their cavities, have been extensively studied in their protonated forms as receptors for anions.

Early in the era of anion coordination, Lehn and coworkers showed that receptor **189** [70, 71], featuring two tren units linked by a diethyl ether link, is able to encapsulate monoatomic anions, such as halides, and the linear three-atomic azide ion, with a marked selectivity for the latter. Synthesis of this receptor was achieved

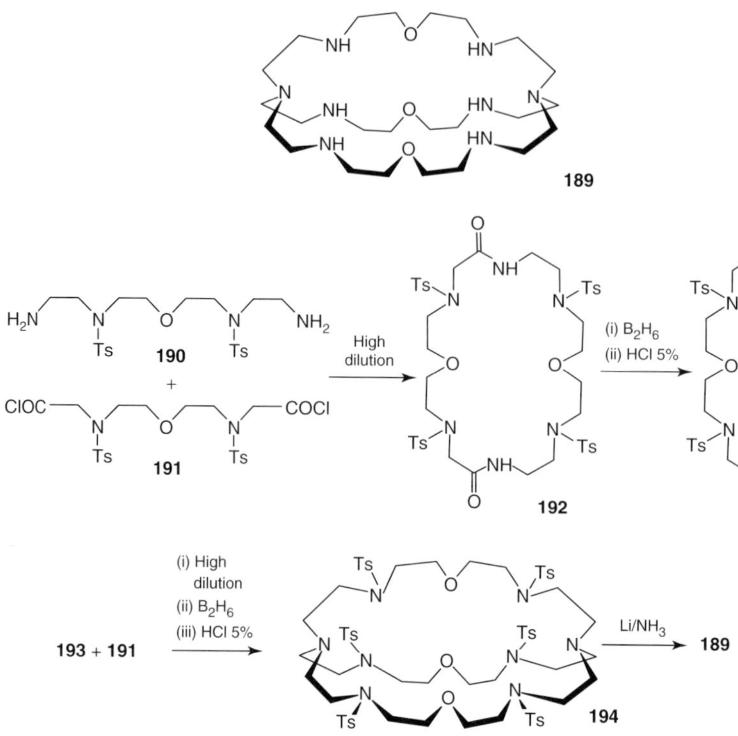

Scheme 4.23

by means of the multistep procedure depicted in Scheme 4.23 [72]. Receptor **189** is obtained starting from the parent tetratosylated macrocycle **193**, which was synthesized via cyclization of diamine (**190**) with acyl chloride (**191**) under high dilution conditions, followed by reduction of the resulting diamide **192** with diborane and hydrolysis of the amine–borane adduct with HCl. Condensation of macrocycle **193** with **191** and subsequent reduction of the resulting diamide affords the tosylated cryptand (**194**), which is finally deprotected with lithium in liquid ammonia to give **189**.

The polyazamacrobicycle (**195**) can be considered the simplest cryptand based on two tren units. In fact, it is composed of two tren units coupled by three short ethylenic chains. Designed for its small cavity, it binds fluoride with exceptionally high affinity and selectivity over larger halide anions [73]. From a synthetic point of view, this receptor can be obtained following a multistep procedure similar to that used for **189** (Scheme 4.24), for example, via assembly of the tetratosylated macrocycle **198** followed by its reaction with the diacyl chloride (**197**). The resulting macrobicyclic diamide (**199**) (28% yield) is reduced with diborane. Final removal of the tosyl groups with lithium in liquid NH_3 gives the desired polyamine macrobicycle (**195**). More interestingly, reaction between **197** and **198** also affords, although in lower yield (9%), the macrotricyclic product **200**, as a result of a competitive 2 + 2 condensation. Compound **200** was then transformed into the

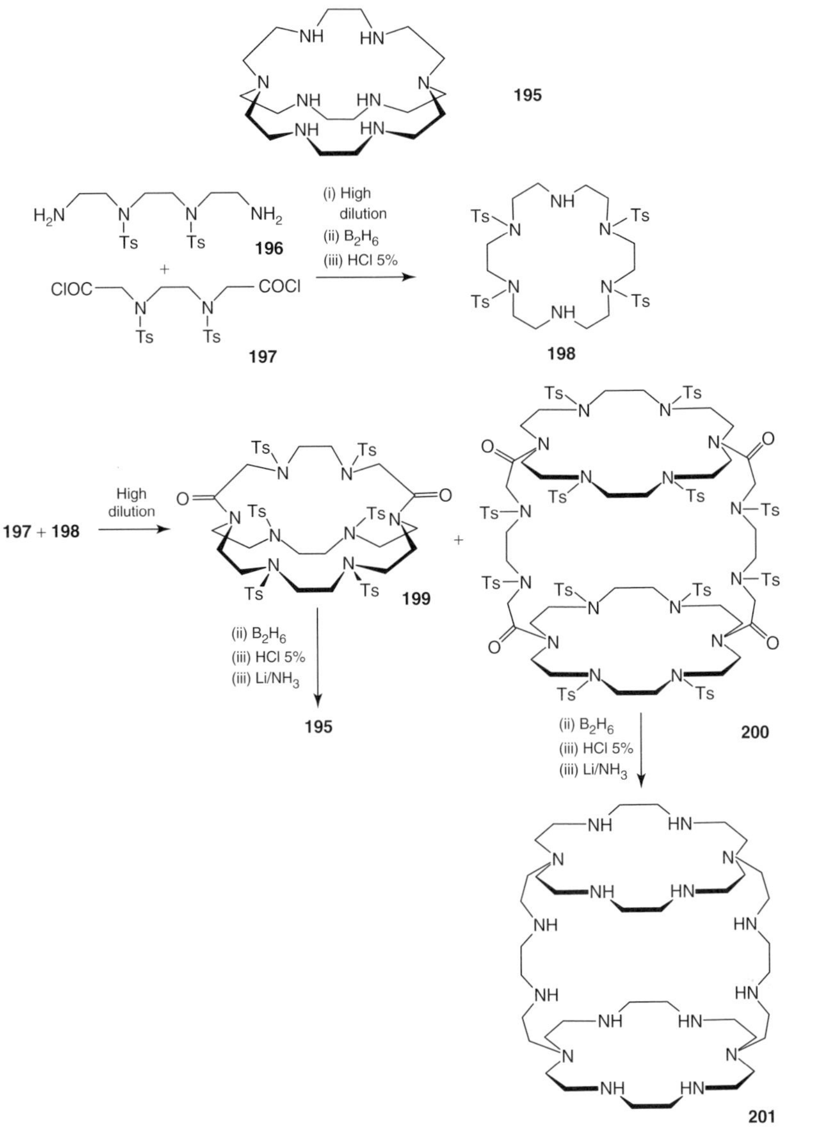

Scheme 4.24

polyamine macrotricycle (**201**), by means of the same procedure used for **195** [71]. The formation of **200** can be considered a representative example of 2 + 2 cyclocondensation resulting in the assembly of macrotricyclic structures. Different factors can contribute to favor the 2 + 2 reaction pathway: (i) the small sizes of the monocycle and of the incoming chain, (ii) the rigidity of the incoming chain, and (iii) steric hindrance, the last two factors probably being the most significant in the case of the formation of **200**.

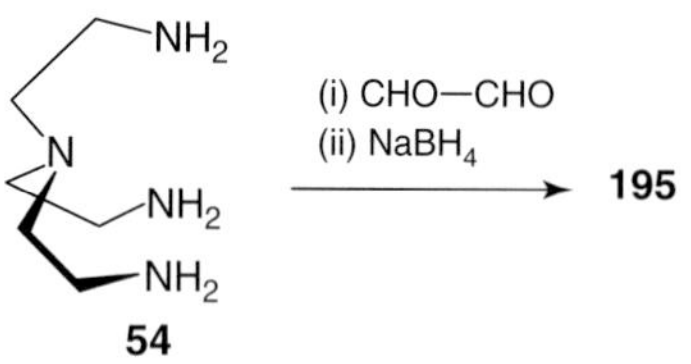

Scheme 4.25

Of note, cryptand (**195**) was also obtained by the multiple condensation method, depicted in Scheme 4.25, involving coupling of two tripodal tetraamine tren (**54**) with glyoxal, followed by reduction of the Schiff base with sodium borohydride [73]. The yield of the condensation (25%) was similar to that found for **199**, and no formation of dimeric products was observed. While this method provides direct access to the final compound, the multistep procedure in Scheme 4.24 is more versatile and makes accessible intermediate products, such as **198**, potentially useful for the synthesis of other receptor molecules.

Coupling of two tren units with aromatic or heteroaromatic dialdehydes, followed by reduction of the resulting Schiff bases, a method originally developed by Nelson and coworkers [6, 74, 75], has produced a variety of cryptands featuring two tripodal tren units separated by aromatic units (**202–206**) (Scheme 4.26). These receptors possess larger cavities than **189** and **195** and therefore are able to encapsulate larger inorganic tetrahedral oxoanions, including dihydrogen phosphate, perchlorate, sulfate, chromate, perrhenate. Furthermore, the particular molecular architecture of these receptors, constituted by three rigid spacers separating two polyamine units with C_3 symmetry, may allow them to encapsulate two inorganic anions, such as F^- or nitrate, each coordinated by a single protonated tren moiety [2, 6, 8], or substrates containing two separated monoanionic functionalities, such as aliphatic dicarboxylates.

Following this synthetic procedure, a number of cryptands containing two tren units linked by aromatic spacer have been synthesized, including receptors containing extended aromatic or heteroaromatic units and capable of binding in an inclusive manner larger anionic substrates. The first and most famous example is **207** [76], which was found to be able to encapsulate the terephthalate anion [1, 8].

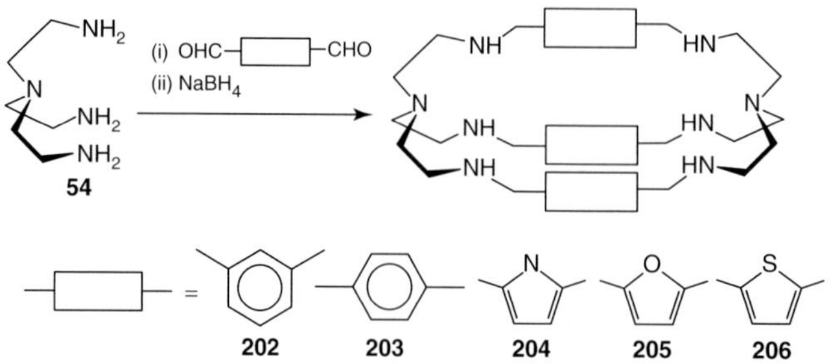

Scheme 4.26

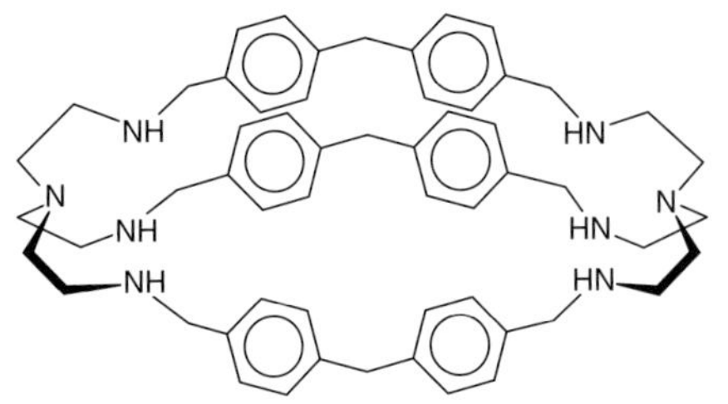

207

Although ligands featuring two tripodal tetraamines linked by aromatic or aliphatic spacers have been probably the most investigated class of cryptands in anion coordination chemistry, other receptors containing a three-dimensional cavity for anion encapsulation have recently been synthesized. Some examples are sketched in **208–212**. The synthetic procedure used is similar to those developed by Lehn and Nelson for the bis-tren cryptands described above.

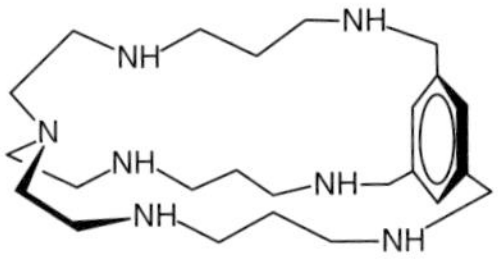

208

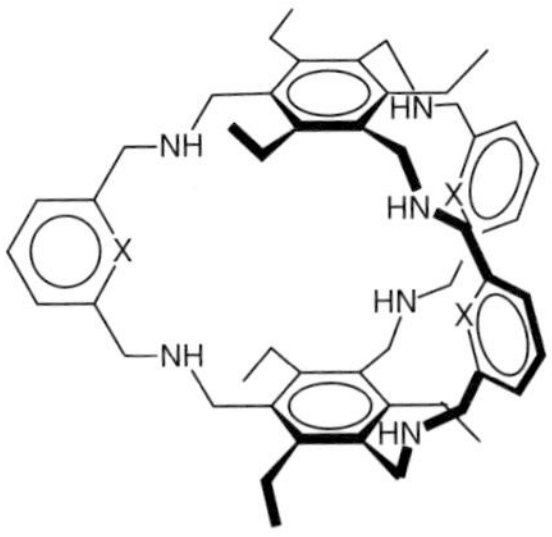

209: X = CH
210: X = N

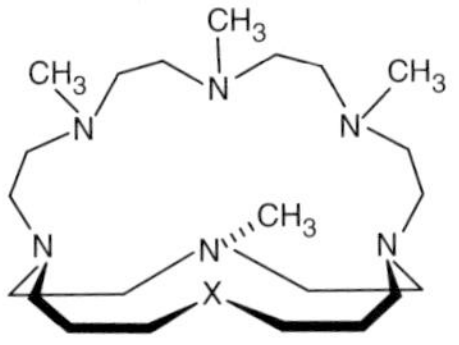

211: X = NH
212: X = O

Cryptand (**208**), designed by Steed and coworkers [77] for the inclusion of halide anions, was synthesized via cyclization of the tosylated tripodal amine (**213**), obtained starting from tren and by adopting already described procedures [11], with 1,3,5-tris(bromomethyl)benzene (**214**) in high dilution conditions in the presence of Cs_2CO_3 as a base. The resulting hexa-tosylated receptor is subsequently deprotected in concentrated HBr in the presence of phenol to give **208** (Scheme 4.27).

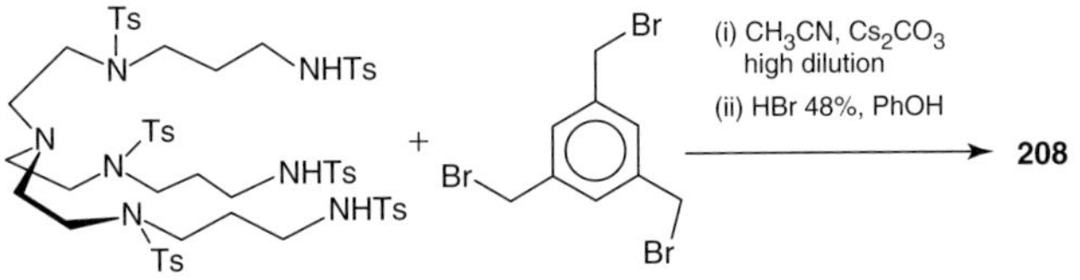

213 **214**

Scheme 4.27

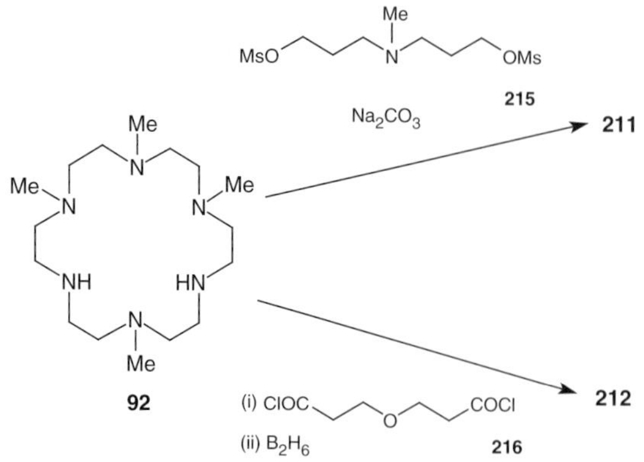

Scheme 4.28

Cryptands **209** [78] and **210** [79] feature a larger cavity than **208**, making them suitable hosts for oxoanions, in particular tetrahedral oxoanions, such as sulfate and monohydrogen phosphate, which are encapsulated within the three-dimensional cavity of these receptors. Cryptands **209** and **210** were obtained, respectively, by condensation of 1,3,5-tris(aminomethyl)-2,4,6-triethylbenzene with isophthaldehyde or 2,6-pyridine-dicarbaldehyde in 2 : 3 ratio, followed by reduction of the resulting imine with $NaBH_4$.

Unlike the tren- or tris(aminomethyl)-benzene-based cryptands described above, **211** and **212** display less symmetric structures. However, they are capable of binding halide anions in an inclusive manner and display different selectivity properties: while **112** selectively binds F^- over Cl^- and Br^- in a wide pH range, **111** preferentially binds Cl^- over F^- in the neutral pH region and F^- over Cl^- at acidic pH values [80]. Their synthesis uses the tetramethylated macrocycle **92** (Scheme 4.12) as starting material. Reaction of **92** with dimesylated *N*-methyl-bis(3-hydroxypropyl)amine (**215**) directly affords the cryptand **111** [37], while condensation of **92** with oxybis(propionyl) chloride (**216**), followed by reduction with diborane of the resulting diamide derivative, gives **212** (Scheme 4.28) [81].

4.3 Design and Synthesis of Amide Receptors

Amide bonds have a paramount relevance in biology because they constitute the building block that ties together amino acids in protein skeletons. However, amide bonds are not only relevant to biological systems since they are also present in a vast array of molecules including many largely marketed drugs. Amide groups have also come to be used in the field of anion coordination chemistry. It is interesting to note that anion coordination in proteins is mostly achieved through neutral amide functions employing the hydrogen bonding donor properties of their NH groups. Mimicking this behavior, many anion receptors have been built incorporating amide motifs. Indeed, amide-based receptors currently constitute a classic class of anionic receptors, and therefore, we have selected this kind of molecules as representative for neutral receptors that make use of hydrogen bonding as the major driving force in anion coordination.

Amide receptors are typically synthesized reacting carboxylic acids with amines. However, this reaction does not proceed readily at room temperature and requires an increase in the temperature to above 200 °C to favor the necessary elimination of water [82]. However, an increase in temperature can be a disadvantage for the stability of the substrates. Therefore, for this reaction to occur, it is usually necessary to first activate the carboxylic acid, transforming its OH group into a good leaving group. Once this first step is carried out, the activated carboxylic acid reacts with the amine group in reasonable yield [83].

4.3.1 Acid Halides as Starting Materials

One of the most popular ways of activating carboxylic acids is their transformation into acid halides. As early as in 1901, Fischer made use of these compounds as starting material to synthesize amides [84]. The general method consisted in using reagents such as thionyl chloride, oxalyl chloride, or phosphorus pentachloride to generate the acid chloride. Bis(trichlorodimethyl)carbonate (BTC) has also been reported to generate amino acid chlorides [85, 86].

4.3.1.1 Acyclic Amide Receptors

Some examples of simple acyclic diamide ligands synthesized by Crabtree [87, 88] are collected in **217–220**.

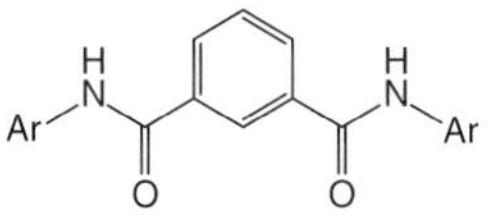

217: Ar = Ph
218: Ar = p-(n-Bu)C_6H_4
219: Ar = 2,4,6-$Me_3C_6H_2$

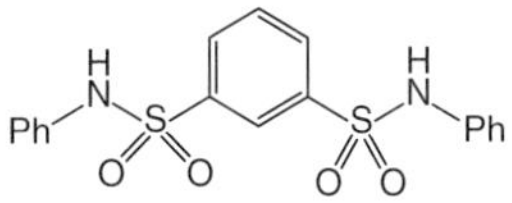

220

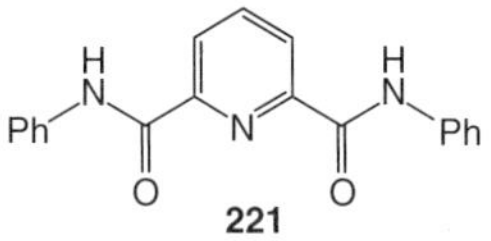

221

These compounds were synthesized in good yield from the acid dichlorides and the corresponding aromatic amines in DMF at room temperature (Scheme 4.29). The purification was carried out by recrystallization. Binding studies conducted by ^{1}H NMR in CD_2Cl_2 showed that the acidity of the amide N–H group has a significant influence on the binding constant; the less acidic receptor **219** shows a sharp decrease in the binding constant for chloride anion as compared to **218**. However, the authors indicated that steric factors can also have a significant effect in the binding affinity. In the same paper, the crystal structure of the Br^- complex of **217** was reported, showing that the amides are in *syn–syn* conformation, allowing therefore for their participation in the hydrogen bond interaction with the anion.

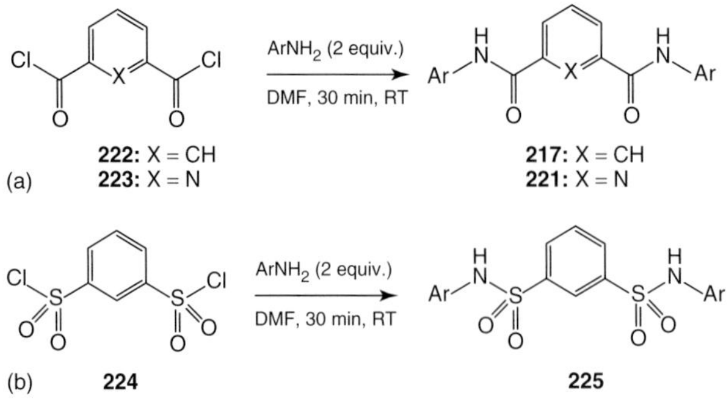

Scheme 4.29

Compounds **226** and **227** were synthesized by reaction of 3-nitroaniline or 3,5-dinitroaniline with isophthaloyl dichloride in dry DMF, with 41 and 39% yields, respectively. In these compounds, the presence of electron-withdrawing nitro- or dinitrophenyl groups enhances the NH acidity. The interaction of these receptors with fluoride was studied both by ^{1}H NMR titrations in 0.5% water-DMSO-d_6 and in the solid state. However, the titration curves resulted to be very complex, showing the presence of multiple equilibria in solution [89].

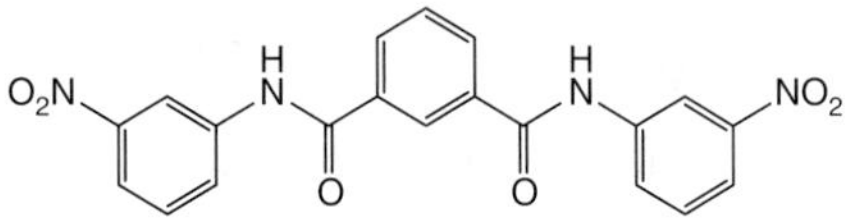

226

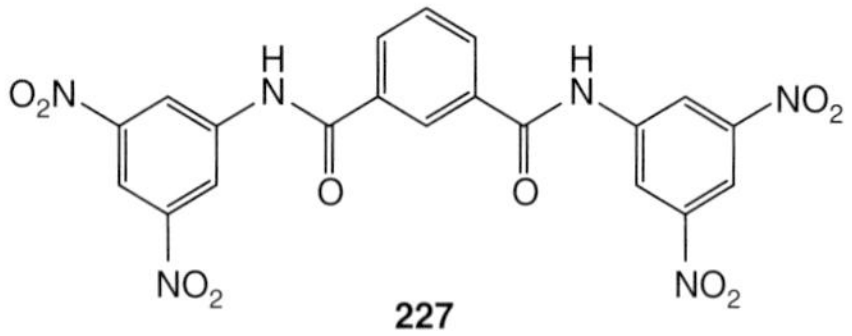

227

Gale and coworkers reported a few years ago the synthesis of several receptors with monoamidopyrrole and diamidopyrrole groups (**228–233**) [90–92]. The precursor used, 5-methyl-3,4-diphenyl-1*H*-pyrrole-2-carboxylic acid ethyl ester, was synthesized via a Paal-Knorr reaction according to a method described by Martell [93]. This synthon was transformed into 5-methyl-3,4-diphenyl-1*H*-pyrrole-2-carboxylic acid butylamide in 29% yield (**228**) by reaction with 50 equiv. of butylamine in the presence of catalytic amounts of cyanide anions [94]. Receptor **230** was obtained in 17% yield by reacting 5-methyl-3,4-diphenyl-1*H*-pyrrole-2-carboxylic acid ethyl ester with an aluminum phenylamine derivative, which was prepared *in situ* by reaction of trimethylaluminum and aniline [95]. 3,4-Diphenyl-1*H*-pyrrole-2,5-dicarboxylic acid was converted into acid chloride by reaction with thionyl chloride heating at reflux. Then, the excess of thionyl chloride was removed in vacuum. The residue was dissolved in a CH_2Cl_2 solution containing either *n*-butylamine or aniline, together with triethylamine and a catalytic amount of 4-dimethylaminopyridine (DMAP), affording the bisamide **229** (18% yield) or **231** (47% yield). While for receptors **228** and **229** the binding constants were calculated in CD_3CN by ^1H NMR, for **230** and **231** they were calculated in 0.5% D_2O/DMSO-d_6 by the same technique. Receptors **228** and **229** show their maximum affinity for benzoate, while **231** has its highest binding constants with $H_2PO_4^-$.

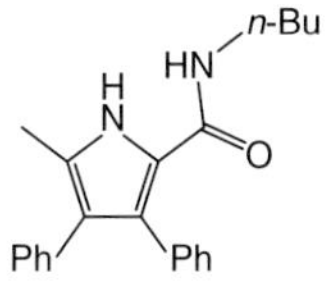

228

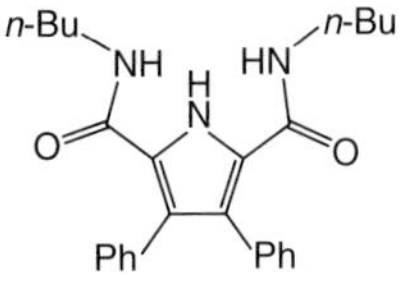

229

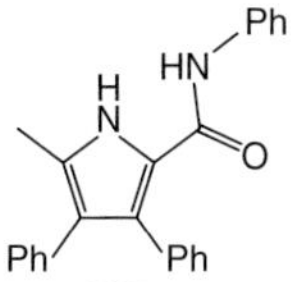

230

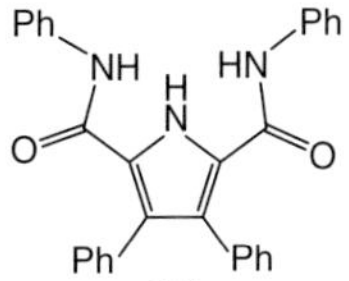

231

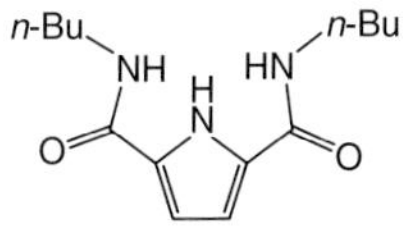

232

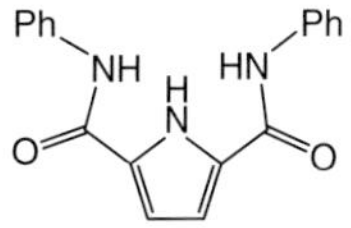

233

Jurczak and Zieliński synthesized the new diamide compounds **232** and **233** (**228–233**) in high yields by reacting 1*H*-pyrrole-2,5-dicarbonyl dichloride and the corresponding amine in dry CH_2Cl_2 at 0 °C [96]. In the case of **233**, a precipitate appeared, which was filtered out and washed with ether, 2 M HCl, and water. The diamide was recrystallized from a MeOH/CH_2Cl_2 mixture. On the other hand, to obtain **232**, the reaction mixture was evaporated to dryness. The crude residue was washed with 2 M HCl, water, and ether. Diamide **232** was purified by recrystallization from ethyl acetate.

3,4-Diphenyl-furan-2,5-dicarbonyl dichloride (**234**) was synthesized by reacting 3,4-diphenyl-furan-2,5-dicarboxylic acid with freshly distilled thionyl chloride heating at reflux overnight [97]. The excess of thionyl chloride was removed, and the solid was dried under high vacuum. Compound **234** was dissolved in dry CH_2Cl_2 under an inert atmosphere and triethylamine (Et_3N), and DMAP in catalytic quantity and either *n*-butylamine (**235** as a product) or aniline (**236** as a product) was added (Scheme 4.30). In a similar way (Scheme 4.31), thiophene-2,5-dicarbonyl dichloride was synthesized by reacting thiophene-2,5-dicarboxylic acid with thionyl chloride. Then, 2,5-dicarbonyl dichloride was dissolved in CH_2Cl_2 and Et_3N, and DMAP (catalytic quantity) and *n*-butylamine (**238** as a product) or aniline (**239** as a product) was added while the reaction mixture was kept under stirring.

The association constants for the interaction of receptors **235–239** with several anions were determined by ^1H NMR in DMSO-d_6–0.5% water. However, in the case of **238** and **239**, addition of fluoride induced a broadening of the signals that impeded obtaining reliable binding constants.

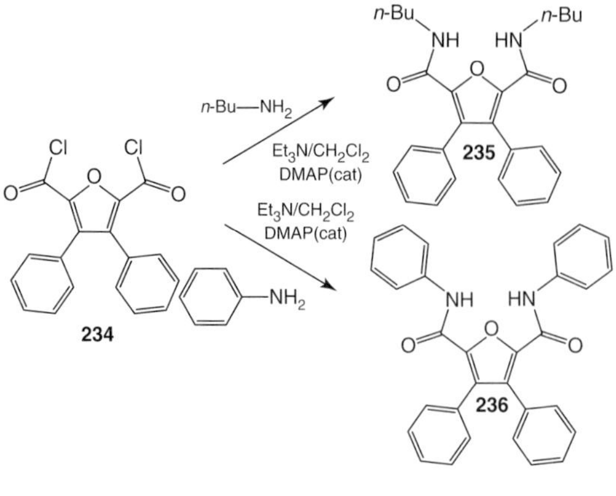

Scheme 4.30

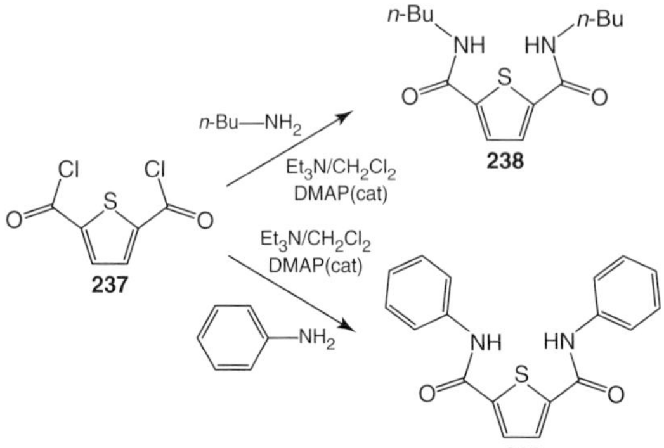

Scheme 4.31

Jeong and Cho [98] synthesized several mono and bipyridinium receptors with the goal of investigating the effect of the CH groups placed at the 2-position of the aromatic ring in hydrogen bond formation in polar solvents. First, the interaction between 3-(acetylamino)pyridine (**240**) with benzoate anions in DMSO-d_6 was studied as a reference, obtaining a value of the association constant of $K_a \sim 16 \pm 1\ M^{-1}$. To prepare a compound with enhanced hydrogen donor ability of the CH group, **240** was reacted with methyl iodide to give the quaternary salt (**241**) in a 57% yield. Benzoate interacts more strongly with **241** ($K_a = 300\ M^{-1}$) than with **240** ($K_a = 16\ M^{-1}$) probably because of the increased hydrogen bond donor ability of the CH group at the 2-position of the aromatic ring in **241** and because of additional electrostatic interaction between opposite charges.

Additionally, the authors described the synthesis of three bipyridinium salts (**242–244**). These receptors were synthesized for studying their binding behavior toward dicarboxylate anions. To achieve the synthesis of compounds **243** and **244**, **240** was reacted with 1,5-dibromopentane or *p*-bis(iodomethyl)benzene in DMF, respectively. On the other hand, receptor **242** was prepared by reaction of

3-aminopyridine with glutaryl dichloride followed by methylation of the pyridine nitrogen with MeI in DMF. All the products were purified by anion exchange column (NH_4PF_6/H_2O).

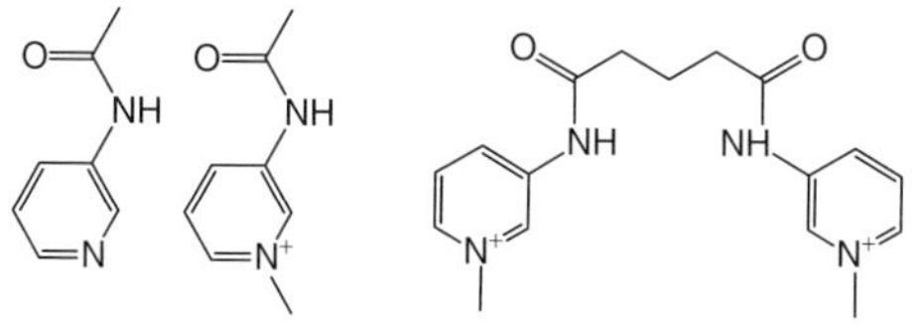

240 241 242

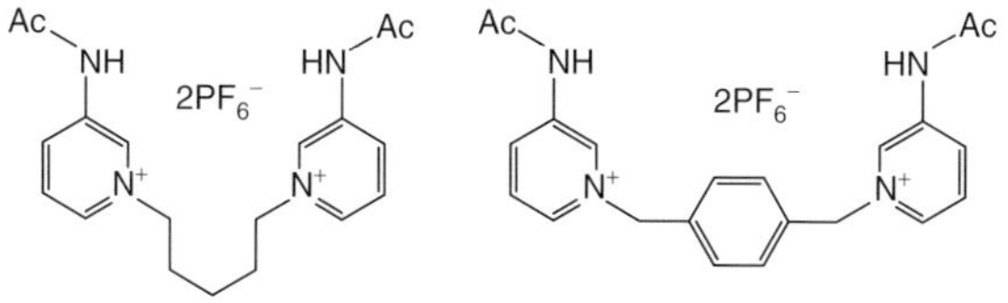

243 244

The ^1H NMR spectra of **242–244** with adipate in DMSO-d_6 showed large downfield shifts of the CH at the 2-position of the aromatic ring and a sharp break in the saturation curves when 1 equiv. of the dianion was added, affording $K_a > 5 \times 10^5\ M^{-1}$. This result leads to the conclusion that two pyridinium units cooperatively bind to adipate through multiple hydrogen bonds and additional electrostatic interactions. Addition of 10% D_2O considerably reduced the K_a by strong solvation of binding partners, and thus, more reliable values could be measured by ^1H NMR. The binding studies of **242–244** with 10% D_2O/DMSO-d_6 showed K_a values with adipate higher than $10^3\ M^{-1}$. These values are much greater than that measured for the monopyridinium salt **241**, with butyrate in the same condition ($K_a = 30 \pm 5\ M^{-1}$).

Prohens, Deyà, Ballester, and Costa described the synthesis and characterization of squaramide-based receptor designed for the recognition of carboxylates anions [99]. Compounds **245–247** were synthesized following the standard pathway described by Schmidt [100]. Receptor **248** was obtained from **245** by methylation with MeI using a refluxing acetone–DMF mixture (2 : 1 v/v) as solvent.

^1H NMR titrations evidenced that addition of a solution of tetramethylammonium acetate (TMAAcO) to the receptors **245–248** shifted the squaramide and methylene proton signals supporting hydrogen bonding formation between the host and guest species. Neutral compounds **245–247** have moderate to high K_a's. As expected, the presence of positively charged groups close to the binding site, as in the **248** receptor, produces about a 10-fold increase in binding constants. These authors synthesized bis squaramide receptors to selectively bind di- and tricarboxylate anions. First, 1,3-bis(aminomethyl)benzene was reacted with diethyl squarate to produce the mixed bis squaramide ester (**249**, in 88% yield). Further condensation of **249** with cyclohexylamine, benzylamine, or *N,N*-dimethylethylenediamine led to bis squaramides **250–252**. Molecular modeling studies showed that compounds **250–253** are appropriate for binding dicarboxylate anions derived from glutaric or glutaconic acids. ^{1}H NMR experiments in DMSO-d_6 evidenced that the K_a values for the interaction of tetrabutylammonium glutarate (TBAGlutarate) and receptors **250–253** were too large ($K_a > 10^4$ M^{-1}) to be measured using NMR techniques. By adding D_2O, the K_a values decrease into a measurable range; K_a(**251**-Glutarate) = $(1.4 \pm 0.2) \times 10^3$ M^{-1} in 10% D_2O/DMSO-d_6 and K_a(**251**·Glutarate) = $(1.5 \pm 0.4) \times 10^2$ M^{-1} in 15% D_2O/DMSO-d_6. Again, the value for the association constant of TBA-Glutarate with **253** in 10% D_2O/DMSO-d_6 was too high to be measured; for this reason it was calculated in 30% H_2O/CD_3CN, obtaining a value of $(5.6 \pm 0.5) \times 10^2$ M^{-1}. To obtain an appropriate receptor for tricarboxylate anions, these authors reacted 1,3,5-tris-(3-aminomethyl-5-butoxy-6-propylphenyl)benzene with diethylsquarate, and the resulting tris squaramide ester (**254**) was condensed with benzylamine to afford tris-squaramide (**255**) in a good yield. ^{1}H NMR titrations for the interaction of **255** with trimesoate and cyclohexentricarboxylate in 10% D_2O/DMSO-d_6 gave $K_a = (3.9 \pm 0.4) \times 10^3$ M^{-1} and $(7.7 \pm 1.3) \times 10^3$ M^{-1}, respectively.

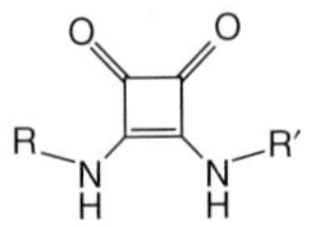

245: R,R′ = *t*Bu
246: R,R′ = Bn
247: R = BnR′ = $CH_2CH_2NMe_2$

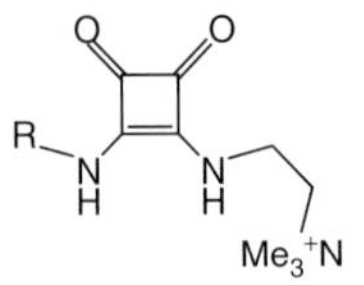

248: R = Bn

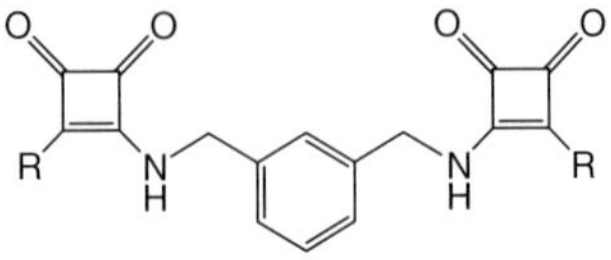

249: R = OEt
250: R = NHC_6H_{11}
251: R = NHBn
252: R = $NHCH_2CH_2NMe_2$
253: R = $NHCH_2CH_2N^+Me_3$

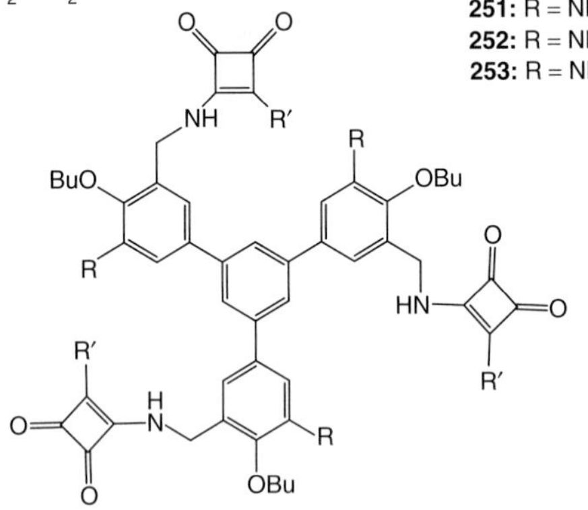

254: R = *n*Pr, R′ = OEt
255: R = *n*Pr, R′ = NHBn

Hughes and Smith described that condensation of the amide carbonyl to a Lewis acidic boronate group greatly improves the binding ability of amide receptors with acetate anions [101]. As a matter of fact, the K_a value for the interaction of **257** with tetrabutylammonium acetate determined by ^{1}H NMR in DMSO-d_6 ($K_a = (2.1 \pm 0.2) \times 10^3$ M^{-1}) resulted to be 10 times higher than that of **256** ($K_a = (1.1 \pm 0.1) \times 10^2$ M^{-1}).

Werner and Schneider reported a few years ago the preparation of mono, di, and triaamide receptors of different molecular topologies (structures **259–265**) [102]. An evaluation of the association constants of these amides with halide, phosphate, and carboxylate anions in $CDCl_3$ showed that in general, all the receptors displayed higher constants than with the reference compound methylformamide (**258**). On the other hand, each one of the receptors displayed its highest binding affinity for the substrate best matching its hydrogen bond pattern and electronic requirements. For instance, while the highest constant for AcO^- was found with receptor **261**, for chloride, the maximum stability was achieved with **265**.

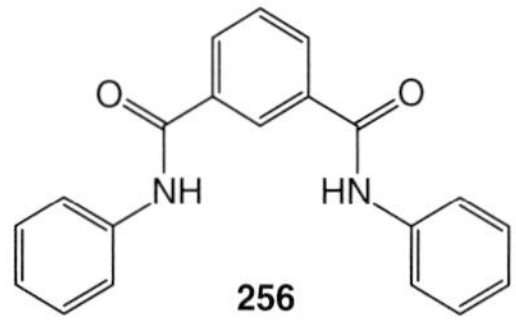

256

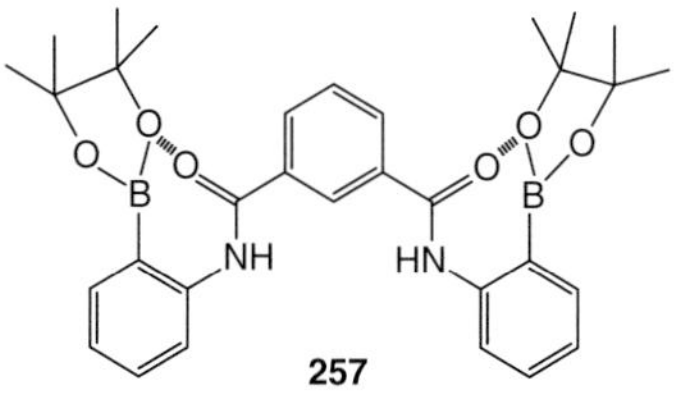

257

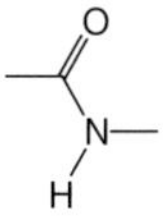

258

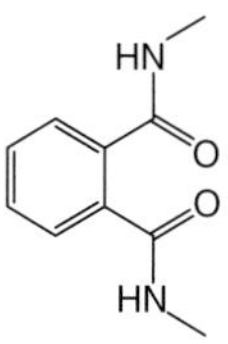

259

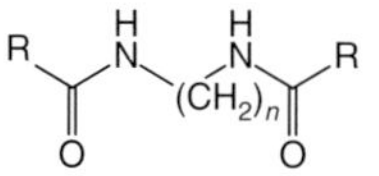

260: $n = 1$, $R = C_6H_5$
261: $n = 2$, $R = CH_3$
262: $n = 3$, $R = CH_3$
263: $n = 4$, $R = CH_3$

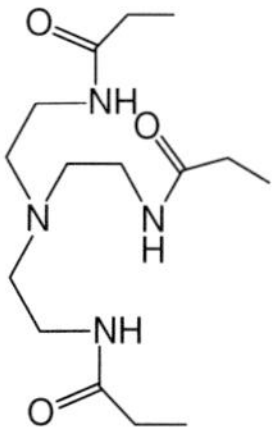

264

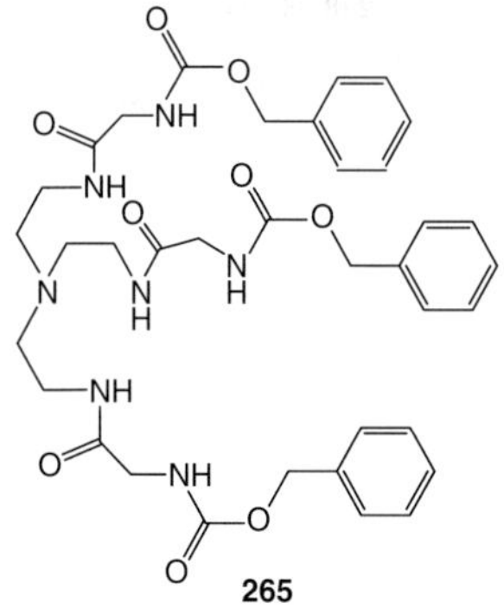

265

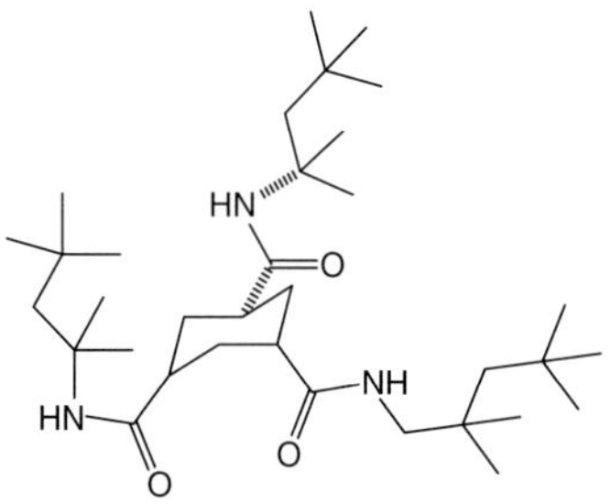

266

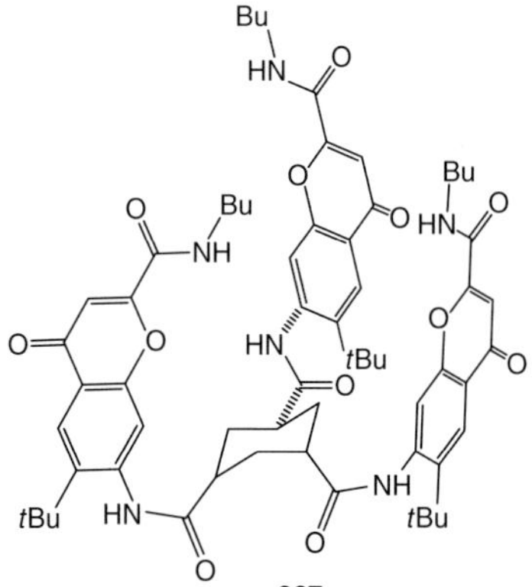

267

Tripodal triamides were also developed by Morán and coworkers (**266** and **267**) [103]. Receptor **266** was prepared from 1,3,5-cyclohexanetricarboxylic acid trichloride and 1,1,3,3-tetramethylbutylamine. Receptor **267** was synthesized from the same triacid chloride and 8-aminochromenone-2-carboxamide. These receptors showed an interesting behavior for the recognition of phosphonic acids in DMSO-d_6.

Reinhoudt and coworkers [104] reported the synthesis of tripodal amide compounds **268–273**, which consisted of the addition of the respective acid chlorides to a solution of tris(aminoethyl)amine and triethylamine in CH_2Cl_2 at 10 °C. The reaction mixture was stirred for 3 h and allowed to warm up to room temperature, maintaining the stirring for another 3 h.

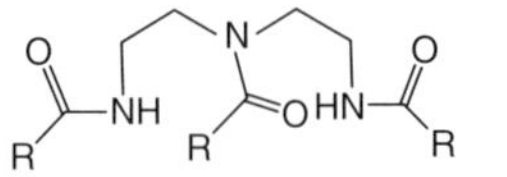

268: R = CH_2Cl
269: R = $(CH_2)_4CH_3$

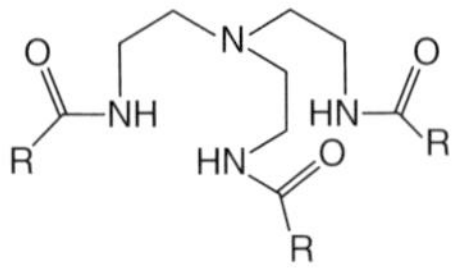

270: R = CH_2Cl
271: R = $(CH_2)_4CH_3$
272: R = C_6H_5
273: R = 4-$CH_3OC_6H_4$

These receptors were prepared trying to mimic the active site of the phosphate-binding protein. ^{1}H NMR titrations in $CDCl_3$ showed that receptors had higher specificity for $H_2PO_4^-$ than for HSO_4^- and Cl^-. The K_a for the interaction of these receptors with $H_2PO_4^-$ in $CDCl_3$ was too high to allow its determination by ^{1}H NMR titration. Therefore, all the K_a values were determined in CH_3CN by a conductometry method.

The interesting tripodal receptor **274** containing quaternary polybipyrimidinium was developed by Beer's group [105]. The synthetic pathway consisted in the addition of a large excess of 4-(aminomethyl)pyridine and triethylamine to a solution of benzene-1,3,5-tricarbonyl trichloride in dry CH_2Cl_2 with a catalytic amount of DMAP. After purification, the product was methylated with CH_3I to obtain receptor **274**. ^{1}H NMR titration studies in DMSO-d_6 showed that these receptors bind chloride and bromide with a 1 : 1 stoichiometry. The association constant for chloride ($K_a = 110\ M^{-1}$) was three times higher than the K_a obtained for a similar ligand without the amide groups, highlighting the contribution of the amide groups to the interaction.

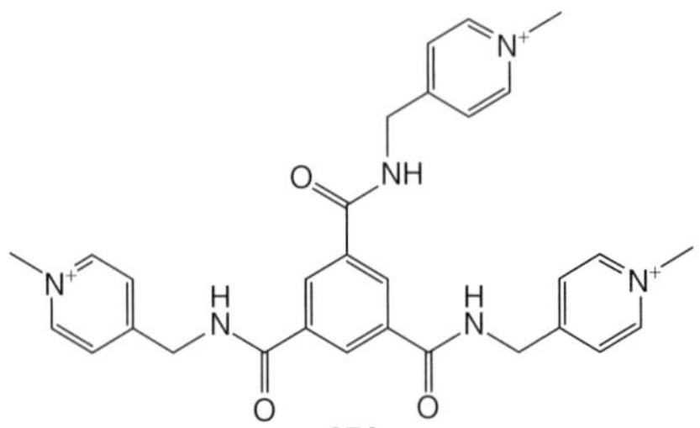

274

4.3.1.2 Macrocyclic Amide Receptors

Bowman-James and coworkers synthesized the new polyamineamide macrocyclic ligand **275** [106] by condensation of *N*′-methyl-2,2′-diaminodiethylamine with isophthaloyl dichloride using triethylamine as a base in toluene. The macrocyclic receptor was isolated in 50% yield after column purification (**275**). ^{1}H NMR titrations conducted in $CDCl_3$ demonstrated that the macrocyclic ligand exhibits a significant selectivity for $H_2PO_4^-$ and HSO_4^- over other anions. Slow evaporation of $CHCl_3$ solution afforded crystals suitable for X-ray of a sandwich complex in which a deprotonated SO_4^{2-} anion was intercalated between two molecules of **275**. It was thought that the presence of the basic amine groups helped to deprotonate the acidic protons of HSO_4^- (see Chapter 3).

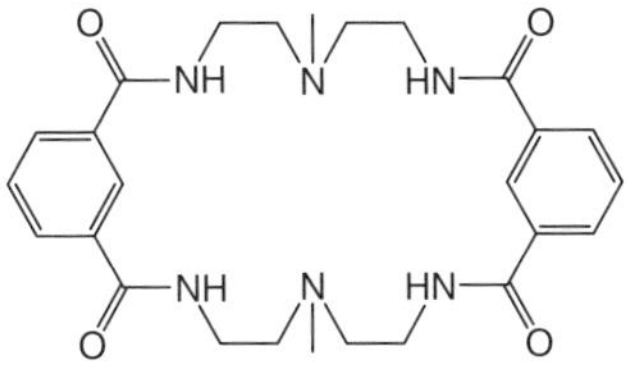

275

The synthesis of receptor **276**, in which two amide functions were appended to a calixarene structure, was reported by Reinhoudt's group several years ago [107]. The synthesis was based on the reaction of 1,3-diaminocalix[4]arene with chloroacetyl chloride in the presence of Et_3N in CH_2Cl_2 to yield the corresponding 1,3-bis(chloroacetamido)calix[4]arene. Years later, Cameron and Loeb employed the same synthetic route to obtain receptor **277** [108]. The binding constants calculated by ^{1}H NMR in CD_3CN indicated preference for Y-shaped carboxylate anions (acetate and benzoate) over tetrahedral anions: HSO_4^-, $H_2PO_4^-$, ReO_4^-, $CH_3CO_2^-$, and $C_6H_5CO_2^-$.

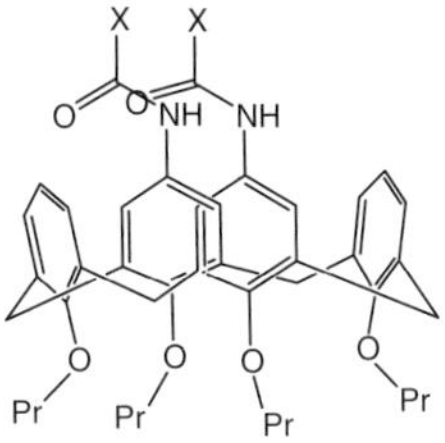

276: X = CH_2Cl
277: X = $CHCl_2$

Smith and coworkers described the synthesis of several anion pair receptors containing amide functionalities [109]. The synthetic pathway begins with a double nitration of dibenzo-18-crown-6. This reaction gives an almost equal mixture of *cis* and *trans* isomers, which can be readily separated by fractional crystallization. The *cis* isomer was chosen to continue the synthesis, and catalytic hydrogenation of the nitro groups afforded the corresponding diamine, which was subsequently coupled with 4-nitrobenzoyl chloride. N-methylation was achieved in DMF using sodium hydride and iodomethane to produce the precursor. Quimioselective reduction with iron powder in HCl produces the diamine with a good yield. The diamine was dissolved in anhydrous CH_2Cl_2 and added to a solution of isophthaloyl dichloride (**222**) in CH_2Cl_2. Product **278** was isolated in 55% yield. A similar procedure but using 5-*tert*-butylisophthaloyl dichloride instead of isophthaloyl dichloride led to **279** in 75% yield.

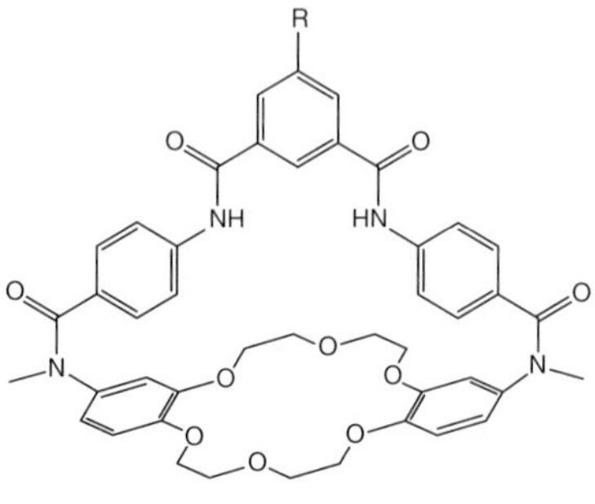

278: R = H
279: R = *t*Bu

The synthesis of the new cryptand **281** (Scheme 4.32) was achieved with a 40% yield by Anslyn and coworkers through the reaction of 2,6-pyridinedicarbonyl dichloride (**223**) with 1,3,5-triaminomethyl-2,4,6-triethylbenzene (**280**) in CH_2Cl_2 and Et_3N [110].

Jurczak reported in 2004 the synthesis of the tetraamide ligand **284** (Scheme 4.33) [111]. The first step consisted in the reaction of 5-*t*-butyl-isophthalic acid (**282**) with thionyl chloride ($SOCl_2$) and catalytic amounts of DMF to obtain 5-*t*-butyl-isophthalic acid chloride. After purification, 5-*t*-butyl-isophthalic acid chloride (**283**) was dissolved in dry CH_2Cl_2 and added dropwise to a mixture of Et_3N and 1,3-diaminepropane in dry CH_2Cl_2. The reaction mixture was concentrated, and the residue was washed with water. Chromatography on SiO_2 with 2.0–2.5% CH_3OH in CH_2Cl_2 gave the desired product (**284**) in 32% yield.

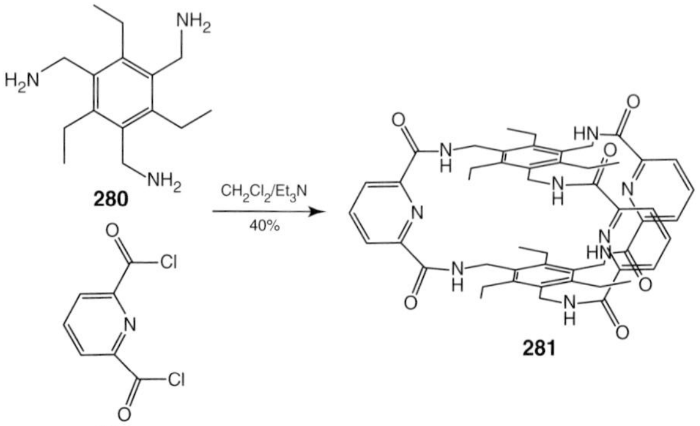

Scheme 4.32

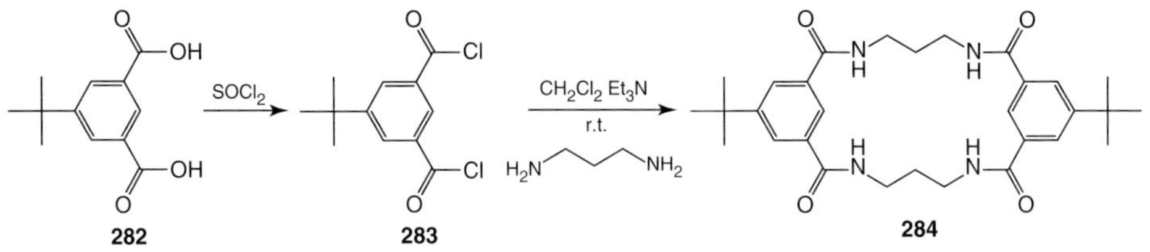

Scheme 4.33

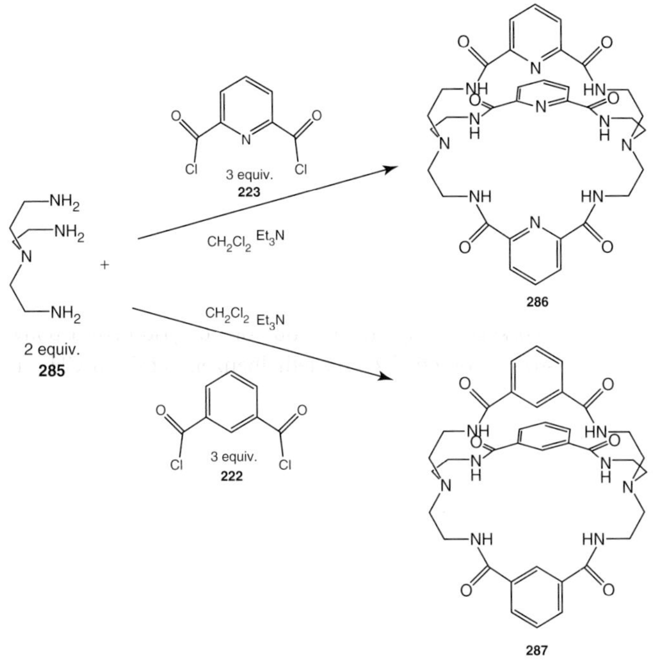

Scheme 4.34

The bicyclic amide cryptands **286** [112] and **287** [113] were synthesized by reacting 2 equiv. of tris(2-aminoethyl)amine with 3 equiv. of 2,6-pyridinedicarbonyl dichloride in CH_2Cl_2 using triethylamine (Et_3N) as a base (Scheme 4.34). The products were isolated in 10% (**286**) and 12% (**287**) yield after column chromatography (basic Al_2O_3, 3% CH_3OH in CH_2Cl_2). In addition, the synthesis of a new cryptand (**289**, Scheme 4.35) was described using the same conditions. Compound **289** was isolated in 10% yield. Chromatographic purification was carried out first with a column of SiO_2 in 15% CH_3OH in CH_2Cl_2 and then with a second column of basic Al_2O_3 in 5% CH_3OH in CH_2Cl_2. Finally, the tertiary nitrogen was quaternized with CH_3I to obtain cryptand **290** in 80% yield [114].

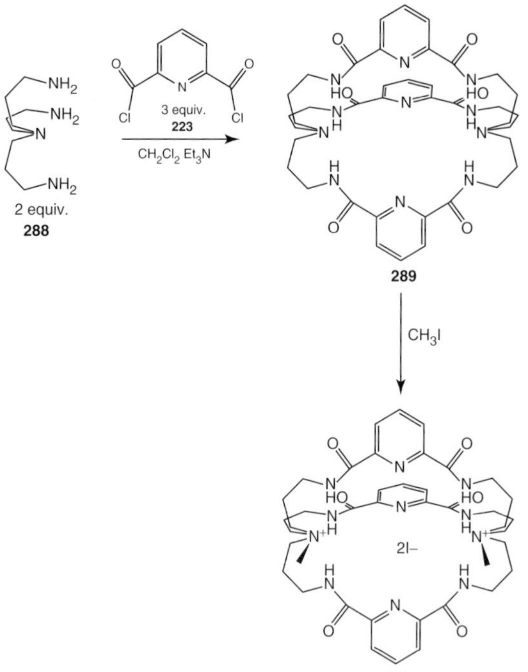

Scheme 4.35

4.3.2
Esters as Starting Materials

Jurczak and coworkers described already in 1991 the synthesis of diamides using esters as starting materials [115]. They found that α,ω-diesters reacted with 1 equiv. of α,ω-diamines under room temperature in methanol during seven days, achieving yields of 60% for **293** and **295**, 70% for **297**, and 75% for **299**.

For this reaction, the yields are particularly high when both the amine and ester reactants have oxygen or nitrogen atoms in γ-position with respect to the reacting group, as occurs in compounds in Scheme 4.36. This method is therefore particularly suited for obtaining diaza crown, an important class of ligands in metal ion recognition.

By the same general procedure but using dimethyl pyridine-2,6-dicarboxylate (**300**) as the ester, Jurczak and coworkers obtained several amidoethers incorporating pyridine rings [116] (Scheme 4.37). Interestingly enough, the condensation degree of the formed amide depended on the length of the α, ω-diamine employed. Short diamines such as **301** (in Scheme 4.37) led to the formation of the larger macrocyclic tetraamide (**305**) resulting from a [2 + 2] condensation reaction, while longer α, ω-diamines such as **303**, **311**, and **312** led to bisamides **308**, **315**, and **317** resulting from a [1 + 1] condensation. However, diamines of intermediate length

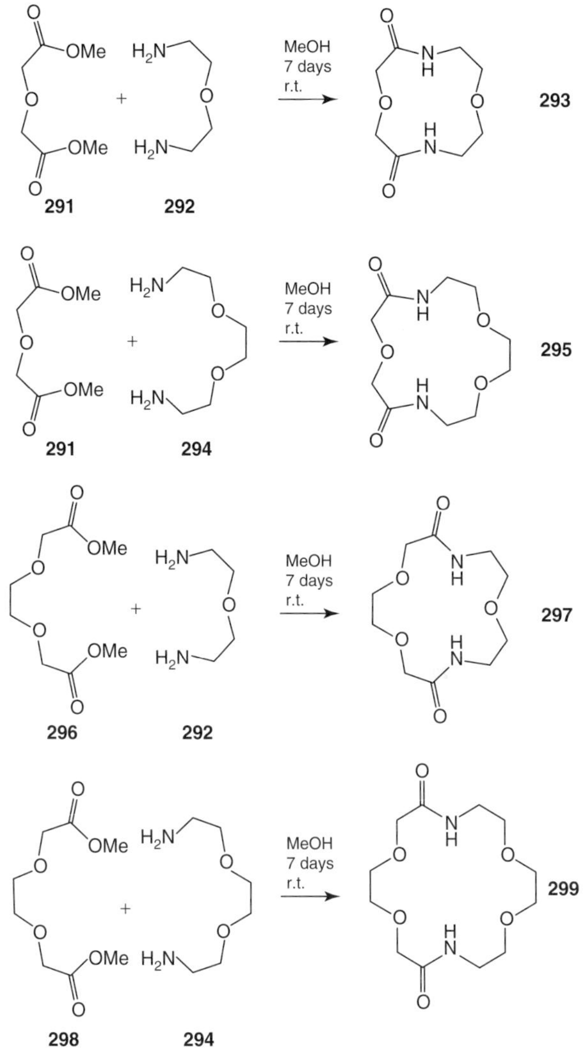

Scheme 4.36

such as **302** and **310** (Scheme 4.37) gave both types of product bisamides (**306** and **313**) and tetraamides (**307** and **314**).

Jurczak and coworkers also synthesized the tetraamide receptors without ether groups **319** and **320** [117]. Compound **319** was prepared by a [2 + 2] condensation implying dimethyl 2,6-pyridinecarboxylate and 1,2-diaminoethane in methanol for seven days. After this time, a precipitate appeared that was filtered off to isolate **319** with a 51% yield. Although the use of aliphatic amines instead of amino acids or amino esters leads to an important decrease in the yield of the process (yield ~10%), the low solubility of the compound that precipitates

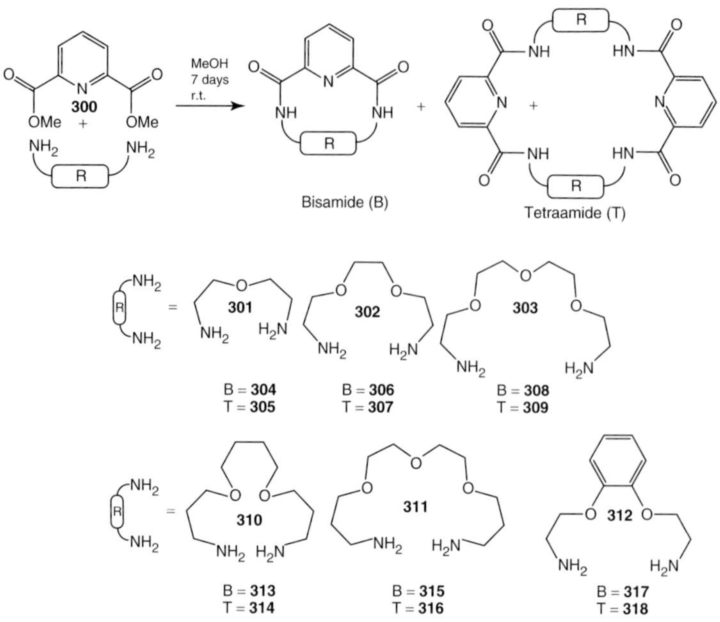

Scheme 4.37 In the scheme B stands for bisamides and T for tetraamides.

from the reaction mixture in an almost pure form makes this procedure very convenient.

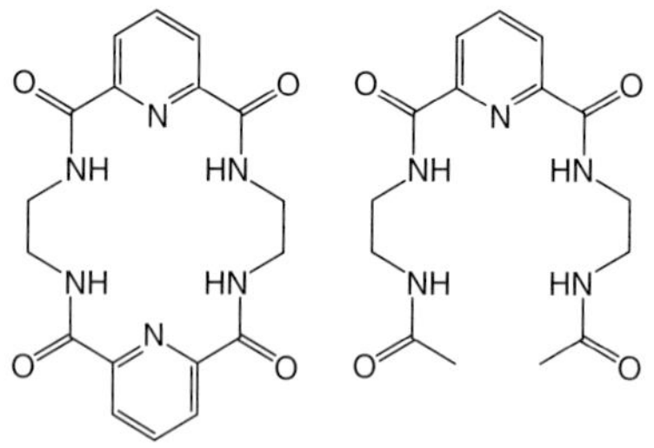

319 **320**

For obtaining **320**, dimethyl 2,6-pyridinecarboxylate and an excess of 1,2-diaminoethane were mixed in methanol and stirred for 2 h. The solvents were removed by coevaporation several times with CH_2Cl_2 and dried under vacuum in order to remove the excess 1,2-diaminoethane. The solid was dissolved in CH_2Cl_2, and pyridine was added. The reaction mixture was cooled to 0 °C, and acetic anhydride was added dropwise. Stirring was continued for 2 h at room temperature. The product was purified by column chromatography.

^{1}H NMR analysis of the interaction of these receptors with anions in DMSO-d_6 showed that the macrocyclic receptor **319** was much better suited for anion recognition than the acyclic counterpart **320**.

In a subsequent publication, Chmielewski and Jurczak reported the synthesis of larger tetraamide macrocycles as well as several hexaamide and octaamide

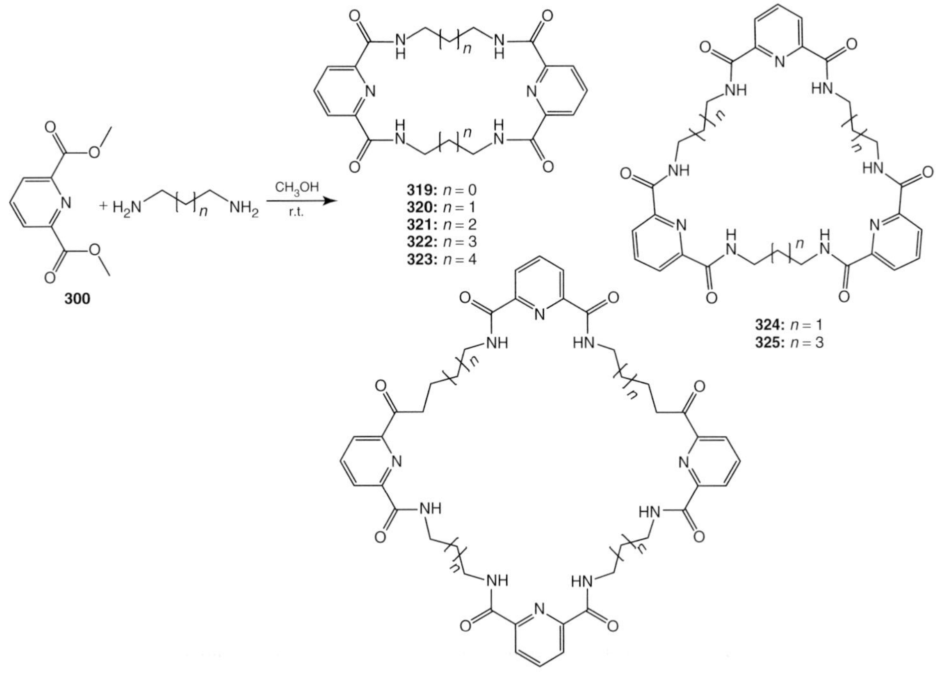

300

319: n = 0
320: n = 1
321: n = 2
322: n = 3
323: n = 4

324: n = 1
325: n = 3

326: n = 0
327: n = 1
328: n = 3

Scheme 4.38

macrocycles (Scheme 4.38) [118]. The same authors performed similar reactions with amines with an odd number of carbon atoms, such as 1,3-diaminopropane and 1,5-diaminopentane. As in the case of **319**, when reacting dimethyl 2,6-pyridinecarboxylate with the diamine 1,3-diaminopropane with an odd number of atoms, a precipitate was obtained. However, in this case, the precipitate obtained was not the tetraamide (**320**) but a mixture of hexaamide (**324**) and octaamide (**327**) in 8 and 6% yields, respectively. The filtrate was concentrated to obtain **320**, which was purified by chromatography (yield 6%).

Analogous products were obtained when reacting dimethyl 2,6-pyridinedicarboxylate with 1,5-diaminopentane. However, in this case, the precipitate was an unidentified product, probably a polymer. The filtrate fraction was purified by chromatography to isolate **322** (7%), **325** (6%), and **328** (2%).

In the same report, Chmielewski and Jurczak established an alternative synthetic strategy to provide macrocyclic tetraamides in high yields (Scheme 4.39). In the first step of this procedure, 1,3-diaminopropane was reacted with 4 equiv. of dimethyl 2,6-pyridinedicarboxylate to obtain **330** in 46% yield after chromatographic purification. In the second step, **330** was reacted with 1 equiv. of 1,3-diaminopropane, giving the tetraamide macrocyclic product **320** in 21% yield as well as the octaamide (**327**) in 17% yield. Using this synthetic pathway, the hexaamide ligands

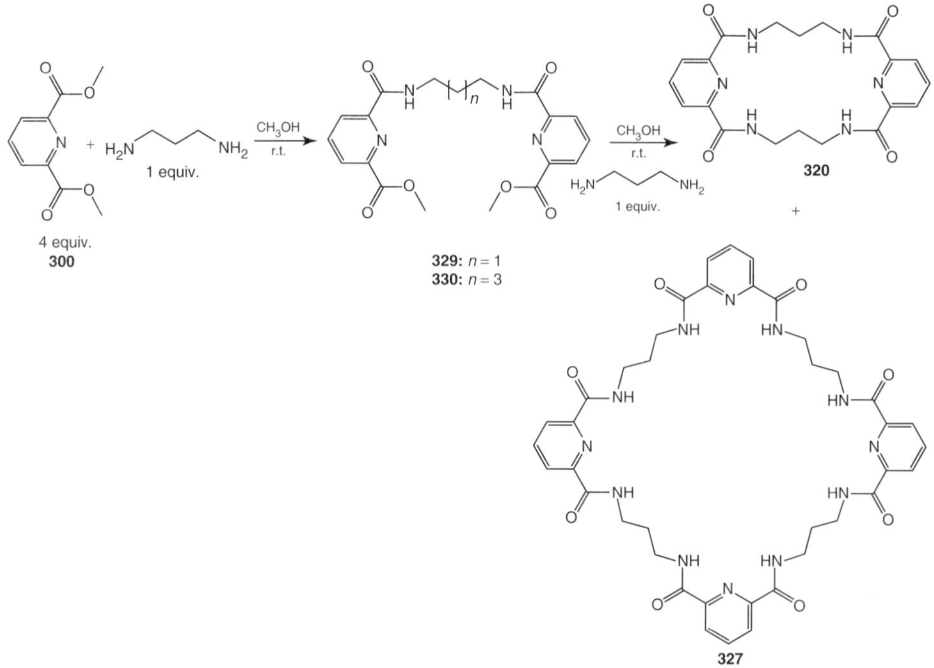

Scheme 4.39

were absent and octaamide (**327**) was obtained directly from the reaction mixture as a precipitate in an almost pure form.

The binding constants (K_a, M^{-1}) for 1 : 1 stoichiometry of **319–322** with several anions in DMSO-d_6 were obtained by ^{1}H NMR titration techniques. All these ligands displayed similar selectivity trends: binding most strongly with $H_2PO_4^-$, acetate, and benzoate, whereas interacting most weakly with cyanide, bromide, *p*-nitrophenolate, and hydrogen sulfate anions.

Szumna and Jurzack described a modification of the synthesis to improve the yield of the [2 + 2] condensation, which consisted in the addition to the reaction mixture of 5 equiv. of tetrabutylammonium hexafluorophosphate ($BuN_4(PF_6)$). This modification gave **319** as a major product in 52% yield, whereas **326** was obtained as a minor product just in 4% (Scheme 4.40) [119].

Chmielewski and Jurczak described the asymmetric synthesis of hybrid macrocyclic tetraamides (**333**) containing both pyridine and phenyl moieties. First, 5-*t*-butyl-isophthalic acid chloride (**283**) was reacted with a monoprotected 1,3-diaminopropane (**331**) to obtain **332** with 74% yield. Then **332** was deprotected with trifluoroacetic acid to afford the ammonium salt, which, without any further purification, was dissolved in a methanolic solution of sodium methoxide and the dimethyl pyridine-2,6-dicarboxylate (**300**) to obtain the macrocyclic tetraamide (**333**) with a 16% yield (Scheme 4.41) [120].

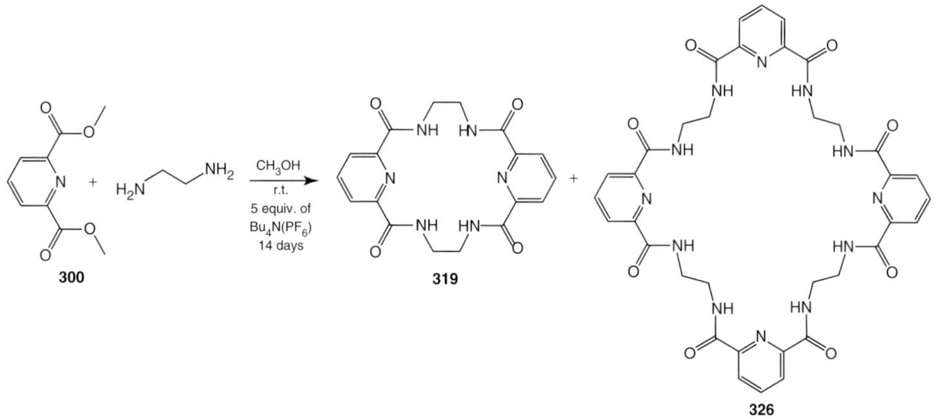

Scheme 4.40

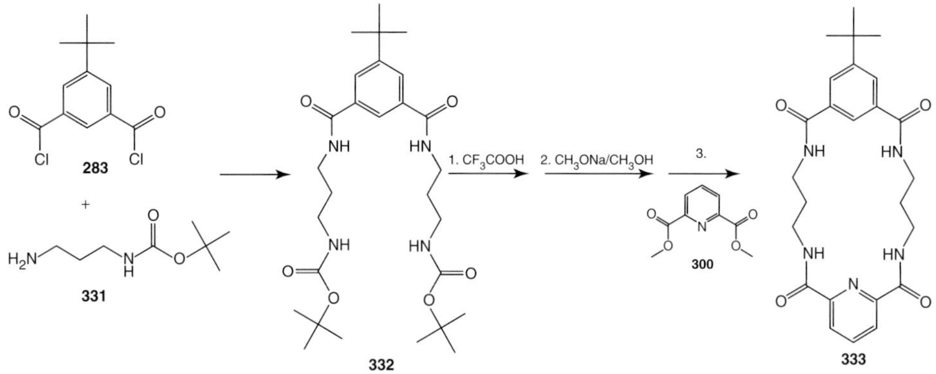

Scheme 4.41

Rybak-Akimova and coworkers followed a modification of the previous procedure to prepare several aminoamide receptors [121]. Reaction of dimethyl pyridine-2,6-dicarboxylate (**300**) with $N^1,N^{1'}$-(ethane-1,2-diyl)bis(ethane-1,2-diamine) overnight in refluxing methanol yielded the [1 + 1] condensation diaminodiamide compound **334**. Similarly, **335** and **336** were obtained using N^1-(3-aminopropyl)propane-1,3-diamine or N^1-(3-aminopropyl)-N^1-methylpropane-1,3-diamine, respectively (Scheme 4.42) [121].

To force the formation of [2 + 2] condensation products, an indirect method, in which first a large excess of dimethyl pyridine-2,6-dicarboxylate (**300**) was mixed with the corresponding amine (1 : 0.15 molar ratio), was used. In this way, [2 + 1] condensation products were obtained (**319** and **320**). Compound **337** was then reacted with a further equivalent of the tetraamine to give the [2 + 2] macrocycle **338** in a 10% overall yield. However, this synthetic procedure did not provide a higher yield of **336** than the procedure previously described, involving the acid chloride [122].

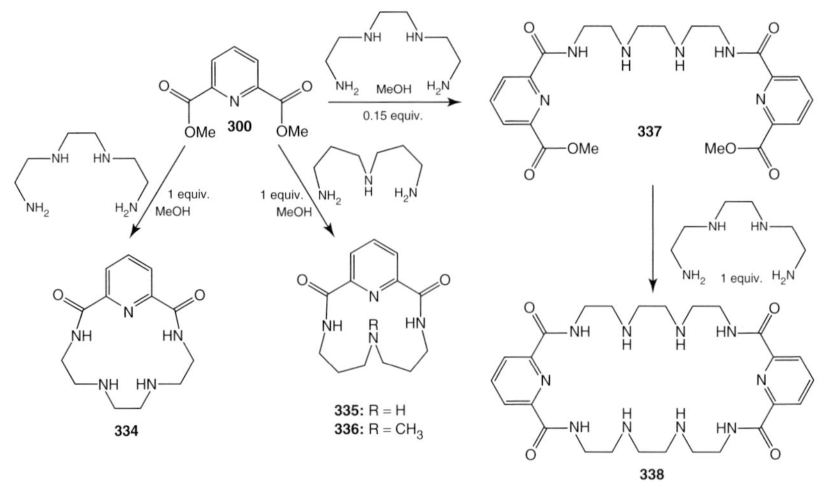

Scheme 4.42

NMR titrations carried out in DMSO-d_6 indicate that [1 + 1] receptor **334** forms with fluoride only complexes of 1 : 1 stoichiometry ($K_a = 5.87(7) \times 10^2$ M^{-1}), whereas [2 + 2] receptor **338** shows the formation of F^- : **338** complexes of 1 : 1 and 2 : 1 stoichiometries with binding constants of $K_1 = 4(7) \times 10^3$ M^{-1} and $K_2 = 3(5) \times 10^2$ M^{-1}, respectively.

4.3.3
Using Coupling Reagents

Another way of activating carboxylic acids is to use coupling reagents, which act as stand-alone reagents to generate compounds such as acid chlorides, anhydrides, carbonic anhydrides, or active esters. In this section, several examples about the use of such coupling reagents for the preparation of cyclic and acyclic amides are described.

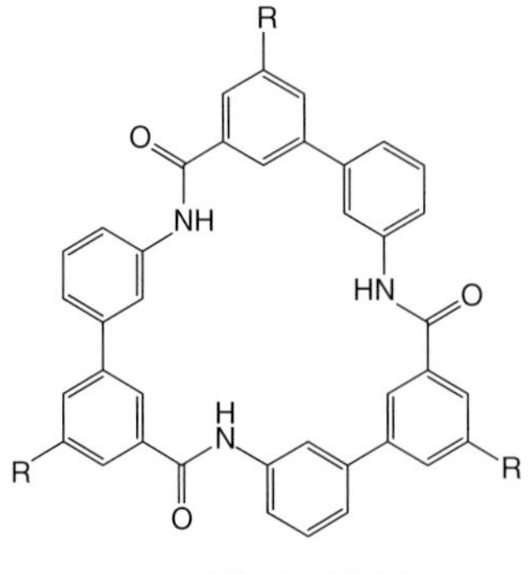

339: R = COOEt
340: R = NHBoc

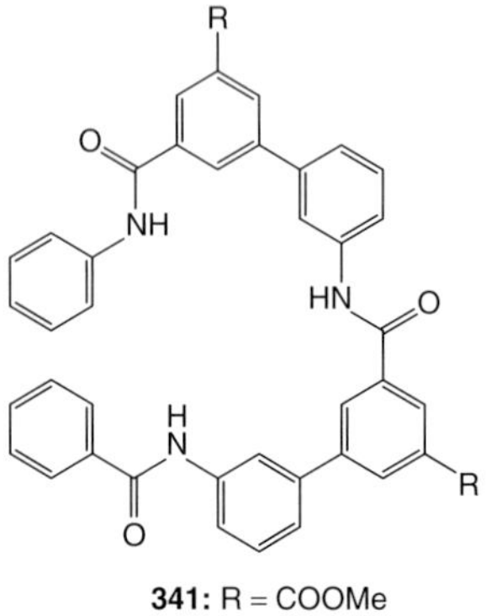

341: R = COOMe

Choi and Hamilton synthesized tetraamides (**339–341**) from 5-substituted-3′-nitro-3-biphenylcarboxylic acid and aniline using THF as a solvent [123]. To the previous mixture, bis(2-oxo-3-oxazolidinyl)phosphinic chloride (BOP-Cl) and diisopropylethylamine (DIEA) were added, and stirring was continued overnight. The mixture was diluted with EtOAc and washed with saturated $NaHCO_3$ and brine. The organic layer was dried over Na_2SO_4 and purified by SiO_2 column chromatography using silica gel (hexane/EtAc). The cyclization of the trimer with the same procedure under dilute condition gave macrocycle **339** in 40–60% yield (**339–341**).

These C_3 symmetric receptors were designed to bind the triangular face of a tetrahedral anion. Binding studies carried out by NMR titration experiments showed a rather complex behavior.

As the anion was added, there was an initial upfield shift of the anion protons. This occurred with up to 0.5 equiv. of added anion, after which the proton signals shifted continuously downfield. The binding stoichiometry fitted to a model including L_2A (2 : 1) and LA (1 : 1) complex formation.

When a more competitive hydrogen bonding solvent was used, such as 100% DMSO-d_6, the K_a values decreased significantly and L_2A complex formation was not observed. The K_a of halide and nitrate anions decreased significantly even in 50% DMSO-d_6/$CDCl_3$. However, tetrahedral anions such as SO_4^{2-}, $H_2PO_4^-$, and pTsO$^-$ retained strong binding constants in DMSO-d_6, even though the $H_2PO_4^-$–L complex showed exchange on the NMR time scale.

Diederich described the synthesis of **342** and **343** [124]. Dicarboxylic acid was cyclized with piperazine-1,4-dipropanamine in the presence of diphenylphosphorylazide (DPPA), triethylamine (Et_3N) in DMF under high dilution conditions to give receptor **342** in 32% yield. It was found that macrocyclization works with other primary diamines, whereas the use of another coupling reagent such as 2-chloro-N-methylpyridinium iodide or conversions with activated esters of dicarboxylic or the corresponding bis(acyl chloride) either failed or gave unreliable results.

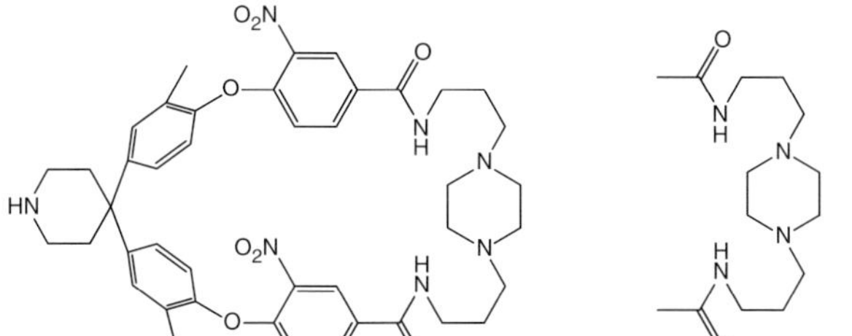

342 **343**

Association constants (K_a, M^{-1}) for **342** with aromatic sulfonates were determined at 300 K by ^{1}H NMR titrations in 0.5 M KCl/DCl buffer in D_2O at pD = 2.0. This pD was employed to form in solution the triprotonated species of the receptor obtaining K_a values for dansyl and benzenesulfonate ions of circa 45 M^{-1}.

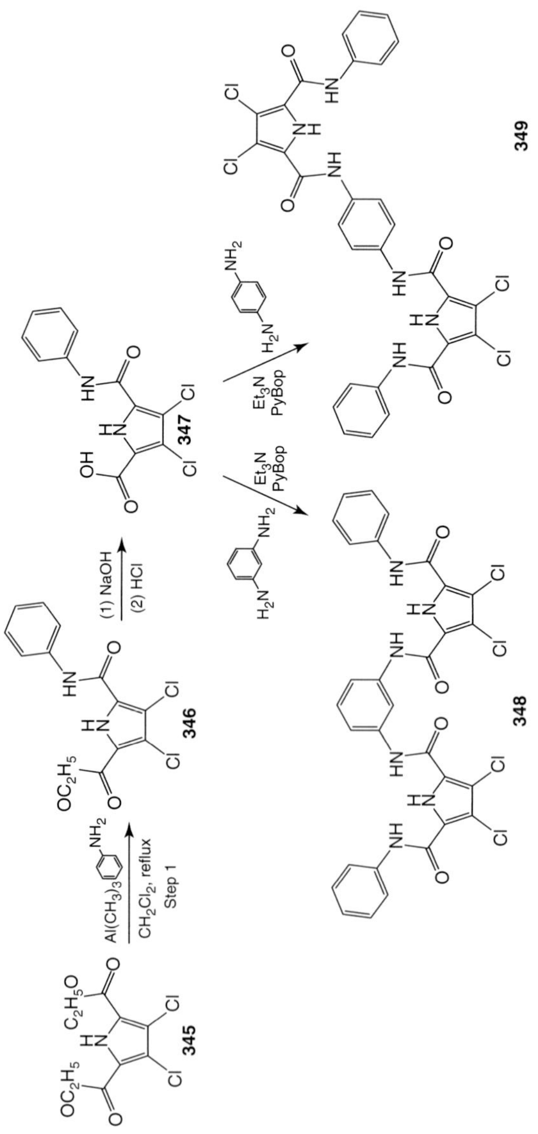
345
$Al(CH_3)_3$
CH_2Cl_2, reflux
Step 1
346
(1) NaOH
(2) HCl
347
Et_3N
PyBop
348
349

Scheme 4.43

Morán synthesized a receptor to recognize chiral molecules [125]. The synthesis of **344** was accomplished by reacting the bisaminomethylspirobifluorene unit with nitrochromenome 2-carboxylic acid chloride followed by reduction of the nitro groups. Receptor **344** showed an ability to resolve racemic mixtures of mandelate ion. Using conventional titrations in DMSO-d_6, **344** showed $K_a((R)$ - mandelate$) = 2.8 \times 10^4$ M^{-1} and $K_a((S)$ - mandelate$) = 1.7 \times 10^3$ M^{-1}.

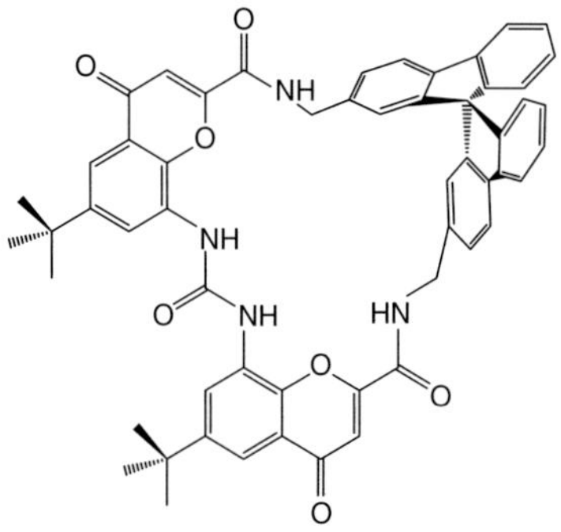

344

Gale described the synthesis of two tetraamide pyrrole ligands (**348** and **349**). Diethyl 3,4-dichloropyrrole-2,5-dicarboxylate (**345**) was reacted with aniline in CH_2Cl_2, and trimethylaluminium solution in hexane was added to obtain 3,4-dichloro-5-phenylcarbamoyl-1*H*-pyrrole-2-carboxylic acid ethyl ester (**346**) [126]. The ester was hydrolyzed in NaOH, and the acid **347** was dissolved in DMF and reacted with benzene-1,3-diamine or benzene-1,4-diamine in the presence of PyBOP/HOBt (cat.) and Et_3N to obtain, respectively, **348** and **349** (Scheme 4.43).

References

1. Fabbrizzi, L., Licchelli, M., Rabaioli, G., and Taglietti, A. (2000) The design of luminescent sensors for anions and ionisable analytes. *Coord. Chem. Rev.*, **205**, 85–108.
2. Llinares, J.M., Powell, D., and Bowman-James, K. (2003) Ammonium based anion receptors. *Coord. Chem. Rev.*, **240**, 57–75.
3. García-España, E., Diaz, P., Llinares, J.M., and Bianchi, A. (2006) Anion coordination in aqueous solution of polyammonium receptors. *Coord. Chem. Rev.*, **250**, 2952–2986.
4. Mateus, P., Bernier, N., and Delgado, R. (2010) Recognition of anions by polyammonium macrocyclic and cryptand receptors: influence of the dimensionality on the binding behaviour. *Coord. Chem. Rev.*, **254**, 1726–1747.
5. Timmons, J.C., and Hubin, T.J. (2010) Preparations and applications of synthetic linked azamacrocyle ligands and complexes. *Coord. Chem. Rev.*, **254**, 1661–1685.
6. McKee, V., Nelson, V., and Town, M.R. (2003) Caged oxoanions. *Chem. Soc. Rev.*, **32**, 309–325.
7. Amendola, V., Bonizzoni, M., Esteban-Gomez, D., Fabbrizzi, L., Licchelli, M., Sancenon, F., and Taglietti, A. (2006) Some guidelines

for the design of anion receptors. *Coord. Chem. Rev.*, **250**, 1451–1470.

8. Bowman-James, K. (2005) Alfred Werner revisited: the coordination chemistry of anions. *Acc. Chem. Res.*, **38**, 671–678.
9. Aragò, J., Bencini, A., Bianchi, A., García-España, E., Micheloni, M., Paoletti, P., Ramirez, J.A., and Paoli, P. (1991) Interaction of "long" open-chain polyazaalkanes with hydrogen and copper(II) ions. *Inorg. Chem.*, **30**, 1843–1849.
10. Dietrich, B., Hosseini, M.W., Lehn, J.-M., and Sessions, R.B. (1983) Synthesis and protonation features of 24-,27- and 32-membered macrocyclic polyamines. *Helv. Chim. Acta*, **66**, 1262–1278.
11. Gampp, H., Haspra, D., Maeder, M., and Zuberbuehler, A.D. (1984) Copper(II) chelates with linear pentadentates chelators. *Inorg. Chem.*, **23**, 3724–3730.
12. Aguilar, J.A., García-España, E., Guerrero, J.A., Luis, S.V., Llinares, J.M., Ramirez, J.A., and Soriano, C. (1996) Synthesis and protonation behaviour of the macrocycle 2,6,10,13,17,21-hexaaza[22]metacyclophane. Thermodynamic and NMR studies on the interaction of 2,6,10,13,17,21-hexaaza[22]metacyclophane and on the-open chain polyamine 4,8,11,15-tetraazaoctadecane-1,18-diamine with ATP, ADP and AMP. *Inorg. Chim. Acta*, **246**, 287–294.
13. Aguilar, J.A., Diaz, P., Escartì, F., García-España, E., Gil, L., Soriano, C., and Verdejo, B. (2002) Cation and anion recognition characteristics of open-chain polyamines containing ethylenic and propylenic chains. *Inorg. Chim. Acta*, **339**, 307–316.
14. Albenda, M.T., Aguilar, J., Alves, S., Aucejo, R., Diaz, P., Lodeiro, C., Lima, J.C., García-España, E., Pina, F., and Soriano, C. (2003) Potentiometric, NMR and fluorescence-emission studies on the binding of adenosine 5'-triphosphate (ATP) by open-chain polyamine receptors containing naphthylmethyl and/or antrylmethyl groups. *Helv. Chim. Acta*, **86**, 3118–3135.
15. Seixas de Melo, J., Albenda, M.T., Diaz, P., García-España, E., Lodeiro, C., Alves, C., Lima, J.C., Pina, F., and Soriano, C. (2002) Ground and excited state properties of polyamine chains bearing two terminal naphthalene units. *J. Chem. Soc., Perkin Trans. 2*, 991–998.
16. Alves, S., Pina, F., Albenda, M.T., García-España, E., Soriano, C., and Luis, S.V. (2001) Open-chain polyamine ligands bearing an anthracene unit – chemosensors for logic operations at the molecular level. *Eur. J. Inorg. Chem.*, **2001** (2), 405–412.
17. Albenda, M.T., Bernardo, M.A., García-España, E., Godino-Salido, M.L., Luis, S.V., Melo, M.J., Pina, F., and Soriano, C. (1999) Thermodynamic and fluorescence emission studies on potential molecular chemosensors for ATP recognition in aqueous solution. *J. Chem. Soc., Perkin Trans. 2*, 2545–2549.
18. Bernardo, M.A., Pina, F., Escuder, B., García-España, E., Godino-Salido, M.L., Latorre, J., Luis, S.V., Ramirez, J.A., and Soriano, C. (1999) Thermodynamic and fluorescence emission studies on chemosensors containing anthracene fluorophores. Crystal structure of {[CuL1Cl]Cl}$_2$·2H_2O [L1 = *N*-(3-aminopropyl)-*N'*-3-(anthracen-9-ylmethyl)aminopropylethane-1,2-diamine]. *J. Chem. Soc., Dalton Trans.*, 915–921.
19. Le Bris, N., Bernard, H., Tripier, R., and Handel, H. (2007) A potentiometric and NMR investigation of triphosphate recognition by linear and bismacrocyclic octaamines possessing a 1,4-xylenyl linker. *Inorg. Chim. Acta*, **360**, 3026–3032.
20. Claudon, G., Le Bris, N., Bernard, H., and Handel, H. (2004) Powerful *N*-monoalkylation of linear tetraamines via bisaminal intermediates. *Eur J. Org. Chem.*, **2004** (24), 5027–5030.
21. Bazzicalupi, C., Bencini, A., Ciattini, S., Giorgi, C., Masotti, A., Paoletti, P., Valtancoli, B., Navon, N., and

Meyerstein, D. (2000) Cu(II) and Cu(I) coordination by hexaamine ligands of different rigidity. A thermodynamic, structural and electrochemical investigation. *J. Chem. Soc., Dalton Trans.*, 2383–2391.

22. Bazzicalupi, C., Bencini, A., Berni, E., Bianchi, A., Fornasari, P., Giorgi, C., Masotti, A., Paoletti, P., and Valtancoli, B. (2001) Cleft-like hexaamine ligands containing large heteroaromatic moieties as receptors for both anions and metal cations. *J. Phys. Org. Chem.*, **14**, 432–439.
23. Nelissen, H.F.M. and Smith, D.K. (2007) Synthetically accessible, high-affinity phosphate anion receptors. *Chem. Commun.*, 3039–3041.
24. Bazzicalupi, C., Bencini, A., Bianchi, A., Danesi, A., Giorgi, C., and Valtancoli, B. (2009) Binding by protonated forms of the tripodal ligand tren. *Inorg. Chem.*, **48**, 2391–2398.
25. Farrell, D., Gloe, K., Gloe, K., Goretzki, G., McKee, V., Nelson, J., Nieuwenhuyzen, M., Pal, I., Stepahn, H., Town, R.M., and Wichmann, K. (2003) Toward promising oxoanion extractants: azacages and open-chain counterparts. *Dalton Trans.*, 1961–1968.
26. Hossain, M.A., Liljegren, J.A., Powell, D., and Bowman-James, K. (2004) Anion binding with a tripodal amine. *Inorg. Chem.*, **43**, 3751–3755.
27. Deroche, A., Morgenstern-Badarau, I., Cesario, M., Guilhem, J., Keita, B., Nadjo, L., and Houée-Levin, C. (1996) A seven-coordinate manganese(II) complex formed with a single tripodal heptadentate ligand as a new superoxide scavenger. *J. Am. Chem. Soc.*, **118**, 4567–4573.
28. Huston, M.E., Akkaya, E.U., and Czarnik, A.W. (1989) Chelation enhanced fluorescence detection of non-metal ions. *J. Am. Chem. Soc.*, **111**, 8735–8737.
29. Vance, D.H., and Czarnik, A.W. (2004) Real-time assay of inorganic pyrophosphatase using a high-affinity chelation-enhanced fluorescence chemosensor. *J. Am. Chem. Soc.*, **116**, 9397–9398.
30. Bencini, A., Bianchi, A., García-España, E., Micheloni, M., and Paoletti, P. (1988) Synthesis and ligational properties of the two very large polyazacycloalkanes [33]aneN_{11} and [36]aneN_{12} forming trinuclear copper(II) complexes. *Inorg. Chem.*, **27**, 176–180, and references therein.
31. Dietrich, B., Hosseini, M.W., Lehn, J.M., and Sessions, R.B. (1981) Anion receptor molecules. Synthesis and anion-binding properties of polyammonium macrocycles. *J. Am. Chem. Soc.*, **103**, 1282–1283.
32. Comarmond, J., Plumere, P., Lehn, J.-M., Agnus, Y., Louis, R., Weiss, R., Kahn, O., and Morgenstern-Badarau, I. (1982) Dinuclear copper(II) cryptates of macrocyclic ligands: synthesis, crystal structure, and magnetic properties. Mechanism of the exchange interaction through bridging azido ligands. *J. Am. Chem. Soc.*, **104**, 6330–6340.
33. Hosseini, M.W. and Lehn, J.-M. (1986) Anion coreceptor molecules. Linear molecular recognition in the selective binding of dicarboxylate substrates by ditopic polyammonium macrocycles. *Helv. Chim. Acta*, **69**, 587–603.
34. Richman, J.E. and Atkins, T.J. (1974) Nitrogen analogs of crown ethers. *J. Am. Chem. Soc.*, **96**, 2268–2270.
35. Hosseini, M.W., Lehn, J.-M., Duff, S.R., Kunjian, G., and Mertes, M.P. (1986) Synthesis of mono- and difunctionalized ditopic [24]aneN_6O_2 macrocyclic receptor molecules. *J. Org. Chem.*, **52**, 1662–1666.
36. Hosseini, M.W., Blacker, A.J., and Lehn, J.-M. (1990) Multiple molecular recognition and catalysis. A multifunctional anion receptor bearing an anion binding site, an intercalating group and a catalytic site for nucleotide binding and hydrolysis. *J. Am. Chem. Soc.*, **112**, 3896–3904.
37. Bencini, A., Bianchi, A., García-España, E., Fusi, V., Micheloni, M., Paoletti, P., Ramirez, J.A., Rodriguez, A., and Valtancoli, B. (1992) Synthesis and protonation behaviour of the macrocyclic ligand

1,4,7,13-tetramethyl-1,4,7,10,13,16-hexaazacyclooctadecane and of its bicyclic derivative 4,7,10, 17,23-pentamethyl-1,4,7,10,13,17,23-heptaazabicyclo[11.7.5]pentacosane. A potentiometric and ^{1}H and ^{13}C NMR study. *J. Chem. Soc., Perkin Trans. 2*, 1059–1065.

38. Andrés, A., Bazzicalupi, C., Bencini, A., Bianchi, A., Fusi, V., García España, E., Giorgi, C., Nardi, N., Paoletti, P., Ramirez, J.A., and Valtancoli, B. (1994) 1,10-Dimethyl-1,4,7,10,13,16-hexa-azacyclcooctadecane L and 1.4.7-trimethyl-1,4,7,10,13,16,19-hepta-azacyclohenicosane L1: two new macro-cyclic receptors for ATP binding. Synthesis, solution equilibria and the crystal structure of $(H_4L)(ClO_4)_4$. *J. Chem. Soc., Perkin Trans. 2*, 2367–2373.

39. Bazzicalupi, C., Bencini, A., Bianchi, A., Fedi, V., Fusi, V., Giorgi, C., Paoletti, P., Tei, L., and Valtancoli, B. (1999) A new function-alized hexaazamacrocycle. Effect of pyridine pendants on cation and anion binding. *J. Chem. Soc., Dalton Trans.*, 1101–1110.

40. Bencini, A., Burguete, M.I., García-España, E., Santiago, V.L., Miravet, J.F., and Soriano, C. (1993) An efficient synthesis of polyaza[*n*] paracyclophanes. *J. Org. Chem.*, **58**, 4749–4753.

41. Albelda, M.T., Frias, J.C., García-España, E., and Santiago, V.L. (2004) Studies on the interaction of phosphate anions with *N*-functionalized polyaza[*n*]paracyclophanes: the role of *N*-methylation. *Org. Biomol. Chem.*, **2**, 816–820.

42. Burguete, M.I., Lopez-Diago, L., García-España, E., Galindo, F., Luis S.V., Miravet, J.F., and Sroczynski D. (2003) New efficient procedure for the use of diethoxyphosphoryl as a protecting group in the synthesis of polyazamacrocycles. Preparation of polyazacyclophanes derived from resor-cinol. *J. Org. Chem.*, **68**, 10169–10171.

43. Bazzicalupi, C., Biagini, S., Bencini, A., Faggi, E., Giorgi, C., Matera, I., and Valtancoli, B. (2006) ATP recognition and sensing with a phenanthroline-containing polyam-monium receptor. *Chem. Commun.*, **39**, 4087–4089.

44. Bazzicalupi, C., Bencini, A., Faggi, E., Giorgi, C., Meini, S., and Valtancoli, B. (2009) Exploring the binding ability of phenanthroline-based polyammo-nium receptors for anions: hints for design of selective chemosensors for nucleotides. *J. Org. Chem.*, **74**, 7349–7363.

45. Anda, C., Llobet, A., Martell, A.E.B., and Parella, T. (2003) Systematic evaluation of molecular recognition phenomena. 3. Selective diphosphate binding to isomeric hexaazamacro-cyclic ligands containing xylylic spacers. *Inorg. Chem.*, **42**, 8545–8550.

46. Chen, D. and Martell, A.E. (1991) The synthesis of new binucleating polyaza macrocyclic and macrobicyclic ligands: dioxygen affinities of the cobalt com-plexes. *Inorg. Chem.*, **34**, 6895–6902.

47. Llobet, A., Reibenspies, J., and Martell, A.E. (1994) Oxydiacetic acid and cop-per(II) complexes of a new hexaaza macrocyclic dinucleating ligand. *Inorg. Chem.*, **33**, 5946–5951.

48. Lindoy, L.F. (1989) *The Chemistry of Macrocyclic Ligand Complexes*, Cam-bridge University Press, Cambridge.

49. Carvalho, S., Delgado, R., Drew, M.G.B., Calisto, V., and Felix, V. (2008) Binding studies of a protonated diox-atetraazamacrocycle with carboxylate substrates. *Tetrahedron*, **64**, 5392–5403.

50. Gibson, D., Dey, K.R., Fronczek, F.R., and Hossain, M.A. (2009) A new hexaaminomacrocycle for ditopic bind-ing of bromide. *Tetrahedron Lett.*, **50**, 6537–6539.

51. Carvalho, S., Delgado, R., Fonseca, N., and Felix, V. (2006) Recognition of dicarboxylate anions by a ditopic hex-aazamacrocycle containing bis-*p*-xylyl spacers. *New J. Chem.*, **30**, 247–257.

52. Li, F., Delgado, R., Costa, J., Drew, M.G.B., and Felix, V. (2005) Ditopic hexaazamacrocycles containing pyri-dine: synthesis, protonation and complexation studies. *Dalton Trans.*, 82–91.

53. Dhaenens, M., Lehn, J.-M., and Vigneron, J.-P. (1993) Molecular recognition of nucleosides, nucleotides and anionic planar substrates by a water-soluble bis-intercaland-type receptor molecule. *J. Chem. Soc., Perkin Trans. 2*, 1379–1381.
54. Cruz, C., Delgado, R., Drew, M.G.B., and Felix, V. (2007) Evaluation of the binding ability of a novel dioxatetraazamacrocyclic receptor that contains two phenanthroline units: selective uptake of carboxylate anions. *J. Org. Chem.*, **72**, 4023–4034.
55. Cruz, C., Calisto, V., Delgado, R., and Felix, V. (2009) Design of protonated polyazamacrocycles based on phenanthroline motifs for selective uptake of aromatic carboxylate anions and herbicides. *Chem. Eur. J.*, **15**, 3277–3289.
56. Bazzicalupi, C., Bencini, A., Bianchi, A., Faggi, E., Giorgi, C., Santarelli, S., and Valtancoli, B. (2008) Polyfunctional binding of thymidine 5'-triphosphate with a synthetic polyammonium receptor containing aromatic groups. Crystal structure of the nucleotide-receptor adduct. *J. Am. Chem. Soc.*, **130**, 2440–2441.
57. Teulade-Fichou, M.-P., Vigneron, J.-P., and Lehn, J.-M. (1995) Molecular recognition of nucleosides and nucleotides by a water-soluble bis-intercaland-type receptor based on acridine subunits. *Supramol. Chem.*, **5**, 139–147.
58. Baudoin, O., Teulade-Fichou, M.-P., Vigneron, J.-P., and Lehn, J.-M. (1997) Cyclobisintercaland macrocycles: synthesis and physicochemical properties of macrocyclic polyamines containing two crescent-shaped dibenzophenanthroline subunits. *J. Org. Chem.*, **62**, 5458–5470.
59. Baudoin, O., Gonnet, F., Teulade-Fichou, M.-P., Vigneron, J.-P., Tabet, J.-C., and Lehn, J.-M. (1999) Molecular recognition of nucleotide pairs by a cyclo-bis-intercaland-type receptor molecule: a spectrophotometric and electrospray mass spectrometry study. *Chem. Eur. J.*, **5**, 2762–2771.
60. Bazzicalupi, C., Bencini, A., Bianchi, A., Fusi, V., Giorgi, C., Paoletti, P., Stefani, A., and Valtancoli, B. (1995) Synthesis and selectivity in metal ion coordination of the new ligands1,4,7-trimethyl-1,7-bis (4-carboxybenzyl)-1,4,7-triazaheptane (L) and 1,4,7,16,19,22-hexamethyl-1,4,7,16, 19,22-hexaaza[9.9]paracyclophane (L1). Crystal structures of $[PdLH_2Cl]NO_3{\cdot}2.6H_2O$ and $[Cu_2L1Cl_2](BPh_4)(ClO_4){\cdot}CH_3CN$. *Inorg. Chem.*, **34**, 552–559.
61. Bazzicalupi, C., Bencini, A., Bianchi, A., Fusi, V., Giorgi, C., Granchi, A., Paoletti, P., and Valtancoli, B. (1997) Basicity properties of two paracyclophane receptors. Their ability in ATP and ADP recognition in aqueous solution. *J. Chem. Soc., Perkin Trans. 2*, 775–781.
62. Granzhan, A. and Teulade-Fichou, M.-P. (2009) Synthesis of mono- and bibrachial naphthalene-based macrocycles with pyrene or ferrocene for anion detection. *Tetrahedron*, **65**, 1349–1360.
63. Develay, S., Tripier, R., Le Baccon, M., Patinec, V., Serratrice, G., and Handel, H. (2005) Cyclen based bis-macrocyclic ligands as phosphates receptors. A potentiometric and NMR study. *Dalton Trans.*, 3016–3024.
64. Delépine, A.S., Tripier, R., and Handel, H. (2008) Cyclen-based bis-macrocycles for biological anion recognition. A potentiometric and NMR study of AMP, ADP and ATP nucleotide complexation. *Org. Biomol. Chem.*, **6**, 1743–1750.
65. Le Baccon, M., Chuburu, F., Toupet, L., Handel, H., Soibinet, M., Déchamps-Olivier, I., Barbier, J.-P., and Aplincourt, M. (2001) Bis-aminals: efficient tools for bis-macrocycle synthesis. *New J. Chem.*, **25**, 1168–1174.
66. Develay, S., Tripier, R., Le Baccon, M., Patinec, V., Serratrice, G., and Handel, H. (2006) Host-guest interaction between cyclen-based macrotricyclic ligands and phosphate anions. A potentiometric investigation. *Dalton Trans.*, 3418–3426.
67. Develay, S., Tripier, R., Chuburu, F., Le Baccon, M., and Handel, H. (2003) A new versatile synthesis of macrotricyclic tetraazacycloalkane-based ligands

for bis-aminal derivatives. *Eur. J. Org. Chem.*, **2003** (16) 3047–3050.
68. Bazzicalupi, C., Bencini, A., Berni, E., Bianchi, A., Ciattini, S., Giorgi, C., Maoggi, S., Paoletti, P., and Valtancoli, B. (2002) Synthesis of new tren-based tris-macrocycles. Anion cluster assembling inside the cavity generated by a bowl-shaped receptor. *J. Org. Chem.*, **67**, 9107–9110.
69. Bazzicalupi, C., Bencini, A., Bianchi, A., Borsari, L., Giorgi, C., and Valtancoli, B. (2005) Tren-based tris-macrocycles as anion hosts. Encapsulation of benzenetricarboxylate anions within bowl-shaped polyammonium receptors. *J. Org. Chem.*, **70**, 4257–4266.
70. Lehn, J.-M., Sonveaux, E., and Willard, A.K. (1978) Molecular recognition. Anion cryptates of a macrobicyclic receptor molecule for linear triatomic species. *J. Am. Chem. Soc.*, **100**, 4914–4915.
71. Dietrich, B., Guilhem, J., Lehn, J.-M., Pascard, C., and Sonveaux, E. (1984) Molecular recognition in anion coordination chemistry. Structure, binding constants and receptor-substrate complementarity of a series of anion cryptates of a macrobicyclic receptor molecule. *Helv. Chim. Acta*, **67**, 91–104.
72. Lehn, J.-M., Pine, S.H., Watanabe, E., and Willard, A.K. (1977) Binuclear cryptates. Synthesis and binuclear cation inclusion complexes of bis-tren macrobicyclic ligands. *J. Am. Chem. Soc.*, **99**, 6766–6769.
73. Dietrich, B., Dilworth, B., Lehn, J.-M., Souchez, J.-P., Cesario, M., Guilheim, J., and Pascard, C. (1996) Anion cryptates: synthesis, crystal structures, and complexation constants of fluoride and chloride inclusion complexes of polyammonium macrobicyclic ligands. *Helv. Chim. Acta*, **79**, 569–587.
74. Hunter, J., Nelson, J., Harding, C., McCann, M., and Mckee, V. (1990) Complexes of a new mononucleating cage ligand; livelier than sepulchrates. *J. Chem. Soc., Chem. Commun.*, 1148–1151.
75. Nelson, J., Mckee, V., and Morgan, G. (1998) Coordination chemistry of azacryptands. *Prog. Inorg. Chem.*, **47**, 167–316.
76. Lehn, J.-M., Meric, R., Vigneron, J.-P., Bkouche-Waksman, I., and Pascard, C. (1987) Molecular recognition of anionic substrates. Binding of carboxylates by a macrobicyclic coreceptor and crystal structure of its supramolecular cryptate with the terephthalate dianion. *J. Chem. Soc., Chem. Commun.*, 1691–1694.
77. Ilioudis, C.A., Tocher, D.A., and Steed, J.W. (2004) A highly efficient, preorganized macrobicyclic receptor for halides based on CH ··· and NH ··· interactions. *J. Am. Chem. Soc.*, **126**, 12395–12402.
78. Mateus, P., Delgado, R., Brandão, P., Carvalho, S., and Felix, V. (2009) Selective recognition of tetrahedral dianions by a hexaaza cryptand receptor. *Org. Biomol. Chem.*, **7**, 4661–4673.
79. Mateus, P., Delgado, R., Brandão, P., and Felix, V. (2009) Polyaza cryptand receptor selective for dihydrogen phosphate. *J. Org. Chem.*, **74**, 8638–8646.
80. Bazzicalupi, C., Bencini, A., Bianchi, A., Danesi, A., Giorgi, C., Martinez Lorente, M.A., and Valtancoli, B. (2006) Inclusive coordination of F^-, Cl^- and Br^- anions into macrobicyclic polyammonium receptors. *New J. Chem.*, **30**, 959–965.
81. Bazzicalupi, C., Bencini, A., Bianchi, A., Fusi, V., Mazzanti, L., Paoletti, P., and Valtancoli, B. (1995) 4,7,10,23-Tetramethyl-17-oxa-1,4,7,10,13,23-hexaazabicyclo[11.7.5]pentacosane (L), a two-binding-site ligand for the assembly of the $[Zn_2(\mu\text{-}OH)_2]^{2+}$ cluster. *Inorg. Chem.*, **34**, 3003–3010.
82. Jursic, B.S. and Zdravkovski, Z. (1993) A simple preparation of amides from acids and amines by heating of their mixture. *Synth. Commun.*, **23** (19), 2761–2770.
83. Valeur, E. and Bradley, M. (2009) Amide bond formation: beyond the myth of coupling reagents. *Chem. Soc. Rev.*, **38** (2), 606–631.
84. Fischer, E. and Armstrong, E.F. (1901) Over isomers of the aceto-halogen

derivative of grape sugar and the synthesis of the glucosides. *Ber. Dtsch. Chem. Ges.*, **34**, 2885–2900.

85. Falb, E., Yechezkel, T., Salitra, Y., and Gi1on, C. (1999) In situ generation of fmoc-amino acid chlorides using bis-(trichloromethyl)carbonate and its utilization for difficult couplings in solid-phase peptide synthesis. *J. Pept. Res.*, **53** (5), 507–517.
86. Thern, B., Rudolph, J., and Jung, G. (2002) Triphosgene as highly efficient reagent for the solid-phase coupling of N-alkylated amino acids – total synthesis of cyclosporin O. *Tetrahedron Lett.*, **43** (28), 5013–5016.
87. Kavallieratos, K., Bertao, C.M., and Crabtree, R.H. (1999) Hydrogen bonding in anion recognition: a family of versatile, nonpreorganized neutral and acyclic receptors. *J. Org. Chem.*, **64** (5), 1675–1683.
88. Kavallieratos, K., de Gala, S.R., Austin, D.J., and Crabtree, R.H. (1997) A readily available non-preorganized neutral acyclic halide receptor with an unusual nonplanar binding conformation. *J. Am. Chem. Soc.*, **119** (9), 2325–2326.
89. Coles, S.J., Frey, J.G., Gale, P.A., Hursthouse, M.B., Light, M.E., Navakhun, K., and Thomas, G.L. (2003) Anion-directed assembly: the first fluoride-directed double helix. *Chem. Commun.*, 568–569.
90. Gale, P.A., Camiolo, S., Tizzard, G.J., Chapman, C.P., Light, M.E., Coles, S.J., and Hursthouse, M.B. (2001) 2-Amidopyrroles and 2,5-diamidopyrroles as simple anion binding agents. *J. Org. Chem.*, **66** (23), 7849–7853.
91. Gale, P.A., Camiolo, S., Chapman, C., Light, M., and Hursthouse, M. (2001) Hydrogen-bonding pyrrolic amide cleft anion receptors. *Tetrahedron Lett.*, **42** (30), 5095–5097.
92. Camiolo, S., Gale, P.A., Hursthouse, M., and Light, M. (2002) Confirmation of a 'cleft-mode' of binding in a 2,5-diamidopyrrole anion receptor in the solid state. *Tetrahedron Lett.*, **43** (39), 6995–6996.
93. Motekaitis, R.J., Heinert, D.H., and Martell, A.E. (1970) 3,4-Dihalopyrroles. *J. Org. Chem.*, **35** (8), 2504–2511.
94. Hoegberg, T., Strom, P., Ebner, M., and Ramsby, S. (1986) Cyanide as an efficient and mild catalyst in the aminolysis of esters. *J. Org. Chem.*, **52**, 2033–2036.
95. Basha, A., Lipton, M., and Weinreb, S.M. (1977) A mild, general method for conversion of esters to amides. *Tetrahedron Lett.*, **18** (48), 4171–4172.
96. Zielinski, T. and Jurczak, J. (2005) Thioamides versus amides in anion binding. *Tetrahedron*, **61**, 4081–4089.
97. Coles, S.J., Gale, P.A., Hursthouse, M.B., Light, M.E., and Warriner, C.N. (2004) Crystallographic and solution anion binding studies of bis-amidofurans and thiophenes. *Supramol. Chem.*, **16** (7), 469–486.
98. Jeong, K. and Cho, Y.L. (1997) Highly strong complexation of carboxylates with 1-alkylpyridinium receptors in polar solvents. *Tetrahedron Lett.*, **38** (18), 3279–3282.
99. Prohens, R., Tomàs, S., Morey, J., Deyà, P.M., Ballester, P., and Costa, A. (1998) Squaramido-based receptors: molecular recognition of carboxylate anions in highly competitive media. *Tetrahedron Lett.*, **39** (9), 1063–1066.
100. Schmidt, A.H. (1980) Reaktionen von quadratsäure und quadratsäure-derivaten. *Synthesis*, **1980** (12), 961–994.
101. Hughes, M.P. and Smith, B.D. (1997) Enhanced carboxylate binding using urea and amide-based receptors with internal Lewis acid coordination: a cooperative polarization effect. *J. Org. Chem.*, **62** (13), 4492–4499.
102. Werner, F. and Schneider, H. (2000) Complexation of anions including nucleotide anions by open-chain host compounds with amide, urea, and aryl functions. *Helv. Chim. Acta.*, **83** (2), 465–478.
103. Raposo, C., Pérez, N., Almaraz, M., Mussons, M.L., Caballero, M.C., and Morán, J.R. (1995) *Tetrahedron Lett.*, **36** (18), 3255–3258.
104. Valiyaveettil, S., Engbersen, J.F.J., Verboom, W., and Reinhoudt, D.N.

(1993) Synthesis and complexation studies of neutral anion receptors. *Angew. Chem. Int. Ed.*, **32** (6), 900–901.

105. Beer, P.D., Fletcher, N.C., Grieve, A., Wheeler, J.W., Moore, C.P., and Wear, T. (1996) Halide anion recognition by new acyclic quaternary polybipyridinium and polypyridinium receptors. *J. Chem. Soc., Perkin Trans. 2*, (8), 1545–1551.
106. Hossain, M.A., Llinares, J.M., Powell, D., and Bowman-James, K. (2001) Multiple hydrogen bond stabilization of a sandwich complex of sulfate between two macrocyclic tetraamides. *Inorg. Chem.*, **40** (13), 2936–2937.
107. Rudkevich, D.M., Verboom, W., and Reinhoudt, D.N. (1994) Calix[4]arene salenes: a bifunctional receptor for NaH_2PO_4. *J. Org Chem.*, **59** (13), 3683–3686.
108. Cameron, B.R. and Loeb, S.J. (1997) Bis(amido)calix[4]arenes in the pinched cone conformation as tuneable hydrogen-bonding anion receptors. *Chem. Commun.*, (6), 573–574.
109. Deetz, M.J., Shang, M., and Smith, B.D. (2000) A macrobicyclic receptor with versatile recognition properties: simultaneous binding of an ion pair and selective complexation of dimethylsulfoxide. *J. Am. Chem. Soc*, **122** (26), 6201–6207.
110. Bisson, A.P., Lynch, V.M., Monahan, M.-K.C., and Anslyn, E.V. (1997) Recognition of anions through NH-π hydrogen bonds in a bicyclic cyclophane – selectivity for nitrate. *Angew. Chem. Int. Ed.*, **36**, 2340–2342.
111. Chmielewski, M.J., Szumna, A., and Jurczak, J. (2004) Anion induced conformational switch of a macrocyclic amide receptor. *Tetrahedron Lett.*, **45**, 8699–8703.
112. Kang, S.O., Llinares, J.M., Powell, D., VanderVelde, D., and Bowman-James, K. (2003) New polyamide cryptand for anion binding. *J. Am. Chem. Soc.*, **125**, 10152–10153.
113. Kang, S.O., VanderVelde, D., Powell, D., and Bowman-James, K. (2004) Fluoride-facilitated deuterium exchange from DMSO-d_6 to polyamide-based cryptands. *J. Am. Chem. Soc.*, **126**, 12272–12273.
114. Kang, S.O., Powell, D., and Bowman-James, K. (2005) Anion binding motifs: topicity and charge in amidocryptands. *J. Am. Chem. Soc.*, **127**, 13478–13479.
115. Jurczak, J., Kasprzyk, S., Salanski, P., and Stankiewicz, T. (1991) A general method for the synthesis of diazacoronands. *J. Chem. Soc. Chem. Commun.*, (14), 956–957.
116. Gryko, D.T., Piatek, P., Pecak, A., Palys, M., and Jurczak, J. (1998) Synthetic and crystallographic studies on pyridinophanes. *Tetrahedron*, **54**, 7505–7516.
117. Szumna, A. and Jurczak, J. (2001) A new macrocyclic polylactam-type neutral receptor for anions – structural aspects of anion recognition. *Eur. J. Org. Chem.*, **2001** (21), 4031–4039.
118. Chmielewski, M.J. and Jurczak, J. (2005) Anion recognition by neutral macrocyclic amides. *Chem. Eur. J.*, **11**, 6080–6094.
119. Szumna, A. and Jurczak, J. (2001) Unusual encapsulation of two anions in the cavity of neutral macrocyclic octalactam. *Helv. Chim. Acta.*, **84**, 3760–3765.
120. Chmielewski, M.J. and Jurczak, J. (2005) A hybrid macrocycle containing benzene and pyridine subunits is a better anion receptor than both its homoaromatic congeners. *Tetrahedron Lett.*, **46**, 3085–3088.
121. Korendovych, I.V., Cho, M., Butler, P.L., Staples, R.J., and Rybak-Akimova, E.V. (2006) Anion binding to monotopic and ditopic macrocyclic amides. *Org. Lett.*, **8**, 3171–3174.
122. Voegtle, F., Weber, E., Wehner, W., Naetscher, R., and Gruetze, J. (1974) Heavy metal complexes with new cyclic ligands. *Chem. Z.*, **98**, 562–563.
123. Choi, K. and Hamilton, A.D. (2001) Selective anion binding by a macrocycle with convergent hydrogen bonding functionality. *J. Am. Chem. Soc.*, **123** (10), 2456–2457.

124. Hinzen, B., Seiler, P., and Diederich, F. (1996) Mimicking the vancomycin carboxylate binding site: synthetic receptors for sulfonates, carboxylates, and N-protected α-amino acids in water. *Helv. Chim. Acta*, **79** (4), 942–960.

125. Tejeda, A., Oliva, A.I., Simón, L., Grande, M., Cruz Caballero, M.A., and Morán, J.R. (2000) A macrocyclic receptor for the chiral recognition of hydroxycarboxylates. *Tetrahedron Lett.*, **41** (23), 4563–4566.

126. Gale, P.A., Navakhun, K., Camiolo, S., Light, M.E., and Hursthouse, M.B. (2002) Anion-anion assembly: a new class of anionic supramolecular polymer containing 3,4-dichloro-2,5-diamido-substituted pyrrole anion dimers. *J. Am. Chem. Soc.*, **124**, 11228–11229.

5
Template Synthesis

Jack K. Clegg and Leonard F. Lindoy

5.1
Introductory Remarks

Although the use of metal ions as templates in the synthesis of metal complexes has long been investigated and, for example, was employed in procedures for generating metal phthalocyanines from the early part of the twentieth century, the use of anion templates in synthesis is more recent. In the 1960s, metal template behavior was systematized by Daryle Busch, who introduced the underlying concepts for what he termed the *kinetic* and *thermodynamic template effects* [1, 2]. Isolated examples of the use of anions as templates were reported from about this time, with a very substantial increase in interest in the area occurring during the last decade or so. Reflecting this, many excellent reviews of developments in the area of anion binding have appeared over recent years [3–22]; in addition, a number of reviews largely focusing on anion template systems have also appeared since 2006 [19, 23–33].

Anions are typically larger than the corresponding isoelectronic cations (Na^+ (116)/F^- (119), K^+ (152)/Cl^- (167), Rb^+ (166)/Br^- (182), Cs^+ (181)/I^- (206 pm)) and hence have a lower charge to radius ratio. This, coupled with the different geometries displayed by particular anions, their often high solvation energies, and in some cases their pH dependence, all result in the design of simple anion receptors being somewhat more complicated than is generally the case for cation receptors. Nevertheless, paralleling the widespread importance of anion binding in biology, many examples of anion binding by a variety of synthetic receptors have now been reported – often in the context of potential applications, with the latter encompassing optical displays, organic gelators, chemosensors as well as anion separation, and/or sensing devices [34–36]. Classically, anion binding has been discussed in terms of electrostatic (ion–ion and ion–dipole) and weak noncovalent interactions such as hydrogen bonding and other van der Waals forces [17], with, more recently, the contributing role of anion–π interactions (see Chapter 6) being postulated in some instances [37–49], although the existence of particular interactions of this type has been contested [50].

One obvious and now increasingly common strategy employed for generating new anion receptors has involved the use of an anion template to yield a

Anion Coordination Chemistry, First Edition. Edited by Kristin Bowman-James, Antonio Bianchi, and Enrique García-España.

cavity-containing system that exhibits a structure complementary to that of the anion (and, in turn, often acts as a selective receptor for that anion and sometimes for other similar anions) [51–54]. Nevertheless, as with metal-ion templates, it is noted that in many instances, the templating anion becomes an integral part of the templated structure, with the latter not existing in the absence of the template.

In this chapter, representative examples of the use of anion templates in the recent literature (largely since 2006) are discussed, with emphasis on discrete systems in which at least one of the components is organic. The use of anion templation in combinatorial systems has recently been reviewed [32] and is not discussed again here. Similarly, in a number of studies, the operation of an anion template effect has been proposed/implied to occur (in some instances because the product generated has been shown to be dependent on the particular anion present) but without the precise templating role of the anion being defined; alternatively, in other cases, the distinction between anion templation and simple anion binding (including anion bridging between metal centers) is not clear cut [55a–e, 56]. In general, studies such as these have not been included in the following discussion, and perhaps "anion-mediated assembly" [56] is a better term for such behavior rather than anion templation.

While the present discussion is chiefly concerned with the anion templation of small discrete systems, it is noted that the role of anions in promoting oligomer folding has also received recent attention. For example, an original study on the role of Cl^- binding in the concomitant folding of a 1,4-diaryl triazole oligomer (through $CH\cdots Cl^-$ interactions) has appeared [57], and the influence of anions on foldamer conformation (where foldamer is defined as a synthetic chain molecule or oligomer that is able to adopt a secondary structure stabilized by noncovalent interactions) has been the subject of a recent tutorial review [58]. In this chapter, for ease of discussion, the templating role of anions is discussed in terms of case studies that are grouped (in some cases somewhat arbitrarily) into five general categories: (i) macrocyclic systems; (ii) bowl-shaped systems; (iii) capsule, cage, and tube-shaped systems; (iv) circular helicates and *meso*-helicates; and (v) mechanically linked systems.

5.2
Macrocyclic Systems

In part based on the results of earlier studies involving imine-linked polypyrrole as well as mixed imine-linked pyrrole–pyridine macrocycles [54, 59], Sessler *et al.* showed that individual systems of the latter type, incorporating both H-donor and acceptor sites within their cyclic frameworks, are excellent receptors for binding phosphate anions [60]. As phosphate and its derivatives can exist as protonated forms (for example, as $H_2PO_4^-$ and HPO_4^{2-}) near physiological pH, the inclusion of imine linkages was seen to be advantageous because of their potential to act as protonation sites when accommodating such ions. The bipyrrole–pyridyl macrocyclic imine receptors **1–3** (Figure 5.1) were isolated using defined reaction conditions involving the Schiff base condensation of the corresponding bipyrroledialdehydes

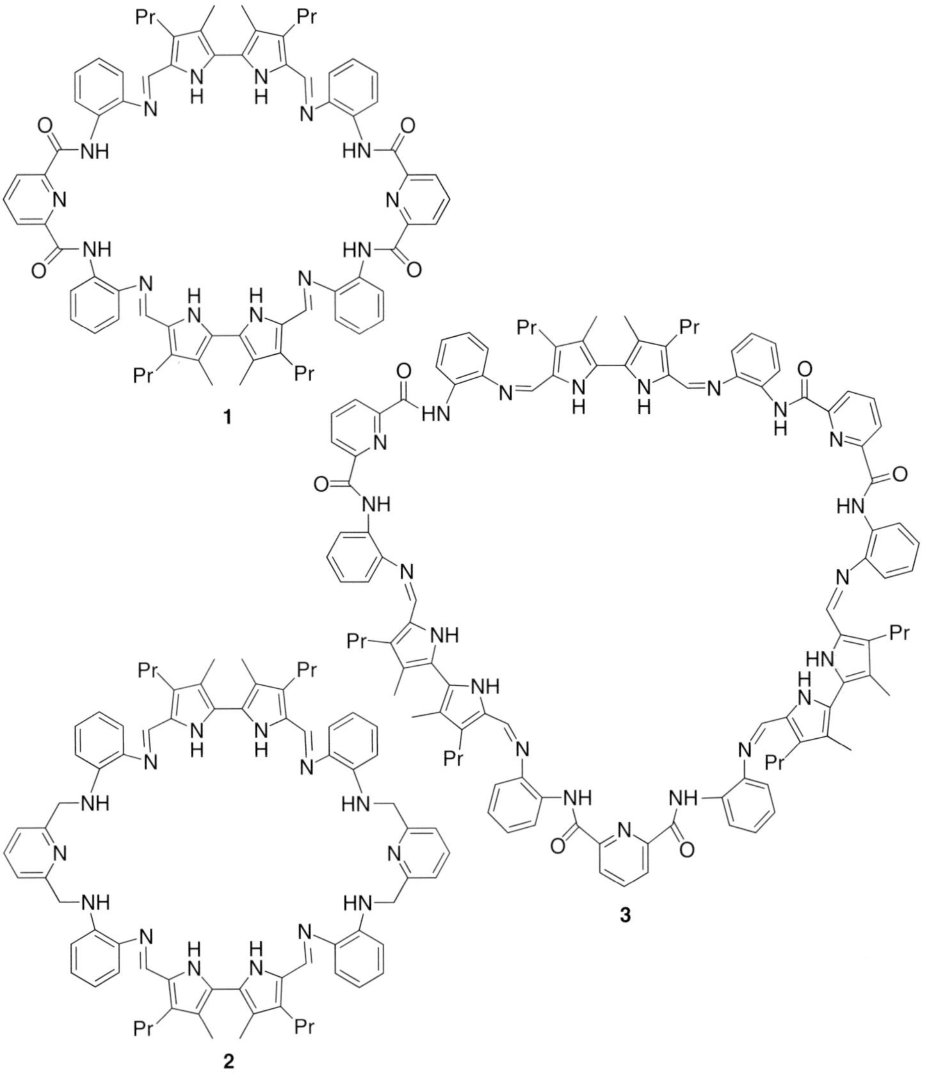

Figure 5.1 Three macrocyclic species incorporating linked bipyrrole units prepared by template synthesis in the presence of phosphoric or sulfuric acid [60].

and bis(aminophenylamino)-substituted pyridine diamines in the presence of phosphoric or sulfuric acid. Neutralization of the acid salts formed with triethylamine in methanol yielded the free receptors [60].

It was demonstrated that the large-ring hexa-imine derivative **3** incorporating three bipyrrole units is selectively prepared by ring expansion from the corresponding macrocyclic species containing two bipyrrole units (**2**) in the presence of tetrabutylammonium hydrogen sulfate or tetrabutylammonium dihydrogenphosphate in acetonitrile. The reaction is postulated to take place under thermodynamic control [54].

Macrocycle **3** forms 1 : 1 complexes with phosphoric and sulfuric acids, whose solid-state structures have been characterized by X-ray diffraction [60]. Anion binding in the respective structures occurs via complementary hydrogen bond networks and resembles the binding occurring in the respective active-site structures of the corresponding phosphate-binding protein (PBP) and sulfate-binding protein (SBP) (see Chapter 1).

UV–vis spectroscopic titrations indicate that the above anions (employed as their tetrabutylammonium salts) interact with **3** in acetonitrile with both high selectivity and affinity, with *K* values for the first binding interaction approaching 10^7 in each case (although the receptor to anion binding ratio is different for each complex: 1 : 1 for HSO_4^- vs 1 : 3 for $H_2PO_4^-$). The binding of 1 equiv. of $H_2PO_4^-$ to **3** was predicted to give rise to a conformational change such that a hydrogen bond network is generated, which again parallels both those observed in the solid state complex salt [$\mathbf{3}H_2^{2+}\cdot HPO_4^{2-}$] and in the active center of the PBP.

With an aim of investigating strategies for preparing larger receptor structures of increased complexity, the reductive amination reaction between a bis(amidoamine) of type **4** (Figure 5.2, where R = *iso*-Pr or CH_2Ph) and 1,4-phenyldialdehyde was investigated, but unfortunately, only a complex mixture of products was obtained, some of which were open chain [61]. However, the formation of the desired [2+2] tetra-imine intermediate in reasonable yield was reasoned to be feasible only if favorable configurational preorganization is present during the reaction. Accordingly, several anions were screened for use as potential anionic templates for the reaction. From Monte Carlo analyses, it was found that the terephthalate

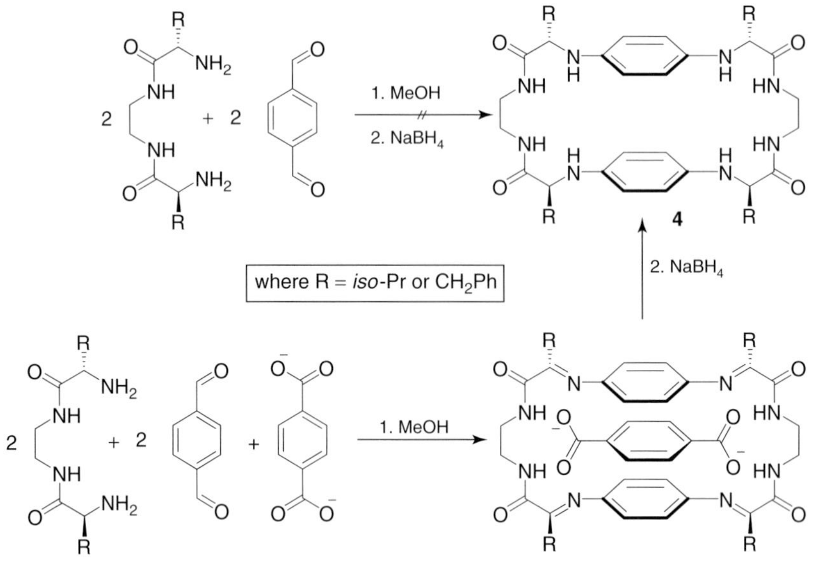

Figure 5.2 The terephthalate-dianion-template-assisted synthesis of macrocycle **4** [61].

dianion appeared to be an excellent choice for use as a template as it seemed to show near-perfect electronic and steric complementarity with the cavity of the target macrocyclic species. In particular, it was seen to be well suited to form four hydrogen bonds with the carboxylate groups and the four amide hydrogen atoms belonging to the two bis-amide domains as well as to participate in π-stacking interactions with each of the macrocycle's aromatic rings. Hence, the macrocyclization reaction was repeated in the presence of the terephthalate dianion (Figure 5.2). The successful formation of the "key" supramolecular (tetra-imine) intermediate complex was probed using ^{1}H NMR (including 1D ROESY and pulse-field gradient spin-echo (PGSE) experiments), ESI-TOF mass spectral studies, and CD spectroscopy. As predicted, these studies confirmed the presence of excellent structural host–guest complementarity between the macrocyclic host and its terephthalate dianion guest. In a further experiment, surprising homochiral self-recognition behavior was also demonstrated for this host–guest system; when the templated reaction was carried out in the presence of both (*R,R*)- and (*S,S*)-bis(amidoamines), only homochiral macrocyclic products were generated – with no "mixed ligand" species being observed within the limits of detection of the NMR experiment used to monitor the results. Finally, the *in situ* reduction of the tetra-imine bonds in the above macrocycles yielded the corresponding tetra-amine derivatives in 60–65% yield (Figure 5.2). Clearly, this procedure represents a one-pot, two-step anion template process of considerable synthetic utility.

In an extension of the above study, the effect of structural variations on each of the components employed for the anion template reaction was investigated [62]. For example, the replacement of the ethylene spacer by a propyl in the bis(amidoamine) reagent led to a significant decrease in the yield of the templated macrocyclization condensation product. This result may, at least in part, be attributed to the increased flexibility working against the required degree of conformational preorganization and "tight" template binding that seems to be a necessary condition for efficient macrocycle formation. Similarly, for the dialdehyde component, it was found that *para* substitution was more effective than *meta* substitution in promoting macrocycle formation, although the meta reagent also gave macrocyclic products in reasonable yields when used with the best fitting template (the isophthalate anion). The effect of variation of the R groups on the bis(amidoamine) precursor reagent was also probed during this study. Overall, it was concluded that it may be possible to obtain analogous larger macrocyclic products by optimizing both the size and the topology of the template employed.

The NH groups of alkyl- and aryl-substituted ureas (and thioureas) have been well established to act as strong hydrogen bonded donors in a wide variety of anion receptor molecules. In one recent study of this type, it has been shown that the sole product of the reaction of the diisocyanate derivative **5** with the diamine derivative **6** in acetonitrile is the macrocyclic [1+1] tri-urea derivative **7** (Figure 5.3, reaction a), although, when the reaction was repeated in dichloromethane (DCM) (Figure 5.3, reaction b), 17% of the corresponding [2+2] hexa-urea macrocycle **8** was generated along with **7** (Table 5.1) [63]. However, there was a dramatic change in the ratio of the products formed when the reaction was carried out in the

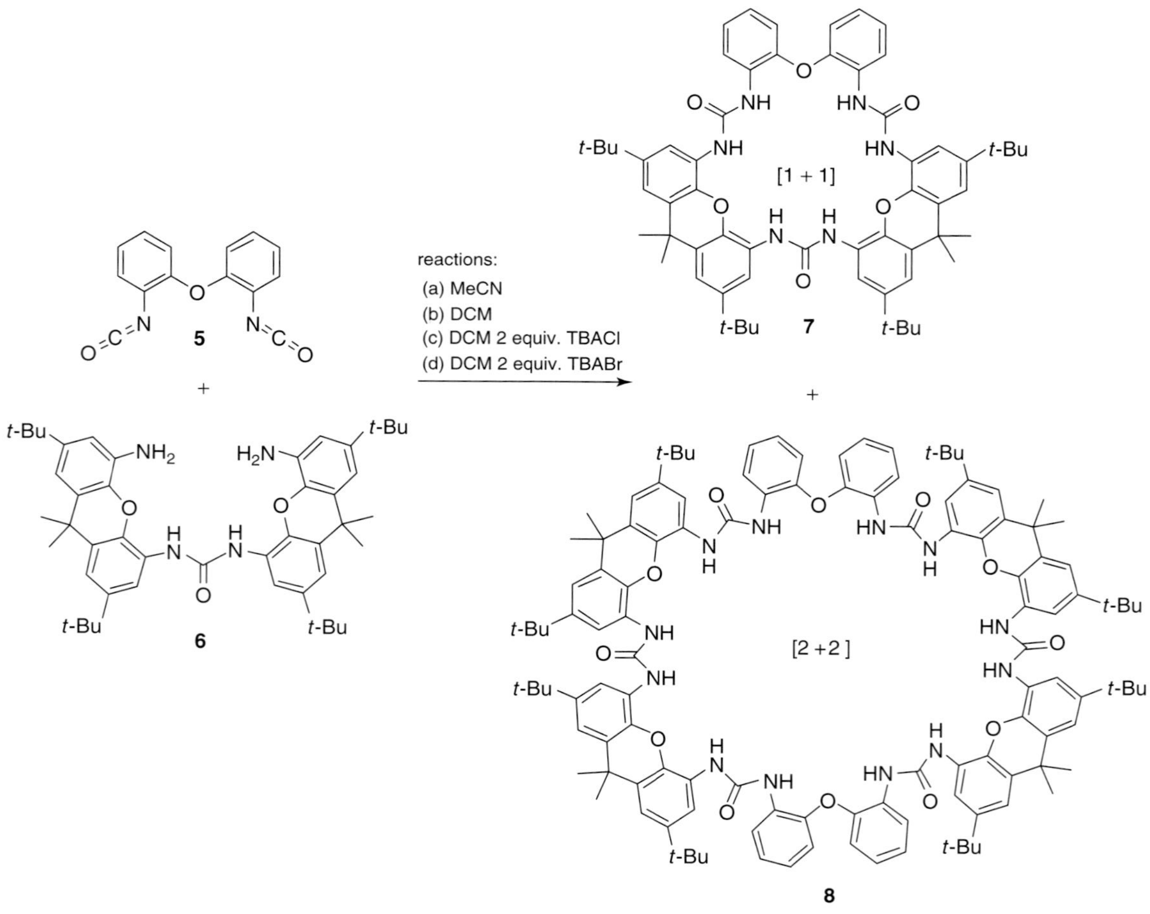

Figure 5.3 Synthesis of macrocycles **7** and **8** under the conditions given by (a)–(d) (also see main text); the respective relative yields for each condition are listed in Table 5.1 [63].

Table 5.1 Relative yields of tri-urea and hexa-urea macrocyclic products **7** and **8**.

Conditions[a]	Mole ratio of [1 + 1] (7) (%)	Mole ratio [2 + 2] (8) (%)
(a)	100	0
(b)	83	17
(c)	17	83
(d)	25	75

[a]See Figure 5.3.

presence of tetrabutylammonium chloride (Figure 5.3, reaction c) – with a similar (but smaller) change when tetrabutylammonium bromide was present (Figure 5.3, reaction d) (see also Table 5.1). These results are best rationalized in terms of two (see below) Cl^- ions acting as a template for the preferential formation of **8** relative to its smaller-ring homolog – as well as being strongly bound within the cavity of **8** once it is formed. Even though the larger bromide ion is expected to bind less strongly to **8**, it is seen that it is still capable of promoting the formation of this larger ring (Table 5.1). However, significantly, when tetrabutylammonium iodide was employed, there was no generation of the hexa-urea derivative; only the [1+1] product **7** was obtained.

X-ray analysis along with solution NMR spectroscopy and ESI-MS showed that a 2 : 1 complex of Cl^- with **8** is generated, with UV spectrophotometric and microcalorimetric data for host–guest formation confirming this assignment. Binding of the first Cl^- is entropically driven, presumably largely reflecting the release of solvent molecules from this receptor on complex formation. In contrast, binding of the second Cl^- is enthalpically driven, possibly reflecting the previously well-documented strong affinity between urea functions and, in this case, the partially prearranged cavity of **8**. The X-ray structure of the dichloride complex is shown in Figure 5.4, which illustrates the "good fit" within the twisted (figure-eight-shaped) configuration of bound **8**. In this arrangement, each Cl^- guest is well positioned to interact with three sequential urea sites in the macrocycle, with these two anions occupying the two "curved" (noncrossing) sections in the "figure eight" arrangement.

The anion binding properties of the pyrrole–pyridine-based macrocyclic polyamides **9** and **10** have also been investigated [64]. Owing to the presence of six NH groups in each macrocycle, it was reasoned that the formation of the latter would be significantly influenced by the presence of anions in the reaction solution. Indeed, the presence of Cl^-, derived initially from the use of 2,6-pyridinedicarbonyl chloride as one component of the respective bis-acylation ring-closing reactions, was demonstrated to serve as a template that controls the [1+1] macrocycle formation in each of these syntheses. The synthesis of **9** is illustrated in Figure 5.5; **10** was prepared using a procedure analogous to the synthesis of **9** from the corresponding diamine precursor. When additional Cl^-

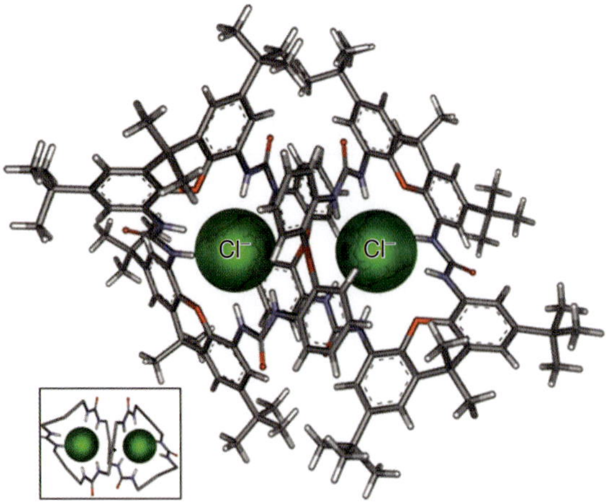

Figure 5.4 A schematic representation of the X-ray structure of the 2 : 1 (Cl^-:**8**) macrocyclic complex. Cations and solvent molecules removed for clarity [63].

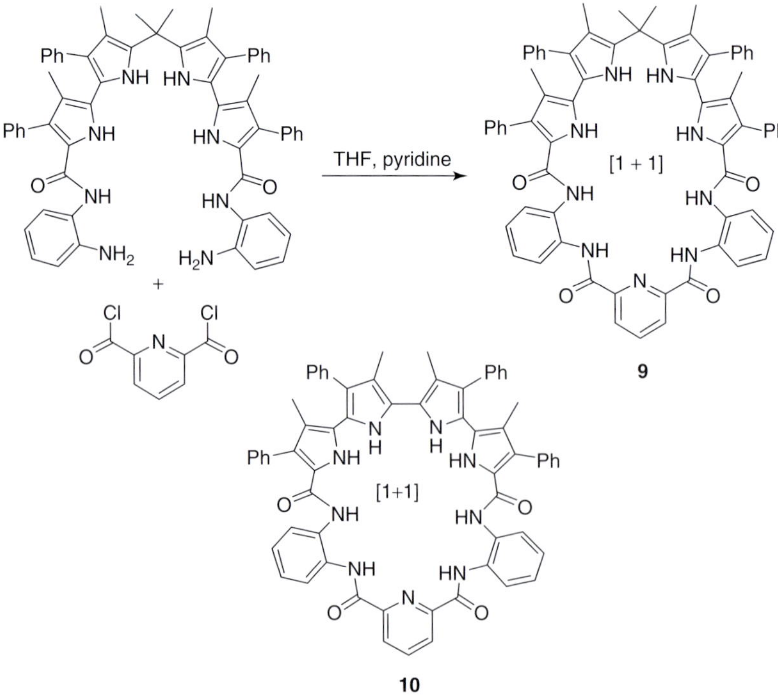

Figure 5.5 The synthesis of **9** together with the structure of **10**, with the latter also prepared by an analogous bis-acylation procedure from the corresponding diamine precursor [64].

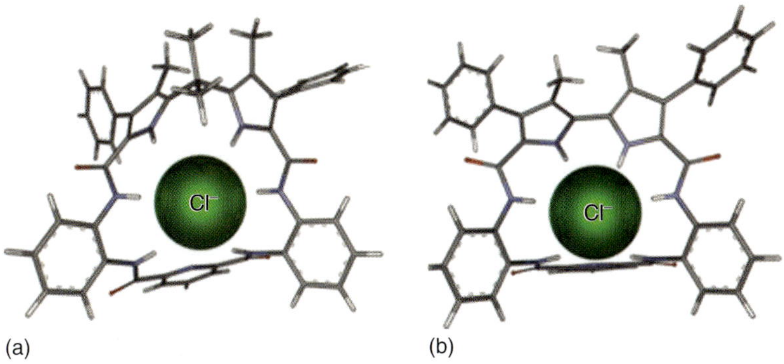

Figure 5.6 Schematic depiction of the X-ray structures of the anionic Cl^- complexes of (a) **9** and (b) **10** [64].

in the form of tetrabutylammonium chloride was added to the reaction mixture, a significant increase in the macrocycle yield was observed in each case. Each bis-acylation (cyclization) reaction was carried out in tetrahydrofuran in the presence of excess pyridine (as base), and the macrocyclic product was observed to be the corresponding Cl^--bound host–guest complex, with a pyridinium ion acting as the required counterion. The X-ray structures of the respective Cl^- complexes are shown in Figure 5.6. When tetrabutylammonium perrhenate was added to the reaction mixture containing Cl^- to probe whether it would act as a larger anion template, a mixture of the [1+1] and [2+2] cyclic condensation products was isolated.

The anion-binding behavior of these cyclic hosts in DMSO has been investigated by spectrophotometric titration. Interestingly, both macrocycles were shown to have higher affinities for the oxyanions HSO_4^-, $H_2PO_4^-$, and OAc^- than for the more symmetric (spherical) Cl^- and F^- anions.

5.3 Bowl-Shaped Systems

Three Cu(I) metallomacrocyclic coordination complexes have been assembled from bridging 4-(2-pyridinyl)pyrimidine (**11**; pprd) and terminal CO and/or 2-ethene co-ligands by anion templation; a fourth polymeric species was also isolated during this study (Figure 5.7) [65]. The X-ray structures of all four products were obtained. The pprd ligand is seen to be useful for generating bridged species since it contains a bidentate site for chelation to one metal center and an exo donor site for bridging to a second center, with the corresponding coordination site vectors mutually orientated at approximately right angles. The unusual metallomacrocyclic species obtained resemble calixarenes in shape, with either C_2H_4 or CO "legs." In each case, an anion is accommodated inside the calixarene-like cavity. Both the type and size of the cavity generated in these cyclic [3+3] and [4+4] species were demonstrated to be controlled by the choice of anion employed, with the smaller tetrahedral perchlorate

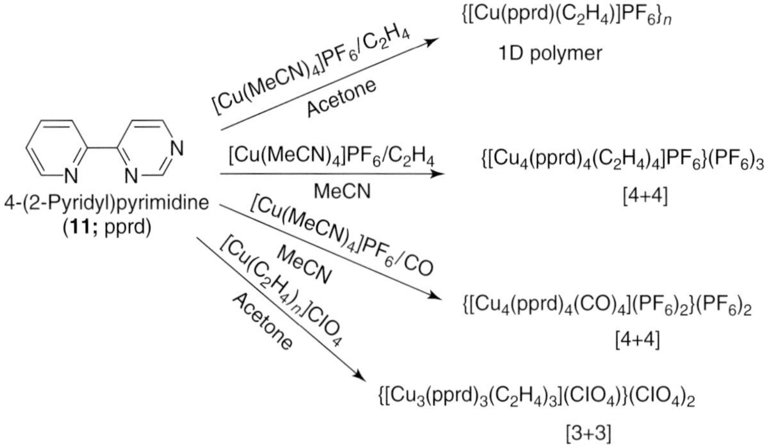

Figure 5.7 Reactions of 4-(2-pyridyl)pyrimidine (pprd) with various Cu(I) salts [65].

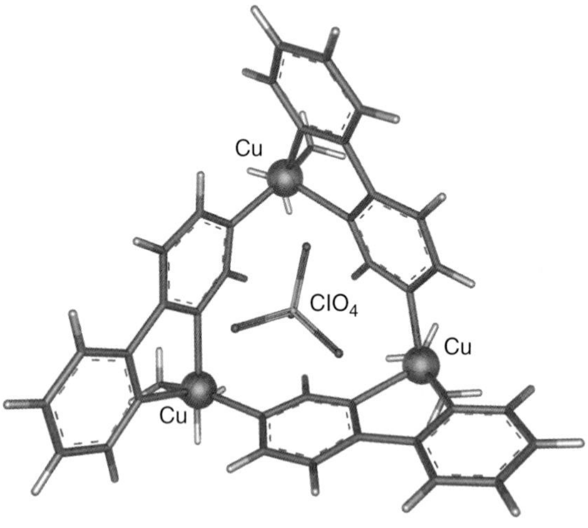

Figure 5.8 The X-ray structure of the [3+3] species, $\{[Cu_3(pprd)_3(C_2H_4)_3](ClO_4)\}^{2+}$, showing the included ClO_4^- ion [65].

anion preferentially inducing formation of a pseudo-metallocalix[3]arene species of type $\{[Cu_3(pprd)_3(C_2H_4)_3](ClO_4)\}^{2+}$ (Figure 5.8), while the larger octahedral hexafluorophosphate anion yielded two such metallocalix[4]arene species, $\{[Cu_4(pprd)_4(CO)_4](PF_6)_2\}^{2+}$ (Figure 5.9) and $\{[Cu_4(pprd)_4)C_2H_4)_4]PF_6\}^{3+}$ (Figure 5.10), with each species forming under different conditions.

In a subsequent study, Cu(I) pprd derivatives incorporating C_2H_4 adducts were assembled under the influence of both solvent and anion type [66]. For example, a [6+6] metallocyclic species of type $\{[Cu_6(pprd)_6(C_2H_4)_6(BF_4)](BF_4)_5]\}_2{\cdot}6H_2O$ was isolated on reaction of $[Cu(MeCN)_4]BF_4$ with pprd (**11**) in methanol under C_2H_4,

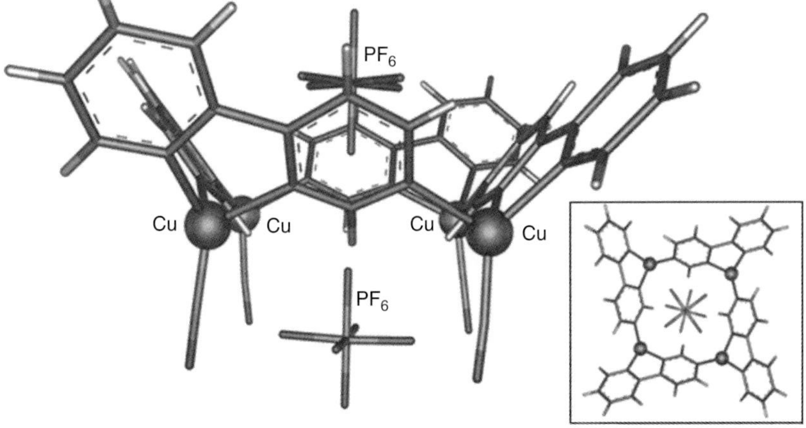

Figure 5.9 X-ray structure (side view) of the [4 + 4] calixarene-shaped cation $\{[Cu_4(pprd)_4(Cl)_4](PF_6)_2\}^{2+}$ showing the two included PF_6^- ions. Inset: view from the top [65].

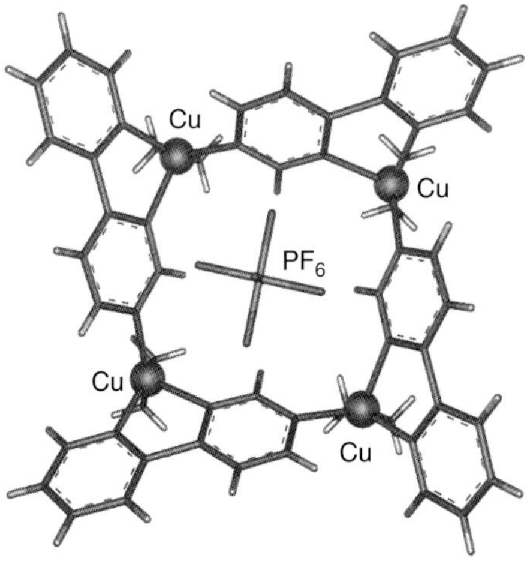

Figure 5.10 X-ray structure of the [4 + 4] calixarene-shaped cation $\{[Cu_4(pprd)_4(C_2H_4)_4]PF_6\}^{3+}$ showing the included PF_6^- ion [65].

while changing the solvent to acetone or the anion to NO_3^-, ClO_4^-, or PF_6^- in each case yielded noncyclic products with various structures. The interesting X-ray structure of the $[Cu_6(pprd)_6(C_2H_4)_6(BF_4)]^{5+}$ cation showing an included BF_4^- anion is illustrated in Figure 5.11. Although the situation is clearly not simple, the authors present evidence for the likely template role of the BF_4^- anion in forming this latter complex.

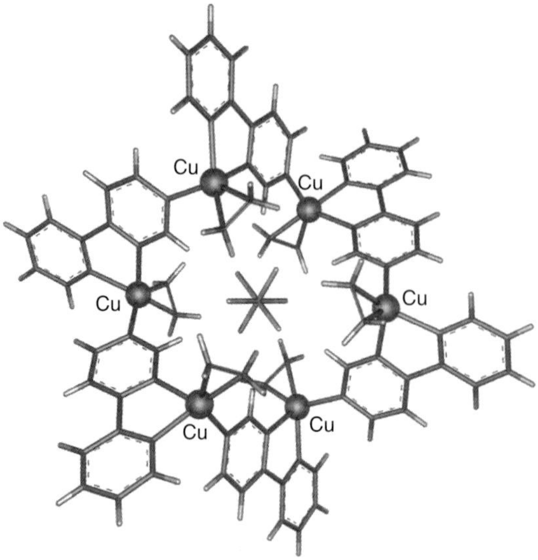

Figure 5.11 X-ray structure of the [6+6] species, $[Cu_6(pprd)_6(C_2H_4)_6(BF_4)]^{5+}$ showing the included disordered BF_4^- anion [66].

5.4 Capsule, Cage, and Tube-Shaped Systems

In an attempt to extend the chemistry of Prussian-blue-type complexes, 3,5-dimethyltris(pyrazolyl)methane (**12**; $HC(3,5\text{-}Me_2Pz)_3$) was employed as a tripodal capping ligand to synthesize new cyanometallate complexes of iron and to investigate their reactivity toward cyanometallate cage formation [67]. Thus, reaction of $K[(HC(3,5\text{-}Me_2Pz)_3)Fe^{II}(CN)_3]$ with $[Fe^{II}(H_2O)_6](BF_4)_2$ yields the air-sensitive cubic cyanometallate cage, $[\{Fe^{II}(H_2O)\}_4(BF_4)\{(HC(3,5\text{-}Me_2Pz)_3Fe^{II}(CN)_3\}_4](BF_4)_3$, incorporating an included BF_4^- anion in its cavity. The reaction was believed to proceed under BF_4^- template control since substitution of ClO_4^- or PF_6^- anion for BF_4^- in the reaction failed to yield a similar cage product. The X-ray structure (Figure 5.12) of the inclusion product showed that it has a cubic box-like structure, with the guest BF_4^- anion disordered over three equivalent positions in the cavity.

The presence of a templating anion can result in a major change in the complexity of the product obtained. For example, reaction of the cytidine nucleoside (**13**, H_2cyd) with an aqueous solution of copper perchlorate in a 1 : 1 metal to ligand ratio yields the "simple" mononuclear complex of type $[Cu(H_2cyd)_4](ClO_4)_2{\cdot}5H_2O$ (Figure 5.13) [68]. However, in the presence of a source of carbonate, deprotonation of the nucleoside occurred to yield a large dodecanuclear Cu(II) globular-shaped assembly of stoichiometry $[Cu_{12}(Hcyd)_{12}(CO_3)_2](ClO_4)_8{\cdot}11H_2O$ (Figure 5.13). In this, two carbonate anions form a key part of the assembly's superstructure, and it was postulated that these trigonal anions play a templating role in formation

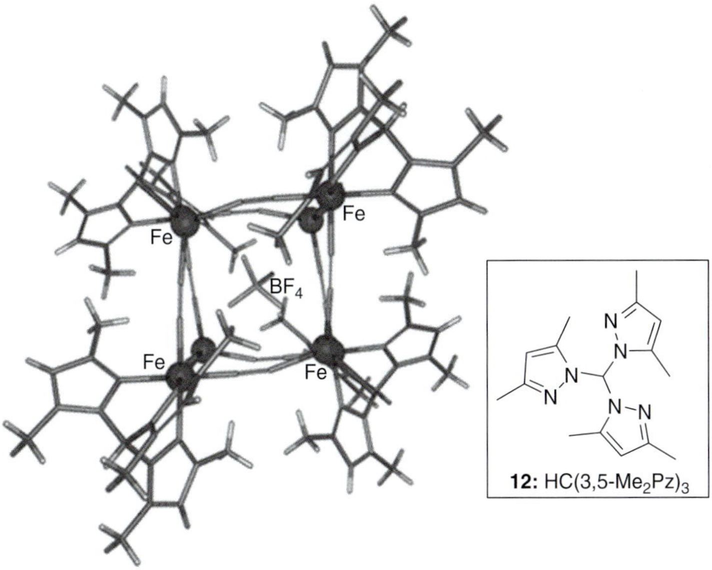

Figure 5.12 X-ray structure of the cage cation in $[\{Fe^{II}(H_2O)\}_4(BF_4)\{(HC(3,5\text{-}Me_2Pz)_3Fe^{II}(CN)_3\}_4](BF_4)_3$ showing one position of the disordered BF_4^- guest in the boxlike cavity [67].

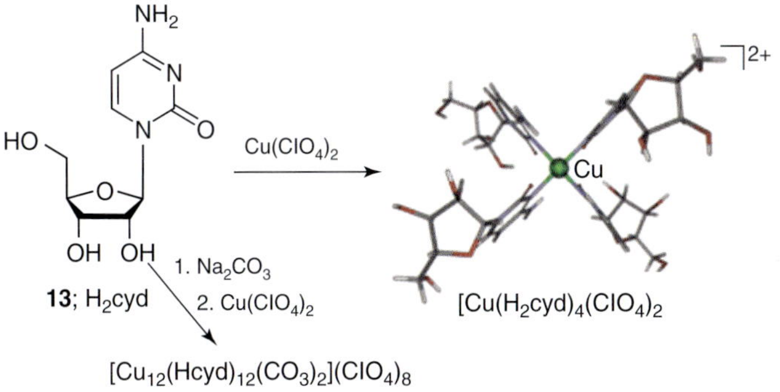

Figure 5.13 Reactions of the cytidine nucleoside (**13**, H_2cyd) with $Cu(ClO_4)_2$ in the absence and presence of carbonate ion (waters of crystallization not shown) [68].

of the capsule. Two views of the X-ray structure of this interesting product are given in Figure 5.14. The globular-shaped dodecanuclear complex ion contains 12 (namely, two hexagonal groups of six) Cu(II) centers. These are connected in turn by two templating carbonate anions and six mono-deprotonated cytidine ligands such that an ellipsoidal capsule is generated, from which six $Hcyd^-$ moieties are appended. The two sets of hexanuclear Cu(II) arrays, which are near planar, are held together by the six $Hcyd^-$ ligands through N- and O-donor coordination and

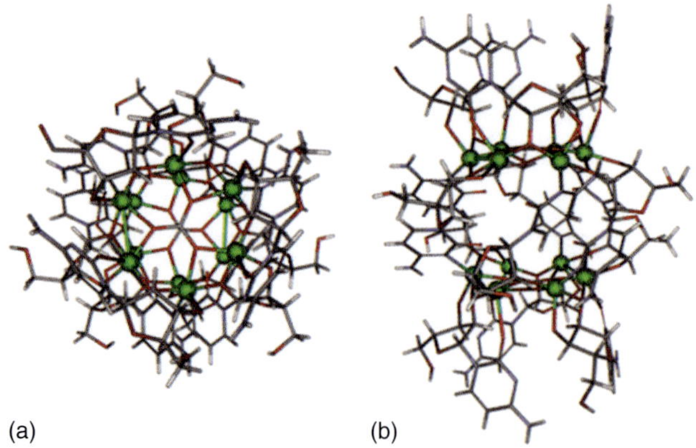

Figure 5.14 X-ray structure of the cation in $[Cu_{12}(Hcyd)_{12}(CO_3)_2](ClO_4)_8 \cdot 11H_2O$; (a) view from top and (b) view from side; copper atoms shown as spheres; lattice waters are not shown [68].

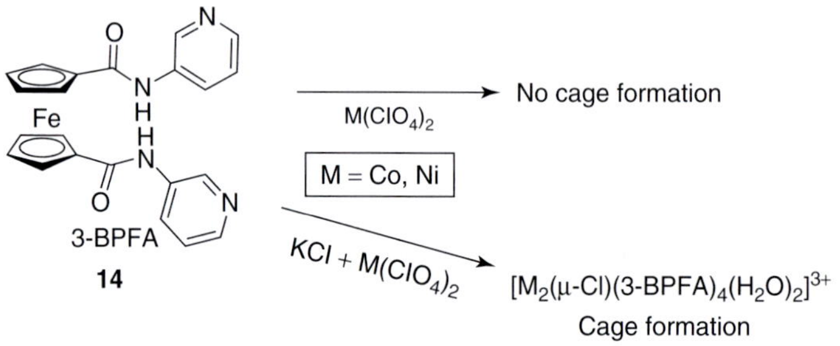

Figure 5.15 Effect of Cl^- addition on the product formed from reaction of 3-BPFA with $M(ClO_4)_2$ (M = Co, Ni) [69].

the bridging carbonate anions. When viewed from the top of the capsule, the two bound carbonate anions are completely staggered (Figure 5.14a).

The interaction of the flexible ferrocenyl-containing, two-armed ligand species, 1,1-bis[(3-pyridylamino)carbonyl]ferrocene (**14**, 3-BPFA – Figure 5.15), with Co(II), Ni(II), and Cu(II) salts has been reported [69]. As part of this study, two discrete cation clusters, $[Ni_2(\mu\text{-}Cl)(3\text{-}BPFA)_4(H_2O)_2](ClO_4)_3$ and $[Co_2(\mu\text{-}Cl)(3\text{-}BPFA)_4(H_2O)_2](ClO_4)_3 \cdot 4CH_3OH$, incorporating μ_2-Cl bridges were reported (Figure 5.15). Each product has been proposed to assemble under the influence of a Cl^- template from a mixture of $Co(ClO_4)_2$ or $Ni(ClO_4)_2$ and KCl in the presence of 3-BPFA. No similar crystalline products were obtained when other halide anions were substituted for Cl^- in both syntheses. In each of the present assemblies, the presence of a bridging chloro ligand (Figure 5.16) is likely driven by the propensities of the high-spin Co(II) and Ni(II) ions to attain stable six-coordinate geometries.

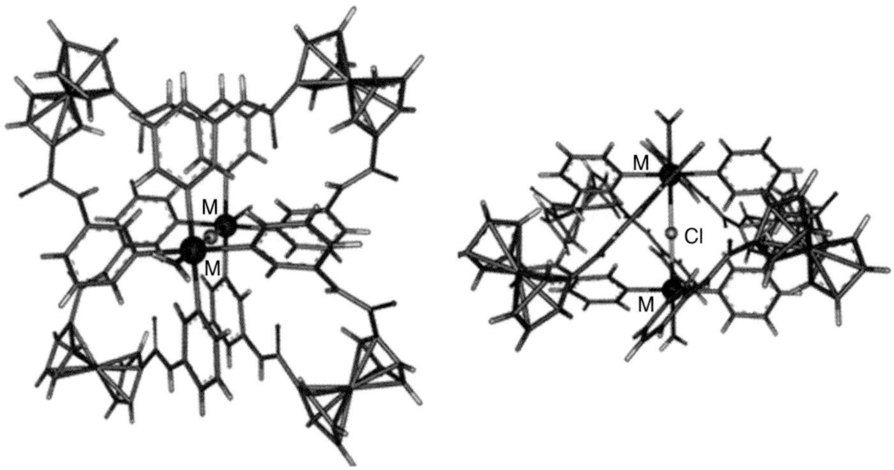

Figure 5.16 Two views of the similar structure of the cationic $[M_2(\mu\text{-Cl})(3\text{-BPFA})_4(H_2O)_2]^{3+}$ cage assemblies (where M = Co or Ni) showing the bridging chloride ligand (small sphere) [69].

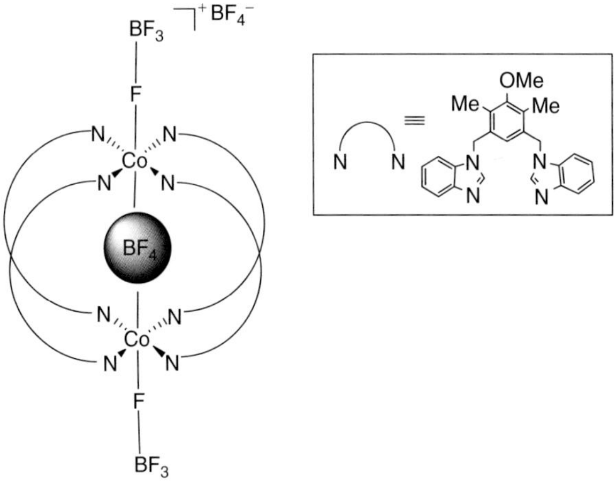

Figure 5.17 The metallo-supramolecular cage assembly of type $[Co_2(L)_4(\textit{endo}\text{-}BF_4)(\textit{exo}\text{-}BF_4)_2]^+$ formed from Co(II) and di(benzimidazole)-1,4-phenylene in the presence of BF_4^- [70].

The anion-templated synthesis of a new supramolecular cage assembly containing Co(II) and di(benzimidazole)-1,4-phenylene (L) of the type $[Co_2(L)_4(\textit{endo}\text{-}BF_4)(\textit{exo}\text{-}BF_4)_2]BF_4$ has been shown to occur in the presence of the weakly coordinating BF_4^- anion (Figure 5.17) [70]. This cage is composed of a tetragonal $[Co_2(L)_4]^{4+}$ unit in which each cobalt center adopts a square-pyramidal geometry. This cavity-containing, three-dimensional species is associated with three weakly coordinated BF_4^- ions, two of which are bound to the two Co(II) centers apically (and *exo* with respect to the cavity), while the third is encapsulated in the cage and is bound to a cobalt center (Co-F, 2.312 Å) (Figure 5.17), with the BF_4^- being disordered over two equivalent sites in the cavity that are displaced toward each cobalt center. In contrast, in the presence of more strongly binding NO_3^- and Cl^- anions, the cage structure is no longer formed. Rather, a one-dimensional coordination polymer of type $[Co(L)(NO_3)_2]_n$ and a discrete metallomacrocycle of type $[Co_2(L)_2Cl_4]$ were obtained – once again emphasizing the importance of the anion in directing the course of the reaction in these cases.

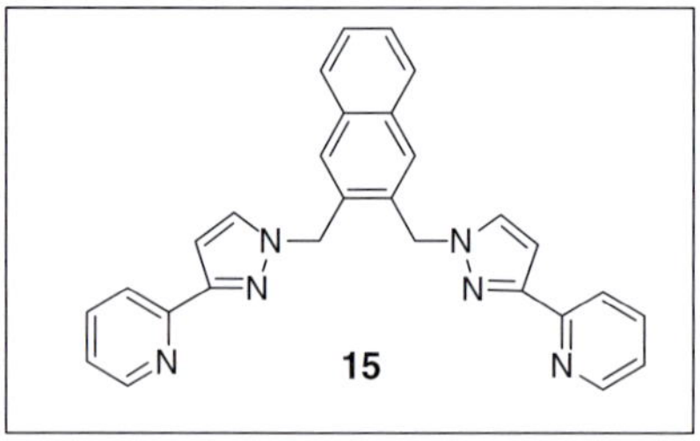

15

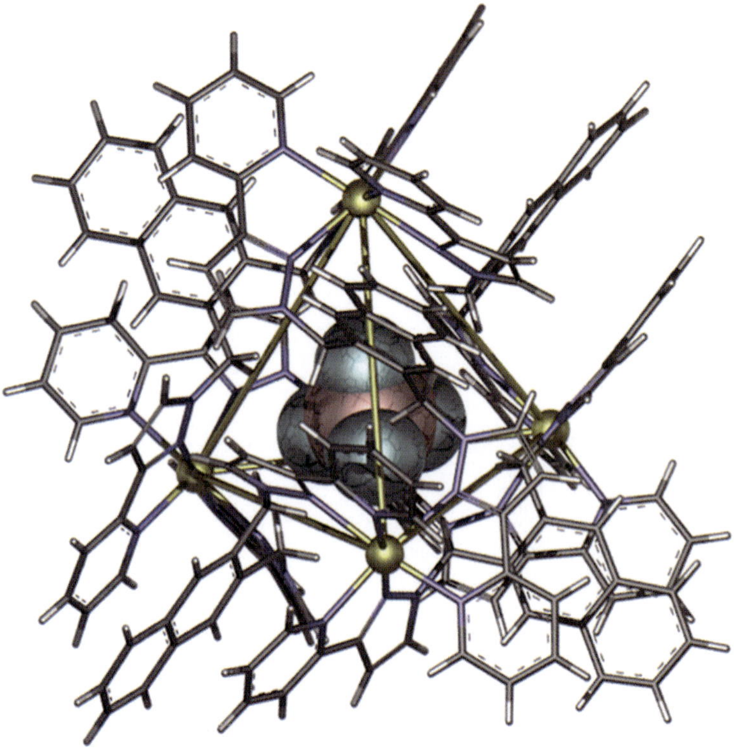

Figure 5.18 X-ray structure of the cationic tetrahedral cage structure of type $[Cd_4L_6(BF_4)]^{7+}$ (where L = **15**) showing the included BF_4^- anion [71].

The Ward group has reported the synthesis of potentially tetradentate ligands incorporating two N-donor bidentate pyrazolyl–pyridine units separated by 1,2-phenyl, 2,3-naphthyl, or 2,3-anthracenyl spacers [71]. Individual ligands of this type were shown to yield new tetrahedral $[M_4L_6]^{8+}$ cage species (where M = Co(II) and Cd(II)) that incorporate an encapsulated anion (BF_4^- or SiF_6^{2-}), for which a possible templating role for the included anion was suggested. The X-ray structure of the cationic cage in one such product, $[Cd_4L_6(BF_4)](BF_4)_7 \cdot 2MeOH \cdot 3H_2O$, derived from the 2,3-naphthyl derivative (**15**) is presented in Figure 5.18.

The possible pitfalls of assigning anion template effects to a given assembly reaction are illustrated by a recent study involving the assembly of a tetrahedral cage of type $[Fe_4L_6(BF_4)]^{7+}$, in which the cationic assembly is formed on reaction of Fe(II) with the quaterpyridine ligand **16** (Figure 5.19) in acetonitrile in the presence of excess BF_4^- added as a precipitating anion [72]. The X-ray structure of the product revealed that a BF_4^- anion occupied the cage's central cavity (Figure 5.19). Initially, it was considered in this study that the tetrahedral anion might have acted as a template for the formation of the tetrahedral cage product. To test this hypothesis, the synthesis was repeated using octahedral PF_6^- substituted for BF_4^-. In this case, the X-ray structure revealed that a PF_6^- anion occupied the central cavity, and clearly, the tetrahedral shape of the BF_4^- anion was not an essential factor in any templating that may have occurred on forming this tetrahedral cage.

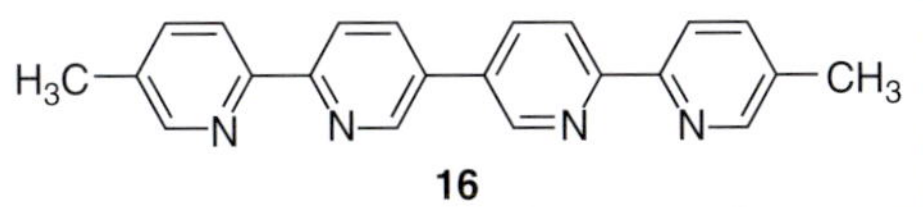

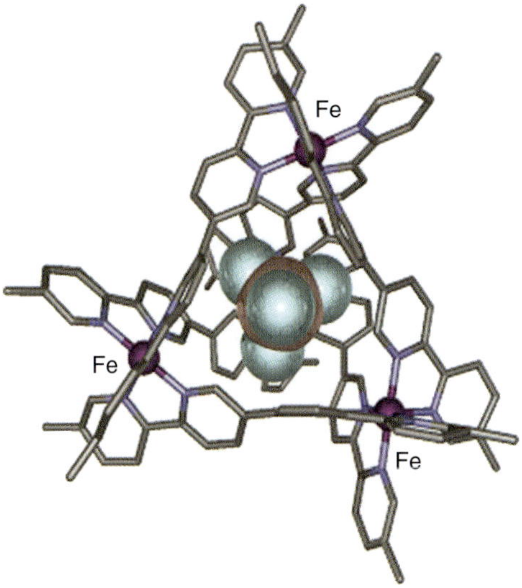

Figure 5.19 Structure of the quaterpyridyl ligand **16** and X-ray structure of the inclusion complex $[Fe_4L_6(BF_4)]^{7+}$ (where L = **16**) [72].

This in turn raised the question of whether a templating anion was necessary at all. Using a microwave synthesis that once again employed Fe(II) and **16**, but with water as the reaction solvent, with subsequent addition of $[ZnCl_4]^{2-}$ as the precipitating anion led to isolation of the corresponding cage free of an included anion. The structure of this product was again confirmed by X-ray diffraction. The success of this procedure was in part proposed to reflect the reluctance of $[ZnCl_4]^{2-}$ to be included in the cage under the (aqueous) conditions of the experiment owing to the energy cost in desolvating this doubly charged anion relative to a monovalent anion. While the results of the above experiments do not prove that an anion templating role does not occur, the ability to isolate the cage free of an included anion suggests that this is likely the case.

5.5 Circular Helicates and *meso*-Helicates

The Ward group has also employed the potentially tetradentate ligand **17** containing two N-donor bidentate pyrazolyl–pyridine units bridged by a 1,8-naphthyl spacer unit as well as two related chiral ligand derivatives bearing pinene groups (as chiral auxiliaries) that are fused to different positions of the terminal pyridyl rings [73]. These ligands give rise to a number of metal complex structural types that include mononuclear species, tetranuclear dodecanuclear coordination cages, and cyclic helicates. For example, **17** forms 1 : 1 metal–ligand complexes with Cu(I) and Ag(I) but 2 : 3 complexes with Co(II), Cu(II), and Cd(II). The 1 : 1 complexes are either a simple mononuclear species in which the ligand acts as a tetradentate chelate or a cyclic helicate in which the ligands are bridging (as illustrated in Figure 5.20), with

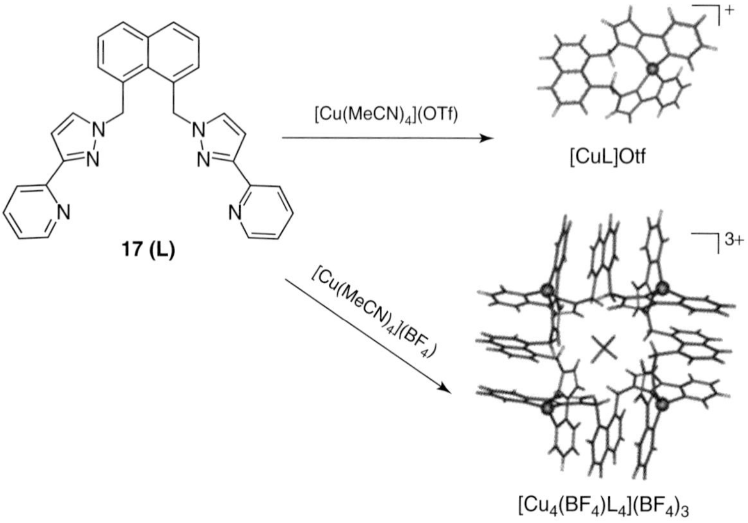

Figure 5.20 Anion-dependent reactions of Cu(II) with **17** [73].

the structure obtained being dependent on whether the anion employed is able to template the formation of the cyclic helicate or not.

An example of the latter type is given by the apparent template formation of $[Cu_4L_4][BF_4]_4$, whose X-ray structure confirmed it to correspond to a cyclic tetranuclear helicate in which a tetrafluoroborate anion is encapsulated in its central cavity (Figure 5.20). Generally, parallel behavior was also observed for the chiral ligand derivatives. Thus, in one case, the use of silver tetrafluoroborate resulted in a cyclic tetranuclear helicate of stoichiometry $[Ag_4L_4](BF_4)_4$, in which one BF_4^- anion presumably acts as a template (and is encapsulated in the $[Ag_4L_4]^{4+}$ cavity), whereas under related conditions, silver perchlorate yielded a simple mononuclear complex of stoichiometry $[AgL]ClO_4$.

Reaction of the bis-pyridylimine ligand derivatives L^1–L^3 (Figure 5.21) with $CuSO_4{\cdot}5H_2O$ in methanol/water/acetonitrile solution yielded uncharged brown or green complexes of type $[CuL(SO_4)]_6{\cdot}24H_2O$ in each case. The X-ray structures of these hexanuclear products indicated that they corresponded to essentially superimposable hexanuclear *meso*-helical circular (toriod) architectures (Figure 5.21). In keeping with their *meso* configurations, the six Cu(II) ions in

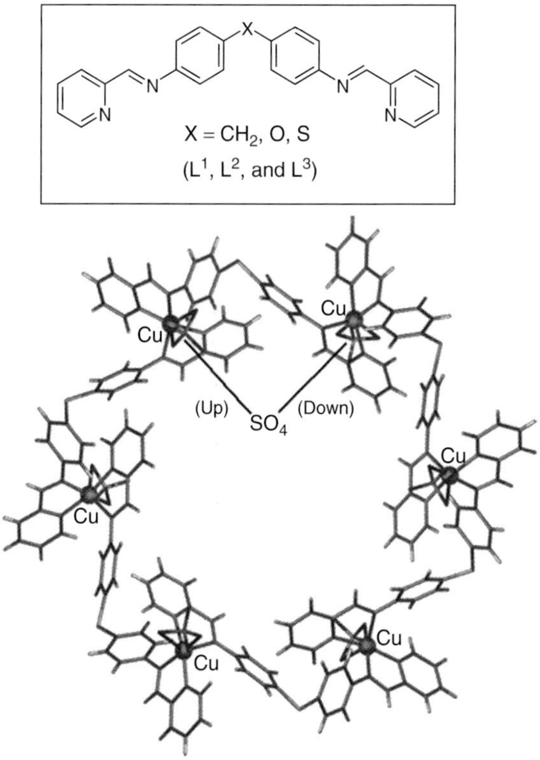

Figure 5.21 X-ray structure of the isostructural circular *meso*-helicates of type $[CuL(SO_4)]_6$ (where $L = L^1$, L^2, or L^3) [73].

each structure have alternately Λ and Δ configurations. The SO_4^{2-} anions are bound in a bidentate manner to the copper centers, with three located on the top rim of the torus and three on the bottom rim. The substitution of ClO_4^- or NO_3^- for SO_4^{2-} in the above procedure resulted in cationic, noncyclic triple helicate products. Overall, the formation of the observed cyclic structure found for the *meso* products was partially assigned to the presence of bidentate coordination of the sulfate groups, which led to the generation of a double- rather than a triple-stranded arrangement about individual octahedral Cu(II) sites. Nevertheless, clearly, interligand hydrogen bonding and $\pi-\pi$ stacking interactions also play their part in the formation of these circular *meso* structures.

5.6 Mechanically Linked Systems

Several informative reviews by Beer *et al.* [23, 25, 27, 28, 31, 74] mainly outlining recent results from their laboratory on the formation of interpenetrated and interlocked architectures have appeared. The above group initially focused on the development of anion templation strategies which employed (spherical) halide anions as templates for directing the assembly of interwoven structures. For example, in early studies, the use of chloride ion in weakly solvating media led to a range of pseudorotaxane, rotaxane, and catenane topologies formed by strong ion-pair bonds that template the insertion of a pyridinium-, imidazolium-, or guanidinium-containing thread through the aperture of an isophthalamide-containing macrocyclic component [52, 53, 75–77]. The assembly process in these cases relied on the rational use of both anion recognition and ion pairing. A further feature of such systems is that, on template removal, the interlocked products commonly act as receptors for selected other anions while typically retaining strong selectivity for Cl^- (although, among the halides, an example of Br^- selectivity has also been documented) [78]. More recently, the use of other (nonspherical) ions such as oxyanions as templates for producing such interlocked assemblies has been investigated.

One strategy employed for [2]catenane formation involved the initial assembly of two identical acyclic, positively charged, anion recognizing units around a Cl^- template into a mutually orthogonal arrangement (Figure 5.22), with subsequent double-ring cyclization resulting in formation of a Cl^--containing [2]catenane. Thus, mixing equimolar amounts of the Cl^- and PF_6^- salts of the positively charged diamide derivative **18** (Figure 5.23) in DCM followed by addition of Grubbs' first-generation catalyst resulted in bis-ring closing to form the corresponding cationic catenane incorporating a bound Cl^- template anion (Figure 5.24) [79]. The PF_6^- acts as the counterion for this species. Solution studies indicated strong selectivity of the guest-free catenane for Cl^- over Br^- or OAc.

A Cl^- templation procedure related to the above has also been employed to generate rotaxanes of the types shown in Figure 5.25 [80]. The procedure involves the initial assembly of the axle and the (unclosed) macrocyclic ring components around the Cl^- template followed by macrocyclic ring-closing metathesis of the required

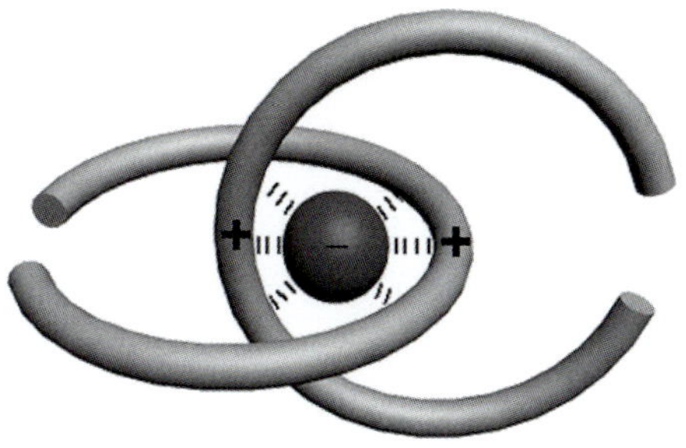

Figure 5.22 Preassembly for catenane formation: orthogonal arrangement of two singly charged cationic components in the presence of a negatively charged spherical anion.

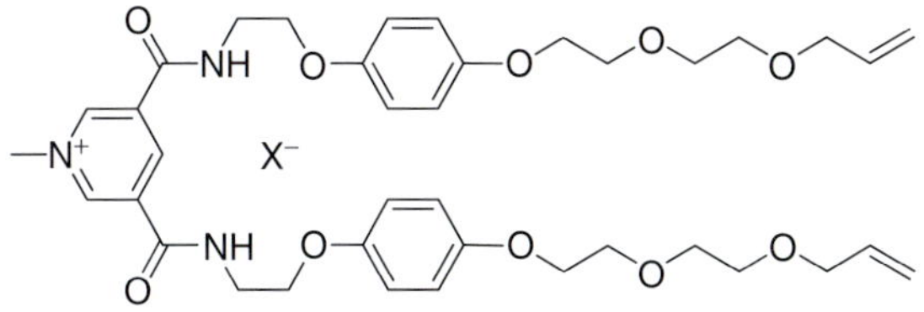

18.X (X = Cl or PF_6)

Figure 5.23 The open chain bis-amide precursor **18** (present as a 1 : 1 mixture of its Cl^- and PF_6^- salts) used for the Cl^- templated catenane shown in Figure 5.23.

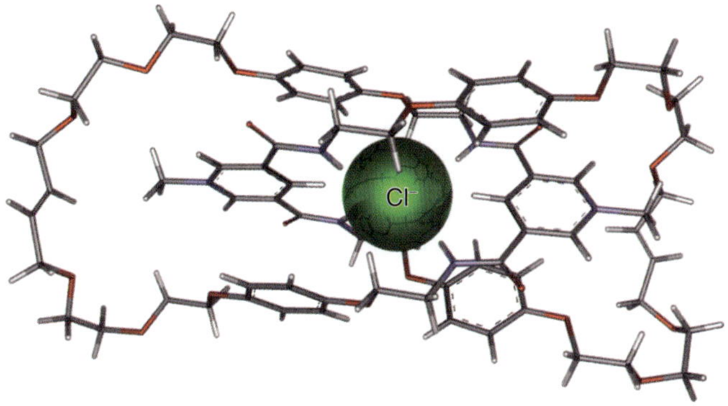

Figure 5.24 X-ray structure of the catenane generated from **18**·Cl and **18**·PF_6 showing the position of the Cl^- template, which is held in position by six hydrogen bonds involving four amide protons and two *para*-pyridinium protons, with all six protons pointing toward the Cl^- anion [79].

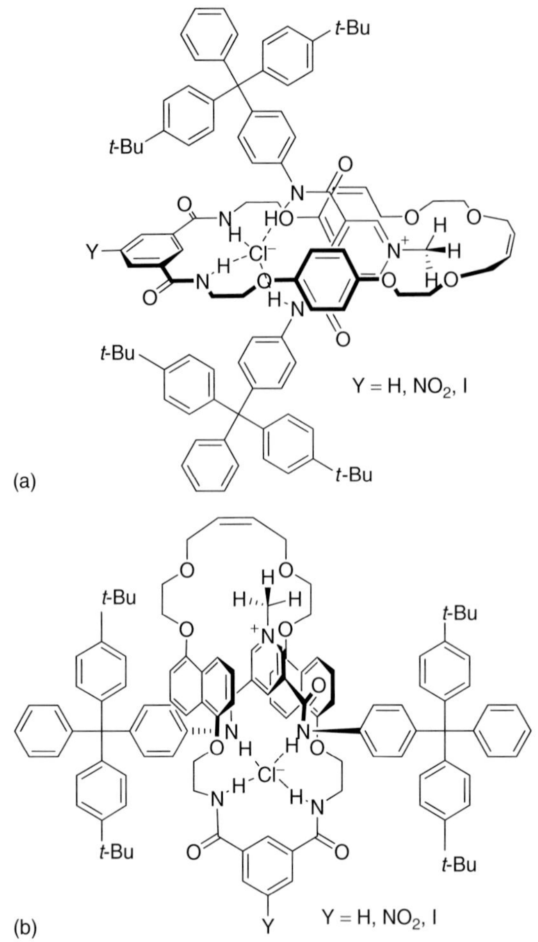

Figure 5.25 Chloride templated [2]rotaxanes consisting of a positively charged pyridinium axle fitted with bulky "stoppers" surrounded by a neutral isophthalamide macrocycle, with the latter incorporating either (a) hydroquinone or (b) naphthalene segments in the respective macrocyclic rings [80].

bis-vinyl acyclic isophthalamide component in the presence of Grubbs' catalyst. The Y-substituted (Y = H, NO_2, I) rotaxane derivatives shown in Figure 5.25 were synthesized. Each consists of a positively charged pyridinium axle fitted with bulky aromatic "stoppers" surrounded by a neutral isophthalamide macrocycle, with the latter incorporating either hydroquinone (Figure 5.25a) or naphthalene (Figure 5.25b) segments in their macrocyclic rings. It was demonstrated that the presence of the electron-withdrawing substituents on the acyclic isophthalamide bis-vinyl precursor unit leads to an increased yield of the resulting Cl^--containing [2]rotaxane. The presence of such substituents was also shown to be reflected in

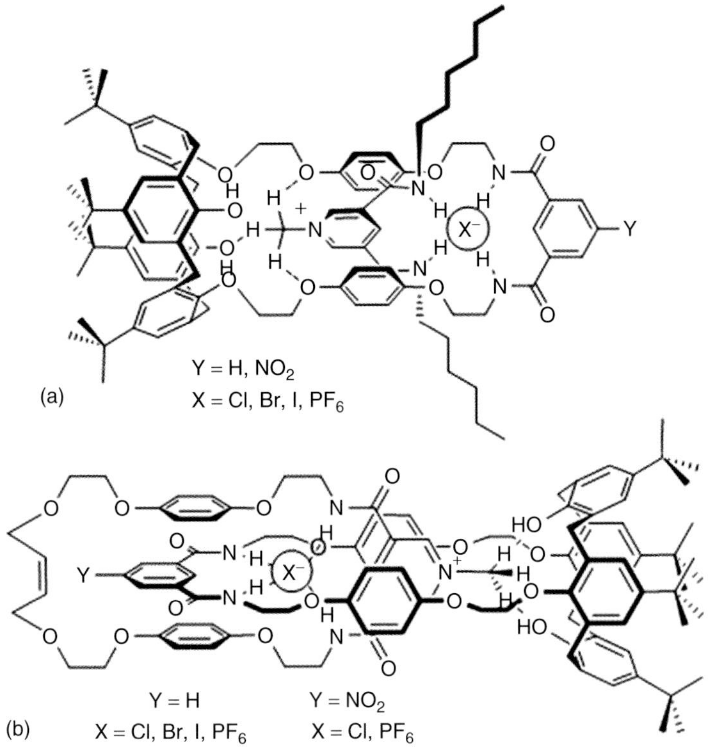

Figure 5.26 Related (a) [2]pseudorotaxane and (b) [2]catenane structures incorporating calix[4]arene scaffolds obtained by Cl^- templation [81].

increased rotaxane anion binding strengths once the Cl^- had been removed from its binding pocket (by precipitation on addition of $AgPF_6$); these studies confirmed, however, that the interlocked anion binding domain maintains selectivity for Cl^- (over $H_2PO_4^-$ and OAc^-).

Examples of anion-templated pseudorotaxanes and catenanes in which a calix[4]arene scaffold unit is employed as part of the cyclic component have been reported [81–83]. For example, using ring-closing metathesis reactions related to those discussed above gave pseudo[2]rotaxane and catenane species of the types shown in Figure 5.26a,b, respectively [81]. Once again, the success of these interlocking reactions was demonstrated to be dependent on the presence of a chloride anion template.

A Cl^- templation strategy has also been employed to form a redox-active bis-ferrocene-functionalized rotaxane of the type shown in Figure 5.27 for use in forming self-assembled monolayers (SAMs) on a gold electrode surface via S–Au bonds [84]. Significantly, the electrochemical anion recognition behavior of the assembled rotaxane SAM is different from its separated wheel-and-axle components. Following removal of the Cl^- template, the binding cavity in the

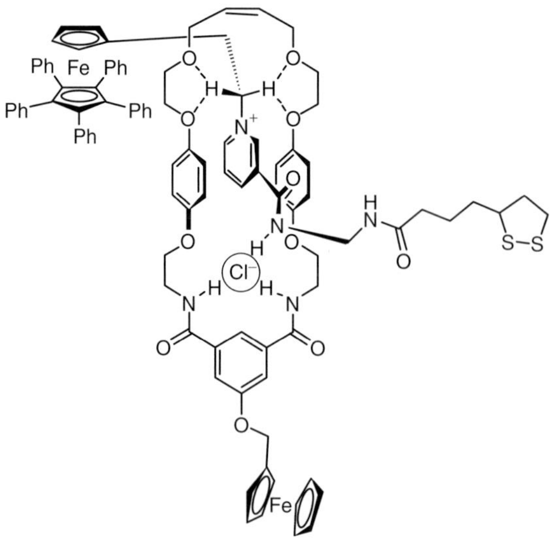

Figure 5.27 A Cl^--templated redox-active pseudo[2]rotaxane incorporating ferrocene groups, subsequently tethered to a gold surface [84].

interlocked assembly again exhibits selective Cl^- sensing (in acetonitrile) compared to larger oxyanions such as $H_2PO_4^-$ and HSO_4^-, which appear too big to access the binding site. An initial experiment indicated that the tethered rotaxane is capable of detecting Cl^- in the presence of 100-fold excess of $H_2PO_4^-$ ion.

More recently, two other examples of anion-templated formation of pseudorotaxane and rotaxane monolayers on gold surfaces have been reported by Beer *et al.* [85], and the potential for the further development of such systems as practical optical and electrochemical sensors has been reviewed [86].

In an extension of the above studies, the use of doubly charged SO_4^{2-} anion as a template has been given attention as a means for obtaining mechanically interlocked structures [87–90]. In general terms, the use of a tetrahedral SO_4^{2-} anion as a template to produce interlocked structures may be seen to parallel the well-documented use of a tetrahedral cation (typically Cu(I) or Ag(I)) for templating the formation of catenane and rotaxane structures [91, 92]. For example, in one such study, a bis-vinyl pyridiniumnicotinamide derivative was prepared as a precursor for use in a SO_4^{2-} template reaction with the aim of generating a [2]catenane via the reaction shown in Figure 5.28 [88]. Thus, the simple addition of Grubbs' second-generation catalyst to the prearranged SO_4^{2-}-containing tetra-vinyl precursor at room temperature was found to yield the corresponding [2]catenane in 80% yield. Attempted parallel reactions in which Cl^-, Br^-, or PF_6^- were substituted for SO_4^{2-} failed to yield a catenane. Clearly, the template process reflects the inherent tetrahedral directionality of the SO_4^{2-} hydrogen-bonding acceptor unit coupled with the ability to achieve complementarity (orthogonal) orientations of the two pairs of hydrogen bond donor amide groups for the binding

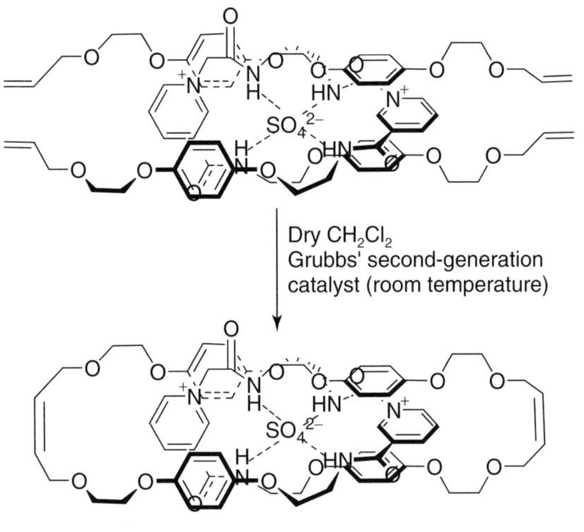

Figure 5.28 An example of a SO_4^{2-}-templated [2]catenane formation [88].

of this ion. Once formed, the anion-free product retains selectivity for sulfate in the presence of Cl^-, Br^-, and OAc^- ions.

As a further extension of the strategies discussed so far in this section, the use of indolocarbazole-containing threading units to achieve new pseudorotaxane [87] and rotaxane [93] assemblies has been reported. For example, the use of such an axle derivative has resulted in the first example of pseudorotaxane assembly via anion templation from two *neutral* components both in solution as well as in a surface-assembled monolayer [87]. In this example, a neutral indolocarbazole unit (**19**) was employed as a strong hydrogen-bonding donor in conjunction with a SO_4^{2-} hydrogen bond acceptor unit. In an initial experiment, solution and solid-state studies confirmed that a 2 : 1 mixture of **19** and tetrabutylammonium sulfate led to isolation of a strong ternary complex of type $[\mathbf{19}\cdot SO_4\cdot \mathbf{19}]^{2-}$. Encouraged by this result, it was then demonstrated that a solution of macrocycle **20** with a 1 : 1 indolocarbazole/tetrabutylammonium sulfate mixture led to the formation of a pseudorotaxane of type $[\mathbf{20}\cdot SO_4\cdot \mathbf{1}]^{2-}$, in which the SO_4^{2-} unit forms a link between the macrocycle **20** and **1**, as shown in Figure 5.29. The occurrence of interpenetration in this species in solution was demonstrated using ^1H NMR spectroscopy, while surface plasmon resonance (SPR) for the corresponding ternary complex incorporating **2**, SO_4^{2-}, and a indolocarbazole derivative incorporating a pendent disulfide group chemisorbed on a gold surface also gave evidence for pseudorotaxane formation. With respect to the above, the specific nature of the SO_4^{2-} templating role is illustrated by the observation that substitution of Cl^- ion for SO_4^{2-} in the above solution studies failed to give evidence for rotaxane formation.

Finally, in this section, it is noted that recent attention has been focused on the potential of anion-templated systems for use as molecular/ionic switches [94–96]. In this context, the Leigh group initially reported the construction of a new pseudorotaxane assembly that involves the binding of a Cl^- anion to a Pd(II)

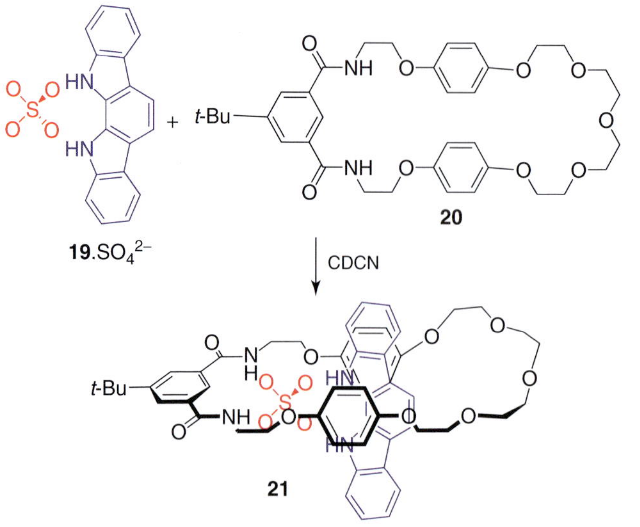

Figure 5.29 Assembly of the pseudorotaxane **21** under the influence of a SO_4^{2-} template [87].

center held in a macrocyclic structure. This structure acted as a receptor for a benzyl pyridinium guest cation, in which ion-pairing between the bound Cl^- and the positively charged guest together with aromatic stacking and CH···O as well as CH···Cl hydrogen-bonding interactions all contribute to the overall stability of this interwoven system [96].

The above benzyl pyridinium/palladium–Cl^- interaction motif was then used as a model for one binding site (site 1) in the construction of a terminally stoppered, switchable "two-site" rotaxane molecular shuttle. The second site (site 2) corresponded to a triazole derivative located at a second position along the linear component ((a) in Figure 5.30). The macrocyclic component was unchanged from that employed in the model assembly. On removal of the Cl^- ligand from site 1 by precipitation with Ag PF_6, the Pd(II)-containing macrocyclic unit was released from this site and translocation to site 2 occurred, with the free coordination position on Pd(II) then being filled by a nitrogen donor from the triazole motif (Figure 5.30b). This switching process is readily reversed by the addition of tetrabutylammonium chloride to the triazole-bound system in chloroform.

5.7
Concluding Remarks

While currently less developed than metal-ion template procedures, the use of anion templation in synthesis is becoming more common – driven in part by the enhanced awareness (and availability) of suitable anion/host interaction types that can be utilized in the rational design of new template strategies. The use of anions

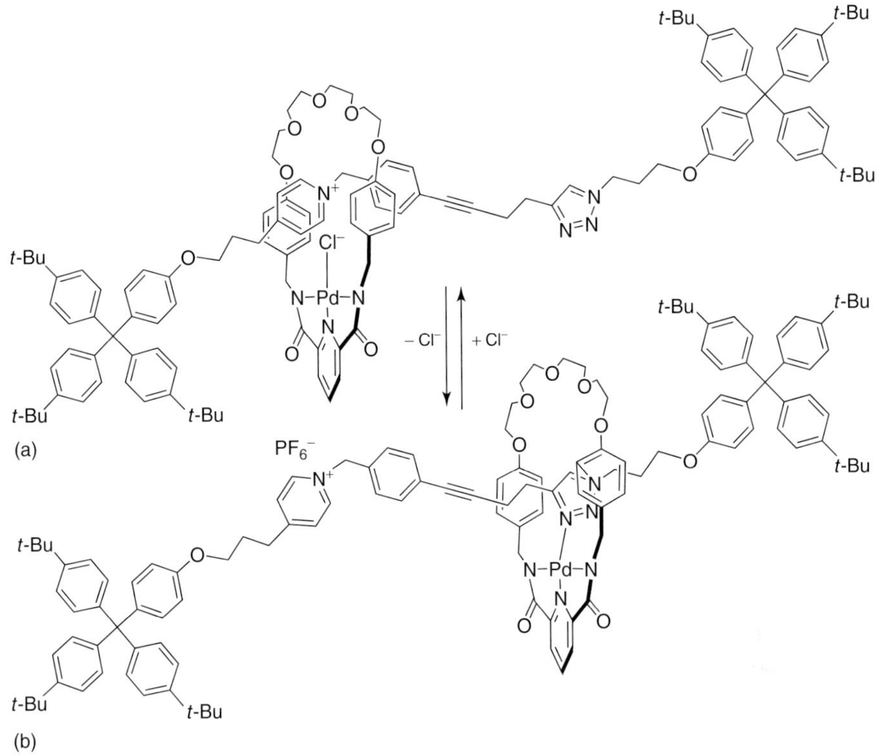

Figure 5.30 A "two-site" rotaxane molecular switch controlled by the addition/removal of Cl^- anion [96].

in this role will undoubtedly continue to lead to new molecular and supramolecular entities that are difficult (or impossible) to obtain by other means. Although the first steps have been taken, the application of anion template synthesis to the design and construction of new practical molecular devices and machines, including sensors, (opto)electronic devices (including electronic components ranging from transistors to logic gates), and enzymelike catalysts all remain significant intellectual and practical challenges for the future that, as yet, remain largely unmet.

References

1. Busch, D.H. (1964) The significance of complexes of macrocyclic ligands and their synthesis by ligand reactions. *Rec. Chem. Progr.*, **25**, 107–126.
2. Lindoy, L.F. and Busch, D.H. (1971) in *Preparative Inorganic Reactions*, vol. 6, (ed. W.J. Jolly), Wiley-Interscience, New York, pp. 1–61.
3. Bowman-James, K. (2005) Alfred Werner revisited: the coordination chemistry of anions. *Acc. Chem. Res.*, **38** (8), 671–678.

4. Hossain, M.D.A., Kang, S.O., and Bowman-James, K. (2005) in *Macrocyclic Chemistry in Current Trends and Future Perspectives* (ed. K. Gloe), Springer, Dordrecht, pp. 173–188.
5. Wichmann, K., Antonioli, B., Söhnel, T., Wenzel, M., Gloe, K., Gloe, K., Price, J.R., Lindoy, L.F., Blake, A.J., and Schröder, M. (2006) Polyamine-based anion receptors: extraction and structural studies. *Coord. Chem. Rev.*, **250** (23–24), 2987–3003.
6. Katayev, E.A., Ustynyuk, Y.A., and Sessler, J.L. (2006) Receptors for tetrahedral oxyanions. *Coord. Chem. Rev.*, **250** (23–24), 3004–3037.
7. Kang, S.O., Hossain, M.A., and Bowman-James, K. (2006) Influence of dimensionality and charge on anion binding in amide-based macrocyclic receptors. *Coord. Chem. Rev.*, **250** (23–24), 3038–3052.
8. Schmuck, C. (2006) How to improve guanidinium cations for oxoanion binding in aqueous solution? The design of artificial peptide receptors. *Coord. Chem. Rev.*, **250** (23–24), 3053–3067.
9. O'Neil, E.J. and Smith, B.D. (2006) Anion recognition using dimetallic coordination complexes. *Coord. Chem. Rev.*, **250** (23–24), 3068–3080.
10. Martinez-Máñez, R., and Sancenón, F. (2006) Chemodosimeters and 3D inorganic functionalized hosts for the fluoro-chromogenic sensing of anions. *Coord. Chem. Rev.*, **250** (23–24), 3081–3093.
11. Gunnlaugsson, T., Glynn, M., Toci, G.M., Kruger, P.E., and Pfeffer, F.M. (2006) Anion recognition and sensing in organic and aqueous media using luminescent and colorimetric sensors. *Coord. Chem. Rev.*, **250** (23–24), 3094–3117.
12. Rice, C.R. (2006) Metal-assembled anion receptors. *Coord. Chem. Rev.*, **250** (23–24), 3190–3199.
13. Filby, M.H. and Steed, J.W. (2006) A modular approach to organic, coordination complex and polymer based podand hosts for anions. *Coord. Chem. Rev.*, **250** (23–24), 3200–3218.
14. Gale, P.A. and Quesada, R. (2006) Anion coordination and anion-templated assembly: highlights from 2002 to 2004. *Coord. Chem. Rev.*, **250** (23–24), 3219–3244.
15. Gale, P.A. (2006) Structural and molecular recognition studies with acyclic anion receptors. *Acc. Chem. Res.*, **39** (7), 465–475.
16. Amendola, V., Esteban-Gmez, D., Fabbrizzi, L., and Licchelli, M. (2006) What anions do to N-H-containing receptors. *Acc. Chem. Res.*, **39** (5), 343–353.
17. Sessler, J.L., Gale, P.A., and Cho, W.-S. (2006) *Anion Receptor Chemistry*, Royal Society of Chemistry, Cambridge.
18. Gamez, P., Mooibroek, T.J., Teat, S.J., and Reedijk, J. (2007) Anion binding involving π-acidic heteroaromatic rings. *Acc. Chem. Res.*, **40** (6), 435–444.
19. Gale, P.A., García-Garrido, S.E., and Garric, J. (2008) Anion receptors based on organic frameworks: highlights from 2005 and 2006. *Chem. Soc. Rev.*, **37** (1), 151–190.
20. Hossain, M.A. (2008) Inclusion complexes of halide anions with macrocyclic receptors. *Curr. Org. Chem.*, **12** (26), 1231–1256.
21. Hudnall, T.W., Chiu, C.-W., and Gabbat, F.P. (2009) Fluoride ion recognition by chelating and cationic boranes. *Acc. Chem. Res.*, **42** (2), 388–397.
22. Hua, Y. and Flood, A.H. (2010) Click chemistry generates privileged CH hydrogen-bonding triazoles: the latest addition to anion supramolecular chemistry. *Chem. Soc. Rev.*, **39** (4), 1262–1271.
23. Lankshear, M.D. and Beer, P.D. (2006) Strategic anion templation. *Coord. Chem. Rev.*, **250** (23–24), 3142–3160.
24. Gimeno, N. and Vilar, R. (2006) Anions as templates in coordination and supramolecular chemistry. *Coord. Chem. Rev.*, **250** (23–24), 3161–3189.
25. Beer, P.D., Sambrook, M.R., and Curiel, D. (2006) Anion-templated assembly of interpenetrated and interlocked structures. *Chem. Commun.*, (20), 2105–2117.
26. Borisova, N.E., Reshetova, M.D., and Ustynyuk, Y.A. (2007) Metal-free methods in the synthesis of macrocyclic Schiff bases. *Chem. Rev.*, **107** (1), 46–79.

27. Vickers, M.S. and Beer, P.D. (2007) Anion templated assembly of mechanically interlocked structures. *Chem. Soc. Rev.*, **36** (2), 211–225.
28. Lankshear, M.D. and Beer, P.D. (2007) Interweaving anion templation. *Acc. Chem. Res.*, **40** (8), 657–668.
29. Bates, G.W. and Gale, P.A. (2008) An introduction to anion receptors based on organic frameworks. *Struct. Bond.*, **129**, 1–44.
30. Vilar, R. (2008) Anion recognition and templation in coordination chemistry. *Eur. J. Inorg. Chem.*, **2008** (3), 357–367.
31. Mullen, K.M. and Beer, P.D. (2009) Sulfate anion templation of macrocycles, capsules, interpenetrated and interlocked structures. *Chem. Soc. Rev.*, **38** (6), 1701–1713.
32. Vilar, R. (2008) Anion templates in synthesis and dynamic combinatorial libraries. *Struct. Bond.*, **129**, 175–206.
33. Gavina, P. and Tatay, S. (2010) Synthetic strategies for the construction of threaded and interlocked molecules. *Curr. Org. Synth.*, **7** (1), 24–43.
34. Fages, F., Vogtle, F., and Zinic, M. (2005) Cholesterol-based gelators. *Top. Curr. Chem.*, **256**, 77–131.
35. Nguyen, B.T. and Anslyn, E.V. (2006) Indicator-displacement assays. *Coord. Chem. Rev.*, **250**, 3118–3127.
36. Ward, M.D. (2006) [Ru(bipy)(CN_4]$^{2-}$ and its derivatives: photophysical properties and its use in photoactive supramolecular assemblies. *Coord. Chem. Rev.*, **250**, 3128–3141.
37. Goodgame, D.M.L., Grachvogel, D.A., and Williams, D.J. (2002) Second-sphere coordination of 'star'-shaped hexakis (*N*-pyridin-4-one)benzene (HPOB) with hexakis(methanol)nickel(II) nitrate; unusual (NO_3^-) (HPOB)NO_3)$^-$ π-stacked 'sandwiching'. *Inorg. Chim. Acta*, **330**, 13–16.
38. Ahuja, R. and Samuelson, A.G. (2003) Non-bonding interactions of anions with nitrogen heterocycles and phenyl rings: a critical Cambridge structural database analysis. *CrystEngComm*, **5**, 395–399.
39. Kim, D., Tarakeshwar, P., and Kim, K.S. (2004) Theoretical investigations of anion-π interactions: the role of anions and the nature of π systems. *J. Phys. Chem. A*, **108** (7), 1250–1258.
40. Maheswari, P.U., Modec, B., Pevec, A., Kozlevcar, B., Massera, C., Gamez, P., and Reedijk, J. (2006) Crystallographic evidence of nitrate-π interactions involving the electron-deficient 1,3,5-triazine ring. *Inorg. Chem.*, **45** (17), 6637–6645.
41. Casellas, H., Massera, C., Buda, F., Gamez, P., and Reedijk, J. (2006) Crystallographic evidence of theoretically novel anion-π interactions. *New J. Chem.*, **30** (11), 1561–1566.
42. Zaccheddu, M., Filippi, C., and Buda, F. (2008) Anion-π and π-π cooperative interactions regulating the self-assembly of nitrate–triazine–triazine complexes. *J. Phys. Chem. A*, **112** (7), 1627–1632.
43. Rotger, C., Soberats, B., Quiñonero, D., Frontera, A., Ballester, P., Benet-Buchholz, J., Deyà, P.M., and Costa, A. (2008) Crystallographic and theoretical evidence of anion- and hydrogen-bonding interactions in a squaramide-nitrate salt. *Eur. J. Org. Chem.*, **2008** (11), 1864–1868.
44. Götz, R.J., Robertazzi, A., Mutikainen, I., Turpeinen, U., Gamez, P., and Reedijk, J. (2008) Concurrent anion$\cdots\pi$ interactions between a perchlorate ion and two π-acidic aromatic rings, namely pentafluorophenol and 1,3,5-triazine. *Chem. Commun.*, (29), 3384–3386.
45. Mooiroek, T.J., Black, C.A., Gamez, P., and Reedijk, J. (2008) What's new in the realm of anion-π binding interactions? Putting the anion-π interaction in perspective. *Cryst. Growth Des.*, **8** (4), 1082–1093.
46. Valencia, L., Pérez-Lourido, P., Bastida, R., and Macías, A. (2008) Dinuclear Zn(II) polymer consisting of channels formed by π,π-stacking interactions with a flow of nitrate anions through the channels. *Cryst. Growth Des.*, **8** (7), 2080–2082.
47. Gural'skiy, I.A., Escudero, D., Frontera, A., Solntsev, P.V., Rusanov, E.B., Chernega, A.N., Krautscheid, H., and Domasevitch, K.V. (2009) 1,2,4,5-Tetrazine: an unprecedented μ_4-coordination that enhances ability for anion$\cdots\pi$ interactions. *Dalton Trans.*, (15), 2856–2864.

48. Antonioli, B., Büchner, B., Clegg, J.K., Gloe, K., Gloe, K., Götzke, L., Heine, A., Jäger, A., Jolliffe, K.A., Kataeva, O., Kataev, V., Klingeler, R., Krause, T., Lindoy, L.F., Popa, A., Seichtere, W., and Wenzel, M. (2009) Interaction of an extended series of *N*-substituted de(2-picolyl)amine derivatives with copper(II). Synthetic, structural, magnetic and solution studies. *Dalton Trans.*, (24), 4795–4805.
49. Hollis, C.A., Hanton, L.R., Morris, J.C., and Sumby, C.J. (2009) 2-D coordination polymers of hexa(4-cyanophenyl)[4]-radialene and silver(I): anion···π-interactions and radialene C-H···anion hydrogen bonds in the solid-state interactions of hexaaryl[3]-radialenes with anions. *Cryst. Growth Des.*, **9** (6), 2911–2916.
50. Hay, B.P., Custelcean, R. (2009) Anion-π interactions in crystal structures: Commonplace or extraordinary? *Cryst. Growth Des.*, **9** (6), 2539–2545.
51. Philp, D. and Stoddart, J.F. (1996) Self-Assembly in natural and unnatural systems. *Angew. Chem. Int. Ed. Engl.*, **35** (11), 1154–1196.
52. Wisner, J.A., Beer, P.D., and Drew, M.G.B. (2001) A demonstration of anion templation and selectivity in pseudorotaxane formation. *Angew. Chem. Int. Ed.*, **40** (19), 3606–3609.
53. Wisner, J.A., Beer, P.D., Drew, M.G.B., and Sambrook, M.R. (2002) Anion-templated rotaxane formation. *J. Am. Chem. Soc.*, **124** (42), 12469–12476.
54. Katayev, E.A., Pantos, G.D., Reshetova, M.D., Khrustalev, V.N., Lynch, V.M., Ustynyuk, Y.A., and Sessler, J.L. (2005) Anion-induced synthesis and combinatorial selection of polypyrrolic macrocycles. *Angew. Chem. Int. Ed.*, **44** (45), 7386–7390.
55. See, for example: (a) Gasperov, V., Galbraith, S.G., Lindoy, L.F., Rumbel, B.R., Skelton, B.W., Tasker, P.A., and White, A.H. (2005) A study of the complexation and extraction of Cu(II) sulfate and Ni(II) sulfate by N_3O_2-donor macrocycles. *Dalton Trans.*, (1), 139–145; (b) Moorthy, J.N., Natarajan, R., Savitha, G., Suchopar, A., and Richards, R.M. (2006) Anion-driven self-assembly of tetrapyridyl ligand with a twist. *J. Mol. Struct.*, **796** (1–3), 216–222; (c) Chand, D.K., Biradha, K., Kawano, M., Sakamoto, S., Yamaguchi, K., and Fujita, M. (2006) Dynamic self-assembly of an M_3L_6 molecular triangle and an M_4L_8 tetrahedron from naked Pd(II) ions and bis(3-pyridyl)-substituted arenes. *Chem. Asian J.*, **1** (1–2), 82–90; (d) Wezenberg, S.J., Escudero-Adán, E.C., Benet-Buchholz, J., and Kleij, A.W. (2009) Anion-templated formation of supramolecular multinuclear assemblies. *Chem. Eur. J.*, **15**, 5695–5700.
56. Pfeffer, F.M., Kruger, P.E., and Gunnlaugsson, T. (2007) Anion recognition and anion-mediated self-assembly with thiourea-functionalised fused [3]polynorbornyl frameworks. *Org. Biomol. Chem.*, **5** (12), 1894–1902.
57. Juwarker, H., Lenhardt, J.M., Pham, D.M., and Craig, S.L. (2008) 1,2,3-Triazole Ch···Cl^- contacts guide anion binding and concomitant folding in 1,4-diaryl triazole oligomers. *Angew. Chem. Int. Ed. Engl.*, **47** (20), 3740–3743.
58. Juwarker, H., Suk, J.-M., and Jeong, K.-S. (2009) Foldamers with helical cavities for binding complementary guests. *Chem. Soc. Rev.*, **38** (12), 3316–3325.
59. Sessler, J.L., Katayev, E., Pantos, G.D., and Ustynyuk, Y.A. (2004) Synthesis and study of a new diamidodipyrromethane macrocycle. An anion receptor with a high sulfate-to-nitrate binding selectivity. *Chem. Commun.*, (11), 1276–1277.
60. Katayev, E.A., Sessler, J.L., Khrustalev, V.N., and Ustynyuk, Y.A. (2007) Synthetic model of the phosphate binding protein: solid-state structure and solution-phase anion binding properties of a large oligopyrrolic macrocycle. *J. Org. Chem.*, **72**, 7244–7252.
61. Bru, M., Alfonso, I., Ignacio, B., Burguete, M.I., and Luis, S.V. (2006) Anion-templated syntheses of pseudopeptidic macrocycles. *Angew. Chem. Int. Ed. Engl.*, **45** (37), 6155–6159.
62. Alfonso, I., Bolte, M., Bru, M., Burguete, M.I., Luis, S.V., and Rubio, J. (2008) Supramolecular control for the modular synthesis of pseudopeptidic macrocycles

through an anion-templated reaction. *J. Am. Chem. Soc.*, **130** (19), 6137–6144.

63. Meshcheryakov, D., Boehmer, V., Bolte, M., Hubscher-Bruder, V., Arnaud-Neu, F., Herschbach, H., Van Dorsselaer, A., Thondorf, I., and Moegelin, W. (2006) Two chloride ions as a template in the formation of a cyclic hexaurea. *Angew. Chem. Int. Ed. Engl.*, **45** (10), 1648–1652.
64. Katayev, E.A., Myshkovskaya, E.N., Boev, N.V., and Khrustalev, V.N. (2008) Anion binding by pyrrole-pyridine-based macrocyclic polyamides. *Supramol. Chem.*, **20** (7), 619–624.
65. Maekawa, M., Konaka, H., Minematsu, T., Kuroda-Sowa, T., Munakata, M., and Kitagawa, S. (2007) Bowl-shaped Cu(I) metallamacrocyclic ethylene and carbonyl adducts as structural analogues of organic calixarenes. *Chem. Commun.*, (48), 5179–5181.
66. Maekawa, M., Nabei, A., Tominaga, T., Sugimoto, K., Minematsu, T., Okubo, T., Kuroda-Sowa, T., Munakata, M., and Kitagawa, S. (2009) A unique chair-shaped hexanuclear Cu(I) metallamacrocyclic C_2H_4 adduct encapsulating a BF_4^- anion. *Dalton Trans.*, (3), 415–417.
67. Shi, C.-C., Chen, C.-S., Hsu, S.C.N., Yeh, W.-Y., Chiang, M.Y., and Kuo, T.-S. (2008) The first anion template cubic cyanometallate cage and its 3,5-dimethyltris(pyrazolyl)methane iron(II,III) tricyanide building blocks. *Inorg. Chem. Comm.*, **11**, 1264–1266.
68. Armentano, D., Marino, N., Mastropietro, T.F., Martinez-Lillo, J., Cano, J., Julve, M., Lloret, F., and De Munno, G. (2008) Self-assembly of a chiral carbonate- and cytidine-containing dodecanuclear copper(II) complex: a multiarm-supplied globular capsule. *Inorg. Chem.*, **47** (22), 10229–10231.
69. Wei, K.-J., Ni, J., Xie, Y.-S., Liu, Y., and Yangzhong, Q.-L. (2007) Self-assembled hetero-bimetallic coordination cage and cation-clusters with μ_2-Cl bridging using a flexible two-arm ferrocene amide linker. *Dalton Trans.*, (31), 3390–3397.
70. Amouri, H., Desmarets, C., Bettoschi, A., Rager, M.N., Boubekeur, K., Rabu, P., and Drillon, M. (2007) Supramolecular cobalt cages and coordination polymers templated by anion guests: self-assembly, structures, and magnetic properties. *Chem. Eur. J.*, **13**, 5401–5407.
71. Tidmarsh, I.S., Taylor, B.F., Hardie, M.J., Russo, L., Clegg, W., and Ward, M.D. (2009) Further investigations into tetrahedral M_4L_6 cage complexes containing guest anions: new structures and NMR spectroscopic studies. *New J. Chem.*, **33** (2), 366–375.
72. Glasson, C.R.K., Meehan, G.V., Clegg, J.K., Lindoy, L.F., Turner, P., Duriska, M.B., and Willis, R. (2008) A new FeII quaterpyridyl M_4L_6 tetrahedron exhibiting selective anion binding. *Chem. Commun.*, (10), 1190–1192.
73. Argent, S.P., Adams, H., Riis-Johannessen, T., Jeffery, J.C., Harding, L.P., Mamula, O., and Ward, M.D. (2006) Coordination chemistry of tetradentate N-donor ligands containing two pyrazolyl-pyridine units separated by a 1,8-naphthyl spacer: dodecanuclear and tetranuclear coordination cages and cyclic helicates. *Inorg. Chem.*, **45** (10), 3905–3919.
74. Chmielewski, M.J. and Beer, P.D. (2008) in *Organic Nanostructures* (eds. J.L. Atwood and J.W. Steed), Wiley Interscience, pp. 63–96.
75. Curiel, D., Beer, P.D., Paul, R.L., Cowley, A., Sambrook, M.R., and Szemes, F. (2004) Halide anion directed assembly of luminescent pseudorotaxanes. *Chem. Commun.*, 1162.
76. Sambrook, M.R., Beer, P.D., Wisner, J.A., Paul, R.L., Cowley, A.R., Szemes, F., and Drew, M.G.B. (2005) Anion-templated assembly of pseudorotaxanes: importance of anion template, strength of ion-pair thread association, and macrocycle ring size. *J. Am. Chem. Soc.*, **127**, 2292–2302.
77. Curiel, D. and Beer, P.D. (2005) Anion directed synthesis of a hydrogensulfate selective luminescent rotaxane. *Chem. Commun.*, (14), 1909.
78. Mullen, K.M., Mercurio, J., Serpell, C.J., and Beer, P.D. (2009) Exploiting the 1,2,3-triazolium motif in anion templated formation of a bromide selective

rotaxane host assembly. *Angew. Chem. Int. Ed.*, **48**, 4781–4784.
79. Ng, K.-Y., Cowley, A.R., and Beer, P.D. (2006) Anion templated double cyclization assembly of a chloride selective [2]catenane. *Chem. Commun.*, (35), 3676.
80. Sambrook, M.R., Beer, P.D., Lankshear, M.D., Ludlow, R.F., and Wisner, J.A. (2006) Anion-templated assembly of [2]rotaxanes. *Org. Biomol. Chem.*, **4**, 1529–1538.
81. Lankshear, M.D., Evans, N.H., Bayly, S.R., and Beer, P.D. (2007) Anion-templated calix[4]arene-based pseudorotaxanes and catenanes. *Chem. Eur. J.*, **13**, 3861–3870.
82. Phipps, D.E. and Beer, P.D. (2009) A [2]catenane containing an upper-rim functionalized calix[4]arene for anion recognition. *Tetrahedron Lett.*, **50**, 3454–3457.
83. McConnell, A.J., Serpell, C.J., Thompson, A.L., Allan, D.R., and Beer, P.D. (2010) Calix[4]arene-based rotaxane host systems for anion recognition. *Chem. Eur. J.*, **16**, 1256–1264.
84. Bayly, S.R., Gray, T.M., Chmielewski, M.J., Davis, J.J., and Beer, P.D. (2007) Anion templated surface assembly of a redox-active sensory rotaxane. *Chem. Commun.*, 2234–2236.
85. Zhao, L., Davis, J.J., Mullen, K.M., Chmielewski, M.J., Jacobs, R.M.J., Brown, A., and Beer, P.D. (2009) Anion templated formation of pseudorotaxane and rotaxane monolayers on gold from neutral components. *Langmuir*, **25**, 2935–2940.
86. Chmielewski, M.J., Davis, J.J., and Beer, P.D. (2009) Interlocked host rotaxane and catenane structures for sensing charged guest species *via* optical and electrochemical methodologies. *Org. Biomol. Chem.*, **7**, 415–424.
87. Chmielewski, M.J., Zhao, L., Brown, A., Curiel, D., Sambrook, M.R., Thompson, A.L., Santos, S.M., Felix, V., Davis, J.J., and Beer, P.D. (2008) Sulfate anion templation of a neutral pseudorotaxane assembly using an indolocarbazole threading component. *Chem. Commun.*, 3154–3156.
88. Huang, B., Santos, S.M., Felix, V., and Beer, P.D. (2008) Sulfate anion-templated assembly of a [2]catenane. *Chem. Commun.*, 4610–4612.
89. Zhao, L., Mullen, K.M., Chmielewski, M.J., Brown, A., Bampos, N., Beer, P.D., and Davis, J.J. (2009) Anion templated assembly of an indolocarbazole containing pseudorotaxane on beads and silica nanoparticles. *New J. Chem.*, **33**, 760–768.
90. Li, Y., Mullen, K.M., Claridge, T.D.W., Costa, P.J., Felix, V., and Beer, P.D. (2009) Sulfate anion templated synthesis of a triply interlocked capsule. *Chem. Commun.*, 7134–7136.
91. Dietrich-Buchecker, C., Jimenez-Molero, M.C., Sartar, V., and Sauvage, J.-P. (2003) Rotaxanes and catenanes as prototypes of molecular machines and motors. *Pure Appl. Chem.*, **75** (10), 1383–1393.
92. Price, J.R., Fenton, R.R., Lindoy, L.F., McMurtrie, J.C., Meehan, G.V., Parkin, A., Perkins, D., and Turner, P. (2009) Copper(I) templated synthesis of a 2,2′-bipyridine derived 2-catenane: synthetic, modelling, and X-ray studies. *Aust. J. Chem.*, **62**, 1014–1019, and references therein.
93. Brown, A., Mullen, K.M., Ryu, J., Chmielewski, M.J., Santos, S.M., Felix, V., Thompson, A.L., Warren, J.E., Pascu, S.I., and Beer, P.D. (2009) Interlocked host anion recognition by an indolocarbazole-containing [2]rotaxane. *J. Am. Chem. Soc.*, **131**, 4937–4952.
94. Mullen, K.M., Davis, J.J., and Beer, P.D. (2009) Anion induced displacement studies in naphthalene diimide containing interpenetrated and interlocked structures. *New J. Chem.*, **33**, 769–776.
95. Curiel, D., Beer, P.D., Tárraga, A., and Molina, P. (2009) Electrochemically induced intermolecular anion transfer. *Chem. Eur. J.*, **15**, 7534–7538.
96. Barrell, M.J., Leigh, D.A., Lusby, P.J., and Slawin, A.M.Z. (2008) An ion-pair template for rotaxane formation and its exploitation in an orthogonal interaction anion-switchable molecular shuttle. *Angew. Chem. Int. Ed. Engl.*, **47**, 8036–8039.

6
Anion–π Interactions in Molecular Recognition

David Quiñonero, Antonio Frontera, and Pere M. Deyà

6.1
Introduction

Weak noncovalent interactions have a constitutive role in many fields. The chemistry derived from noncovalent binding is the intelligent and, in most cases, elegant utilization of interactions between molecules [1]. In biology, these interactions are the basis of a great deal of processes, the efficiency of which is always impressive. Interactions between predesigned binding centers are able to achieve complex functions in highly organized molecular systems, which is one fundamental aspect of supramolecular chemistry [2–20]. A deep understanding and accurate description of interactions between organic molecules, even in the condensed phase is needed, in particular, the mechanisms of molecular recognition. Aromatic interactions play a vital role in chemistry and biology [12]. The role of aromatic interactions becomes prominent in drug–receptor interactions, crystal engineering, and protein folding [21]. Moreover, the role of stacking interactions in DNA/RNA is of utmost importance [22, 23]. With the emergence of supramolecular chemistry, dendrimers [24, 25], molecular tweezers, rotaxanes, catenanes [26–29], and several supramolecular aggregates are associated with aromatic interactions [30–34]. Molecular recognition, which is of central importance in chemistry and biology, relies on noncovalent interactions such as H-bonding, stacking interaction [35], cation–π interaction, ionic interaction, and hydrophobic interaction [1–20]. The role of noncovalent interactions in Nature was fully recognized only in the past two decades and is of key importance in biodisciplines [21]. The side chains of the aromatic amino acid residues, Phe, Tyr, and Trp, provide a surface of negative electrostatic potential that can bind to a wide range of cations through a predominantly electrostatic interaction. Several examples of cation–π interactions involving aromatic amino acids have been described, including contributions to protein secondary structure, in which Phe/Tyr/Trp···Lys/Arg interactions are common [36–39].

The importance of anion–π interactions has been widely demonstrated, that is, noncovalent forces between electron-deficient aromatic systems and anions, by a great deal of theoretical [40–42] and experimental investigations [18, 43–51]. Several

Anion Coordination Chemistry, First Edition. Edited by Kristin Bowman-James, Antonio Bianchi, and Enrique García-España.

pioneering theoretical studies revealed that these interactions are energetically favorable [40–42]. Anion–π interactions are gaining significant recognition, and their pivotal role in many key chemical and biological processes is being increasingly appreciated [52]. The design of highly selective anion receptors and channels [53–57] represent important advances in this nascent field of supramolecular chemistry. The closely related lone pair–π interaction has been observed in several biological systems. For instance, Egli and Sarkhel have reported an interesting case of O–π interactions involving an RNA pseudoknot [58]. There are several excellent reviews [52, 59–61] that describe several aspects of the anion–π interaction.

This chapter focuses on providing information necessary to understand the physical nature of the interaction to a wider readership, which allows choosing intuitively the best aromatic molecule and the more appropriate ion to design a given system with a given function. This topic is treated in Sections 6.2 and 6.3. The understanding of these points also allows predicting how the strength of the interaction is affected by the presence of other interactions. This interesting topic is treated in Section 6.4. Finally, in Section 6.5, we describe several selected examples of the literature where the relevance of anion–π interaction in molecular recognitions is illustrated, both in the solid state and in solution.

6.2
Physical Nature of the Interaction

The physical nature of the anion–π interaction has been analyzed in a great deal of manuscripts [40–42, 62–65]. From them, it can be concluded that, essentially, electrostatic and ion-induced polarization are the main forces that contribute to the interaction energy [66, 67]. The electrostatic term is explained by means of the permanent quadrupole moment of arene. The negative quadrupole moment value of benzene can be turned positive by attaching electron-withdrawing substituents to the ring (Figure 6.1). Thus the *a priori* electrostatically unfavorable anion–π interaction can be turned favorable. The polarization of the π-electron system by the anion inducing a dipole moment has a significant contribution, whereas the reverse effect is expected to be sensibly lower. Therefore, the contribution of

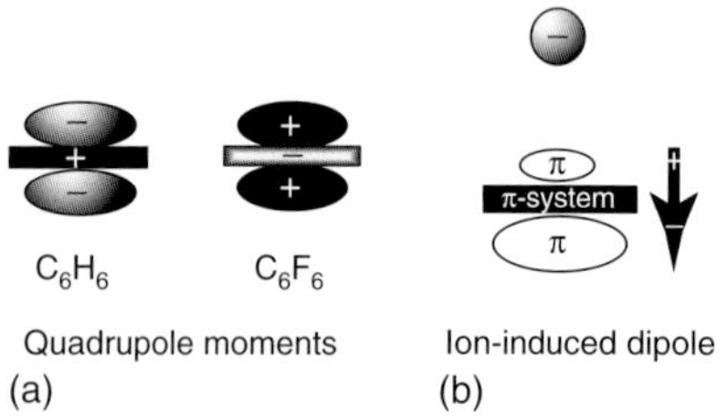

Figure 6.1 (a) Schematic representation of the quadrupole (Q_{zz}) moments of benzene and hexafluorobenzene ($Q_{zz} = -8.45$ and $Q_{zz} = 9.50$ Buckinghams, respectively). (b) Schematic representation of the anion-induced dipole.

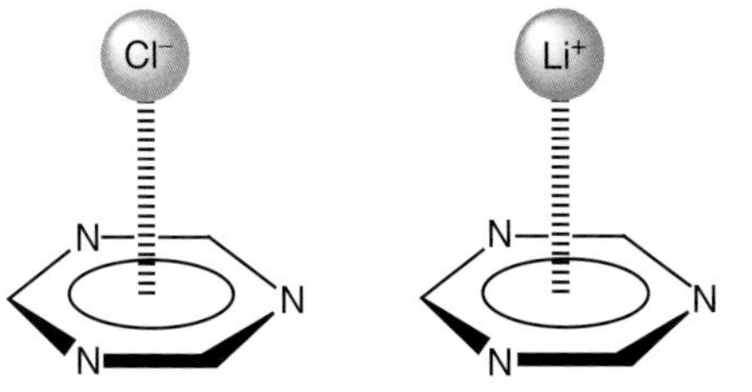

Figure 6.2 Interaction energies of the complexes of s-triazine with sodium cation and chloride. See Ref. [68] for computational details.

polarization to the total interaction energy basically comes from the interaction of the anion with the induced dipole in the π-system. The dispersion contribution to the total interaction energy in anion–π interactions is modest [62–65].

This knowledge of the physical nature of the interaction has allowed us to predict the dual behavior exhibited by aromatic rings with negligible quadrupole moments [68, 69]. Since both anion–π and cation–π interactions are dominated by electrostatic and polarization effects, these particular molecules (e.g., 1,3,5-trifluorobenzene, *s*-triazine) must be able to interact with cations and anions because the polarization term is always favorable. This has been demonstrated using theoretical calculations; see Figure 6.2 [68, 69].

Intuitively, the interaction of anions with electron-rich aromatic rings such as benzene is expected to be repulsive. However, this assumption is false because of a compensating effect of electrostatic and ion-induced polarization forces. As a matter of fact, the interaction energy of either benzene with chloride or hexafluorobenzene with sodium is negligible [70]. An interesting example of this compensating effect is observed in complexes of anions with isocyanuric acid, when the oxygen atoms are replaced by sulfur atoms to form thiocyanuric acids (Figure 6.3). The binding energy of the complexes of chloride with the four possible (thio)cyanuric acids is constant ($\sim$15 kcal mol^{-1}). In this case, the quadrupole moment progressively decreases on going from isocyanuric acid to trithiocyanuric acid, while the molecular polarizability increases [70].

6.3
Energetic and Geometric Features of the Interaction Depending on the Host (Aromatic Moieties) and the Guest (Anions)

Regarding the host and from the considerations explained in Section 6.2, two issues arise. First, to guarantee a strong anion–π interaction, the aromatic ring must have a large and positive quadrupole moment and a large molecular polarizability ($\alpha_{||}$). Second, depending on the magnitudes of these two physical properties, the interaction can be dominated by either electrostatic or polarization effects. For example,

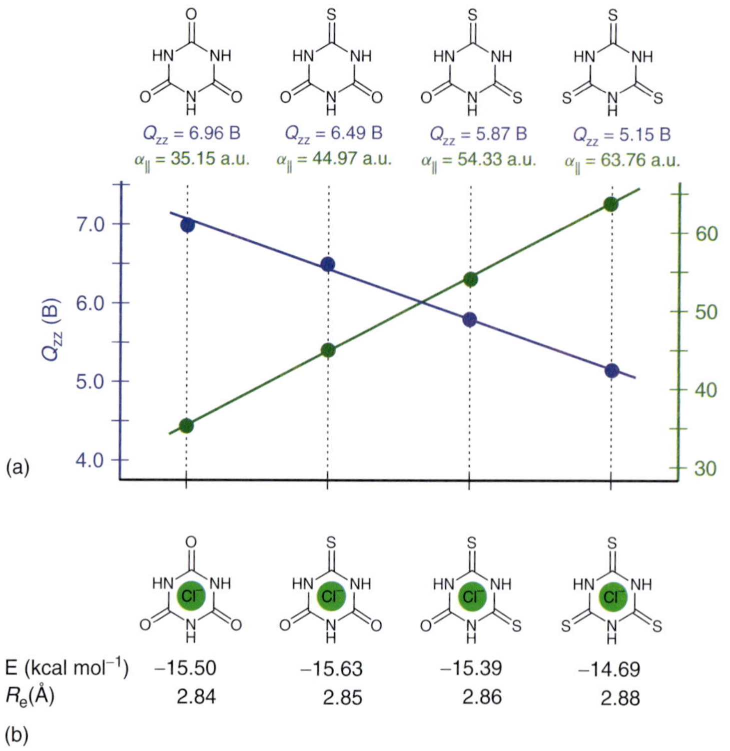

Figure 6.3 Figure 6.3(a) Variation of the Q_{zz} (blue line) and molecular polarizability ($\alpha_{||}$, green line) in isocyanuric acid derivatives depending on the number of sulfur atoms.(b) Interaction energies and equilibrium distances of several chloride anion–π complexes. See Ref. [70] for computational details.

aromatic molecules with negative permanent quadrupole moments and extended π-systems, such as triphenylene derivatives, interact favorably with anions because of the large magnitude of their molecular polarizability. The molecular polarizability of the example illustrated in Figure 6.4 (1,4,5,8,9,12-hexaazatriphenylene) is almost three times the value of benzene, and the quadrupole moment is similar.

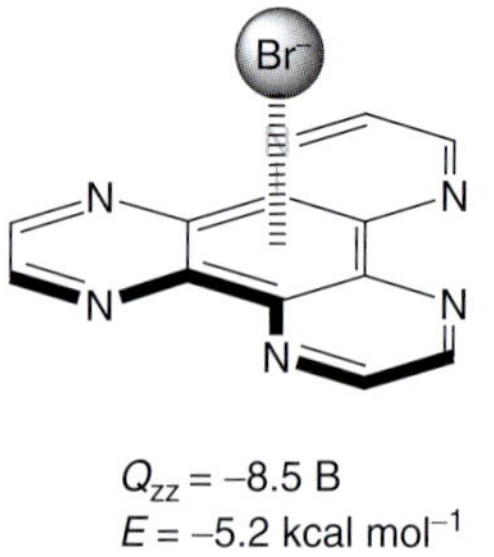

Figure 6.4 1,4,5,8,9,12-Hexaazatriphenylene complex with bromide. The interaction energy of the complex and the quadrupole moment of the aromatic ring are also indicated.

Table 6.1 Interaction energies (E, kcal mol^{-1}) with basis set superposition error (BSSE) and zero point energy (ZPE) corrections and equilibrium distances measured from the anion to the center of the ring at the MP2/6–31++G** level of theory in several hexafluorobenzene complexes.

Complex	E	R_e
$H^-@C_6F_6$	−12.1	2.693
$F^-@C_6F_6$	−18.2	2.570
$Cl^-@C_6F_6$	−12.6	3.148
$Br^-@C_6F_6$	−11.6	3.201
$NO_3^-@C_6F_6$	−12.2	2.917
$CO_3^{2-}@C_6F_6$	−34.7	2.720

As a result, the interaction energy of its complex with bromide is negative (the interaction energy of bromide with benzene is +1.9 kcal mol^{-1}) [71].

Thus, in these cases, induction effects dominate the interaction. Both the electrostatic and the polarization contributions to the total interaction energy strongly depend on the ion–arene distance. Therefore, not only should the properties of the arene be considered but the characteristics of the guest (anion) are also important. Small anions are more polarizing and present short equilibrium distances, and consequently, they exhibit more negative interaction energies (Table 6.1 for some energetic and geometric data). In addition, planar anions or even linear anions, such as NO_3^- or N_3^-, interact with the aromatic ring, resembling a $\pi-\pi$ stacking interaction. This theoretically predicted binding mode was confirmed experimentally for nitrate in pyrimidine salts (Figure 6.5) [72].

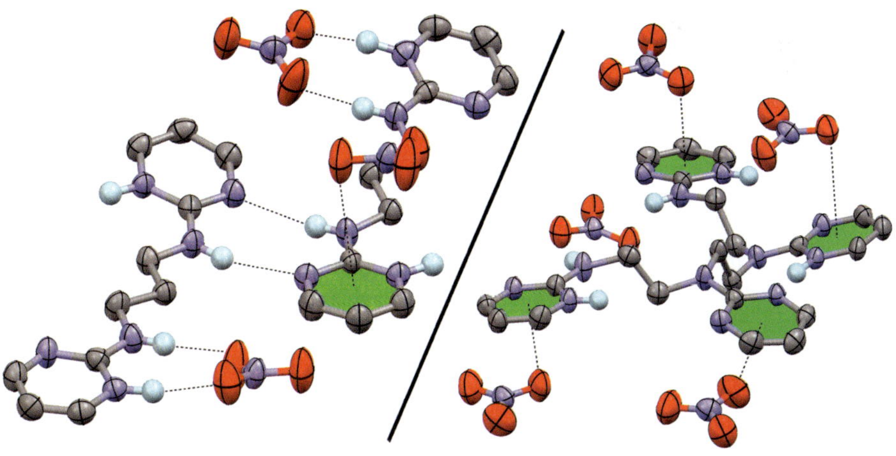

Figure 6.5 X-ray structures of amino-pyrimidine salts exhibiting anion–π interactions.

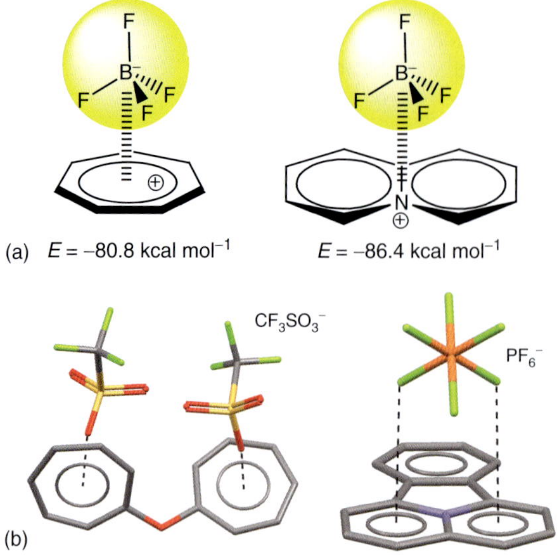

Figure 6.6 (a) Schematic representation of $[BF_4]^-$ complexes with positively charged aromatic ring. The interaction energies are also indicated. (b) Related X-ray structures retrieved from the CSD where anion–π interactions are established.

Energetically, in relation to the cation–π interaction, the anion–π interaction is less favorable because the van der Waals radii of anions are larger than those of cations, and consequently, the equilibrium distances are larger in anion–π than in cation–π complexes [69]. Therefore, the potential use of electron-deficient aromatic rings as building blocks for constructing anion receptors is reduced with respect to the wide use of cation receptors based on cation–π interactions. A solution to overcome this disadvantage has been proposed inspired by guanidinium salts that combine both hydrogen-bonding and electrostatic forces to bind anions. Therefore, charged aromatic compounds are used to increment the anion–π binding ability of the ring. The geometric and energetic features of anion–π complexes between several aromatic cations (tropylium, quinolizinylium, protonated 2-aminopyrimidine) and various anions have been reported, and some crystallographic structures support the theoretical findings (Figure 6.6) [49, 72–76].

Another important aspect that should be considered for the potential use of electron-deficient rings as binding blocks to construct anion receptors is the additivity of the interaction. Using *s*-triazine and trifluoro-*s*-triazine as electron-deficient aromatic rings, the additivity of the anion–π interactions has been examined (Figure 6.7) [77]. In Table 6.2, the energetic and geometric features of some complexes are summarized. It can be observed that the interaction energies of the ternary complexes ($\pi - X^- - \pi$) are approximately twice the interaction energies of the corresponding binary complexes ($X^- - \pi$). For the quaternary complexes ($X^- - \pi_3$), the interaction energies are not additive,

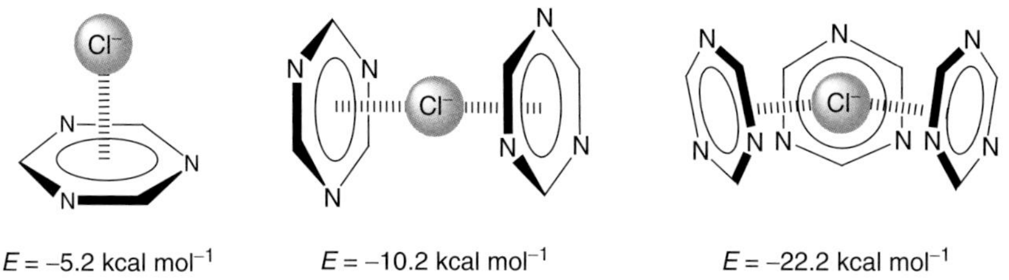

Figure 6.7 Schematic representation of the anion–π_n(n = 1–3) of s-triazine with chloride, indicating the computed interaction energies.

Table 6.2 Binding energies (E, kcal mol^{-1}) with the basis set superposition error correction and equilibrium distances (R_e, Å) at the RI-MP2/6–31++G** level of theory for several complexes.

Complex	E	R_e
$Cl^-@C_3N_3H_3$	−5.2	3.220
$Br^-@C_3N_3H_3$	−5.0	3.338
$Cl^-@(C_3N_3H_3)_2$	−10.4	3.213
$Br^-@(C_3N_3H_3)_2$	−10.2	3.370
$Cl^-@(C_3N_3H_3)_3$	−22.2	3.015
$Br^-@(C_3N_3H_3)_3$	−21.7	3.372
$Cl^-@C_3N_3F_3$	−15.1	3.008
$Br^-@C_3N_3F_3$	−14.2	3.176
$Cl^-@(C_3N_3F_3)_2$	−28.6	3.006
$Br^-@(C_3N_3F_3)_2$	−26.8	3.170
$Cl^-@(C_3N_3F_3)_3$	−41.0	3.019
$Br^-@(C_3N_3F_3)_3$	−38.6	3.172

because additional interactions take place. In triazine, additional N···H–C interactions are established between the rings stabilizing the $X^-@(C_3N_3H_3)_3$ complexes. In contrast, the interaction energies of the trifluoro-s-triazine quaternary complexes are less than three times the binding energy of the respective binary complexes because unfavorable C–F···F–C contacts are present.

The "atoms-in-molecules" (AIM) analysis [78] gives hints about the strength of the anion–π interaction [42, 79]. AIM is based on those critical points (CPs) at which the gradient of the density, $\nabla\rho$, vanishes. Such points are classified by the curvature of the electron density, for example, a bond CP has one positive curvature (in the internuclear direction) and two negative ones (perpendicular to the bond). Two bonded atoms are then connected with a bond path through the bond CP. The properties evaluated at such bond CPs characterize the bonding interactions. They

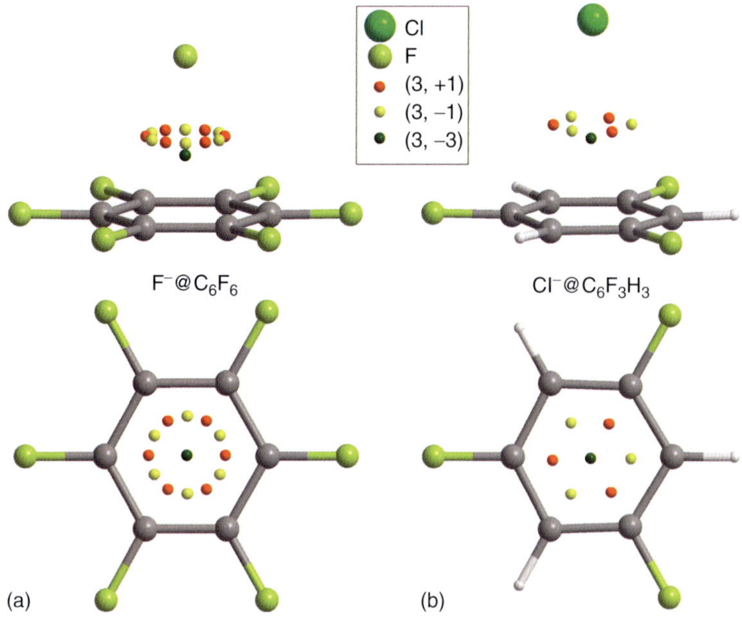

Figure 6.8 Representation of the critical points that characterize the anion–π interaction in $F^-@C_6F_6$ (a) and $Cl^-@C_6F_3H_3$ (b) complexes. The anion has been omitted for clarity in the on-top representations. See the legend for the color code.

have been widely used to study a great variety of molecular interactions. In anion–π complexes, it has been shown that the value of the charge density computed at the cage CP that appears on complexation correlates with the binding energy, and therefore, it can be used as a measure of bond order [66, 67]. In Figure 6.8, we show the typical distribution of CPs in two exemplifying complexes, $F^-@C_6F_6$ and $Cl^-@C_6F_3H_3$. In the fluoride–hexafluorobenzene complex, the anion–π interaction is characterized by six bond (3, +1) CPs that connect the anion with the six carbon atoms of the ring and six ring CPs that connect the anion with the middle of the six CC bonds of the ring. Finally, the interaction is characterized by a cage CP located along the main symmetry axis that connects the anion with the center of the aromatic ring. In the chloride-1,3,5-trifluorobenzene complex, the anion–π interaction is characterized by three bond (3, +1) CPs that connect the anion with the three carbon atoms of the ring bonded to fluorine and three ring CPs that connect the anion with the three carbon atoms of the ring bonded to hydrogen. Finally, the interaction is also characterized by a cage CP located along the main symmetry axis that connects the anion with the center of the aromatic ring.

It is also interesting to know how the aromatic ring is affected when it is interacting with ions in terms of aromaticity. It has been reported that when the

Table 6.3 Mulliken, Merz-Kollman (MK), ChelpG, and AIM charges of the ion in several ion–π complexes.

Anion@C_6F_6	q (Mull)	q (MK)	q (CHelpG)	q (AIM)
F^-@C_6F_6	−0.934	−0.805	−0.813	−0.989
Cl^-@C_6F_6	−0.921	−0.895	−0.878	−0.946
Br^-@C_6F_6	−0.868	−0.902	–	−0.954
NO_3^-@C_6F_6	−0.923	−0.859	−0.867	−0.973

aromatic ring participates in cation–π interactions, the aromaticity of the ring decreases [69, 80]. In contrast, it increases in anion–π complexes. This variation in the aromaticity of the ring has been related to a weakening or strengthening of the C-C bonds of the aromatic ring. Regarding charge-transfer effects on anion–π interactions, the results are somewhat contradictory. The theoretically predicted values strongly depend on the method used to derive the atomic charges. In general, Natural Population Analysis (NPA) and AIM charges predict negligible charge-transfer effects, and Merz-Kollman and CHelpG charges predict values ranging 0.1–0.25 (e) (Table 6.3 for some data). It is also interesting to emphasize the differences observed between the cation–π and the anion–π interactions from the molecular orbital point of view. The differences in two isoelectronic ion–π complexes have been studied, namely, 1,3,5-trifluorobenzene interacting with either fluoride anion or sodium cation [69, 81]. From the orbital analysis, it can be deduced that, while the atomic orbitals of the cation do not participate in the molecular orbitals of the complex, the atomic orbitals of the anion have an active participation in the molecular orbitals of the anion–π complexes (Figure 6.9).

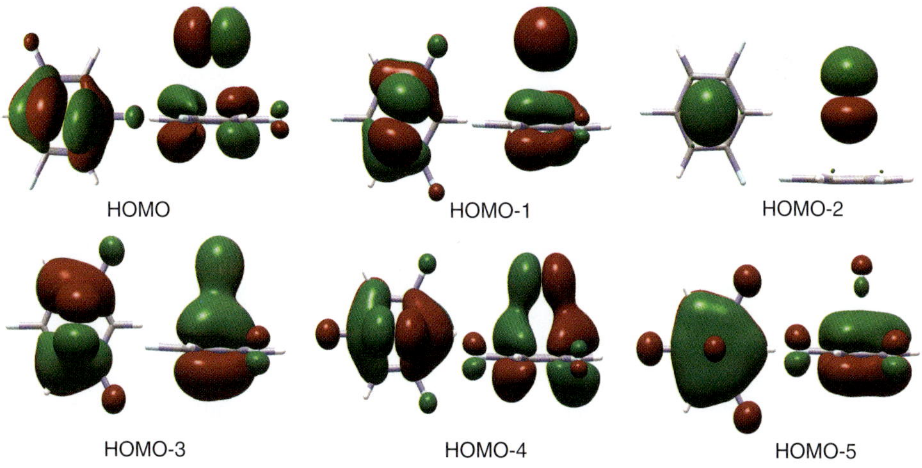

Figure 6.9 Representation of the π-type orbitals in the F^-@$C_6F_3H_3$ complex.

6.4
Influence of Other Noncovalent Interactions on the Anion–π Interaction

Weak noncovalent interactions are important in many areas of modern chemistry [1–20]. For instance, they are important in chemical reactions, in molecular recognition, in controlling the conformation of many molecules, and in regulating biochemical processes [9–13]. These chemical processes are accomplished with specificity and efficiency, taking advantage of intricate combinations of different intermolecular interactions. Noncovalent interactions such as hydrogen bonding, ion–π and π–π interactions, and other weak forces govern the organization of multicomponent supramolecular assemblies [1–39, 82]. A deep understanding of these interactions is of outstanding importance for the rationalization of effects observed in several fields, such as biochemistry and materials science. A quantitative description of these interactions can be performed by means of quantum chemical calculations on small model systems [83–85]. In complex biological systems and in the solid state, a multitude of these noncovalent interactions may operate simultaneously, giving rise to interesting cooperativity effects. For instance, it is well known that hydrogen bonding shows a highly cooperative behavior. In general, the cumulative strength of networks of hydrogen bonds is larger than the sum of the individual bond strengths when they work simultaneously [86–89]. Similar observations have been made for the interplay between stacking and hydrogen-bonding interactions [90]. This combination is of great importance for structural control in oligonucleotides. A recent review has covered the study of a multitude of combinations of weak interactions [91]. The aim of this section is to show the influence of a variety of interactions on the anion–π interaction. An important application of this knowledge is the fine tuning of the strength of the anion–π interaction as needed to accomplish a given function.

6.4.1
Interplay between Cation–π and Anion–π Interactions

Previously in this chapter, we have learned that the utilization of an aromatic ring with a large and positive quadrupole moment guarantees a strong anion–π interaction. However, it is also possible to establish a strong anion–π interaction between anions and benzene. This is achieved whenever the aromatic ring simultaneously interacts with a cation on the other side of the ring. The simultaneous interaction of anions and cations on different faces of the same π-system has been studied [92–95]. The aromatic system can act as a charge insulator or as an intermediate in the transfer of *information* between the charged systems. For these ternary anion–π–cation (A–π–C) complexes, the interaction energies are large and negative (Table 6.4), and the equilibrium distances are shorter than the corresponding distances in the binary ion–π complexes. In addition to the binding energies, the cooperativity energies (E_{coop}) are also given in Table 6.4. They are computed as the difference between the energy of the A–π–C complex and the sum of the energies $A - \pi + \pi - C + AC$ (Figure 6.10). For A–π and π–C, the

Table 6.4 Interaction (BSSE corrected) and cooperativity energies (E and E_{coop}, respectively in kcal mol^{-1}) and equilibrium distances (R_e, Å) for several cation–π–anion complexes at the MP2/6–31++G** level of theory.

Complex	$E_{A-\pi-C}$	E_{coop}	R_e (cation–π)	R_e (anion–π)
$Na^+@C_6H_6@F^-$	−93.10	−3.37	2.280	2.482
$Na^+@C_6H_6@Cl^-$	−85.11	−2.78	2.304	3.049
$Na^+@C_6H_6@Br^-$	−84.16	−2.80	2.313	3.157
$Na^+@C_6F_3H_3@F^-$	−90.64	−3.43	2.353	2.368
$Na^+@C_6F_3H_3@Cl^-$	−80.35	−3.67	2.389	2.925
$Na^+@ C_6F_3H_3@Br^-$	−78.89	−4.50	2.399	3.006
$Na^+@C_6F_6@F^-$	−88.42	−4.24	2.437	2.286
$Na^+@C_6F_6@Cl^-$	−75.63	−4.85	2.488	2.835
$Na^+@C_6F_6@Br^-$	−74.04	−4.74	2.495	2.913

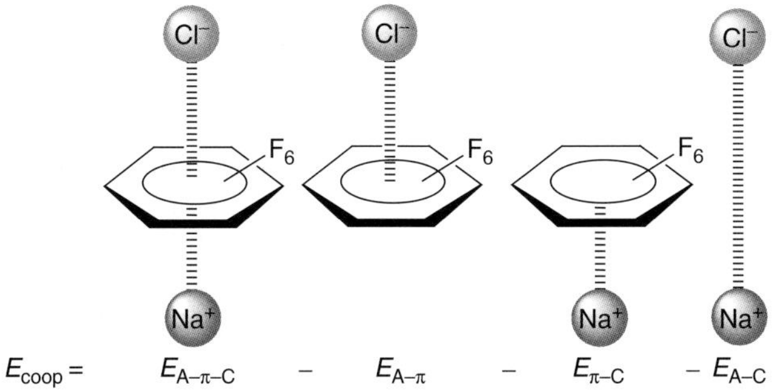

Figure 6.10 Explanation of the different terms used to measure the cooperativity energy (E_{coop}) in A–π–C complexes.

energy of the isolated dimers within their corresponding minima configurations is used, and AC is the interaction energy of the ions A and C in the positions they have in the trimer. All E_{coop} values are negative, indicating that the simultaneous interaction of the ions with the same aromatic ring implies a reinforcement of both cation–π and anion–π interactions. This is confirmed by the shortening of the equilibrium distances in the ternary complexes in relation to the binary complexes, which ranges from 0.15 to 0.40 Å. In addition, this mutual and favorable influence does not depend on the nature of the arene, since negative E_{coop} values are obtained for three aromatic rings with different quadrupole moments (C_6H_6, $Q_{zz} = -8.45$ B; $C_6F_3H_3$, $Q_{zz} = 0.19$ B; C_6F_6, $Q_{zz} = 9.50$ B).

An elegant experimental evidence for these theoretical findings can be obtained from the interesting work of Atwood's group [96–99]. They have demonstrated using X-ray analysis and ^1H NMR titration experiments that the host–guest

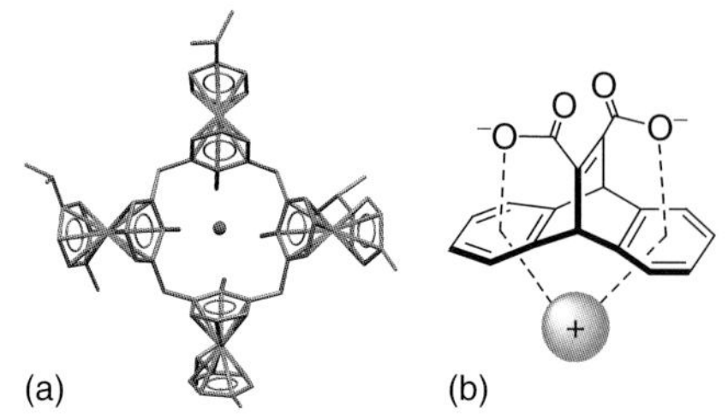

Figure 6.11 (a) X-ray of an organometallic calixarene with included iodide. (b) Dougherty's cyclophanic receptor.

behavior of calixarenes and cyclotriveratrylenes is drastically altered in the presence of transition metal centers (Ru, Ir, Rh), allowing anionic guest species (instead of cationic) to be included within the molecular cavity. One of these examples is shown in Figure 6.11a. In addition, experimental evidence of the interplay between anion–π and cation–π interactions in solution comes from Dougherty's group [100]. Using ^{1}H NMR spectroscopy, they have demonstrated that cyclophanic receptors that present carboxylate groups close to the external face of the phenylic rings have higher cation-binding affinities than those without the anionic groups. This effect was attributed to the induced dipole generated in the aromatic ring by the presence of the carboxylate anion, see Figure 6.11b.

6.4.2
Interplay between π–π and Anion–π Interactions

Since π–π interactions are very important and omnipresent in a great variety of systems, the study of the mutual influence between anion–π and π–π interactions is a topic of interest. It has been recently demonstrated that the interplay between ion–π and π–π interactions can lead to strong cooperativity effects [101–103]. Three aromatic systems (benzene, trifluorobenzene, and hexafluorobenzene) and three halide ions (F^-, Cl^-, and Br^-) have been used in the theoretical study of cooperativity effects in the anion–π–π' complexes. Some of them are shown in Figure 6.12. While all ternary complexes involving trifluorobenzene exhibit favorable cooperativity [102], for the other two arenes, the cooperativity is favorable in all complexes apart from X^-@C_6F_6@C_6H_6 complexes, see Table 6.5.

The theoretical results obtained for anion–π–π complexes have been used to explain an unexpected experimental finding regarding the face-to-face stacking of pentafluorophenyl groups in substituted ferrocenes. In Figure 6.13, we show the three ferrocene derivatives studied by Blanchard *et al.* [104] in the conformation observed in their crystal structures. While the conformation adopted in the crystal of the substituted ferrocenes **A** and **B** can be explained intuitively, the conformation of **C** is counterintuitive, as stated by the authors. We also show in Figure 6.13 a fragment of the crystal structure of **C** where the unexpected face-to-face π-stacking of two pentafluorophenyl groups is observed. As noticed by the authors, the cyclopentadienyl ring of a neighboring molecule forms an intermolecular stacking

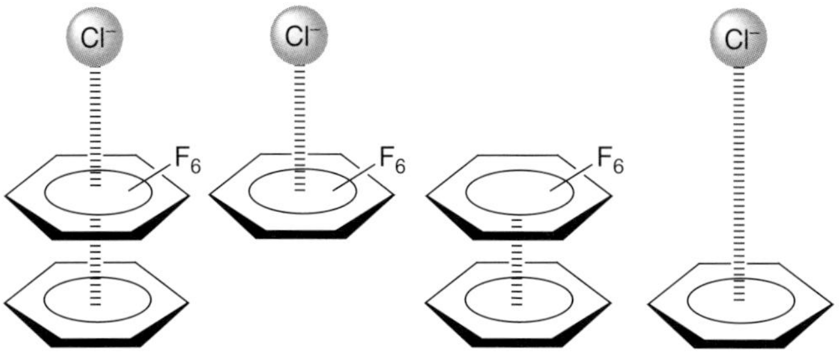

$$E_{coop} = E_{A-\pi-\pi'} - E_{A-\pi} - E_{\pi-\pi'} - E_{A-\pi'}$$

Figure 6.12 Explanation of the different terms used to measure the cooperativity energy (E_{coop}) in A–π–π′ complexes.

Table 6.5 Interaction (BSSE corrected) and cooperativity energies (E and E_{coop} respectively in kcal mol^{-1}) and equilibrium distances of anion–π–π complexes at the MP2/6–31++G** level of theory.

Complex	$E_{A-\pi-\pi'}$	E_{coop}	R_e (ion–π)	$R_e(\pi - \pi')$
$F^-@C_6H_6@C_6F_6$	−6.5	−2.6	3.35	3.00
$Cl^-@C_6H_6@C_6F_6$	−6.0	−2.6	3.36	3.59
$Br^-@C_6H_6@C_6F_6$	−6.2	−2.5	3.36	3.59
$F^-@C_6F_6@C_6H_6$	−21.8	0.5	3.48	2.59
$Cl^-@C_6F_6@C_6H_6$	−16.3	0.2	3.46	3.16
$Br^-@C_6F_6@C_6H_6$	−16.1	0.4	3.44	3.31
$F^-@C_6F_6@C_6F_6$	−24.4	−0.8	3.40	2.53
$Cl^-@C_6F_6@C_6F_6$	−17.4	−1.5	3.38	3.08
$Br^-@C_6F_6@C_6F_6$	−17.1	−1.0	3.39	3.24

with the pentafluorophenyl ring. The distance between the ring planes of both molecules is 3.35 Å and between the cyclopentadienyl ring centroid and the pentafluoropheyl centroid is 3.61 Å. This unexpected face-to-face stacking of the pentafluorophenyl groups can be explained considering the cooperativity energies of the anion–π–π complexes where both π-systems are hexafluorobenzene rings (Table 6.5). In the three examples studied, the E_{coop} is favorable and the equilibrium distances are shorter than in the related 1 : 1 complexes, indicating that cooperativity effects are found between anion–π and π–π interactions, whereas in the ternary complexes both π–systems are hexafluorobenzene. In the crystal structure **C** (Figure 6.13), the presence of the cyclopentadienyl anion interacting with a pentafluorophenyl group induces the face-to-face stacking interaction with the other pentafluorophenyl unit. A similar behavior is observed in related ferrocene compounds, which have been extensively studied by Deck *et al.* [105–109].

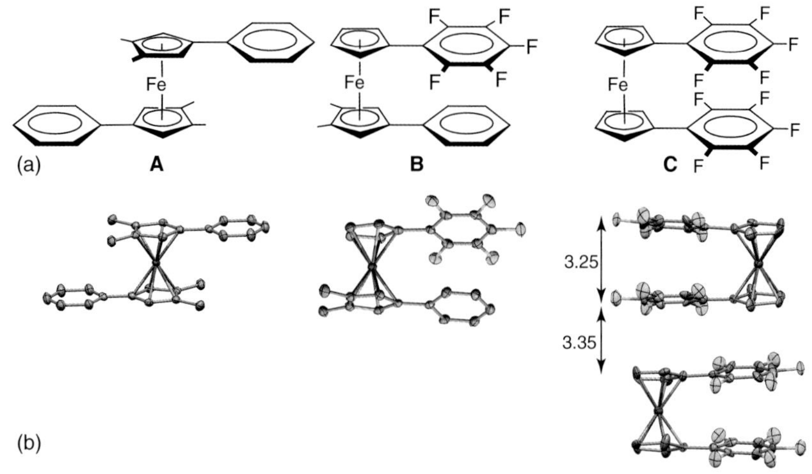

Figure 6.13 (a) Schematic representation of the ferrocene derivatives synthesized by Blanchard *et al.* (b) X-ray structures of compounds A–C are shown.

Although it was either not discussed or not noticed by the authors, an important number of structures with face-to-face stacking interactions between pentafluorophenyl groups are present, and in all cases, a cyclopentadienyl anion interacts with the pentafluorophenyl group. To finish this topic, it should be mentioned that several recently reported works [110, 111] have given experimental evidence of the interplay between anion–π and π–π interactions in π-acidic rings, in particular, between two *s*-tetrazine rings and anions. This interplay may influence self-assembly reactions.

6.4.3
Interplay between Anion–π and Hydrogen-Bonding Interactions

The interplay between the anion–π and H-bonding interactions has also been studied [112–114]. It has been demonstrated that a reinforcement of both interactions is obtained when they act simultaneously in complexes in which the aromatic rings are hydrogen bond acceptors, for instance, pyrazine or pyridazino[4,5-*d*]pyridazine. In contrast, a weakening of both interactions is observed in complexes where the aromatic ring is a hydrogen bond donor, for instance, pyromellitic imide. From the partitioning of the interaction energy, it has been demonstrated that this interplay is basically due to electrostatic effects [112]. These mutual effects were first published in relatively small aromatic systems, where the distance between the interacting anion and the water molecules used to generate the hydrogen-bonding interactions is not large. However, in a second report, the aromatic systems studied were very large and the distance from the anion to the water molecules is as long as 11 Å [114]. Even in these cases, a remarkable interplay between both interactions has been observed. Some of the systems studied are shown in Figure 6.14. Interestingly

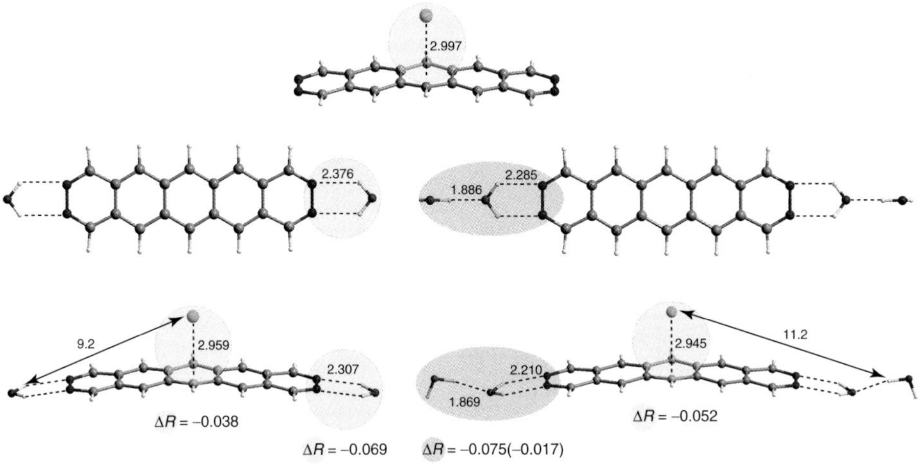

Figure 6.14 RI-MP2(full)/6–31++G** optimized complexes ($n = 3$). The chloride anion is represented in light grey. Distances in angstrom.

(Table 6.6), for a given series of complexes, E_{coop} is almost independent of the number of aromatic rings. That is to say, it is not affected by the water–anion distance. Therefore, the aromatic ring is able to transmit the synergetic effect from the anion (anion–π interaction) through the aromatic π-system to the water molecule (H-bonding interaction), even to a second set of water molecules, and vice versa.

All these findings can be rationalized by inspecting the shortening of both equilibrium distances (R_{HB} and R_e in Table 6.6) with respect to the complexes where only one interaction is present. It is important to note that this mutual reinforcement of both interactions is maintained in systems where the H-bonding interaction is very distant from the anion–π interaction. To exemplify this, the largest system is illustrated in Figure 6.14. In the anion–π complex with two water molecules, the anion–water distance is 9.2 Å. In spite of this long distance, a shortening of the anion–π interaction ($\Delta R_e = -0.038$ Å) with respect to the anion–π complex without water is observed. Moreover, the addition of a second set of water molecules further strengthens the anion–π interaction. The equilibrium distance of this complex shortens 0.052 Å with respect to the anion–π complex without water and 0.015 Å with respect to the anion–π complex with two water molecules. These variations of the equilibrium distances of the complexes are remarkable, taking into account that the second set of water molecules are located at 11.2 Å from the chloride anion.

It is worth noting that the behavior is totally different when the aromatic ring participates in H-bonding interaction as donor. The cooperativity energies are positive, and a lengthening of the equilibrium distances is observed (Table 6.6), indicating that both interactions weaken.

Table 6.6 Interaction and cooperativity energies with the BSSE correction (E_{BSSE} and E_{coop}, respectively in kcal mol^{-1}) and equilibrium distances (R_{HB} and R_e, Å) at the RI-MP2(full)/6–31++G** level of theory for anion–π and H-bonding multi-component complexes. The variation of the equilibrium distances in the multicomponent system with respect to the complexes with only one type of interaction (HB or anion–π) are also summarized (ΔR_{HB} and ΔR_e, Å).

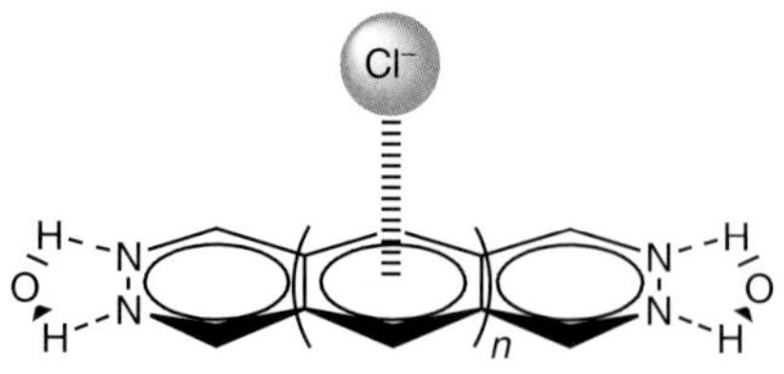

n	E_{BSSE}	E_{coop}	R_{HB}	R_e	ΔR_{HB}	ΔR_e
0	−30.6	−2.1	2.316	2.943	−0.073	−0.098
1	−31.8	−2.6	2.305	3.044	−0.088	−0.051
2	−32.5	−3.1	2.303	2.917	−0.078	−0.042
3	−32.6	−3.4	2.307	2.959	−0.069	−0.038

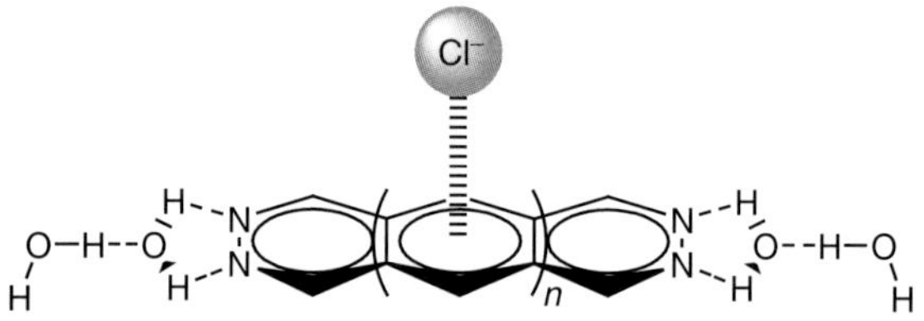

n	E_{BSSE}	E_{coop}	R_{HB}	R_e	ΔR_{HB}	ΔR_e
0	−46.1	−3.9	2.224 (1.879)	2.909	−0.111 (−0.017)	−0.107
1	−47.2	−4.3	2.215 (1.875)	3.022	−0.103 (−0.017)	−0.073
2	−47.6	−4.9	2.207 (1.867)	2.900	−0.085 (−0.020)	−0.059
3	−47.8	−5.0	2.210 (1.869)	2.945	−0.075 (−0.017)	−0.052

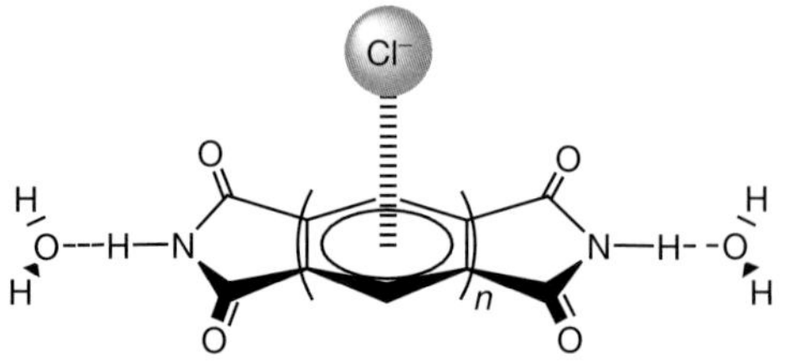

n	E_{BSSE}	E_{coop}	R_{HB}	R_e	ΔR_{HB}	ΔR_e
1	−20.7	1.6	1.974	3.149	0.092	0.042
2	−20.8	2.2	1.962	2.990	0.076	0.011
3	−21.2	2.2	1.954	3.059	0.064	0.013

n, number of benzene rings (see the figures in the table).

6.4.4
Influence of Metal Coordination on the Anion–π Interaction

The influence of metal coordination of the heteroaromatic ring on the ability to establish anion–π interactions has been studied in two recent works [74, 115]. In the first one, we have studied theoretically how the coordination of both pyridine and pyrazine with Ag(I) dramatically changes their anion–π binding properties (Figure 6.15) [115]. Using high-level *ab initio* calculations, it has been demonstrated that the anion–π complexes of pyridine and pyrazine with Cl^- are considerably more favorable when Ag(I) is coordinated to the nitrogen atom since the π-acidity of the aromatic rings increases. In fact, the interaction energy of pyridine with chloride is unfavorable (2 kcal mol^{-1}), indicating that pyridine is not π-deficient enough to form stable anion–π complexes. However, the interaction energy is favorable (−14.8 kcal mol^{-1}) when the nitrogen atom of pyridine is coordinated to silver. A similar behavior is observed in pyrazine complexes. The interaction energy of pyrazine with Cl^- is very modest (−2.6 kcal mol^{-1}); however, it becomes considerably more negative when it is coordinated to two silver atoms (−35.8 kcal mol^{-1}).

The second work [74] combines theory and experiment to demonstrate that *s*-tetrazine is a very strong anion–π acceptor when it is coordinated to four Ag(I) atoms. As a matter of fact, it is the first experimental work where the unprecedented μ^4-coordination of 1,2,4,5-tetrazine, the most nitrogen-rich 6-membered aromatic, is reported. The computed interaction energy of *s*-tetrazine with nitrate ion is −9.6 kcal mol^{-1} (Figure 6.15), and when it is tetracoordinated to Ag(I), the

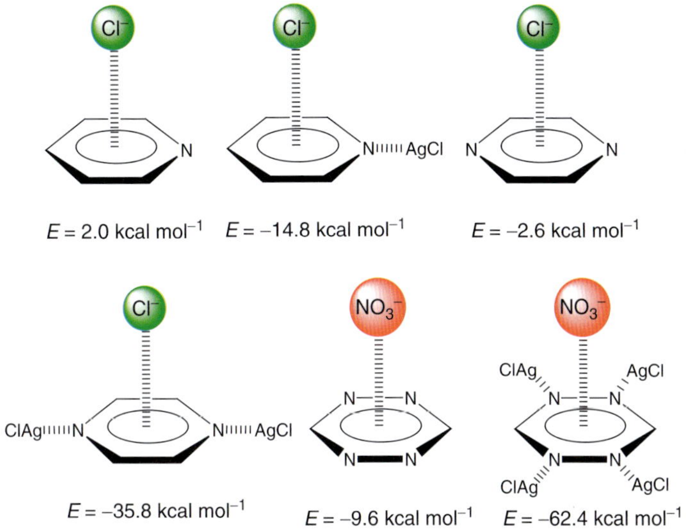

Figure 6.15 Schematic representation of the pyridine, pyrazine, and *s*-tetrazine anion–π complexes. The computed interaction energies are also indicated.

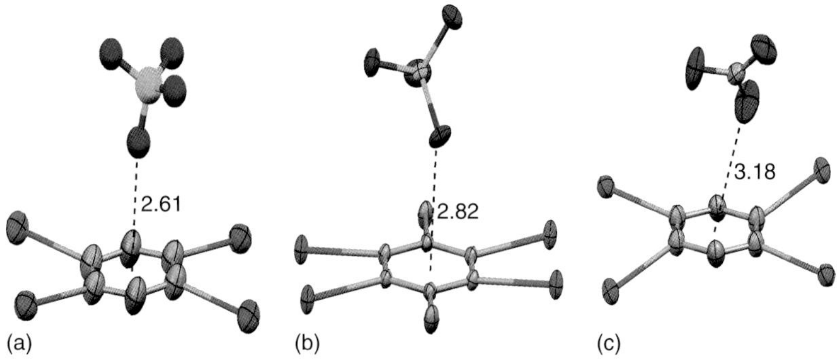

Figure 6.16 (a–c) Fragments of the X-ray structures containing the unprecedented μ^4-coordination of 1,2,4,5-tetrazine and relevant anion–π interactions.

interaction energy is −62.4 kcal mol^{-1}. In addition, a significant shortening of the equilibrium distance is predicted by the theoretical study. Experimentally, very short equilibrium distances between the anion and the *s*-tetrazine ring are observed in the X-ray structures (Figure 6.16), indicating strong anion–π interactions, in agreement with the theoretical predictions. Actually, for the perchlorate anion (Figure 6.16a), it is the shortest anion–π contact reported to date.

6.5
Experimental Examples of Anion–π Interactions in the Solid State and in Solution

Molecular modeling would be highly speculative without the validity provided by experimental results. Crystal structures are so rich in information and often reveal effects that had not been noticed by the original authors. The Cambridge Structural Database (CSD) is a convenient and reliable storehouse for geometrical information [116, 117]. Taking advantage of this, one of the pioneering theoretical works on anion–π interactions provided experimental evidence retrieved from a search in the CSD [118]. After that, a great deal of experimental work has provided further evidence of the importance of this nascent noncovalent interaction. This weight of evidence notwithstanding, a recent analysis of the CSD published by Hay and Custelcean [119] states that the CSD completely fails to provide any convincing crystal structure evidence for anion–π interactions involving charge-neutral, 6-membered rings with Cl^-, Br^-, I^-, NO_3^-, ClO_4^-, BF_4^-, and PF_6^- anions, in marked contrast with previous studies, which had found them to be ubiquitous. However, a subsequent analysis [120] demonstrates that the conclusions arrived at by Hay and Custelcean [119] were the result of inappropriate search criteria, leading to an unfortunate misinterpretation of the data in the CSD, and that, indeed, anion–π interactions in the solid state are commonplace and, to those with interests in noncovalent design elements for anion recognition, extraordinary [120].

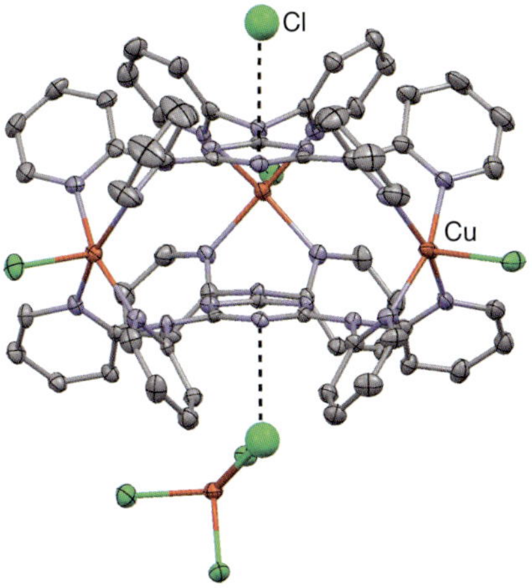

Figure 6.17 Fragment of the X-ray structure of the s-triazine-based host interacting with Cl and $[CuCl_4]^{2-}$ ions by means of anion–π interactions.

In this section, we describe some selected experimental examples where anion–π interactions are particularly relevant. We have not sorted the examples by relevance since it is subjective; we just present the results chronologically. The primary example is the first experimental work [43] devoted to the study of anion–π interactions after the original theoretical manuscripts [40–42]. The synthesis and X-ray characterization of a host based on electron-deficient s-triazine ring is reported. Interestingly, in the host, both triazine rings are arranged in an almost perfect face-to-face arrangement (Figure 6.17), with the N atoms of one s-triazine ring located over the N atoms of the second ring. The most interesting feature of the crystal structure concerns the position of the charge-compensating chloride and $[CuCl_4]^{2-}$ ions. One chloride anion resides above one of the triazine rings (Figure 6.17), where the distance between the centroid of this ring and the chlorine atom is 3.17 Å. The angle of the $Cl^- \cdots$centroid axis to the plane of the ring (87°) gives evidence that the chlorine atom is almost perfectly located on the C_3-axis above the ring. Both values are in excellent agreement with values obtained from *ab initio* molecular orbital calculations for the parent chloride-1,3,5-triazine complex [42]. In a similar manner, the opposite triazine face of the host is capped by the second counteranion, with an even shorter distance of 3.11 Å observed between one chlorine atom of the $[CuCl_4]^{2-}$ ion and the ring centroid. Anion–π interactions between chloride and the host (H) appear to be persistent also in solution and the gas phase. A stable $H{\cdot}Cl_2^+$ ion is detected as a prominent peak in the ESI mass spectrum of $H{\cdot}[CuCl_4]{\cdot}Cl$ in methanol.

Another pioneering experimental work was published some months later by de Hoog *et al.* [46]. They published the synthesis and X-ray characterization of a cationic tetranuclear copper moiety formed by one dendritic ligand, where the four s-triazinyl groups stack two by two in a parallel manner and the copper ions are coordinated by two 2,2′-dipyridylamino units belonging to two different s-triazine

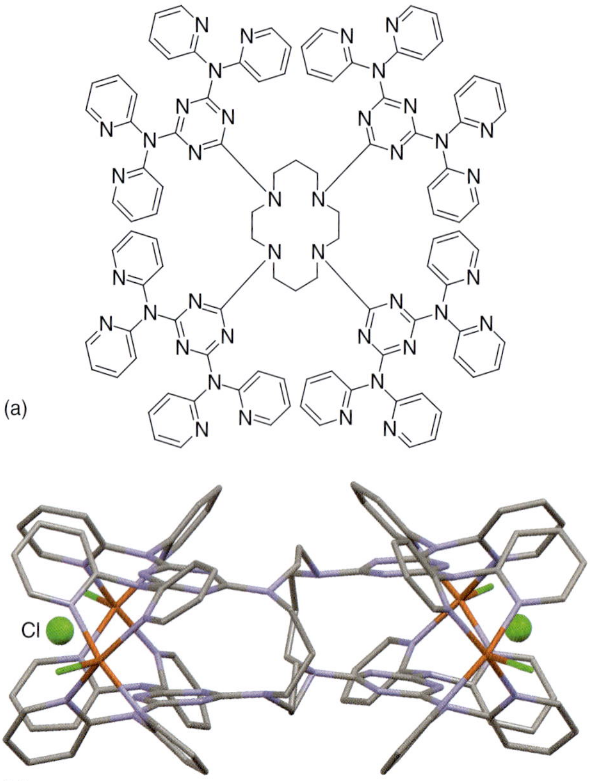

(a)

(b)

Figure 6.18 (a) Representation of the dendritic ligand. (b) Fragment of the X-ray structure of the complex showing the encapsulation of two chloride ions.

rings. The coordination spheres of the metal centers are completed by a chloride anion at the apical position. The most important feature of this supramolecular assembly is its ability to encapsulate two chloride anions, which are the guests of two host cavities formed by four pyridine rings of the ligand (Figure 6.18). The anion–π interactions observed here are favored by the fact that the pyridine rings involved are coordinated to copper ions, which enhances their electron-poor character.

After the publication of these experimental works, where relevant anion–π interactions are observed mainly in the solid state, Maeda *et al.* [121, 122] reported two manuscripts devoted to the study of anion receptors based on tetraperfluorophenyl-substituted N-confused porphyrin (**FNCP**) in solution. The magnitudes of the association constant (K_a) values of divalent metal complexes of **FNCP** (M(II)-**FNCP**) for respective anions in CH_2C1_2 are nearly equal and increase in the order $F^- > Cl^- > Br^- > I^-$ in each complex (Figure 6.19). The K_a values of the M(II)-**FNCP** complexes with F^- in CH_2C1_2 are $> 3 \times 10^5 M^{-1}$. The small

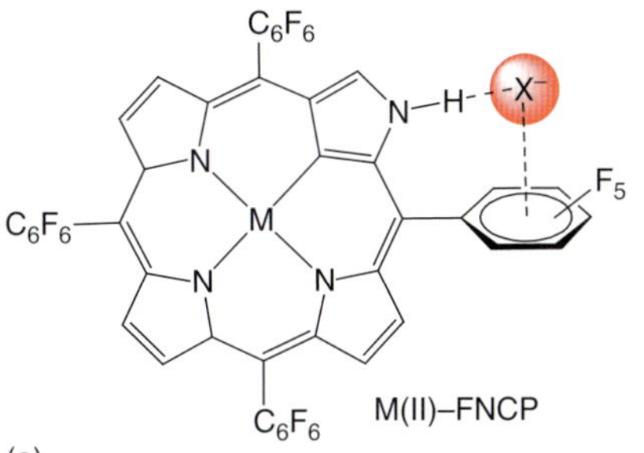

(a)

K_a (M^{-1})	Ni(II)	Pd(II)	Cu(II)
F^-	$> 3 \times 10^5$	$> 3 \times 10^5$	$> 3 \times 10^5$
Cl^-	5.7×10^4	4.6×10^4	4.9×10^4
Br^-	8.4×10^3	1.4×10^4	6.9×10^3
I^-	1.2×10^3	3.0×10^2	1.2×10^3

(b)

Figure 6.19 (a) Representation of the tetraperfluorophenyl-substituted N-confused porphyrin divalent metal complex (M(II)-**FNCP**) interacting with a halide. (b) The magnitudes of the K_a values of divalent metal complexes (M(II)-**FNCP**) for halide ions in CH_2Cl_2 are indicated in the table.

differences among the metal complexes have been attributed to the total charge included in the core cation, which affects the acidity of the outer NH through the confused pyrrole. Interestingly, the association constants of tetraphenyl instead of tetraperfluorophenyl-substituted N-confused porphyrin (**HNCP**) divalent metal complexes (M(II)-**HNCP**) with Cl^- are extremely small (<10 M in Ni(II)), indicating that the influence of the pentafluorophenyl ring is important. Partially, the anion-binding ability of M(II)-**FNCP** and M(II)-**HNCP** complexes is ascribable to the zwitterionic resonance form. Moreover, higher K_a values of M(II)-**FNCP** than of M(II)-**HNCP** are ascribable not only to the enhancement of acidity of the outer NH group resulting from electron-withdrawing effects but also to the additional anion–π interaction. Interaction of the neighboring $-C_6F_5$ group with anions was suggested by ^{19}F NMR experiments.

In 2004, Kochi [45] and collaborators experimentally characterized direct anion–π interactions in the most unambiguous way using a series of neutral organic π-acceptors with electron-deficient olefinic and aromatic centers (Figure 6.20). The recognition of halide anions X^- (X = Cl, Br, I) was established by both isolation and X-ray structure determination of a series of well-defined 1 : 1 salt-admixed complexes and definitive spectral assignments of each of their diagnostic charge-transfer absorption bands. Therefore, the formation constants of the halide complexes with neutral π-acceptors, together with the intense absorption and compression of the intermolecular separations found by X-ray structural analysis, were used to demonstrate the existence of substantial anion–π interactions. The authors proposed that charge-transfer effects are the origin of the seminal anion–π interactions. In addition, they propose that the formation of relatively strong complexes together with the distinctive colorations of various anion–π interactions encourage their use in the design of anion-sensing receptors.

An elegant and fascinating work was reported by Dunbar and collaborators in 2005 [123] in which they accomplished a comprehensive investigation of an anion-templated self-assembly reaction, namely, that of solvated first-row transition

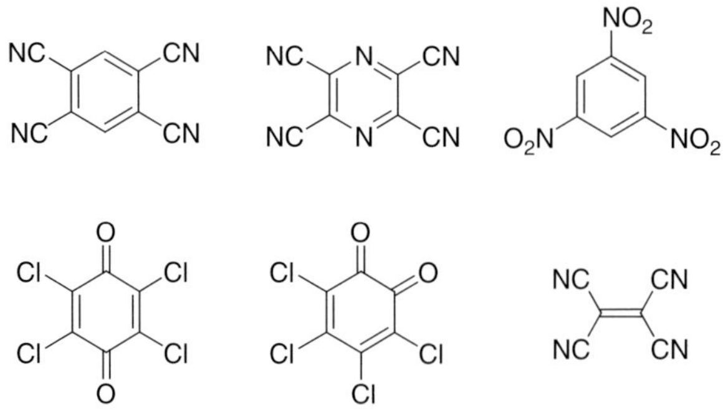

Figure 6.20 Representation of the electron-deficient aromatic centers used by Kochi and collaborators.

metal ions M(II) (M = Ni, Zn, Mn, Fe, Cu) with the divergent bis-bipyridine ligand 3,6-bis(2-pyridyl)-1,2,4,5-tetrazine (bptz; Figure 6.21). The formation of several molecular polygons in the presence of the appropriate anions is reported. They give evidence that the anions play a decisive role in the formation of a particular cyclic structure, both in the solid state and in solution. Considering the divergent nature of bptz and the absence of protective ligands around metal ions such as Ni(II) and Zn(II), the isolation of the discrete molecular squares and pentagons in high yields is quite remarkable. The formation of molecular squares is dominant in the presence of both $[BF_4]^-$ and $[ClO_4]^-$ ions, regardless of the presence of $[SbF_6]^-$ ions. The single-crystal X-ray structural determination of $[\{Ni_5(bptz)_5(CH_3CN)_{10}\} \subset SbF_6]^{9+}$ revealed that the shape and size of the $[SbF_6]^-$ anion render it appropriate to occupy a cavity slightly larger than that of the square, namely, that of a pentagon (Figure 6.21). Furthermore, ESI-MS studies in CH_3CN established that the $[Ni_5]^{10+}$ moiety persists in solution for long periods, with no evidence of a $[Ni_4]^{8+}$ species. Therefore, the anion $[SbF_6]^-$ leads to the exclusive formation of the molecular pentagon both in the solid state and in solution. The competing influence of the anions in stabilizing the different cyclic entities was also studied by mass spectrometry and X-ray crystallography. The $[Ni_5]^{10+}$ pentagon was found to be less stable compared to its $[Ni_4]^{8+}$. Ion signals corresponding to the molecular square begin to appear in the ESI-MS spectra after addition of $[n\text{-}Bu_4N][BF_4]$ or $[n\text{-}Bu_4N][ClO_4]$ to a solution of $[\{Ni_5(bptz)_5(CH_3CN)_{10}\} \subset SbF_6]^{9+}$, and complete conversion of the Ni(II) pentagon to the square is achieved by adding an excess of either tetrahedral anion (Figure 6.21). It can be concluded from this study that the nuclearity of the cyclic products is dictated by the identity of the anion present in solution during the self-assembly process. The latter can be attributed to a template effect that stabilizes one particular cyclic molecule over another because of favorable anion–π interactions between the anion inside the cavity and the bptz ligands.

In an attempt to probe the efficacy of the anion–π interaction to bind anions in solution, Berryman *et al.* [124] prepared two receptor molecules (Figure 6.22). The design of the receptor that incorporates the pentafluorophenyl ring was

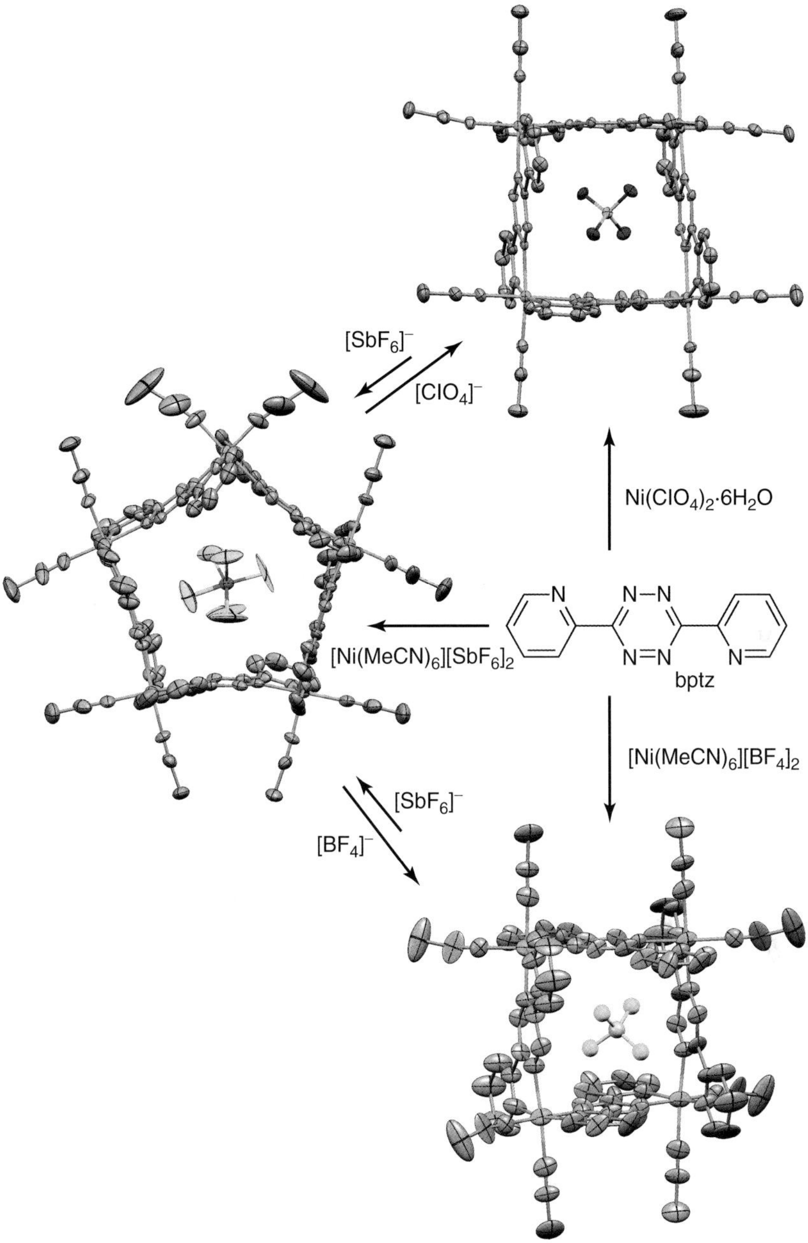

Figure 6.21 Representations of the cationic units $[\{Ni_5(bptz)_5(CH_3CN)_{10}\} \subset SbF_6]^{9+}$ (middle), $[\{Ni_4(bptz)_4(CH_3CN)_8\} \subset ClO4]^{7+}$ (top), and $[\{Ni_4(bptz)_4(CH_3CN)_8\} \subset BF_4]^{7+}$ (bottom) and their scheme of interconversion.

	K_a (M^{-1}) X = Cl	Br	I
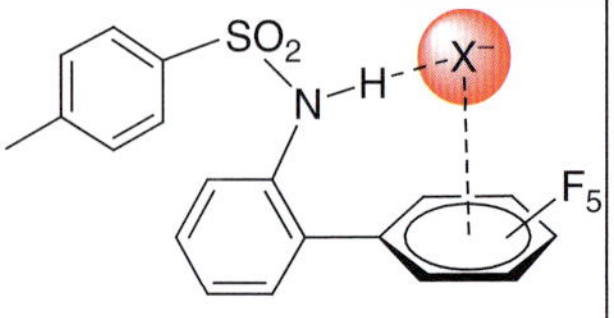	30	20	34
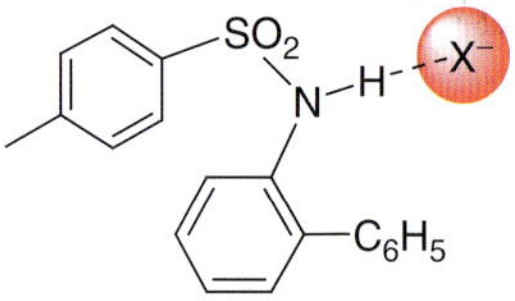	<1	<1	<1

Figure 6.22 Anion–receptor synthesized by Berryman *et al.* and their association constants in $CDCl_3$.

focused on a two-point recognition motif utilizing both a hydrogen bond and an electron-deficient aromatic ring. In contrast, the other receptor lacked the electron-deficient aromatic substituent required for the anion–π interaction. Any enhanced association for anions that the former receptor exhibits over the latter should be a result of the favorable anion–π interaction present in the anion complex. This was the first neutral receptor molecule designed to incorporate the anion–π interaction to bind anions in solution. ^{1}H NMR spectroscopic titration experiments were performed for each receptor with the tetra-*n*-butylammonium salts of chloride, bromide, and iodide in $CDCl_3$. The reported K_a is included in Figure 6.22, which depicts a stark contrast between the association constants of both receptors with a given halide. The receptor that incorporated the electron-deficient ring binds all the halides with a measurable, albeit modest, association constant. However, in the case of the receptor where an electron-deficient aromatic ring is not present, there is no measurable association with any of the halides. These association constants measured for both receptors provide strong support demonstrating the anion–π interaction in solution, highlighting the possibility of utilizing the anion–π interaction to bind anions by design.

Anion–π interactions have been found crucial as controlling elements in self-assembly reactions of Ag(I) complexes with π-acidic aromatic rings [110]. Using the bptz ligand, Dunbar's group has further investigated the role of anion–π interactions in Ag(I) complexes. Since silver is a suitably flexible transition metal, it can adopt a variety of coordination numbers and geometries, which makes it an attractive candidate for the study of "self-healing" thermodynamic systems. This work was the first example of a comprehensive investigation of anion–π interactions as controlling elements in self-assembly reactions. The reactions of bptz with the appropriate AgX salts (X = $[PF_6]^-$, $[AsF_6]^-$, $[SbF_6]^-$) afford complexes of three different structural types, depending on the experimental conditions and the anion used. Reactions of Ag(I) and bptz, in a 1 : 1 ratio, in the presence of $[PF_6]^-$ ions afford either a polymer or a discrete molecular compound, whereas

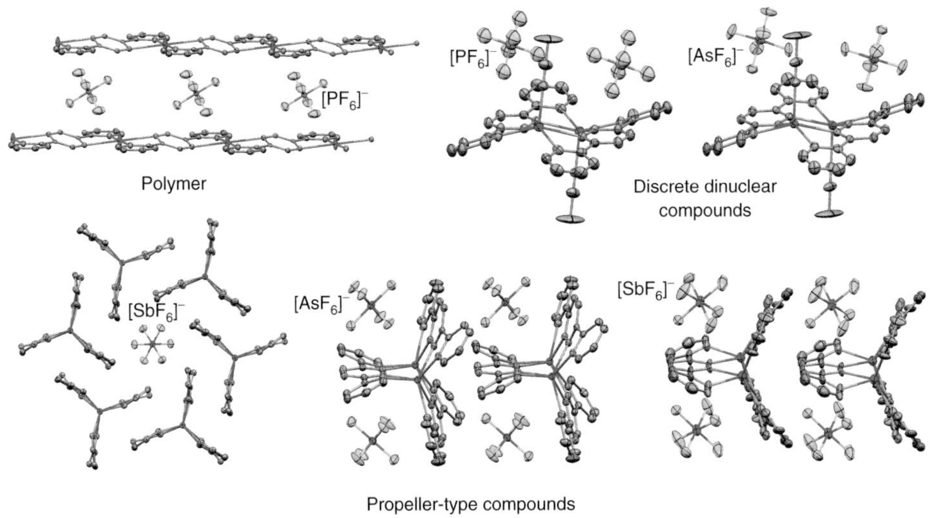

Figure 6.23 Thermal ellipsoid plots of fragments of crystal structures of several bptz complexes exhibiting different structural types.

in the presence of $[AsF_6]^-$ ions, the reaction produces only a discrete molecular compound (dinuclear product). When the ratio of Ag(I):bptz is 2 : 3, a propeller-type compound $[Ag_2\text{-(bptz)}_3][AsF_6]_2$ is formed in high yield in the presence of $[AsF_6]^-$ ions but not in the presence of $[PF_6]^-$ ions. At a $[AgPF_6]$:bptz ratio of 2 : 3, only the polymer or the dimer is obtained in low yield. Reactions of Ag(I):bptz in a 1 : 1 ratio in the presence of $[SbF_6]^-$ ions, however, yield the propeller-type compounds (Figure 6.23), as indicated by the single-crystal X-ray structural determinations. Apart from the anion size, which plays a role in the packing of the resulting structures, anion–π interactions are a dictating factor in determining the preferred structural motif. This conclusion was corroborated by the fact that reactions of a similar ligand (bppn), where the central tetrazine ring has been substituted by a pyridazine, with AgX salts (X = $[PF_6]^-$, $[AsF_6]^-$, $[SbF_6]^-$, $[BF_4]^-$) in a 1 : 1 ratio lead only to the grid-type structures, regardless of the anion present; see Ref. 110 for a more comprehensive treatment.

Another interesting example of self-assembly of Ag(I) coordination networks directed by anion–π interactions has been published by Zhou *et al.* [125]. In their study of Ag(I) metal complexes with 2,4,6-tri(2-pyridyl)-1,3,5-triazine (tpt), they found that multiatomic anions ($[ClO_4]^-$, $[BF_4]^-$, and $[PF_6]^-$) determine the self-assembly of Ag-tpt coordination polymers through anion–π interactions. It has been proved that factors such as reaction temperature, ratio of reactants, counterions, and solvents influence the self-assembly of resulting supramolecular compounds [110, 126, 127]. In this work, the authors provide systematically synthetic evidence for the determination of anion–π interactions on directing the self-assembly of metal complexes. Reactions of AgX with tpt yielded three 3D highly

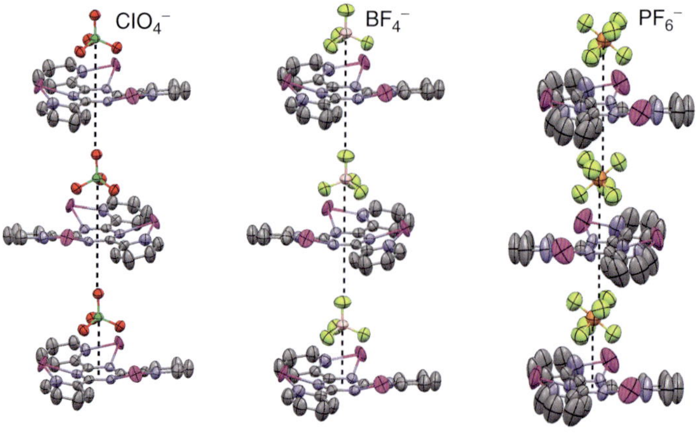

Figure 6.24 Thermal ellipsoid plots of fragments of the isostructural coordination polymers reported by Zhou *et al.* [125].

symmetric isostructural coordination polymers ($X^- = [ClO_4]^-$, $[BF_4]^-$, and $[PF_6]^-$). In the coordination polymers, anion–π interactions were found between the multiatomic anions and tpt ligands. Systematic variation including synthetic methods, ratios of reactants, and solvents gives synthetic evidence proving that anion–π interactions play a decisive role in the assembly of the coordination polymers. Single-crystal X-ray measurements revealed that all complexes are isostructural and crystallize in cubic space groups. Notably, all the multiatomic anions in the structures reside above the central triazine rings of the tpt ligands and are perfectly located on the C_3-axis above the rings (Figure 6.24).

Structurally directing isomorphous (4, 4) nets formed from Ag(I) salts and bis(4-pyrimidylmethyl)sulfide have been elegantly used to encapsulate anions by a uniform mode of anion–π binding. The strategy used by Black *et al.* [128] was to design a flexible multimodal ligand, bis(4-pyrimidylmethyl)sulfide, based on a 4-substituted pyrimidine moiety. This moiety is able to provide both suitably strong N-donors and sufficiently π-acidic ring centers. Furthermore, the particular arrangement of N-donors was considered conducive to the formation of more open coordination polymer networks. The reaction of the ligand with AgX ($X^- = [BF_4]^-$, $[ClO_4]^-$, and $[PF_6]^-$) in a 1 : 1 molar ratio afforded three isomorphous complexes with two-dimensional sheet structures characterized by a (4, 4) topology. The open nature of the two-dimensional structures and the stabilization provided by anion–π interactions allowed the anions to be situated in cavities formed within the sheets rather than being located between adjacent sheets. The cavities were bounded by pyrimidine rings, and the embedded anions were all held in place by four complementary π–anion–π sandwich interactions with two pyrimidine rings (Figure 6.25). Despite containing anions of different volume, the three structures are isomorphous. As stated by the authors [128], this invariance in structure indicated that either the sheets were robust enough to accommodate the guest

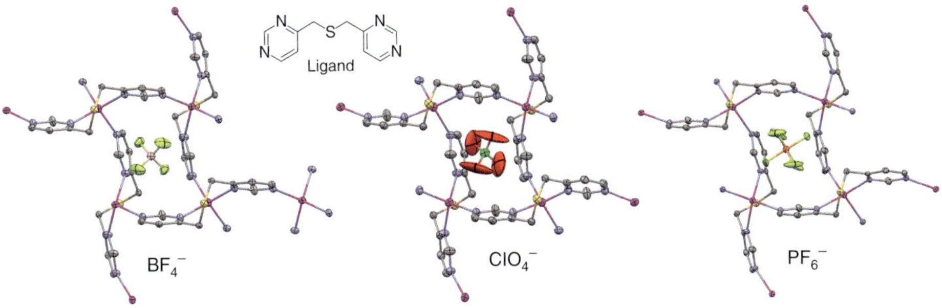

Figure 6.25 Thermal ellipsoid plots of fragments of the isomorphous structures reported by Black *et al.* [128].

anions or that the anion–π interactions were strong enough to engender the same overall structural arrangement. If space filling was the major determinant, different overall network structures might be expected, as such Ag–heterocyclic systems are especially difficult to direct. Therefore, it appeared in this case that the anion–π interactions were structurally directing. This important investigation provided further experimental evidence for the usefulness of diazines in the design of anion receptors by demonstrating the ability of pyrimidine to interact with anions through multiple anion–π interactions. In addition, both the structural consistency displayed by these networks and the uniform mode of anion binding demonstrate the potential for the use of anion–π interactions in a structurally directing role.

An interesting work by Mascal *et al.* [48] that validated a previous theoretically proposed host [129] for the encapsulation of fluoride was published in 2007. In the theoretical treatment of the receptor (Figure 6.26), the complexation of fluoride ion was shown to benefit from intrinsically stronger π–X–π and $^{+}NH\cdots X^{-}$ interactions than for chloride, in addition to acute steric issues associated with accommodating the larger chloride anion in the cavity. Subsequently, the first practical application of anion–π bonding in a purpose-designed macrocyclic anion host was published. The cyanuric-acid-based cylindrophane was synthesized and the triprotonated receptor forms an inclusion complex with fluoride by a combination of anion–π interactions and ion-pair-reinforced hydrogen bonding. The receptor showed no affinity for chloride in the electrospray mass spectrum and indeed was predicted by comparative binding energetics in models to be completely fluoride selective. Although a number of halide complexing agents have been described [130], this receptor introduces a new genre of anion binding, wherein anion–π interactions operate alongside with conventional ion pairing, hydrogen bonding, and the classic "preorganization" effect. The X-ray crystal structure shows the fluoride ion occupying the center of the cavity, in very close agreement with the previously published theory.

Berryman *et al.* [131] have reported very interesting experimental and theoretical results on a series of neutral tripodal receptors that utilize only electron-deficient arenes to bind halides in solution (Figure 6.27). These receptors employ steric

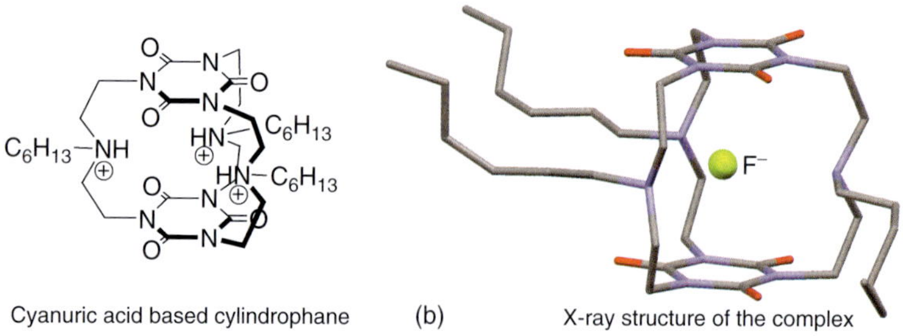

Figure 6.26 (a) Representation of the theoretically proposed host for fluoride. (b) Experimental X-ray structure of the complex.

gearing to preorganize electron-deficient arenes. This study represents the first designed receptor to quantitatively measure weak σ contacts between anions and arenes utilizing only electron-deficient aromatic rings. The ^{1}H NMR spectroscopy was used to measure the magnitude of anion-binding constant (K_a) and to determine the structure (anion–π contacts vs hydrogen bonding). It should be emphasized that this work represents the first observation of receptors binding anions in solution using only electron-deficient aromatic rings and it provides the first quantitative data that allows the comparison of the relative stabilities of weak σ anion–π or C–H$\cdots$X$^-$ hydrogen bonding interactions in solution. Some representative receptors used by Berryman *et al.* [131] are shown in Figure 6.27. They are 2,4,6-trisubstituted 1,3,5-triethylbenzene derivatives, which provide access for monatomic anions to interact with the electron-deficient dinitroarenes via the π-system or hydrogen bonding (Figure 6.27a,b). These receptors are structural isomers composed of two electron-deficient arenes differing only in the position of their nitro substituents, which grants understanding of the effect of substitution pattern on receptor function. The key feature used by the authors in their design strategy is that the middle receptor of Figure 6.27 cannot form hydrogen bonds to anions (due to the bulky nitro groups being positioned ortho to each acidic aryl hydrogen), allowing the study of the interaction between the anion and the π-system. The last receptor (Figure 6.27c) lacks electron-deficient arenes, and was used as a control. The association constants determined for the receptors based on electron-deficient arenes measured 11–53 M^{-1}, while control receptor distinctly lacking electron-deficient arenes exhibited no measurable binding by NMR (Figure 6.27). These results support the hypothesis that electron-deficient aromatic rings are required to bind anions in this neutral system. Additional solution studies yielded important results that further support the conclusions. Significantly larger chemical shift changes were observed for the left receptor over those of the middle receptor, again consistent with the fact that the former can form aryl CH$\cdot\cdot$X$^-$H-bonds while the latter cannot. With two NO_2 and one ester substituent, these highly electron-deficient arenes adopt binding motifs of "weak σ" anion–π contact and aryl H-bonding. The isomeric receptors exhibit the strongest interactions with Cl$^-$ followed by Br$^-$ and I$^-$, and larger association constants are

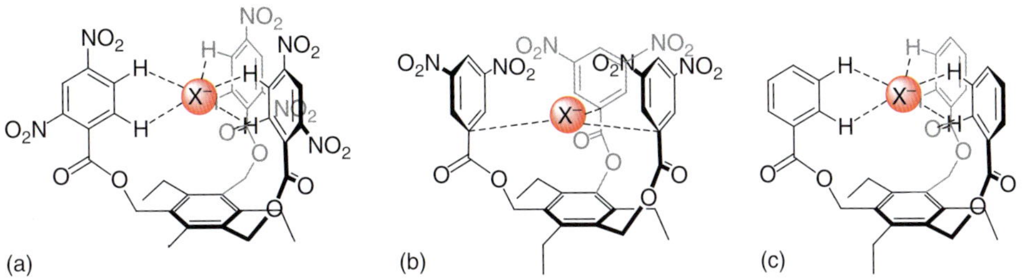

K_a (M^{-1}, C_6D_6)			
Cl^-	26	53	<1
Br^-	18	35	<1
I^-	11	26	<1

(d)

Figure 6.27 (a–c) Representation of the receptors used by Berryman *et al.* interacting with a halide. (d) The magnitudes of the K_a values for halide ions in benzene are indicated in the table.

observed when the halide is restricted to interact solely through contacts to the π-system (Figure 6.27b).

A related work devoted to the study of the anion–π interaction on neutral receptors attempting to measure the energy of the interaction has been published by Ballester and collaborators [18]. They have shown that a series of mesotetraaryl calix[4]pyrrole receptors can be used as model systems to quantify chloride–π interactions in solution. By means of ^{1}H NMR spectroscopy and X-ray crystallography they demonstrated that the chloride–arene interactions observed in these complexes are established exclusively with the π-aromatic system. The quantitative Hammett free-energy relationship derived was used to demonstrate that the detected chloride–π interactions are dominated by electrostatic effects. The observation, in the solid state and in solution, of chloride ions placed next to aromatic surfaces does not relate directly to the existence of attractive interactions between them. The values of the free energies estimated for the interaction of chloride with the π-aromatic systems were modest (about 1.1 kcal·mol^{-1} per aromatic ring) and they are probably not transferable to other model systems.

Han *et al.* [132] have extensively and comprehensively studied the facile halide/π-arene associations to afford supramolecular structures. They raise an interesting question of cocrystal growth between highly disparate partners, such as an ionic salt and an oleaginous arene, in which the usual tendency is for structurally distinctive partners to segregate themselves into separate crystalline stacks. Therefore, they fully examined the nature and scope of such a cocrystallization problem. In their study they examine the behavior of a series of planar π-acids of different shapes and sizes toward various types of polyatomic anions (e.g., thiocyanate, nitrite/nitrate, sulfite/sulfate, and tetrahalometallates) that are considerably more complex than the monatomic halides with regard to size and geometric shape as well as the number of negative atomic sites (hapticity). Four prototypical π-acids (cyano- and nitrosubstituted pyrazine and benzene, as well as tetracyanoethylene) and four classes of anions were used to probe the effect of quite distinctive hapticities (η) and thus to allow the possibility of different multisite interactions with the π-acids. From the viewpoint of supramolecular assembly, this two-component system consisting of the anion (D^-) salt and the π-acid (A) is relatively straightforward to the extent that anion/solution will simply lead to separate crystals of the (D^-) salt and of (A) or cocrystals of (D^-, A) which are readily distinguished by either the superimposed electronic absorption spectra of the local bands of the (D^-) salt plus that of (A) or the hybrid CT band of (D^-, A). Since the absorption band of the latter commonly occurs in the visible spectral region, the formation of cocrystals should be visually apparent by the immediate color change. The most important aspect of this study is the X-ray crystallographic analysis to elucidate the intimate (bonding) nature of the anion–π interaction leading to the usually colored cocrystals. The self-assembly of the anions with the prototypical π-acids occurred rapidly and selectively to yield a series of novel one-dimensional structures. The wirelike molecular chains all consist of parallel stacks of π-acids and alternate anions of different sizes and shapes that define these unique structures. Analogy of such linear arrays to nanoscopic wires is reinforced by counterions that are completely arrayed around the linear cores. The critical feature of these self-assemblies is shown to derive from anion–π recognitions via charge-transfer forces between the electron-rich anions acting as electron donors and the electron-poor π-acids acting as electron acceptors to spontaneously generate synthons according to Mulliken theory [133, 134].

Heteroatom-bridged heteroaromatic calixarenes are an emerging type of novel macrocyclic molecules and they have been recently utilized as versatile host molecules in supramolecular chemistry [135, 136]. Because of their electronic nature, nitrogen atoms can adopt different configurations to form varying degrees of conjugation with adjacent heteroaromatic rings, resulting in macrocyclic heteroatom-linked heteroaromatic calixarenes with conformations and sizes different from those of conventional calixarenes. A typical example of heteroatom-bridged heteroaromatic calixarene is tetraoxacalix[2]arene[2]triazine (Figure 6.28), which adopts a preorganized 1,3-alternate conformation, yielding a cleft formed by two π-electron deficient triazine rings. Recently, Wang *et al.* [137] envisioned that this π-electron-deficient cavity would act as a receptor to interact with anions through anion–π interactions. They have reported halide recognition by tetraoxacalix[2]arene[2]triazine host molecules, and they have shown considerable substituent effects of the triazine on the halide–π interaction. Most astonishingly, X-ray crystallography revealed the concurrent formation of noncovalent halide–π and lone-pair–π electron interactions between water, chloride (or bromide), and the dichloro-substituted host (Figure 6.28). The interaction of halides with the host was examined by means of spectrophotometric measurements. They studied the effect of the triazine substituent of the host on the anion–π interaction, the macrocyclic compounds containing bis(*N*,*N*-dimethylamino) groups showed no

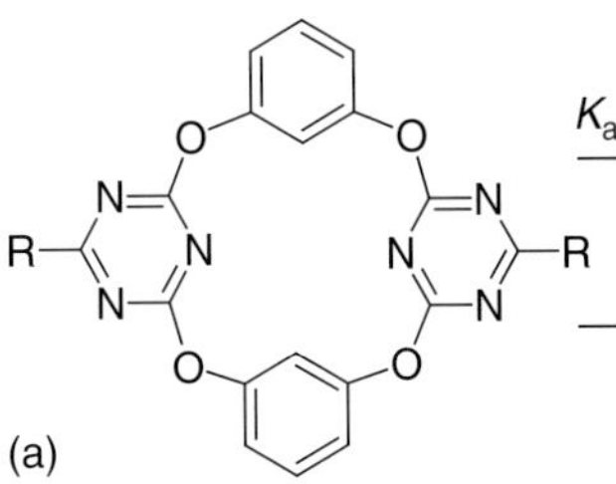

K_a (M^{-1}, MeCN)	R = Cl	H	NMe_2
F^-	4036	68	—
Cl^-	4246	–	—
Br^-	–	–	—

(a)

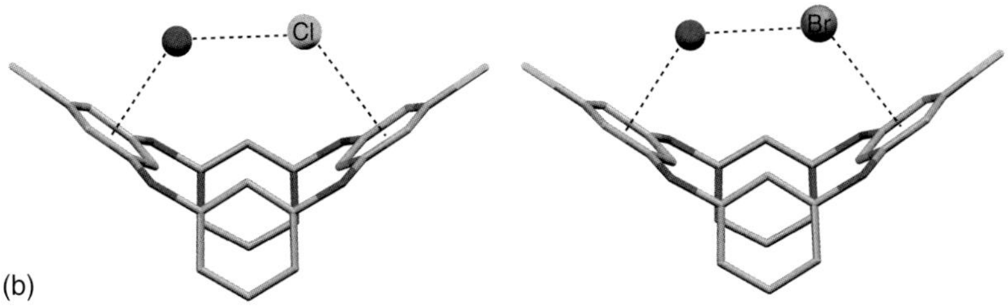

(b)

Figure 6.28 (a) Representation of the hosts used by Wang *et al.* The magnitudes of the K_a values for halide ions in acetonitrile are indicated in the table. (b) Fragments of the X-ray structures of the complexes of the host (R = Cl) with chloride and bromide establishing relevant anion–π interactions.

change in either the absorption or emission spectrum when titrated with fluoride, chloride, or bromide. In addition, whereas both the absorption and emission spectra of the unsubstituted calixarene host remained unchanged when it was treated with chloride and bromide, the chlorine-substituted macrocyclic host molecule was able to form a 1 : 1 complex with fluoride and chloride, with binding constants of $4036 \pm 36\ M^{-1}$ and $4246 \pm 83\ M^{-1}$, respectively. They represent the strongest halide–π interactions in solution reported to date. The results shown in Figure 6.28 clearly indicate a considerable effect of the substituent on the π-deficient triazine ring on the halide–π interaction. The dichloro-substituted host molecule exhibited much stronger binding affinity toward the same anion than the unsubstituted host molecule which contains no electron-withdrawing substituents on either triazine ring. This effect was attributed by the authors to the electron-withdrawing nature of the chloro substituent, which renders the triazine ring more electron deficient. Conversely, the calixarene containing bis(*N*,*N*-dimethylamino) groups did not act as the π-deficient host to interact with halide species in solution, probably as a result of the electron-donating nature of the amino substituent increasing the electron density of the triazine ring. To further clarify the halide–π interaction on the molecular level, the authors obtained single crystals of the Cl^- and Br^- complexes (Figure 6.28), which reveals the formation of very similar ternary complexes incorporating the host, a halide ion, and a water molecule. Some interesting structural features are that in both complexes, the calixarene moiety adopts a 1,3-alternate conformation, with the two benzene rings being nearly face-to-face, whereas the two π-deficient triazine rings form a V-shaped cleft. In addition, both chloride and bromide in complexes form typical noncovalent anion–π interactions with the triazine rings. In both cases, no arene hydrogen bonding occurred between the calixarene host and the halide guest. Furthermore, both host–halide complexes co-crystallized with water molecules, and one water molecule was found to form a ternary complex with halide and host (Figure 6.28), as indicated by the short distances between halogen atom and oxygen. Moreover, the hydrogen-bonded water molecule in both cases forms an intriguing lone-pair–π interaction between water oxygen atom and the triazine ring, as evidenced by the location of the water molecule virtually above the triazine centroid with a very short distance of 2.833–2.849 Å. Such a short distance excluded the other possible water–arene interaction model, namely an O–H ··· π interaction.

Finally, the extensive work of Matile's group on anion transport deserves a special mention [53–57]. They have used anion–π interactions for selective and efficient anion transport across lipid bilayer membranes. In biology, the selectivity of ion transport is of vital importance. In natural cation channels, selectivity often originates from the ion coordination to preorganized arrays of oxygen lone pairs, whereas cation–π interactions are much less important than expected [138]. In biological anion channels, hydrogen bonding, ion pairing, and anion–dipole interactions contribute to selectivity. Hitherto, anion–π interactions are not known in biological anion channels. The question if anion–π interactions could be used to create significant function such as transmembrane transport has been to some extent answered by Matile's group. To achieve anion transport, anion recognition by means of anion–π interactions must be combined with anion translocation. This combination is not easy, because tight binding of ions tends to decelerate rather than accelerate translocation. In biological anion and cation channels, the alignment of several ion-binding sites one after another is a good solution that combines selectivity with speed [139, 140]. The cooperative transport of ions along such strings of partially occupied ion-binding sites is referred to as *multi-ion hopping* [53]. To achieve significant transmembrane transport with anion–π interactions, it is therefore essential to align multiple π-acidic aromatic binding sites in series. Anions could then move fast and selectively along these anion–π "slides." Inspired in a complementary approach, which was explored almost a decade ago to achieve transmembrane potassium transport with cation–π interactions, Matile's group have used π-acidic, rigid-rod oligonaphthalenediimides (O-NDIs) as anion–π slides (Figure 6.29). These shape-persistent O-NDI rods have been introduced as anion–π slides for chloride-selective multi-ion hopping across lipid bilayers. Results from end-group engineering and covalent capture as O-NDI hairpins have suggested that self-assembly into transmembrane O-NDI bundles is essential for activity. The reported halide topology implies strong anion binding along the anion–π slides with relatively weak contributions from size exclusion. Anomalous mole fraction effects supported the occurrence of multi-ion hopping along the π-acidic O-NDI rods. The existence of anion–π interactions was corroborated by high-level *ab initio* and density functional theory (DFT) calculations.

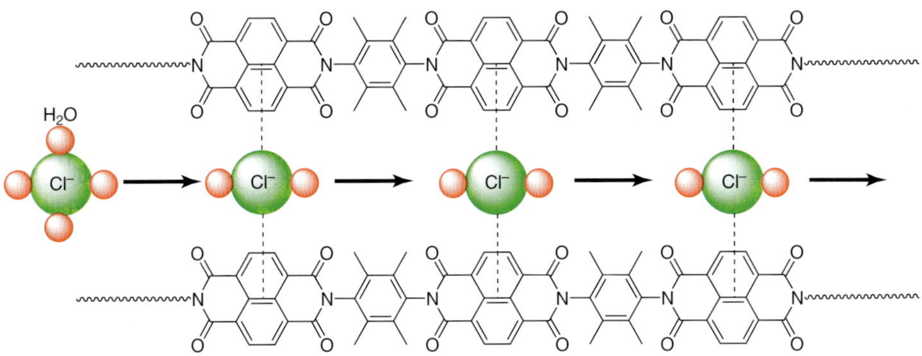

Figure 6.29 Schematic representation of the anion–π "slide" based on O-NDIs used by Matile's group.

6.6 Concluding Remarks

In this chapter, we have described the interaction of anions with aromatic rings from several points of view, that is, energetic and geometric features, aromatic aspects, charge-transfer effects, and orbitalic analysis. We have described the main forces involved in the anion–π interactions, which are electrostatic and polarization. The electrostatic force is explained as an anion–quadrupole interaction where the arene uses its permanent quadrupole moment. The polarization force is described as an anion–dipole interaction where the dipole is generated by the π-cloud polarization induced by the anion. The comprehension of these forces allows a rationalization of the requirements that should present both the anion and the aromatic compound to improve the interaction. Some physical properties of the aromatic rings such as Q_{zz} and $\alpha_{||}$ are directly related to the strength of the interaction. This knowledge also allows predicting the behavior of some special rings. For instance, the duality of arenes with negligible quadrupole moment or the unexpected anion-binding affinity of electron-rich molecules with greatly extended π-systems (large polarizability).

An attractive part of this chapter is the study of the interplay of the anion–π interactions with other noncovalent forces such as cation–π, hydrogen bonding, π–π stacking interactions. For instance, the strength of the anion–π interaction is considerably influenced by the presence of hydrogen bonding. Either a weakening or strengthening of the anion–π interaction is observed depending on the donor/acceptor hydrogen bonding behavior of the arene. Regarding the interplay between the anion–π and π–π stacking interactions, it is interesting to remark that the anion–π interaction is reinforced in the complexes of the hexafluorobenzene dimer with anions with respect to the complexes of hexafluorobenzene alone. Curiously, in the benzene-hexafluorobenzene dimer complexes with anions, the anion–π interaction is reinforced when the anion interacts with the dimer via the electron-rich benzene and it is weakened when it interacts with the electron-deficient hexafluorobenzene.

Finally, selected examples retrieved from the literature clearly establish the importance of this nascent noncovalent interaction in the solid state, in solution and in the gas phase. The final challenge would be to transfer the knowledge gained from those investigations to the utilization of attractive anion–π interactions in the design of advanced anion receptors for different purposes, including separation, sensing, or catalysis.

References

1. Schneider, H.J. (2009) Binding mechanisms in supramolecular complexes. *Angew. Chem. Int. Ed. Engl.*, **48** (22), 3924–3977.
2. Lehn, J.M. (1995) *Supramolecular Chemistry. Concepts and Perspectives*, Wiley-VCH Verlag GmbH, Weinheim.
3. Vögtle, F. (1993) *Supramolecular Chemistry: An Introduction*, John Wiley & Sons, Inc., New York.
4. Beer, P.D., Gale, P.A., and Smith, D.K. (1999) *Supramolecular Chemistry*, Oxford University Press, Oxford.
5. Schneider, H.-J. and Yatsimirski, A. (2000) *Principles and Methods in Supramolecular Chemistry*, John Wiley & Sons, Ltd, Chichester.
6. Steed, J.W. and Atwood, J.L. (2000) *Supramolecular Chemistry*, John Wiley & Sons, Ltd, Chichester.
7. Inoue, Y., Gokel G.W. (eds) (1990) *Cation Binding by Macrocycles*, Marcel Dekker New York.
8. Bianchi, A., Bowman-James, K., and García-España, E. (eds) (1997) *The Supramolcular Chemistry of Anions*, Wiley-VCH Verlag GmbH, Weinheim.
9. Oshovsky, G.V., Reinhoudt, D.N., and Verboom, W. (2007) Supramolecular chemistry in water. *Angew. Chem. Int. Ed.*, **46** (14), 2366–2393.
10. Kruppa, M., and Konig, B. (2006) Reversible coordinative bonds in molecular recognition. *Chem. Rev.*, **106** (9), 3520–3560.
11. Paulini, R., Muller, K., and Diederich, F. (2005) Orthogonal multipolar interactions in structural chemistry and biology. *Angew. Chem. Int. Ed.*, **44** (12), 1788–1805.
12. Meyer, E.A., Castellano, R.K., and Diederich, F. (2003) Interactions with aromatic rings in chemical and biological recognition. *Angew. Chem. Int. Ed.*, **42** (11), 1210–1250.
13. Saalfrank, R.W., Maid, H., and Scheurer, A. (2008) Supramolecular coordination chemistry: the synergistic effect of serendipity and rational design. *Angew. Chem. Int. Ed.*, **47** (46), 8794–8824.
14. Nassimbeni, L.R. (2003) Physicochemical aspects of host-guest compounds. *Acc. Chem. Res.*, **36** (8), 631–637.
15. Hunter, C.A. (2004) Quantifying intermolecular interactions: guidelines for the molecular recognition toolbox. *Angew. Chem. Int. Ed. Engl.*, **43** (40), 5310–5324.
16. Schneider, H.J. (1994) Linear free-energy relationships and pairwise interactions in supramolecular chemistry. *Chem. Soc. Rev.*, **23** (4), 227–234.
17. Schneider, H.J. and Yatsimirsky, A.K. (2008) Selectivity in supramolecular host-guest complexes. *Chem. Soc. Rev.*, **37** (2), 263–277.
18. Gohlke, H. and Klebe, G. (2002) Approaches to the description and prediction of the binding affinity of small-molecule ligands to macromolecular receptors. *Angew. Chem. Int. Ed.*, **41** (15), 2645–2676.
19. Arvizo, R.R., Verma, A., and Rotello, V.M. (2005) Biomacromolecule surface recognition using nanoparticle receptors. *Supramol. Chem.*, **17** (1–2), 155–161.

20. Williams, D.H., Stephens, E., O'Brien, D.P., and Zhou, M. (2004) Understanding noncovalent interactions: ligand binding energy and catalytic efficiency from ligand-induced reductions in motion within receptors and enzymes. *Angew. Chem. Int. Ed.*, **43** (48), 6596–6616.
21. Muller-Dethlefs, K. and Hobza, P. (2000) Noncovalent interactions: a challenge for experiment and theory. *Chem. Rev.*, **100** (1), 143–167.
22. Burley, S.K. and Petsko, G.A. (1985) Aromatic-aromatic interaction-a mechanism of protein-structure stabilization. *Science*, **229** (4708), 23–28.
23. Li, S., Cooper, V.R., Thonhauser, T., Lundqvist, B.I., and Langreth, D.C. (2009) Stacking interactions and DNA intercalation. *J. Phys. Chem. B*, **113** (32), 11166–11172.
24. Lovinger, A.J., Nuckolls, C., and Katz, T.J. (1998) Structure and morphology of helicene fibers. *J. Am. Chem. Soc.*, **120** (2), 264–268.
25. Larsen, M., Krebs, F.C., Jorgensen, M., and Harrit, N. (1998) Synthesis, structure, and fluorescence properties of 5,17-distyryl-25,26,27,28-tetrapropoxy-calix[4]arenes in the cone conformation. *J. Org. Chem.*, **63** (13), 4420–4424.
26. Anelli, P.L., Ashton, P.R., Spencer, N., Slawin, A.M.Z., Stoddart, J.F., and Williams, D.J. (1991) Self-assembling [2]pseudorotaxanes. *Angew. Chem. Int. Ed. Engl.*, **30** (8), 1036–1039.
27. Gatti, F.G., Leigh, D.A., Nepogodiev, S.A., Slawin, A.M.Z., Teat, S.J., and Wong, J.K.Y. (2001) Stiff, and sticky in the right places: the dramatic influence of preorganizing guest binding sites on the hydrogen bond-directed assembly of rotaxanes. *J. Am. Chem. Soc.*, **123** (25), 5983–5989.
28. Williams, D.H. and Bardsley, B. (1999) The vancomycin group of antibiotics and the fight against resistant bacteria. *Angew. Chem. Int. Ed. Engl.*, **38** (9), 1173–1193.
29. Anelli, P.L., Ashton, P.R., Ballardini, R., Balzani, V., Delgado, M., Gandolfi, M.T., Goodnow, T.T., Kaifer, A.E., Philp, D., Pietraszkiewicz, M., Prodi, L., Reddington, M.V., Slawin, A.M.Z., Spencer, N., Stoddart, J.F., Vicent, C., and Williams, D.J. (1992) Molecular Meccano 1. [2]Rotaxanes and a [2]catenane made to order. *J. Am. Chem. Soc.*, **114** (1), 193–218.
30. Gellman, S.H. (1998) Foldamers: a manifesto. *Acc. Chem. Res.*, **31** (4), 173–180.
31. Masu, H., Sakai, M., Kishikawa, K., Yamamoto, M., Yamaguchi, K., and Kohmoto, S. (2005) Aromatic foldamers with iminodicarbonyl linkers: their structures and optical properties. *J. Org. Chem.*, **70** (4), 1423–1431.
32. Han, J.J., Wang, W., and Li, A.D.Q. (2006) Folding and unfolding of chromophoric foldamers show unusual colorful single molecule spectral dynamics. *J. Am. Chem. Soc.*, **128** (3), 672–673.
33. Heemstra, J.M. and Moore, J.S. (2004) Helix stabilization through pyridinium–π interactions. *Chem. Commun.*, (13), 1480–1481.
34. Hamuro, Y., Geib, S.J., and Hamilton, A.D. (1997) Novel folding patterns in a family of oligoanthranilamides: non-peptide oligomers that form extended helical secondary structures. *J. Am. Chem. Soc.*, **119** (44), 10587–10593.
35. Hunter, C.A. and Sanders, J.K.M. (1990) The nature of π–π interactions. *J. Am. Chem. Soc.*, **112** (14), 5525–5534.
36. Ma, J.C. and Dougherty, D.A. (1997) The cation–π interaction. *Chem. Rev.*, **97** (5), 1303–1324.
37. Zacharias, N. and Dougherty, D.A. (2002) Cation–π interactions in ligand recognition and catalysis. *Trends Pharmacol. Sci.*, **23** (6), 281–287.
38. Scrutton, N.S. and Raine, A.R.C. (1996) Cation–π bonding and amino-aromatic interactions in the biomolecular recognition of substituted ammonium ligands. *Biochem. J.*, **319** (1), 1–8.
39. Gallivan, J.P. and Dougherty, D.A. (1999) Cation–π interactions in structural biology. *Proc. Natl. Acad. Sci. U.S.A.*, **96** (17), 9459–9464.
40. Mascal, M., Armstrong, A., and Bartberger, M.D. (2002) Anion-aromatic

bonding: a case for anion recognition by π-acidic rings. *J. Am. Chem. Soc.*, **124** (22), 6274–6276.

41. Alkorta, I., Rozas, I., and Elguero, J. (2002) Interaction of anions with perfluoro aromatic compounds. *J. Am. Chem. Soc.*, **124** (29), 8593–8598.
42. Quiñonero, D., Garau, C., Rotger, C., Frontera, A., Ballester, P., Costa, A., and Deyà, P.M. (2002) Anion–π interactions: do they exist? *Angew. Chem. Int. Ed.*, **41** (18), 3389–3392.
43. Demeshko, S., Dechert, S., and Meyer, F. (2004) Anion–π interactions in a carousel copper(II)-triazine complex. *J. Am. Chem. Soc.*, **126** (14), 4508–4509.
44. Schottel, B.L., Bacsa, J., and Dunbar, K.R. (2005) Anion dependence of Ag(I) reactions with 3,6-bis(2-pyridyl)-1,2,4,5-tetrazine (bptz): isolation of the molecular propeller compound $[Ag_2(bptz)_3][AsF_6]_2$. *Chem. Commun.*, (1), 46–47.
45. Rosokha, Y.S., Lindeman, S.V., Rosokha, S.V., and Kochi, J.K. (2004) Halide recognition through diagnostic "anion–π" interactions: molecular complexes of Cl^-, Br^-, and I^- with olefinic and aromatic π receptors. *Angew. Chem. Int. Ed.*, **43** (35), 4650–4652.
46. de Hoog, P., Gamez, P., Mutikainen, H., Turpeinen, U., and Reedijk, J. (2004) An aromatic anion receptor: anion–π interactions do exist. *Angew. Chem. Int. Ed.*, **43** (43), 5815–5817.
47. Estarellas, C., Rotger, M.C., Capó, M., Quiñonero, D., Frontera, A., Costa, A., and Deyà, P.M. (2009) Anion–π interactions in four-membered rings. *Org. Lett.*, **11** (9), 1987–1990.
48. Mascal, M., Yakovlev, I., Nikitin, E.B., and Fettinger, J.C. (2007) Fluoride-selective host based on anion–π interactions, ion pairing, and hydrogen bonding: synthesis and fluoride-ion sandwich complex. *Angew. Chem. Int. Ed.*, **46** (46), 8782–8784.
49. García-Raso, A., Albertí, F.M., Fiol, J.J., Tasada, A., Barceló-Oliver, M., Molins, E., Escudero, D., Frontera, A., Quiñonero, D., and Deyà, P.M. (2007) Anion–π interactions in bisadenine derivatives: a combined crystallographic and theoretical study. *Inorg. Chem.*, **46** (25), 10724–10735.
50. Domasevitch, K.V., Solntsev, P.V., Gural'skiy, I.A., Krautscheid, H., Rusanov, E.B., Chernega, A.N., and Howard, J.A. (2007) Silver(I) ions bridged by pyridazine: doubling the ligand functionality for the design of unusual 3D coordination frameworks. *Dalton Trans.*, (35), 3893–3905.
51. Gotz, R.J., Robertazzi, A., Mutikainen, I., Turpeinen, U., Gamez, P., and Reedijk, J. (2008) Concurrent anion···π interactions between a perchlorate ion and two π-acidic aromatic rings, namely pentafluorophenol and 1,3,5-triazine. *Chem. Commun.*, (29), 3384–3386.
52. Schottel, B.L., Chifotides, H.T., and Dunbar, K.R. (2008) Anion–π interactions. *Chem. Soc. Rev.*, **37** (1), 68–83.
53. Mareda, J. and Matile, S. (2009) Anion–π slides for transmembrane transport. *Chem. Eur. J.*, **15** (1), 28–37.
54. Gorteau, V., Bollot, G., Mareda, J., and Matile, S. (2007) Rigid-rod anion–π slides for multiion hopping across lipid bilayers. *Org. Biomol. Chem.*, **5** (18), 3000–3012.
55. Gorteau, V., Bollot, G., Mareda, J., Perez-Velasco, A., and Matile, S. (2006) Rigid oligonaphthalenediimide rods as transmembrane anion–π slides. *J. Am. Chem. Soc.*, **128** (46), 14788–14789.
56. Gorteau, V., Julliard, M.D., and Matile, S. (2008) Hydrophilic anchors for transmembrane anion–π slides. *J. Membr. Sci.*, **321** (1), 37–42.
57. Perez-Velasco, A., Gorteau, V., and Matile, S. (2008) Rigid oligoperylenediimide rods: anion–π slides with photosynthetic activity. *Angew. Chem. Int. Ed.*, **47** (5), 921–923.
58. Egli, M. and Sarkhel, S. (2007) Lone pair-aromatic interactions: to stabilize or not to stabilize. *Acc. Chem. Res.*, **40** (3), 197–205.
59. Gamez, P., Mooibroek, T.J., Teat, S.J., and Reedijk, J. (2007) Anion binding involving π-acidic heteroaromatic rings. *Acc. Chem. Res.*, **40** (6), 435–444.

60. Hay, B.P. and Bryantsev, V.S. (2008) Anion-arene adducts: C-H hydrogen bonding, anion-pi interaction, and carbon bonding motifs. *Chem. Commun.*, (21), 2417–2428.
61. Ballester, P. (2008) Recognition of anions. *Struct. Bond.*, **129**, 127.
62. Kim, D., Tarakeshwar, P., and Kim, K.S. (2004) Theoretical investigations of anion–π interactions: the role of anions and the nature of π systems. *J. Phys. Chem. A*, **108** (7), 1250–1258.
63. Kim, D., Lee, E.C., Kim, K.S., and Tarakeshwar, P. (2007) Cation–π–anion interaction: a theoretical investigation of the role of induction energies. *J. Phys. Chem. A*, **111** (32), 7980–7986.
64. Kim, D.Y., Singh, N.J., Lee, J.W., and Kim, K.S. (2008) Solvent-driven structural changes in anion–π complexes. *J. Chem. Theory Comput.*, **4** (7), 1162–1169.
65. Kim, D.Y., Singh, N.J., and Kim, K.S. (2008) Cyameluric acid as anion–π type receptor for ClO_4 and NO_3: π-stacked and edge-to-face structures. *J. Chem. Theory Comput.*, **4** (8), 1401–1407.
66. Garau, C., Frontera, A., Quiñonero, D., Ballester, P., Costa, A., and Deyà, P.M. (2003) A topological analysis of the electron density in anion–π interactions. *ChemPhysChem*, **4** (12), 1344–1348.
67. Quiñonero, D., Garau, C., Frontera, A., Ballester, P., Costa, A., and Deyà, P.M. (2002) Counterintuitive interaction of anions with benzene derivatives. *Chem. Phys. Lett.*, **359** (5-6), 486–492.
68. Garau, C., Quiñonero, D., Frontera, A., Ballester, P., Costa, A., and Deyà, P.M. (2003) Dual binding mode of s-triazine to anions and cations. *Org. Lett.*, **5** (13), 2227–2229.
69. Garau, C., Frontera, A., Quiñonero, D., Ballester, P., Costa, A., and Deyà, P.M. (2004) Cation–π versus anion–π interactions: energetic, charge transfer, and aromatic aspects. *J. Phys. Chem. A*, **108** (43), 9423–9427.
70. Frontera, A., Saczewski, F., Gdaniec, M., Dziemidowicz-Borys, E., Kurland, A., Deyà, P.M., Quiñonero, D., and Garau, C. (2005) Anion–π interactions in cyanuric acids: a combined crystallographic and computational study. *Chem. Eur. J.*, **11** (22), 6560–6567.
71. Garau, C., Frontera, A., Ballester, P., Quiñonero, D., Costa, A., and Deyà, P.M. (2005) A theoretical ab initio study of the capacity of several binding units for the molecular recognition of anions. *Eur. J. Org. Chem.*, (1), 179–183.
72. García-Raso, A., Albertí, F.M., Fiol, J.J., Tasada, A., Barceló-Oliver, M., Molins, E., Estarellas, C., Frontera, A., Quiñonero, D., and Deyà, P.M. (2009) 2-Aminopyrimidine derivatives exhibiting anion–π interactions: a combined crystallographic and theoretical study. *Cryst. Growth Des.*, **9** (5), 2363–2376.
73. Quiñonero, D., Frontera, A., Escudero, D., Ballester, P., Costa, A., and Deyà, P.M. (2007) A theoretical study of anion–π interactions in seven-membered rings. *ChemPhysChem*, **8** (8), 1182–1187.
74. Gural'skiy, I.A., Escudero, D., Frontera, A., Solntsev, P.V., Rusanov, E.B., Chernega, A.N., Krautscheid, H., and Domasevitch, K.V. (2009) 1,2,4,5-Tetrazine: an unprecedented μ^4-coordination that enhances ability for anion–π interactions. *Dalton Trans.*, (15), 2856–2864.
75. García-Raso, A., Albertí, F.M., Fiol, J.J., Tasada, A., Barceló-Oliver, M., Molins, E., Escudero, D., Frontera, A., Quiñonero, D., and Deyà, P.M. (2007) A combined experimental and theoretical study of anion–π interactions in bis(pyrimidine) salts. *Eur. J. Org. Chem.*, (35), 5821–5825.
76. Estarellas, C., Frontera, A., Quiñonero, D., and Deyà, P.M. (2008) Theoretical and crystallographic study of the dual σ/π anion binding affinity of quinolizinylium cation. *J. Chem. Theory Comput.*, **4** (11), 1981–1989.
77. Garau, C., Quiñonero, D., Frontera, A., Ballester, P., Costa, A., and Deyà, P.M. (2005) Approximate additivity of anion–π interactions: an ab initio study on anion–π anion–π_2 and anion–π_3 complexes. *J. Phys. Chem. A*, **109** (41), 9341–9345.

78. Bader, R.F.W. (1991) A quantum-theory of molecular-structure and its applications. *Chem. Rev.*, **91** (5), 893–928.
79. Cubero, E., Orozco, M., and Luque, F.J. (1999) A topological analysis of electron density in cation–π complexes. *J. Phys. Chem. A*, **103** (2), 315–321.
80. Garau, C., Frontera, A., Quiñonero, D., Ballester, P., Costa, A., and Deyà, P.M. (2004) Cation–π versus anion–π interactions: a comparative ab initio study based on energetic, electron charge density and aromatic features. *Chem. Phys. Lett.*, **392** (1-3), 85–89.
81. Garau, C., Frontera, A., Quiñonero, D., Ballester, P., Costa, A., and Deyà, P.M. (2004) Cation–π vs anion–π interactions: a complete π-orbital analysis. *Chem. Phys. Lett.*, **399** (1–3), 220–225.
82. Dougherty, D.A. (1996) Cation–π interactions in chemistry and biology: a new view of benzene, Phe, Tyr, and Trp. *Science*, **271** (5246), 163–168.
83. Rappe, A.K. and Bernstein, E.R. (2000) Ab initio calculation of nonbonded interactions: are we there yet? *J. Phys. Chem. A*, **104** (26), 6117–6128.
84. Hesselmann, A., Jansen, G., and Schutz, M. (2006) Interaction energy contributions of H-bonded and stacked structures of the AT and GC DNA base pairs from the combined density functional theory and intermolecular perturbation theory approach. *J. Am. Chem. Soc.*, **128** (36), 11730–11731.
85. Piacenza, M. and Grimme, S. (2005) Van der Waals interactions in aromatic systems: structure and energetics of dimers and trimers of pyridine. *ChemPhysChem*, **6** (8), 1554–1558.
86. Parthasarathi, R., Subramanian, V., and Sathyamurthy, N. (2005) Hydrogen bonding in phenol, water, and phenol-water clusters. *J. Phys. Chem. A*, **109** (5), 843–850.
87. Ludwig, R. (2001) Water: from clusters to the bulk. *Angew. Chem. Int Ed.*, **40** (10), 1808–1827.
88. Jorgensen, W.L. and Pranata, J. (1990) Importance of secondary interactions in triply hydrogen-bonded complexes guanine-cytosine vs uracil-2,6-diaminopyridine. *J. Am. Chem. Soc.*, **112** (5), 2008–2010.
89. Pranata, J., Wierschke, S.G., and Jorgensen, W.L. (1991) OPLS potential functions for nucleotide bases-relative association constants of hydrogen-bonded base-pairs in chloroform. *J. Am. Chem. Soc.*, **113** (8), 2810–2819.
90. Estarellas, C., Escudero, D., Frontera, A., Quiñonero, D., and Deyà, P.M. (2009) Theoretical ab initio study of the interplay between hydrogen bonding, cation–π and π–π interactions. *Theor. Chem. Acc.*, **122** (5–6), 325–332.
91. Alkorta, I., Blanco, F., Deyà, P.M., Elguero, J., Estarellas, C., Frontera, A., and Quiñonero, D. (2010) Cooperativity in multiple unusual weak bonds. *Theor. Chem. Acc.*, **126** (1–2), 1–14.
92. Garau, C., Quiñonero, D., Frontera, A., Ballester, P., Costa, A., and Deyà, P.M. (2003) Anion–π interactions: must the aromatic ring be electron deficient? *New J. Chem.*, **27** (2), 211–214.
93. Alkorta, I. and Elguero, J. (2003) Aromatic systems as charge insulators: their simultaneous interaction with anions and cations. *J. Phys. Chem. A*, **107** (44), 9428–9433.
94. Alkorta, I., Quiñonero, D., Garau, C., Frontera, A., Elguero, J., and Deyà, P.M. (2007) Dual cation and anion acceptor molecules. The case of the $(\eta^6\text{-}C_6H_6)(\eta^6\text{-}C_6F_6)Cr(0)$ complex. *J. Phys. Chem. A*, **111** (16), 3137–3142.
95. Quiñonero, D., Frontera, A., Deyà, P.M., Alkorta, I., and Elguero, J. (2008) Interaction of positively and negatively charged aromatic hydrocarbons with benzene and triphenylene: towards a model of pure organic insulators. *Chem. Phys. Lett.*, **460** (4–6), 406–410.
96. Steed, J.W., Juneja, R.K., and Atwood, J.L. (1995) A water-soluble bear trap exhibiting strong anion complexation properties. *Angew. Chem. Int. Ed.*, **33** (23–24), 2456–2457.
97. Staffilani, M., Hancock, K.S.B., Steed, J.W., Holman, K.T., Atwood, J.L., Juneja, R.K., and Burkhalter, R.S. (1997) Anion binding within the cavity of π-metalated calixarenes. *J. Am. Chem. Soc.*, **119** (27), 6324–6335.
98. Holman, K.T., Halihan, M.M., Jurisson, S.S., Atwood, J.L., Burkhalter,

R.S., Mitchell, A.R., and Steed, J.W. (1996) Inclusion of neutral and anionic guests within the cavity of π-metalated cyclotriveratrylenes. *J. Am. Chem. Soc.*, **118** (40), 9567–9576.

99. Staffilani, M., Bonvicini, G., Steed, J.W., Holman, K.T., Atwood, J.L., and Elsegood, M.R.J. (1998) Bowl vs saddle conformations in cyclononatriene-based anion binding hosts. *Organometallics*, **17** (9), 1732–1740.

100. Ngola, S.M., Kearney, P.C., Mecozzi, S., Russell, K., and Dougherty, D.A. (1999) A selective receptor for arginine derivatives in aqueous media. Energetic consequences of salt bridges that are highly exposed to water. *J. Am. Chem. Soc.*, **121** (6), 1192–1201.

101. Frontera, A., Quiñonero, D., Garau, C., Costa, A., Ballester, P., and Deyà, P.M. (2006) MP2 study of cation–π_n–π interactions ($n = 1$–4). *J. Phys. Chem. A*, **110** (30), 9307–9309.

102. Quiñonero, D., Frontera, A., Garau, C., Ballester, P., Costa, A., and Deyà, P.M. (2006) Interplay between cation–π, anion–π and π–π interactions. *ChemPhysChem*, **7** (12), 2487–2491.

103. Frontera, A., Quiñonero, D., Costa, A., Ballester, P., and Deyà, P.M. (2007) MP2 study of cooperative effects between cation–π, anion–π and π–π interactions. *New J. Chem.*, **31** (4), 556–560.

104. Blanchard, M.D., Hughes, R.P., Concolino, T.E., and Rheingold, A.L. (2000) π-Stacking between pentafluorophenyl and phenyl groups as a controlling feature of intra- and intermolecular crystal structure motifs in substituted ferrocenes. Observation of unexpected face-to-face stacking between pentafluorophenyl rings. *Chem. Mater.*, **12** (6), 1604–1610.

105. Deck, P.A. and Fronczek, F.R. (2000) Tricarbonylrhenium(I) complexes of pentafluorophenyl-substituted indenyl ligands. *Organometallics*, **19** (3), 327–333.

106. Deck, P.A., Lane, M.J., Montgomery, J.L., Slebodnick, C., and Fronczek, F.R. (2000) Synthesis and structural trends in pentafluorophenyl-substituted ferrocenes, 1,4-tetrafluorophenylene-linked diferrocenes, and 1,1'-ferrocenylene-1,4-tetrafluorophenylene co-oligomers. *Organometallics*, **19** (6), 1013–1024.

107. Thornberry, M., Slebodnick, C., Deck, P.A., and Fronczek, F.R. (2000) Structural and electronic effects of pentafluorophenyl substituents on cyclopentadienyl complexes of Fe, Co, Mn, and Re. *Organometallics*, **19** (25), 5352–5369.

108. Thornberry, M.P., Slebodnick, C., Deck, P.A., and Fronczek, F.R. (2001) Synthesis and structure of piano stool complexes derived from the tetrakis(pentafluorophenyl)cyclopentadienyl ligand. *Organometallics*, **20** (5), 920–926.

109. Deck, P.A., Kroll, C.E., Hollis, W.G., and Fronczek, F.R. (2001) Conformational control of intramolecular arene stacking in ferrocene complexes bearing *tert*-butyl and pentafluorophenyl substituents. *J. Organomet. Chem.*, **637**, 107–115.

110. Schottel, B.L., Chifotides, H.T., Shatruk, M., Chouai, A., Perez, L.M., Bacsa, J., and Dunbar, K.R. (2006) Anion–π interactions as controlling elements in self-assembly reactions of Ag(I) complexes with π-acidic aromatic rings. *J. Am. Chem. Soc.*, **128** (17), 5895–5912.

111. Barrios, L.A., Aromi, G., Frontera, A., Quiñonero, D., Deyà, P.M., Gamez, P., Roubeau, O., Shotton, E.J., and Teat, S.J. (2008) Coordination complexes exhibiting anion$\cdots\pi$ interactions: synthesis, structure, and theoretical studies. *Inorg. Chem.*, **47** (13), 5873–5881.

112. Escudero, D., Frontera, A., Quiñonero, D., and Deyà, P.M. (2009) Interplay between anion–π and hydrogen bonding interactions. *J. Comput. Chem.*, **30** (1), 75–82.

113. Alkorta, I., Blanco, F., Elguero, J., Estarellas, C., Frontera, A., Quiñonero, D., and Deyà, P.M. (2009) Simultaneous interaction of tetrafluoroethene with anions and hydrogen-bond donors: a cooperativity study. *J. Chem. Theory Comput.*, **5** (4), 1186–1194.

114. Lucas, X., Estarellas, C., Escudero, D., Frontera, A., Quiñonero, D., and

Deyà, P.M. (2009) Very long-range effects: cooperativity between anion–π and hydrogen-bonding interactions. *ChemPhysChem*, **10** (13), 2256–2264.
115. Quiñonero, D., Frontera, A., and Deyá, P.M. (2008) High-level ab initio study of anion–π interactions in pyridine and pyrazine rings coordinated to Ag1. *ChemPhysChem*, **9** (3), 397–399.
116. Nangia, A., Biradha, K., and Desiraju, G.R. (1996) Correlation of biological activity in beta-lactam antibiotics with Woodward and Cohen structural parameters-A Cambridge database study. *J. Chem. Soc. Perkin Trans. 2*, (5), 943–953.
117. Desiraju, G.R. (1989) *Crystal Engineering. The Design of Organic Solids*, Elsevier, Amsterdam.
118. Allen, F.H. (2002) The Cambridge structural database: a quarter of a million crystal structures and rising. *Acta Crystallogr.*, **B58** (1), 380–388.
119. Hay, B.P. and Custelcean, R. (2009) Anion–π interactions in crystal structures: commonplace or extraordinary? *Cryst. Growth Des.*, **9** (6), 2539–2545.
120. Frontera, A., Gamez, P., Mascal, M., Mooibroek, T.J., and Reedijk J. (2011) Putting Anion–π interactions into Perspective Angew. *Chem. Int. Ed.*, (Accepted).
121. Maeda, H. and Furuta, H. (2004) N-confused porphyrins as new scaffolds for supramolecular architecture. *J. Porphyrins Phthalocyanines*, **8** (1–3), 67–75.
122. Maeda, H., Osuka, A., and Furuta, H. (2004) Anion binding properties of N-confused porphyrins at the peripheral nitrogen. *J. Inclusion Phenom. Macrocyclic Chem.*, **49** (1–2), 33–36.
123. Campos-Fernandez, C.S., Schottel, B.L., Chifotides, H.T., Bera, J.K., Bacsa, J., Koomen, J.M., Russell, D.H., and Dunbar, K.R. (2005) Anion template effect on the self-assembly and interconversion of metallacyclophanes. *J. Am. Chem. Soc.*, **127** (37), 12909–12923.
124. Berryman, O.B., Hof, F., Hynes, M.J., and Johnson, D.W. (2006) Anion-pi interaction augments halide binding in solution. *Chem. Commun.*, (5), 506–508.
125. Zhou, X.P., Zhang, X.J., Lin, S.H., and Li, D. (2007) Anion–π-interaction-directed self-assembly of Ag(I) coordination networks. *Cryst. Growth Des.*, **7** (3), 485–487.
126. Beer, P.D. and Gale, P.A. (2001) Anion recognition and sensing: the state of the art and future perspectives. *Angew. Chem. Int. Ed.*, **40** (3), 486–516.
127. Vilar, R. (2003) Anion-templated synthesis. *Angew. Chem. Int. Ed.*, **42** (13), 1460–1477.
128. Black, C.A., Hanton, L.R., and Spicer, M.D. (2007) A coordination polymer strategy for anion encapsulation: anion–π interactions in (4,4) nets formed from Ag(I) salts and a flexible pyrimidine ligand. *Chem. Commun.*, (30), 3171–3173.
129. Mascal, M. (2006) Precedent and theory unite in the hypothesis of a highly selective fluoride receptor. *Angew. Chem. Int. Ed.*, **45** (18), 2890–2893.
130. Bowman-James, K. (2005) Alfred Werner revisited: the coordination chemistry of anions. *Acc. Chem. Res.*, **38** (8), 671–678.
131. Berryman, O.B., Sather, A.C., Hay, B.P., Meisner, J.S., and Johnson, D.W. (2008) Solution phase measurement of both weak sigma and C-H···X^- hydrogen bonding interactions in synthetic anion receptors. *J. Am. Chem. Soc.*, **130** (33), 10895–10897.
132. Han, B., Lu, J.J., and Kochi, J.K. (2008) Anion recognitions via cocrystallizations with organic π-acids in the efficient self-assembly of nanoscopic one-dimensional molecular chains (wires). *Cryst. Growth Des.*, **8** (4), 1327–1334.
133. Mulliken, R.S. (1952) Molecular compounds and their spectra 2. *J. Am. Chem. Soc.*, **74** (3), 811–824.
134. Mulliken, R.S. and Person, W.B. (1969) *Molecular Complexes*, John Wiley & Sons, Inc., New York.
135. Wang, M.X., Zhang, X.H., and Zheng, Q.Y. (2004) Synthesis, structure, and [60]fullerene complexation properties

of azacalix[*m*]arene[*n*]pyridines. *Angew. Chem. Int. Ed.*, **43** (7), 838–842.

136. Gong, H.Y., Zheng, Q.Y., Zhang, X.H., Wang, D.X., and Wang, M.X. (2006) Methylazacalix[4]pyridine: En route to Zn^{2+}-specific fluorescence sensors. *Org. Lett.*, **8** (21), 4895–4898.

137. Wang, D.X., Zheng, Q.Y., Wang, Q.Q., and Wang, M.X. (2008) Halide recognition by tetraoxacalix[2]arene[2]triazine receptors: concurrent noncovalent halide–π and lone-pair–π interactions in host-halide-water ternary complexes. *Angew. Chem. Int. Ed.*, **47** (39), 7485–7488.

138. Kumpf, R.A. and Dougherty, D.A. (1993) A mechanism for ion selectivity in potassium channels-computational studies of cation–π interactions. *Science*, **261** (5129), 1708–1710.

139. Doyle, D.A., Cabral, J.M., Pfuetzner, R.A., Kuo, A.L., Gulbis, J.M., Cohen, S.L., Chait, B.T., and MacKinnon, R. (1998) The structure of the potassium channel: molecular basis of K^+ conduction and selectivity. *Science*, **280** (5360), 69–77.

140. Dutzler, R., Campbell, E.B., and MacKinnon, R. (2003) Gating the selectivity filter in ClC chloride channels. *Science*, **300** (5616), 108–112.

7
Receptors for Biologically Relevant Anions

Stefan Kubik

7.1
Introduction

A clear majority, namely, 70–75%, of substrates and cofactors engaged in biological processes are negatively charged. Examples are adenosine-5′-triphosphate (ATP), the most important energy source involved in biochemical processes, including biosynthetic reactions, motility, and cell division. In signal transduction pathways, ATP is used as a substrate by kinases that phosphorylate proteins and lipids, as well as by adenylate cyclase, which uses ATP to produce the second messenger molecule, cyclic adenosine monophosphate. The products formed in the last two reactions act as anionic substrates themselves in subsequent biosynthetical pathways. Phosphorylated monosaccharides are intermediates in glycolysis and gluconeogenesis, isopentenyl pyrophosphate is involved in terpene biosynthesis, thiamine pyrophosphate is a coenzyme for several enzymes that catalyze the decarboxylation of α-keto acids, ATP as well as the other nucleoside triphosphates are substrates for RNA polymerase during mRNA synthesis, and the genetic code is stored in the polyanion DNA. Besides phosphate esters, another large group of biochemically relevant anionic substrates are carboxylates, which include, for example, fatty acids and the intermediates involved in the citric acid cycle. Of somewhat smaller importance in biochemistry are sulfate and sulfate esters (sulfated sugars such as heparin), chloride (chloride channels), and other anions.

These examples demonstrate that the course and interplay of biological processes strongly rely on the selective recognition of anionic species, either of small inorganic anions or of structurally more complex organic anions or polyanions. With the rapid progress of supramolecular chemistry, the field of anion coordination chemistry also advanced substantially, a trend that included the development of abiotic receptors possessing high binding affinity and selectivity for natural anionic substrates, sometimes under physiological conditions. These receptors are not only useful to elucidate general principles of anion recognition but also serve to address questions in more applied areas. Can a synthetic receptor that interacts with a biologically relevant anionic substrate interfere in the biological process in which this compound is involved, for example? In this case, medicinal applications may

Anion Coordination Chemistry, First Edition. Edited by Kristin Bowman-James, Antonio Bianchi, and Enrique García-España.

be envisaged. Alternatively, such receptors could be useful to detect trace amounts of the respective substrate in a mixture, or, if also catalytically active, they could induce transformations in a natural substrate.

In this chapter, current and past research in the development of synthetic receptors for anions of biological relevance is presented, with focus on receptors for phosphates and carboxylates. Because of the vast amount of literature available, this review cannot be comprehensive. Care was therefore taken to choose instructive examples of receptors whose properties reveal general principles in anion coordination and to provide an overview of the different receptor types relevant in the area. In this context, receptors with potential applications are particularly highlighted. This overview is structured according to substrate types, a classification system that has the advantage of allowing comparison of the different approaches with which binding of a specific anion can be achieved. Receptors that interact with more than one substrate type are therefore mentioned in different contexts throughout the text.

7.2 Phosphate Receptors

7.2.1 Introduction

To a first approximation, decisive factors determining the efficiency and selectivity of an artificial receptor are its structural and electronic complementarity to the targeted substrate. Receptors for orthophosphate should therefore possess appropriately positioned binding sites to host a tetrahedral anion with an ionic radius of 1.52 Å. An additional aspect particularly important in orthophosphate recognition is that this anion is involved in protonation equilibria in aqueous solution. Thus, completely deprotonated PO_4^{3-} is only existent in strongly basic media, while according to the pK_a values of H_3PO_4 (2.12), $H_2PO_4^-$ (7.20), and HPO_4^{2-} (10.9), the monoprotonated and the diprotonated forms predominate at neutral pH (Figure 7.1). These two anions differ in charge and in the number of hydrogen bond donors and acceptors around the central atom, characteristics that affect their affinity to a host. Receptors whose binding modes are based on attractive coulombic interactions and/or the presence of hydrogen bond donors should, for example, possess a higher affinity to HPO_4^{2-} than to $H_2PO_4^-$ because of the higher charge and larger number of acceptors in the former anion. Receptor design more often aims at systems, however, that favor binding of both anions. Phosphate recognition is less complex in organic aprotic media where protonation equilibria can usually be neglected as long as acidity or basicity of the binding partners is not too high. Binding of a receptor to either HPO_4^{2-} or $H_2PO_4^-$ can therefore be studied independently.

Diphosphate $P_2O_7^{4-}$ formally derives from condensing two HPO_4^{2-} anions, and triphosphate $P_3O_{10}^{5-}$ from condensing two HPO_4^{2-} and one $H_2PO_4^-$ anion. The

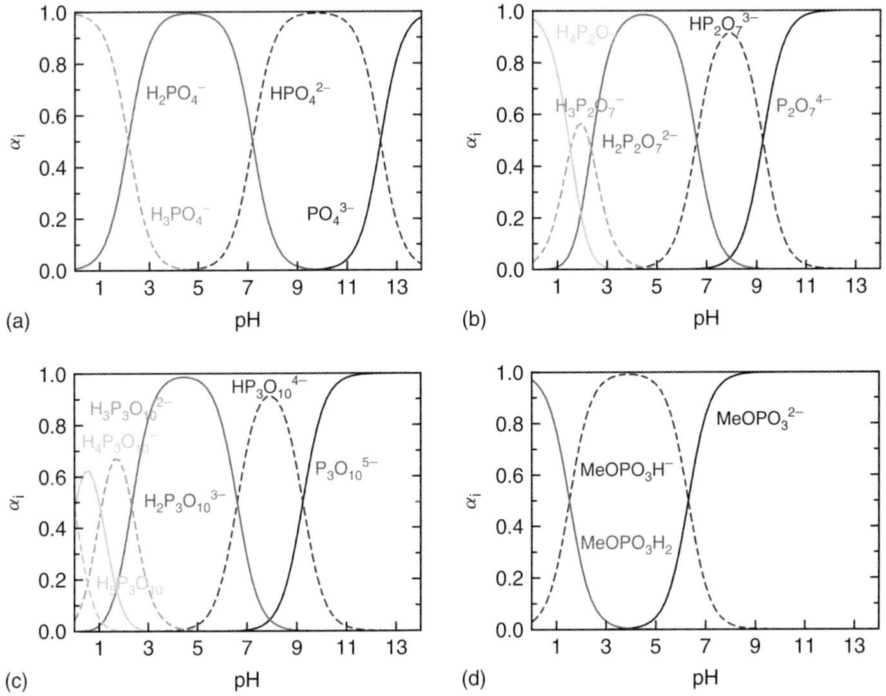

Figure 7.1 Distribution diagrams for (a) H_3PO_4, (b) $H_4P_2O_7$, (c) $H_5P_3O_{10}$, and (d) $MeOPO(OH)_2$.

pK_a values of the corresponding acids show that each subunit in $H_4P_2O_7$ and $H_5P_3O_{10}$ contains one strongly acidic proton, while two additional more weakly acidic protons are located at the terminal residues ($H_4P_2O_7$: pK_{a1} = 1.52, pK_{a2} = 2.36, pK_{a3} = 6.60, pK_{a4} = 9.25; $H_5P_3O_{10}$: pK_{a1} < 0.5, pK_{a2} = 1.06, pK_{a3} = 2.30, pK_{a4} = 6.60, pK_{a5} = 9.24). $P_2O_7^{4-}$ and $P_3O_{10}^{5-}$ therefore also predominantly exist in aqueous solution as a mixture of the respective monoprotonated and diprotonated forms (Figure 7.1). In terms of structure, these anions can be described as two- and three- corner-linked tetrahedra. Their interaction with a receptor requires, of course, a larger cavity than binding of monophosphate, but both anions also possess a higher negative charge and more oxygen atoms with which hydrogen bond donors in a receptor can interact.

Inorganic phosphates are of relevance in several biochemical processes; orthophosphate is, for example, consumed during ATP synthesis from ADP, and pyrophosphate plays a role in energy transduction in organisms. More important are, however, derivatives of these anions containing additional organic residues. Examples are nucleotides in which the D-ribose unit contains a phosphate, diphosphate, or triphosphate group in the 5′-position or proteins with phosphorylated serine, threonine, or tyrosine residues. A recent review analyzes the binding motifs found in proteins for recognition of such phosphate-containing substrates [1].

Of the 3003 protein structures identified in the Research Collaboratory for Structural Bioinformatics (RCSB) protein database as phosphate binders, 547, hence about 20%, contain a metal coordinating to the anionic group in the substrate. In 1386 of the remaining 2456 structures, the positively charged side chains of lysine and arginine residues contribute to binding, which leaves a surprisingly high number of 1070 structures in which neither a metal nor a positively charged residue is involved in anion recognition. A comparison of the different subsets shows that the lack of positively charged residues is compensated for by an increase in polar residues such as Ser, Thr, His, and Asn, as well as apolar residues such as Gly, Ala, or Ile.

The different types of interactions identified in proteins for coordinating to the anionic residue in nucleotides or phosphorylated alcohols have all been realized in synthetic receptors. In the design of such receptors, one has to consider that, on conversion of orthophosphate into a phosphate ester, one potential contact point to which the receptor can bind is removed, which could cause a reduction in complex stability. Secondary interactions between the organic part of a phosphate ester and appropriate binding sites in the receptor can, however, compensate for this effect. For example, abiotic nucleotide receptors generally contain aromatic residues that contribute to complex stability by binding to the nucleobase residue of the substrate via aromatic and/or hydrophobic interactions. Secondary interactions also provide a means to achieve binding selectivity. A ditopic receptor with a binding site for a specific nucleobase in addition to the phosphate recognition site should allow differentiation of the different nucleotides, for example. Another type of binding selectivity in nucleotide recognition involves discrimination of nucleotide monophosphates, diphosphates, or triphosphates, which can be realized by adapting the anion-binding site of the receptor. Finally, it should be noted that binding of phosphate esters in aqueous solution involves less anionic species than that of orthophosphate because the protonation equilibrium consists in only two steps and the predominant species at neutral pH is the completely deprotonated dianion as indicated by the pK_a values of phosphoric acid methyl ester ($MeOPO(OH)_2$: pK_{a1} = 1.52, pK_{a2} = 6.31) (Figure 7.1).

Abiotic receptors for the polyanions DNA or RNA usually target more than one phosphodiester group along the polynucleotide chain. This is similar to the natural DNA ligands spermine or spermidine, RNA-binding proteins in which arginine-rich regions contact the phosphodiesters of the polynucleotide, or histones in which basic residues along the peptide chain induce the folding of the DNA chain around these proteins.

7.2.2
Phosphate, Pyrophosphate, Triphosphate

Inorganic phosphates are probably the most widely studied substrates in anion coordination chemistry, which is partly due to their participation in numerous biological processes. Since practically every large class of anion receptors has been studied with respect to phosphate recognition, this chapter also provides the opportunity to introduce the various types of receptors relevant in the field [2]. It is appropriate to start the chapter with polyaza macrocycles that were among the first receptors whose anion affinity was characterized. A detailed review covering all aspects of the anion coordination chemistry of polyaza macrocycles has recently been published [3].

Polyaza macrocycles are a structurally widely diverse class of receptors, most of which contain several, usually at least six, nitrogen atoms in the form of secondary amines appropriately distributed along the cavity of a macrocyclic or bicyclic molecular framework. The nitrogen atoms are held apart by linking units that can differ in length and rigidity. Studies on anion complexation using polyaza macrocycles were inspired by the anion recognition properties of some naturally occurring polyamines, such as spermidine **1** and spermine **2**.

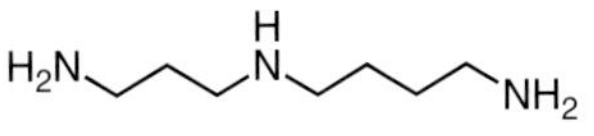

1

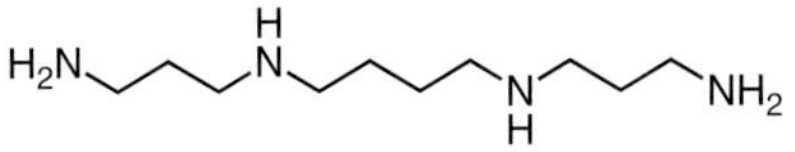

2

These compounds precipitate DNA and protect this polyanion from denaturation by heat or damage by shearing; stabilizing effects were attributed to neutralization of the negative charges on the DNA phosphate groups by polycationic forms of **1** and **2** [4]. Also, the effects of spermidine and spermine on RNA biosynthesis and protein kinase activity have been demonstrated, both of which involve interactions of the polycations with anionic substructures of the binding partners [5].

Anion binding of polyaza macrocycles benefits from the macrocyclic effect but is otherwise closely related to that of **1** and **2** in that it also takes place in water and involves protonated forms of the hosts. The driving force of complex formation therefore mainly comes from electrostatic interactions between the oppositely charged binding partners, although additional types of interactions, particularly hydrogen bonding, can play a decisive role in determining substrate selectivity (*vide infra*). In this respect, the advantage of the nitrogen atoms in polyaza macrocycles is their ability to act as hydrogen bond donors or acceptors depending on whether they are protonated or not.

Although the behavior of polyaza macrocycles in anion coordination is often complex, involving, for example, several protonation equilibria, several rules apply. The strongest anion binding is, for example, usually observed for the fully protonated host with a maximum number of ammonium groups because anion affinity correlates with the positive charge density around the cavity. Analogously, the higher the negative charge on the anion, the higher its affinity to a specific protonation state of a host. The pH required to achieve full protonation of a polyaza macrocycle depends on host structure. The smaller the distance between two nitrogen atoms, for example, the more difficult the protonation of both becomes because of unfavorable charge accumulation [6, 7]. It was therefore suggested that receptors containing either 1,3-propylenediammonium subunits as **1** or **2** or 1,2-ethylenediamine units isolated by longer intervening chains would represent effective phosphate ligands [8]. At a pH at which the host is not completely protonated, multiple equilibria have to be considered, and, as a consequence, several host species exist simultaneously in solution, all behaving differently. Quantitative evaluation of the concentration of each compound present at a defined pH and its contribution to binding is often performed by using potentiometric measurements during which the change in the protonation constants of the host caused by the presence of an anionic guest is determined. The binding properties of a macrocyclic polyamine are controlled not only by the degree of protonation but also by the position where protonation occurs, particularly when hydrogen bonds contribute to complex stabilization. In general, nonadjacent amino groups along the host are protonated first and the least basic ones such as tertiary amines last, if at all. Finally, protonation has an effect on host conformation because charge repulsion causes ammonium groups to move away from each other, causing the molecular framework of a polyamine to expand and become more rigid on protonation.

There are several useful strategies to affect affinity and selectivity of a polyaza macrocycle, the most simple of which is variation of ring size. A means to control the protonation pattern and, as a consequence, anion selectivity, involves introduction of tertiary amines in defined positions along the ring, which usually require a much lower pH for protonation than secondary amines. Rigidity of the host can be controlled by introducing aromatic linkers, and the introduction of a strap across the ring that converts a macrocyclic into a bicyclic host also affects binding properties. Finally, quaternization of nitrogen atoms introduces a permanent charge that is independent of pH. The disadvantage of quaternization is, however, that quaternized nitrogen atoms lose the ability to interact with anions via hydrogen bond formation.

Numerous polyaza macrocycles exhibiting affinity for inorganic phosphates have been described [9–22], representative examples of which are compounds **3**–**7** [23, 24].

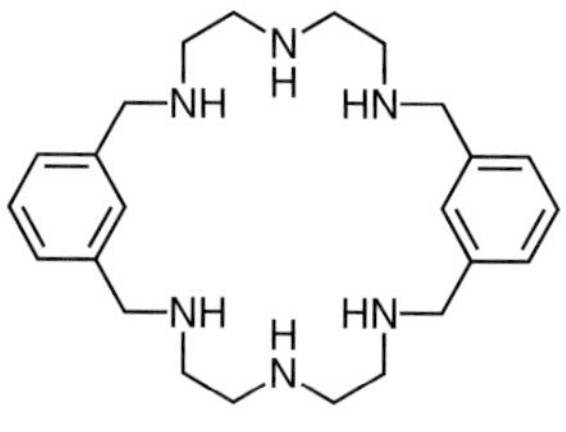

3

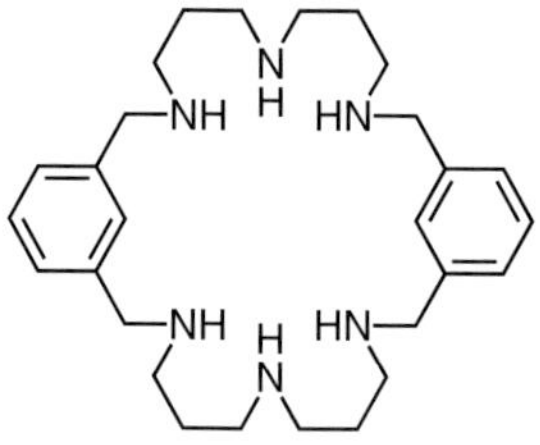

4

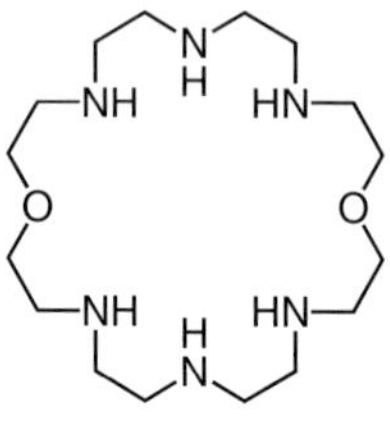

5

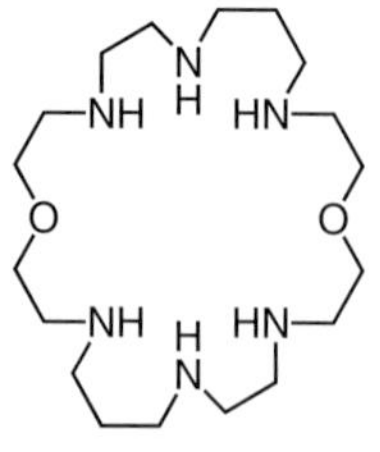

6

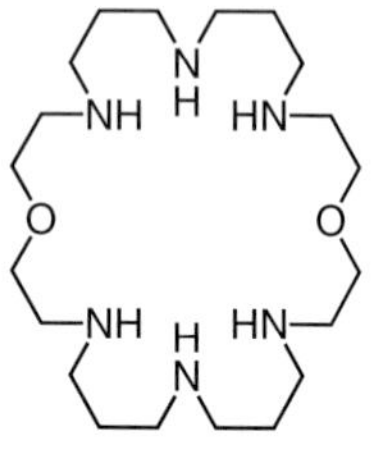

7

Quantitative evaluation of the affinity of receptors **3**–**7** toward orthophosphate, pyrophosphate, and triphosphate using potentiometric measurements revealed some general trends. One is that the binding affinity of each receptor usually increases in the order phosphate < pyrophosphate < triphosphate. The stability constants log K_a of the hexaprotonated complexes $PO_4^{3-} \subset$ **3**, $P_2O_7^{4-} \subset$ **3**, and $P_3O_{10}^{5-} \subset$ **3** amount to, respectively, 7.36, 13.07, and 14.19, for example [23]. This order can easily be rationalized by the increasing negative charge on the anion that causes a strengthening of coulombic attraction. The electrostatic interactions between host and guest are, however, modulated by additional contributions from hydrogen-bonding interactions as indicated by the effects of ligand structure on complex stability. The gradual increase in pyrophosphate affinity on going from 28-membered receptor **7** (log K_a of hexaprotonated $P_2O_7^{4-} \subset 7 = 8.55$) to 26-membered **6** (log K_a of hexaprotonated $P_2O_7^{4-} \subset 6 = 10.19$) to 24-membered **5** (log K_a of hexaprotonated $P_2O_7^{4-} \subset 5 = 12.56$) indicates, for example, a better fit of the anion into the cavity of the smallest host [23].

There are, however, receptors for which deviations from these trends have been observed. For instance, the stability of the complexes formed by HPO_4^{2-} with the mono-, di-, and triprotonated forms of **8** decrease with increasing charge on the ligand, while the stability of the complexes formed by **9** with $HP_2O_7^{3-}$, $H_2P_2O_7^{2-}$, and $H_3P_2O_7^{-}$ increases with decreasing charge of the anion [19].

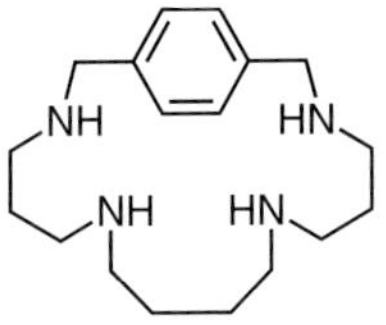

8

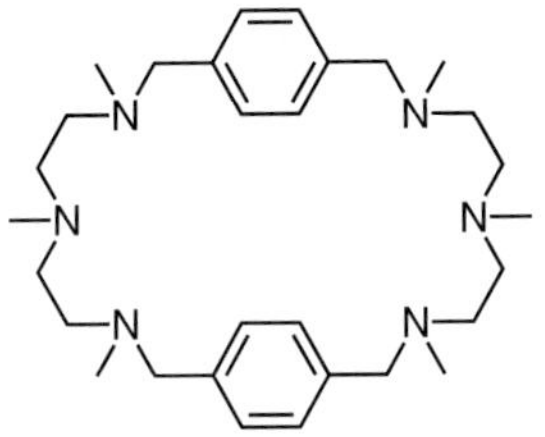

9

Microcalorimetric investigations showed that these unusual stability trends can be rationalized by assuming the formation of different types of hydrogen bonds between host and guest, whose contribution to binding strength can be a decisive component in receptor–substrate interactions, even in a highly competitive environment such as water [19].

Changing receptor topology from macrocyclic to macrobicyclic often significantly improves binding affinity or selectivity. This is not so pronounced in phosphate recognition when using macrobicyclic compounds such as **10**, however, as demonstrated by the lower phosphate and pyrophosphate affinity of **10** (log K_a of hexaprotonated $P_2O_7^{4-} \subset 10 = 10.30$ and hexaprotonated $PO_4^{3-} \subset 10 = 2.75$) with respect to the monocyclic analog **5** (log K_a of hexaprotonated $P_2O_7^{4-} \subset 5 = 12.56$ and hexaprotonated $PO_4^{3-} \subset 5 = 6.97$). This result was ascribed to the fact that these substrates can only be partially included into the cavity of cryptand **10** [10]. The cages **11a** and **11b**, on the other hand, were shown to efficiently interact with tetrahedral oxoanions [25, 26]. These receptors engage in aqueous solution in the typical protonation equilibria of polyaza macrocycles, and the protonated host species were shown to bind anions such as Cl^-, I^-, NO_3^-, ClO_4^-, AcO^-, $H_2PO_4^-$, SO_4^{2-}, and SeO_4^{2-} [25]. Potentiometry revealed remarkable selectivity of **11a** for the dianionic tetrahedral anions of this series, with association constants ranging from

5.03 to 5.3 log units for the dianionic substrates and from 1.49 to 2.97 log units for the monoanionic ones. Receptor **11a** possesses significantly lower affinity for dihydrogenphosphate in acidic media than for sulfate, which was attributed to the lack of hydrogen bond acceptor sites inside the receptor cavity. Selectivity reverses for **11b**, a result that was explained by favorable contributions of hydrogen bonds on complex stability between the $H_2PO_4^-$ protons and the ring nitrogen atoms [26]. Unfortunately, X-ray crystallography of the sulfate and the dihydrogenphosphate complexes of **11b** did not reveal inclusion of these anions into the macrocyclic cavity of the host or involvement of the ring nitrogen atoms in anion binding. In contrast, an anion is situated inside the receptor cavity in the crystal structure of the sulfate complex of **11a** [25].

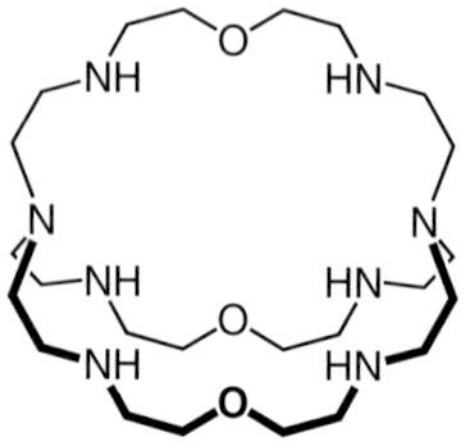

10

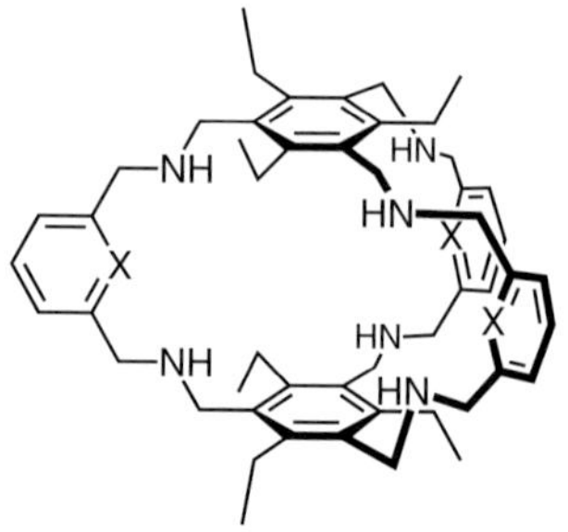

11a; X = CH

11b; X = N

Changing the cavity shape of the receptor from almost spherical to cleftlike as in **12** can cause interesting selectivity effects. The stability constants log K_a of complexes between fully deprotonated phosphate, diphosphate, and triphosphate anions and the tetraprotonated form of **12** amount to 5.32, 5.52, and 4.34, for example, demonstrating that despite its higher negative charge, triphosphate cannot be bound as efficiently inside the receptor cleft as the other anions [21]. The thermodynamic parameters determined for binding of the phosphate anions in water demonstrate that complex formation is invariably promoted by a favorable entropic contribution, indicating that it is accompanied by the release of water molecules from the receptor cavity. More recently, tris-macrocyle **13** was described containing three reinforced cyclic tetraamine ligands connected by two 2,6-pyridine diylbis(methylene) linkers [27]. This compound behaves as a "double" proton sponge in that the two lateral macrocycles are each protonated even in highly basic aqueous media. Further protonation affords a tetraprotonated species, where each lateral tetraamine macrocycle is diprotonated. The central macrocycle is involved in proton binding only in species with a higher protonation degree. The protonated forms of **12** interact with inorganic phosphate anions in aqueous solution, with a clear selectivity for triphosphate over diphosphate and monophosphate over the pH range of 2–10. Complex formation involves electrostatic interactions and hydrogen bond formation.

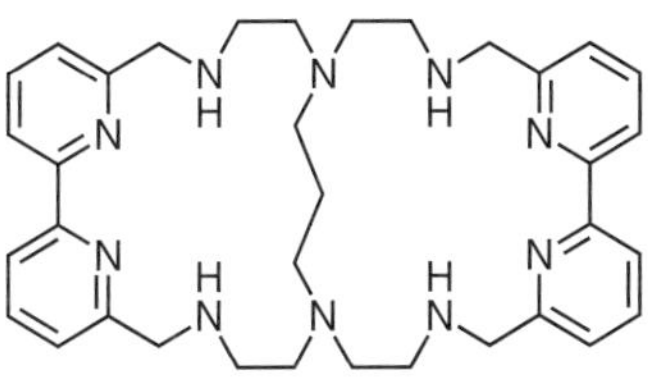

12

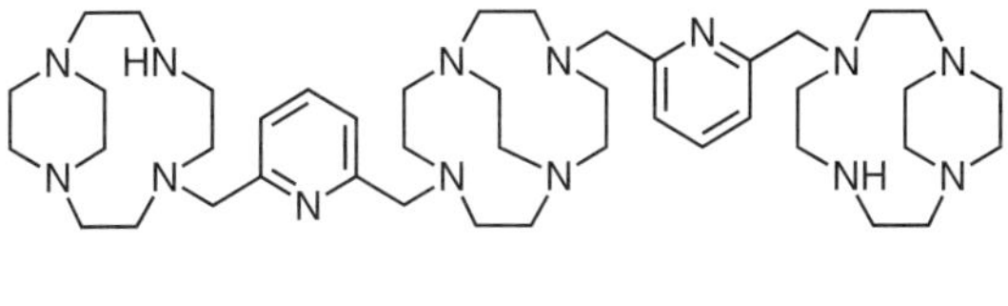

13

Another strategy to modulate anion binding of polyaza macrocycles involves introduction of two metal ions in opposing positions of the ring. If coordination of the metal centers to the ring nitrogens leaves one coordination site unsaturated (or saturated with a weakly bound solvent molecule), anions can bridge these binding sites, leading to the formation of so-called cascade complexes. The dicopper(II) complex of **5**, receptor **14**, has been shown to bind pyrophosphate in water (0.1 M KCl) with an association constant log K_a of 8.5, for example [28]. In comparison with an equally charged mononuclear derivative in which two amino groups are protonated, **14** shows stronger binding for pyrophosphate by 2.25 log units. A mononuclear species in which three amino groups are protonated binds the anion 1.43 log units more strongly than **14**, which was attributed to the organizing effect of the three ammonium groups on the ligand that causes a better fit of the pyrophosphate anion inside the host cavity [28]. Another example of a host that strongly interacts with pyrophosphate in water is **15** (logK_a = 7.2) [29]. Complex formation involves bridging of the two copper ions of the host by oxygens of both phosphate moieties. To allow for binding of the smaller orthophosphate anion, the macrocycle has to contract, which causes a strained situation and, as a consequence, a reduced stability of the corresponding complex. Expansion of the cavity of **15** leads to host **16** that tightly binds triphosphate (logK_a = 8.0) [30]. As in the case of metal-free ligands, the cavities of macrobicyclic ligands containing metal centers are usually too small to allow efficient phosphate binding.

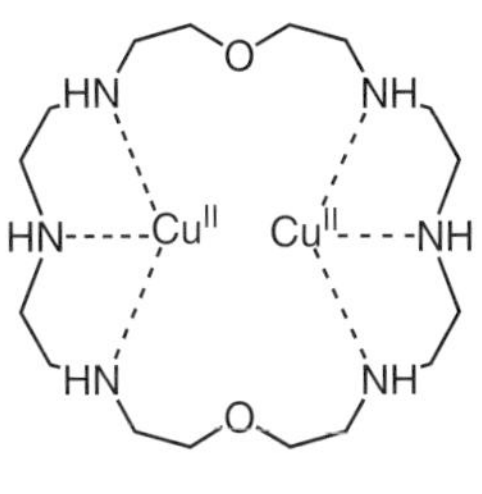

14

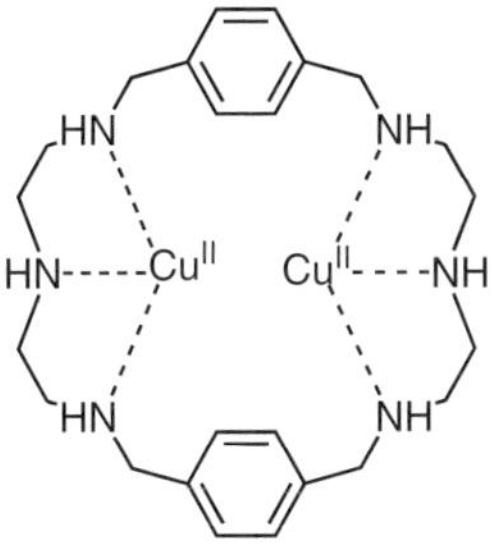

15

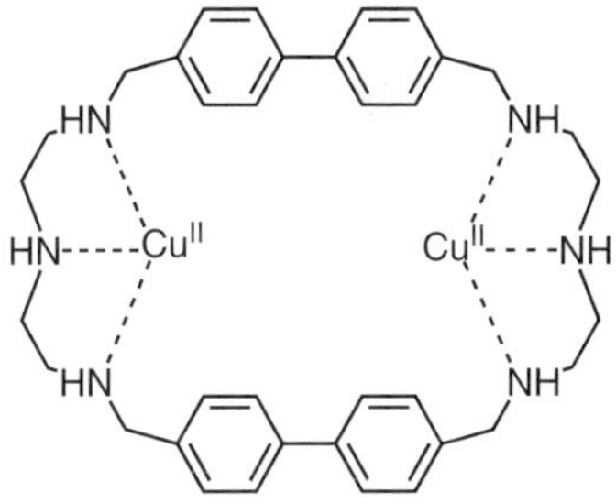

16

By the careful choice of an anionic fluorescent indicator whose complex with the host is almost as stable as that of the target substrate but more stable than complexes of competing substrates, a fluorescent sensor for pyrophosphate employing the dye displacement strategy was devised on the basis of **15** [29]. Electrochemical anion sensors derived from polyaza macrocycles were devised by the Beer group by appending ferrocene units to or incorporating them into the ring [31]. Examples are receptors **17** and **18**, whose tetraprotonated complexes with orthophosphate have stability constants log K_a of, respectively, 15.55 and 11.01 (0.1 mM aqueous tetrabutylammonium perchlorate) [32]. Phosphate binding results in a cathodic shift of the redox potential, which allows complex formation to be followed qualitatively as well as quantitatively. Comparative studies involving a series of structurally related hosts showed that the values of the oxidation potential do not directly correlate with the observed binding affinities, which indicates that structural parameters of the host and the anion affect both binding selectivity and the observed electrochemical responses. Optical sensors for phosphate on the basis of polyammonium-based receptors were developed by the Czarnik group [33].

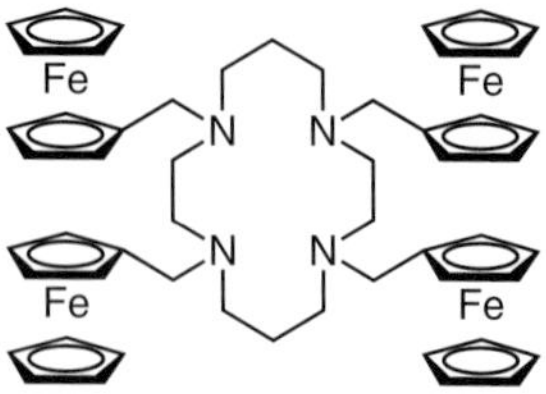

17

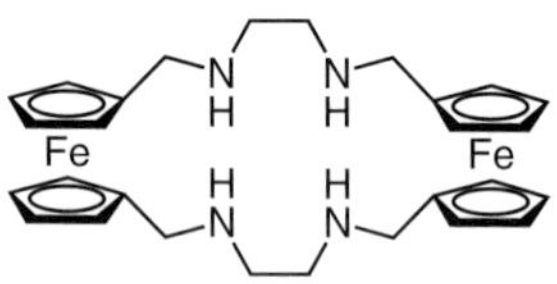

18

Macrotricyclic receptors **19a,b** were introduced by Schmidtchen into the field of anion coordination chemistry [34, 35]. These receptors contain four quaternary nitrogen atoms; their overall charge is therefore independent of pH. Since quaternization eliminates the hydrogen-bonding ability of (protonated) nitrogen, anion binding of **19a,b** solely relies on electrostatic interactions and is therefore usually weaker than that of conventional polyaza macrocycles. Phosphate complexation is, however, still clearly detectable and associated with stability constants of 345 M^{-1} for the HPO_4^{2-} complex of **19a** and 125 M^{-1} for the $H_2PO_4^-$ complex in water (0.1 M aqueous sodium tosylate), consistent with the expected correlation between complex stability and negative charge on the anion [36].

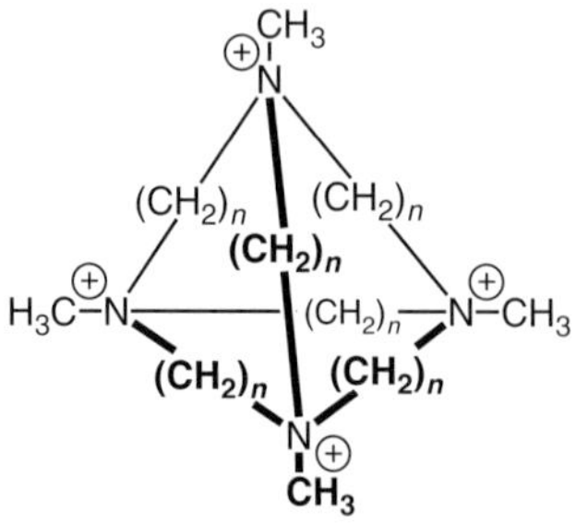

19a, $n = 6$
19b, $n = 8$

The use of guanidinium groups (or amidinium groups) as binding sites in synthetic receptors is a strategy to mimic the anion-binding properties of the side chain of arginine [37–39]. Guanidinium moieties are important elements in the interaction of many biological receptors with anions. They operate, for example, in substrate recognition of phosphate-binding proteins or RNA-binding proteins. The advantage of guanidinium and amidinium moieties lies in their ability to combine coulombic attraction in the binding of oxoanions such as carboxylates or phosphates with the formation of two parallel hydrogen bonds. Moreover, both groups are strongly basic with pK_a values typically ranging between 11 and 13, ensuring that they remain protonated over a wide pH range. A disadvantage of guanidinium ions is their strong solvation, which is so efficient that, despite the favorable binding pattern, ion pairing with carboxylates or phosphates in aqueous solution is practically negligible ($K_a < 5\,M^{-1}$) [40]. Hydrophilic anion hosts therefore often make use of two or more guanidinium moieties or the cooperative effect of other anion-binding sites such as pyrroles to be effective.

The first examples of guanidinium-based anion receptors were described by Lehn and coworkers [41, 42]. Complex formation was followed by means of pH titrations between a number of cyclic and noncyclic hosts such as **20–22** and phosphates (PO_4^{3-}, HPO_4^{2-}, $P_2O_7^{4-}$, $HP_2O_7^{3-}$, $H_2P_2O_7^{2-}$) in water. Stability constants log K_a typically range between 1 and 3 and increase with increasing charge on host or guest. This and the fact that the investigated guanidinium-based hosts form less stable complexes than ammonium-based hosts of equivalent structure and charge, a result that is consistent with the lower charge density of the guanidinium moiety, led to the conclusion that electrostatic interactions dominate anion binding.

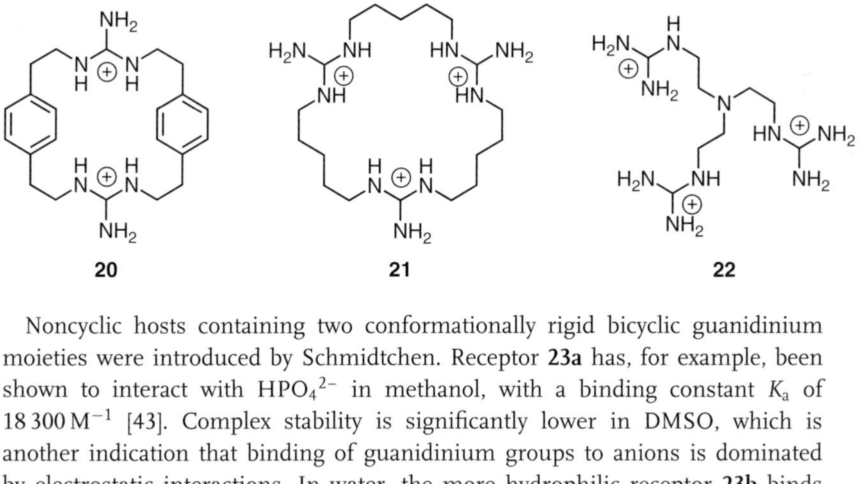

Noncyclic hosts containing two conformationally rigid bicyclic guanidinium moieties were introduced by Schmidtchen. Receptor **23a** has, for example, been shown to interact with HPO_4^{2-} in methanol, with a binding constant K_a of 18 300 M^{-1} [43]. Complex stability is significantly lower in DMSO, which is another indication that binding of guanidinium groups to anions is dominated by electrostatic interactions. In water, the more hydrophilic receptor **23b** binds HPO_4^{2-} with an appreciable K_a of 970 M^{-1}.

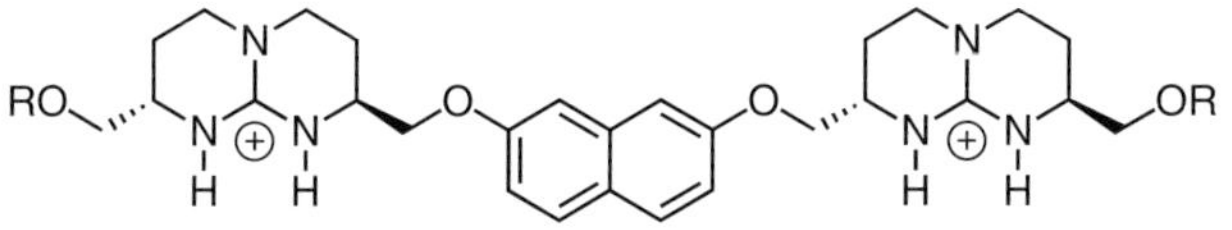

23a; R = SiPh$_2$^{t}Bu
23b; R = H

As in the case of polyaza macrocycles, anion binding of guanidinium-derived receptors can be improved by the cooperative effects of simultaneously bound metal ions. Two examples should illustrate this approach. The first is receptor **24**, whose zinc affinity increases in aqueous solution (10 mM HEPES, pH 6.8) from $< 100\,M^{-1}$ in the absence of phosphate to 4300 M^{-1} in the presence of 1 equiv. of

Na_2HPO_4, which was ascribed to the combined stabilizing effects of coordination of one phosphate oxygen to the zinc, ion pairing, and hydrogen-bonding interactions between the phosphate and the guanidinium moieties of the host [44].

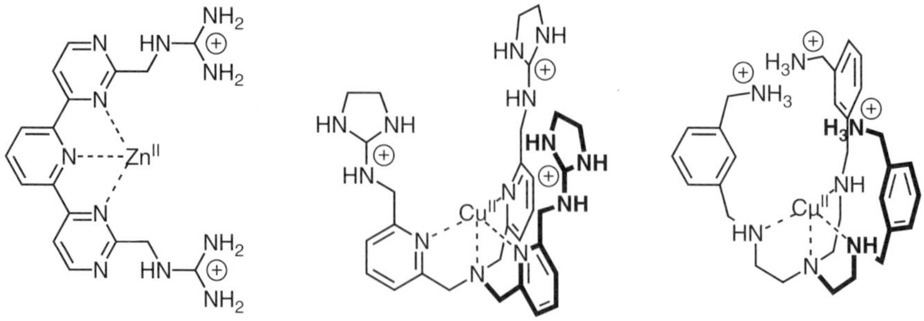

24 **25** **26**

The second example involves the structurally related receptors **25** and **26**, which were also developed by the Anslyn group. Anion complexation of these hosts is achieved by a combination of coordinative interactions to the copper(II) center, coulombic attraction, and hydrogen bonding to protonated aminoimidazoline moieties or amino groups arranged around the cavity. In addition, the shape of the cavity of these hosts provides an optimal environment for the inclusion of tetrahedral anions. As a consequence, **25** and **26** strongly bind to, for example, HPO_4^{2-} in 2% methanol/water (5 mM HEPES, pH 7.4), with stability constants K_a of, respectively, 1.5×10^4 and $2.5 \times 10^4\ M^{-1}$ [45, 46]. Very weak or no interactions of both receptors with anions such as acetate, sulfate, nitrate, hydrogencarbonate, and chloride were detected. Arsenate is bound by both receptors comparably well as phosphate, whereas the perrhenate complex of **26** has an about 1 order of magnitude smaller stability constant than the phosphate complex, most probably because of the larger size and the reduced charge of this anion. Interestingly, binding of **25** to perrhenate is almost negligible ($K_a < 100\ M^{-1}$), demonstrating that this host has a lower affinity but a higher selectivity for phosphate in comparison to **26**. The lower selectivity of **26** was initially attributed to the higher flexibility of this receptor [45], but a subsequent thermodynamic characterization of the complexation equilibria revealed that differences in the solvation of the positively charged binding sites of the two hosts are the main cause for the selectivity pattern [46]. The high phosphate affinity and selectivity of **25** allowed assembly of a dye displacement assay in combination with the indicator carboxyfluorescein for the quantitative determination of phosphate in horse serum and saliva [47].

Optical sensing of pyrophosphate [48] could be achieved with metal-free pyrene derivative **27** [49]. Complex formation between **27** and pyrophosphate involves coordination of two receptor molecules to the anion. This positions the two pyrene moieties in a sandwich-type arrangement so that complex formation could be observed by following the change in the ratio of excimer to monomer emission. The overall stability of the complex amounts to $9.8 \times 10^7\ M^{-2}$, with a K_a of the 1 : 1 complex of $1.3 \times 10^4\ M^{-1}$. No emission change was observed on binding of **27** to orthophosphate.

The guanidiniocarbonyl pyrrole moiety that was introduced into the field of anion coordination by Schmuck mainly for carboxylate recognition (*vide infra*) forms the basis for pyrophosphate sensor **28**. Although this compound only weakly interacts with pyrophosphate in aqueous solution (10% v/v DMSO/water), the K_a of the complex amounts only to $78\,M^{-1}$, complex formation strongly increases the fluorescence of the receptor, an effect that is not observed with any of the 10 other anions tested, including orthophosphate, several nucleotides, halides, sulfate, and acetate [50]. Receptor **28** therefore allows the highly selective optical sensing of pyrophosphate in water. Similar to the related compounds **17** and **18**, the ferrocene-containing receptor **29** developed by the Beer group allowed pyrophosphate complexation to be studied by cyclic voltammetry [51].

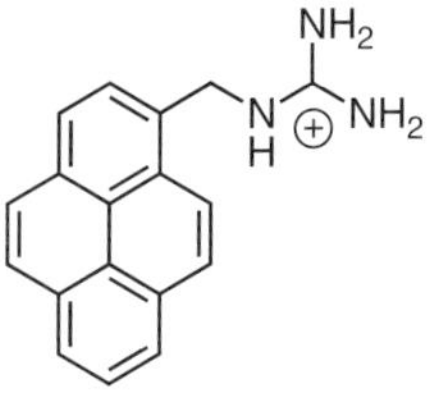

27

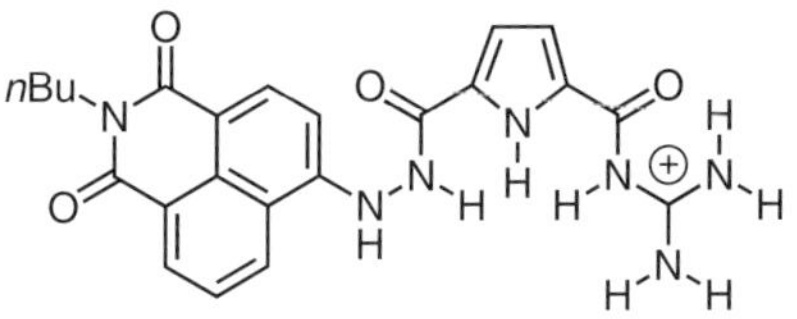

28

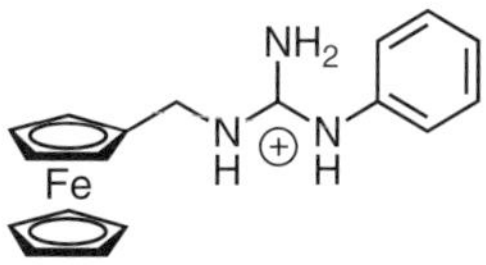

29

Somewhat related to receptors containing guanidinium groups are those in which imidazolinium groups are responsible for anion recognition. These groups bind anions via a combination of electrostatic interactions and direct hydrogen bonds between the imidazolinium C–H groups and the anion. Examples are receptors **30** and **31**, which were developed by Yoon *et al.* and whose emission of the anthracene moiety is quenched on binding to $H_2PO_4^-$ in acetonitrile. Although **30** and **31** possess the same affinity for this anion ($K_a = 1.3 \times 10^6\ M^{-1}$), selectivity of **31** for $H_2PO_4^-$ over other anions, particularly over fluoride, is significantly higher presumably because the cavity dimensions of the cyclic receptor are better defined and no complexes of higher stoichiometry can be formed with fluoride anions [52, 53].

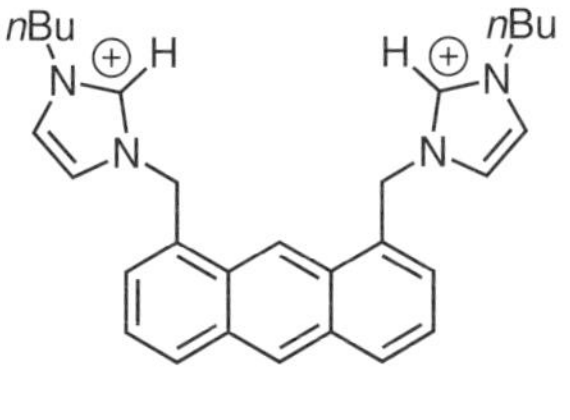

30

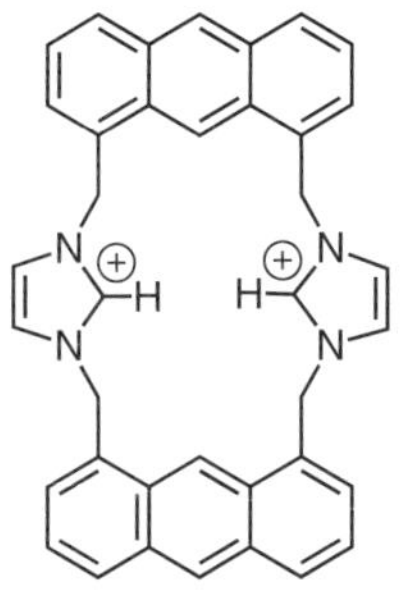

31

Anion hosts containing hydrogen bond donors such as the NH groups of amide, urea, or thiourea moieties rely on hydrogen-bonding interactions for anion recognition, thereby mimicking the interaction of amide NH groups along the protein backbone with the substrate in anion-binding proteins, a type of interaction termed "*nest*" by Watson and Milner-White [54, 55]. One advantage of such receptors is their electroneutrality that prevents competition in solution of counterions for the binding sites of the host. Another is the directionality of hydrogen bonds that allows a well-defined arrangement of binding sites in the molecular framework of a host to translate into predictable substrate selectivity. One major disadvantage of hydrogen-bonding interactions is, however, their weakness in polar and particularly in protic solvents, which is probably the reason why there are only few examples of neutral anion hosts active in aqueous media [56].

In a series of systematic investigations, the Bowman-James group established general rules that determine the structure–anion affinity relationship of amide-containing receptors [57]. Among the receptors studied in this respect are the monocyclic, bicyclic, and tricyclic derivatives **33a–d**, **34a,b**, **35a,b**, and **36a,b** [58–60]. Closely related to these receptors are the macrocyclic lactams **32a–c** investigated by the Jurczak group [61–65]. Recently, Rybak-Akimova's group also contributed to this field by studying the anion-binding properties of analogs of **33b** with an extended cavity [66].

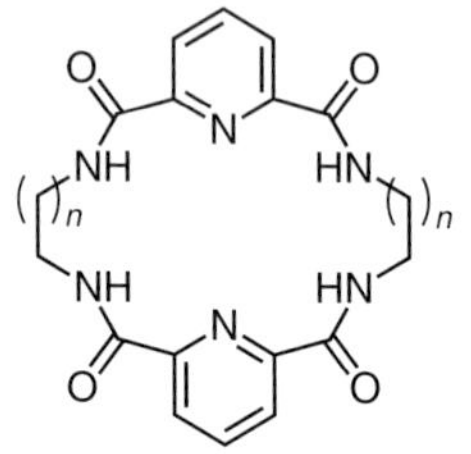

32a, $n = 1$
32b, $n = 2$
32c, $n = 4$

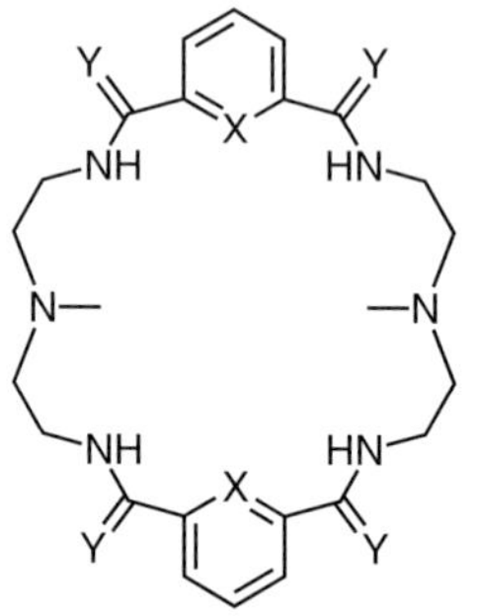

33a; X = CH, Y = O
33b; X = N, Y = O
33c; X = CH, Y = S
33d; X = N, Y = S

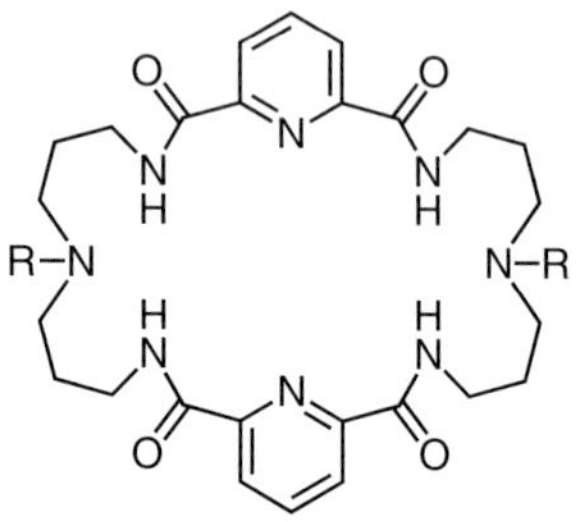

34a; R = H
34b; R = CH_3

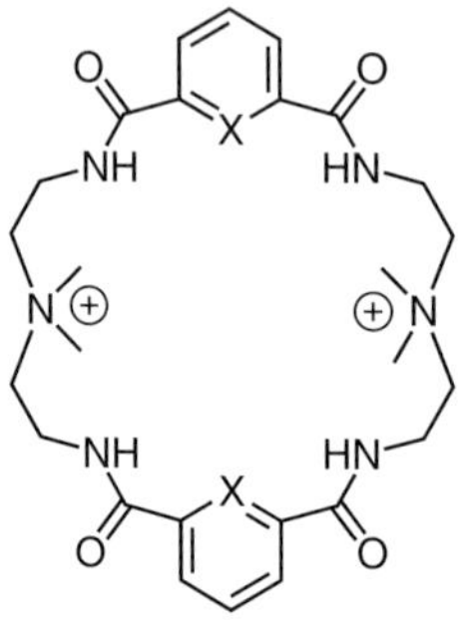

35a; X = CH
35b; X = N

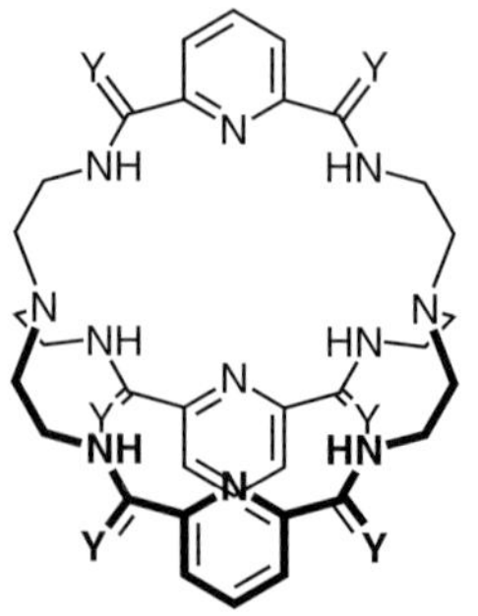

36a; Y = O
36b; Y = S

These investigations showed that replacement of isophthalic diamide subunits with 2,6-pyridine diamide subunits increases anion affinity, which can be attributed to the better preorganization of the pyridine-containing receptors caused by the intramolecular hydrogen-bonding interactions between the amide NH groups with the ring nitrogen atoms. Expectedly, introduction of a positive charge also improves anion affinity as does replacement of the amides by thioamides and the change in receptor topology from monocyclic to bicyclic. While these rules nicely apply to halide binding, structure–affinity relationships are not so clear in phosphate recognition. The thioamide-containing receptors **33c** and **33d**, for example, indeed possess higher $H_2PO_4^-$ affinity in DMSO than analogs **33a** and **33b** ($H_2PO_4^- \subset$ **33a**: log $K_a = 2.92$; $H_2PO_4^- \subset$ **33b**: log $K_a = 4.04$; $H_2PO_4^- \subset$ **33c**: log $K_a = 4.97$; $H_2PO_4^- \subset$ **33d**: log $K_a = 4.63$) [60], but replacement of the isophthalic subunits in thioamide-containing receptor **33c** with pyridine subunits in **33d** leads to a reduction in phosphate affinity. In comparison, the $H_2PO_4^-$ affinities of hosts **32a–c** lacking the central amino groups are of a similar order of magnitude in DMSO ($H_2PO_4^- \subset$ **32a**: log $K_a = 3.23$; $H_2PO_4^- \subset$ **32b**: log $K_a = 3.87$; $H_2PO_4^- \subset$ **32c** : log$K_a = 2.65$) [63]. The corresponding investigations indicate that the observed order of complex stabilities cannot be interpreted exclusively in terms of matching between anion diameter and the size of macrocyclic cavity because **32b** forms the most stable complexes with all anions studied, irrespective of their sizes. Instead, the results suggest that anion binding by this family of macrocycles is governed by the competitive interplay between their ability to adjust to a guest, requiring longer aliphatic spacers, and preorganization, calling for shorter spacers. Obviously, the 20-membered receptor **32b** represents the optimal compromise between these factors. Analogs of **32a–c** containing isophthalic acid subunits possess lower anion affinity, in agreement with the results from the Bowman-James group [65]. Analogs of receptors **33a–c**, such as compound **34a**, were shown to bind sulfate with higher affinity than dihydrogenphosphate in DMSO ($H_2PO_4^- \subset$ **34a**: $K_a = 4.4 \times 10^3$ M^{-1}; $HSO_4^- \subset$ **34a**: $K_a = 6.4 \times 10^4$ M^{-1}) [67]. Methylation of the secondary amide groups in **34a** reverses this selectivity while simultaneously causing a substantial drop in complex stability ($H_2PO_4^- \subset$ **34b**: $K_a = 500$ M^{-1}; $HSO_4^- \subset$ **34b**: $K_a = 73$ M^{-1}). Crystal structures of SO_4^{2-}, HPO_4^{2-}, $H_2PO_4^-$, and $H_2P_2O_7^{2-}$ complexes indicated different macrocyclic conformations depending on the substituents on the amino groups.

As expected, the charged derivatives **35a** and **35b** have a higher phosphate affinity than the neutral analogs **34a** and **34b** ($H_2PO_4^- \subset$ **35a**: log $K_a = 4.06$; $H_2PO_4^- \subset$ **35b**: log $K_a = 5.32$) [59]. As also observed in other cases, phosphate affinity decreases on going from monocyclic to bicyclic hosts, reflecting the fact that the phosphate anion cannot be efficiently included into the more rigid cavity of the latter ($H_2PO_4^- \subset$ **36a**: log $K_a = 3.40$; $H_2PO_4^- \subset$ **36b**: log $K_a = 3.31$) [58, 60]. Consistent with this interpretation is that tricyclic host **37a** is highly selective for the small bifluoride anion in DMSO ($FHF^- \subset$ **37a**: $K_a = 5500$ M^{-1}; $H_2PO_4^- \subset$ **37a** : $K_a = 740$ M^{-1}) [68]. Derivative **37b**, having an extended cavity, has been shown by X-ray crystallography to bind $H_2PO_4^-$, $H_2P_2O_7^{2-}$, and $H_3P_3O_{10}^{2-}$ in the side pockets (and not between the aromatic rings) via hydrogen bonds to the amide NH groups [69]. Complex stability in DMSO of the $H_2PO_4^-$ complex of **37b** amounts to 2300 M^{-1}.

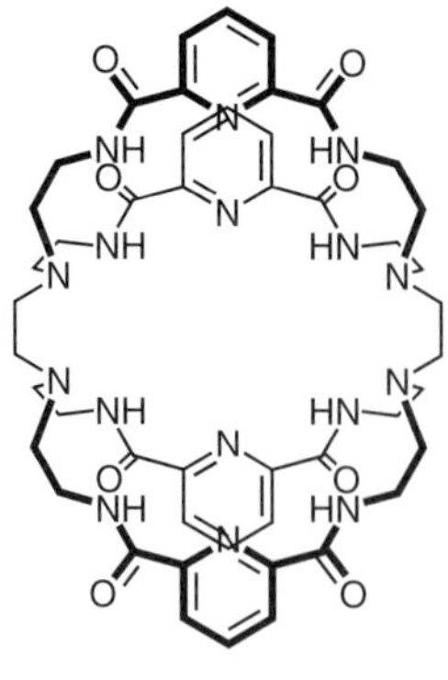

37a

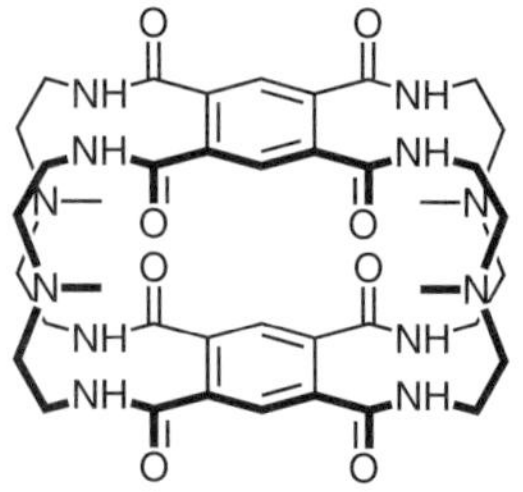

37b

Acyclic phosphate receptors containing two converging thiourea units were described by Pfeffer and coworkers. The norbornane-derived receptors **38a–c** form 2 : 1 complexes with dihydrogenphosphate in DMSO, for example, in which two anions interact with one receptor molecule [70]. Thus, instead of acting cooperatively in anion complexation, each receptor arm interacts individually with an anionic guest. NMR spectroscopic investigations indicated that this lack of cooperativity is most likely due to steric constraints imposed by the rigid bicyclic scaffold. In contrast, the [3]polynorbornyl-derived host **39** forms a 1 : 1 complex with $H_2PO_4^{2-}$ in DMSO whose stability constant log K_a amounts to 3.9 [71]. Pyrophosphate $H_2P_2O_7^{2-}$ is, however, bound with 2 : 1 stoichiometry by two host molecules. A structure is proposed for this complex, in which the anion is sandwiched between two molecules of **39** and surrounded by four thiourea moieties that interact with the guest by hydrogen bonding. Anthracene-containing receptor **40** forms a 1 : 1 complex with pyrophosphate in DMSO (log K_a = 3.40), whose formation can be monitored by the PET quenching of the anthracene moiety [72].

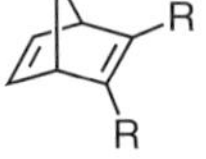

38a

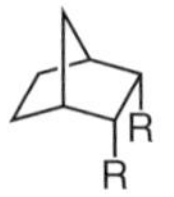

38b

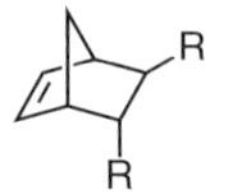

38c

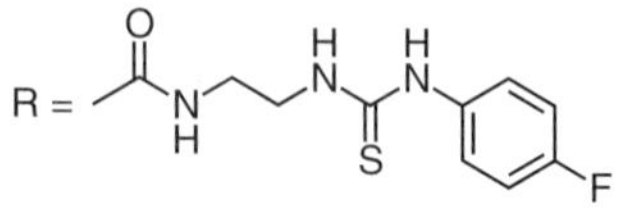

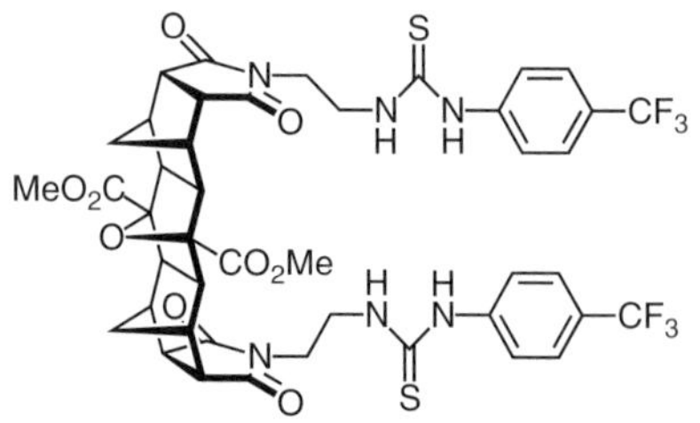

39

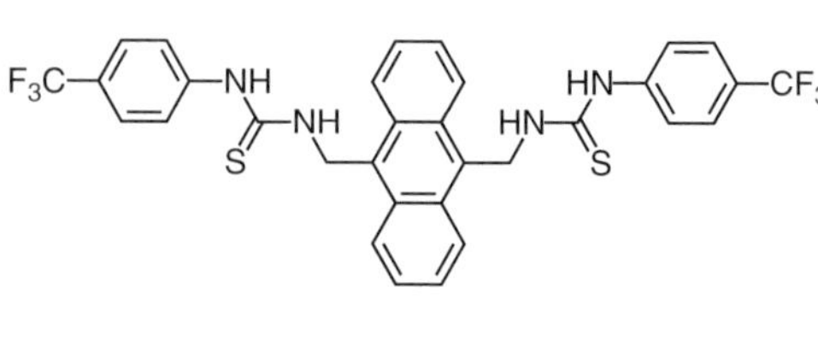

40

The Sessler group was the first to demonstrate the anion-binding ability of protonated sapphyrins [73]. The mode of interaction between sapphyrins and phosphate is clearly visible in the single-crystal X-ray structure of the HPO_4^{2-} complex of diprotonated sapphyrin **41** [74]. In this complex, one oxygen atom from each phosphate oxyanion interacts via hydrogen bonds with the pyrrolic NHs of the protonated sapphyrin core. This motif was also detected in several 2 : 1 phosphate anion–sapphyrin complexes, in which the two anions were found to be coordinated on opposite faces of the sapphyrin macrocycle [74]. Evidence for binding in solution includes the observation of upfield shifts in the ^{31}P NMR signals of phosphoric acid on addition of sapphyrin [74]. These measurements allowed quantitative evaluation of the stability of the HPO_4^{2-} 1 : 1 complex of diprotonated **41a**, which amounts to 12 600 M^{-1} in methanol.

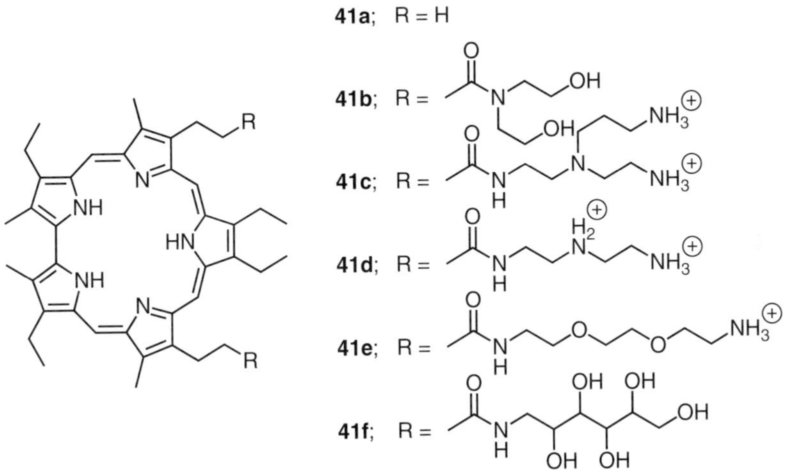

Hydrophilic sapphyrins such as **41b–f** were prepared for investigations on anion binding in water [74, 75]. Although these compounds are monoprotonated at pH 7, which should enable them to interact with anions in water, they are also highly aggregated. Addition of anions to aqueous solutions of sapphyrins does, however, cause a gradual shift of the aggregation equilibrium to the monomeric state, the only sapphyrin state in aqueous solution that is fluorescent, allowing complex formation to be followed by monitoring the change in fluorescence. This method showed that stability constants of the phosphate complexes of sapphyrin derivatives **41b–f** range between 6 and 19 M^{-1} in buffered solution (25 mM PIPES, pH 7.0) in the presence of 150 mM NaCl [75]. Although these constants may seem small, the strong increase of fluorescence emission that is observed on complex formation allows even small changes in phosphate concentration to be detected visually. Moreover, chloride does not interfere with phosphate recognition even at high concentrations, which is advantageous for the use of hydrophilic sapphyrins as selective anion sensors in water.

The finding that pyrrole units in sapphyrins can act as potent ligands for anionic species prompted the search for other pyrrole-derived anion receptors [76]. One class of receptors studied in detail in this respect is based on calixpyrroles, with calix[4]pyrrole **42a** representing the best studied derivative [77, 78]. This compound, whose name derives from its resemblance, not only structurally but also in terms of conformational flexibility, to calixarenes, contains four pyrrole subunits linked between the 2 and 5 positions via disubstituted methylene groups. In solution, **42a** preferentially adopts a conformation with the pyrrole units alternately pointing upward and downward (1,3-alternate conformation), while in the complex with, for example, chloride, a *cone* conformation is observed, with all NH groups pointing to one direction, thus allowing for four simultaneous hydrogen-bonding interactions with the anion [79]. Characterization of the anion affinity of **42a** mainly involved halide and phosphate complexation in organic solvents such as dichloromethane or acetonitrile. Complex stability of the $H_2PO_4^-$ complex of **42a** in 0.5% water/acetonitrile amounts to $1300\,M^{-1}$, for example. The complex of fluorinated calix[4]pyrrole **42b** is seven times more stable ($K_a = 9100\,M^{-1}$), which can be explained by the better hydrogen-bonding donor ability of the NH groups in this receptor [80]. Calix[4]pyrrole derivative **43** binds $H_2PO_4^-$ and HPO_4^{2-} in 4% water/acetonitrile with high association constants of, respectively, 6.8×10^5 and $> 2 \times 10^6\ M^{-1}$, most probably due to two-point interactions of the anions with the calix[4]pyrrole and the thiourea moiety of the receptor [81]. Fluorescence of **43** is quenched upon complex formation, allowing this receptor to be used as an optical anion sensor. Also, electrochemical sensors were devised on the basis of calixpyrroles by appending a ferrocene unit to the macrocyclic core [82].

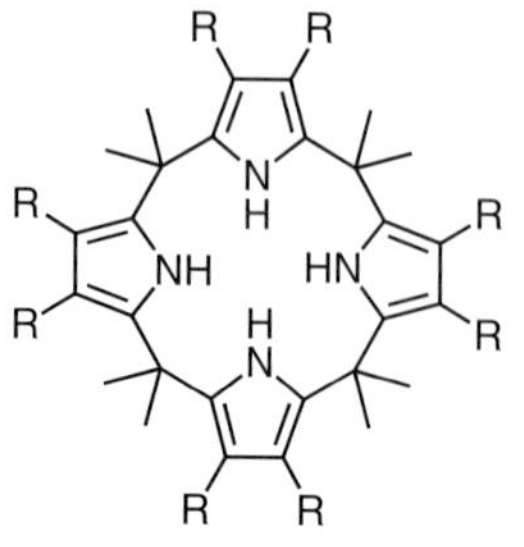

42a; R = H

42b; R = F

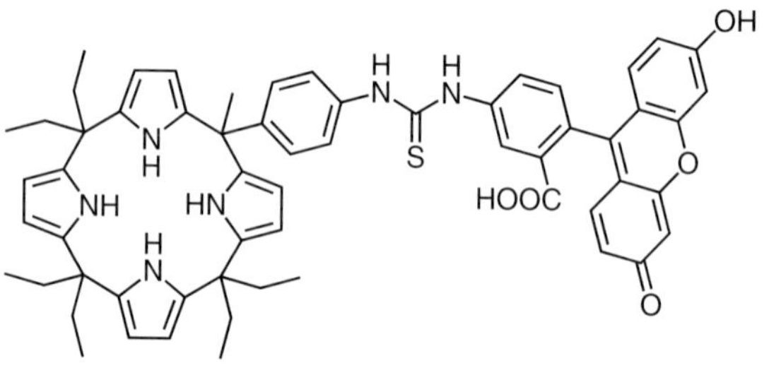

43

Macrocyclic anion receptors containing pyrrole units are also accessible by Schiff-base formation between appropriately functionalized polypyrrole building blocks, for example, diformyldimethyldipyrrolylmethane, and aromatic diamines. This elegant approach gives rise to a whole family of receptors, examples of which are compounds **44–46**. Receptor **44** was shown to bind $H_2PO_4^-$ with a stepwise 2 : 1 stoichiometry in acetonitrile. Binding of the first anion is associated with a K_a of $342\,000\,M^{-1}$, while the second anion is bound with a more

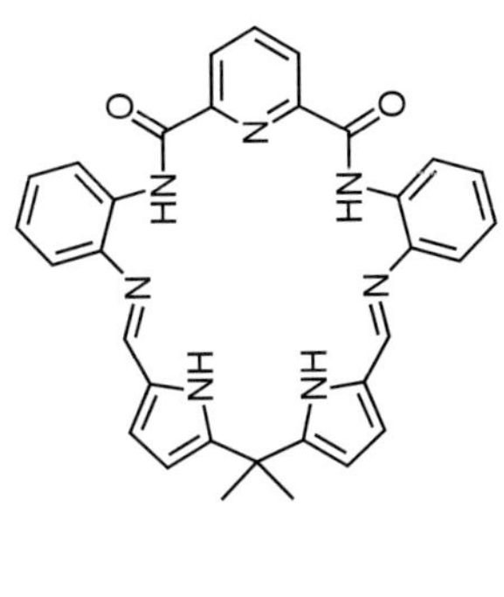

44

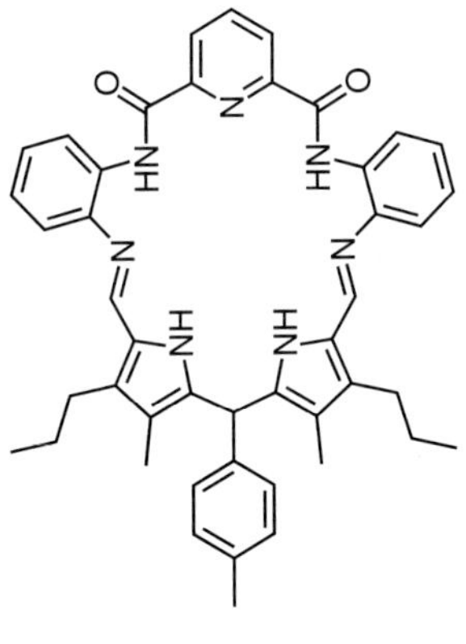

45

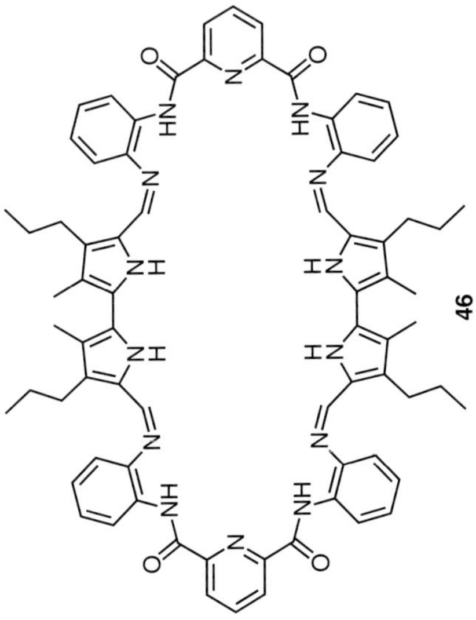

46

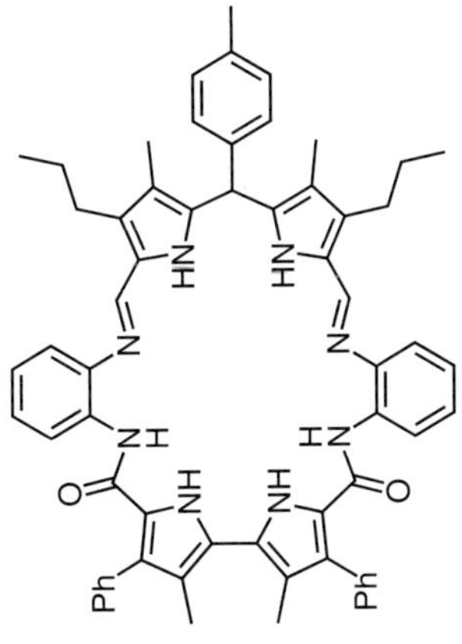

47

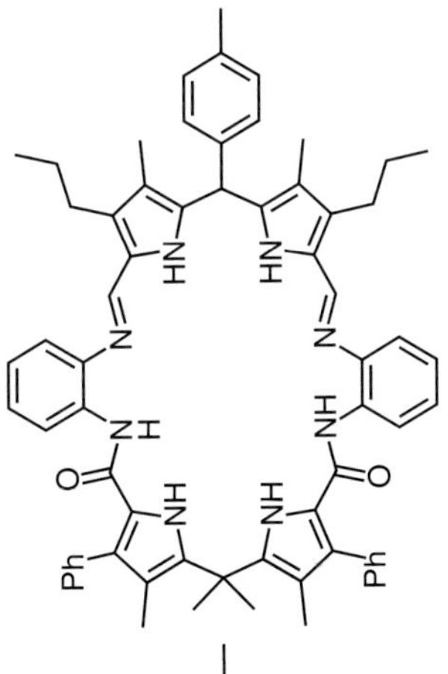

48

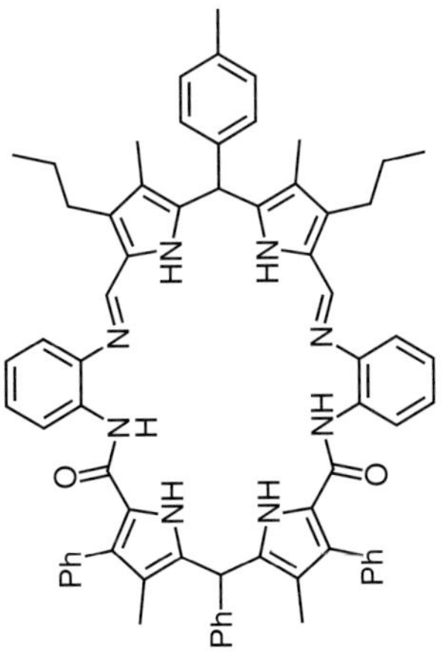

49

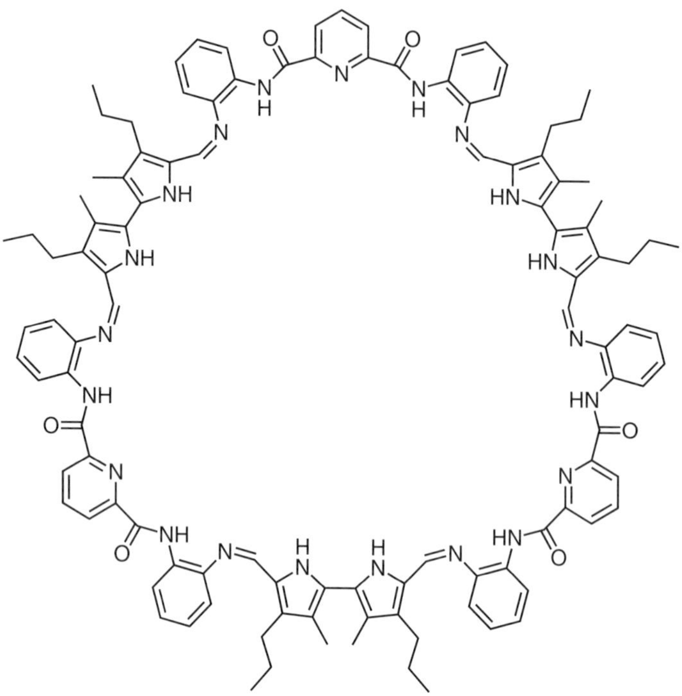

50

than 10 times lower affinity ($K_a = 26\,000\ M^{-1}$) [83]. For comparison, sulfate is bound by **44** in the form of a 1 : 1 complex, whose stability amounts to $64\,000\ M^{-1}$ in acetonitrile. Introduction of a more rigid pyrrole building block reverses phosphate versus sulfate selectivity, leading to receptor **42** that binds HSO_4^- about four times more strongly than $H_2PO_4^-$ ($HSO_4^- \subset$ **45** : $K_a = 1\,08\,000\ M^{-1}$; $H_2PO_4^- \subset$ **45** : $K_a = 29\,000\ M^{-1}$) [84]. Interestingly, receptor synthesis is strongly affected by the presence of oxoanions that can bind to the products, indicating that it is influenced by thermodynamic template effects [85]. For example, reaction between diformyl bipyrrole and bis(2-aminophenyl)pyridine-2,6-dicarboxamide cleanly produces the [2+2] macrocycle **46**, but only in the presence of acids containing tetrahedral oxoanions, such as sulfuric acid or phosphoric acid. Binding properties of **46** with these anions are similar to those of **44**, with the exception that no evidence for chloride binding is seen ($HSO_4^- \subset$ **46** $K_a = 63\,500\ M^{-1}$; $H_2PO_4^- \subset$ **46**:$K_{a1} = 1\,91\,000\ M^{-1}$; $K_{a2} = 60\,200\ M^{-1}$). In the presence of tetrabutylammonium hydrogensulfate or tetrabutylammonium dihydrogenphosphate, receptor **46** rearranges to give a larger [3 + 3] analog in acetonitrile, provided that the reaction mixture is left without stirring [85]. Stirring causes precipitation of the anion complexes of **46**. Derivatives **47–49** have been prepared using a similar strategy. They all form very stable 1 : 1 complexes with dihydrogenphosphate (and other tetrahedral oxoanions) in acetonitrile ($H_2PO_4^- \subset$ **47** : $K_a = 124\,000\ M^{-1}$; $H_2PO_4^- \subset$ **48** : $K_a = 850\,000\ M^{-1}$; $H_2PO_4^- \subset$ **49** : $K_a = 127\,000\ M^{-1}$) [86]. The analysis of the crystal structure of the HPO_4^{2-} complex with diprotonated receptor

50 provided information about the mode of interaction between tetrahedral oxoanions and this type of macrocyclic ligands, also revealing a strong resemblance to the hydrogen-bonding pattern found inside the active center of the phosphate-binding protein [87]. The repertoire of building blocks has recently been extended to obtain new members of this family of macrocyclic anion receptors [88].

Gale showed that acyclic pyrrole derivatives, particularly those containing additional hydrogen bond donors in the form of amide NH groups, can also possess high phosphate affinity [89, 90]. Receptor **51**, for example, binds to $H_2PO_4^-$ in 5% water/DMSO with an affinity too high to be estimated quantitatively by NMR titrations [91]. In 25% water/DMSO, the affinity for the same anion was still substantial, amounting to $234\,M^{-1}$. Unfortunately, **51** was found to be unstable over time in solution, oxidizing to the corresponding dipyrrolylmethene. To overcome this problem, the hydrogen atoms in the *meso*-methylene group were substituted by methyl groups. The corresponding receptor **52** again showed high affinity for dihydrogen phosphate, amounting to $1092\,M^{-1}$ in 5% water/DMSO [92]. DFT calculations suggest that the two pyrrole-amide groups bind to different oxygen atoms in the $H_2PO_4^{2-}$ anion.

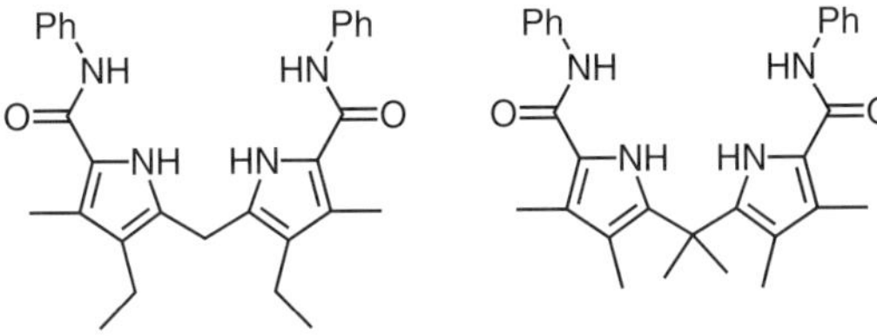

51 **52**

Recently, several groups have shown that indole derivatives can also be useful building blocks for phosphate receptors [93]. Examples include 2,7-disubstituted indole **53** [94], macrocyclic [95] or acyclic indolocarbazole derivatives (**54** and **55**) [96–98], benzodipyrrole **56** [99], indolocarbazole-quinoxaline **57** [100, 101], or bis(indoles) **58** and **59** in which the two heterocyclic units are connected via appropriate linkers [102–104]. Phosphate affinity of these compounds varies, also strongly depending on the solvent. Some of the more potent phosphate receptors of this type are macrocycle **54**, whose dihydrogenphosphate complex has a K_a of $6.5 \times 10^6\ M^{-1}$ in acetonitrile [95], and 1,3-diindolylurea **59**, which exhibits appreciable affinity to the same anion even in 25% water/DMSO ($H_2PO_4^- \subset$ **59** : $K_a = 160\,M^{-1}$) [104]. The benzodipyrrole-derived receptor **56** was shown to possess good selectivity for $HP_2O_7^{3-}$ in acetone containing 5% (v/v) water ($HP_2O_7^{3-} \subset$ **56** : $K_a = 9081\ M^{-1}$) [99]. Molecular modeling indicated that complex formation involves a tight network of hydrogen bonds between the donors of **56** and oxygen atoms of the substrate.

Several metal-containing receptors for phosphate anions have already been discussed (*vide supra*). These examples show that, in general, introduction of one or more coordinatively unsaturated metal ions into a receptor framework can result in high anion affinity even in aqueous media. The Reinhoudt group made use of this concept when developing receptors **60** and **61** [105].

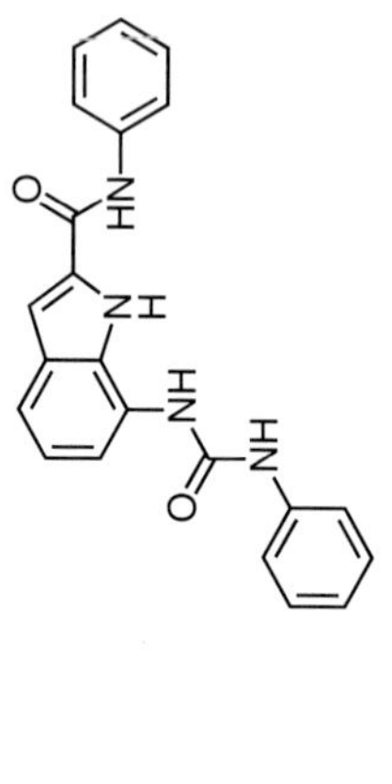

53

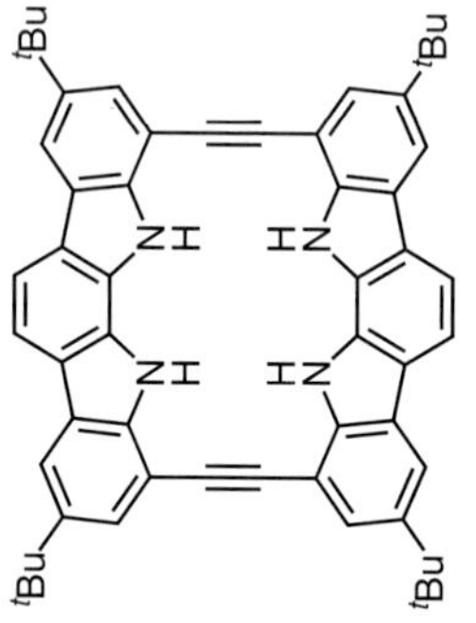

54

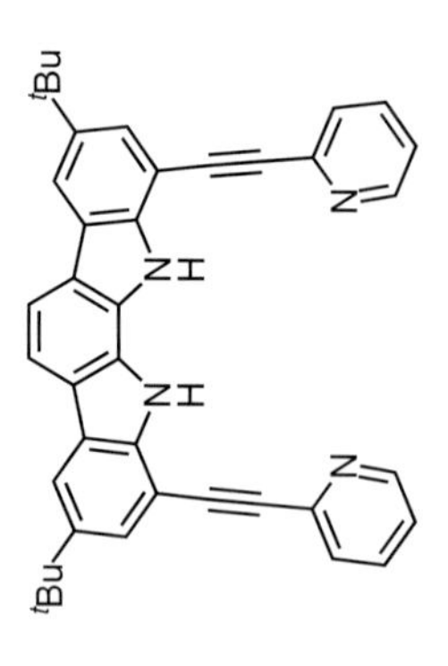

55

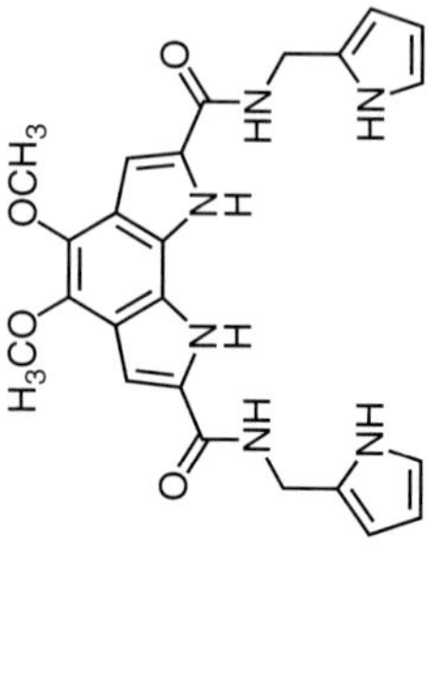

56

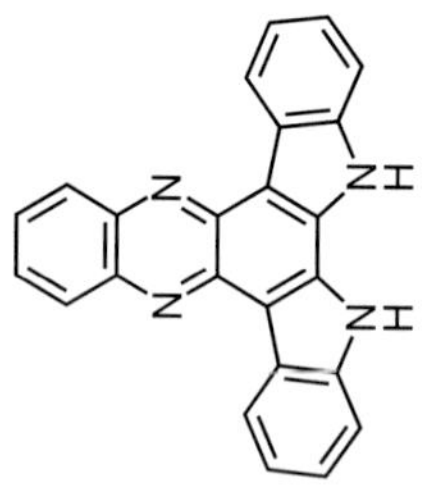

57

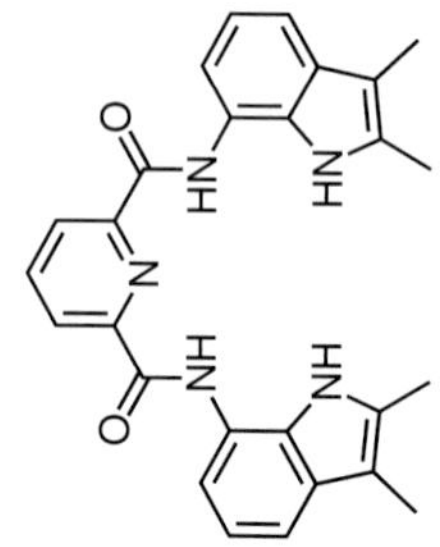

58

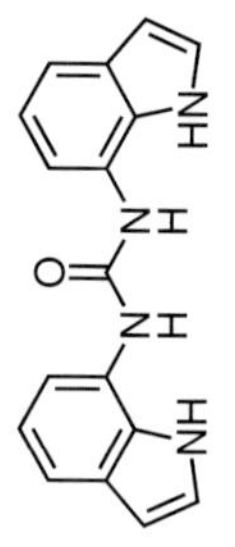

59

These receptors combine the uranyl salen complex as a Lewis-acid-binding site with amide groups that cooperatively contribute to anion binding. Binding studies showed that such receptors are highly selective for dihydrogenphosphate. Stability of the dihydrogen phosphate complexes of receptor **60** amounts to $>10^5$ M^{-1} in 1% DMSO/acetonitrile, for example, about 100 times higher than chloride affinity and even 1000 times higher than that for nitrate.

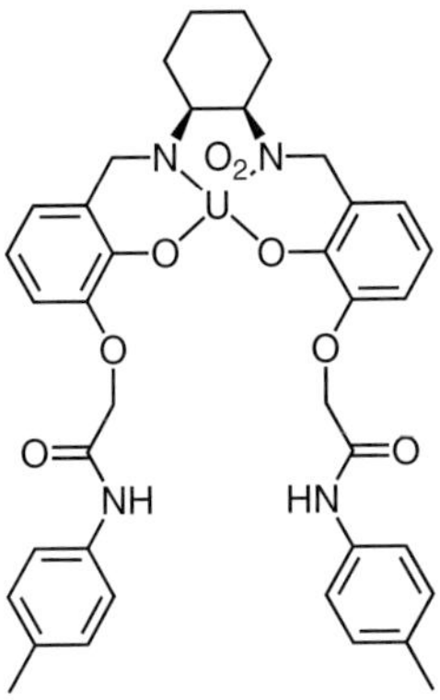

60

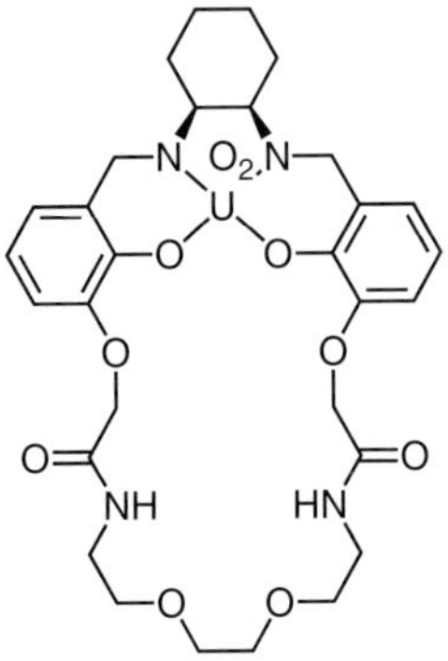

61

Examples of dimetallic coordination complexes [106] for phosphate include compounds **62**–**64**. In a systematic study, zinc(II) dipicolylamine-derived receptor **62** served to elucidate the influence of substituents R^1 and R^2 on phosphate affinity [107]. In 50 mM HEPES buffer at pH 7 and 25°C, the parent compound **62a** binds the dihydrogenphosphate anion with an affinity of 19.3×10^4 M^{-1}. Affinity increases by a factor of about 10 if amino groups are introduced on one side of the receptor as in **62b**. Isothermal titration calorimetry indicated that the adverse effect the introduction of the amino groups has on binding enthalpy is more than compensated for by more favorable binding entropy. This result was rationalized by assuming that the increased enthalpic penalty is required to break the solvation sphere of the hydrophilic amino groups in **62b**, but this cost is more than repaid by the associated liberation of several ordered water molecules. The even more pronounced unfavorable binding enthalpy and more favorable binding entropy observed for **62c** is consistent with this assumption. For this receptor, enthalpy and entropy compensate each other, however, leading to a binding constant that is similar to that of **62a**.

Phosphate sensing was realized by using an indicator displacement assay comprising receptor **63** and pyrocatechol violet as indicator [108]. In the absence of phosphate, an equimolar mixture of the host and the indicator in water (10 mM HEPES, pH 7.0) gives rise to a blue solution. Addition of phosphate to this solution causes the color to change to yellow, whereas no color change is observed in the presence of other anions such as sulfate, halides, carbonate, acetate, azide, or nitrate. Stability of the HPO_4^{2-} complex of **63** amounts to 11.2×10^4 M^{-1}. A fluorescent version of this indicator displacement assay was devised by Smith and coworkers employing a coumarine derivative as indicator [109].

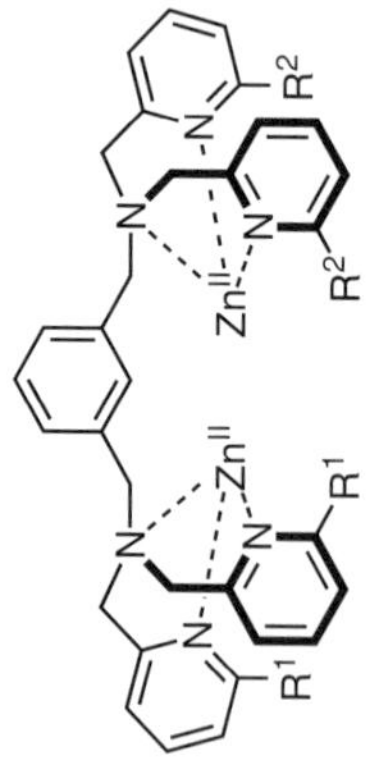

62a; $R^1 = R^2 = H$
62b; $R^1 = NH_2$, $R^2 = H$
62c; $R^1 = R^2 = NH_2$

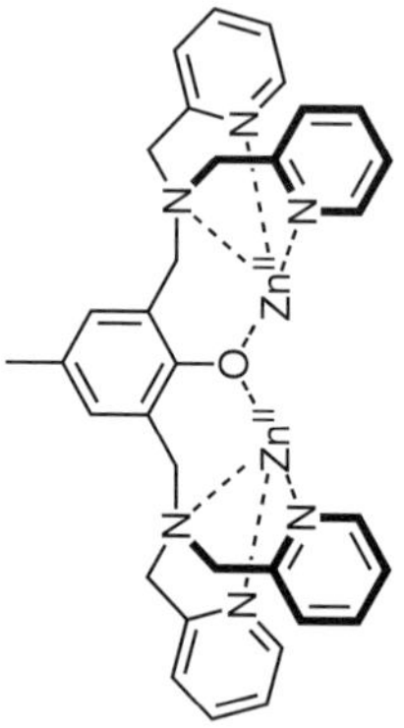

63

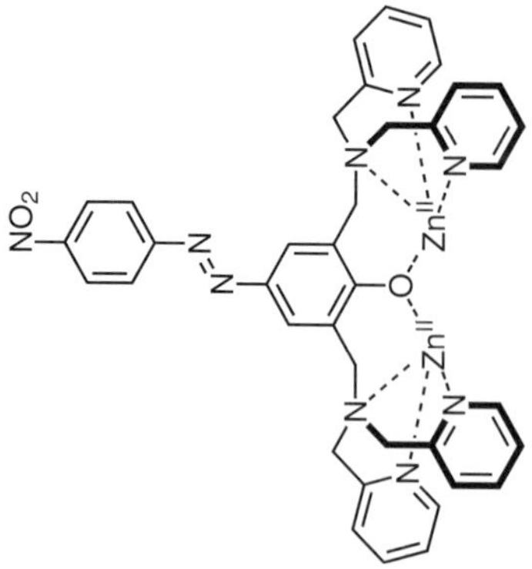

64

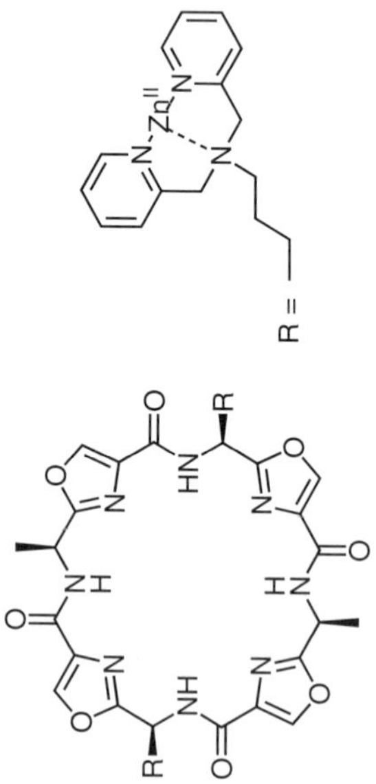

65

The azophenol-based derivative **64** is selective for pyrophosphate [110]. Binding can be detected by the color change of an aqueous solution (10 mM HEPES, pH 7.4) of this host from yellow to orange on addition of $Na_4P_2O_7$, an effect not observed on addition of the sodium salts of hydrogenphosphate, citrate, acetate, or fluoride. Binding studies demonstrated that the stability of the pyrophosphate complex of **64** is 3 orders of magnitude larger than that of the HPO_4^{2-} complex of **64** ($P_2O_7^{4-} \subset$ **64** : $K_a = 6.6 \times 10^8\ M^{-1}$).

The Jolliffe group demonstrated that C_2 symmetrical cyclic octapeptide with two appended dipicolylamino groups complexed to zinc(II) **65** can also serve as a highly selective sensor for pyrophosphate [111]. Of the 15 anions tested, only pyrophosphate and, to a lesser extent, ATP, ADP, and citrate were able to displace a coumarine derivative from its complex with **65**, causing a recovery of the coumarine fluorescence. Pyrophosphate affinity of **65** in aqueous solution (pH 7.2, HEPES buffer) was determined to amount to a log K_a of 8.0, at least 2 orders of magnitude larger than ADP or ATP affinity. The large selectivity for pyrophosphate is due to the distance of the primary binding sites in **65** that allows optimal contacts with this anion. In addition, NMR spectroscopic investigations indicated that the amide groups along the peptidic scaffold of **65** could provide a secondary binding site for the pyrophosphate anion.

7.2.3
Nucleotides

Early investigations concerned with complexes between protonated polyaza macrocycles and nucleotides usually addressed the question as to what extent complex stability is affected by the degree of protonation of the receptor or ring size. These studies focused on binding of AMP, ADP, or ATP as model nucleotides because no specific interactions between receptor and the hydrogen bond acceptors and donors in the nucleobase residue of the substrate were expected. Variation of substrate structure therefore only involved the phosphate residue. Ring size of polyaza macrocycle **66** turned out to be optimal for ATP binding. The complex between the heptaprotonated form of **66** and the fully deprotonated ATP anion has a stability constant log K_a of 12.93, for example (0.15 M $NaClO_4$), while those of larger receptor analogs (having the same degree of protonation) become progressively less stable with increasing ring size [112]. Complex stabilities of the ATP, ADP, and AMP complexes of **66** decrease in the order ATP > ADP > AMP, consistent with the general correlation between complex stability and charge on host or, in this case, substrate (log K_a of heptaprotonated $ADP^{3-} \subset$ **66** $= 10.04$; log K_a of heptaprotonated $AMP^{2-} \subset$ **66** $= 8.87$).

Control over ATP affinity of polyaza macrocycles can be achieved by methylation of secondary amino groups along the ring. The tetraprotonated form of **67**, in which the protons are preferentially located on the secondary nitrogens, binds ATP^{4-} in 0.15 M $NaClO_4$ at 298 K more than 1 order of magnitude more strongly than the tetraprotonated form of **68** (log K_a of tetraprotonated $ATP^{4-} \subset$ **67** $= 7.39$; log K_a of tetraprotonated $ATP^{4-} \subset$ **68** $= 5.91$) [16]. No large effect is, however, observed on introduction of two other methyl groups in polyaza macrocycle **69** (log K_a of tetraprotonated $ATP^{4-} \subset$ **69** $= 7.48$) [14].

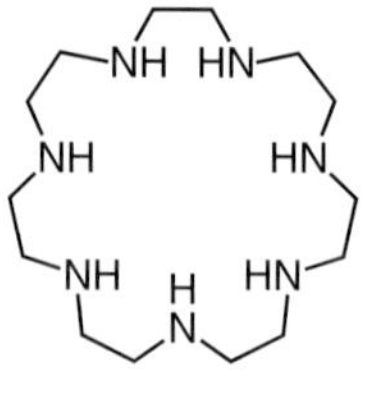

66

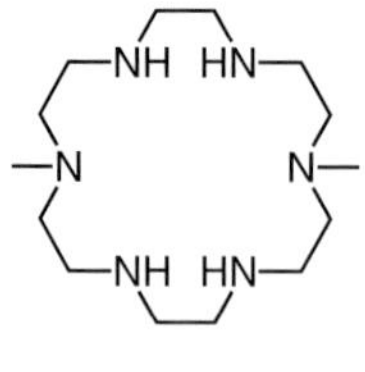

67

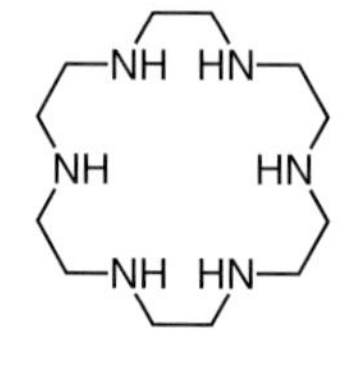

68

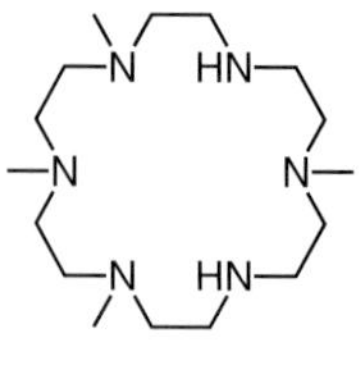

69

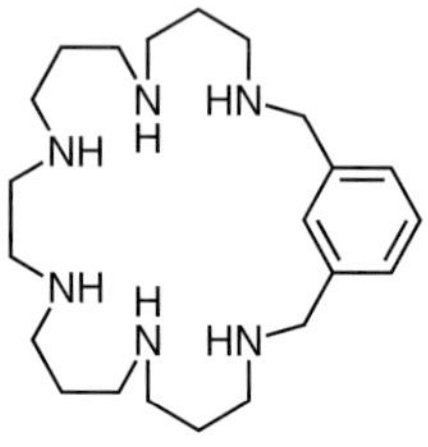

70

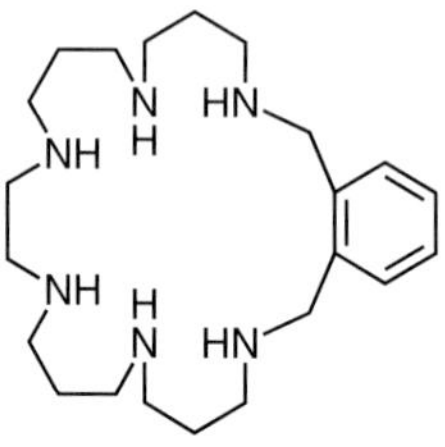

71

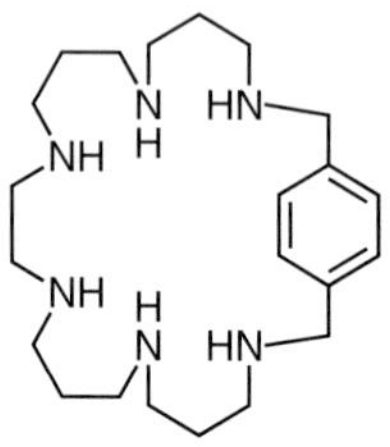

72

An unusual feature of several polyaza macrocycles that goes beyond simple phosphate recognition is their ability to catalyze both dephosphorylation of ATP and, under suitable conditions, phosphorylation of ADP. Also, hydrolysis of acetyl phosphate and activation of formate can be achieved with such compounds. Work in this area has been reviewed and is not considered further here [2, 3, 113, 114].

Another successful approach to improve the affinity of polyaza macrocycles for nucleotides involves introduction of aromatic subunits. These subunits not only impose a higher rigidity on the receptors but also allow for π-stacking interactions with the nucleobase residues of the guests, thus cooperatively contributing to complex stability. Complex formation between **70** and ATP, ADP, and AMP, for example, was found to combine electrostatic and hydrogen-bonding interactions between the ammonium groups of the host and the guest's phosphate moiety as well as aromatic interactions. The latter induced characteristic shifts of guest and host signals in the ^1H NMR spectra of the complexes with respect to the spectra of the free components, demonstrating that the adenine group of the guests and the aromatic subunit of the host are in close contact [115–117]. Comparison of the ATP affinities of structurally closely related receptors **70–72** revealed that ortho-disubstitution at the aromatic moiety in **71** causes the highest affinity, while the lowest is observed for the para-disubstituted derivative **72**, indicating that receptor structure has a decisive effect on complex stability (log K_a of hexaprotonated $ATP^{4-} \subset$ **70** = 7.6; log K_a of hexaprotonated $ATP^{4-} \subset$ **71** = 8.1; log K_a of hexaprotonated $ATP^{4-} \subset$ **72** = 6.7) [118].

High affinity for ATP, ADP, AMP, and inorganic phosphate anions has also been detected for host **73**. Comparison of the nucleotide affinity of **73** with that of **74** lacking the aromatic subunits revealed that while the more basic host **74** forms more stable complexes with phosphate than **73**, the opposite is true for nucleotide binding, which demonstrates a stabilizing effect of $\pi-\pi$ interactions in complex formation [24]. Spectroscopic evidence for a close contact in the complexes between the aromatic subunits of **73** and the nucleotide adenine moiety came again from NMR investigations. Complex formation between **12** and ATP involves partial inclusion of the triphosphate moiety into the cleft of this receptor, while the adenine moiety remains outside the cavity, forming π-stacking interactions with one bipyridine group [21]. Also, receptor **13** has been shown to strongly interact with nucleotides (log K_a of hexaprotonated $ATP^{4-} \subset$ **13** = 6.5; log K_a of hexaprotonated $ADP^{3-} \subset$ **13** = 5.8) [27]. In this case, molecular dynamic calculations and ^{31}P NMR spectroscopy were used to elucidate structural aspects of the complexes formed. These methods showed that binding involves interaction between the lateral macrocyclic units in **13** and the terminal and central phosphate unit of ATP or the terminal phosphate unit of ADP, while the adenine units of these nucleotides seem to play a minor role in complex stabilization.

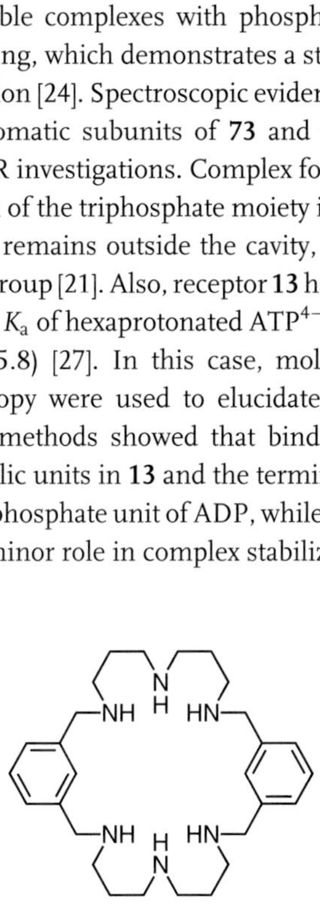

73

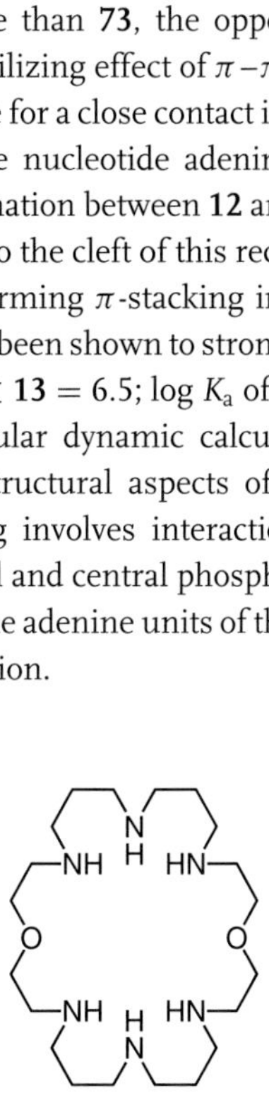

74

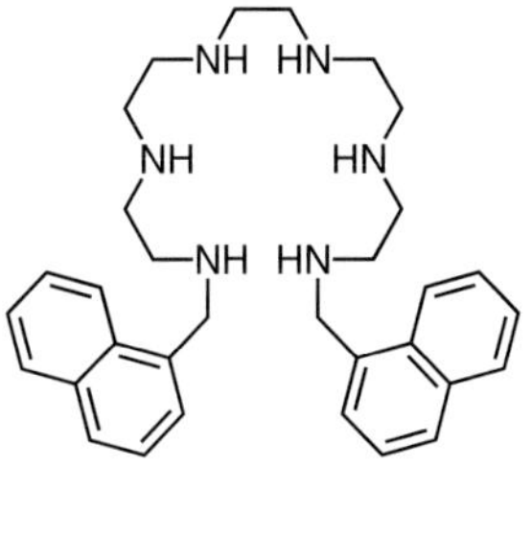

75

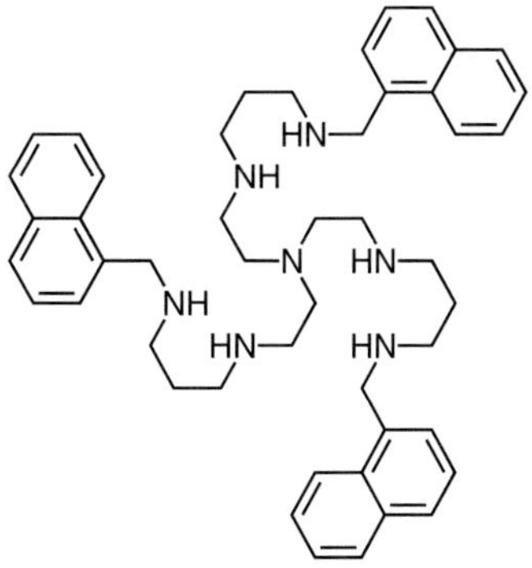

76

Also, acyclic receptors such as **75** or **76** are able to interact with nucleotides. For example, the triprotonated form of **75** binds to fully deprotonated ATP with a log K_a of 5.2 (0.15 M NaCl) [119]. There is NMR spectroscopic evidence of a close proximity of the adenine moiety of the substrate and the aromatic substituents of the receptor, and steady-state fluorescence measurements revealed quenching of the emission of the receptor fluorophores at acidic pH following the protonation of the adenine ring. Receptor **76** can bind up to three AMP anions simultaneously [120]. This compound is also an example of a receptor, whose copper(II) complex was characterized with respect to nucleotide binding. Other metal-containing nucleotide receptors are zinc complexes **77–80** that form complexes with ATP [121]. Association constants of the ATP complexes increase in the order **77** (log $K_a = 3.25$) < **78** (log $K_a = 3.74$) < **79** (log $K_a = 4.35$) < **80** (log $K_a = 5.18$), which was rationalized in terms of a more "open" coordination sphere at the zinc ion in hosts **79** and **80**. ^{31}P NMR spectroscopy showed that the terminal phosphate group of ATP and not the ring nitrogens in the adenine moiety interacts with the metal in the complexes. Contributions to complex stability from the adenine system were detected in the form of π-stacking interactions. Protonation of amino groups of the host not involved in metal coordination results in a further increase in complex stability due to the formation of salt bridges. Because protonated metal-free ligands form much less stable complexes, the high ATP affinity of protonated species of **77–80** was ascribed to synergetic effects of the metal ion and of the ammonium groups [121]. A fluorescent sensor employing the dye displacement strategy was devised on the basis of macrocyclic dicopper(II) complex **16** for the detection of ATP in water [30]. More recently, the dicopper(II) cryptand complex **81** was described, which was shown to preferentially recognize guanosine monophosphate (GMP) over other nucleotide monophosphates in methanol/water 1 : 1 (v/v) mixtures at pH 7 [122]. Thus, compound **81** represents one of the few receptors that also selectively recognize the nucleobase residue on the substrate. This discriminating behavior of **81** was ascribed to the capability of GMP to bridge, with its phosphate and carbonyl oxygen atoms, the distance between the two copper centers inside the cryptate, leading to a stable 1 : 1 complex (log $K_a = 4.7$). Increase in mononucleotide concentration then leads to the formation of a 2 : 1 complex consisting of two cryptates bridged via one nucleotide molecule. Only in the case of GMP, this consecutive binding equilibrium is associated with a smaller binding constant than the first one. Thus, inclusion of other nucleotides into the cavity of **81** seems to be significantly less favorable than that of GMP, indicating good size complementarity between this substrate and the receptor. Recognition of GMP is effectively signaled through the displacement of the indicator 6-carboxyfluorescein, which resulted in the recovery of the indicator's yellow emission.

Sensing of guanosine triphosphate over other nucleotide triphosphates was achieved by using a fluorescence assay comprising metal-free receptor **82** and 8-hydroxy-1,3,6-pyrene trisulfonate (HPTS) as the indicator [123]. Fluorescence of HPTS is effectively quenched on binding to **82** and can be recovered when displacing the dye from the receptor with GTP. This effect is negligible on addition of adenosine, AMP, ADP, CTP, or UTP (uridine triphosphate) to the aqueous solution (phosphate buffer, pH 7.4) of HTPS ⊂ **82** and much smaller when ATP is used. For UTP and uridine diphosphate (UDP) sensing, zinc(II) dipicolylamine-derived receptor **83** was developed [124]. Fluorescence of the perylene platform in this receptor is turned on in CH_3CN/HEPES buffer (0.01 M, pH 7.4) in the presence of these nucleotides, which was ascribed to simultaneous binding of the phosphate moiety and the nucleobase (via the deprotonated imide nitrogen) to the two metal centers. Other nucleotides, for example, ATP, CTP, GTP, ADP, AMP, or UMP, did not cause a similar effect. This allowed real-time monitoring of the enzymatic dephosphorylation of UTP in a fluorescence assay.

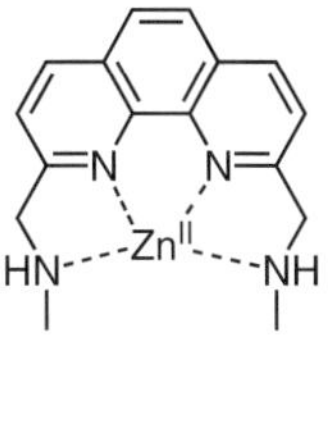

77

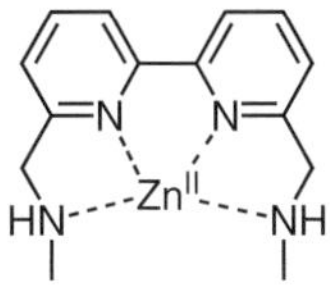

78

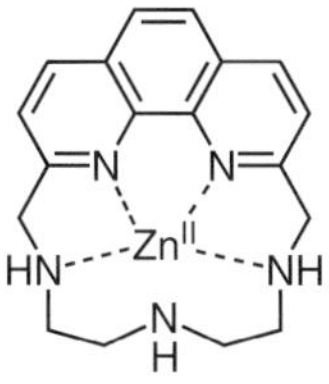

79

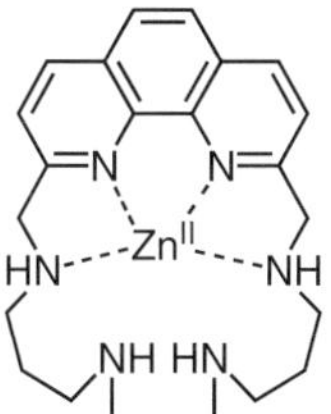

80

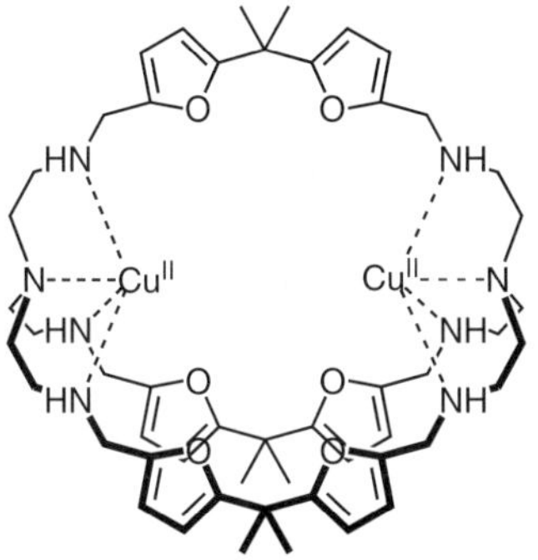

81

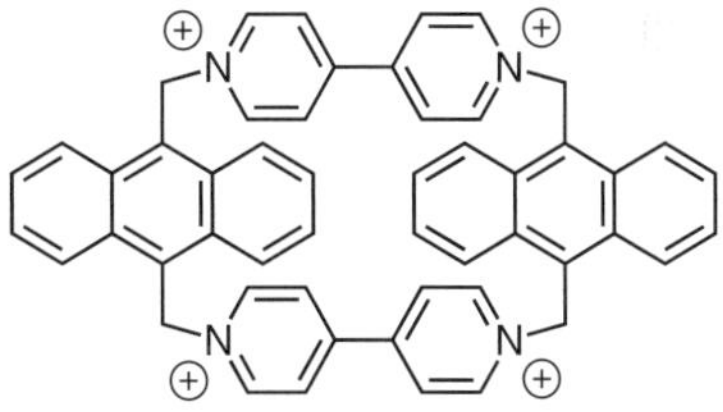

82

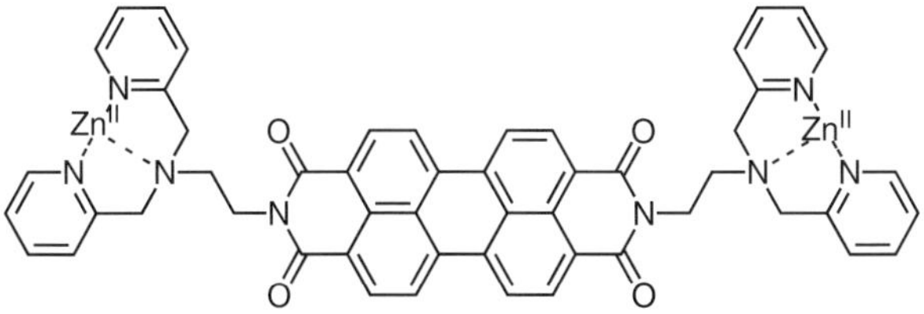

83

Host **66** has been shown to interact with the dinucleotides NAD and NADP. Of these two substrates, NADP is bound with higher affinity most probably because the complex is stabilized by interactions between the phosphate moiety of NADP and two adjacent ammonium groups of the host (log K_a of heptaprotonated NAD ⊂ **66** = 7.59; log K_a of heptaprotonated NADP ⊂ **66** = 11.84) [125]. Another example of a dinucleotide receptor, compound **84**, has been obtained by appending two acridine moieties to **5**. This receptor exhibits a remarkable selectivity for NADPH, which is bound by a factor of about 10^3 better than NADP and by a factor of $>10^6$ better than NAD(H) [126]. Moreover, binding can be followed by optical methods as complex formation causes an enhancement of the acridine fluorescence. It should be noted that NAD recognition can also be achieved with the so-called "molecular clips" developed recently in a collaborative approach of the Klärner and the Schrader groups. However, these receptors target only the aromatic moieties of the substrate without directly interacting with the phosphate residues [127].

Structurally somewhat related to **84** is the pincerlike imidazolium-derived receptor **85** [128]. This compound senses ATP in buffered water (20 mM HEPES, pH 7.4) by a decrease in the excimer band of the two pyrene residues in the fluorescence spectrum and the concomitant enhancement of the monomer pyrene emission. Other nucleotide triphosphates only induce quenching of the excimer band, thus allowing selective ATP sensing. The observed effects of complex formation on the fluorescence spectrum of **85** were attributed to different complex geometries: the adenine residue of ATP is sandwiched between the two pyrene residues, while other nucleobases can stack on one pyrene residue only from the outside. The selectivity of **85** for ATP is also sufficiently high to selectively detect ATP in the presence of ADP or AMP. A ratiometric fluorescence assay could thus be established to monitor hydrolytic dephosphorylation of ATP by the enzyme apyrase. Other imidazolium-derived receptors for the selective nucleotide recognition have also been described [129, 130].

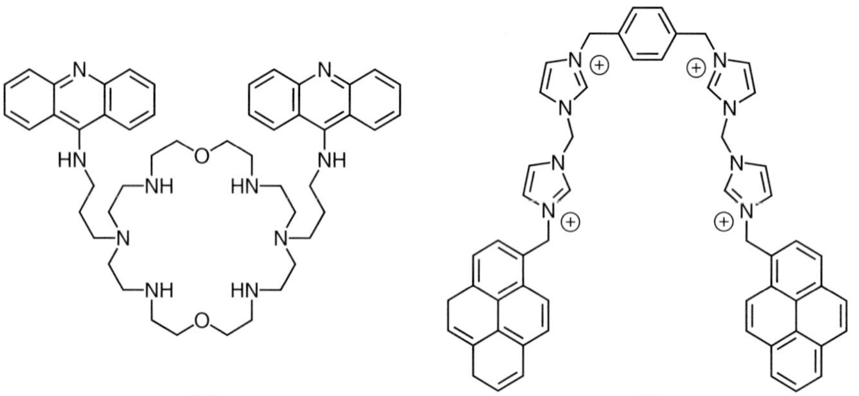

84 **85**

Another important class of receptors for nucleotides relies on guanidinium or amidinium groups for substrate binding. Receptor **23b** has, for example, been shown to bind AMP, ATP, and NAD in water (K_a of AMP $\subset$ **23b** $= 204\ M^{-1}$; K_a of ATP $\subset$ **23b** $= 840\ M^{-1}$; K_a of NAD $\subset$ **23b** $= 140\ M^{-1}$) [43]. The Diederich group has shown that arranging four amidinium groups around the cavity of a resorcin[4]arene also furnishes nucleotide receptors [131]. The corresponding cavitand (**86**) forms 1 : 1 complexes with *c*AMP, AMP, ADP, and ATP in water (2.5 mM TRIS, pH 8.3), with stability constants ranging from $1.4 \times 10^3\ M^{-1}$ (for *c*AMP) to 6.6 × 10^5 M^{-1} (for ATP). The increase in binding strength with increasing charge of the substrate indicates a large contribution of electrostatic interactions to binding. Additionally, a selectivity for AMP over nucleotide monophosphates containing other bases was also detected, which was explained, on the basis of detailed ^{1}H NMR studies, by an inclusion of the nucleobase part of AMP into the bowl-shaped cavity of **86**. The major driving force of complex formation is believed to come from ion pairing and hydrogen-bonding interaction between the substrate and the host. Apolar interactions and hydrophobic desolvation most probably do not make a large contribution to complex stability.

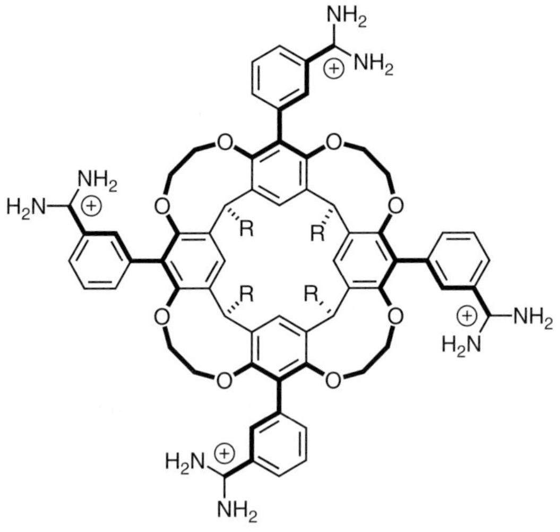

R = $(CH_2)_3(OCH_2CH_2)_3OCH_3$

86

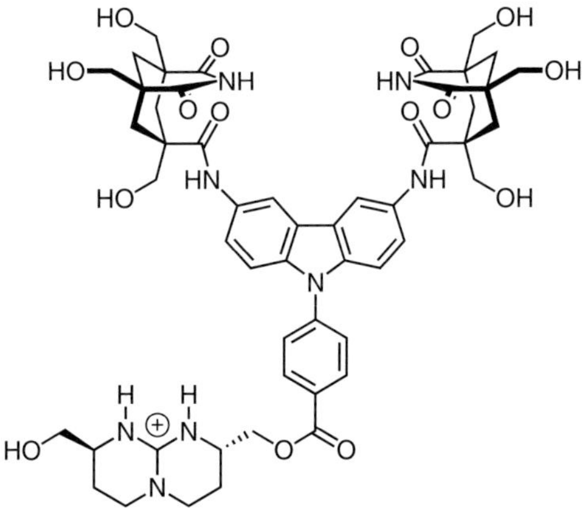

87

Host **87**, developed by Rebek and coworkers, arranges a bicyclic guanidinium moiety and two Kemp's triacid imides around a 3,5-diaminocarbazole platform.

In water (10°C, pH 6.0, 51 mM NaCl), **87** binds 2′3′-*c*AMP and 3′,5′-*c*AMP with association constants of, respectively, 660 and 600 M^{-1} [132]. The slight preference for 2′3′-*c*AMP vanishes, however, on increasing the ionic strength of the solution. Adenosine or 9-ethyladenine is bound significantly less strongly by the host, clearly demonstrating the cooperative effect of the guanidinium group in nucleotide recognition. Since complex formation also involves specific Watson–Crick and Hoogsteen hydrogen-bonding interactions with the nucleobase, **87** is selective for adenine-containing substrates [133]. Host **87** and a derivative containing two carbazole moieties linked by a bicyclic guanidinium have also been shown to allow extraction of nucleotides and dinucleotides from water into dichloromethane and nucleotide transport across liquid membranes [133–135].

Compound **87** is another example of a ditopic receptor that combines a binding site for a phosphate moiety with a specific nucleobase recognition motif to achieve selectivity in nucleotide recognition. Two other nucleotide receptors of this type should be mentioned. One was developed by the Anslyn group by using a combinatorial approach. This receptor is built on a 1,3,5-trisubstituted 2,4,6-triethylbenzene core, whose usefulness as receptor scaffold derives from the fact that steric repulsion between adjacent aromatic substituents cause every other substituent to be oriented on the same face of the benzene ring [136]. Two identical peptide chains of variable sequence terminated by fluorophore F^2 are attached to positions 1 and 3 of the central benzene unit, while position 5 is attached to a resin via a lysine linker containing fluorophore F^1 in the side chain. The corresponding compound **88** is based on the structurally much simpler host **89**,

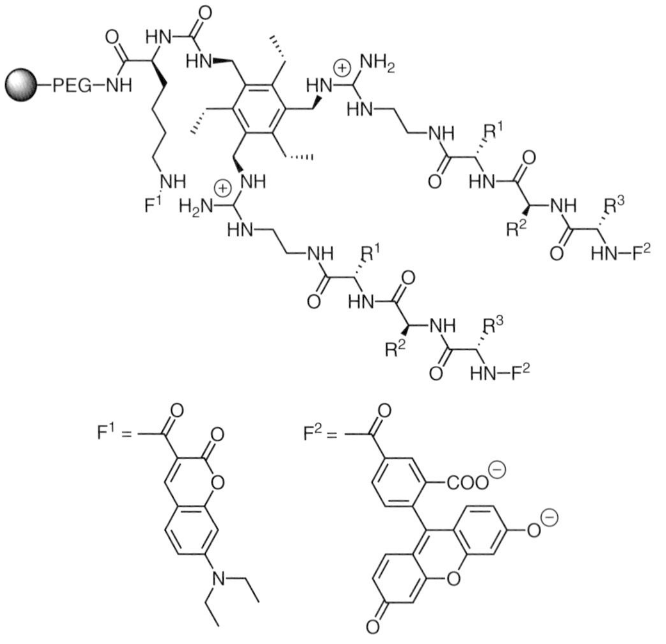

88

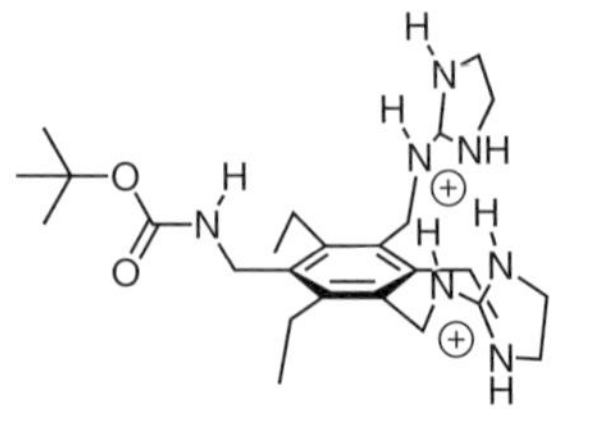

89

which has been shown to interact with ATP with a binding constant of $3.5 \times 10^2\ M^{-1}$ in water [137]. Screening of a library containing 4913 derivatives of **88** differing in the sequence of the side chains furnished a host that binds ATP about 10 times stronger than **89** ($K_a = 3.4 \times 10^3\ M^{-1}$ in 200 mM HEPES, pH 7.4) [137]. In addition, this host turned out to be selective for ATP over GTP and AMP. In a slightly different approach, a selection of 12 beads of the host library (before attachment of the fluorophores) was chosen to construct a chip-based array for the optical differentiation between structurally similar anions such as AMP, GTP, and ATP using an indicator displacement assay [138].

Cytosine- and guanine-functionalized sapphyrin derivatives **90a,b** were designed to selectively recognize mononucleotides containing the complementary nucleobase, namely, GMP and CMP. These receptors indeed enhance transport through a liquid membrane of substrates that form complementary Watson–Crick base pairs with respect to mismatched nucleotides by a factor of about 10 [139, 140]. Interestingly, transport rate is also sensitive to the position of the phosphate group on the ribose moiety, as 2′, 3′, and 5′-cytidine monophosphates are transported by **90b** at different rates.

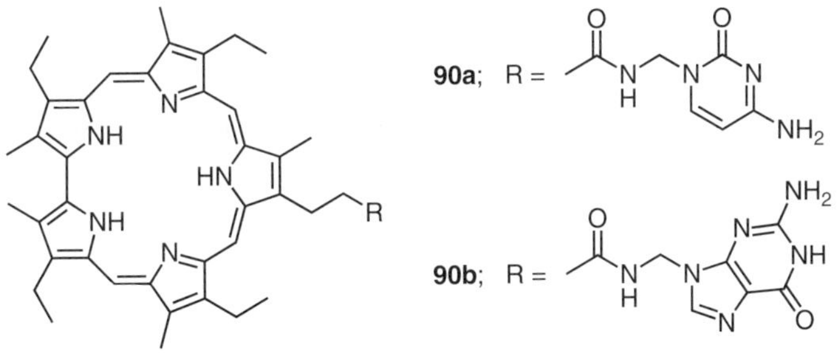

7.2.4
Phosphate Esters

Various anion receptors have been shown to interact with phosphate esters such as 4-nitrophenylphosphate or dibenzylphosphate. These substrates serve as model compounds for the biologically relevant phosphate esters of carbohydrates or amino acids with hydroxy groups in the side chain or for phosphate diesters in polynucleotides. Examples are bicyclic-guanidinium-containing receptor **23b** [43], sapphyrin **91** [74], tripodal-amide-containing ligand **92** [141], and cyclic hexapeptide **93** [142]. The last compound has been shown to bind 4-nitrophenyl phosphate with a K_a of $1.2 \times 10^6\ M^{-1}$ in DMSO. Replacement of alanine in **93** with serine causes a slight reduction in complex stability. A much larger drop was, however, observed for acyclic and larger analogs of **93**, for example, the corresponding cyclic octapeptide, demonstrating that the hexapeptide has the correct size and shape for phosphate ester recognition.

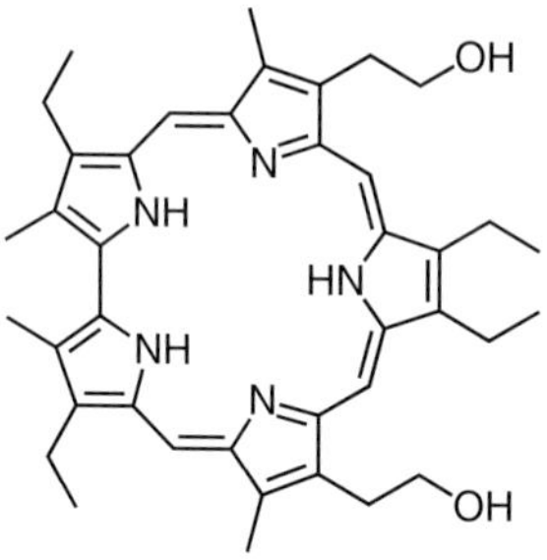

91

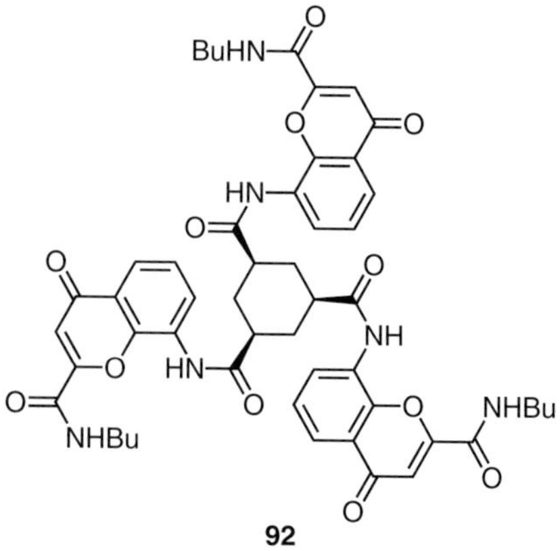

92

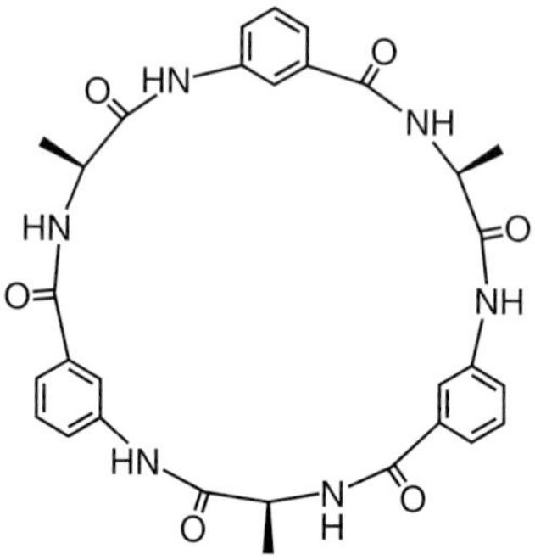

93

Further neutral receptors for phosphate esters have been developed by the Smith and Miller groups. Smith's group is interested in the development of molecular mimics of translocases, enzymes that flip phospholipids in biological membranes from one side to the other [143–145]. In this context, it was demonstrated that tren sulfonamide **94** facilitates phosphatidylcholine translocation across synthetic vesicles and erythrocyte membranes [146, 147]. Activity of **95** proved to be even higher and includes the translocation of phosphatidylserine [148]. Mechanistic studies indicate that both hosts form a complex with the phosphatidyl headgroup of the lipid by hydrogen bond formation between the sulfonamide and urea NH groups and the phosphate oxygens, thus reducing headgroup hydrophobicity and promoting diffusion through the nonpolar interior of the bilayer membrane. Such complexes are most stable in a nonpolar environment such as in chloroform [149]. In collaboration with the Davis group, Smith showed that translocation of phospholipids can also be promoted with cholate esters such as **96** [150, 151]. The urea groups at the 7α- and 12α-positions of **96** are essential for strong binding of the phosphate group in the phosphocholine headgroup of the substrate and cannot be replaced by amide, hydroxy, or amino groups. Interestingly, these synthetic translocases have very weak affinity for phosphatidylethanolamine and phosphatidylserine.

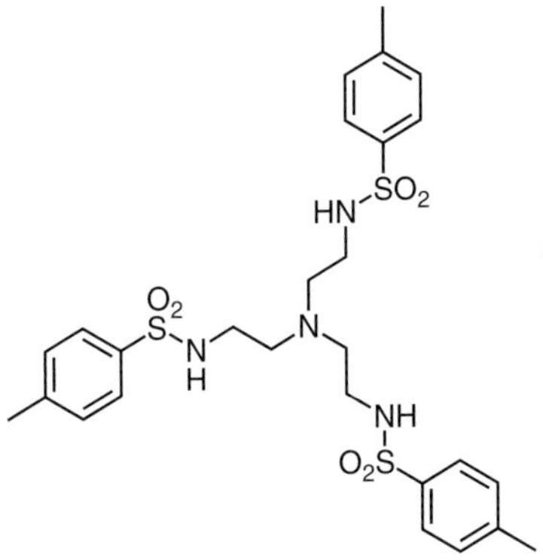

94

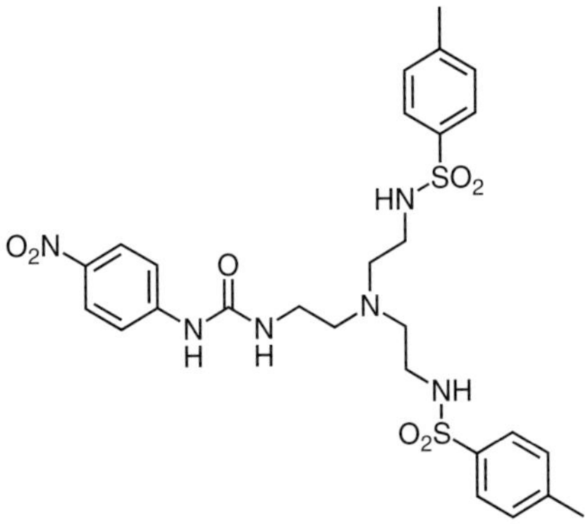

95

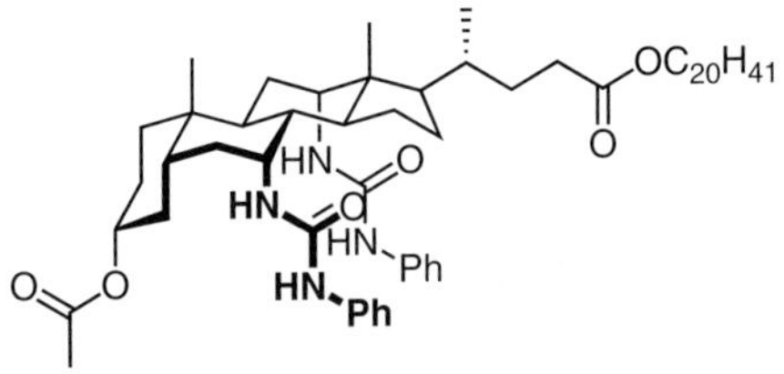

96

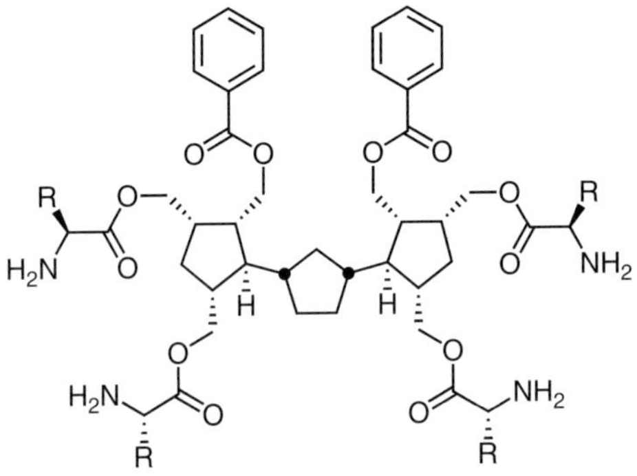

97a; R = CH_2Ph

97b; R =

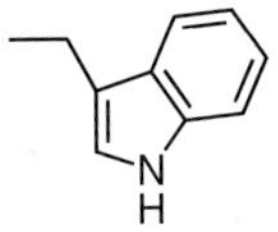

97c; R = H

Interactions between the neutral *ter*-cyclopentanes **97a** or **97b** and the anionic bisphosphate lipid A in phosphate-buffered saline at pH 7.4 were demonstrated by Miller and coworkers [152]. Lipid A constitutes the innermost of the three regions of the lipopolysaccharide (LPS) endotoxin of gram-negative bacteria. Its hydrophobic nature allows it to anchor the LPS to the outer membrane. While its toxic effects can be damaging, sensing of lipid A by the human immune system could also trigger immune responses to bacterial infection. Job plots indicate that defined 1 : 1 complexes are formed, with dissociation constants of 587 and 592 nM for the complexes of, respectively, **97a** and **97b**. Because only weak binding was detected between **97c** and lipid A, complexation is obviously promoted by the hydrophobic substituents of **97a** and **97b**.

A polymeric receptor for an oligopeptide containing a phosphorylated tyrosine moiety has recently been prepared by using the molecular imprinting technique [153]. This polymer showed selective binding of the peptide that was present during polymerization over the corresponding nonphosphorylated analog or a peptide containing a phosphorylated serine residue.

Two structurally related hosts, both containing a 1,3,5-trisubstituted 2,4,6-triethylbenzene core, were designed for sugar–phosphate recognition. Host **98**, developed by the Schmuck group, contains three guanidiniocarbonyl-pyrrole-binding sites [154]. Binding affinity for glucose-1-phosphate in 30% water/DMSO (pH 7.4) amounts to $12\,940\,M^{-1}$, more than twice as high as the affinity for methyl phosphate (K_a methyl phosphate $\subset$ **98** $= 4850\ M^{-1}$), indicating that complexation of monosaccharides also involves hydrogen-bonding interactions between the receptor and hydroxy groups of the substrates. However, selectivity of **98** in binding of different monosaccharide-1-phosphates is moderate.

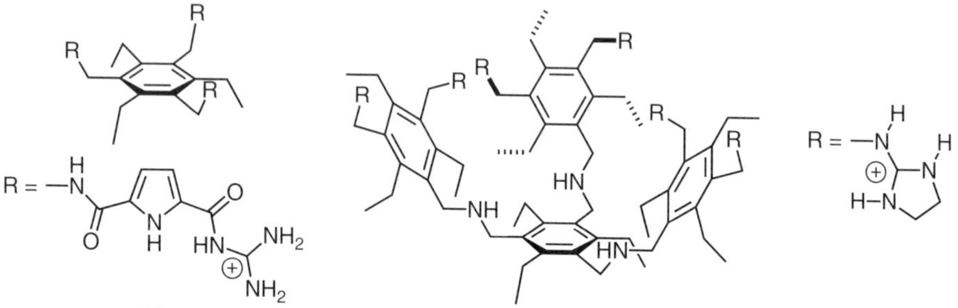

98 **99**

The structurally somewhat more elaborate receptor **99** was developed by Anslyn and coworkers for recognition of inositol-1,4,5-triphosphate (IP_3). Binding affinity and selectivity were characterized by using a dye displacement assay, which yielded a K_a of $4.7 \times 10^5\ M^{-1}$ for the IP_3 complex of **99** in water (10 mM HEPES, pH 7.4), slightly smaller than the K_a of the complex between the ammonium analog of this host and IP_3 ($5.0 \times 10^5\ M^{-1}$) [155]. In the presence of 50 mM NaCl, IP_3 complex stability of **99** remains high ($K_a = 8.2 \times 10^4\ M^{-1}$), while that of the ammonium analog decreased ($K_a < 1 \times 10^4\ M^{-1}$), clearly demonstrating specific interactions in the complex between **99** and the guest. In methanol and in the presence of 5-carboxyfluorescein as the indicator, complex formation can easily be detected by a color change of the solution from fluorescent yellow to nonfluorescent colorless [155]. The same color change is also observed at pH 4.0 in 40% methanol/water or in water in the presence of 2% of the surfactant Triton X-100 [156]. Under these conditions, complex stability amounts to $5.0 \times 10^6\ M^{-1}$ (in 40% methanol/water) and $1.2 \times 10^6\ M^{-1}$ (in 2% Triton X-100/water). Complex formation in the presence of Triton-X-100 is thus believed to occur to a significant extent in the lower dielectric environment of the micelles.

Another type of interaction frequently used in receptors for phosphate esters is coordination to a metal ion. Monometallic receptor **100**, for example, has been investigated by Mareque-Rivas and coworkers. This receptor forms a complex with phenyl phosphate in water (50 mM HEPES, pH 7.0), whose stability constant log K_a amounts to 3.6 [157]. Interestingly, the phenyl phosphate complex of a derivative of **100** containing two additional amino groups on the ligand is about 1 order of magnitude more stable (log $K_a = 4.4$) despite the fact that steric and electronic effects of the amino groups should reduce phosphate affinity with respect to **100**. The increase in complex stability has therefore been attributed to NH···OP hydrogen bonds between the amino groups and the guest, an assumption that was supported by the crystal structure of the nitrate complex of **101**.

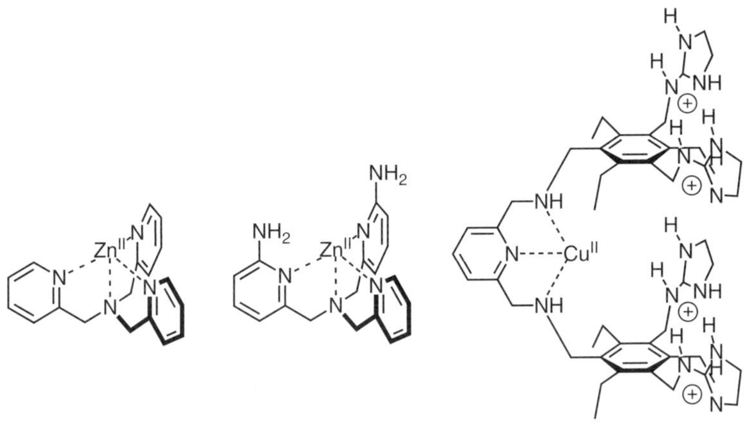

100 **101** **102**

Receptor **102** combining a metal ion and aminoimidazolinium groups for anion recognition was developed by the Anslyn group to selectively recognize 2,3-bisphosphoglycerate (2,3-BPG), an allosteric effector that modulates the oxygenation level of hemoglobin, in aqueous solution [158]. The K_a of 8×10^8 M^{-1} determined for the complex between 2,3-BPG and **102** in 50% methanol/water (10 mM HEPES, pH 7.4) is indeed remarkable. It only slightly decreases to 4×10^7 M^{-1} on changing the solvent to 100% water (pH 6.8). Moreover, substrates with only one phosphate group are bound much less tightly to the host under the same conditions. This allowed the use of **102** to modulate the oxygenation affinity of hemoglobin in horse red cell hemolyzate in 20 mM phosphate buffer solution at pH 7.2. Thus, decrease in oxygen affinity of hemoglobin resulting from interactions with 2,3-BPG could be reversed on addition of **102** to the solution, which indicates that the synthetic host is able to strip the natural effector from the protein. One reason for this ability is the significantly higher affinity of **102** to 2,3-BPG in comparison to hemoglobin [158].

Work by the Kimura group centers on the development of synthetic hosts and enzyme mimics on the basis of metal complexes of macrocyclic polyamines, for example, the zinc(II) complex of 1,4,7,10-tetraazadodecane (cyclen) **103** [159–161]. Compound **103** was shown to reversibly form a 1 : 1 complex with various anions including phenyl phosphate and *p*-nitrophenyl phosphate in aqueous solution, whereas the corresponding diprotonated metal-free macrocycle only weakly interacts with such substrates under the same conditions [162, 163]. The stability constant log K_a of, for example, the *p*-nitrophenyl phosphate complex of **103** amounts to 3.3. Linking two zinc(II)-cyclen subunits together via an *m*-xylylene spacer, as in **104**, increases the *p*-nitrophenyl phosphate affinity with respect to **103** by about 1 order of magnitude [164]. A much more pronounced boost in phosphate affinity could be achieved, however, by arranging three zinc(II)-cyclen complexes around an aromatic benzene scaffold [165]. The corresponding tripodal host **105** that was inspired by the crystal structures of the PO_4^{3-} complex of a derivative of **103** with an ethanolic residue on one cyclen nitrogen [162] and by the crystal structure of the *p*-nitrophenyl phosphate complex of **103** [165], in both of which three oxygens of the anion coordinate to the metal center of a zinc(II)-cyclen unit, indeed forms 1 : 1 complexes with phosphates in slightly acidic solutions (pH $<$ 6), of which the *p*-nitrophenyl phosphate has a stability constant log K_a of 5.8. Thus, phosphate ester affinity of **103–105** increases with the number of binding sites available for anion binding, and the cooperative action of all three binding sites in **105** is obviously the reason for the highest phosphate affinity of this host. Phosphate affinity seems to parallel the basicity of the substrate, as *p*-nitrophenyl phosphate (pK_a = 5.2; log K_a = 5.8), phenyl phosphate (pK_a = 5.8; log K_a = 6.6), α-D-glucose-1-phosphate (pK_a = 6.1; log K_a = 7.0), and phenyl phosphonate (pK_a = 7.0; log K_a = 7.9) are bound increasingly more strongly [165].

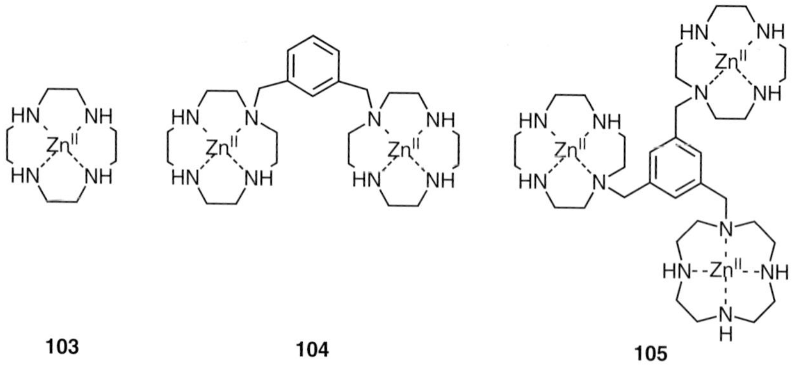

103 **104** **105**

Hamachi and coworkers developed a series of zinc(II) dipicolylamine-based receptors **106–110** that interact with phosphate in aqueous solution and with phosphorylated peptides [166–168]. Addition of inorganic phosphate or monophosphate esters such as phenyl phosphate, *o*-phospho-L-tyrosine, or methyl phosphate to a solution of the binuclear hosts **106** or **107** (10 mM HEPES, pH 7.2), for example, causes an increase in fluorescence intensity, thus allowing a determination of complex stoichiometry and stability by fluorescence spectroscopy. Both hosts form 1 : 1 complexes with different phosphate esters, whose stability constants range between 10^4 and 10^5 M^{-1} [169, 170]. Binding affinity toward pyrophosphate, ADP, or ATP is higher [170, 171], most probably because of the higher charge of these

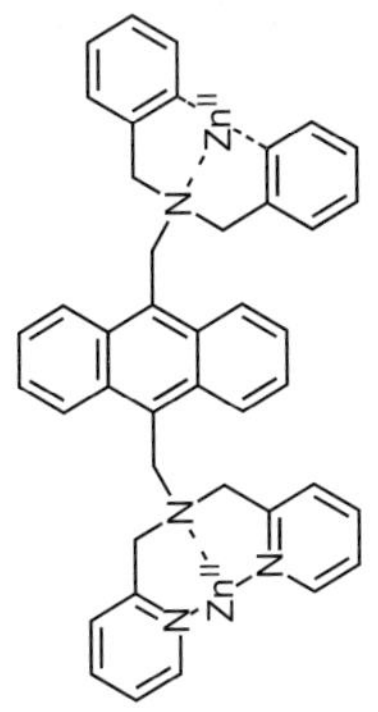

106

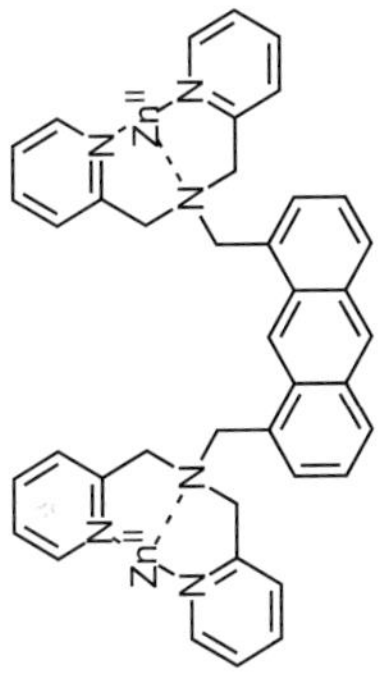

107

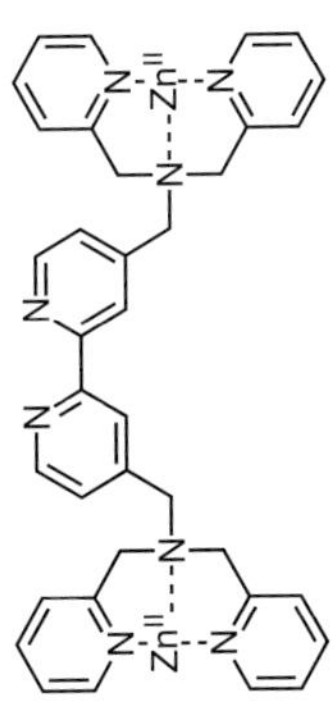

108

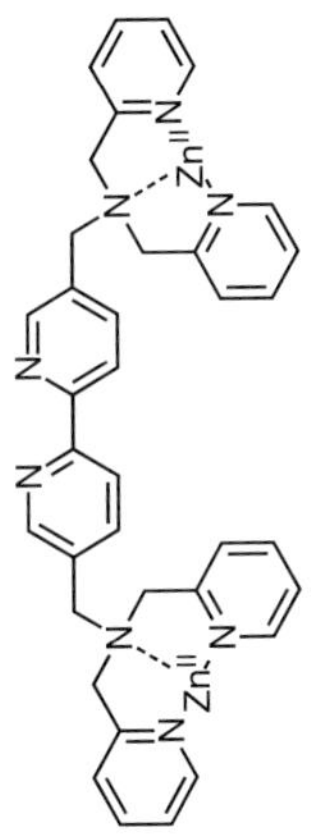

109

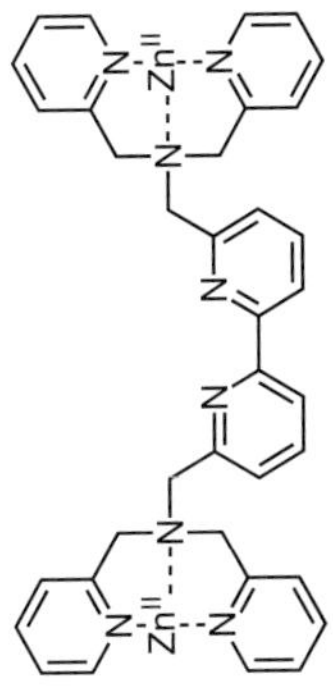

110

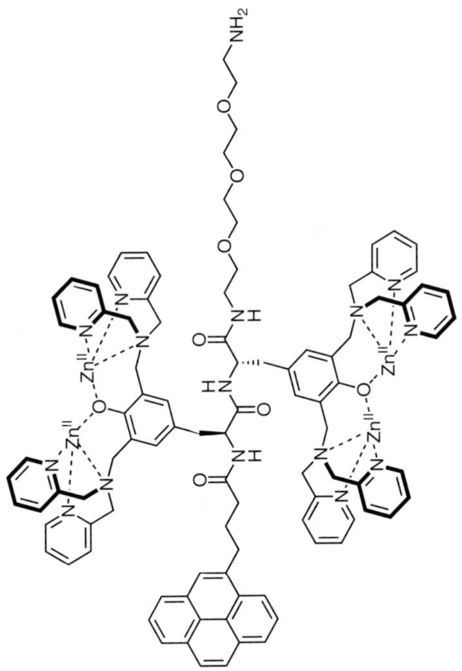

111

anions. No interaction could be detected with diphosphate esters, *c*AMP, or other anions such as carbonate, sulfate, nitrate, and acetate, demonstrating the selectivity of the hosts for phosphate anions.

Binding of **106** and **107** to several phosphorylated peptides containing optimal consensus sequences that are phosphorylated by certain kinases was also studied [169, 170]. These investigations showed that binding strength becomes stronger with increasing negative charge on the substrate. Thus, both receptors scarcely sense peptides bearing net charges of 0 or +2 even at 10^{-4} M concentrations, whereas they form highly stable complexes with an eightfold negatively charged peptide, whose stability constants amount to 9.5×10^5 M^{-1} for **106** and 8.9×10^6 M^{-1} for **107**. Moreover, binding of phosphorylated peptides is much stronger than of nonphosphorylated analogs, showing that the hosts can distinguish phosphorylated peptides from nonphosphorylated ones. This ability and the fact that the interactions with a negatively charged phosphorylated peptide are much stronger than those with inorganic phosphate allowed the phosphatase-catalyzed dephosphorylation of a model peptide to be monitored in real time in the presence of these receptors [170].

Circular dichroism spectroscopy demonstrated that **108** and **109** but not **110** induce α-helical conformations in peptides containing two appropriately spaced phosphorylated serine residues [172]. Since no effect of the receptors on the conformation of a monophosphorylated peptide was observed, two-point interactions seem to be required to induce helix stabilization. A similar binding mechanism was found to effectively disrupt phosphoprotein–protein interactions at a micromolar level [173].

The larger differences detected in the interactions of **109** with monophosphorylated and diphosphorylated derivatives of a naturally occurring peptide fragment of the insulin receptor kinase indicate that such zinc(II) dipicolylamine-based receptors are promising leads for the recognition and sensing of phosphorylated peptide surfaces in aqueous solution [172]. Anthracene-containing receptor **106** could indeed be used as a fluorescent staining dye for the selective detection of phosphoproteins in SDS-PAGE [174]. Moreover, selective recognition and sensing of phosphorylated peptides has recently been achieved by hybridization of a zinc(II) dipicolylamine-derived artificial fluorescent chemosensor with a natural phosphoprotein-binding domain [175]. Another interesting development is receptor **111**, which has been shown to transport phosphorylated peptides across cell membranes [176]. Somewhat related to this work are investigations carried out by the Smith group, which revealed that **106** is also an effective fluorescent sensor for phosphate residues on the surface of bilayer membranes such as the ones on phosphatidylserine residues [177, 178]. Since phosphatidylserine is enriched on the cell surface during apoptosis, **106** can be used in a fluorescent and flow cytometry assay to identify mammalian cells undergoing cell death [179].

Phosphorylated serine residues in oligopeptides can also be targeted by using bis(zinc(II)–cyclen)-triazine-derived receptors **112–114** [180]. In addition to the bis(zinc(II)–cyclen) triazine residue, all of these receptors contain a second binding site in an appropriate distance to recognize a second side-chain functionality in the peptidic substrate in *i*-3 position. Binding is particularly strong if this functionality is the carboxylate group of a glutamate or the imidazole ring in a histidine residue. In comparison, a model receptor consisting of only the bis(zinc(II)–cyclen) triazine moiety binds to peptides containing a phosphorylated serine with much weaker affinity.

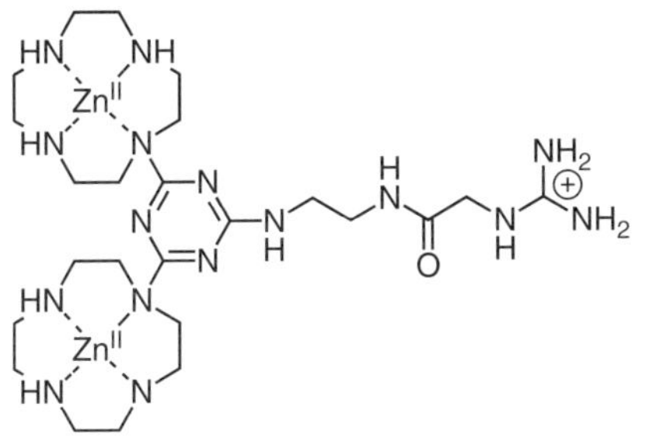

112

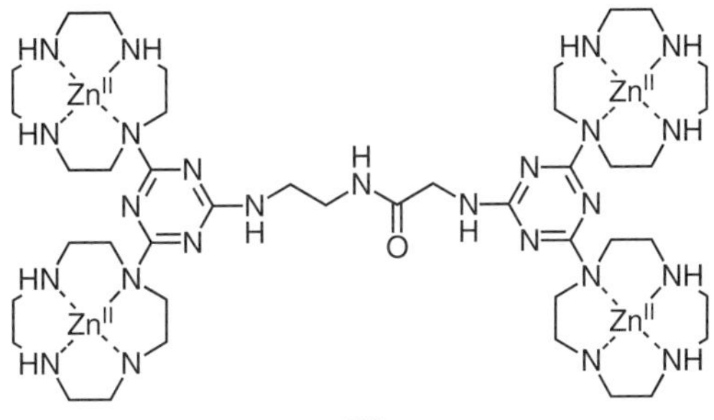

113

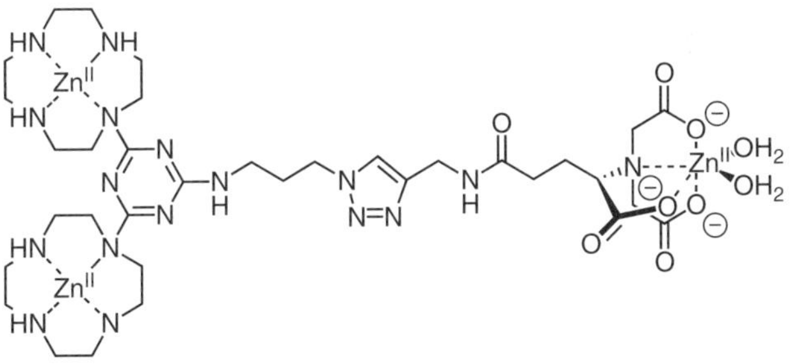

114

A strategy to differentiate phosphorylated peptides not only with respect to degree of phosphorylation but also with respect to primary structure has recently been described by the Ansyln group [181]. This technique involves a series of receptors **115** produced in a combinatorial manner by appending peptide fragments to an analog of **25**. On the basis of these receptors, a family of indicator displacement assays containing 45 members was created by combining five receptors differing in the appended peptide sequences, three metal ions, and three pH indicators. Addition of peptidic substrates varying in sequence and phosphorylation state to the assays resulted in characteristic signatures in the UV spectra. Collecting the results for the 45 sensing ensembles followed by linear discriminant analysis gave patterns that allowed 100% classification of the substrates. This approach impressively illustrates the potential of pattern-based recognition protocols involving synthetic anion receptors for the detection and highly selective analysis of protein phosphorylation.

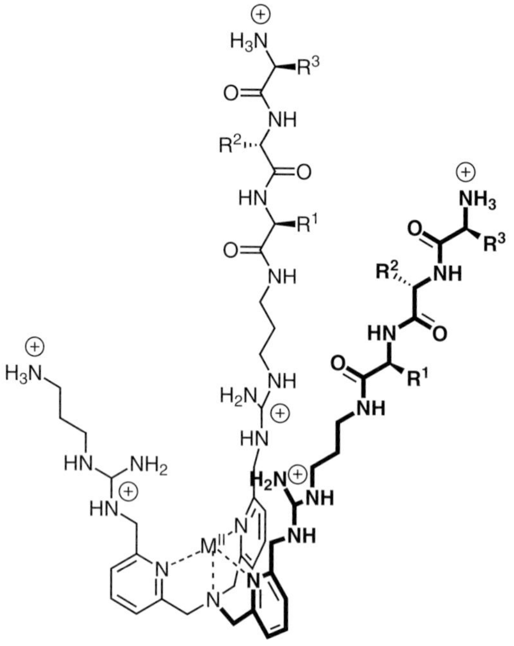

115

Receptors for phosphodiesters serve as models for polynucleotide-binding systems or, if also catalytically active, for nucleases. A receptor of the first type is macrocycle **116** containing a bicyclic guanidinium moiety as the anion-binding site [182]. This compound binds to phosphodiesters such as diphenylphosphate (diphenylphosphate ⊂ **116** $K_a = 10^3\ M^{-1}$ in $CDCl_3$), which are, however, not included into the cavity of the host but reside outside rapidly exchanging at room temperature between both sides of the ring.

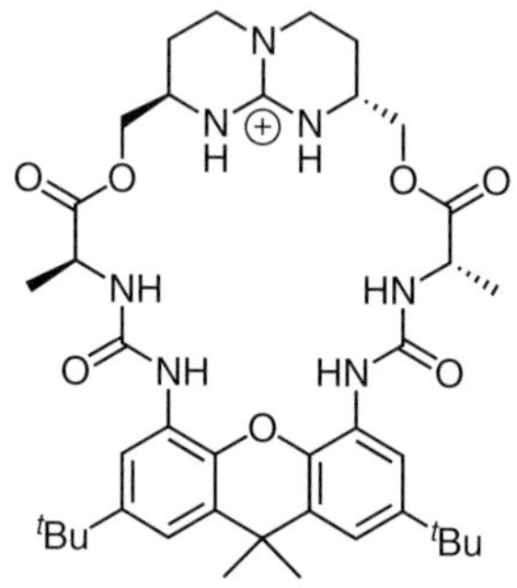

116

Hosts **117** and **118** containing a rigid octahydroacridine linker with two aminoimidazoline-binding sites were designed by Anslyn and coworkers to mimic phosphate binding inside the active center of the staphylococcal nuclease that

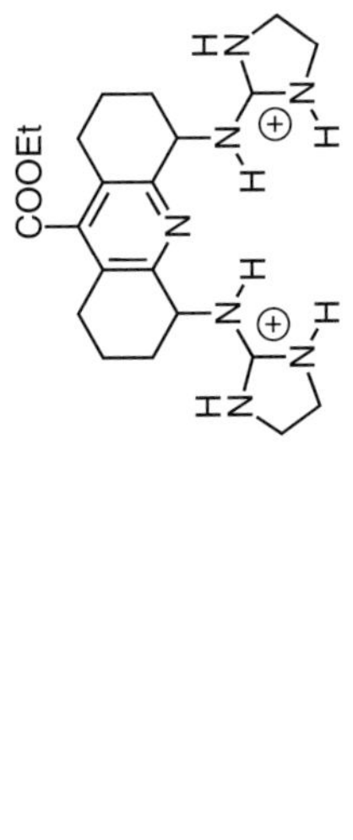

117

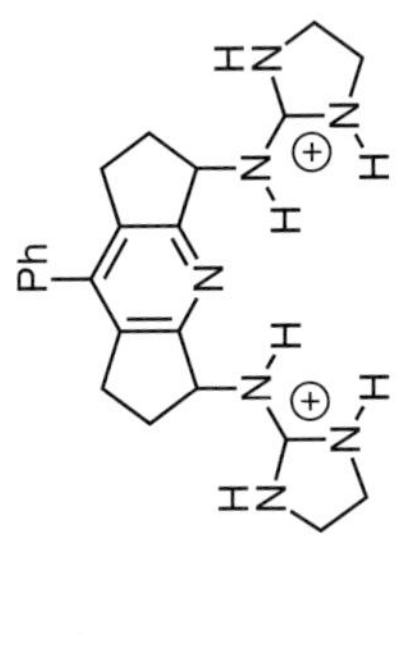

118

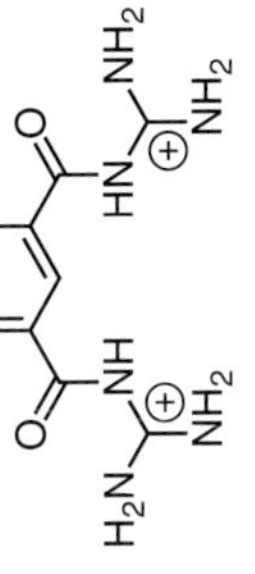

119

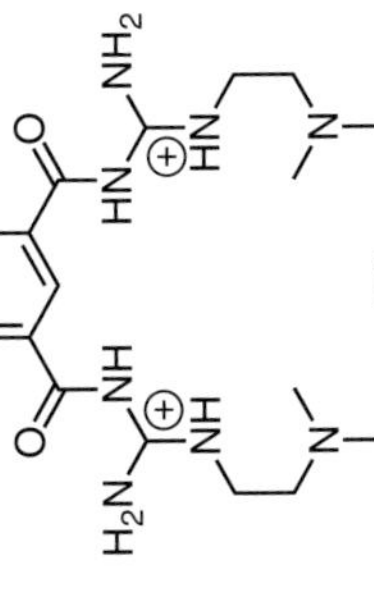

120

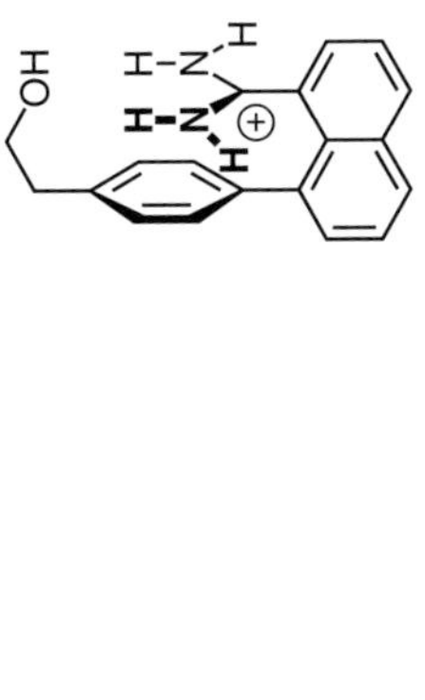

121

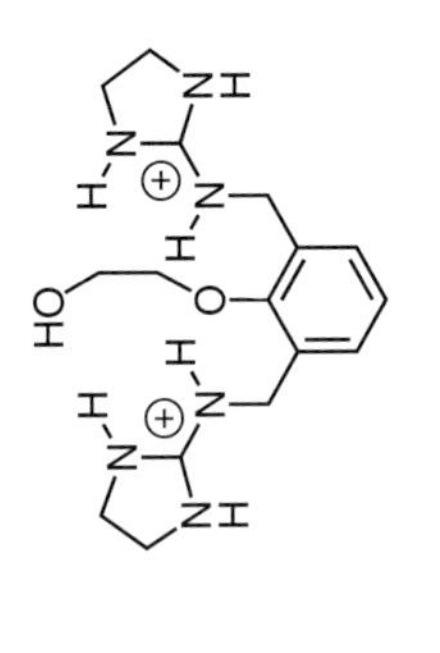

122

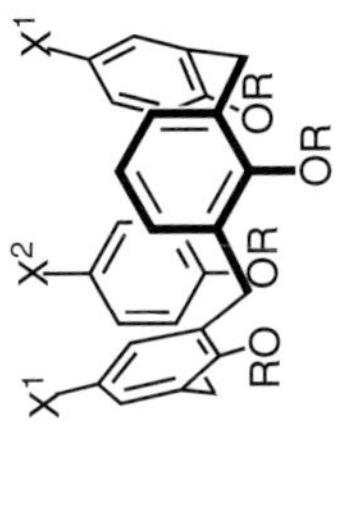

R = CH_2CH_2OEt

123a; X^1 =

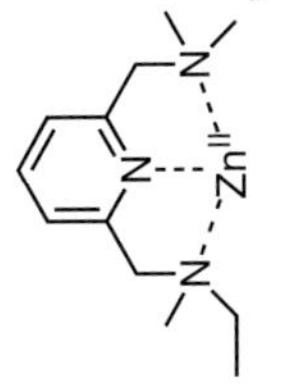

, X^2 = H

123b; X^1 = X^2 =

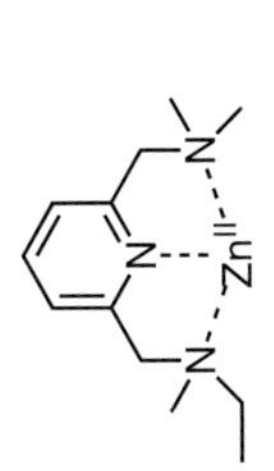

contains two arginine residues [183–185]. Initial binding studies using a mixture of all stereoisomers of **117** showed that this host can bind up to two dibenzyl phosphate ions in DMSO [183]. Formation of complexes of higher stoichiometry can be suppressed by increasing the solvent polarity, and in 2 : 1 DMSO/water, only the more stable 1 : 1 complexes are formed. Not surprisingly, complex stability is reduced in more polar solvent mixtures with respect to DMSO, but addition of chloride reverses this effect. The origin of this unexpected behavior was elucidated by characterizing the anion affinity of the individual stereoisomers of **117** and **118** independently. This investigation revealed that in the presence of tetraphenylborate as counterion, the meso forms of the hosts are the best receptors because of preorganization of the imidazoline moieties on the same face of the octahydroacridine scaffold [184]. Chloride, on the other hand, causes a stabilization of the complexes of the *d,l* receptors, thus improving anion affinity. In accordance with the design principle, **117** was shown to catalyze the cleavage of mRNA with a 20-fold rate increase over the uncatalyzed reaction [186, 187].

The two guanidinium moieties in related receptor **119** developed by the Hamilton group are also well suited to stabilize the trigonal-bipyramidal intermediate during phosphodiester cleavage via four hydrogen bonds with concomitant charge neutralization. Accordingly, **119** acts as an efficient catalyst in this reaction. The intramolecular transesterification of 2-hydroxypropyl 4-nitrophenyl phosphate is, for example, accelerated 700-fold over the uncatalyzed reaction [188]. The binding arrangement provided by the two guanidinium groups in **119** is crucial for this large rate enhancement because an analog of **119** containing only one guanidinium is a significantly less potent catalyst. Subsequently, the system was improved by appending substituents to **119** containing amino groups that can act as general bases in the transesterification reaction [189]. The most efficient receptor **120** accelerates the cleavage of 2-hydroxypropyl 4-nitrophenyl phosphate by a factor of 45 relative to the rate at which **119** catalyzes this reaction under the same conditions.

A similar strategy was followed by Göbel and coworkers by using receptors **121** and **122** [190, 191]. Both receptors bind to a cyclic phosphodiester in DMF in the absence of a base; the affinity of **122** is significantly higher, however. On addition of a tertiary amine to the solution, the hydroxy groups in both receptors are phosphorylated by the substrate in a pseudointramolecular reaction. Phosphorylation of **122** is 800 times faster than that of **121** and an impressive 380 000 times faster than that of 2-phenylethanol lacking any anion-binding site [191].

Models for RNA nucleases have also been devised on the basis of dimetallic receptors such as the calix[4]arene derivatives **123a** and **123b**, which were developed by the groups of Reinhoudt and Engbersen. These receptors promote the release of 4-nitrophenol from the RNA model substrate 2-hydroxypropyl 4-nitrophenyl phosphate (HPNP) as well as hydrolysis of RNA dinucleotides [192–196]. Interestingly, large differences were found in the rate of hydrolysis of different dinucleotides. Compound **123a**, for example, cleaves GpG at least 8.5 times faster than any other RNA dinucleotide investigated [195]. This and related work on metal-containing systems catalyzing phosphoester or phosphodiester hydrolysis have been reviewed [197].

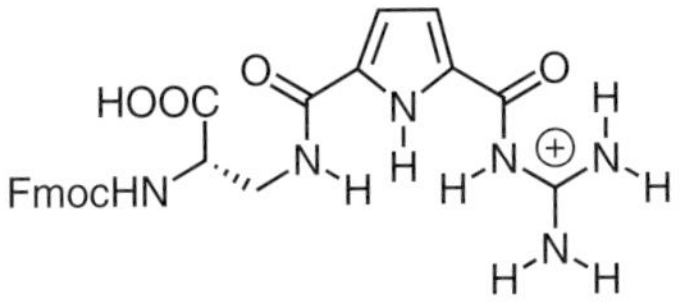

124

Schmuck and Geiger have developed the arginine analog **124** containing a 2-guanidiniocarbonyl pyrrole moiety in the side chain [198]. Extensive investigations using a variety of guanidiniocarbonyl pyrrole derivatives have revealed high affinity toward oxoanions, in particular carboxylate (*vide infra*) [199, 200], but this group can also participate in phosphoester recognition. Incorporation of **124** into a library of octapeptides containing 625 individual members has furnished, for example, peptides that accelerate phosphoester hydrolysis in water by a factor of 175 over the uncatalyzed background reaction [201]. The most active peptides all contain serine or histidine besides the artificial arginine analog **124**. It is therefore postulated that the guanidiniocarbonyl pyrrole moiety is responsible for substrate binding by interacting with the negatively charged subunit of the substrate, while the nucleophilic amino acids histidine or serine induce the actual bond cleavage. Catalytic activity also depends on the amino acid sequence in the octapeptide.

This work has recently been extended to the cleavage of RNA model compounds [202]. A polypeptide that was used in this investigation adopts a characteristic helix–loop–helix conformation and accelerates the hydrolysis of the RNA model HPNP by about 2 orders of magnitude. Two histidine and two arginine units in this peptide were believed to be responsible for catalytic activity. Analogs of this peptide were prepared, in which the two histidine, the two arginine, or all four amino acid residues were replaced by **124**. Interestingly, the last peptide proved to possess the highest activity, accelerating the hydrolysis of HPNP by a factor of 150 over the rate of the reaction catalyzed by the parent peptide and by a factor of 1500 over the rate of the imidazole catalyzed reaction. It was thus concluded that **124** can not only serve as a binding site for the substrate but can also provide general base catalysis in the reaction investigated.

7.2.5
Polynucleotides

DNA- and RNA-binding molecules possess a variety of potential medicinal applications, which explains the intensive search for efficient (sequence-selective) polynucleotide-binding agents. In nature, DNA recognition mainly follows four different paths: recognition of the phosphodiester backbone, intercalation between base pairs, or binding to the minor or the major groove. Of these, only the first mechanism involves ligands with affinity for anionic phosphate residues. Examples of natural ligands whose DNA affinity is due to interactions with the phosphodiester backbone are spermidine **1** and spermine **2** (*vide supra*). Considering the close structural relationship of **1** and **2** with polyaza macrocycles or acyclic analogs thereof, it is not surprising that a number of abiotic polyamines have been shown to possess affinity for polynucleotides [203–208]. Examples are receptors **72** and **125–128**.

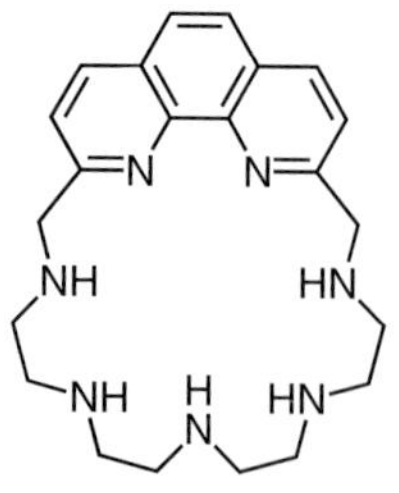

125

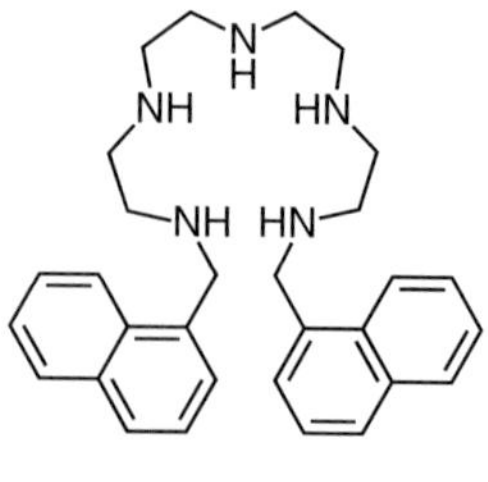

126

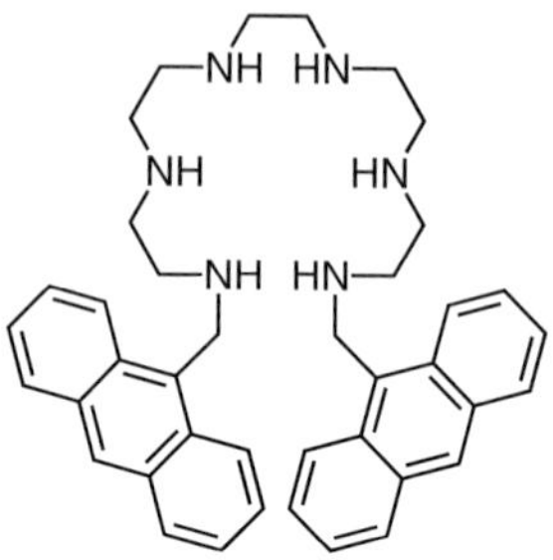

127

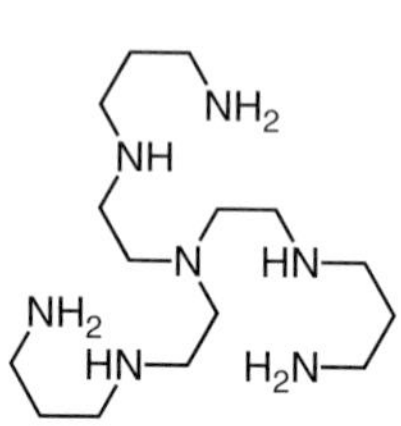

128

Association of these compounds with nucleic acids is primarily due to salt bridges with the groove phosphates. Affinity can be studied by measuring the change in melting temperature ΔT_M of the nucleic acid. These investigations showed that the affinity of macrocyclic ligands toward double-stranded DNA or RNA is generally lower than that of their linear analogs bearing the same charge. Phenanthroline-containing ligand **125**, however, produces a distinctly larger increase in the melting temperature of DNA, which reverses in favor of RNA on metalation to the dinuclear copper(II) complex [203]. The fact that phenanthroline-containing polyaza macrocycles have higher affinity for DNA than acyclic diethylenetriamine has been interpreted in terms of an intercalation of the phenanthroline moiety between base pairs along the DNA double helix. The macrocyclic complex between 72 and $CuCl_2$ exhibits a particularly strong discrimination between RNA and DNA, with a $\Delta\Delta T_M$ of 41°C, and a small destabilization of DNA [205].

Also, polyamines such as **126** and **127** with pendant aromatic moieties have been shown to interact with polynucleotides [206, 208]. While polyamines without aromatic residues possess stronger affinity to RNA because binding is dominated by ion pairing of the ammonium centers with the RNA phosphates, appropriately spaced aromatic substituents cause preferential binding to DNA because of bisintercalation, a result that demonstrates how selectivity for binding RNA rather than DNA can be achieved with rather simple compounds. Bisintercalation is possible only if the aromatic moieties are separated by at least 12 atoms, and preferential binding to DNA decreases with increasing distance of the aromats. Interestingly, interaction of polyamine **126** with DNA is disrupted in the presence copper(II) ions, whose coordination to the ligand's nitrogen atoms causes a conformational distortion thus preventing the naphthyl groups from efficiently intercalating between DNA base pairs simultaneously. Anthracene-containing ligand **127**, on the other hand, intercalates so strongly that no decrease in DNA affinity is observed on addition of copper(II) salts. Of all polyamines studied so far, tripodal ligand **128** shows by far the strongest preference for RNA ($\Delta T_{M(RNA)}/\Delta T_{M(DNA)} = 40$) [208].

Rotello's group has studied binding of mixed monolayer-protected gold nanoparticles containing trimethylammonium end groups to DNA [209, 210]. The cationic groups on the surface of these nanoparticles are believed to bind to the negatively charged phosphate backbone through electrostatic complementarity. In solution, interaction of 37mer duplex DNA with four nanoparticles leads to the formation of discrete DNA–nanoparticle clusters that have a diameter of about 20 nm. These aggregates are sufficiently stable to inhibit DNA transcription by T7 RNA polymerase, indicating that either the nanoparticles bind with higher affinity to DNA than the T7 RNA polymerase or the altered conformation of the nanoparticle-bound DNA prevents the binding of the polymerase. In further studies, such nanoparticles were shown to induce gene delivery into cells [211]. Strong binding of cationic calixarene dimers to DNA was demonstrated by Schrader and coworkers. Molecular modeling and experimental evidence suggested, however, that this dimer does not interact with the DNA phosphodiester backbone but rather inserts into the major grove [212, 213].

Amidinium- or guanidinium-based receptors not only strongly bind to DNA but also exhibit sequence specificity. The tetraguanidinium derivative **129** has, for example, been shown to bind to the minor groove of calf thymus DNA, with an association constant of about 10^7 M^{-1}, preferentially interacting with 3′-GAA-5′ regions [214]. Affinity of the tricationic receptor **130** is even larger ($K_a = 7 \times 10^7$ M^{-1}), but this receptor binds strongly to AATT DNA [215, 216]. Strong interactions were also found with an alternating AT sequence but very weak association with an alternating GC sequence. The mode of interaction between **130** and DNA was elucidated by X-ray crystallography.

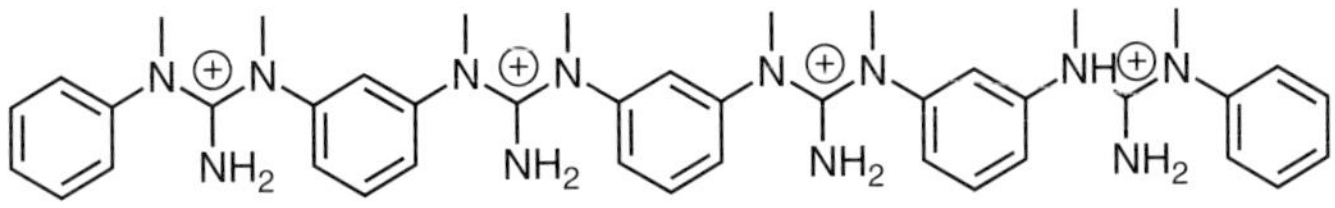

129

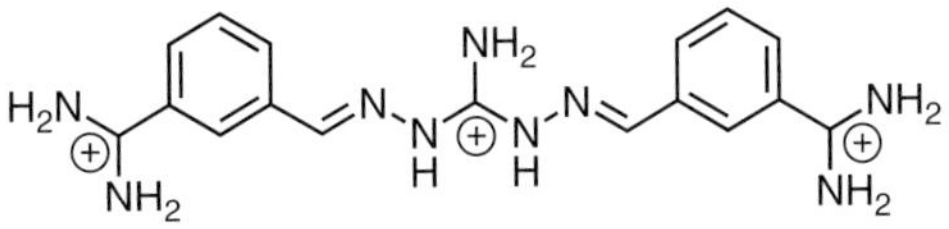

130

Interaction of calixarenes bearing guanidinium groups with DNA was studied by Ungaro and coworkers. Calix[*n*]arenes with guanidinium groups directly attached to the aromatic calixarene subunits along the upper rim have been shown to condense plasmid and linear DNA and perform cell transfection in a way that is strongly dependent on the size of the calixarene ring, on lipophilicity, and on conformation [217]. These compounds also exhibit low transfection efficiency and high cytotoxicity, however. These undesired properties could be much improved by using calixarenes with guanidinium groups attached to the phenol OH groups along the lower rim via an appropriate spacer [218]. Importantly, one of the calixarenes studied, when formulated with dioleoylphosphatidylethanolamine (DOPE), surpasses a commercial transfection agent in terms of transfection efficiency as well as toxicity. The mode of interaction between these calixarene derivatives and DNA is believed to be due to a combination of hydrophobic interactions and electrostatic interactions between the guanidinium groups and the DNA phosphate esters.

Also, sapphyrins have been shown to bind to DNA. First, evidence for these interactions came from the observation that addition of **41b** to double-stranded DNA at pH 7 caused precipitation of green fibers [219–221]. The formation of these fibers was ascribed to charge neutralization arising from the interaction between the DNA phosphodiesters and the protonated sapphyrin. These interactions, termed "*phosphate chelation*," were proposed to involve specific contacts between phosphate oxygens and sapphyrin NH groups similar to those found in some crystal structures. They thus differ from other mechanisms of DNA recognition, namely groove binding, intercalation, or simple electrostatic interactions, an assumption that was confirmed by a number of independent experiments [219–221]. Interestingly, sapphyrins were also found to catalyze the photocleavage of DNA, a property that could be suppressed, for example, by addition of sodium dodecyl sulfate (SDS) a reagent known to inhibit sapphyrin–phosphate interactions [222]. Appending nucleotide conjugates to the sapphyrin core even allowed site-specific photocleavage of complementary DNA strands, which suggests a potential value of sapphyrins in medicinal applications [223].

7.3 Carboxylate Receptors

7.3.1 Introduction

The prototype substrate for carboxylate receptors is the acetate anion. Its anionic headgroup has a trigonal planar structure with one carbon and two oxygen atoms located in the three corners of the triangle, the latter of which can function as hydrogen bond acceptors. Recognition of acetate groups therefore usually involves two appropriately positioned hydrogen bond donors in a receptor or two coordinatively unsaturated metal ions. In both cases, binding can be assisted by electrostatic interactions.

Two examples of natural systems containing binding sites for the carboxylate group are the enzyme carboxypeptidase A and the antibiotic vancomycin. Carboxypeptidase A is responsible for the hydrolytic release of a C-terminal hydrophobic (aromatic) amino acid from a polypeptide chain [224, 225]. To recognize the terminal carboxylate group of the substrate, carboxypeptidase A contains the guanidinium group of an arginine residue in its active center that binds to the substrate's anionic headgroup via a combination of coulombic attractions and hydrogen bond formation. Similar guanidinium/carboxylate interactions also occur in the active centers of the enzymes thrombine and trypsin, both of which cleave peptides directly behind an arginine residue [226, 227]. In the latter cases, the guanidinium group is, however, part of the substrate and the carboxylate part of the protein. The antibiotic activity of vancomycin is due to the inhibitory effect of this compound on the mechanical stabilization of the bacterial cell wall during cell wall biosynthesis [228–230]. This process involves the transpeptidase-catalyzed crosslinking of the peptidoglycan cell wall precursors by connecting pendant peptide fragments of which one contains a terminal D-Ala–D-Ala dipeptide unit. By specifically binding to the carboxylate group of this unit, vancomycin prevents crosslinking, eventually leading to the death of the bacterium. The crystal structure of vancomycin complexing a peptide ligand shows that complex formation involves formation of four intermolecular hydrogen bonds, three of which are formed between amide NH groups in the vancomycin molecule and the carboxylate of the substrate's terminal D-Ala residue [231].

Because of the substantial acidity of carboxylic acids, the pK_a of acetic acid amounts to 4.75, they are fully deprotonated in water at concentrations relevant in biochemistry, thus rendering binding equilibria less complex than those involving, for example, phosphate anions (Figure 7.2). This is true not only for monocarboxylic acids such as acetate but also for di- or tricarboxylic acids or amino acids. To demonstrate this property the distribution diagrams of succinic acid, citric acid, and glutamic acids are depicted in Figure 7.2 besides that of acetic acid. All diagrams clearly show that the predominant species at pH 7 are those with the carboxylate group(s) fully deprotonated.

The simplest carboxylic acids, formic acid and acetic acid, are essential components in a number of biochemical processes. N^{10}-Formyl-tetrahydrofolate is a coenzyme derived from formic acid, which serves as a donor of formyl groups, for example, in purine biosynthesis. Acetic acid, when bound to the coenzyme A, is consumed in the biosynthesis of fatty acids or fuels the citric cycle. Other biologically relevant monocarboxylic acids are fatty acids and several ketocarboxylic acids or hydroxycarboxylic acids. Di- and tricarboxylic acids such as citric acid, isocitric acid, succinic acid, malic acid, and α-ketoglutaric acid are central intermediates in the citric acid cycle. Two amino acids contain carboxylic acids in the side chain, namely, glutamic acid and aspartic acid. Incorporation of these subunits into a polypeptide backbone causes a defined arrangement of side-chain carboxylate groups along secondary structure motifs or on the surface of the corresponding protein.

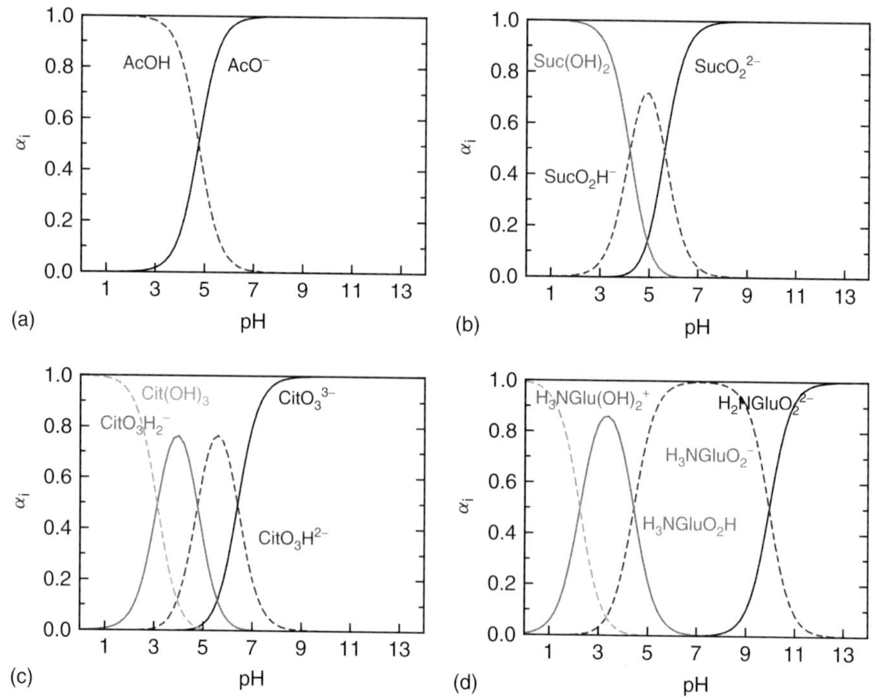

Figure 7.2 Distribution diagrams for (a) acetic acid AcOH, (b) succinic acid $Suc(OH)_2$, (c) citric acid $Cit(OH)_3$, and (d) glutamic acid $H_2NGlu(OH)_2$.

Binding to only the carboxylate group in all of these substrates (or to only one carboxylate group in a substrate containing more than one) is not sufficient to achieve binding selectivity. Synthetic receptors for biologically relevant carboxylates therefore usually contain a combination of binding sites, a primary binding site that interacts with the anionic carboxylate group and additional ones that recognize other functional groups of the substrate. Receptors for substrates with more than one carboxylate group contain the corresponding number of (not necessarily identical) carboxylate binding sites appropriately positioned in space.

7.3.2
Acetate

Receptors for acetate often serve as models to study carboxylate binding [232]. They are useful to demonstrate the ability of certain functional groups or receptor structures to interact with the carboxylate group and can form the basis for further developments aiming at, for example, binding selectivity.

Binding studies involving monocarboxylates such as acetate and polyaza macrocycles are complicated by the fact that at pH values required to produce host species with a high degree of protonation and, as a consequence, high anion affinity, carboxylates are also protonated and hence not available for interactions with the host. Monocyclic polyaza receptors have therefore only rarely been investigated with respect to acetate recognition; binding studies mainly targeted di- or tricarboxylates (*vide infra*). Bicyclic polyaza receptors **131–133** have, however, been shown to possess affinity also for acetate (and lactate) [233]. The corresponding investigations revealed that binding constants of the acetate complexes of penta-, tetra-, or triprotonated receptors **131–133** are even larger than those of the complexes of tetrahedral or trigonal inorganic monoanions because of the greater charge density in the carboxylate group (log K_a of pentaprotonated $AcO^- \subset$ **131** = 4.21, pentaprotonated $AcO^- \subset$ **132** = 3.99, and pentaprotonated $AcO^- \subset$ **133** = 5.73). Affinity decreases at lower pH, which favors the neutral acetic acid species. The unusually high stability of the acetate complex of **133** has been rationalized by assuming additional interactions, presumably ones involving stacking or hydrogen-bonding.

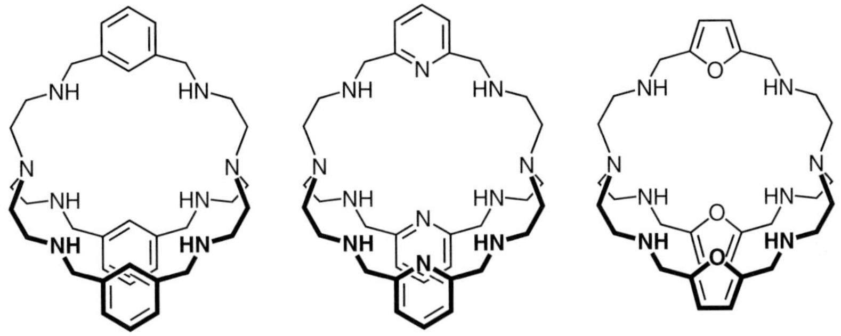

131 **132** **133**

The dicopper(II) complex of **131** has been shown to bind acetate with a stability constant log K_a of 2.97 [234]. However, other anions such as azide (log K_a = 4.78), cyanate (log K_a = 4.60), or hydrogen carbonate (log K_a = 4.56) are bound much more efficiently, a result that was rationalized by the almost perfect fit of azide and, to a lesser extent, hydrogencarbonate and cyanate between the two copper centers. An example of a monometallic acetate receptor is compound **26**, whose affinity amounts to 900 M^{-1} in 5 mM aqueous HEPES buffer (pH 7.4) according to a UV–vis titration [235].

Because of the importance of carboxylate/guanidinium interactions in natural carboxylate binding systems, the underlying interactions of which are two parallel hydrogen bonds complemented by attractive electrostatic interactions as depicted in Figure 7.3, acetate complexes of several guanidinium derivatives have been studied intensively. These investigations addressed, for example, the dependence of complex stability on the structure of the guanidinium-derived host, on the counterion, or on the solvent [236, 237].

The Hamilton group, for example, has studied acetate binding of a series of monotopic hosts **134–141** by using isothermal titration calorimetry [237]. They showed that anion affinity is higher in guanidinium derivatives that can form bidentate hydrogen bonds. Host **139** therefore only weakly interacts with acetate, while no formation of a complex could be observed for hosts **135** and **136**. Bicyclic

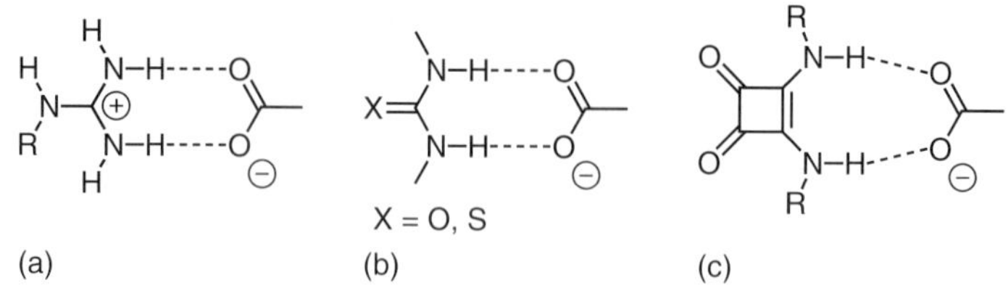

Figure 7.3 Schematic representation of (a) the guanidinium–carboxylate interaction, (b) the (thio)urea–carboxylate interaction, and (c) the squaramide–carboxylate interaction.

guanidinium derivative **134** forms a defined 1 : 1 complex, whose stability strongly depends on the counterion in the salt of **134** used for the binding studies. The least stable complex in DMSO is formed in the presence of chloride ($K_a = 2900\ M^{-1}$), while tetraphenylborate less pronouncedly competes for the binding sites of **134**, causing complex stability to increase to $5600\ M^{-1}$. Derivatives **137** and **140** can interact with up to two acetate anions simultaneously; the second acetate is bound significantly weaker than the first, however. The fact that complex formation in DMSO is exothermic suggests that discrete hydrogen bonding is the driving force of complex formation in this solvent. Binding in methanol is endothermic presumably because of solvent reorganization during bimolecular association.

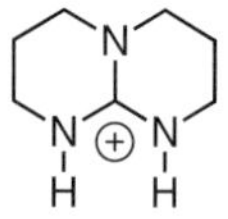
134

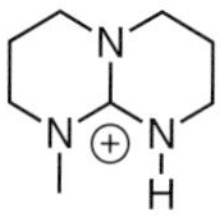
135

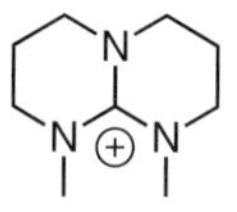
136

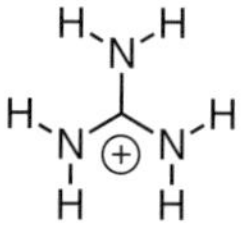
137

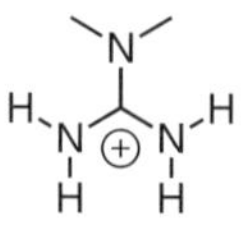
138

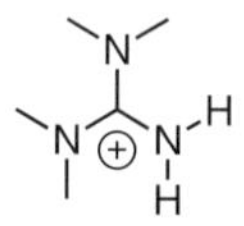
139

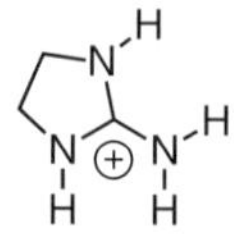
140

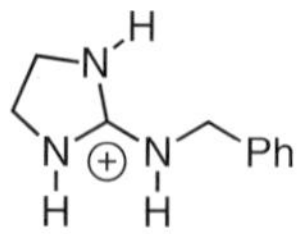
141

Systematic investigations by Schmuck demonstrated that the 2-guanidiniocarbonyl pyrrole group combines the anion-binding properties of guanidinium cations with that of pyrrole rings [199, 200]. The potential of this strategy was first demonstrated by using the simple guanidiniocarbonyl pyrrole derivative **142** [238].

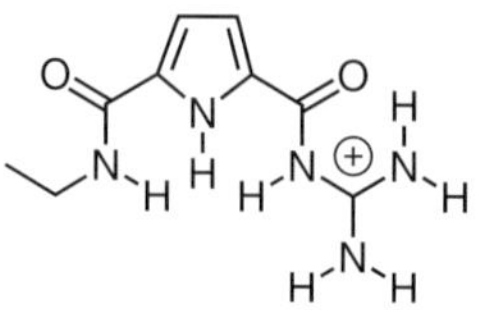
142

This compound was shown to interact with various carboxylates in highly competitive solvent mixtures such as 40% water/DMSO-d_6. Strongest binding was observed for acetate, with the stability of the complex amounting to $2700\,M^{-1}$, much higher than the stability of the corresponding acetate complex of *N*-acetyl guanidinium ($50\,M^{-1}$) in the same solvent. This result was ascribed to cooperative effects of the pyrrole and amide NH groups in carboxylate recognition, an assumption consistent with the complexation-induced shifts observed in the ^{1}H NMR spectrum of **142** on complex formation and with molecular modeling studies. In the crystal structure of 2-guanidiniocarbonyl pyrrole acetate, no discrete host–guest entities were observed, but an extended two-dimensional network in which the acetate carboxylate group simultaneously interacts with the guanidinium moiety of one molecule of **142** and with the pyrrole NH of another [239].

Interaction of urea groups with carboxylates is very similar to the guanidinium/carboxylate interaction (Figure 7.3b), however, lacking the electrostatic component. Complex stability is therefore usually considerably lower, and binding studies involving urea-based receptors are restricted to organic solvents and often involve receptors containing more than one urea moiety. Because of the higher acidity of thiourea NH protons (pK_a about 21) compared to urea (pK_a about 27), replacement of urea by thiourea groups generally improves carboxylate affinity. ^{1}H NMR spectroscopy was used to systematically study the influence of substituents on the urea moiety on anion affinity [240]. These investigations showed that anion affinity correlates with the number of aromatic substituents. Thus, bisphenyl-based ureas bind anions more strongly than ureas containing a single phenyl group. Although it is generally believed that electron-withdrawing substituents on the aromatic receptor give rise to stronger binding than electron-donating ones, this correlation is not straightforward. Urea **143b**, for example, does indeed bind acetate more strongly than **143a**; a further reduction of the electron density of the aromatic moieties in **143c** causes anion affinity to substantially drop, however (acetate $\subset$ **143a** : $\log K_a = 3.06$; acetate $\subset$ **143b** : $\log K_a = 3.70$; acetate $\subset$ **143c** : $\log K_a = 2.30$). This result is presumably due to repulsive electrostatic interactions between the ortho-substituents and the guest. Furthermore, solid-state data from X-ray crystallography has shown that electron-withdrawing groups can give rise to changes in the dihedral angle between the urea and the aryl groups, suggesting that such substituents can influence binding in an unpredictable manner. In line with this assumption is that ureas **144a–c** were found to bind

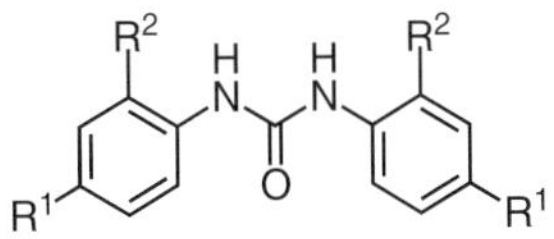

143a; $R^1 = R^2 = H$
143b; $R^1 = F$, $R^2 = H$
143c; $R^1 = R^2 = F$

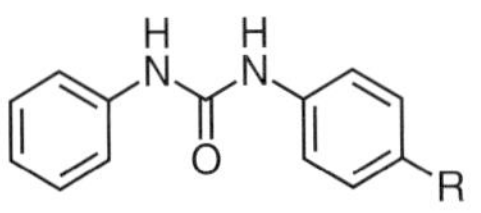

144a; $R = CH_3$
144b; $R = CF_3$
144c; $R = F$

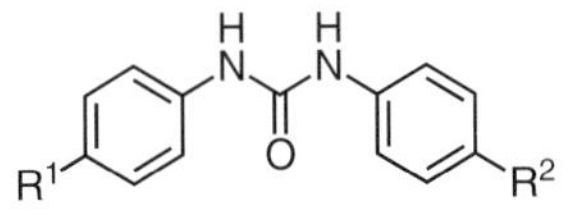

145a; $R^1 = R^2 = NCH_2CH_2OCH_3$
145b; $R^1 = NO_2$, $R^2 = NCH_2CH_2OCH_3$

acetate with log K_a values of 3.63, 2.59, and 3.61, respectively, which demonstrates that the nature of the electron-withdrawing group has a significant influence on the receptor's anion affinity and sometimes affects it in a counter-intuitive manner.

Redox-active ureas **145a,b** were synthesized to investigate the effect of anion binding on the oxidation potential of the *p*-phenylenediamine subunit [241]. These compounds bind acetate with stability constants approaching $10^6\ M^{-1}$ in acetonitrile (acetate $\subset$ **145a** : $\log K_a = 4.4$; acetate $\subset$ **145b**: $\log\ K_a = 5.7$). As expected, anion binding lowers the oxidation potentials of these hosts. Because of the high stability of the acetate complexes, a two-wave behavior was observed in electrochemical studies, with the appearance of a second oxidation wave cathodic to the one of the free receptor signifying the presence of the complexed state. A binding enhancement factor could be calculated from the voltammetric data, which provides a measure of the increase in complex stability on oxidation of the host. This enhancement factor correlates with anion affinity of the neutral host.

Examples of macrocyclic urea-based receptors for acetate are tristhioureas **146a,b** developed by Lee and Hong [242]. The 1,3,5-triethylbenzene group in **146b** was chosen as a linker to preorganize the thiourea-binding sites as in several other previously mentioned receptors (**88**, **89**, **98**, **99**, **102**). Interestingly, **146a** possesses affinity for dihydrogenphosphate in DMSO similar to acyclic bisthioureas (*vide infra*), while **144b** preferentially binds acetate. Stability of the acetate complexes of **146a** and **146b** amount to, respectively, 320 and 5300 M^{-1}.

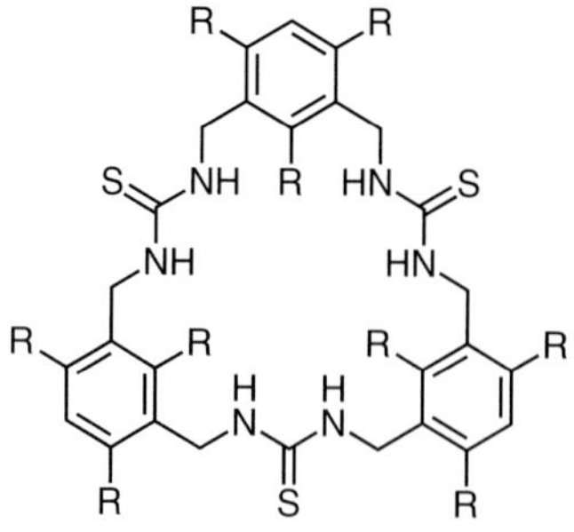

146a; R = H
146b; R = Et

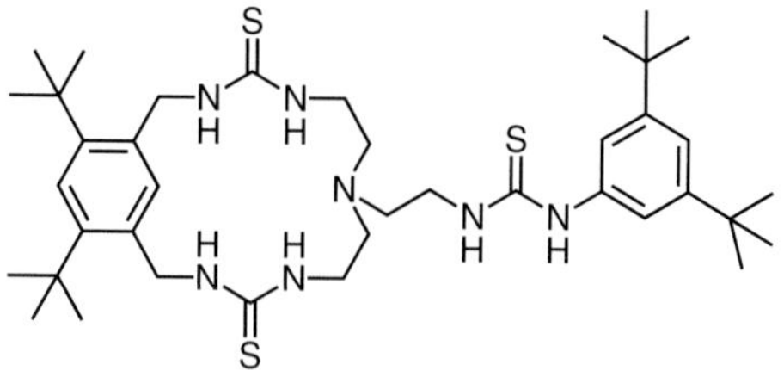

147

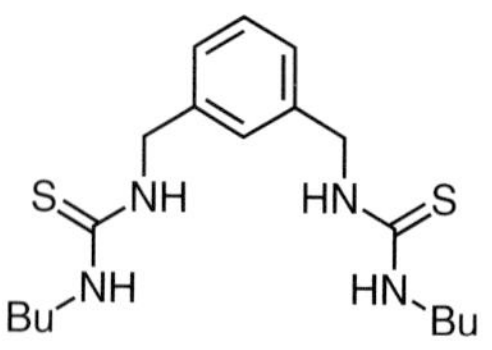

148

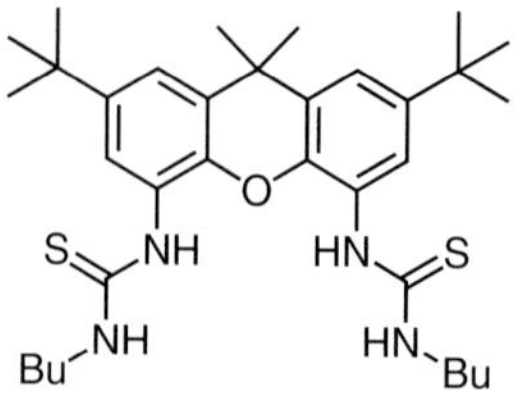

149

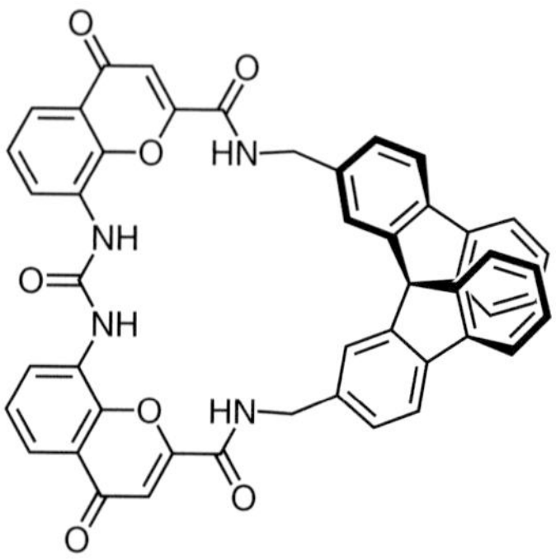

150

The acetate complex of tristhiourea **147**, developed by the Tobe group, is even more stable, possessing a K_a of 8300 M^{-1}, which is also significantly higher than that of several other cyclic bisthioureas described in the same publication [243]. All of these compounds bind more strongly to dihydrogenphosphate in DMSO, which has also been observed for **146a** and the acyclic receptors **148** and **149**. For comparison, acetate affinity of **148** in DMSO amounts to 470 M^{-1} and an impressive K_a of 38 000 M^{-1} has been determined for the acetate complex of **149**, while the dihydrogenphosphate complexes of both receptors have a K_a of 820 and 55 000 M^{-1}, respectively [244, 245]. The high anion affinity of **149** was attributed to the rigidity of the xanthene spacer, which leads to an optimal preorganization of this receptor for anion binding.

Enantioselective recognition of α-hydroxycarboxylates such as mandelate or lactate has been achieved with chiral macrocycle **150** containing a spirofluorene unit. Compound **150** shows preferential binding of the *R*-enantiomers of mandelate and lactate in DMSO, which most likely has steric reasons (*R*-mandelate $\subset$ **150** $K_a = 2.8 \times 10^4$ M^{-1}; *S*-mandelate $\subset$ **150** $K_a = 1.7 \times 10^3$ M^{-1}; *R*-lactate $\subset$ **150** $K_a = 3.5 \times 10^4$ M^{-1}; *S*-lactate $\subset$ **150** $K_a = 3.5 \times 10^3$ M^{-1}) [246]. Assuming that the mandelate phenyl group or the lactate methyl group points outward from the receptor cavity, the α-hydrogen must reside close to the spirobifluorene upper aromatic residue in the sterically most congested region of the receptor, while the hydroxy group lies over the spirobifluorene lower aromatic rings in the complex of the *R*-enantiomers of the guests. In the complex of the *S*-enantiomers, on the other hand, the α-hydrogen and hydroxy group exchange positions, which brings the larger hydroxy group in the most congested receptor region, causing a destabilization of the corresponding complexes.

In a systematic binding study, Davis, Smith and coworkers have investigated the anion affinity of a series of cholapod anion receptors in chloroform, for example, compounds **151a–c**, which are closely related to receptor **96** [247]. These receptors contain between three and six H-bond donors residing in urea, thiourea, sulfonamide, carbamate, trifluoroacetamide, or isophthalamide moieties. Because the anion affinity of these receptors proved to be extremely high in chloroform, binding constants were determined by using Cram's extraction method. The results show that high binding constants ($K_a > 10^{10}$ M^{-1}) are achievable with several types of cholapods, especially those bearing five or six H-bond donors. The binding constants of the acetate complexes of receptors **151a–c** amount to, respectively,

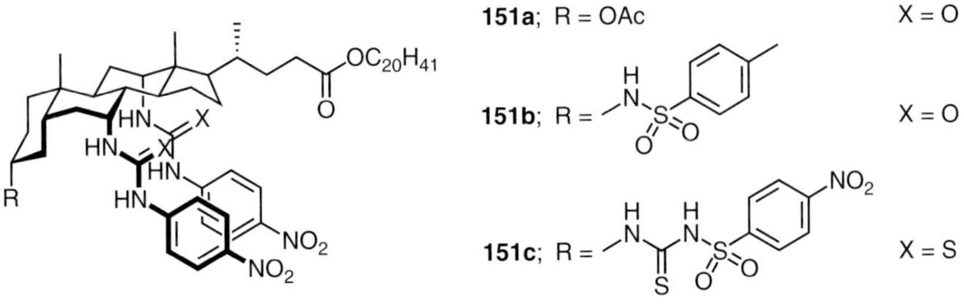

1.2×10^8, 1.5×10^9, and 1.3×10^{11} M^{-1}, for example; complex stability thus correlates with the number of hydrogen bond donors in these receptors.

In addition, the arrangement of H-bond donors also affects selectivity, although the rationalization of the observed effects proved to be not straightforward. Interestingly, selectivities are strongly affected by the inherent binding strength of the receptor. Thus, as affinities increase, the spread of binding constants also increases so that selectivities for the more strongly bound guests tend to increase.

Subsequently, the Davis group also included in their studies cholapods containing a quaternary ammonium group in the 3α-position. These receptors were designed to preferentially extract small lipophilic anions into organic solvents contrary to what is expected from the Hofmeister series. Anion selectivity was assessed by comparing the ability of such cholapod receptors to partition an anion between the organic and the aqueous phase with that of the tetraoctylammonium cation, whose extraction properties are expected to reflect the Hofmeister series. The investigations showed that cholapod **152** shows selectivity for chloride [248]. The urea appendages in **152** are rather flexible, however, allowing them to accommodate other anions as well, which causes acetate extraction of **152** to be very similar to chloride extraction. Receptors **153** and **154** containing more rigid urea-binding sites were therefore designed. While the macrocyclic unit in **153** causes an improvement in chloride selectivity, **154** turned out to show a distinct preference for acetate [249]. It should be noted that work by the Davis group on cholapod anion receptors mainly focused on developing systems able to selectively transport chloride across lipid membranes [145, 250].

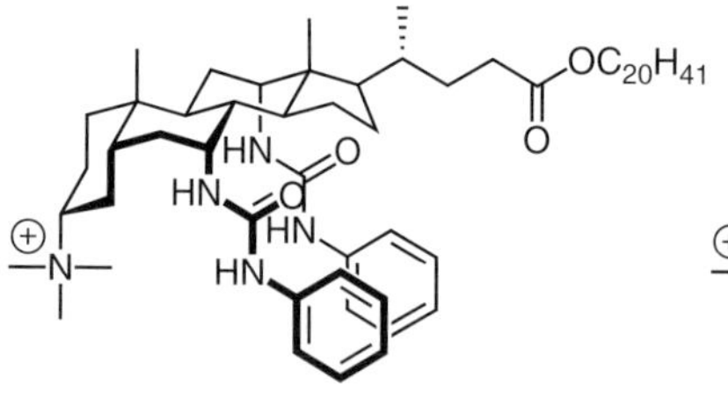

152

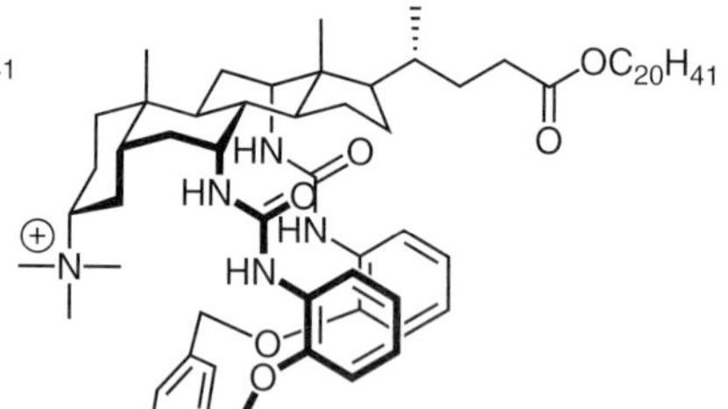

153

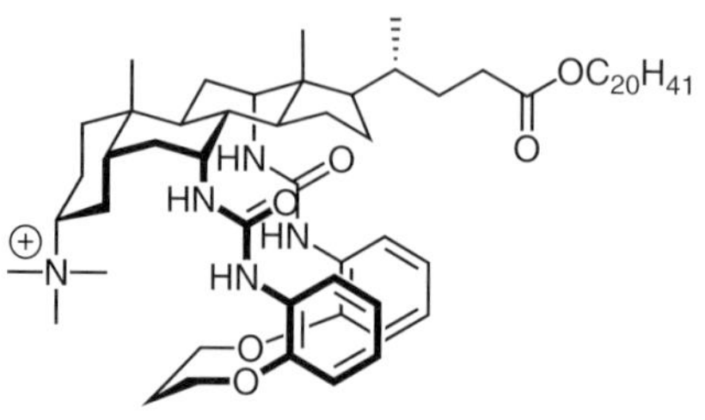

154

Several groups have investigated the anion-binding properties of calixarene derivatives containing urea moieties. Receptor **155** has, for example, been synthesized by Lhoták, Stibor, and coworkers [251]. This compound binds acetate in $CDCl_3/CD_3CN$ 4 : 1 with a K_a of 3940 M^{-1}. Interestingly, benzoate affinity is more than 1 order of magnitude higher (benzoate $\subset$ **155** : $K_a = 161\,000\ M^{-1}$), indicating that the latter complex is stabilized by interactions between the aromatic part of the guest and the calixarene moiety. Calixarenes containing two urea or thiourea moieties along the lower rim were prepared by Nam *et al.* [252]. Contrary to the expectations, thiourea derivative **156c** forms less stable complexes than bisurea **156a**. Moreover, **156b**, which differs from **156a** in two additional methyl groups, forms significantly more stable complexes than **156a**. The acetate complex of **156b**, for example, has a K_a of 1150 M^{-1} in $CDCl_3$ versus 500 M^{-1} for the complex of **156a**, while acetate affinity of **156c** amounts to only 50 M^{-1}. The lower affinity of **156a** with respect to **156b** was ascribed to hydrogen bonds between the free hydroxyl groups and the urea moieties, which inhibit anion binding.

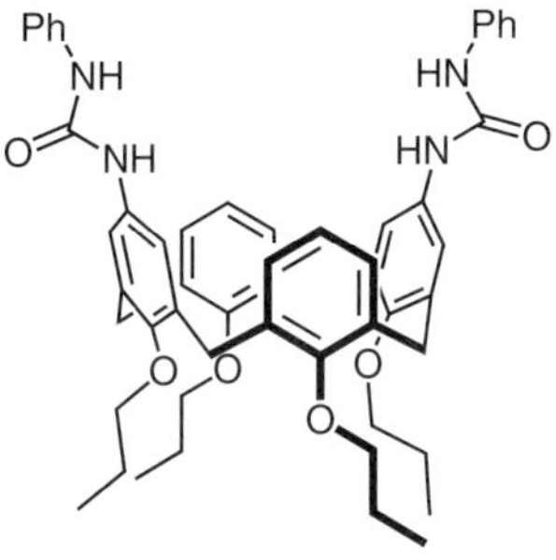

155

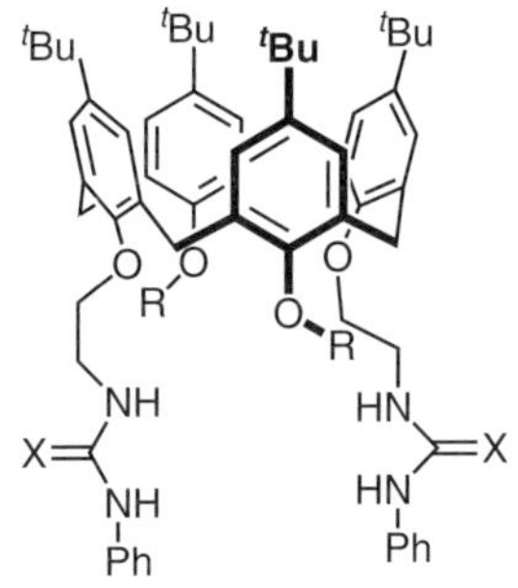

156a; R = H, X = O

156b; R = CH_3, X = O

156c; R = H, X = S

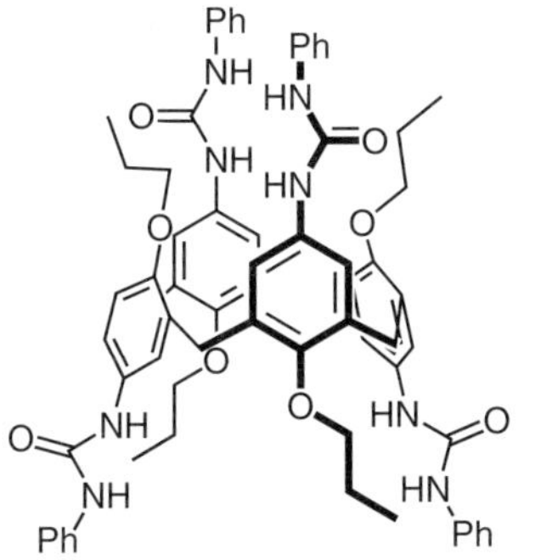

157

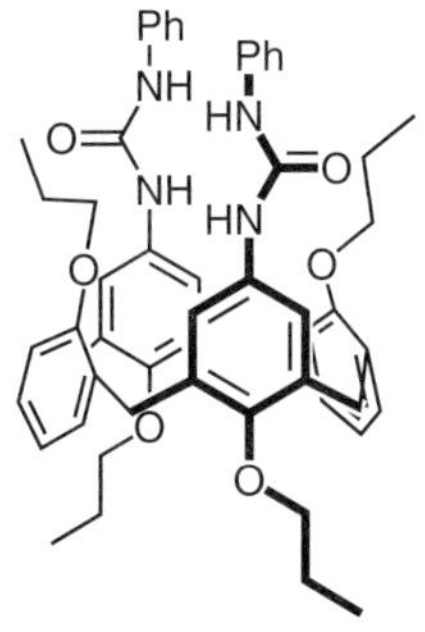

158

Tetraurea derivative **157** that derives from a calixarene with 1,3-alternate conformation was designed as a ditopic anion receptor with two binding sites at each side of the ring [251]. Surprisingly, **157** only forms 1 : 1 complexes with several anions including acetate, which was ascribed to a change of receptor conformation after binding of the first anion, resulting in an arrangement of the urea moieties on the opposite side of the receptor unsuitable for interactions with a second guest. An analog of **157**, receptor **158** containing only two urea moieties, has been shown to possess exceptional anion affinity [253]. The stability constant of the acetate complex of **158** in $CDCl_3/CD_3CN$ 4 : 1 is in fact too large to be determined by using ^1H NMR titrations, while it amounts to 920 M^{-1} in $CDCl_3$/DMSO 4 : 1.

Acetate binding to the anthracene-containing receptor **40** can be monitored by the PET quenching of the anthracene moiety [72].

Crabtree and coworkers investigated binding of simple acyclic diamides such as **159a** and **160** to anions including acetate [254]. Complex formation of **159a** is accompanied by a conformational reorganization of the receptor from the thermodynamically preferred *syn–anti* conformation to the *syn–syn* conformation, allowing both NH groups to simultaneously interact with the anion. Despite the lack of preorganization of **159a**, anion affinity is high; the acetate complex has, for example, a K_a of 19 800 M^{-1} in CD_2Cl_2. In contrast to **159a**, receptor **160** adopts the *syn–syn* conformation also in the absence of anions. This better preorganization does not translate into higher anion affinity, however, because hydrogen bonds have to be broken to allow interactions of the NH groups with the guest (acetate ⊂ **160** $K_a = 525\ M^{-1}$).

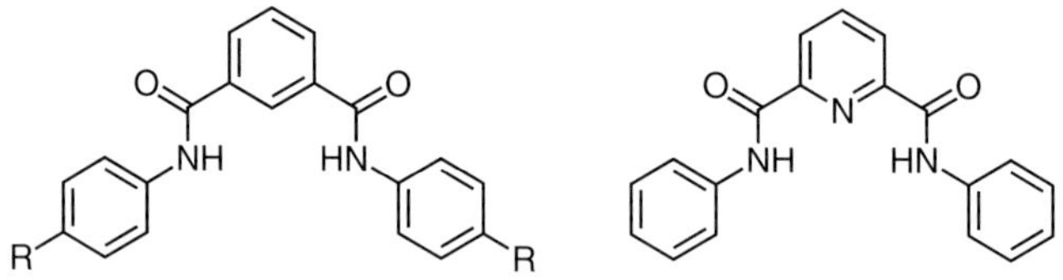

159a; R = ^{n}Bu
159b; R = H

160

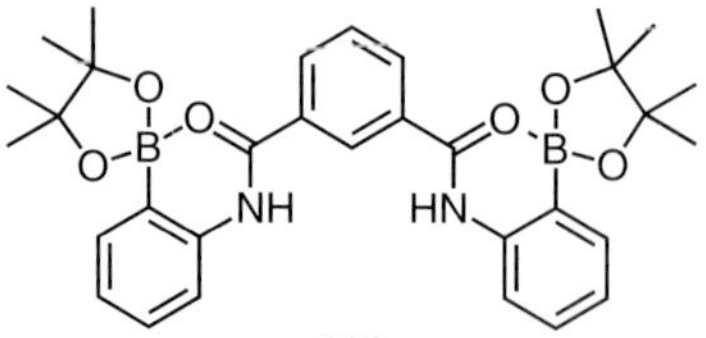

161

A significant increase in anion affinity was observed on introduction of boronate groups into **159b**; the acetate complex of **161** has a K_a of $2.1 \times 10^3\ M^{-1}$ in DMSO, for example, about 20 times higher than **159b** in the same solvent (acetate ⊂ **159b** $K_a = 1.1 \times 10^3\ M^{-1}$) [255]. This effect was explained by the intramolecular coordination of the boron atoms to the receptor carbonyl group, which (i) induces a larger host dipole moment thus strengthening dipole interactions and (ii) increases the positive surface potential at the urea NH groups, which strengthens short-range coulombic interactions with the anionic substrate.

The binding motif of receptor **160** was subsequently incorporated into macrocyclic receptor **32a–c**, which were shown to bind various anions in DMSO in the form of 1 : 1 complexes [61, 63]. While **32a** possesses the largest affinity for acetate (acetate ⊂ **32a** $K_a = 2640\ M^{-1}$), **32b** binds more strongly to dihydrogenphosphate (*vide supra*), although increase in ring size also results in an increase in acetate affinity (acetate ⊂ **32a** $K_a = 3240\ M^{-1}$). The crystal structure of the acetate complex of **32a** shows that only one oxygen atom of the anion interacts with all four receptor NH groups in the solid state. The second oxygen binds either to the four NH groups of a second receptor molecule or to a water molecule, depending on the counterion used for crystallization [61]. Hybrid tetraamide (**162**) benefits from both suitable preorganization provided by the 2,5-diamidopyridine unit and good anion-binding properties of the 1,3-diamidobenzene moiety and therefore proved to be a better anion receptor than the homoaromatic analogs **32b** and **163** [64].

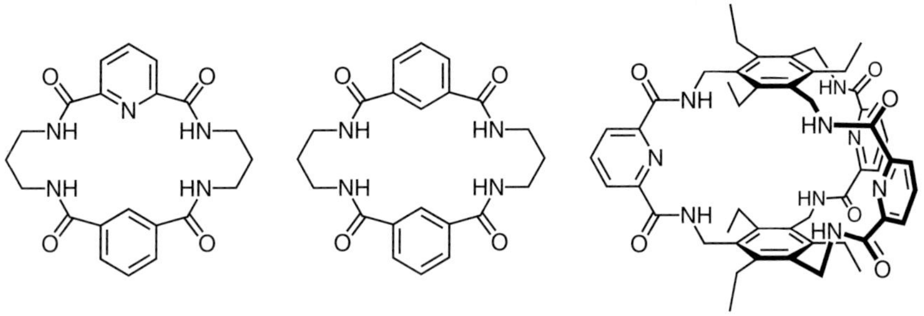

162 **163** **164**

Also, macrocyclic hosts **44–46** [83, 84], macrobicyclic host **36a** [58], and macrotricyclic host **37a** [68] have been shown to possess affinity for acetate. A structurally unusual acetate receptor consisting of a [2]catenane with two identical interlocked amido bipyrrole-based macrocycles was developed in a collaborative approach between the Sessler and Vögtle groups. This compound binds acetate with a K_a of $9.63 \times 10^5\ M^{-1}$ in 1,1,2,2-tetrachloroethane [256].

Bicyclic host **164** was designed by the Anslyn group as a receptor selective for the complexation of planar trigonal anions [257]. The capping aromats in this compound are separated by 7.0 Å, and the size of the cavity amounts to 78.3 Å^3. In agreement with the design strategy, strong binding was indeed detected in 25% CD_2Cl_2/CD_3CN for acetate and nitrate. The K_a of the acetate complex amounts to $770\ M^{-1}$, for example. The crystal structure of the acetate complex of **164** revealed that the anion resides inside the bicyclic cavity, where each of its oxygen atoms is hydrogen bonded to two amide hydrogen atoms from the acyl pyridines.

Calixarenes have been shown to be useful scaffolds for the construction of acetate receptors. Loeb designed calix[4]arene **165a** containing two opposing amide groups on the upper rim for the selective recognition of "Y-shaped" carboxylate anions [258]. It was proposed that binding of the carboxylates causes a pinched cone conformation of **165a**, which allows two linear hydrogen bonds to be formed between the receptor and the anion. Acetate affinity of **165a** amounts to $88\ M^{-1}$

in acetonitrile. Introduction of the stronger electron-withdrawing group in **165b** improves anion affinity without reducing selectivity.

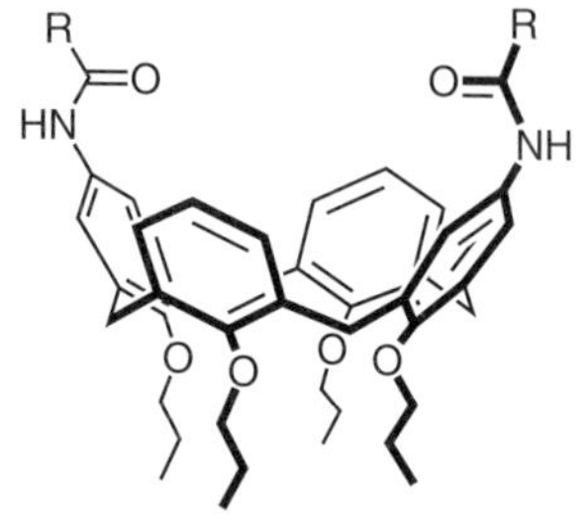

165a; R = CH_2Cl
165b; R = $CHCl_2$

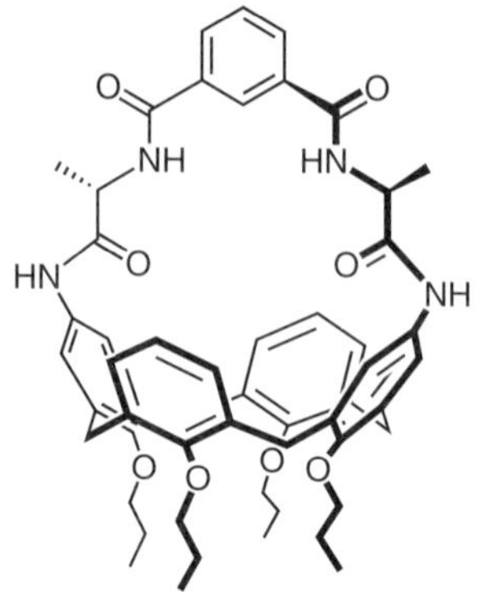

166a

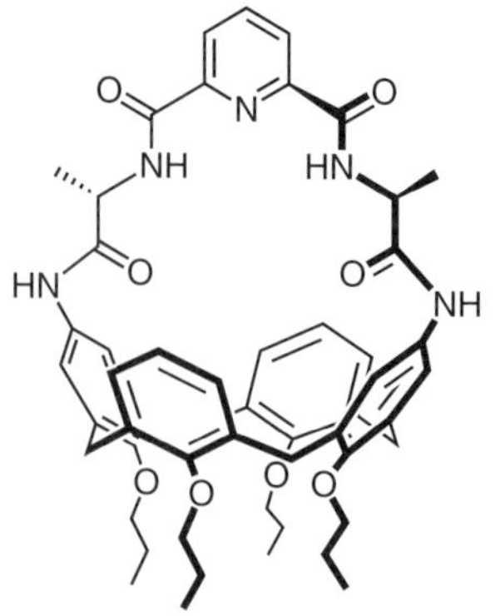

166b

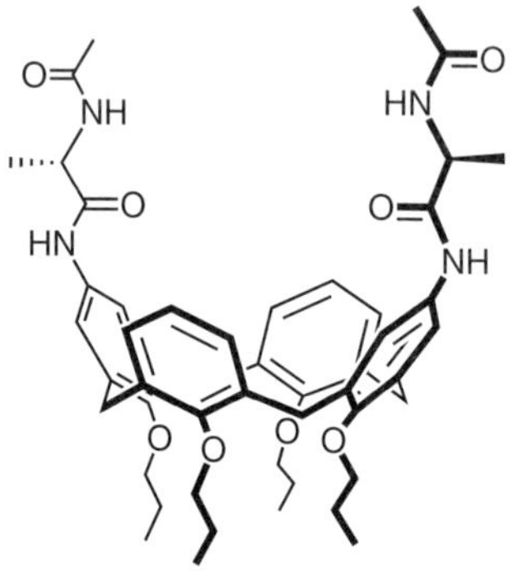

167

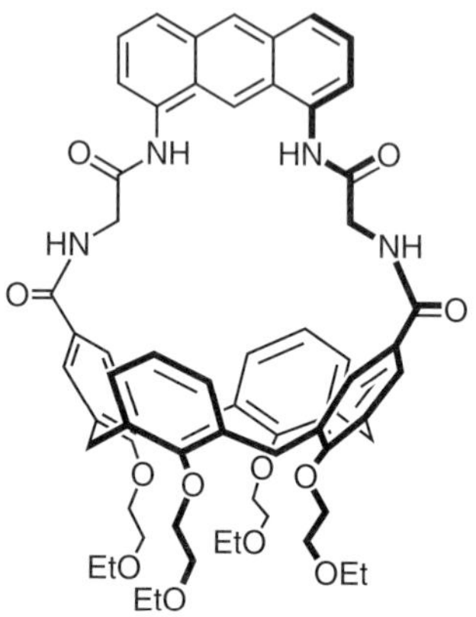

168

The group around Ungaro showed that bridged calixarenes such as **166a** and **166b** possess much higher anion affinities than the unstrapped analogs because the bridge reduces conformational flexibility and the tendency of the unstrapped analog **167** to form intermolecular and intramolecular hydrogen bonds [259]. Peptido[4]calixarenes **166a** and **166b** were shown to bind carboxylates but not the corresponding carboxylic acids. High affinity was observed for aromatic carboxylates; the K_a of the benzoate complex of **166b**, for example, amounts to 40 100 M^{-1} in acetone, while that of the acetate complex is only 10 500 M^{-1}. Benzoate affinity is about 100 times higher than that of the unstrapped analog **167**, whose complex with benzoate has a K_a of 680 M^{-1}. The preference of receptors **166a** and **166b** for aromatic over aliphatic amino acids has been rationalized by stabilizing $\pi-\pi$ stacking interactions between the phenyl group of the guest and the bridging aromat of the host. NMR spectroscopic investigations indicated that complexation is mainly due to hydrogen bonding of the anions to the NH amide groups of the receptors. Interactions of structurally related hosts with the terminal carboxylate group in peptides are discussed in Section 7.3.5.

Structurally related to receptor **166** is **168** containing an anthracene moiety in the bridge [260]. The close proximity of this chromophore to the anion-binding site efficiently couples fluorescence of the system to its binding properties. Consequently, fluorescence of **168** is quenched on addition of basic anions such as acetate or fluoride. Stability constant of the acetate complex of **168** amounts to 3200 M^{-1}.

Prohens *et al.* showed that squaramide groups can act as potent binding sites for carboxylates including acetate even in competitive solvents since the hydrogen-bonding pattern, featuring two hydrogen donor atoms, parallels those in ureas, thioureas, and guanidinium cations (Figure 7.3c) [261]. Receptor **169a** was shown to bind acetate with a K_a of 1980 M^{-1} in DMSO. Combining hydrogen bonding with electrostatic interactions in receptor **169b** improves acetate affinity by a factor of about 7 (acetate ⊂ **169b** $K_a = 14200\ M^{-1}$), while arranging several squaramide moieties on suitable scaffolds produces receptors for di- or tricarboxylates.

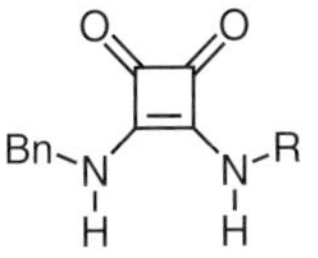

169a; R = Bn
169b; R = $CH_2CH_2\overset{\oplus}{N}(CH_3)_3$

A variety of metal complexes have been described as acetate receptors. Two examples should illustrate the different approaches. The Beer group synthesized ruthenium(II) and rhenium(II) calix[4]arene and calix[4]diquinone receptors **170a,b** and **171a,b**, both of which somewhat resemble compounds **156a–c**, for the recognition and sensing of anions. No direct interaction of anions with the metal coordinated to the 2,2′-biphenyl moiety, whose role is primarily to act as a sensing group, is expected. The investigations revealed that, among the different anions studied, all receptors favor binding of acetate [262]. Acetate affinity is the highest for the calix[4]diquinone derivatives (**171a,b**), which also allow optical sensing of this anion (acetate ⊂ **170a** $K_a = 4060\ M^{-1}$, acetate ⊂ **171a** $K_a = 9990\ M^{-1}$). Addition of acetate to a solution of **171a**, for example, increases the emission intensity of the receptor by 500%, concomitant with a slight blueshift of the emission maximum. The structurally related receptor **172**, also developed by the Beer group, binds acetate in DMSO with a binding constant of 3600 M^{-1} [263]. Its ruthenium(II) bipyridyl redox system undergoes significant cathodic perturbations on addition of anionic guests, allowing complex formation to be followed not only by NMR spectroscopy but also electrochemically.

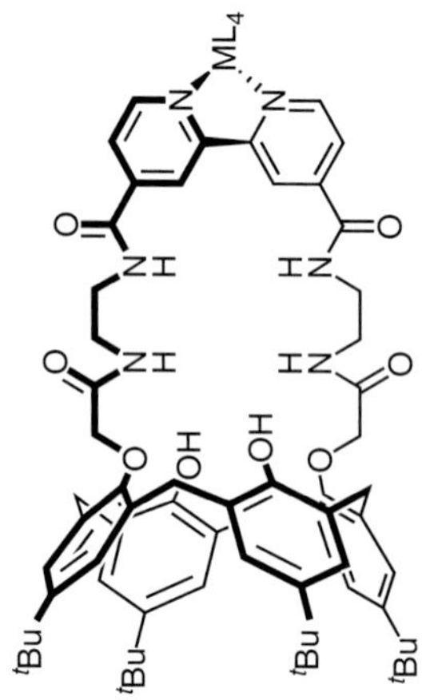

170a; $ML_4 = [Ru(bipy)_2]^{2+}$
170b; $ML_4 = ReCl(CO)_3$

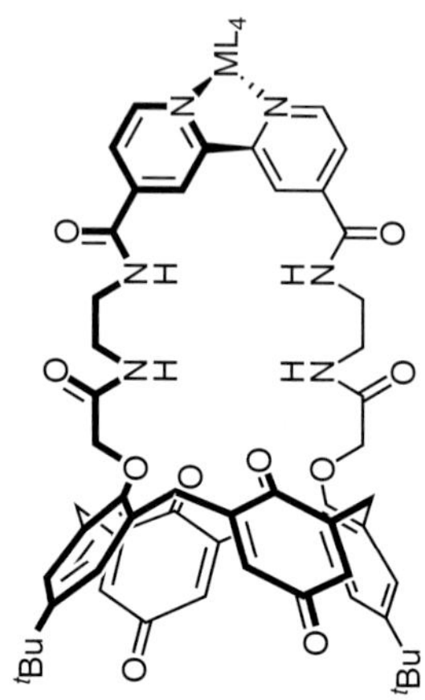

171a; $ML_4 = [Ru(bipy)_2]^{2+}$
171b; $ML_4 = ReCl(CO)_3$

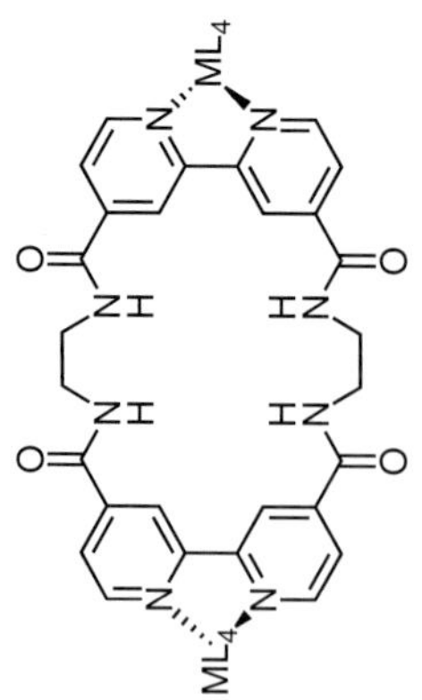

172; $ML_4 = [Ru(bipy)_2]^{2+}$

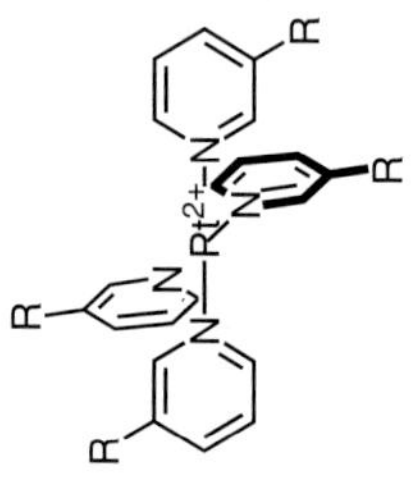

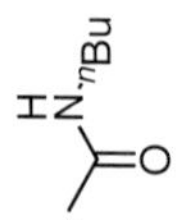

173a; R =

173b; R =

The second approach to designing metal-containing acetate receptors involves the use of a metal as a structure-determining element on which the anion-binding sites are assembled. This approach was pursued by Loeb, who investigated the anion-binding properties of platinum(II) complexes **173a,b** [264, 265]. These twofold positively charged complexes were isolated as the hexafluorophosphate salts because this anion only weakly interacts with the anion-binding sites. ^{1}H NMR titrations showed that **173a** efficiently recognizes various oxoanions. Affinity was the highest for planar bidentate anions such as nitrate and acetate, which was attributed to the shape match between these anions and two *cis*-oriented amide groups in **173a**. Complexation of two acetate anions was observed in CD_3CN/DMSO 1 : 9. The stability constants associated with the two binding steps are $K_{11} = 230\ M^{-1}$ and $K_{12} = 491\ M^{-1}$, demonstrating that binding of the first anion has a positive allosteric effect, favoring complexation of the second. A similar behavior was detected for receptor **173b** [265].

7.3.3
Di- and Tricarboxylates

Common strategies to obtain synthetic receptors for di- or tricarboxylates include incorporation of two or more carboxylate-binding sites into a ring or attachment of two or more monotopic carboxylate receptors to a suitable molecular scaffold. Subsequent binding studies then usually address the question as to what extent receptor selectivity is controlled by the mutual orientation of the binding sites in the receptor. To assess selectivity, substrates differing in the distance of carboxylate groups along an aliphatic chain or in the arrangement of carboxylate groups at a double bond (maleate vs fumarate) or along an aromatic or aliphatic ring are used. Although some investigations relevant in this context are presented in this chapter, its main focus lies on receptors that target biologically relevant anions such as malate, tartrate, or citrate.

In a systematic investigation, Hossain and Schneider have evaluated the strength of ion-pairing interactions in a series of salt pairs containing an acyclic bisquaternary ammonium cation and a structurally matching α,ω-dicarboxylate [266]. This work not only yielded an estimation of the strength of a single carboxylate–ammonium interaction in water (8 kJ mol^{-1}) but also showed that with every additional flexible bond in the ion pair, complex stability decreases by only 0.5 kJ mol^{-1}. According to the authors, this surprisingly small value indicates that the importance of conformational preorganization for effective molecular recognition may have been overestimated. Similar studies were carried out by characterizing the binding of dicarboxylates with varying distances of the two carboxylate groups to structurally matching bisamides [267].

While there are only few examples of binding studies involving polyaza macrocycles and monocarboxylates (*vide supra*), interaction of such receptors with di- or tricarboxylates has been studied extensively [3]. An early investigation addressed binding of various anions including di- and tricarboxylates to receptors **174–176**, which were shown to interact with oxalate, malonate, succinate, tartrate, maleate, fumarate, squarate, and citrate in water [8]. The results indicate that electrostatic interactions play a major role in binding strength. Structural effects and size complementarity do affect complex stability, but these effects do not translate into significant differences in anion selectivity, which is most probably due to the conformational flexibility of these receptors that allows them to easily adapt to different anion geometries. Thus, although **174–176** do possess different affinities toward various dicarboxylates, hexaprotonated **176** binds oxalate more strongly than hexaprotonated **174**, which was attributed to higher local charge density in an ethylenediammonium group than in a propylenediammonium group; affinities of the three receptors toward a given dicarboxylate differ by a factor of 16 at most, however. Appreciable binding selectivity was observed only in the complexation of citrate, which is bound by **175** up to 800 times more strongly than by the other two hosts. Related work was carried out by the Kimura group [268].

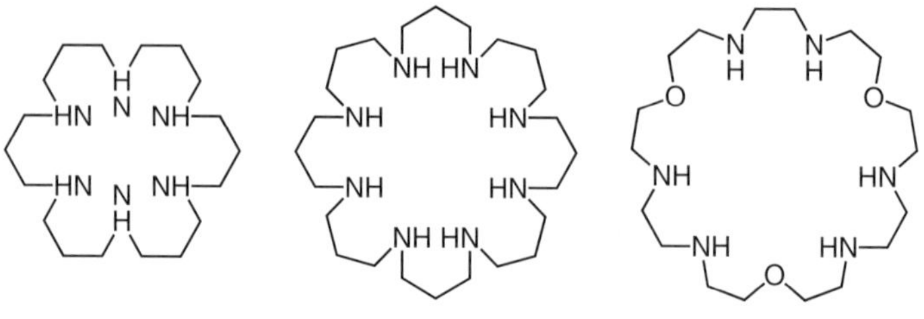

174 **175** **176**

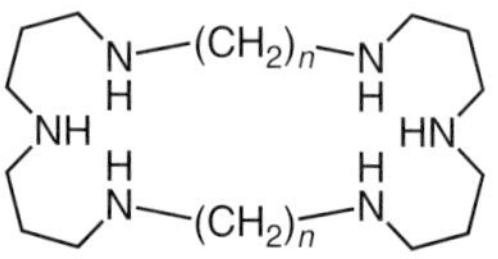

177a, $n = 3$
177b, $n = 7$
177c, $n = 10$

Receptors **177a–c** have been shown to exhibit selectivity for dicarboxylates matching the distance of the two 1,5,9-triazanonane moieties [269, 270]. Thus, **177b** has the highest affinity for succinic acid (log K_a of hexaprotonated $^{-}OOC(CH_2)_2COO^{-} \subset$ **177b** = 4.30) and glutaric acid (log K_a of hexaprotonated $^{-}OOC(CH_2)_3COO^{-} \subset$ **177b** = 4.40), while longer dicarboxylic acids such pimelic acid and suberic acid are bound best by **177c** (log K_a of hexaprotonated $^{-}OOC(CH_2)_5COO^{-} \subset$ **177c** = 4.40; log K_a of hexaprotonated $^{-}OOC(CH_2)_6COO^{-}$**177 c** = 4.25). Receptor **177a** exhibits selectivity for oxalate (log K_a of hexaprotonated $^{-}OOCCOO^{-} \subset$ **177a** = 3.80).

One major problem in assessing binding selectivity of polyammonium-based receptors toward di- and tricarboxylates arises from the fact that both receptor and substrate are involved in pH-dependent protonation equilibria. A comparison of receptor affinity toward different substrates has to take into account the degree of protonation of receptor and substrate, but although potentiometric titrations provide information about the total degree of protonation of the complex, it is rarely possible to identify the exact locations where protonation occurs. It has therefore been proposed to derive a receptor's selectivity pattern from the distribution diagram of the mixed system containing all interacting species or by calculating the analytical apparent stability constants of the complexes for every pH value as the quotient between the summation of the complexed species and the summation of the free reagents [271, 272].

High affinity for dicarboxylic acids has also been detected for cryptands **131–133** and **178**. Azacryptand **178** binds dicarboxylates in aqueous solution at pH 6 [273]. The highest affinity was observed for adipate ($K_a = 2.6 \times 10^3$ M^{-1}), while shorter or longer dicarboxylates are complexed less efficiently. Again, this behavior was attributed to an optimal size and shape complementarity of the cavity of **178** to adipate. A very stable complex is also formed between **178** and terephthalate ($K_a = 2.5 \times 10^4$ M^{-1}), and for this anion, a crystal structure confirmed that binding occurs inside the cavity of the host.

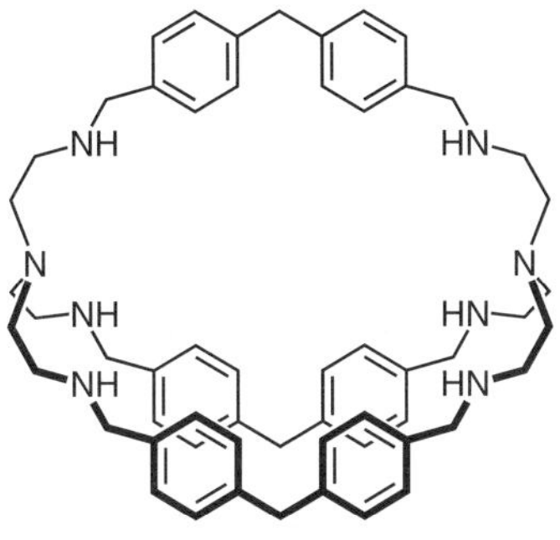

178

Oxalate possesses excellent steric complementarity with protonated cryptands **131** and **133** [274]. At each amine-derived cap, only one of the carboxylate oxygen atoms is involved in direct hydrogen-bonding interactions to these hosts, three in each case. This strong interaction has the effect of polarizing the carboxylate function, generating at each end of the dicarboxylate one partly single and one partly double C–O bond. The partly double C–O lies parallel to an aromatic ring, suggesting that π–π stacking also plays a part in stabilizing the aggregate and explaining the large complexation constants seen for the oxalate cryptates of **131** and **133**; the log K_a of the oxalate complex of hexaprotonated **131** amounts to 10.71, for example. In contrast, the protonated host **132** prefers a cleftlike conformation and includes dicarboxylates such as oxalate or malonate between two of the straps [275]. Oxalate and malonate affinities of hexaprotonated **132** are very similar, amounting to a log K_a of about 8.3.

Host **179** was prepared by the Kimura group in an attempt to obtain dicarboxylate receptors by tethering two monotopic carboxylate hosts [11]. Compound **179** was shown to form a complex with citrate when tetraprotonated presumably by sandwiching the anion between the two polyammonium rings. This complex has a log K_a of 3.57 in water, about 1 order of magnitude higher than the stability of the complex between citrate and the triprotonated monotopic parent host, demonstrating the cooperative effect of the two receptor subunits of **179** in anion recognition.

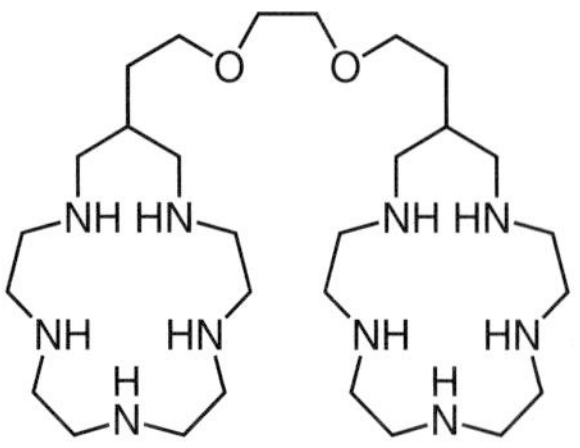

179

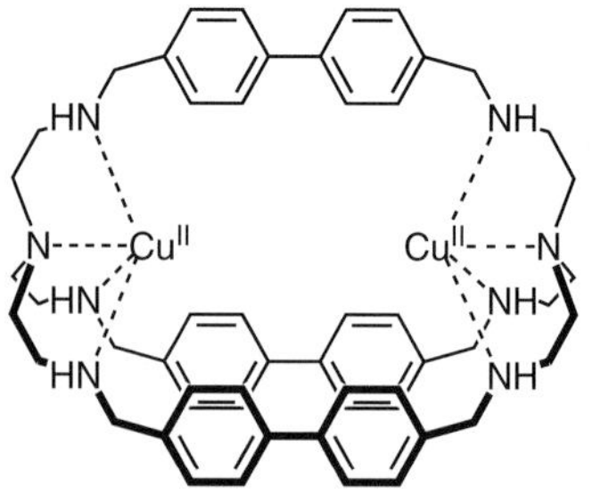

180

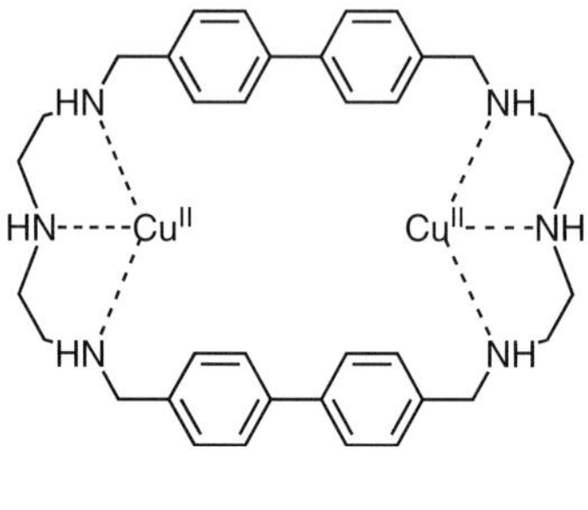

181

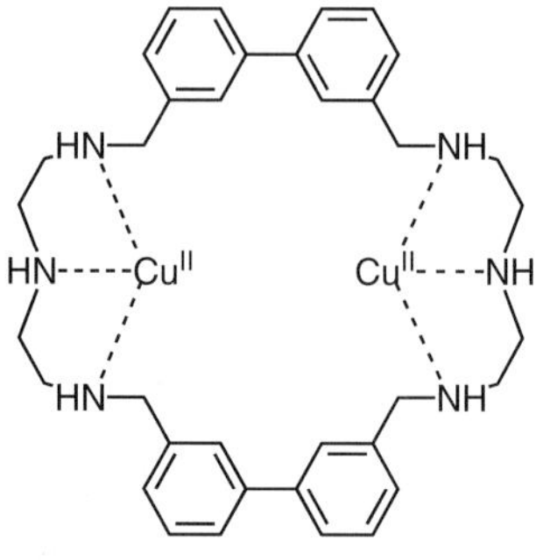

182

Shape-selective dicarboxylate recognition could also be achieved by using dicopper(II) cryptate **180** [276]. Complex formation can be detected visually by combining **180** with carboxyrhodamine, a fluorescent dye containing a terephthalate subunit, whose interaction with **180** leads to a nonfluorescent complex in aqueous HEPES-buffered solution at pH 7. Of the isomeric benzene dicarboxylic acids, only terephthalic acid is able to compete with carboxyrhodamine in binding to the host thus causing a fluorescence enhancement, while almost no effect on fluorescence was observed in the presence of phthalic or isophthalic acid. In the family of aliphatic dicarboxylates $^{-}OOC{-}(CH_2)_n{-}COO^{-}$ with $n = 0$–5, the receptor–dye ensemble can be used for glutarate ($n = 3$) and adipate ($n = 4$) sensing. No displacement of the dye from the host occurs in the presence of the longer or the shorter dicarboxylic acids, with the exception of oxalate ($n = 0$), which was ascribed to the formation of a 2 : 1 complex between **180** and two guest molecules.

In a similar manner, dicarboxylate binding to the two monocyclic dicopper(II) complexes **181** and **182** was investigated [277]. Despite the close structural relationship of both hosts, characteristic differences in binding selectivity were detected. Thus, while **181** exhibits preferential binding of terephthalate among the three benzene dicarboxylates, host **182** prefers the 1,3-isomer. Selectivity of **181** in binding to aliphatic dicarboxylates differing in chain length is similar to that of **180**, whereas host **182** more strongly binds to shorter dicarboxylates such as succinate ($n = 2$). Moreover, affinities for the best matching dicarboxylates of the monocyclic hosts are distinctly lower than those of cryptate **180**.

The tricopper(II) complex of receptor **183** has been shown to selectively bind and, in combination with 5-carboxy-fluorescein, sense tricarboxylic acids in aqueous solution [278]. The association constants log K_a of the complexes with citrate, 1,3,5-benzenetricarboxylate, or 1,3,5-cyclohexanetricarboxylate range between 5.00 and 5.81. Dicarboxylates are bound with a significantly lower affinity. Similarly, the dicopper(II) complex of receptor **184** efficiently and selectively binds to oxalate in aqueous solution ($K_a = 1.3 \times 10^5\ M^{-1}$) [279]. Binding of malonate, succinate, or glutarate anion is weaker by at least 1 order of magnitude. Complex formation can be visualized by establishing a dye displacement assay using eosine as the indicator.

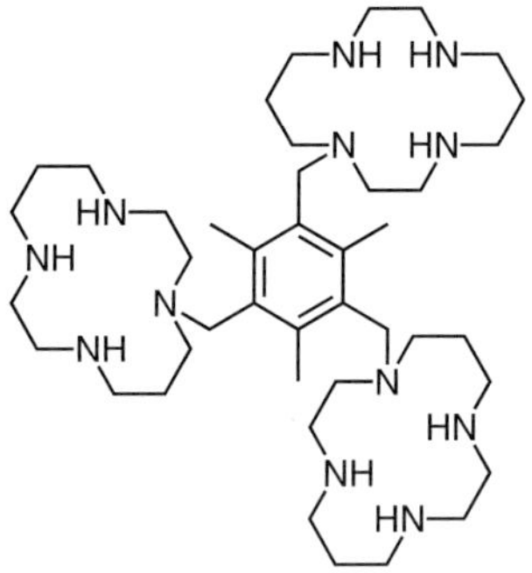

183

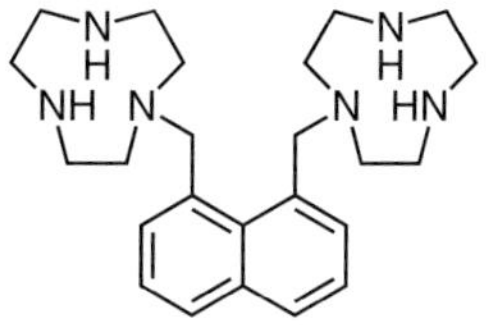

184

Receptors for dicarboxylates based on the guanidinium/carboxylate-binding motif usually contain two or more guanidinium groups for simultaneous interactions with both carboxylate groups of the substrate. An early example is host **23a**, which has been shown to bind various dicarboxylates in methanol [280]. This receptor exhibits preference for malonate ($K_a = 1.6 \times 10^4\ M^{-1}$) over shorter or longer dicarboxylates. The fact that even aromatic dicarboxylates are bound with appreciable affinity indicates that **23a** can adapt well to host structure, which is detrimental for binding selectivity.

A number of di- and tricarboxylate receptors containing aminoimidazoline moieties attached to the 1,3,5-triethylbenzene scaffold have been developed by the Anslyn group. Receptor **185** has, for example, been shown to interact with citrate and tricarballate in water. The binding constant of the citrate complex amounts to $6.9 \times 10^3\ M^{-1}$, for example [281]. Dicarboxylates, monocarboxylates, or phosphates are bound with considerably reduced affinity. Moreover, derivatives of **185** lacking the ethyl groups or containing ammonium instead of imidazoline groups possess inferior receptor properties, illustrating the importance of receptor preorganization and of hydrogen-bonding interactions between the substrate and the guanidinium moieties. The high affinity and selectivity of **185** allowed this receptor to be used in a dye displacement assay to quantitatively determine the citrate content of various sports drinks [282].

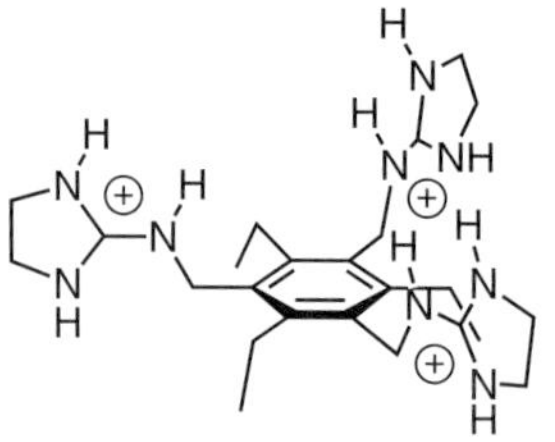

185

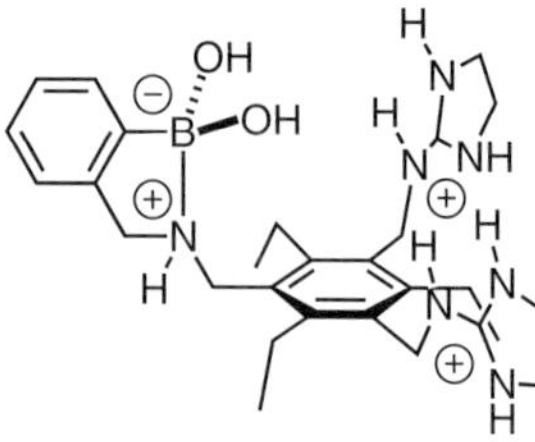

186

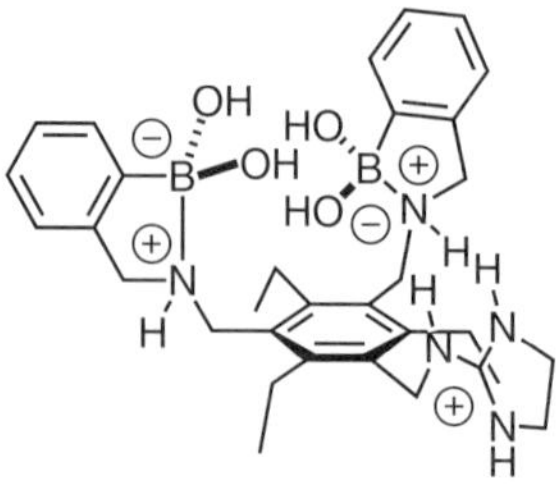

187

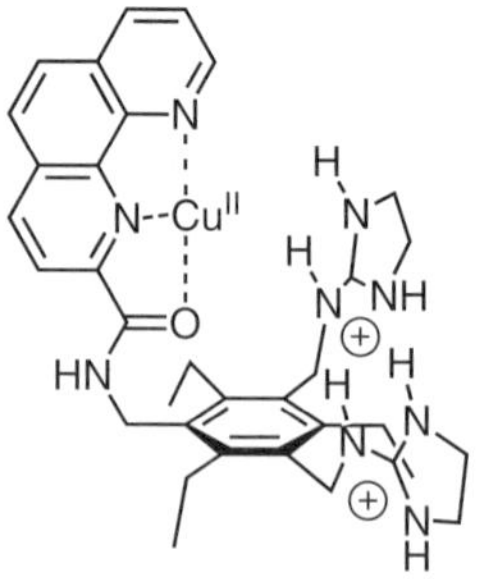

188

Subsequent structural modifications of **185** were carried out with the aim of inducing binding selectivity for other anions. In this context, boronic acid residues were introduced into **186** and **187** to induce binding selectivity for anions containing vicinal hydroxy groups. Receptor **186** binds tartrate with a high binding constant of 5.5×10^4 M^{-1} in 25% water/methanol, for example [283]. With the exception of malate, every other anion investigated is bound considerably less tightly, thus allowing a quantitative determination of the tartrate/malate content of several beverages by dye displacement. A mathematical method was derived for the analysis of indicator displacement assay isotherms, which was employed for the determination of malate in Pinot Noir using host **186** and alizarine complexone as the indicator [284]. Host **187** was shown to bind gallate and phenolic acids of similar structure in 25% water/methanol, a property that was used for the development of a colorimetric assay for the aging of scotch [285]. This host also binds tartrate and malate, but in contrast to **186**, it has a higher affinity for tartrate. Thus, a multicomponent sensing ensemble could be devised consisting of **186**, **187**, and two indicators that, in combination with pattern-recognition analysis of UV–vis spectra, allowed the simultaneous quantitative estimation of the malate and the tartrate concentration of an aqueous solution [286]. On the basis of host **187**, a simple colorimetric test strip for tartrate that makes use of an indicator displacement could also be designed [287]. A systematic comparison of the interaction of receptors **185**–**187** and a derivative containing three boronic acid subunits with carboxylates, α-hydroxycarboxylates, and diols revealed that receptor **185** is selective for citrate, while receptors that incorporate boronic acids have higher affinities for guests containing α-hydroxycarboxylate or catechol moieties over guests containing only carboxylate or hydroxy groups [288].

Copper(II)-containing receptor **188** also interacts with citrate [289]. Its metal-free form has a citrate affinity of $3.9 \times 10^6\ M^{-1}$ in 15% water/methanol (0.1 mM HEPES, 0.1 mM NaCl, pH 7.4), whereas a stability constant of at least $8.3 \times 10^6\ M^{-1}$ was estimated for the complex between citrate and **188**. Moreover, the fluorescence of the phenanthroline moiety that is quenched in **188** returns on citrate binding, thus allowing optical sensing of the anion. The cooperative effect of copper indicates a participation of the metal center in citrate complexation, and detailed studies involving model compounds showed that the return in fluorescence is indeed due to metal–anion interaction and not to a displacement of the metal from the host. Also in this case, the sensing properties of **188** allowed the quantification of the citrate content of various beverages [289].

In a somewhat related approach, citrate receptors have been devised by Schmuck and Schwegmann by assembling three 2-guanidiniocarbonyl pyrrole moieties along the 1,3,5-triethylbenzene scaffold [290]. Receptor **189** indeed binds citrate and other tricarboxylates such as 1,3,5-benzene tricarboxylate with high association constants of $K_a > 10^5\ M^{-1}$ in water. Complex stabilities are so high that even the presence of a large excess of competing anions or buffer salts does not significantly impair binding. The association constant K_a of the citrate complex, for example, changes only from $1.6 \times 10^5\ M^{-1}$ in water to $8.6 \times 10^4\ M^{-1}$ in the presence of a 170-fold excess of bis–tris buffer and a 1000-fold excess of chloride. Structural investigations indicated that the tricarboxylates are bound within the inner cavity of the receptor.

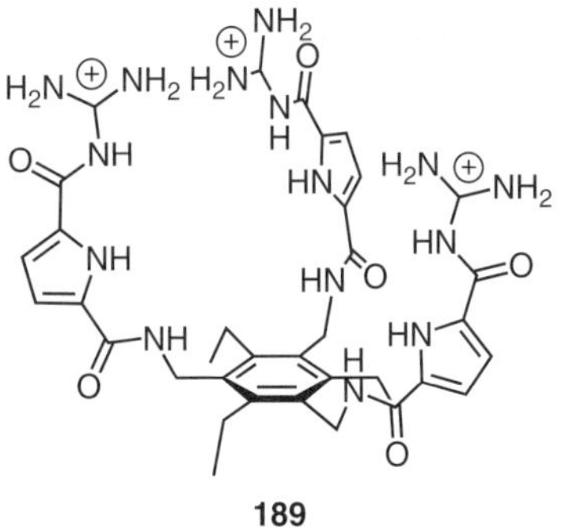

189

In combination with carboxyfluorescein, an indicator displacement assay could be established on the basis of **189**, which allowed the fluorimetric detection of citrate in aqueous solution [291]. Competing anions such as malate or tartrate were shown to be unable to replace the dye from the host.

Neutral receptors for dicarboxylates have been obtained by attaching two urea, thiourea, or amide moieties to a suitable molecular scaffold. Examples are tweezer-type receptors **190a,b**, described by Hamilton and coworkers, which bind the bis(tetrabutylammonium) salt of glutaric acid with stability constants K_a of, respectively, 6.4×10^2 and $1.0 \times 10^4\ M^{-1}$ [292]. As usual, the thiourea-derived receptor proved to be more effective. It should be noted that the combination of a hydrogen bond donor and a hydrogen bond acceptor in an amidopyridine moiety gives rise to receptors for free carboxylic acids [232]. Receptor **191**,which is structurally closely related to **190a**, has been shown to bind glutaric acid with a K_a of $6.4 \times 10^2\ M^{-1}$ in 5% THF/$CDC1_3$, for example, [293].

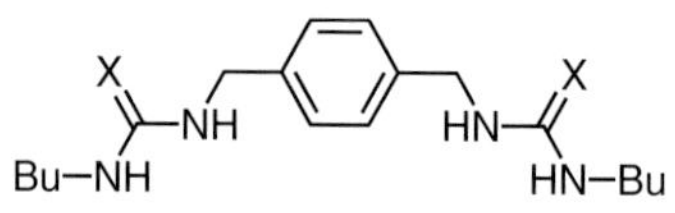

190a; X = O
190b; X = S

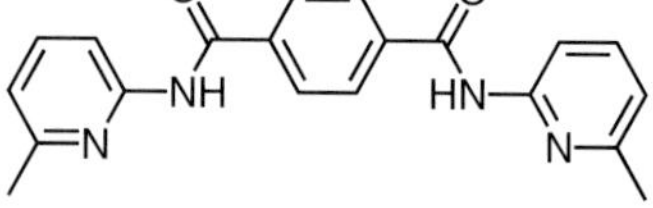

191

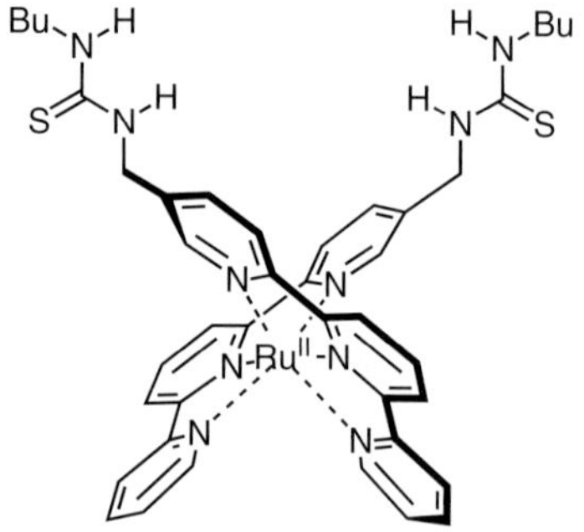

192

Similar strategies were used by Kelly [294], Jeong [295], and Reinhoudt [296] for the construction of di- and tricarboxylate receptors.

In a combinatorial approach, the Hamilton group has synthesized a series of receptors by assembling two functionalized terpyridine moieties into a ruthenium(II) center [297]. Compound **192** containing two thiourea moieties proved to be an efficient receptor for dicarboxylates. Dicarboxylate affinity of **192** is in fact too high to be measured accurately in DMSO by NMR titrations ($K_a > 10^4\ M^{-1}$) presumably because electrostatic interactions between the substrate and the positively charged metal ion in **192** cooperatively contribute to complex stability. In the more competitive solvent 5% water/DMSO, affinity of **192** to the best substrate glutarate amounts to $8.3 \times 10^3\ M^{-1}$ [298].

A recent example of a dicarboxylate receptor containing pyrrole units is ditopic host **193** [299]. This receptor consists of two 1,3-dipyrrolyl-1,3-propanedione boron complexes assembled on a *p*-phenylene-bridged dicatechol scaffold. Boron complexes of 1,3-dipyrrolyl-1,3-propanedione containing a BF_2 group been shown to preferentially adopt a conformation with divergent NH groups. Reorganization occurs on binding of carboxylates, leading to a conformation that allows both pyrrole NH groups to interact with the carboxylate oxygen atoms simultaneously. The cooperative action of two of these binding sites in **193** induces affinity for dicarboxylates as well as selectivity for substrates, which best spans the distance between them. Thus, affinity of **193** in CH_2Cl_2 is the highest for the bis(tetrabutylammonium) salt of pimelate (K_a $^-OOC(CH_2)_5COO^- \subset$ **193** $= 8.1 \times 10^5\ M^{-1}$) and decreases for longer and shorter dicarboxylates.

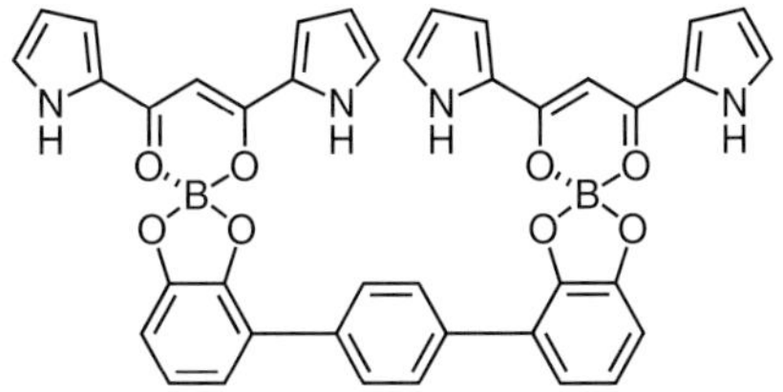

193

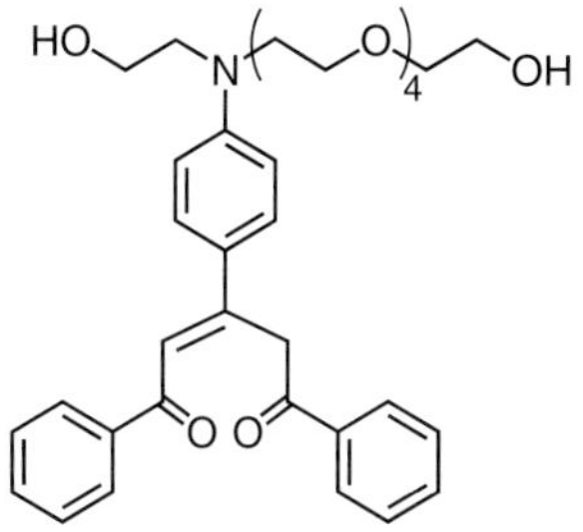

194

A highly selective colorimetric sensor for oxalate, malonate, and maleate, compound **194**, has been described by Martínez-Mañez [300]. Addition of carboxylates (acetate and benzoate) or dicarboxylates (terephthalate, succinate, glutarate, adipate, pimelate, and suberate) to a solution of **194** in water (buffered with 0.1 M HEPES) gives rise to a yellow solution. Only oxalate, malonate, and maleate cause the color of the solution to turn to red-magenta. This color change is due to a cyclization of the pent-2-en-1,5-dion moiety in **194** to give a 3,4,6-trisubstituted pyrylium cation, a reaction that is only efficiently catalyzed by dicarboxylic acids in which the two carboxylate groups are appropriately arranged to allow for simultaneous hydrogen-bonding interactions with the OH proton in the enol form of **194**.

7.3.4 Amino Acids

Synthetic amino acid receptors should possess one or preferably both of the following properties: they should be able to distinguish different amino acid derivatives and/or the enantiomers of a chiral amino acid. Either property requires the receptor to make contact with the substrate in more than one position. There is a rule of thumb, for example, that chiral recognition can be achieved only if there are three simultaneous interactions between the receptor and at least one of the substrate's enantiomers, not all of which have to be attractive, however, with at least one of these interactions being stereochemically independent [301]. Selective recognition of a specific amino acid necessarily requires the receptor to interact with the amino acid side chain.

Not all amino acid receptors are also anion receptors. There are, for example, many receptors that target the protonated cationic amino group of the substrate while its carboxylate group is protected, exerting steric effects on the binding process at best. A classic example is the binol-derived crown ether developed by Cram, which is able to differentiate the enantiomers of chiral amino acid esters [302]. In keeping with the general topic of this monograph, these receptors are not considered here but only ones that contain a characteristic structural element for carboxylate recognition. An additional binding site for an ammonium group is allowed but not required. The types of substrates covered in this chapter are therefore N-protected amino acids as well as zwitterionic unprotected amino acids containing one or two carboxylate groups.

Somewhat complementary to Cram's approach of achieving enantioselective amino acid recognition by using chiral crown ethers that bind to the ammonium group of the substrate is the use of chiral polyaza macrocycles with which the carboxylate group is targeted. Compounds **195–197** are examples of such receptors.

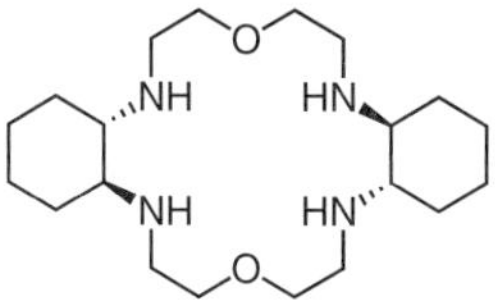

195

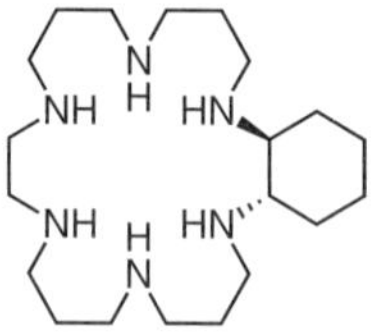

196

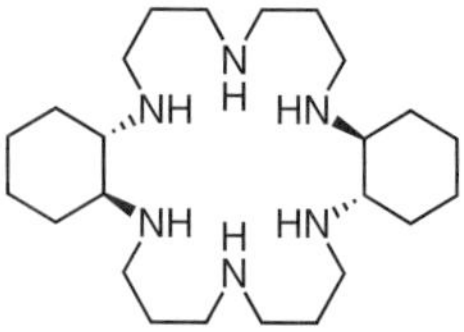

197

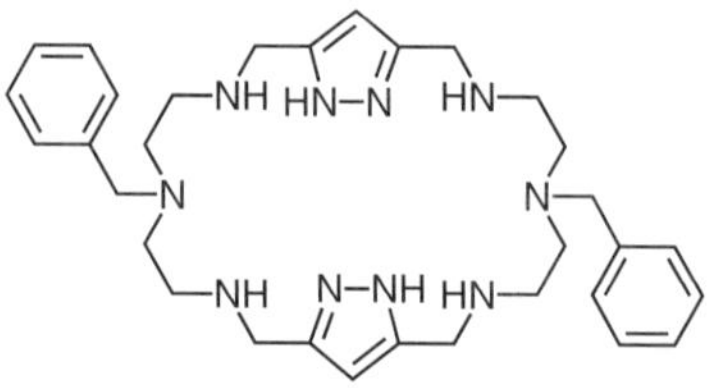

198

Tetraprotonated **195**, for example, exhibits good enantioselectivity in binding to *N*-acetyl aspartate; the log K_a values of the complexes with the L- and the D-enantiomer of the substrate amount to, respectively, 3.54 and 4.21 [303]. On the basis of the experimental results, the authors propose a structural model of the complexes formed involving a combination of coulombic interactions and hydrogen bond formation. Subsequently, the same group compared the binding properties of receptors **196** and **197** [304]. Measurement of the protonation constants revealed particularly large differences in the fifth protonation step of these receptors, in which **196** is 10 times more basic than **197**. According to the authors' interpretation, the fifth protonation involves the ethylenediamine fragment in **196**, which causes the macrocycle to relax avoiding electrostatic repulsion. This is not possible in **197**, which explains its reluctance to form the fully protonated species.

The tetraprotonated form of **197** has a lower affinity and enantioselectivity in binding *N*-acetyl asparate than **195**. Affinity as well as enantioselectivity substantially increases after full protonation, reaching a log K_a of 5.34 for the complex of the D-enantiomer and a log K_a of 4.57 for that of the L-enantiomer of the substrate. Enantioselectivity of **196** is somewhat lower, but binding of D-aspartate is also preferred. Interestingly, the hexaprotonated form of **197** binds *N*-acetyl D-glutamate in the form of a 1 : 2 receptor–substrate complex, while the L-enantiomer of this substrate is bound with 1 : 1 stoichiometry.

Pyrazole-containing macrocyclic polyamine **198** has been shown to interact with L-glutamate in water [305]. NMR investigations suggested that, in addition to electrostatic and hydrogen-bonding interactions, cation–π interactions between the benzyl groups of **198** and the L-glutamate ammonium group contribute to complex stability. This assumption is supported by the fact that an analog of **198** lacking the appended benzyl groups forms less stable complexes under the same conditions.

The chiral ditopic receptor **199** containing a guanidinium group for carboxylate recognition and a crown ether moiety for binding to ammonium groups was introduced as a host for zwitterionic amino acids by de Mendoza [306]. The two noncomplementary binding sites were introduced because they are unable to cause internal collapse of the receptor, while the naphthoyl group should induce selectivity for aromatic amino acids, which could bind to this binding site via π-stacking interactions. Because of the low solubility of **199** in water, binding properties were evaluated by using extraction experiments involving evaluation of the efficiency with which amino acids were extracted from water into CD_2Cl_2. In a competition experiment with a mixture of L-phenylalanine, L-tryptophane, and L-valine, the organic solvent contained these amino acids after extraction in a ratio of 100 : 97 : 6 Phe/Trp/Val, confirming the expected higher affinity of **199** for aromatic amino acids. Chiral recognition was demonstrated by showing that the D-enantiomers of these amino acids were practically not extracted. Subsequently, the ability of a series of altogether 14 structurally similar receptors was evaluated to act as carriers for zwitterionic aromatic amino acids across bulk liquid membranes [307]. These studies showed that receptors in which the subunits were linked to the bicyclic guanidinium scaffold through amide groups are better carriers than those containing esters, although the latter are more enantioselective. In addition, molecular modeling indicated that the previously proposed three-point binding involving the participation of the aromatic side chain of the amino acid guest is unnecessary to explain the enantioselectivities

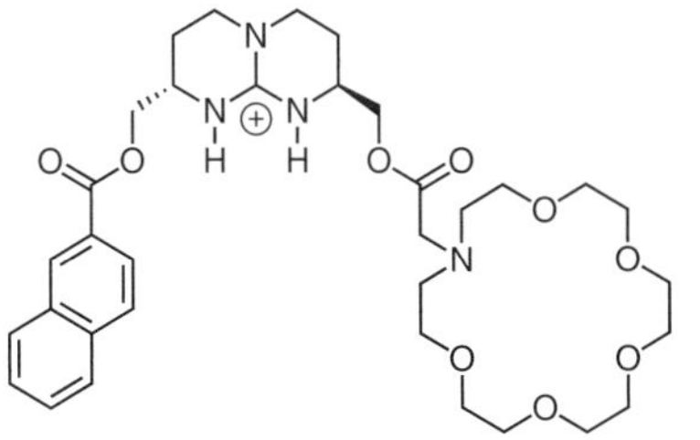

199

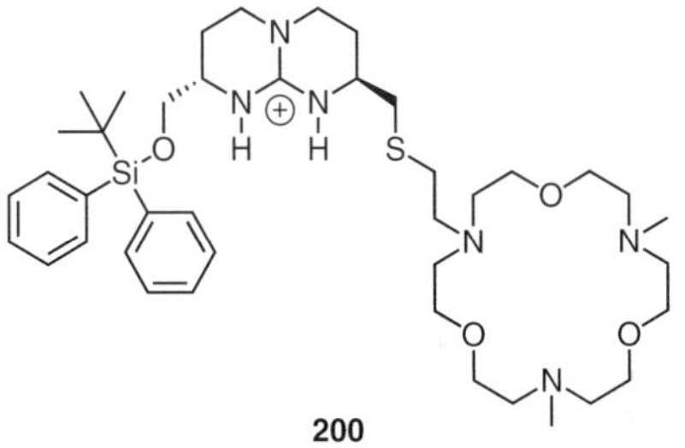

200

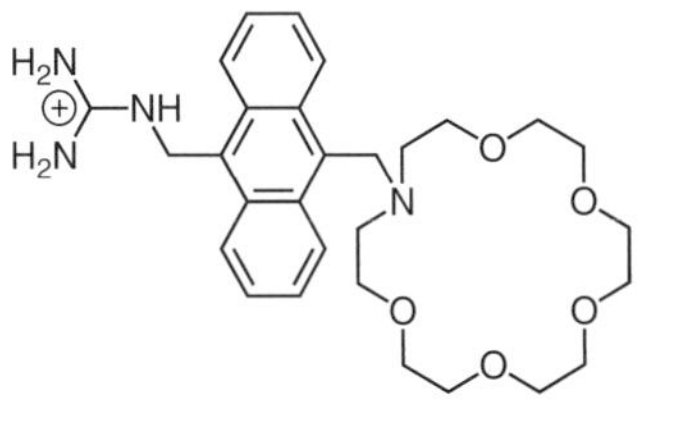

201

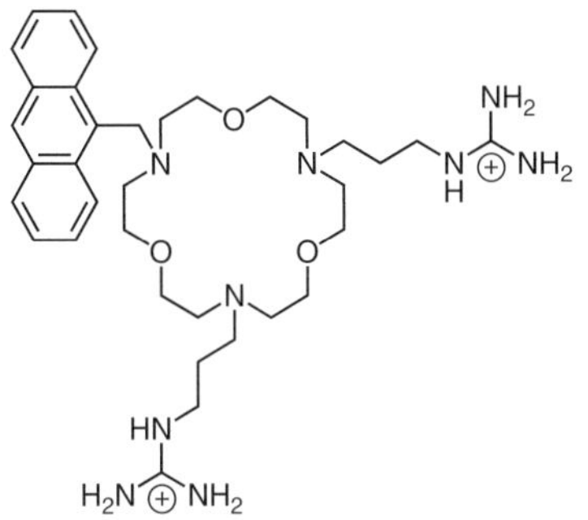

202

observed. Instead, a two-point model involving only the guanidinium and crown ether moieties is fully adequate.

In a similar approach, receptor **200** containing a diphenyl-*tert*-butylsilyl group as well as a triaza crown ether was shown to extract L-phenylalanine into chloroform with 40% e.e. in a wide pH range [308]. Moreover, even small hydrophilic (Ser, Gly) but no charged amino acids were extracted by factors up to 3000-fold better than by **199**. Ditopic hosts **201** and **202** allow the optical sensing of γ-aminobutyric acid (GABA) in water/methanol mixtures [309, 310].

Several cholapod derivatives developed by the Davis group, examples of which are compounds **203a,b** and **204a**, have been shown to extract *N*-acetyl-α-amino acids from aqueous phosphate buffer into chloroform [311, 312]. Extraction efficiency of **203b** is higher than that of **203a**, possibly due to the greater acidity of the dichlorophenylcarbamoyl NH groups, and **203b** is also more sensitive to the structure of the side chain in the substrate. While the substrate with the most bulky side chain, *N*-Ac-*tert*-leucine, gave the lowest enantioselectivity, good L/D selectivities of 9 : 1 were observed for phenylalanine, methionine, and valine. ^{1}H NMR spectroscopy combined with molecular modeling suggested that the carboxylate group of the substrate interacts with the 7-carbamoyl NH group and the two guanidinium NH groups, while the acetyl oxygen binds to the 12-carbamoyl NH. The lipophilic analog of **204a**, compound **204b**, has been used for the enantioselective transport of *N*-Ac-phenylalanine across dichloromethane and dichloroethane bulk liquid membranes and across 2.5% (v/v) octanol/hexane via hollow fiber membrane contactors. Significant enantioselectivities and multiple turnovers were observed for both types of transport systems, indicating the ability of **204b** to function as an agent for enantioselective separation [312].

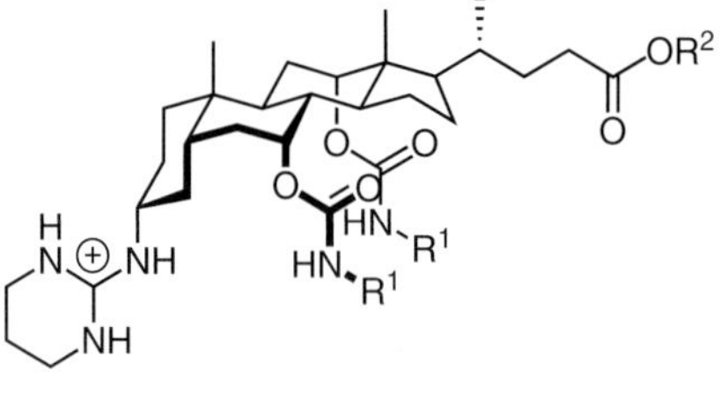

203a; R^1 = phenyl, R^2 = Me

203b; R^1 = 2,6-dichlorophenyl, R^2 = Me

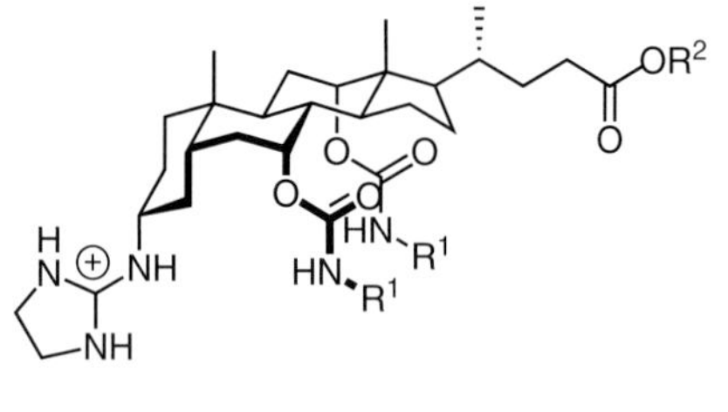

204a; R^1 = 4-trifluoromethylphenyl, R^2 = Me

204b; R^1 = 4-trifluoromethylphenyl, R^2 = $C_{20}H_{41}$

In a series of papers, the group around Schmuck demonstrated that 2-guanidiniocarbonyl pyrrole **142** and related compounds are able to bind not only acetate in competitive solvent mixtures but also N-protected amino acids. The affinity of **142** toward *N*-acetylated amino acid carboxylates is somewhat lower than acetate affinity, with stability constants ranging between 360 M^{-1} for the Ac-L-Lys complex and 1700 M^{-1} for the Ac-L-Phe complex, which has been ascribed to steric effects of the amino acid side chains that prevent the carboxylate group from adopting an optimal coplanar arrangement with the receptor [238]. The higher binding constant of the phenylalanine complex is most probably due to stabilizing

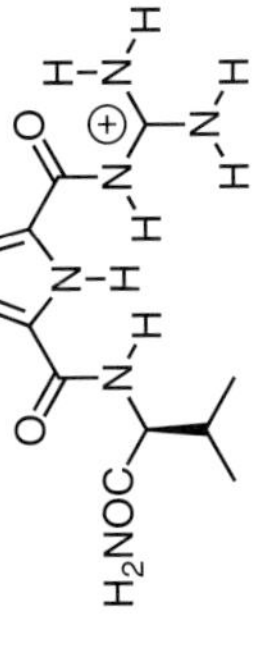

205a

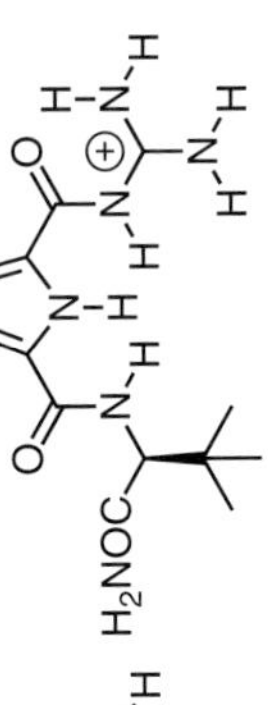

205b

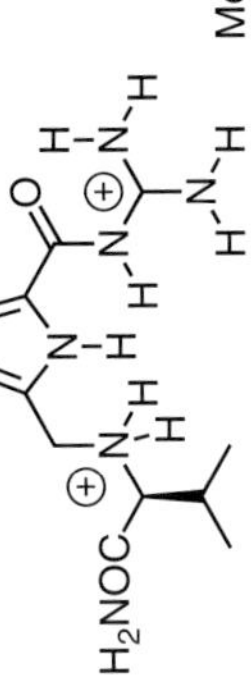

206

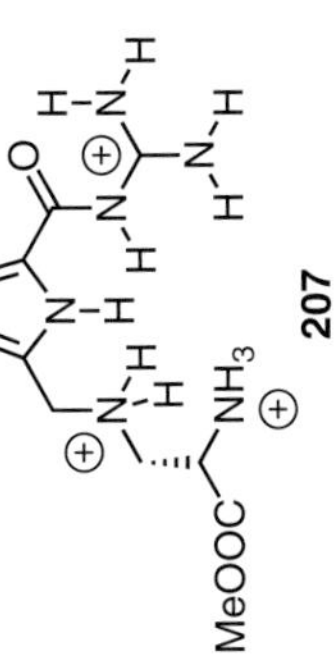

207

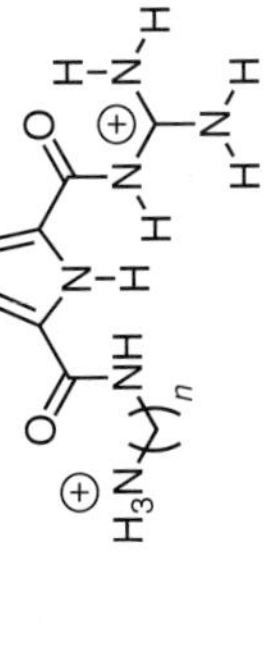

208a, $n = 2$
208b, $n = 3$
208c, $n = 4$
208d, $n = 5$

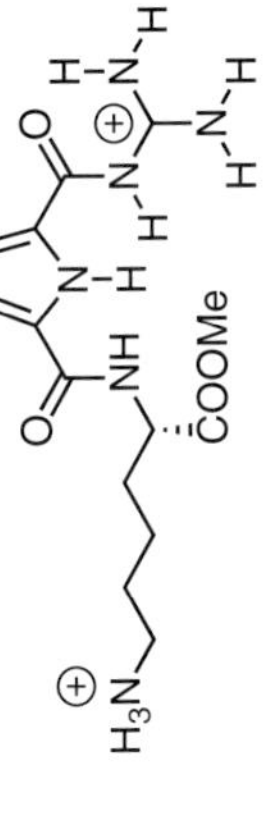

208e

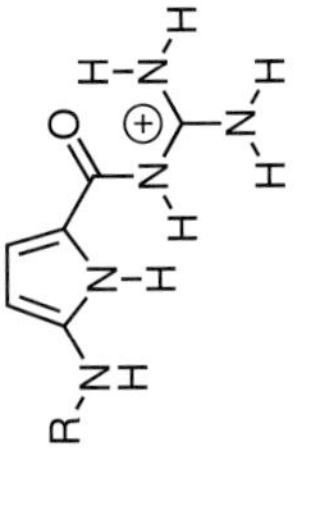

209a; R = Ac
209b; R = *N*-Ac-L-Val

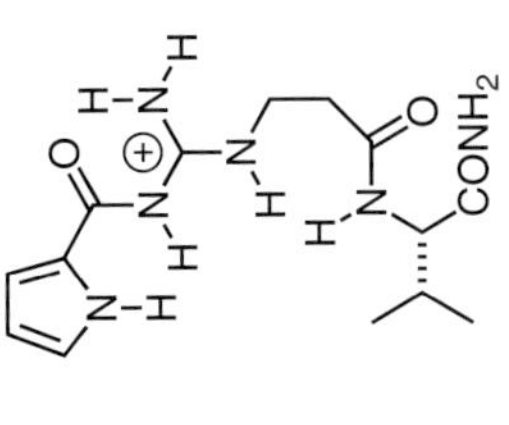

210

cation–π interactions between the guest's aromatic side chain and the receptor's pyrrole subunit.

Substituents in the 5-position of the pyrrole ring have a distinct effect on amino acid selectivity [313]. L-Valine-derived receptor **205a**, for example, binds Ac-L-Ala significantly more strongly ($K_a = 1610\ M^{-1}$) than **142** ($K_a = 770\ M^{-1}$). Moreover, the chirality in the side chain of **205a** induces enantioselectivity in complex formation: Ac-L-Ala, for example, is preferred over the corresponding D-enantiomer. Unfortunately, further increase in the steric bulk in the amino acid substituent, as in receptor **205b**, causes a complete loss of stereoselectivity [200].

Another structural modification that increases carboxylate affinity with respect to **142** by a factor of about 2 in aqueous solution is the introduction of a positive charge as in receptor **206** [314]. Systematic variation of the distance between the ammonium ion and the guanidiniocarbonyl-binding site in receptors **208a–e** showed that carboxylate affinity of such bis-cations increases until an optimum is reached for bis-cation **208c** containing the C4-linker [315]. The best substrate for **208c** is the *N*-acetyl phenylalanine anion, most likely due to favorable contributions of cation–π interactions to complex stability. Significantly improved binding properties were observed for receptor **207** containing a third positive charge in the form of a primary ammonium cation [316]. This receptor binds amino acid carboxylates efficiently with affinities generally exceeding $1000\ M^{-1}$ in 10% DMSO/H_2O. Interestingly, complexation of *N*-acetyl glutamate involves two receptor molecules the second one being bound significantly stronger than the first one. In contrast, a 1 : 1 complex is formed with the structurally closely related substrate *N*-acetyl aspartate.

Reversal of the direction of the amide group in **142** causes a change in substrate selectivity and binding affinity [317]. Thus, receptor **209a** binds Ac-L-Ala more strongly than Ac-L-Lac in 40% D_2O/DMSO, while **142** possesses a larger affinity for the ester. Another effect of the reversal of amide direction is that the isopropyl group in **205b** has an adverse effect on alanine affinity while it improved binding of alanine in **205a**. Both effects can be rationalized on the basis of the calculated complex structures.

An alternative approach to strengthen interactions between guanidiniocarbonyl pyrroles and carboxylates involves introduction of substituents into the guanidinium group. *N*′-Alkylated guanidinium derivative **210**, for example, was shown to possess high affinity toward carboxylates with a remarkable K_a of $1750\ M^{-1}$ for the Ac-L-Val complex in water (3 mM bis–TRIS, pH 6.1) [318].

Bis(sapphyrins) **211–213** were used to bind N-protected aspartate or glutamate [319]. Receptors **211** and **212** were found to display a preference for glutamate over aspartate in 19 : 1 (v/v) dichloromethane–methanol (*N*-Cbz-L-Asp ⊂ **211** $K_a = 45\,000\ M^{-1}$; *N*-Cbz-L-Glu ⊂ **211** $K_a = 112\,700\ M^{-1}$; *N*-Cbz-L-Asp ⊂ **212** $K_a = 20\,600\ M^{-1}$; *N*-Cbz-L-Glu ⊂ **212** $K_a = 324\,500\ M^{-1}$), with **212** showing a modest level of enantiomeric selectivity (*N*-Cbz-D-Asp ⊂ **212** $K_a = 43\,500\ M^{-1}$; *N*-Cbz-D-Glu ⊂ **212** $K_a = 217\,100\ M^{-1}$) and preferential binding of the D-enantiomer of aspartate and the L-enantiomer of glutamate. The cyclic receptor **213** binds these anions less effectively but displays excellent chiral discrimination between the D- and L-enantiomers of *N*-Cbz-protected glutamate (*N*-Cbz-L-Asp $\subset$ **213** $K_a = 16\,700\ M^{-1}$; *N*-Cbz-D-Asp $\subset$ **213** $K_a = 9700\ M^{-1}$; *N*-Cbz-L-Glu $\subset$ **213** $K_a = 3800\ M^{-1}$; *N*-Cbz-D-Glu $\subset$ **213** $K_a = 16\,200\ M^{-1}$).

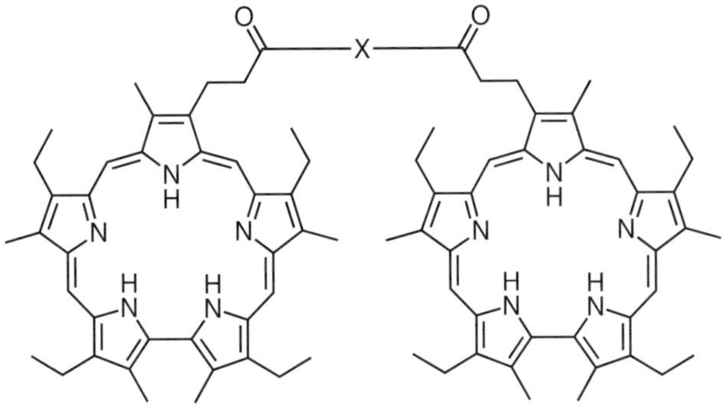

211; X = 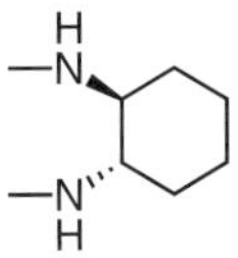 **212**; X = 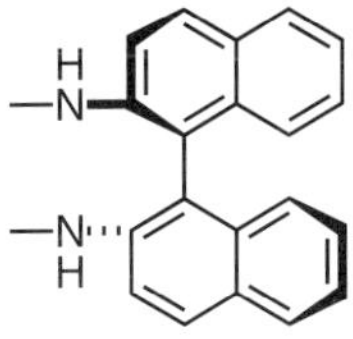

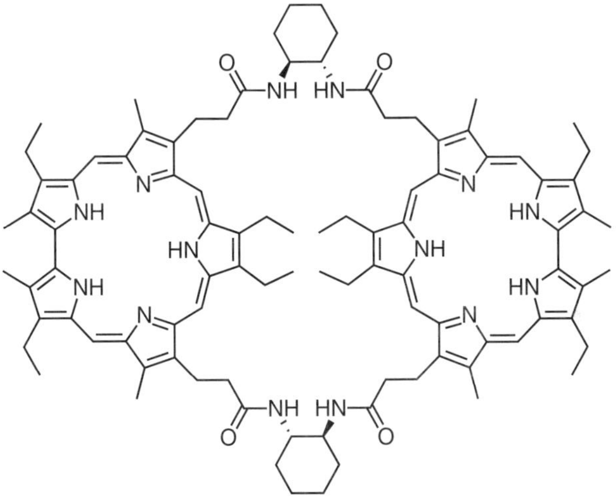

213

Several thiourea-containing amino acid receptors were developed by the Kilburn group. Noncyclic receptor **214** was designed to create a cleft with four hydrogen bond donors appropriately arranged for binding to the amino acid carboxylate group [320]. Binding studies involving a range of *N*-acetylated amino acid tetrabutylammonium salts in $CDCl_3$ demonstrated that complex formation involves not only hydrogen bond formation from the carboxylate oxygens of the substrates to the thiourea secondary amide and to the primary NH's but also breaking of an intramolecular hydrogen bond to the other primary amide NH. Receptor **214** shows the highest affinity for amino acids with side chains incorporating a hydrogen-bonding functionality (*N*-Ac-L-Gln $\subset$ **214** $K_a = 9000\ M^{-1}$) or an electron-rich aromatic side chain (*N*-Ac-L-Trp $\subset$ **214** $K_a = 12\,400\ M^{-1}$); enantioselectivity is, however, modest with a general preference for the L-enantiomer of the substrate.

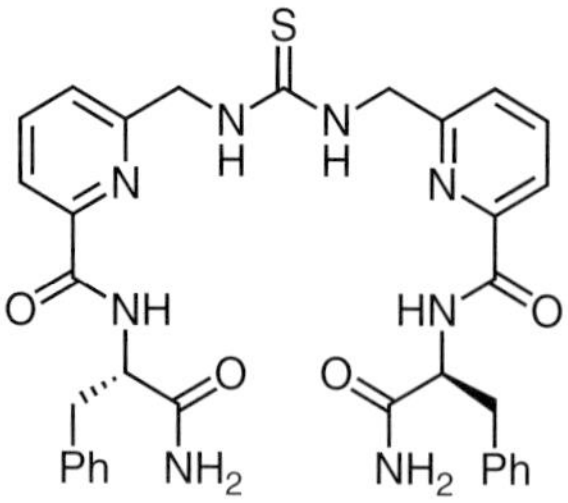

214

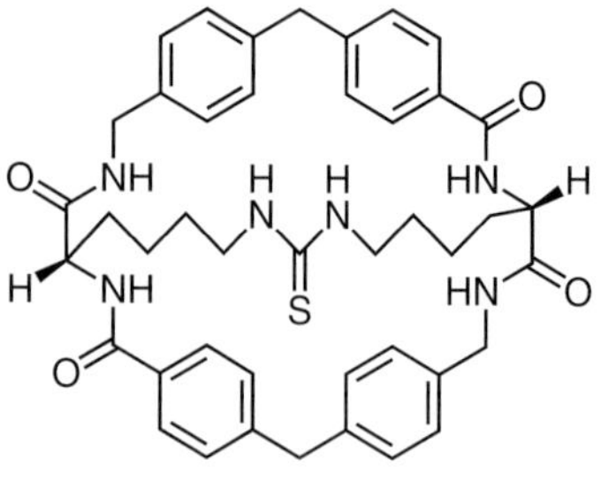

215

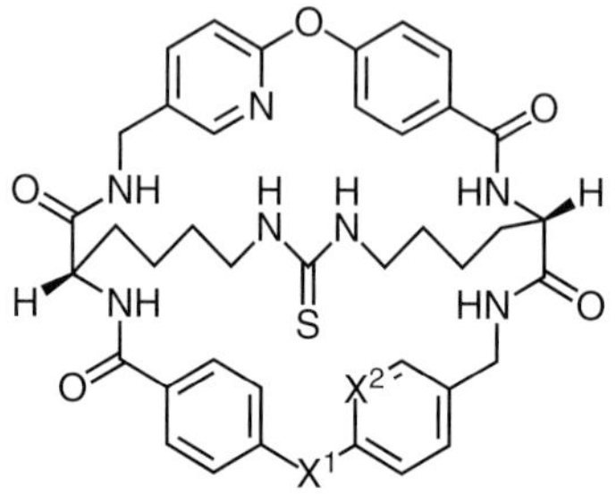

216a; $X^1 = CH_2$, $X^2 = CH$
216b; $X^1 = O$, $X^2 = N$

The bowl-shaped lysine-containing macrobicyclic receptor **215** binds to chiral amino acid anions albeit with almost no enantioselectivity [321, 322]. Affinity to *N*-Ac-L-Ala amounts to $16\,900\ M^{-1}$ in $CDCl_3$, for example, while the complex of *N*-Ac-D-Ala has a K_a of $14\,600\ M^{-1}$. Detailed NMR spectroscopic investigations revealed profound differences in the modes with which both enantiomers of this substrate interact with **215**. While the D-amino acid is bound predominantly on the outside of the macrobicycle cavity via carboxylate–thiourea interaction, the L-amino acid is included into the cavity, also establishing a strong carboxylate–thiourea interaction, but with the acetyl amide in the *cis* amide conformation. Molecular modeling studies suggested that the energetic penalty associated with the guest adopting this conformation is compensated for by intermolecular hydrogen bonds between the *cis* amide and macrocycle amide functionality. Inclusion of the substrate into the receptor cavity therefore leads to a situation that is energetically similar to the one with the substrate bound on the receptor outside.

Subsequently, receptor **215** was modified by replacing one or two benzene rings along the upper rim by pyridine units leading to **216a,b** that contain additional hydrogen-bonding acceptors along the rim of the cavity [323]. Binding studies showed that **216b** is an efficient receptor for the tetrabutylammonium salt of the *N*-Ac-L-Asn anion in CH_2Cl_2 ($K_a = 5.5 \times 10^4\ M^{-1}$).

The Kilburn group also described macrocyclic bis(thiourea) receptor **217**, which was anticipated to serve as an enantioselective receptor for glutamate or aspartate [324, 325]. Interestingly, **217** forms 1 : 1 complexes with *N*-Boc-L-Glu in acetonitrile, with the guest accommodated within the macrocyclic cavity (*N*-Boc-L-Glu ⊂ **217** $K_a = 2.83 \times 10^4\ M^{-1}$), while binding of *N*-Boc-D-Glu predominantly leads to the formation of a 1 : 2 host–guest complex with a weak binding constant K_{11} of the 1 : 1 complex and a significantly stronger binding constant K_{12} describing the formation of the 1 : 2 complex from the 1 : 1 complex (*N*-Boc-D-Glu ⊂ **217** $K_{11} = 38\ M^{-1}$, $K_{12} = 4.92 \times 10^4\ M^{-1}$). Interestingly, behavior of **217** toward Boc-protected L- and D-aspartate is very similar, while the acetyl-protected forms of L- and D-glutamate are both bound in the form of 1 : 2 complexes. Not unexpectedly, binding of glutamate is weaker in the more competitive solvent DMSO, but the different complex stoichiometries of the complexes of *N*-Boc-L-Glu and *N*-Boc-D-Glu are retained in this solvent. Complexation of aspartate in DMSO is complicated by the fact that complex formation is slow on the NMR time scale at room temperature. No binding of carboxylates was observed in the less polar solvent chloroform because the energy required to break the intramolecular hydrogen bonds that stabilize a folded conformation of **217** in this solvent is not compensated for by the binding interactions that would thereby be established.

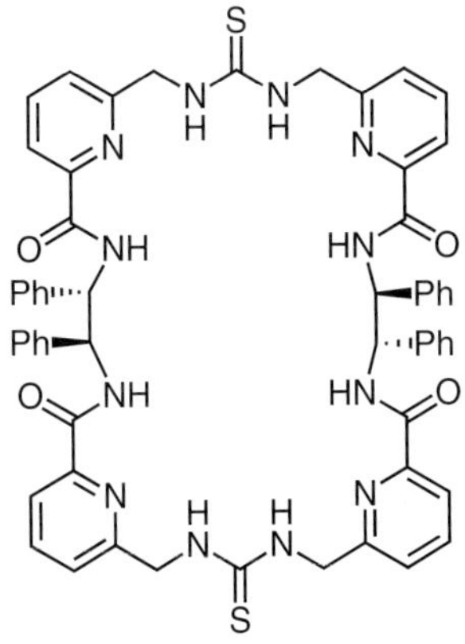

217

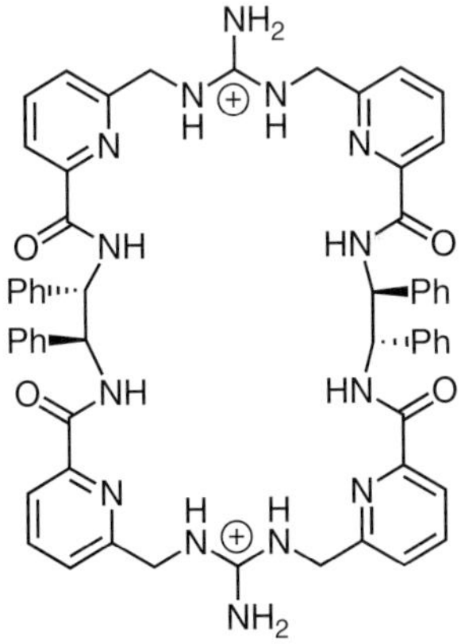

218

The observed cooperativity in the interaction between **217** and certain carboxylates in DMSO is explained by assuming that solvation of the intramolecular hydrogen bonds weakens the preference for a tightly wrapped conformation of **217** in this solvent. However, binding of the first carboxylate group seems to require a significant structural reorganization, which is energetically costly, reflecting the weak binding of the first guest. This energetic penalty has not to be paid when a second carboxylate is bound. The strong preference of *N*-Boc-L-Glu to form a 1 : 1 complex with **217** demonstrates the optimal fit of this substrate into the binding pocket of this receptor.

Systematic binding studies showed that small structural changes such as replacement of the pyridine rings in **217** with phenylene moieties or the variation of the substituents in the ethylene diamine subunits have a distinct effect on the binding properties of such macrocycles [325]. Exchange of the thiourea moieties with guanidinium subunits gave the charged analog **218**, which allowed glutamate binding to be studied in aqueous solvent mixtures. As already seen for **217**, this receptor forms a strong 1 : 1 complex with *N*-Boc-L-Glu ($K_a = 3.8 \times 10^4\ M^{-1}$) in 1 : 1 DMSO/$H_2O$ and a 1 : 2 complex with *N*-Boc-D-Glu [326]. In this case, 1 : 2 complexation has also been observed for either enantiomer of *N*-Boc-Asp. Binding studies in 1 : 1 DMSO/H_2O using unprotected zwitterionic L-Glu as guest revealed very weak binding ($K_a < 10^2\ M^{-1}$).

Metal centers in amino acid receptors can serve to bind the carboxylate groups and/or the amino group. On the basis of this idea, receptor **24** containing two guanidinium-binding sites as well as a zinc ion was developed by the Anslyn group [327]. A comparison of the amino acid affinity of **24** with the affinity for model substrates containing only an amino or a carboxylate group shows that amino acid binding indeed involves simultaneous coordination of the carboxylate and the amino group to the zinc center of the host. Affinity is largest for aspartate in 50% methanol/water (10 mM HEPES, pH 7.4) (**24** ⊂ L-Asp $K_a = 1.5 \times 10^5\ M^{-1}$), which was attributed to cooperative ion-pairing interactions between a guanidinium group of the host and the side-chain carboxylate group of the guest. Moreover, the fact that the glutamic acid complex is much less stable (**24** ⊂ L-Glu $K_a = 2.2 \times 10^4\ M^{-1}$) indicates that an optimal complementarity between the interacting groups contributes to the high stability of the aspartate complex. In combination with pyrocatechol violet, a dye displacement assay was established on the basis of **24**, which allowed the detection of α-amino acids by a color change of the solution from blue to yellow.

Zinc porphyrins decorated with appropriate substituents have also been shown to interact with amino acids. Examples are compounds **219–221** described by Imai and coworkers [328]. These compounds interact with α-amino carboxylates in aqueous $NaHCO_3/Na_2CO_3$ buffer (pH 10.4) by a combination of coordinative, coulombic, and hydrophobic interactions. The fact that butylamine is bound significantly more weakly than amino carboxylates clearly shows that the carboxylate group in the substrates is specifically recognized by these receptors. Hydrophobic interactions between the hosts and side-chain functional groups cause the complexes with, for example, phenylalanine and tryptophan to be significantly more stable than those with glycine or aspartate.

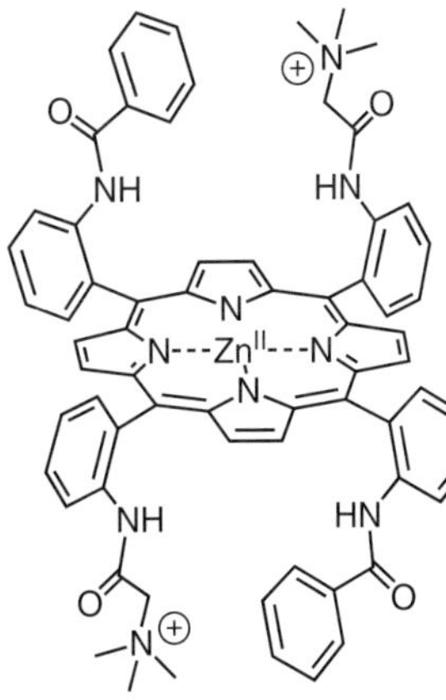

219

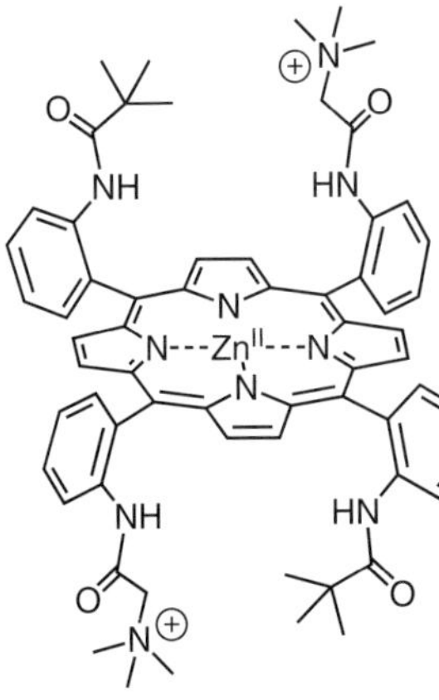

220

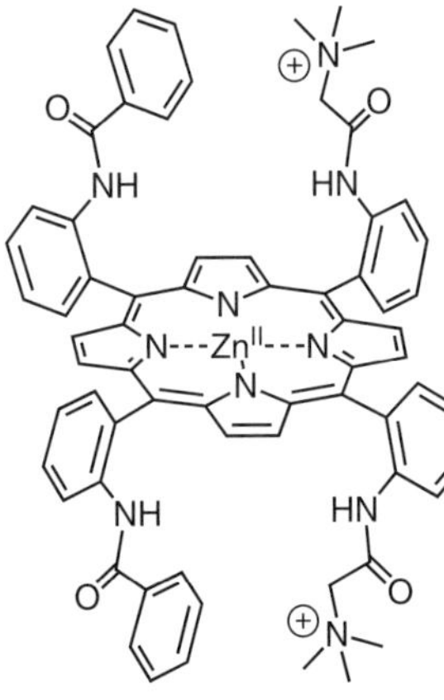

rac-**221**

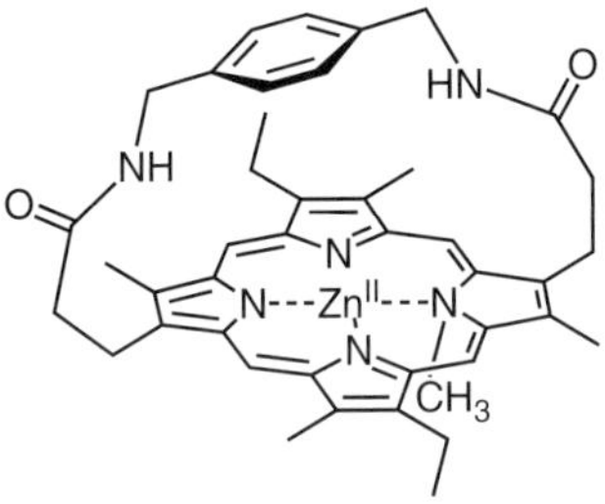

222

Amino acid binding to chiral porphyrin derivatives, an example of which is **222**, was investigated by the Inoue group [329]. Receptor **222** contains a *p*-xylene-based strap on one face of the porphyrin ring, while a substituent on one pyrrole residue blocks coordination of the substrates to the metal center from the unstrapped face. Extraction experiments using racemic N-protected amino acids showed that the (+)-enantiomer of **222** preferentially binds to L-amino acids, while the (−)-enantiomer prefers D-amino acids. Among the amino acids studied, the highest enantioselectivity was observed for *N*-(3,5-dinitrobenzoyl)-Phe(NBP) (L-NBP ⊂ (+)-**222** + D-NBP ⊂ (−)-**222**/D-NBP ⊂ (+)-**222** + L-NBP ⊂ (−)-**222** = 96/4). *N*-Cbz-Pro and *N*-Cbz-*N*-methylalaninate are bound with no enantioselectivity, indicating that NH groups in the substrate are required for enantioselective recognition. Structural studies showed that receptor–substrate interactions involve a combination of hydrogen-bonding interactions between the substrate and the amide groups in the strap as well as coordination of the substrate's carboxylate group to the metal center.

Complex **223** forms well-defined ternary adducts with α-amino acids in aqueous solution at pH 7 [330]. Of the 20 common α-amino acids studied by means of, for example, ^{1}H NMR spectroscopy, all form complexes with **223**, in which the ytterbium has a mono-capped square antiprismatic coordination environment with the amine nitrogen of the substrate oriented axially and the carboxylate oxygen occupying the equatorial position. Competitive chelation through amino acid side-chain functionalities has not been observed for simple amino acids. Enantioselectivity in complex formation depends on the amino acid side chain. Receptor **223** can, for example, not discriminate the enantiomers of Ala, while it binds preferentially to the *S*-enantiomers of Asp, His (2 : 1), and Ser, Thr (4 : 1).

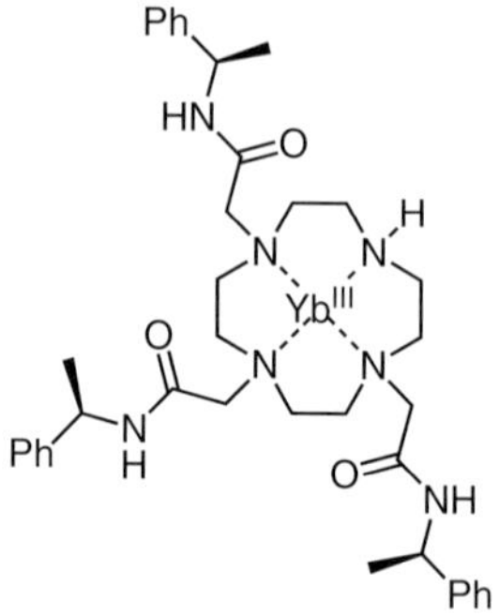

223

7.3.5
Peptide C-Terminal Carboxylates

Pioneering work in peptide recognition that involves binding of the receptor to the terminal carboxylate group was performed by the Schneider group [331]. The ditopic receptor **224** developed in this context contains a quaternary ammonium group for electrostatic interactions with the carboxylate group of the substrate, a crown ether moiety for binding to the N-terminal ammonium group, and an aromatic residue between these two binding sites that should induce selectivity for substrates containing aromatic amino acids. Pronounced length selectivity was detected for **224**; this receptor preferentially binds to tripeptides, while affinity for dipeptides or tetrapeptides is significantly lower. Binding is strong even in water, with the association constants of the tripeptide complexes ranging between about

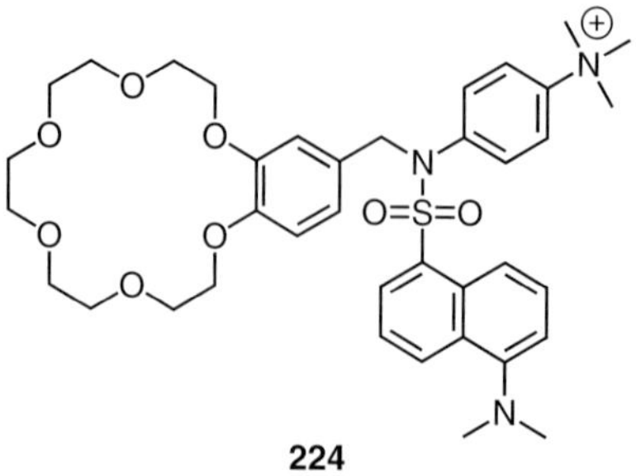

224

200 and 2200 M^{-1}. In accordance with the design principle, the highest affinity was detected for peptides containing an aromatic residue in the second position (Gly-Phe-Gly $\subset$ **224** $K_a = 1770\ M^{-1}$; Gly-Trp-Gly $\subset$ **224** $K_a = 2150\ M^{-1}$). Shifting the aromatic residue to position 1 or 3 of the substrate reduces complex stability by about 1 order of magnitude. High tripeptide affinity that relies on similar types of interactions was also detected for receptors containing ammonium residues and a crown ether moiety on a porphyrin core [332, 333].

Tweezer-type receptor **225** was developed by Srinivasan and Kilburn for the sequence-selective recognition of short peptides [334]. Complex formation was

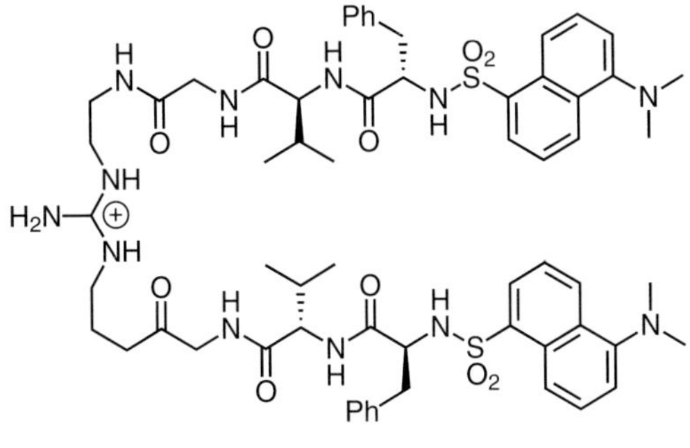

225

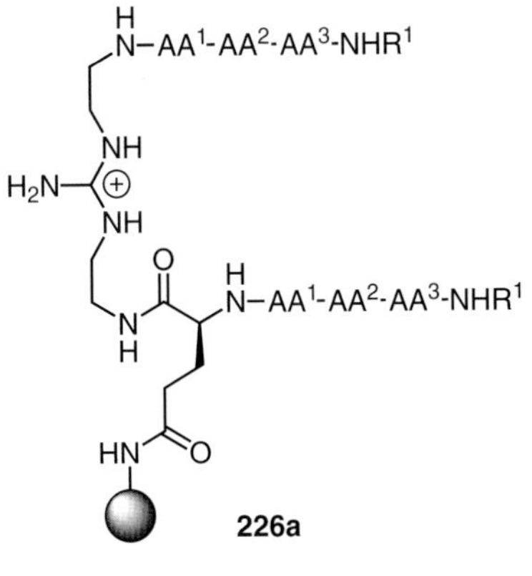

226a

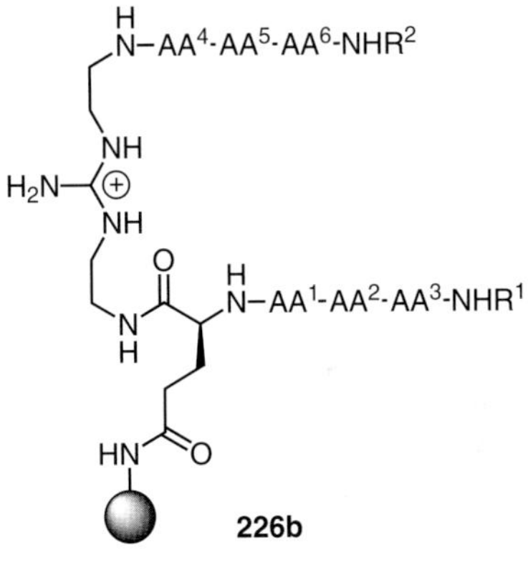

226b

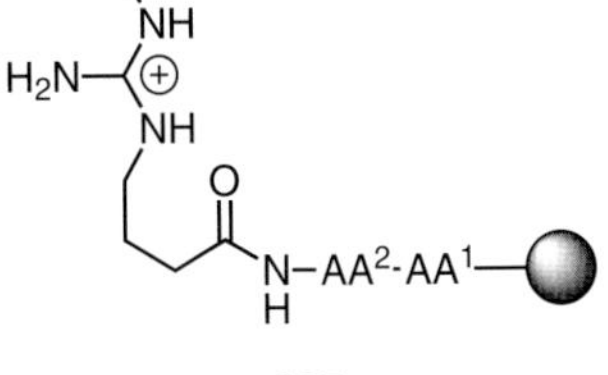

227

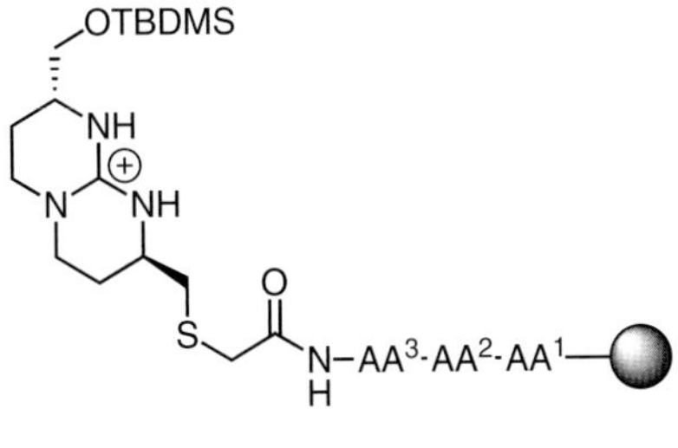

228

anticipated to involve a salt bridge between the *C*-terminal carboxylate group of a peptidic guest and the guanidinium moiety of the host. To test this idea, **225** was incubated in aqueous sodium borate buffer (pH 9.2, 16.7% DMSO) with a 1000-member library of tripeptides attached to a TentaGel resin via the amino terminus. Mainly, hydrophobic amino acid residues were incorporated into these tripeptides to ensure that receptor–substrate interactions are largely due to hydrophobic interactions. Receptor **225** was found to bind to about 3% of the library members and showed 95% selectivity for Val at the carboxy terminus of the tripeptides and 40% selectivity for Glu(O*t*Bu) at the amino terminus [335]. A stability constant of 4×10^5 M^{-1} was determined for the complex between **225** and Z-Glu(O*t*Bu)-Ser(O*t*Bu)-Val-OH in 16.7% DMSO/water by means of isothermal titration microcalorimetry.

The reverse experiment involved a resin-bound library of symmetrical tweezer-type receptors **226a** with identical peptide fragments appended to both sides of the guanidinium scaffold. Screening with a dye-labeled tripeptide as substrate allowed the identification of a host that was shown to bind this tripeptide in 15% DMSO/water with appreciable selectivity over the corresponding enantiomer [336]. This approach was subsequently extended to "unsymmetrical" receptor **226b**, whose arms were synthesized independently [337]. Libraries of **226b** were screened to identify receptors for the *N*-Ac-Lys-D-Ala-D-Ala tripeptide sequence, the target of vancomycin. A receptor thus identified (AA^1 = Gly; AA^2 = Val; AA^3 = Val; AA^4 = Met; AA^5 = His; AA^6 = Ser; R^1 = Ac; R^2 = H) was re-synthesized and used to determine the association constant with *N*-Ac-Lys-D-Ala-D-Ala. In free solution, although UV and NMR binding studies provided evidence for an association between the receptor and the tripeptide, the data did not allow the estimation of complex stability using a 1 : 1 binding model. When attached to the solid support, however, the receptor binds to the tripeptide in an aqueous buffered solution with a millimolar association constant. The failure of binding on the solid phase to translate into equivalent binding in free solution was ascribed to the environment created by the resin that could affect binding affinity by excluding or organizing solvent molecules within the resin matrix or by suppressing aggregation of the tweezer receptor molecules, which occurs readily in free solution. The same group also developed a rapid sequencing method for libraries of guanidinium-based peptide receptors **227** and **228** [338]. Both libraries were successfully screened to identify binding partners for Val-Val-Ile-Ala, the *C*-terminal tetrapeptide sequence of the amyloid-β protein. Again, significantly different behaviors were detected for resin-bound receptors and free receptors in solution.

The Schmuck group showed that remarkably efficient receptors can be obtained for the selective recognition of peptides in water by appending suitable residues to the guanidiniocarbonyl pyrrole moiety. Receptor **229**, which efficiently recognizes the deprotonated dipeptide Ac-D-Ala-D-Ala in buffered water with a K_a of 33 100 M^{-1}, was prepared, for example, by linking a cyclotribenzylene-substituted alanine derivative and the guanidiniocarbonyl pyrrole unit [339]. No stereoselectivity in complex formation was observed, as the enantiomer of the substrate is bound with almost the same affinity. Dipeptides with smaller (Ac-Gly-Gly) or larger (Ac-D-Val-D-Val) substituents in the side chains are, however, bound significantly less efficiently by factors of more than 10, indicating that the bowl-shaped cyclotribenzylene moiety has an influence on binding selectivity.

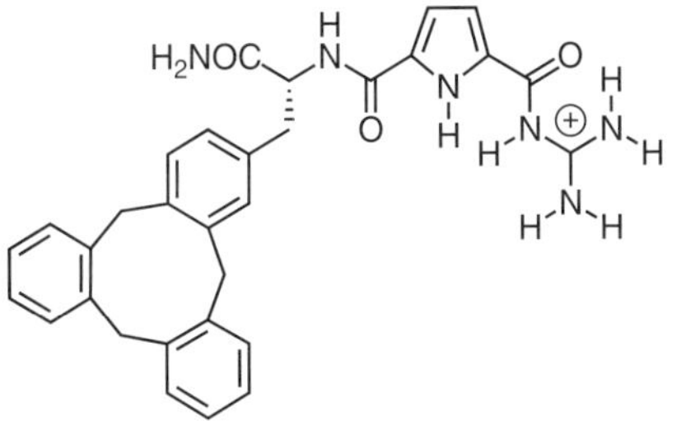

229 R = 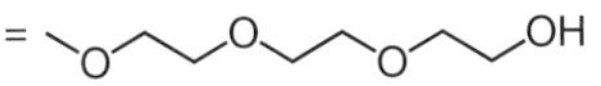

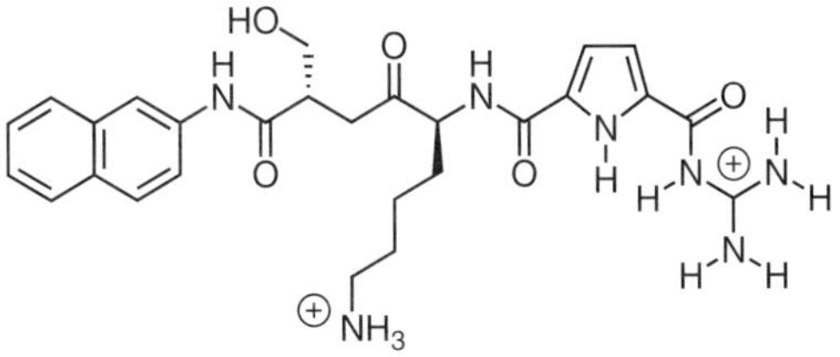

230

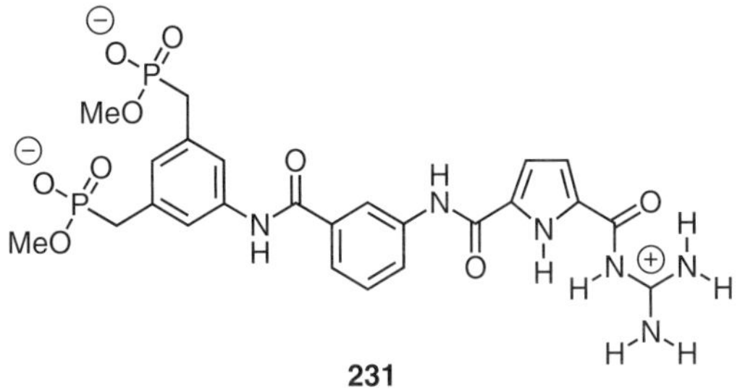

231

Bis-cationic host **230** contains a lysine residue for improved anion affinity, serine for water solubility, and a naphthyl group to allow for favorable hydrophobic contacts with the substrate. Binding studies showed that **230** possesses mM affinities ($K_a = 2\text{–}5 \times 10^5\ M^{-1}$) for dipeptide carboxylates in 20% DMSO/water (pH = 6.0) and exhibits some preference for alanine in the C-terminal position [340]. A receptor for the Arg-Gly-Asp tripeptide was obtained by attaching the guanidiniocarbonyl pyrrole to a bisphosphonate moiety, a known recognition motif for the guanidinium group in the arginine side chain, through a linker of sufficient length and rigidity that prevents intramolecular and intermolecular self-association of the two complementary recognition sites. UV and fluorescence titrations showed that receptor **231** containing 3-aminobenzoic acid as linker binds the protected tripeptide Ac-Arg-Gly-Asp-NH_2 with an association constant of 2700 M^{-1} [341].

Besides a rational receptor design based on molecular modeling studies, the Schmuck group also made use of combinatorial chemistry to identify potent guanidiniocarbonyl-pyrrole-derived peptide receptors [200, 342, 343]. To this end, a library of hosts of the general structure **232** was prepared on solid support, and binding affinity was screened toward the N-protected tetrapeptide Ac-L-Val-L-Val-L-Ile-L-Ala, the C-terminal sequence of the amyloid-β peptide [344, 345]. A first investigation involved a library of 125 receptors of which about 7% showed selective binding to the target substrate [344]. After selection of the most efficient receptors, an on-bead binding assay was used to determine relative affinities for L-Val-L-Val-L-Ile-L-Ala in methanol in the presence of formate as counterion. Substrate affinities of the best two receptors **232a** and **232b** amounted to 9800 and 9300 M^{-1}, respectively. Subsequently, a larger library containing 512 potential hosts was synthesized and screened in water (5 mM bis-TRIS, pH 6) [345]. In this solvent, binding affinities of the most potent receptors toward the target is less than half of those observed in methanol. More important is, however, that the best receptors in water are structurally quite different from the ones in methanol; largest affinity for the target peptide in bis-TRIS buffer at pH 6.1 ($K_a = 6025\ M^{-1}$) and on solid support ($K_a = 8800\ M^{-1}$) was detected, respectively, for receptor **232c** and the soluble analog Gua-Lys-Leu-Lys-NH_2, for example. This result clearly demonstrates the important influence of the solvent on supramolecular complex formation both in terms of binding affinity and selectivity. Notably, the best receptors that have thus been identified inhibit the formation of amyloid plaques *in vitro* but only for Aβ(1–42) and not Aβ(1–40), which has the wrong *C*-terminal sequence [346].

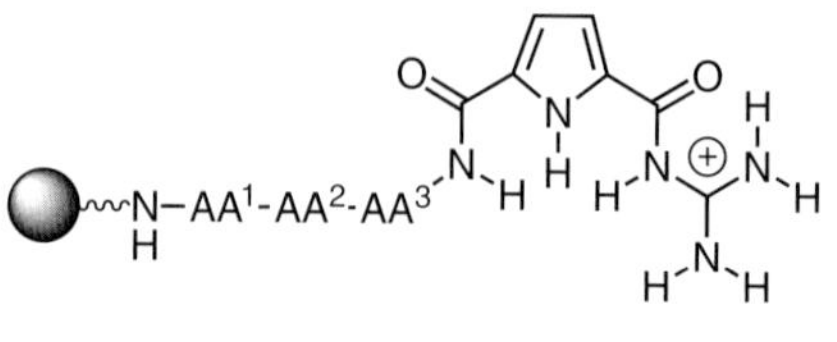

232

232a; AA^1 = Phe; AA^2 = Val; AA^3 = Val
232b; AA^1 = Val; AA^2 = Val; AA^3 = Val
232c; AA^1 = Lys; AA^2 = Leu; AA^3 = Lys
232d; AA^1 = Phe; AA^2 = Lys; AA^3 = Lys

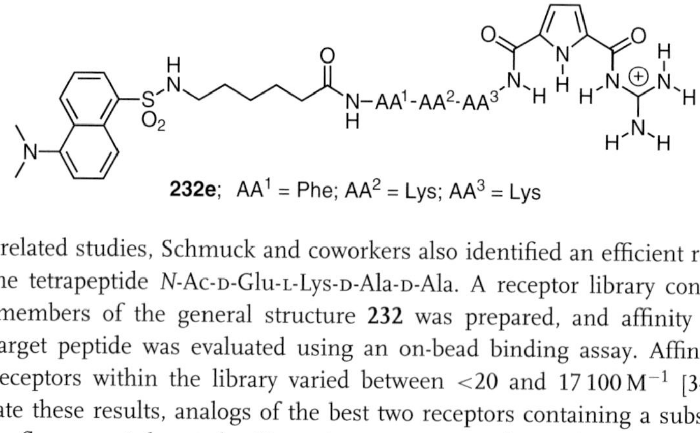

232e; AA^1 = Phe; AA^2 = Lys; AA^3 = Lys

In related studies, Schmuck and coworkers also identified an efficient receptor for the tetrapeptide *N*-Ac-D-Glu-L-Lys-D-Ala-D-Ala. A receptor library containing 512 members of the general structure **232** was prepared, and affinity toward the target peptide was evaluated using an on-bead binding assay. Affinities of the receptors within the library varied between <20 and 17 100 M^{-1} [347]. To validate these results, analogs of the best two receptors containing a substituent with a fluorescent dye at the *N*-terminus were re-synthesized and their binding properties in solution characterized. These experiments confirmed that both receptors form stable complexes also in solution. Stability of the complex between *N*-Ac-D-Glu-L-Lys-D-Ala-D-Ala and the receptor with the same amino acid sequence as **232d** amounts to 15 400 M^{-1}, for example.

To probe substrate selectivity of the receptors thus identified, binding of a tetrapeptide with the inverse sequence, *N*-Ac-D-Ala-D-Ala-D-Lys-D-Glu, was studied. This inverse substrate is bound by **232d** and the soluble analog of this receptor, compound **232e**, albeit with a significantly lower affinity (K_a about 6000 M^{-1}) than the peptide used in the screening experiments [348]. Affinity of **232e** is sensitive not only to the amino acid sequence in the substrate but also to the absolute configuration of the subunits. Stability of the complex with TentaGel (TEG) immobilized tetrapeptide TEG-D-Glu-D-Ala-D-Glu-D-Glu amounts to 5200 M^{-1}, for example, while the complex with a tetrapeptide that differs only in the configuration of the alanine residue (TEG-D-Glu-L-Ala-D-Glu-D-Glu) has a K_a of only 600 M^{-1} [349].

Another receptor for the tripeptide substrate vancomycin was identified by the Ungaro group, namely, the N-linked peptidocalix[4]arene derivative (**233**) containing a 1,3,5-triethylenetriamine moiety in the bridge [350]. This receptor binds to carboxylic acids, but since complex formation is initiated by protonation of the secondary amine in the spacer, the protonated form of **233** can be regarded as a carboxylate receptor. Binding studies demonstrated that **233** forms 1 : 1 complexes with *N*-Ac-D-Ala-D-Ala as well as with simple α-amino acids or carboxylic acids. Stabilities of the complexes increase in $CDCl_3$ from lauric acid (log $K_a = 3.00$) over *N*-lauroyl-D-Ala (log $K_a = 4.10$) or *N*-lauroyl-L-Ala (log $K_a = 4.05$) to *N*-lauroyl-D-Ala-D-Ala (log $K_a > 5$). NMR diffusion studies performed in 3% DMSO-d_6/$CDCl_3$ confirmed that the *N*-Ac-L-Ala-L-Ala dipeptide is bound more strongly (log $K_a = 3.4$) than amino acid derivative *N*-Ac-L-Ala (log $K_a = 2.4$) [351].

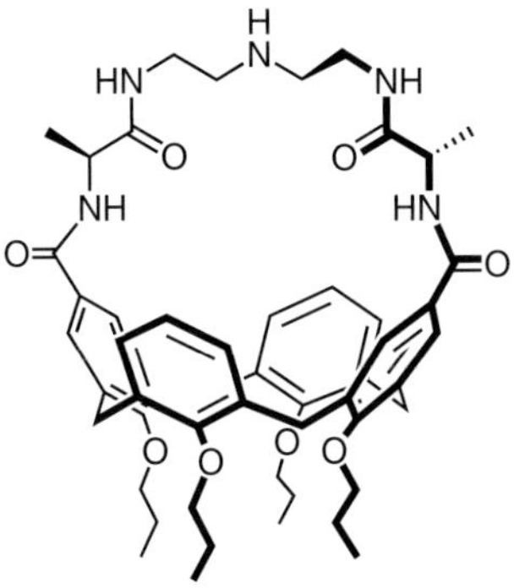

233

On the basis of these experimental findings, a structure for the complex between **233** and *N*-Ac-D-Ala-D-Ala was proposed in which the guest is threaded under the bridge of the receptor to allow for electrostatic interactions between the ammonium ion and the carboxylate anion together with H-bonds between the NH donor and CO acceptor units of host and guest. Additional stabilizing interactions could come from insertion of one side-chain methyl group of the guest into the hydrophobic cavity of the host.

Biological activity of **233** and derivatives thereof has also been studied. Interestingly, activity of this receptor against several gram-positive bacterial strains is close to that of vancomycin [350]. Increasing the length of the bridge in **232** by using the L-Ala-L-Ala dipeptide chain segments instead of a simple L-Ala causes a pronounced drop in activity, while protecting the central NH group with a Boc group or substituting it with a methylene group completely inhibits biological activity. The active compounds show a behavior very close to that of vancomycin: no activity is observed against gram-negative bacteria (*Escherichia coli*), yeast (*Saccharomyces* *cerevisiae*) or cell-wall-lacking bacteria (*Acholeplasma laidlawii*), indicating that the biological target of this class of calixarene-based antimicrobials is, as for vancomycin, the terminal D-Ala-D-Ala part of the peptidoglycan that, after crosslinking, constitutes the bacterial cell wall.

7.3.6
Peptide Side-Chain Carboxylates

Interaction of synthetic receptors with anionic groups located in the side chains of oligopeptides or proteins can cause a change or stabilization of the preferred peptide conformation. Anionic groups that are usually targeted are phosphoesters of tyrosine, serine, or threonine or carboxylate groups in glutamate or aspartate. Receptors for side-chain phosphate groups in peptides are described in Section 3.4. Typical examples are Hamachi's zinc(II) dipicolylamine-based receptors **106–110**. The ditopic receptors **112–114** developed by the König group simultaneously interact with a phosphate group and a carboxylate group (*vide supra*). Guanidinium-derived receptors **234** and **235**, developed by Hamilton, de Mendoza, and coworkers, on the other hand, interact *only* with carboxylate groups.

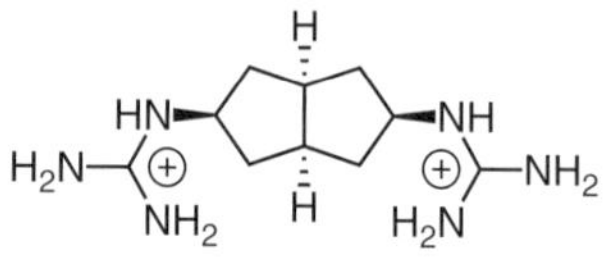

234

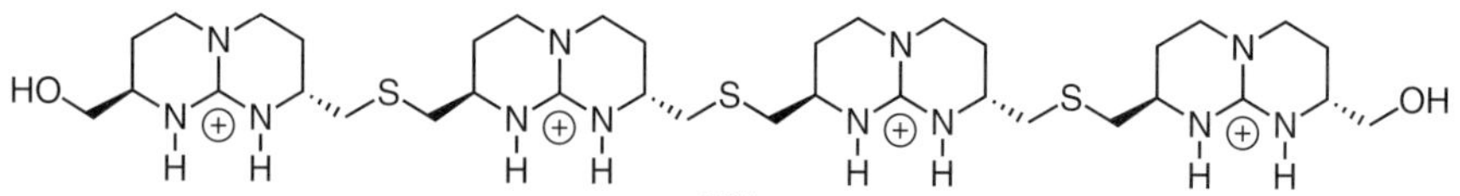

235

Bis(guanidinium) receptor **234** binds glutarate with an association constant of 3900 M^{-1} in 10% water/methanol [352]. Investigation of the interaction between **234** and different 16-mer α-helical peptides containing two aspartate residues separated by a variable number of other amino acids revealed a preference for the binding of helices with aspartates in the i and the $i + 3$ position [352, 353]. Circular dichroism spectroscopy indicated that the addition of **234** to such a peptide in 15% water/methanol at 25°C causes a 23% enhancement of helicity, while the increase in helicity of peptides with aspartate residues located farther apart is significantly lower. In buffered aqueous solution, similar but weaker binding and helix induction were observed. These results suggest that the two guanidinium moieties of **234** can simultaneously interact with appropriately spaced carboxylate groups on the surface of a helical peptide.

In a similar approach, the linear tetraguanidinium host **235** was synthesized, and its interactions with aspartate-containing oligopeptides were tested [354, 355]. Molecular modeling studies indicated that **235** is able to wrap around an ideal right-handed α-helix of a peptide containing aspartate residues at positions $i +$

$3n$ ($n = 0$–3) with an almost perfect match of each guanidinium moiety with the corresponding aspartate carboxylate group. Consistent with these calculations, an increase in the helicity of such a peptide from 21 to 45% was observed in 10% water/methanol in the presence of 2 equiv. of **235**, with the association constant of the aggregate amounting to $3.4 \times 10^5\ M^{-1}$. Interestingly, spermine 2, which has been shown to enhance the helical content of a peptide containing glutamate residues in an i, $i+4, i+7, i+11$ arrangement from 19 to 38% [356], had no effect on the helicity of the peptide used in these studies, which demonstrates the importance of the geometrical fit between the binding sites in the host and in the substrate. In water, the peptide is conformationally much more flexible, and although interactions seem to occur, no significant induction of a helical structure was found in the presence of **235**.

7.3.7 Sialic Acids

There are several biologically relevant mono- and oligosaccharides containing carboxylate groups. An example is *N*-acetylneuraminic acid (NeuAc), the most commonly occurring sialic acid, which plays a key role in a wide range of biological processes, including cell-recognition events. The hemagglutinins of a number of viruses such as the influenza virus exhibit specificity for NeuAc-containing ligands. In addition, NeuAc is overexpressed on the cell surfaces of tumor cells.

Receptors for NeuAc were developed by the Mazik group. Neutral ligands **236** and **237**, for example, bind NeuAc in DMSO [357]. Complex formation presumably involves multiple hydrogen-bonding interactions, including ones to the substrate's carboxylate group, in addition to CH–π interactions. Comparison of NeuAc affinity with the affinity toward glucuronic acids showed that both receptors are able to distinguish the two anionic sugars possessing strong preference for NeuAc.

Cationic receptors **238–240** bind NeuAc even in competitive solvents such as water/DMSO mixtures. Interactions of **238** with NeuAc in water/DMSO 1 : 9 proved to be too strong to be evaluated quantitatively by NMR titration [358]. Microcalorimetric titrations showed that this receptor can bind up to two NeuAc molecules, with the stepwise binding constants of the 1 : 1 and the 1 : 2 complex amounting to, respectively, $K_{11} = 1.5 \times 10^5\ M^{-1}$ and $K_{12} = 3.2 \times 10^4\ M^{-1}$. Benzimiazolium-based receptor **239** forms a very stable 1 : 1 complex with NeuAc in DMSO ($K_{11} > 10^6\ M^{-1}$), followed by the formation of weaker 1 : 2 and 1 : 3 complexes at higher NeuAc concentrations [358]. Receptor **240** binds NeuAc in aqueous media [359]. Binding constants determined by isothermal titration calorimetry in water/DMSO 1 : 9 amount to $K_{11} = 5860\ M^{-1}$ for the 1 : 1 complex and $K_{12} = 10\,600\ M^{-1}$ for the binding of a second NeuAc molecule to the 1 : 1 complex.

Recognition of sialylated oligosaccharides with the anthracene-derived receptors **241** was investigated by Nilsson and coworkers [360]. The anthracene moiety of **241** was expected to interact with hydrophobic regions of the oligosaccharide guests, while the amino and guanidinium groups should provide binding sites for

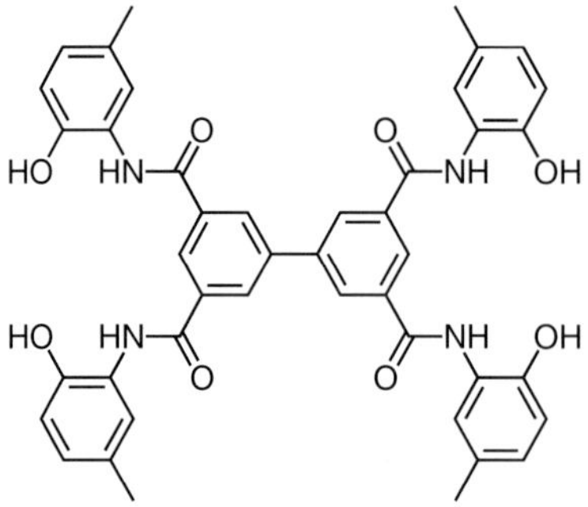

236

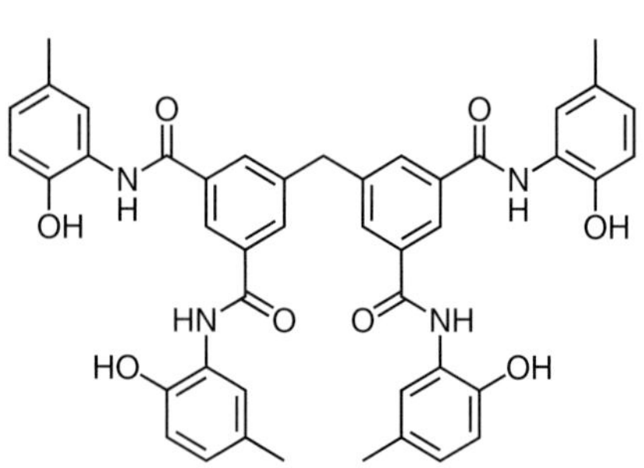

237

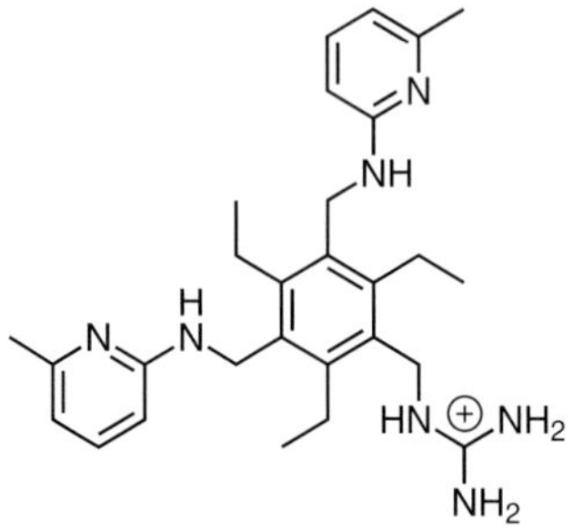

238

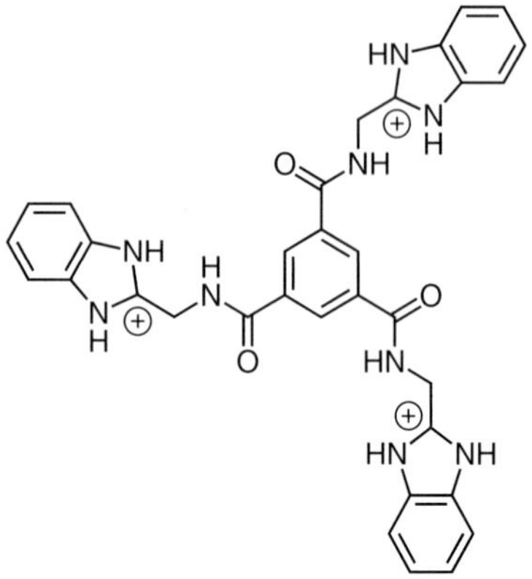

239

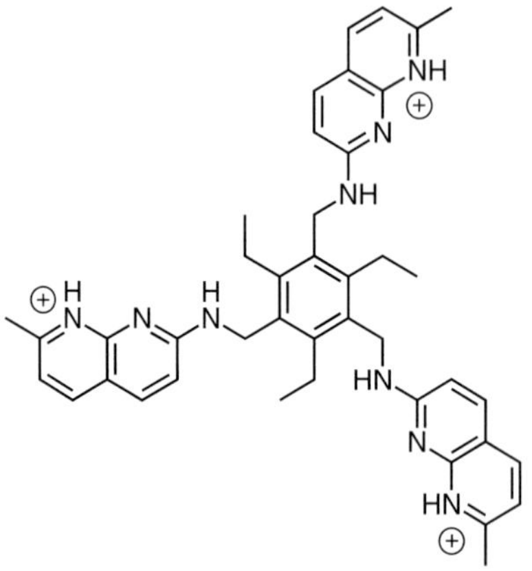

240

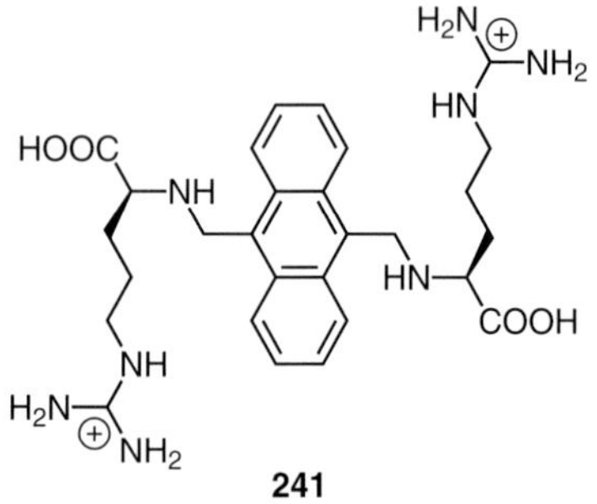

241

carboxylates. Binding in water proved to be weak, approaching stability constants of about 100 M^{-1}.

7.4 Conclusion

Anion coordination chemistry has reached a state where binding studies are no longer mainly concerned with elucidating general principles of molecular recognition phenomena but increasingly with tackling applied areas. The development of receptors for biorelevant anions is a good example. Such receptors usually derive from structural elements that have been shown to exhibit affinity for typical anionic groups found in biological substrates, for example, phosphate or carboxylate. Secondary interactions between additional functional groups in the receptors and the organic part of the substrate then serve to induce binding selectivity for the biorelevant target. This chapter demonstrates that this general strategy has furnished numerous potent anion receptors, some of which possess remarkable binding properties even under physiological conditions, rivaling natural systems with respect to affinity and selectivity. The future will show whether these investigations will pave the way for applications of synthetic anion receptors in medicinal chemistry or for the *in vitro* or *in vivo* sensing of biorelevant anions in diagnostics.

References

1. Hirsch, A.K.H., Fischer, F.R., and Diederich, F. (2007) *Angew. Chem.*, **119**, 342–357; (2007) *Angew. Chem. Int. Ed.*, **46**, 338–352.
2. Katayev, E.A., Ustynyuk, Y.A., and Sessler, J.L. (2006) *Coord. Chem. Rev.*, **250**, 3004–3037.
3. García-España, E., Díaz, P., Llinares, J.M., and Bianchi, A. (2006) *Coord. Chem. Rev.*, **250**, 2952–2986.
4. Tabor, C.W. and Tabor, H. (1976) *Ann. Rev. Biochem.*, **45**, 285–306.
5. Tabor, C.W. and Tabor, H. (1984) *Ann. Rev. Biochem.*, **53**, 749–790.
6. Leugger, A.P., Hertli, L., and Kaden, T.A. (1978) *Helv. Chim. Acta*, **61**, 2296–2306.
7. Suet, E., Laouenan, A., Handel, H., and Guglielmetti, R. (1984) *Helv. Chim. Acta*, **67**, 441–449.
8. Dietrich, B., Hosseini, M.W., Lehn, J.-M., and Sessions, R.B. (1981) *J. Am. Chem. Soc.*, **103**, 1282–1283.
9. Kimura, E., Kodama, M., and Yatsunami, T. (1982) *J. Am. Chem. Soc.*, **104**, 3182–3187.
10. Dietrich, B., Guilhem, J., Lehn, J.-M., Pascard, C., and Sonveaux, E. (1984) *Helv. Chim. Acta*, **67**, 91–104.
11. Kimura, E., Kuramoto, Y., Koike, T., Fujioka, H., and Kodama, M. (1990) *J. Org. Chem.*, **55**, 42–46.
12. Bencini, A., Bianchi, A., Burguete, M.I., Doménech, A., García-España, E., Luis, S.V., Niño, M.A., and Ramírez, J.A. (1991) *J. Chem. Soc., Perkin Trans. 2*, 1445–1451.
13. Motekaitis, R.J. and Martell, A.E. (1992) *Inorg. Chem.*, **31**, 5534–5542.
14. Andrés, A., Aragó, J., Bencini, A., Bianchi, A., Doménech, A., Fusi, V., García-España, E., Paoletti, P., and Ramírez, J.A. (1993) *Inorg. Chem.*, **32**, 3418–3424.
15. Lu, Q., Motekaitis, R.J., Reibenspies, J.J., and Martell, A.E. (1995) *Inorg. Chem.*, **34**, 4958–4964.

16. Bencini, A., Bianchi, A., Giorgi, C., Paoletti, P., Valtancoli, B., Fusi, V., García-España, E., Llinares, J.M., and Ramírez, J.A. (1996) *Inorg. Chem.*, **35**, 1114–1120.
17. Nation, D.A., Reibenspies, J., and Martell, A.E. (1996) *Inorg. Chem.*, **35**, 4597–4603.
18. English, J.B., Martell, A.E., Motekaitis, R.J., and Murase, I. (1997) *Inorg. Chim. Acta*, **258**, 183–192.
19. Bazzicalupi, C., Bencini, A., Bianchi, A., Cecchi, M., Escuder, B., Fusi, V., García-España, E., Giorgi, C., Luis, S.V., Maccagni, G., Marcelino, V., Paoletti, P., and Valtancoli, B. (1999) *J. Am. Chem. Soc.*, **121**, 6807–6815.
20. Bazzicalupi, C., Bencini, A., Berni, E., Bianchi, A., Fornasari, P., Giorgi, C., Masotti, A., Paoletti, P., and Valtancoli, B. (2001) *J. Phys. Org. Chem.*, **14**, 432–443.
21. Anda, C., Bazzicalupi, C., Bencini, A., Berni, E., Bianchi, A., Fornasari, P., Llobet, A., Giorgi, C., Paoletti, P., and Valtancoli, B. (2003) *Inorg. Chim. Acta*, **356**, 167–178.
22. Albelda, M.T., Frías, J.C., García-España, E., and Luis, S.V. (2004) *Org. Biomol. Chem.*, **2**, 816–820.
23. Anda, C., Llobet, A., Salvado, V., Reibenspies, J., Motekaitis, R.J., and Martell, A.E. (2000) *Inorg. Chem.*, **39**, 2986–2999.
24. Anda, C., Llobet, A., Salvado, V., Martell, A.E., and Motekaitis, R.J. (2000) *Inorg. Chem.*, **39**, 3000–3008.
25. Mateus, P., Delgado, R., Brandão, P., and Félix, V. (2009) *J. Org. Chem.*, **74**, 8638–8646.
26. Mateus, P., Delgado, R., Brandão, P., Carvalho, S., and Félix, V. (2009) *Org. Biomol. Chem.*, **7**, 4661–4673.
27. Bencini, A., Biagini, S., Giorgi, C., Handel, H., Le Baccon, M., Mariani, P., Paoletti, P., Paoli, P., Rossi, P., Tripier, R., and Valtancoli, B. (2009) *Eur. J. Org. Chem.*, 5610–5621.
28. Jurek, P.E., Martell, A.E., Motekaitis, R.J., and Hancock, R.D. (1995) *Inorg. Chem.*, **34**, 1823–1829.
29. Fabbrizzi, L., Marcotte, N., Stomeo, F., and Taglietti, A. (2002) *Angew. Chem.*, **114**, 3965–3968; (2002) *Angew. Chem. Int. Ed.*, **41**, 3811–3814.
30. Marcotte, N. and Taglietti, A. (2003) *Supramol. Chem.*, **15**, 617–625.
31. Beer, P.D. (1998) *Acc. Chem. Res.*, **31**, 71–80.
32. Beer, P.D., Cadman, J., Lloris, J.M., Martínez-Máñez, R., Padilla, M.E., Pardo, T., Smith, D.K., and Soto, J. (1999) *J. Chem. Soc., Dalton Trans.*, 127–133.
33. Czarnik, A.W. (1994) *Acc. Chem. Res.*, **27**, 302–308.
34. Schmidtchen, F.P. (1977) *Angew. Chem. Int. Ed. Engl.*, **89**, 751–752; (1977) *Angew. Chem. Int. Ed. Engl.*, **16**, 720–721.
35. Schmidtchen, F.P. (1980) *Chem. Ber.*, **113**, 864–874.
36. Schmidtchen, F.P. (1981) *Chem. Ber.*, **114**, 597–607.
37. Anslyn, E.V. and Hannon, C.L. (1993) in *Bioorganic Chemistry Frontiers*, vol. **3** (eds H. Dugas and F.P. Schmidtchen), Springer, Berlin, pp. 193–255.
38. Best, M.D., Tobey, S.L., and Anslyn, E.V. (2003) *Coord. Chem. Rev.*, **240**, 3–15.
39. Schug, K.A. and Lindner, W. (2005) *Chem. Rev.*, **105**, 67–113.
40. Springs, B. and Haake, P. (1977) *Bioorg. Chem.*, **6**, 181–190.
41. Dietrich, B., Fyles, T.M., Lehn, J.-M., Pease, L.G., and Fyles, D.L. (1978) *J. Chem. Soc., Chem. Commun.*, 934–936.
42. Dietrich, B., Fyles, D.L., Fyles, T.M., and Lehn, J.-M. (1979) *Helv. Chim. Acta*, **62**, 2763–2787.
43. Schiessl, P. and Schmidtchen, F.P. (1994) *J. Org. Chem.*, **59**, 509–511.
44. Folmer-Andersen, J.F., Aït-Haddou, H., Lynch, V.M., and Anslyn, E.V. (2003) *Inorg. Chem.*, **42**, 8674–8681.
45. Tobey, S.L., Jones, B.D., and Anslyn, E.V. (2003) *J. Am. Chem. Soc.*, **125**, 4026–4027.
46. Tobey, S.L. and Anslyn, E.V. (2003) *J. Am. Chem. Soc.*, **125**, 14807–14815.
47. Tobey, S.L. and Anslyn, E.V. (2003) *Org. Lett.*, **5**, 2029–2031.
48. Kim, S.K., Lee, D.H., Hong, J.-I., and Yoon, J. (2009) *Acc. Chem. Res.*, **42**, 23–31.

49. Nishizawa, S., Kato, Y., and Teramae, N. (1999) *J. Am. Chem. Soc.*, **121**, 9463–9464.
50. Sun, Y., Zhong, C., Gong, R., and Fu, E. (2008) *Org. Biomol. Chem.*, 3044–3047.
51. Beer, P.D., Drew, M.G.B., and Smith, D.K. (1997) *J. Organomet. Chem.*, **543**, 259–261.
52. Kim, S.K., Singh, N.J., Kim, S.J., Kim, H.G., Kim, J.K., Lee, J.W., Kim, K.S., and Yoon, J. (2003) *Org. Lett.*, **5**, 2083–2086.
53. Yoon, J., Kim, S.K., Singh, N.J., Lee, J.W., Yang, Y.J., Chellappan, K., and Kim, K.S. (2004) *J. Org. Chem.*, **69**, 581–583.
54. Watson, J.D. and Milner-White, E.J. (2002) *J. Mol. Biol.*, **315**, 171–182.
55. Watson, J.D. and Milner-White, E.J. (2002) *J. Mol. Biol.*, **315**, 183–191.
56. Kubik, S., Reyheller, C., and Stüwe, S. (2005) *J. Inclusion Phenom. Macrocyclic Chem.*, **52**, 137–187.
57. Kang, S.O., Hossain, M.D., and Bowman-James, K. (2006) *Coord. Chem. Rev.*, **250**, 3038–3052.
58. Kang, S.O., Llinares, J.M., Powell, D., VanderVelde, D., and Bowman-James, K. (2003) *J. Am. Chem. Soc.*, **125**, 10152–10153.
59. Hossain, M.A., Kang, S.O., Powell, D., and Bowman-James, K. (2003) *Inorg. Chem.*, **42**, 1397–1399.
60. Hossain, M.A., Kang, S.O., Llinares, J.M., Powell, D., and Bowman-James, K. (2003) *Inorg. Chem.*, **42**, 5043–5045.
61. Szumna, A. and Jurczak, J. (2001) *Eur. J. Org. Chem.*, 4031–4039.
62. Chmielewski, M.J. and Jurczak, J. (2004) *Tetrahedron Lett.*, **45**, 6007–6010.
63. Chmielewski, M.J. and Jurczak, J. (2005) *Chem. Eur. J.*, **11**, 6080–6094.
64. Chmielewski, M.J. and Jurczak, J. (2005) *Tetrahedron Lett.*, **46**, 3085–3088.
65. Chmielewski, M.J. and Jurczak, J. (2006) *Chem. Eur. J.*, **12**, 7652–7667.
66. Korendovych, I.V., Cho, M., Makhlynets, O.V., Butler, P.L., Staples, R.J., and Rybak-Akimova, E.V. (2008) *J. Org. Chem.*, **73**, 4771–4782.
67. Kang, S.O., Day, V.W., and Bowman-James, K. (2009) *Org. Lett.*, **11**, 3654–3657.
68. Kang, S.O., Powell, D., Day, V.W., and Bowman-James, K. (2006) *Angew. Chem.*, **118**, 1955–1959; (2006) *Angew. Chem. Int. Ed.*, **45**, 1921–1925.
69. Kang, S.O., Day, V.W., and Bowman-James, K. (2008) *Org. Lett.*, **10**, 2677–2680.
70. Lowe, A.J., Dyson, G.A., and Pfeffer, F.M. (2008) *Eur. J. Org. Chem.*, 1559–1567.
71. Pfeffer, F.M., Kruger, P.E., and Gunnlaugsson, T. (2007) *Org. Biomol. Chem.*, **5**, 1894–1902.
72. Gunnlaugsson, T., Davis, A.P., O'Brien, J.E., and Glynn, M. (2002) *Org. Lett.*, **4**, 2449–2452.
73. Sessler, J.L. and Davis, J.M. (2001) *Acc. Chem. Res.*, **34**, 989–997.
74. Král, V., Furuta, H., Shreder, K., Lynch, V., and Sessler, J.L. (1996) *J. Am. Chem. Soc.*, **118**, 1595–1607.
75. Sessler, J.L., Davis, J.M., Král, V., Kimbrough, T., and Lynch, V. (2003) *Org. Biomol. Chem.*, **1**, 4113–4123.
76. Sessler, J.L., Camiolo, S., and Gale, P.A. (2003) *Coord. Chem. Rev.*, **240**, 17–55.
77. Gale, P.A., Anzenbacher, P. Jr., and Sessler, J.L. (2001) *Coord. Chem. Rev.*, **222**, 57–102.
78. Gale, P.A., Sessler, J.L., and Král, V. (1998) *Chem. Commun.*, 1–8.
79. Gale, P.A., Sessler, J.L., Král, V., and Lynch, V. (1996) *J. Am. Chem. Soc.*, **118**, 5140–5141.
80. Anzenbacher, P. Jr., Try, A.C., Miyaji, H., Jursíková, K., Lynch, V.M., Marquez, M., and Sessler, J.L. (2000) *J. Am. Chem. Soc.*, **122**, 10268–10272.
81. Anzenbacher, P. Jr., Jursíková, K., and Sessler, J.L. (2000) *J. Am. Chem. Soc.*, **122**, 9350–9351.
82. Sessler, J.L., Gebauer, A., and Gale, P.A. (1997) *Gazz. Chim. Ital.*, **127**, 723–726.
83. Sessler, J.L., Katayev, E., Pantos, G.D., and Ustynyuk, Y.A. (2004) *Chem. Commun.*, 1276–1277.
84. Sessler, J.L., Katayev, E., Pantos, G.D., Scherbakov, P., Reshetova, M.D., Khrustalev, V.N., Lynch, V.M., and

Ustynyuk, Y.A. (2005) *J. Am. Chem. Soc.*, **127**, 11442–11446.
85. Katayev, E.A., Pantos, G.D., Reshetova, M.D., Khrustalev, V.N., Lynch, V.M., Ustynyuk, Y.A., and Sessler, J.L. (2005) *Angew. Chem.*, **117**, 7552–7556; (2005) *Angew. Chem. Int. Ed.*, **44**, 7386–7390.
86. Katayev, E.A., Boev, N.V., Khrustalev, V.N., Ustynyuk, Y.A., Tananaev, I.G., and Sessler, J.L. (2007) *J. Org. Chem.*, **72**, 2886–2896.
87. Katayev, E.A., Sessler, J.L., Khrustalev, V.N., and Ustynyuk, Y.A. (2007) *J. Org. Chem.*, **72**, 7244–7252.
88. Katayev, E.A., Boev, N.V., Myshkovskaya, E., Khrustalev, V.N., and Ustynyuk, Y.A. (2008) *Chem. Eur. J.*, **14**, 9065–9073.
89. Gale, P.A. (2005) *Chem. Commun.*, 3761–3772.
90. Gale, P.A. (2006) *Acc. Chem. Res.*, **39**, 465–475.
91. Vega, I.E.D., Camiolo, S., Gale, P.A., Hursthouse, M.B., and Light, M.E. (2003) *Chem. Commun.*, 1686–1687.
92. Vega, I.E.D., Gale, P.A., Hursthouse, M.B., and Light, M.E. (2004) *Org. Biomol. Chem.*, **2**, 2935–2941.
93. Gale, P.A. (2008) *Chem. Commun.*, 4525–4540.
94. Bates, G.W., Triyanti, Light, M.E., Albrecht, M., and Gale, P.A. (2007) *J. Org. Chem.*, **72**, 8921–8927.
95. Chang, K.-J., Moon, D., Lah, M.S., and Jeong, K.-S. (2005) *Angew. Chem.*, **117**, 8140–8143; (2005) *Angew. Chem. Int. Ed.*, **44**, 7926–7929.
96. Curiel, D., Cowley, A., and Beer, P.D. (2005) *Chem. Commun.*, 236–238.
97. Chang, K.-J., Chae, M.K., Lee, C., Lee, J.-Y., and Jeong, K.-S. (2006) *Tetrahedron Lett.*, **47**, 6385–6388.
98. Kwon, T.H. and Jeong, K.-S. (2006) *Tetrahedron Lett.*, **47**, 8539–8541.
99. Curiel, D., Espinosa, A., Más-Montoya, M., Sánchez, G., Tárraga, A., and Molina, P. (2009) *Chem. Commun.*, 7539–7541.
100. Sessler, J.L., Cho, D.-G., and Lynch, V. (2006) *J. Am. Chem. Soc.*, **128**, 16518–16519.
101. Wang, T., Bai, Y., Ma, L., and Yan, X.-P. (2008) *Org. Biomol. Chem.*, **6**, 1751–1755.
102. Bates, G.W., Gale, P.A., and Light, M.E. (2007) *Chem. Commun.*, 2121–2123.
103. Pfeffer, F.M., Lim, K.F., and Sedgwick, K.J. (2007) *Org. Biomol. Chem.*, **5**, 1795–1799.
104. Caltagirone, C., Gale, P.A., Hiscock, J.R., Brooks, S.J., Hursthouse, M.B., and Light, M.E. (2008) *Chem. Commun.*, 3007–3009.
105. Rudkevich, D.M., Verboom, W., Brzozka, Z., Palys, M.J., Stauthamer, W.P.R.V., van Hummel, G.J., Franken, S.M., Harkema, S., Engbersen, J.F.J., and Reinhoudt, D.N. (1994) *J. Am. Chem. Soc.*, **116**, 4341–4351.
106. O'Neil, E.J. and Smith, B.D. (2006) *Coord. Chem. Rev.*, **250**, 3068–3080.
107. Drewry, J.A., Fletcher, S., Hassan, H., and Gunning, P.T. (2009) *Org. Biomol. Chem.*, **7**, 5074–5077.
108. Han, M.S. and Kim, D.H. (2002) *Angew. Chem.*, **114**, 3963–3965; (2002) *Angew. Chem. Int. Ed.*, **41**, 3809–3811.
109. Hanshaw, R.G., Hilkert, S.M., Jiang, H., and Smith, B.D. (2004) *Tetrahedron Lett.*, **47**, 8721–8724.
110. Lee, D.H., Im, J.H., Son, S.U., Chung, Y.K., and Hong, J.-I. (2003) *J. Am. Chem. Soc.*, **125**, 7752–7753.
111. McDonough, M.J., Reynolds, A.J., Lee, W.Y.G., and Jolliffe, K.A. (2006) *Chem. Commun.*, 2971–2973.
112. Bencini, A., Bianchi, A., García-España, E., Scott, E.C., Morales, L., Wang, B., Deffo, T., Takusagawa, F., Mertes, M.P., Mertes, K.B., and Paoletti, P. (1992) *Bioorg. Chem.*, **20**, 8–29.
113. Mertes, M.P. and Mertes, K.B. (1990) *Acc. Chem. Res.*, **23**, 413–418.
114. Hosseini, M.W. (1997) in *Supramolecular Chemistry of Anions* (eds A. Bianchi, K. Bowman-James, and E. García-España), Wiley-VCH Verlag GmbH, New York, pp. 421–448.
115. Aguilar, J.A., García-España, E., Guerrero, J.A., Luis, S.V., Llinares, J.M., Miravet, J.F., Ramírez, J.A., and Soriano, C. (1995) *J. Chem. Soc., Chem. Commun.*, 2237–2239.
116. Aguilar, J.A., García-España, E., Guerrero, J.A., Luis, S.V., Llinares, J.M., Ramírez, J.A., and

Soriano, C. (1996) *Inorg. Chim. Acta*, **246**, 287–294.

117. Aguilar, J.A., Descalzo, A.B., Díaz, P., Fusi, V., García-España, E., Luis, S.V., Micheloni, M., Ramírez, J.A., Romani, P., and Soriano, C. (2000) *J. Chem. Soc., Perkin Trans. 2*, 1187–1192.

118. Aguilar, J.A., Celda, B., Fusi, V., García-España, E., Luis, S.V., Martínez, C., Ramírez, J.A., Soriano, C., and Tejero, R. (2000) *J. Chem. Soc., Perkin Trans. 2*, 1323–1328.

119. Albelda, M.T., Aguilar, J., Alves, S., Aucejo, R., Díaz, P., Lodeiro, C., Lima, J.C., García-España, E., Pina, F., and Soriano, C. (2003) *Helv. Chim. Acta*, **86**, 3118–3135.

120. Albelda, M.T., García-España, E., Jiménez, H.R., Llinares, J.M., Soriano, C., Sornosa-Ten, A., and Verdejo, B. (2006) *Dalton Trans.*, 4474–4481.

121. Bazzicalupi, C., Bencini, A., Berni, E., Bianchi, A., Fornasari, P., Giorgi, C., Marinelli, C., and Valtancoli, B. (2003) *Dalton Trans.*, 2564–2572.

122. Amendola, V., Bergamaschi, G., Buttafava, A., Fabbrizzi, L., and Monzani, E. (2010) *J. Am. Chem. Soc.*, **132**, 147–156.

123. Neelakandan, P.P., Hariharan, M., and Ramaiah, D. (2006) *J. Am. Chem. Soc.*, **128**, 11334–11335.

124. Chen, X., Jou, M.J., and Yoon, J. (2009) *Org. Lett.*, **11**, 2181–2184.

125. Doménech, A., García-España, E., Ramírez, J.A., Celda, B., Martínez, M.C., Monleón, D., Tejero, R., Bencini, A., and Bianchi, A. (1999) *J. Chem. Soc., Perkin Trans. 2*, 23–32.

126. Fenniri, H., Hosseini, M.W., and Lehn, J.-M. (1997) *Helv. Chim. Acta*, **80**, 786–803.

127. Fokkens, M., Jasper, C., Schrader, T., Koziol, F., Ochsenfeld, C., Polkowska, J., Lobert, M., Kahlert, B., and Klärner, F.-G. (2005) *Chem. Eur. J.*, **11**, 477–494.

128. Xu, Z., Singh, N.J., Lim, J., Pan, J., Kim, H.N., Park, S., Kim, K.S., and Yoon, J. (2009) *J. Am. Chem. Soc.*, **131**, 15528–15533.

129. Kwon, J.Y., Singh, N.J., Kim, H.N., Kim, S.K., Kim, K.S., and Yoon, J. (2004) *J. Am. Chem. Soc.*, **126**, 8892–8893.

130. Wang, S. and Chang, Y.-T. (2006) *J. Am. Chem. Soc.*, **128**, 10380–10381.

131. Sebo, L., Diederich, F., and Gramlich, V. (2000) *Helv. Chim. Acta*, **83**, 93–113.

132. Kato, Y., Conn, M.M., and Rebek, J. Jr. (1994) *J. Am. Chem. Soc.*, **116**, 3279–3284.

133. Andreu, C., Galán, A., Kobiro, K., de Mendoza, J., Park, T.K., Rebek, J. Jr., Salmerón, A., and Usman, N. (1994) *J. Am. Chem. Soc.*, **116**, 5501–5502.

134. Galán, A., de Mendoza, J., Toiron, C., Bruix, M., Deslongchamps, G., and Rebek, J. Jr. (1991) *J. Am. Chem. Soc.*, **113**, 9424–9425.

135. Deslongchamps, G., Galán, A., de Mendoza, J., and Rebek, J. Jr. (1992) *Angew. Chem.*, **104**, 58–60; (1992) *Angew. Chem. Int. Ed. Engl.*, **31**, 61–63.

136. Hennrich, G. and Anslyn, E.V. (2002) *Chem. Eur. J.*, **8**, 2219–2224.

137. Schneider, S.E., O'Neil, S.N., and Anslyn, E.V. (2000) *J. Am. Chem. Soc.*, **122**, 542–543.

138. McCleskey, S.C., Griffin, M.J., Schneider, S.E., McDevitt, J.T., and Anslyn, E.V. (2003) *J. Am. Chem. Soc.*, **125**, 1114–1115.

139. Sessler, J.L., Furuta, H., and Král, V. (1993) *Supramol. Chem.*, **1**, 209–220.

140. Král, V. and Sessler, J.L. (1995) *Tetrahedron*, **51**, 539–554.

141. Raposo, C., Pérez, N., Almaraz, M., Mussons, M.L., Caballero, M.C., and Morán, J.R. (1995) *Tetrahedron Lett.*, **36**, 3255–3258.

142. Ishida, H., Suga, M., Donowaki, K., and Ohkubo, K. (1995) *J. Org. Chem.*, **60**, 5374–5375.

143. Boon, J.M. and Smith, B.D. (2002) *Curr. Opin. Chem. Biol.*, **6**, 749–756.

144. Lambert, T.N. and Smith, B.D. (2003) *Coord. Chem. Rev.*, **240**, 129–141.

145. Smith, B.D. and Lambert, T.N. (2003) *Chem. Commun.*, 2261–2268.

146. Boon, J.M. and Smith, B.D. (1999) *J. Am. Chem. Soc.*, **121**, 11924–11925.

147. Boon, J.M. and Smith, B.D. (2001) *J. Am. Chem. Soc.*, **123**, 6221–6226.

148. Sasaki, Y., Shukla, R., and Smith, B.D. (2004) *Org. Biomol. Chem.*, **2**, 214–219.

149. Boon, J.M., Lambert, T.N., Smith, B.D., Beatty, A.M., Ugrinova, V., and Brown, S.N. (2002) *J. Org. Chem.*, **67**, 2168–2174.
150. Boon, J.M., Lambert, T.N., Sisson, A.L., Davis, A.P., and Smith, B.D. (2003) *J. Am. Chem. Soc.*, **125**, 8195–8201.
151. Lambert, T.N., Boon, J.M., Smith, B.D., Pérez-Payán, M.N., and Davis, A.P. (2002) *J. Am. Chem. Soc.*, **124**, 5276–5277.
152. Hubbard, R.D., Horner, S.R., and Miller, B.L. (2001) *J. Am. Chem. Soc.*, **123**, 5810–5811.
153. Emgenbroich, M., Borrelli, C., Shinde, S., Lazraq, I., Vilela, F., Hall, A.J., Oxelbark, J., De Lorenzi, E., Courtois, J., Simanova, A., Verhage, J., Irgum, K., Karim, K., and Sellergren, B. (2008) *Chem. Eur. J.*, **14**, 9516–9529.
154. Schmuck, C. and Schwegmann, M. (2005) *Org. Lett.*, **7**, 3517–3520.
155. Niikura, K., Metzger, A., and Anslyn, E.V. (1998) *J. Am. Chem. Soc.*, **120**, 8533–8534.
156. Niikura, K. and Anslyn, E.V. (2003) *J. Org. Chem.*, **68**, 10156–10157.
157. Mareque-Rivas, J.C., de Rosales, R.T.M., and Parsons, S. (2004) *Chem. Commun.*, 610–611.
158. Zhong, Z. and Anslyn, E.V. (2003) *Angew. Chem.*, **115**, 3113–3116; (2003) *Angew. Chem. Int. Ed.*, **42**, 3005–3008.
159. Kimura, E. (1992) *Tetrahedron*, **48**, 6175–6217.
160. Kimura, E. (1994) *Prog. Inorg. Chem.*, **41**, 443–491.
161. Kimura, E. (2001) *Acc. Chem. Res.*, **34**, 171–179.
162. Koike, T., Kajitani, S., Nakamura, I., Kimura, E., and Shiro, M. (1995) *J. Am. Chem. Soc.*, **117**, 1210–1219.
163. Koike, T., Takamura, M., and Kimura, E. (1994) *J. Am. Chem. Soc.*, **116**, 8443–8449.
164. Fujioka, H., Koike, T., Yamada, N., and Kimura, E. (1996) *Heterocycles*, **42**, 775–787.
165. Kimura, E., Aoki, S., Koike, T., and Shiro, M. (1997) *J. Am. Chem. Soc.*, **119**, 3068–3076.
166. Ojida, A., Miyahara, Y., Kohira, T., and Hamachi, I. (2004) *Biopolymers*, **76**, 177–184.
167. Ojida, A. and Hamachi, I. (2006) *Bull. Chem. Soc. Jpn.*, **79**, 35–46.
168. Sakamoto, T., Ojida, A., and Hamachi, I. (2009) *Chem. Commun.*, 141–152.
169. Ojida, A., Mito-oka, Y., Inoue, M., and Hamachi, I. (2002) *J. Am. Chem. Soc.*, **124**, 6256–6258.
170. Ojida, A., Mito-oka, Y., Sada, K., and Hamachi, I. (2004) *J. Am. Chem. Soc.*, **126**, 2454–2463.
171. Ojida, A., Park, S., Mito-oka, Y., and Hamachi, I. (2002) *Tetrahedron Lett.*, **43**, 6193–6195.
172. Ojida, A., Inoue, M., Mito-oka, Y., and Hamachi, I. (2003) *J. Am. Chem. Soc.*, **125**, 10184–10185.
173. Ojida, A., Inoue, M., Mito-oka, Y., Tsutsumi, H., Sada, K., and Hamachi, I. (2006) *J. Am. Chem. Soc.*, **128**, 2052–2058.
174. Ojida, A., Kohira, T., and Hamachi, I. (2004) *Chem. Lett.*, **33**, 1024–1025.
175. Anai, T., Nakata, E., Koshi, Y., Ojida, A., and Hamachi, I. (2007) *J. Am. Chem. Soc.*, **129**, 6232–6239.
176. Kohira, T., Honda, K., Ojida, A., and Hamachi, I. (2008) *ChemBioChem*, **9**, 698–701.
177. Koulov, A.V., Stucker, K.A., Lakshmi, C., Robinson, J.P., and Smith, B.D. (2003) *Cell Death Differ.*, **10**, 1357–1359.
178. Koulov, A.V., Hanshaw, R.G., Stucker, K.A., Lakshmi, C., and Smith, B.D. (2005) *Isr. J. Chem.*, **45**, 373–379.
179. Hanshaw, R.G. and Smith, B.D. (2005) *Bioorg. Med. Chem.*, **13**, 5035–5042.
180. Grauer, A., Riechers, A., Ritter, S., and König, B. (2008) *Chem. Eur. J.*, **14**, 8922–8927.
181. Zhang, T., Edwards, N.Y., Bonizzoni, M., and Anslyn, E.V. (2009) *J. Am. Chem. Soc.*, **131**, 11976–11984.
182. Alcázar, V., Segura, M., Prados, P., and de Mendoza, J. (1998) *Tetrahedron Lett.*, **39**, 1033–1036.
183. Ariga, K. and Anslyn, E.V. (1992) *J. Org. Chem.*, **57**, 417–419.
184. Kneeland, D.M., Ariga, K., Lynch, V.M., Huang, C.-Y., and Anslyn, E.V. (1993) *J. Am. Chem. Soc.*, **115**, 10042–10055.
185. Perreault, D.M., Chen, X., and Anslyn, E.V. (1995) *Tetrahedron*, **51**, 353–362.

186. Smith, J., Ariga, K., and Anslyn, E.V. (1993) *J. Am. Chem. Soc.*, **115**, 362–364.

187. Perreault, D.M., Cabell, L.A., and Anslyn, E.V. (1997) *Bioorg. Med. Chem.*, **5**, 1209–1220.

188. Jubian, V., Dixon, R.P., and Hamilton, A.D. (1992) *J. Am. Chem. Soc.*, **114**, 1120–1121.

189. Jubian, V., Veronese, A., Dixon, R.P., and Hamilton, A.D. (1995) *Angew. Chem.*, **107**, 1343–1345; (1995) *Angew. Chem. Int. Ed. Engl.*, **34**, 1237–1239.

190. Göbel, M.W., Bats, J.W., and Dürner, G. (1992) *Angew. Chem.*, **104**, 217–218; (1992) *Angew. Chem. Int. Ed. Engl.*, **31**, 207–209.

191. Muche, M.-S. and Göbel, M.W. (1996) *Angew. Chem.*, **108**, 2263–2265; (1996) *Angew. Chem. Int. Ed. Engl.*, **35**, 2126–2129.

192. Molenveld, P., Kapsabelis, S., Engbersen, J.F.J., and Reinhoudt, D.N. (1997) *J. Am. Chem. Soc.*, **119**, 2948–2949.

193. Molenveld, P., Engbersen, J.F.J., Kooijman, H., Spek, A.L., and Reinhoudt, D.N. (1998) *J. Am. Chem. Soc.*, **120**, 6726–6737.

194. Molenveld, P., Engbersen, J.F.J., and Reinhoudt, D.N. (1999) *Eur. J. Org. Chem.*, **12**, 3269–3275.

195. Molenveld, P., Engbersen, J.F.J., and Reinhoudt, D.N. (1999) *Angew. Chem.*, **111**, 3387–3390; (1999) *Angew. Chem. Int. Ed.*, **38**, 3189–3192.

196. Molenveld, P., Stikvoort, W.M.G., Kooijman, H., Spek, A.L., Engbersen, J.F.J., and Reinhoudt, D.N. (1999) *J. Org. Chem.*, **64**, 3896–3906.

197. Molenveld, P., Engbersen, J.F.J., and Reinhoudt, D.N. (2000) *Chem. Soc. Rev.*, **29**, 75–86.

198. Schmuck, C. and Geiger, L. (2005) *Chem. Commun.*, 772–774.

199. Schmuck, C. and Geiger, L. (2003) *Curr. Org. Chem.*, **7**, 1485–1502.

200. Schmuck, C. (2006) *Coord. Chem. Rev.*, **250**, 3053–3067.

201. Schmuck, C. and Dudaczek, J. (2007) *Org. Lett.*, **9**, 5389–5392.

202. Lindgren, N.J.V., Geiger, L., Razkin, J., Schmuck, C., and Baltzer, L. (2009) *Angew. Chem.*, **121**, 6850–6853; (2009) *Angew. Chem. Int. Ed.*, **48**, 6722–6725.

203. Chand, D.K., Schneider, H.-J., Bencini, A., Bianchi, A., Giorgi, C., Ciattini, S., and Valtancoli, B. (2000) *Chem. Eur. J.*, **6**, 4001–4008.

204. Chand, D.K., Bharadwaj, P.K., and Schneider, H.-J. (2001) *Tetrahedron*, **57**, 6727–6732.

205. Chand, D.K., Schneider, H.-J., Aguilar, J.A., Escarti, F., García-España, E., and Luis, S.V. (2001) *Inorg. Chim. Acta*, **316**, 71–78.

206. Lomadze, N., Gogritchiani, E., Schneider, H.-J., Albelda, M.T., Aguilar, J., García-España, E., and Luis, S.V. (2002) *Tetrahedron Lett.*, **43**, 7801–7803.

207. Bencini, A., Berni, E., Bianchi, A., Giorgi, C., Valtancoli, B., Chand, D.K., and Schneider, H.-J. (2003) *Dalton Trans.*, 793–800.

208. Lomadze, N., Schneider, H.-J., Albelda, M.T., García-España, E., and Verdejo, B. (2006) *Org. Biomol. Chem.*, **4**, 1755–1759.

209. McIntosh, C.M., Esposito, E.A. III, Boal, A.K., Simard, J.M., Martin, C.T., and Rotello, V.M. (2001) *J. Am. Chem. Soc.*, **123**, 7626–7629.

210. Verma, A. and Rotello, V.M. (2005) *Chem. Commun.*, 303–312.

211. Sandhu, K.K., McIntosh, C.M., Simard, J.M., Smith, S.W., and Rotello, V.M. (2002) *Bioconjug. Chem.*, **13**, 3–6.

212. Zadmard, R. and Schrader, T. (2006) *Angew. Chem.*, **118**, 2769–2772; (2006) *Angew. Chem. Int. Ed.*, **45**, 2703–2706.

213. Breitkreuz, C.J., Zadmard, R., and Schrader, T. (2008) *Supramol. Chem.*, **20**, 109–115.

214. Fukutomi, R., Tanatani, A., Kakuta, H., Tomioka, N., Itai, A., Hashimoto, Y., Shudo, K., and Kagechika, H. (1998) *Tetrahedron Lett.*, **39**, 6475–6478.

215. Nguyen, B., Lee, M.P.H., Hamelberg, D., Joubert, A., Bailly, C., Brun, R., Neidle, S., and Wilson, W.D. (2002) *J. Am. Chem. Soc.*, **124**, 13680–13681.

216. Nguyen, B., Hamelberg, D., Bailly, C., Colson, P., Stanek, J., Brun, R., Neidle, S., and Wilson, W.D. (2004) *Biophys. J.*, **86**, 1028–1041.

217. Sansone, F., Dudic, M., Donofrio, G., Rivetti, C., Baldini, L., Casnati, A., Cellai, S., and Ungaro, R. (2006) *J. Am. Chem. Soc.*, **128**, 14528–14536.
218. Bagnacani, V., Sansone, F., Donofrio, G., Baldini, L., Casnati, A., and Ungaro, R. (2008) *Org. Lett.*, **10**, 3953–3956.
219. Iverson, B.L., Shreder, K., Král, V., and Sessler, J.L. (1993) *J. Am. Chem. Soc.*, **115**, 11022–11023.
220. Iverson, B.L., Shreder, K., Král, V., Smith, D.A., Smith, J., and Sessler, J.L. (1994) *Pure Appl. Chem.*, **66**, 845–850.
221. Iverson, B.L., Shreder, K., Král, V., Sansom, P., Lynch, V., and Sessler, J.L. (1996) *J. Am. Chem. Soc.*, **118**, 1608–1616.
222. Magda, D., Wright, M., Miller, R.A., Sessler, J.L., and Sansom, P.I. (1995) *J. Am. Chem. Soc.*, **117**, 3629–3630.
223. Sessler, J.L., Sansom, P.I., Král, V., O'Connor, D., and Iverson, B.L. (1996) *J. Am. Chem. Soc.*, **118**, 12322–12330.
224. Christianson, D.W. and Lipscomb, W.N. (1989) *Acc. Chem. Res.*, **22**, 62–69.
225. Mangani, S. and Ferraroni, M. (1997) in *Supramolecular Chemistry of Anions* (eds A. Bianchi, K. Bowman-James, and E. García-España), Wiley-VCH Verlag GmbH, New York, pp. 63–78.
226. Keil, B. (1971) in *The Enzymes, Hydrolysis: Peptide Bond*, vol. **3** (ed. P.D. Boyer), Academic Press, New York, pp. 249–275.
227. Magnusson, S. (1971) in *The Enzymes, Hydrolysis: Peptide Bond*, vol. **3** (ed. P.D. Boyer), Academic Press, New York, pp. 278–321.
228. Rao, A.V.R., Gurjar, M.K., Reddy, K.L., and Rao, A.S. (1995) *Chem. Rev.*, **95**, 2135–2167.
229. Nicolaou, K.C., Boddy, C.N.C., Bräse, S., and Winssinger, N. (1999) *Angew. Chem.*, **111**, 2230–2287; (1999) *Angew. Chem. Int. Ed.*, **38**, 2097–2152.
230. Williams, D.H. and Bardsley, B. (1999) *Angew. Chem.*, **111**, 1264–1286; (1999) *Angew. Chem. Int. Ed.*, **38**, 1173–1193.
231. Loll, P.J., Miller, R., Weeks, C.M., and Axelsen, P.H. (1998) *Chem. Biol.*, **5**, 293–298.
232. Fitzmaurice, R.J., Kyne, G.M., Douheret, D., and Kilburn, J.D. (2002) *J. Chem. Soc., Perkin Trans. 1*, 841–864.
233. McKee, V., Nelson, J., and Town, R.M. (2003) *Chem. Soc. Rev.*, **32**, 309–325.
234. Fabbrizzi, L., Pallavicini, P., Parodi, L., and Taglietti, A. (1995) *Inorg. Chim. Acta*, **238**, 5–8.
235. Tobey, S.L. and Anslyn, E.V. (2003) *J. Am. Chem. Soc.*, **125**, 10963–10970.
236. Berger, M. and Schmidtchen, F.P. (1999) *J. Am. Chem. Soc.*, **121**, 9986–9993.
237. Linton, B. and Hamilton, A.D. (1999) *Tetrahedron*, **55**, 6027–6038.
238. Schmuck, C. (1999) *Chem. Commun.*, 843–844.
239. Schmuck, C. and Lex, J. (1999) *Org. Lett.*, **1**, 1779–1781.
240. Gunnlaugsson, T., Glynn, M., Tocci, G.M., Kruger, P.E., and Pfeffer, F.M. (2006) *Coord. Chem. Rev.*, **250**, 3094–3117.
241. Clare, J.P., Statnikov, A., Lynch, V., Sargent, A.L., and Sibert, J.W. (2009) *J. Org. Chem.*, **74**, 6637–6646.
242. Lee, K.H. and Hong, J.-I. (2000) *Tetrahedron Lett.*, **41**, 6083–6087.
243. Sasaki, S., Mizuno, M., Naemura, K., and Tobe, Y. (2000) *J. Org. Chem.*, **65**, 275–283.
244. Nishizawa, S., Bühlmann, P., Iwao, M., and Umezawa, Y. (1995) *Tetrahedron Lett.*, **36**, 6483–6486.
245. Bühlmann, P., Nishizawa, S., Xiao, K.P., and Umezawa, Y. (1997) *Tetrahedron*, **53**, 1647–1654.
246. Tejeda, A., Oliva, A.I., Simón, L., Grande, M., Caballero, M.C., and Morán, J.R. (2000) *Tetrahedron Lett.*, **41**, 4563–4566.
247. Clare, J.P., Ayling, A.J., Joos, J.-B., Sisson, A.L., Magro, G., Peréz-Payán, M.N., Lambert, T.N., Shukla, R., Smith, B.D., and Davis, A.P. (2005) *J. Am. Chem. Soc.*, **127**, 10739–10746.
248. Sisson, A.L., Clare, J.P., Taylor, L.H., Charmant, J.P.H., and Davis, A.P. (2003) *Chem. Commun.*, 2246–2247.
249. Sisson, A.L., Clare, J.P., and Davis, A.P. (2005) *Chem. Commun.*, 5263–5265.

250. Davis, A.P. (2006) *Coord. Chem. Rev.*, **250**, 2939–2951.
251. Budka, J., Lhoták, P., Michlová, V., and Stibor, I. (2001) *Tetrahedron Lett.*, **42**, 1583–1586.
252. Nam, K.C., Kang, S.O., and Ko, S.W. (1999) *Bull. Korean Chem. Soc.*, **20**, 953–956.
253. Stibor, I., Budka, J., Michlová, V., Tkadlecová, M., Pojarová, M., Cuřínová, P., and Lhoták, P. (2008) *New J. Chem.*, **32**, 1597–1607.
254. Kavallieratos, K., Bertao, C.M., and Crabtree, R.H. (1999) *J. Org. Chem.*, **64**, 1675–1683.
255. Hughes, M.P. and Smith, B.D. (1997) *J. Org. Chem.*, **62**, 4492–4499.
256. Andrievsky, A., Ahuis, F., Sessler, J.L., Vögtle, F., Gudat, D., and Moini, M. (1998) *J. Am. Chem. Soc.*, **120**, 9712–9713.
257. Bisson, A.P., Lynch, V.M., Monahan, M.-K.C., and Anslyn, E.V. (1997) *Angew. Chem.*, **109**, 2435–2437; (1997) *Angew. Chem. Int. Ed. Engl.*, **36**, 2340–2342.
258. Cameron, B.R. and Loeb, S.J. (1997) *Chem. Commun.*, 573–574.
259. Sansone, F., Baldini, L., Casnati, A., Lazzarotto, M., Ugozzoli, F., and Ungaro, R. (2002) *Proc. Natl. Acad. Sci. U.S.A.*, **99**, 4842–4847.
260. Miao, R., Zheng, Q.-Y., Chen, C.-F., and Huang, Z.-T. (2005) *Tetrahedron Lett.*, **46**, 2155–2158.
261. Prohens, R., Tomàs, S., Morey, J., Deyà, P.M., Ballester, P., and Costa, A. (1998) *Tetrahedron Lett.*, **39**, 1063–1066.
262. Beer, P.D., Timoshenko, V., Maestri, M., Passaniti, P., and Balzani, V. (1999) *Chem. Commun.*, 1755–1756.
263. Beer, P.D., Szemes, F., Balzani, V., Salà, C.M., Drew, M.G.B., Dent, S.W., and Maestri, M. (1997) *J. Am. Chem. Soc.*, **119**, 11864–11875.
264. Bondy, C.R., Gale, P.A., and Loeb, S.J. (2001) *Chem. Commun.*, 729–730.
265. Vega, I.E.D., Gale, P.A., Light, M.E., and Loeb, S.J. (2005) *Chem. Commun.*, 4913–4915.
266. Hossain, M.A. and Schneider, H.-J. (1999) *Chem. Eur. J.*, **5**, 1284–1290.
267. Eblinger, F. and Schneider, H.-J. (1998) *Angew. Chem.*, **110**, 821–824; (1998) *Angew. Chem. Int. Ed.*, **37**, 826–829.
268. Kimura, E., Sakonaka, A., Yatsunami, T., and Kodama, M. (1981) *J. Am. Chem. Soc.*, **103**, 3041–3045.
269. Hosseini, M.W. and Lehn, J.-M. (1982) *J. Am. Chem. Soc.*, **104**, 3525–3527.
270. Hosseini, M.W. and Lehn, J.-M. (1986) *Helv. Chim. Acta*, **69**, 587–603.
271. Bencini, A., Bianchi, A., Burguete, M.I., Dapporto, P., Doménech, A., García-España, E., Luis, S.V., Paoli, P., and Ramírez, J.A. (1994) *J. Chem. Soc., Perkin Trans. 2*, 569–577.
272. Bianchi, A. and García-España, E. (1999) *J. Chem. Educ.*, **76**, 1727–1732.
273. Lehn, J.-M., Méric, R., Vigneron, J.-P., Bkouche-Waksman, I., and Pascard, C. (1991) *J. Chem. Soc., Chem. Commun.*, 62–64.
274. Nelson, J., Nieuwenhuyzen, M., Pál, I., and Town, R.M. (2002) *Chem. Commun.*, 2266–2267.
275. Nelson, J., Nieuwenhuyzen, M., Pál, I., and Town, R.M. (2004) *Dalton Trans.*, 229–235.
276. Boiocchi, M., Bonizzoni, M., Fabbrizzi, L., Piovani, G., and Taglietti, A. (2004) *Angew. Chem.*, **116**, 3935–3940; (2004) *Angew. Chem. Int. Ed.*, **43**, 3847–3852.
277. Boiocchi, M., Bonizzoni, M., Moletti, A., Pasini, D., and Taglietti, A. (2007) *New J. Chem.*, **31**, 352–356.
278. Fabbrizzi, L., Foti, F., and Taglietti, A. (2005) *Org. Lett.*, **7**, 2603–2606.
279. Tang, L., Park, J., Kim, H.-J., Kim, Y., Kim, S.J., Chin, J., and Kim, K.M. (2008) *J. Am. Chem. Soc.*, **130**, 12606–12607.
280. Schießl, P. and Schmidtchen, F.P. (1993) *Tetrahedron Lett.*, **34**, 2449–2452.
281. Metzger, A., Lynch, V.M., and Anslyn, E.V. (1997) *Angew. Chem.*, **109**, 911–914; (1997) *Angew. Chem. Int. Ed. Engl.*, **36**, 862–865.
282. Metzger, A. and Anslyn, E.V. (1998) *Angew. Chem.*, **110**, 682–684; (1998) *Angew. Chem. Int. Ed.*, **37**, 649–652.
283. Lavigne, J.J. and Anslyn, E.V. (1999) *Angew. Chem.*, **111**, 3903–3906; (1999) *Angew. Chem. Int. Ed.*, **38**, 3666–3669.

284. Piątek, A.M., Bomble, Y.J., Wiskur, S.L., and Anslyn, E.V. (2004) *J. Am. Chem. Soc.*, **126**, 6072–6077.

285. Wiskur, S.L. and Anslyn, E.V. (2001) *J. Am. Chem. Soc.*, **123**, 10109–10110.

286. Wiskur, S.L., Floriano, P.N., Anslyn, E.V., and McDevitt, J.T. (2003) *Angew. Chem.*, **115**, 2116–2118; (2003) *Angew. Chem. Int. Ed.*, **42**, 2070–2072.

287. Nguyen, B.T., Wiskur, S.L., and Anslyn, E.V. (2004) *Org. Lett.*, **6**, 2499–2501.

288. Wiskur, S.L., Lavigne, J.J., Metzger, A., Tobey, S., Lynch, V., and Anslyn, E.V. (2004) *Chem. Eur. J.*, **10**, 3792–3804.

289. Cabell, L.A., Best, M.D., Lavigne, J.J., Schneider, S.E., Perreault, D.M., Monahan, M.-K., and Anslyn, E.V. (2001) *J. Chem. Soc., Perkin Trans. 2*, 315–323.

290. Schmuck, C. and Schwegmann, M. (2005) *J. Am. Chem. Soc.*, **127**, 3373–3379.

291. Schmuck, C. and Schwegmann, M. (2006) *Org. Biomol. Chem.*, **4**, 836–838.

292. Fan, E., Van Arman, S.A., Kincaid, S., and Hamilton, A.D. (1993) *J. Am. Chem. Soc.*, **115**, 369–370.

293. Garcia-Tellado, F., Goswami, S., Chang, S.K., Geib, S.J., and Hamilton, A.D. (1990) *J. Am. Chem. Soc.*, **112**, 7393–7394.

294. Kelly, T.R. and Kim, M.H. (1994) *J. Am. Chem. Soc.*, **116**, 7072–7080.

295. Jeong, K.-S., Park, J.W., and Cho, Y.L. (1996) *Tetrahedron Lett.*, **37**, 2795–2798.

296. Scheerder, J., Engbersen, J.F.J., Casnati, A., Ungaro, R., and Reinhoudt, D.N. (1995) *J. Org. Chem.*, **60**, 6448–6454.

297. Goodman, M.S., Jubian, V., Linton, B., and Hamilton, A.D. (1995) *J. Am. Chem. Soc.*, **117**, 11610–11611.

298. Goodman, M.S., Jubian, V., and Hamilton, A.D. (1995) *Tetrahedron Lett.*, **36**, 2551–2554.

299. Maeda, H., Fujii, Y., and Mihashi, Y. (2008) *Chem. Commun.*, 4285–4287.

300. Sancenón, F., Martínez-Máñez, R., Miranda, M.A., Seguí, M.-J., and Soto, J. (2003) *Angew. Chem.*, **115**, 671–674; (2003) *Angew. Chem. Int. Ed.*, **42**, 647–650.

301. Pirkle, W.H. and Pochapsky, T.C. (1989) *Chem. Rev.*, **89**, 347–362.

302. Peacock, S.C., Domeier, L.A., Gaeta, F.C.A., Helgeson, R.C., Timko, J.M., and Cram, D.J. (1978) *J. Am. Chem. Soc.*, **100**, 8190–8202.

303. Alfonso, I., Rebolledo, F., and Gotor, V. (2000) *Chem. Eur. J.*, **6**, 3331–3338.

304. Alfonso, I., Dietrich, B., Rebolledo, F., Gotor, V., and Lehn, J.-M. (2001) *Helv. Chim. Acta*, **84**, 280–295.

305. Miranda, C., Escarti, F., Lamarque, L., Yunta, M.J.R., Navarro, P., García-España, E., and Jimeno, M.L. (2004) *J. Am. Chem. Soc.*, **126**, 823–833.

306. Galán, A., Andreu, D., Echavarren, A.M., Prados, P., and de Mendoza, J. (1992) *J. Am. Chem. Soc.*, **114**, 1511–1512.

307. Breccia, P., Van Gool, M., Pérez-Fernández, R., Martín-Santamaría, S., Gago, F., Prados, P., and de Mendoza, J. (2003) *J. Am. Chem. Soc.*, **125**, 8270–8284.

308. Metzger, A., Gloe, K., Stephan, H., and Schmidtchen, F.P. (1996) *J. Org. Chem.*, **61**, 2051–2055.

309. de Silva, A.P., Gunaratne, H.Q.N., McVeigh, C., Maguire, G.E.M., Maxwell, P.R.S., and O'Hanlon, E. (1996) *Chem. Commun.*, 2191–2192.

310. Sasaki, S., Hashizume, A., Citterio, D., Fujii, E., and Suzuki, K. (2002) *Tetrahedron Lett.*, **43**, 7243–7245.

311. Davis, A.P. and Lawless, L.J. (1999) *Chem. Commun.*, 9–10.

312. Baragaña, B., Blackburn, A.G., Breccia, P., Davis, A.P., de Mendoza, J., Padrón-Carrillo, J.M., Prados, P., Riedner, J., and de Vries, J.G. (2002) *Chem. Eur. J.*, **8**, 2931–2936.

313. Schmuck, C. (2000) *Chem. Eur. J.*, **6**, 709–718.

314. Schmuck, C. and Graupner, S. (2005) *Tetrahedron Lett.*, **46**, 1295–1298.

315. Schmuck, C. and Bickert, V. (2007) *J. Org. Chem.*, **72**, 6832–6839.

316. Schmuck, C. and Geiger, L. (2005) *J. Am. Chem. Soc.*, **127**, 10486–10487.

317. Schmuck, C. and Dudaczek, J. (2005) *Tetrahedron Lett.*, **46**, 7101–7105.

318. Schmuck, C. and Bickert, V. (2003) *Org. Lett.*, **5**, 4579–4581.

319. Sessler, J.L., Andrievsky, A., Král, V., and Lynch, V. (1997) *J. Am. Chem. Soc.*, **119**, 9385–9392.
320. Kyne, G.M., Light, M.E., Hursthouse, M.B., de Mendoza, J., and Kilburn, J.D. (2001) *J. Chem. Soc., Perkin Trans. 1*, 1258–1263.
321. Pernía, G.J., Kilburn, J.D., and Rowley, M. (1995) *J. Chem. Soc., Chem. Commun.*, 305–306.
322. Pernía, G.J., Kilburn, J.D., Essex, J.W., Mortishire-Smith, R.J., and Rowley, M. (1996) *J. Am. Chem. Soc.*, **118**, 10220–10227.
323. Jullian, V., Shepherd, E., Gelbrich, T., Hursthouse, M.B., and Kilburn, J.D. (2000) *Tetrahedron Lett.*, **41**, 3963–3966.
324. Rossi, S., Kyne, G.M., Turner, D.L., Wells, N.J., and Kilburn, J.D. (2002) *Angew. Chem.*, **114**, 4407–4409; (2002) *Angew. Chem. Int. Ed.*, **41**, 4233–4236.
325. Ragusa, A., Rossi, S., Hayes, J.M., Stein, M., and Kilburn, J.D. (2005) *Chem. Eur. J.*, **11**, 5674–5688.
326. Bartoli, S., Mahmood, T., Malik, A., Dixon, S., and Kilburn, J.D. (2008) *Org. Biomol. Chem.*, **6**, 2340–2345.
327. Aït-Haddou, H., Wiskur, S.L., Lynch, V.M., and Anslyn, E.V. (2001) *J. Am. Chem. Soc.*, **123**, 11296–11297.
328. Imai, H., Misawa, K., Munakata, H., and Uemori, Y. (2001) *Chem. Lett.*, 688–689.
329. Konishi, K., Yahara, K., Toshishige, H., Aida, T., and Inoue, S. (1994) *J. Am. Chem. Soc.*, **116**, 1337–1344.
330. Dickins, R.S., Batsanov, A.S., Howard, J.A.K., Parker, D., Puschmann, H., and Salamano, S. (2004) *Dalton Trans.*, 70–80.
331. Hossain, M.A. and Schneider, H.-J. (1998) *J. Am. Chem. Soc.*, **120**, 11208–11209.
332. Sirish, M. and Schneider, H.-J. (1999) *Chem. Commun.*, 907–908.
333. Sirish, M., Chertkov, V.A., and Schneider, H.-J. (2002) *Chem. Eur. J.*, **8**, 1181–1188.
334. Srinivasan, N. and Kilburn, J.D. (2004) *Curr. Opin. Chem. Biol.*, **8**, 305–310.
335. Davies, M., Bonnat, M., Guillier, F., Kilburn, J.D., and Bradley, M. (1998) *J. Org. Chem.*, **63**, 8696–8703.
336. Jensen, K.B., Braxmeier, T.M., Demarcus, M., Frey, J.G., and Kilburn, J.D. (2002) *Chem. Eur. J.*, **8**, 1300–1309.
337. Shepherd, J., Gale, T., Jensen, K.B., and Kilburn, J.D. (2006) *Chem. Eur. J.*, **12**, 713–720.
338. Shepherd, J., Langley, G.J., Herniman, J.M., and Kilburn, J.D. (2007) *Eur. J. Org. Chem.*, 1345–1356.
339. Schmuck, C., Rupprecht, D., and Wienand, W. (2006) *Chem. Eur. J.*, **12**, 9186–9195.
340. Schmuck, C. and Hernandez-Folgado, L. (2007) *Org. Biomol. Chem.*, **5**, 2390–2394.
341. Schmuck, C., Rupprecht, D., Junkers, M., and Schrader, T. (2007) *Chem. Eur. J.*, **13**, 6864–6873.
342. Schmuck, C. and Wich, P. (2006) *New J. Chem.*, **30**, 1377–1385.
343. Schmuck, C. and Wich, P. (2007) *Top. Curr. Chem.*, **277**, 3–30.
344. Schmuck, C. and Heil, M. (2003) *Org. Biomol. Chem.*, **1**, 633–636.
345. Schmuck, C. and Heil, M. (2003) *ChemBioChem*, **4**, 1232–1238.
346. Schmuck, C., Frey, P., and Heil, M. (2005) *ChemBioChem*, **6**, 628–631.
347. Schmuck, C., Heil, M., Scheiber, J., and Baumann, K. (2005) *Angew. Chem.*, **117**, 7374–7379; (2005) *Angew. Chem. Int. Ed.*, **44**, 7208–7212.
348. Schmuck, C. and Heil, M. (2006) *Chem. Eur. J.*, **12**, 1339–1348.
349. Schmuck, C. and Wich, P. (2006) *Angew. Chem.*, **118**, 4383–4387; (2006) *Angew. Chem. Int. Ed.*, **45**, 4277–4281.
350. Casnati, A., Fabbi, M., Pelizzi, N., Pochini, A., Sansone, F., Ungaro, R., Di Modugno, E., and Tarzia, G. (1996) *Bioorg. Med. Chem. Lett.*, **6**, 2699–2704.
351. Frish, L., Sansone, F., Casnati, A., Ungaro, R. and Cohen, Y. (2000) *J. Org. Chem.*, **65**, 5026–5030.
352. Albert, J.S., Peczuh, M.W., and Hamilton, A.D. (1997) *Bioorg. Med. Chem.*, **5**, 1455–1467.
353. Albert, J.S., Goodman, M.S., and Hamilton, A.D. (1995) *J. Am. Chem. Soc.*, **117**, 1143–1144.
354. Peczuh, M.W., Hamilton, A.D., Sánchez-Quesada, J., de Mendoza,

J., Haack, T., and Giralt, E. (1997) *J. Am. Chem. Soc.*, **119**, 9327–9328.

355. Haack, T., Peczuh, M.W., Salvatella, X., Sánchez-Quesada, J., de Mendoza, J., Hamilton, A.D., and Giralt, E. (1999) *J. Am. Chem. Soc.*, **121**, 11813–11820.

356. Tabet, M., Labroo, V., Sheppard, P., and Sasaki, T. (1993) *J. Am. Chem. Soc.*, **115**, 3866–3868.

357. Mazik, M. and König, A. (2007) *Eur. J. Org. Chem.*, 3271–3276.

358. Mazik, M. and Cavga, H. (2007) *J. Org. Chem.*, **72**, 831–838.

359. Mazik, M. and Cavga, H. (2007) *Eur. J. Org. Chem.*, 3633–3638.

360. Billing, J. Grundberg, H., and Nilsson, U.J. (2002) *Supramol. Chem.*, **14**, 367–372.

8
Synthetic Amphiphilic Peptides that Self-Assemble to Membrane-Active Anion Transporters

George W. Gokel and Megan M. Daschbach

8.1
Introduction and Background

The development of synthetic ion channels has taken place over parallel pathways during the past two decades. We refer to these here as the *synporins* and *synthetic channels*. Both strategies involve synthetic compounds but differ in the approach. Minimalist synporin structures resulted from an analysis of peptides and proteins followed by synthesis of peptide sequences thought to be appropriate. Examples may be found in the studies by DeGrado and coworkers [1] and the efforts of Montal *et al.* [2]. DeGrado and their coworkers referred to their novel peptides as "*minimalist peptides*" and Montal *et al.* designated their novel transporters "*synporins*" [2b].

An example of the structures developed by DeGrado *et al.* is the 21-amino acid peptide having the sequence $H_2N(Leu\text{-}Ser\text{-}Leu_3\text{-}Ser\text{-}Leu)_3CONH_2$. This compound was found to insert into phospholipid bilayers and form proton-conducting channels [1b]. Montal and coworkers modeled their "synthetic proteins" on the sequence of the known *Torpedo californica* acetylcholine receptor [2b]. The 23-amino acid peptide has the sequence GKMST AISVL LAEAV FLLLT SER. As with the 21-mer above, this compound forms a channel in bilayers. In this case, planar bilayer conductance evidence that confirmed cation selectivity was presented.

Pioneering work describing completely abiotic synthetic channels was reported by Lehn [3] and by Fyles [4]. The results of our own work in this area appeared shortly thereafter [5]. The initial structure, reported by Lehn and Jullien, was referred to as a "*chundle*." The name derived from "channel formed from a bundle of fibers." The name was likely inspired by the English Channel Tunnel, the "Chunnel," then under construction. The original chundle compound was built on a central crown ether that had dendrimer-like chains radiating from it. The chains were capped by carboxyl terminated oligoethyleneoxy residues. The structure was complex and intriguing, but no ion-transport data were disclosed in the original paper. These appeared later, and a variant of the channel that incorporated a β-cyclodextrin (β-CD) central element was introduced. The latter, called a "*bouquet*" molecule, had more, but less complex, pendant chains than the original crown variant [6].

Anion Coordination Chemistry, First Edition. Edited by Kristin Bowman-James, Antonio Bianchi, and Enrique García-España.

Fyles developed a family of ion transporters [7] based on the same tartaric-acid-derived crown ether used in some of Lehn's studies. In the Fyles family of compounds, bolaamphiphiles (bolytes) constructed from succinic acids were linked to the central macrocycles. Four bolytes, the distal ends of which were polar, were linked to the central macrocycle to form a tunnel-like structure. Ion transport was assayed by using a proton-sensitive fluorescent dye to follow metal-ion-dependent changes in H^+ concentration.

Our own efforts to develop an ion-selective channel also used a crown ether central unit. It differed from the approaches of Lehn and Fyles because it was a diaza crown rather than a tetracarboxycrown. Further, two additional crown ethers were incorporated as amphiphilic headgroups. We, like Fyles, undertook extensive studies to verify membrane insertion and functional and selectivity properties. Our studies with the cation-conducting compounds we called "*hydraphiles*" were summarized in 2000 [8]. Figure 8.1 shows the structures of the three synthetic channels described above.

As noted above, the work on synporins or minimalist peptides and synthetic channels has generally occurred in parallel, with little convergence. An exception is Voyer's use of a DeGrado peptide [1b] substituted by regularly placed macrocyclic rings to obtain a functional cation channel [9]. Still, there have been relatively few intersections. This is important because the protein ion channel community is large and active. The biochemist wishes to know the ion-transport efficacy, the ion selectivity, the evidence for membrane insertion, the evidence for conformation within the membrane, whether the transporter is a carrier or a channel, and whether the channel is a unimolecular or an oligomeric pore. The answers to these questions are not always available for natural peptides and proteins or for biologically active synthetic peptides. Even so, it is important for the chemical community to be aware of and to address these questions.

A number of other synthetic channels, carriers, or pore formers that transport cations have been developed. Several reviews summarize these cation channel studies [7e, 8, 10]. In recent years, the complexation of anions has received increasing attention [11, 12]. This has naturally led to the study of anion transport, which is an area of considerable current interest and the focus of this chapter.

8.2
Biomedical Importance of Chloride Channels

It is known that the cystic fibrosis transport regulator (CFTR) is a chloride ion transporter as well as a regulator of other transport systems. Mutations in this protein that result in cystic fibrosis are part of a complex pathology involved in the most common fatal genetic disease among Caucasians. In addition to cystic fibrosis, myotonia congenita, Dents' disease, and Bartter's syndrome also appear to involve dysfunction in chloride transport. The possibility that synthetic,

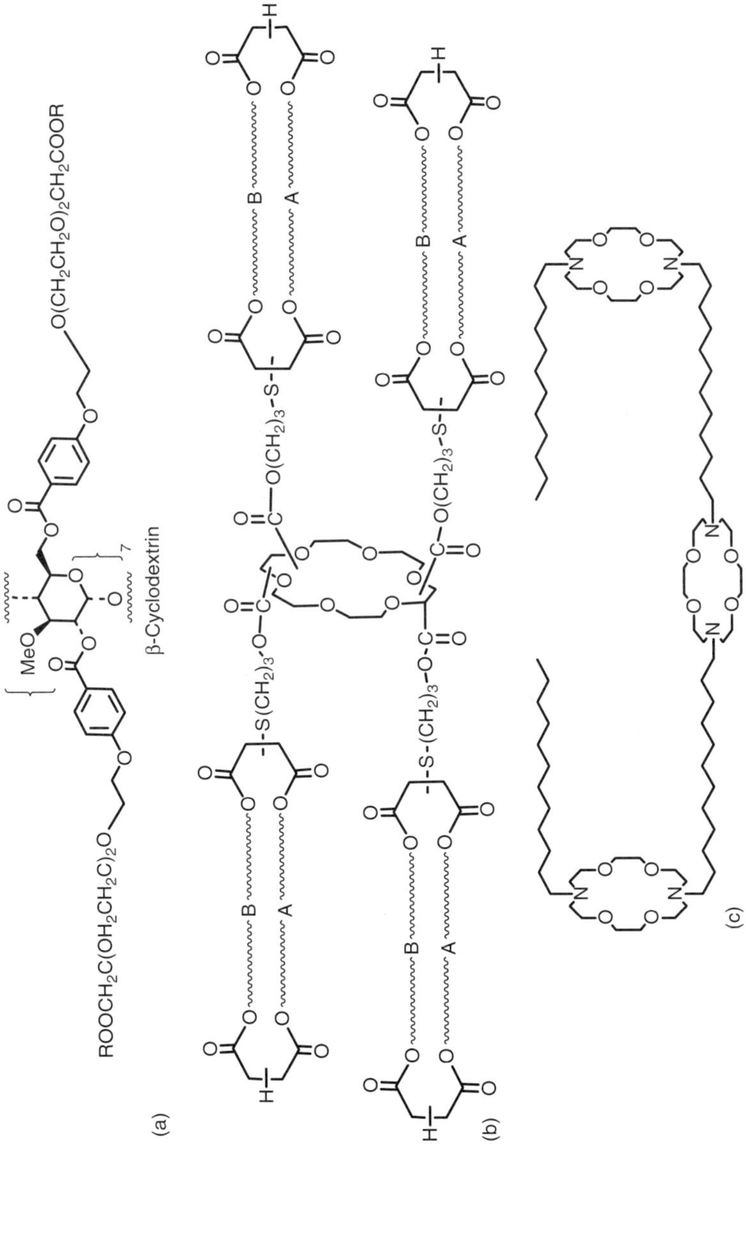

Figure 8.1 (a) Lehn's "bouquet." (b) Fyles' bolaamphiphile channels. (c) A "hydraphile."

chloride-selective transporters can assist in ameliorating these pathologies is a powerful motivation for their development [13]. Of course, numerous obstacles must be overcome to do so. Some of these efforts are described in the sections below.

8.2.1
A Natural Chloride Complexing Agent

Prodigiosin is the tris(pyrrole) compound shown as **1**. It is a natural substance produced by the gram-negative bacterium *Serratia marcescens*, which grows on bread and other starches [14, 15]. Rapoport and Holden established the chemical structure by total synthesis in 1962 [16]. The name prodigiosin or prodigiosins is often used to describe the molecule or molecules within a mixture, as the *n*-pentyl group shown in **1** varies in other derivatives.

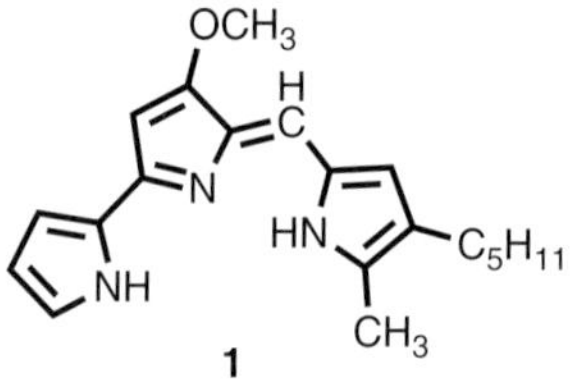

1

The prodigiosins show good antimicrobial activity but generally poor selectivity for microbes over mammalian cells [17]. Prodigiosins have also shown cytotoxicity to human cancer cell lines with selectivity over nonmalignant cells. These molecules bind chloride ion and have the ability to decrease the pH of certain cellular organelles [18]. Prodigiosin analogs have recently been reported to transport both chloride and bicarbonate ions, presumably by a carrier mechanism [19].

8.3
The Development of Synthetic Chloride Channels

The emergence of chloride channels paralleled the course of cation channel development. Several key steps were involved. First, anion receptors were designed and prepared. The ability of these receptor molecules to bind anions, particularly fluoride and chloride, was confirmed by solution and solid-state structural methods. The accumulation of structure–activity data led to attempts to mediate anion transport.

8.3.1
Cations, Anions, Complexation, and Transport

The ions that are common in seawater are also those that are common in vital biological entities. In order of diminishing ion concentration in the oceans, the common ions are $Na^+ \gg Mg^{2+} > Ca^{2+} \sim K^+$. The amounts of ions in cells are typically regulated in mammals so that the concentration of Na^+ is ~135 mM in extracellular fluid and much lower (~12 mM) within cells. In contrast, the concentrations of $K_{inside}{}^+$ and $K_{outside}{}^+$ are 140 and 4 mM, respectively. The concentrations of Mg^{2+} and Ca^{2+} are far lower. This presents a problem in terms of selectivity since all four of these cations are spherical and Na^+ and Ca^{2+} are nearly identical in diameter (~2 Å) [20].

In the early consideration of synthetic ion channel design, it seemed a simple matter to select a smaller ion over a larger one, although the question of the larger ion blocking the smaller conductance pathway loomed. The far more difficult question was how to select the larger cation over a smaller one. One clever suggestion was that control resulted from cation–π interaction(s) between ions and amino acid side chains in the channel's so-called selectivity filter [21]. This explanation was shown by site-directed mutagenesis not to account for the ion selectivity [22], and the mechanism remained unclear until 1998, when Mackinnon and coworkers reported the solid-state structure of the voltage-gated potassium channel isolated from *Streptomyces lividans* [23]. In fact, main chain amide carbonyl groups, rather than the arenes attached to the respective amino acids, formed the selectivity filter.

Recognition of K^+ involves donor–acceptor interactions between oxygen and the cation. The donor elements are appropriately sized to replace transient potassium ion's waters of solvation. It is clear that the carbonyl–cation interaction must be strong enough to permit recognition but not so strong as to impede transport. Similar issues are involved in the recognition of Cl^-. The matter is somewhat less complex, however, owing to the general lack of competing anions in nature. The anions commonly present *in vivo* are Cl^-, HCO_3^-, $CH_3CO_2^-$, NO_3^-, HSO_4^-, and $H_2PO_4^-$ [24]. Of course, the potentially multivalent ions will exhibit charge states that vary with pH. Even so, carbonate, nitrate, sulfate, and phosphate also differ significantly in geometry from spherical chloride.

Several synthetic cation channels have used crown ethers in the selectivity filter [3–7]. The ether oxygens of crowns can interact as Lewis bases with Lewis acidic Na^+ or K^+. It is interesting to note that the equilibrium complexation constant between Na^+ and K^+ with 18-crown-6 in water favors the latter by about 10-fold [25]. Both ions are bound at similar rates, but Na^+ is released about 10-fold faster than K^+. Thus, a crown ether serving as a selectivity filter should pass (release, transport) Na^+ more rapidly (effectively, easily) than it will pass K^+.

Much of the information concerning Cl^- complexation derives from solution-state studies of complexation or solid-state structures of Cl^- complexes. Both types of information are clearly valuable, but the latter studies reveal nothing dynamic. Solution complexation studies are valuable in that they quantify the position of the equilibrium, but the equilibrium constants *per se* do not reveal the rates of ion complexation and release. An additional issue that remains unresolved is the hydration state of the anion. The same issue persists in studies of cation transport.

8.3.2
Anion Complexation Studies

The report in 1967 that crown ethers complex alkali metal ions [26] was the seminal discovery from which much of modern supramolecular chemistry arose. Shortly after this initial publication, legions of chemists synthesized a vast number of crown ethers and analogs that varied in ring size and contained a range of substituents, subcyclic units, and heteroatoms different from oxygen. By the 1980s, reviews and monographs had begun to summarize the structures [27] and their binding properties with cations [28], anions [29], and neutral molecules [30].

Although cation and anion complexation studies developed in parallel, the former vastly exceeded the latter in extent. Pioneering anion complexation studies were reported by Lehn [31], Newcomb [32], Schmidtchen [33], De Mendoza [34], and others [35]. Extensive recent work has been summarized in reviews by Beer and Gale [36], Beer and Hayes [37], Gale [38], Bowman-James [39], and Kubik [40]. Monographs coauthored by Bianchi *et al.* [11] and by Sessler *et al.* [41] have also appeared.

8.3.3
Transport of Ions

The design and preparation of a functional ion receptor is a significant achievement. The characterization of its transport function requires additional study and is also of great importance. Ion transport may be analyzed and quantified in several ways. One of the earliest methods was to use a U-tube device [42], also called a "*Pressman cell*" [43]. Indeed, the U-tube and concentric tube [44] variant have been used extensively to demonstrate carrier transport through hydrophobic solvent barriers such as $CHCl_3$ or CH_2Cl_2.

Evaluation of ion transport through a phospholipid bilayer is more complex than measuring ion flux using a bulk membrane system. To use this technique at all, one must prepare liposomes (vesicles) from phospholipid monomers [45] and characterize their average size and size distribution. In doing so, one must consider whether analysis of transport will involve ions entering or departing the liposomes. If the latter, the ions (salts) must be encapsulated within the vesicles during formation. The analytical method to be used obviously bears on how the liposome will be prepared. If fluorescence is chosen to detect transport, lower concentrations of ions can often be used than if the anion is to be detected directly. Several of the methods used to characterize transport are described in subsequent sections.

8.3.4
Synthetic Chloride Transporters

Thus far, relatively few examples of Cl^- transport mediated by small molecule binders have appeared, at least compared to analogous cation transport studies. The compounds reported to date embrace receptor molecules that appear to be carriers and those that appear to form pores. Since ion release from, or entry into, vesicles will be apparent in either case, the planar bilayer conductance voltage clamp experiment is valuable for confirmation of pore formation or channellike behavior.

8.4
Approaches to Synthetic Chloride Channels

From the organic or bioorganic chemist's perspective, the design of an anion channel will parallel the requirements for a cation channel. Clearly, the element or elements that define selectivity must differ. Still, the requirements that a putative channel must insert in the bilayer and form some type of conductance pathway persist. The approaches taken are as diverse as the laboratories from which they emerged. A common problem, however, is the lack of detailed structural information about natural chloride channels and the complexity of these remarkable transport or exchange proteins [46–49]. Nevertheless, a number of designs have emerged. In addition, a number of natural lipids such as ceramide and sphingosine have been found to form pores in bilayers [50]. In the following sections, several approaches to artificial anion channels are described.

Some commentary is in order concerning the design of synthetic ion channels. First, a tubular molecule may appear to have a structure suitable for a channel, but the compound must at least function as a transporter to actually qualify as a channel. There is more than one analytical method that can detect ion transport, pore formation, or open–close behavior. However, in the absence of a clear indication of function, a tubular molecule is simply a tubular molecule. Such a compound may fairly be referred to as having a *channellike structure*. Molecules that stack in such a way that a pathway among the monomers is created may be porelike [51]. Whatever such structures are called or however they are described, only experimentally determined function can validate channel behavior.

Second, there is no rule that requires channel detection experiments to be conducted in a single, inflexible way. In the work that was done in our laboratory, planar or vesicular membranes were created and the channel was added to the bulk phase. Prior to addition of the transporter, no channel activity was observed. Detection of channel function after the compound in question has been added to the buffer suggests autonomic insertion. The distinction is significant because not all compounds that may function as channels do so in a manner that presumably mimics nature. When a protein is synthesized *in vivo*, it must still fold appropriately and insert into the bilayer in order to function. An interesting example illustrates this issue.

Triton X-100 is an inexpensive, commercially available detergent. The phenyloctylene fragment is hydrophobic, and the chain of approximately nine ethyleneoxy units comprises the polar end of this amphiphile. In 1977, it was reported that Triton X-100 (structure shown) functions as a channel in phospholipid bilayers [52]. This experiment was done by embedding the detergent in the bilayer membrane as it was formed. Channel activity was then detected by the planar bilayer conductance technique (see Section 8.5.5). To some, the fact that this simple detergent exhibited channel function trivialized the pursuit of synthetic channels.

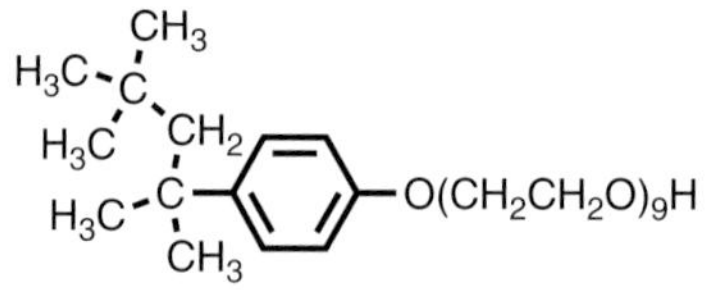

Those involved in the development of synthetic, ion-conducting channels are obviously obliged to confirm the structure and purity of the compounds they prepare. In our view, there are two other criteria that should be met to sustain the credibility of the field. First, channel function must be demonstrated by experimental measurements. The transporter may be embedded in the bilayer or added to the bulk phase. There is no regulation concerning the approach. At least in the case of Triton X-100, embedding in the membrane during its formation suggested activity that was not apparent when the compound was added to the bulk phase. It seems more appropriate to us to mimic Nature in this respect and add the compound to a membrane-containing mixture to confirm autoinsertion.

Second, control experiments are especially important in studies such as these. As the complexity of science has increased along with the multiplicity of analytical techniques used, it has become nearly impossible for results to be checked by others. Honest mistakes and simple oversights can remain undetected because checking the work requires too much time, an expertise unique to very few labs, or equipment that is not accessible. It is therefore incumbent on the reporting lab to undertake as many control experiments as feasible.

8.4.1
Tomich's Semisynthetic Peptides

Tomich and coworkers [53] have pursued peptide-based chloride-selective transporters. They have modified the 23-amino acid sequence of the M2 segment of the α-subunit of the glycine-gated Cl^- channel. The sequence is H_2N-PARVG LGITT VLTMT TQSSG SRA-COOH. They have extended this sequence by the addition of four lysines at the carboxy terminus. The resulting sequence is H_2N-PARVG LGITT VLTMT TQSSG SRAKK KK-COOH. They have given this peptide the abbreviated name C-K4-M2GlyR for C-terminus modified tetralysyl M2 glycine receptor. The work reported by Tomich and coworkers is focused on the modification of natural peptide sequences. As such, it is tangential to the focus of this article, notwithstanding its importance. In summary, Tomich *et al.* have demonstrated (i) Cl^- channel activity, (ii) enhanced solubility by dint of the tetralysyl terminus, and (iii) increased capacity for epithelial anion secretion when studied in Madin-Darby canine kidney and T84 cell monolayers.

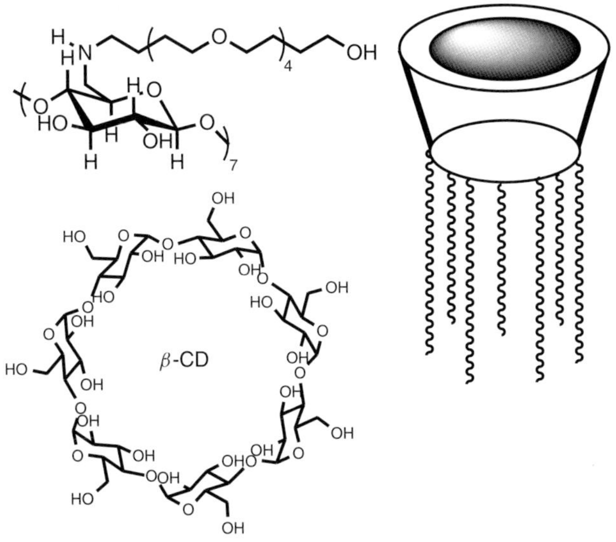

Figure 8.2 Cyclodextrin-based chloride transporter. β-Cyclodextrin (β-CD) is shown at the lower left and a schematic representation of the transporter is shown at the right.

8.4.2
Cyclodextrin as a Synthetic Channel Design Element

The CD residue was used as an element in early synthetic channel designs by both Tabushi [54] and Lehn [55] and their coworkers. Transport of cations was studied in both cases. More recently, Gin and coworkers appended oligohydroxybutyl groups to the seven primary hydroxyl groups of β-CD. The linkage was accomplished by converting the primary hydoxyls to amines prior to attachment [56]. Figure 8.2 illustrates the resulting structure in both chemical and schematic forms. The authors have referred to this transporter as hydrobutylated aminocyclodextrin or hbaCD.

Several assays evaluated transport by β-CD. Changes in the fluorescence emission spectra of the widely used dye 8-hydroxypyrene-1,3,6-trisulfonic acid (HTPS) occur with changes in pH. Proton transport mediated by β-CD was demonstrated by using this fluorescence technique. Further, the ^{23}Na exchange method was used to quantitate Na^+ transport. The latter method, developed for a variety of cations by Riddell and Hayer [57], relies on different ^{23}Na-NMR signals when the ion is within a liposome. Internal Na^+ is protected from a shift reagent present in the bulk phase and is observed at a different chemical shift than the remainder of the Na^+. When a transport mechanism is available so that Na^+ within the vesicles can exchange with Na^+ outside the liposomes, the change in linewidth gives a measure of the exchange rate through the bilayer. This excellent analytical method is tedious to perform and requires careful comparison of the results obtained with controls.

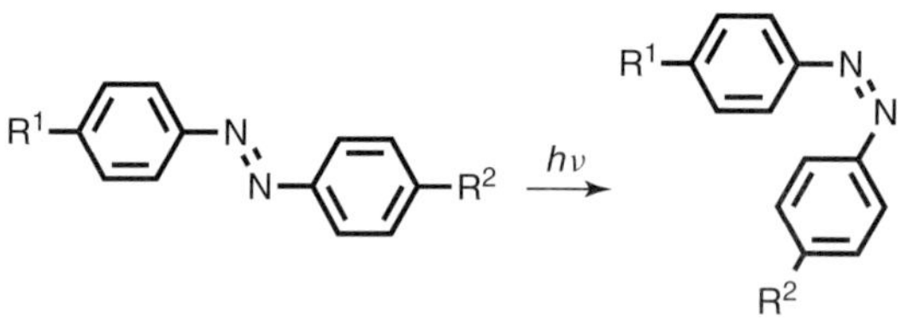

Figure 8.3 Photoswitched trans to cis photoisomerization of azobenzene.

In the present case, transport of Na^+ mediated by β-CD was 36% of that mediated by the known channel former, gramicidin A. When Na^+ transport was evaluated using NaCl, NaBr, and NaI, the Na^+ exchange rates varied in the order $I^- > Br^- > Cl^-$. The success of I^- transport may be aided by the ability of this anion to traverse bilayer membranes unaided. Additional studies [58] using the ^{23}Na-NMR method revealed that transport was pH dependent. Using the pH values 5.6, 6.3, and 7–10, the observed Na^+ exchange rate was higher when the bulk phase was more acidic or more basic than neutral. In additional studies, higher exchange rates were observed for both Cl^- and Br^- at higher pH but iodide was too membrane permeables to give comparable results [59].

8.4.3 Azobenzene as a Photo-Switchable Gate

Azobenzene possesses the property of "photo switchability." In the ground state *E*- or *trans*-form, it is a nearly linear molecule that has a span of ~9 Å between the two para-positions. When irradiated at the appropriate wavelength, it undergoes trans to cis ($E \rightarrow Z$) isomerization, which alters its geometry and the spatial relationships of attached units. Azobenzene has been incorporated into gramicidin and several other peptides by Woolley and coworkers [60, 61]. Trauner and collaborators have used azobenzene as a photoactivatable cap on channel proteins [62, 63]. In further work with the CD scaffold, Jog and Gin attached azobenzene to one of the secondary hydroxyl groups on the larger rim of CD [64]. In the ground-state trans conformation, the CD cavity was blocked, presumably by insertion of the diarene. After photoswitching, the more compressed structure was less able to block ion entry, which improved ion flux. As is often the case with azobenzene molecules, the relatively slow and incomplete switching from *E* to *Z* failed to give a binary result, even though the improved transport rates correlate well with the expected geometrical changes (Figure 8.3).

8.4.4 Calixarene-Derived Chloride Transporters

DeMendoza and coworkers used the hydraphile design [8] to probe the value of calix[4]arenes as "central relay" [65] units within a potential channel former [66]. In this case, substitution of a calix[4]arene for a diaza-18-crown-6 residue was intended to probe two questions. First, can a calix[4]arene act as a central relay unit in a

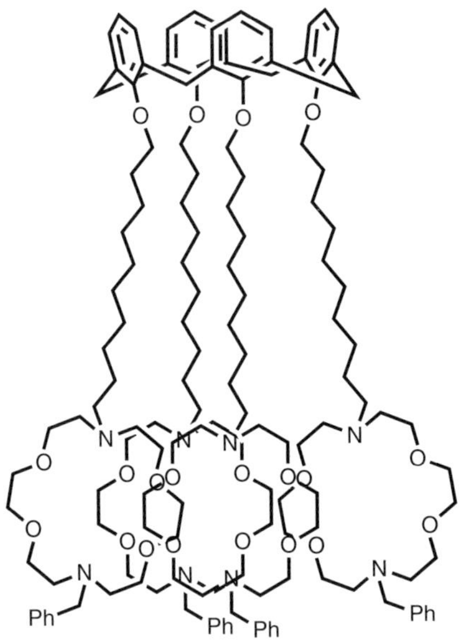

2

Figure 8.4 A channel control compound in which the central relay calixarene is in the cone conformation.

structural entity known to function as a cation channel? Second, if ion transport is observed, do the ions pass through the central calix?

Three molecules were constructed and studied in an effort to answer these questions. The first such compound is shown as **2**. In this material, the calixarene central relay is in the "cone" arrangement (Figure 8.4). This places the four dodecyl chains in register. The molecule is too short to span the bilayer and is expected not to show open–close behavior when studied by the planar bilayer conductance method. Indeed, no transport activity was observed even when **2** was added to the aqueous buffer in concentrations as high as 100 pM.

The compounds that were expected to be functional are shown as **3** and **4**. They differ only in the presence (in **4**) of four *t*-butyl groups in the calix[4]arene para-positions. It was hypothesized that the *t*-butyl groups would prevent cations from passing through the calix. If both **3** and **4** were active transporters and showed similar transport behavior, it would demonstrate the formation of a conductance pathway in the bilayer and that cations were circumventing, rather than passing through, the calix. In fact, both compounds showed "bursting" or "spiking" behavior. Such behavior is characteristic of ion transport by membrane disruption rather than by the formation of an ordered conductance pathway. The more hindered compound (**4**) showed a lower level of this type of activity than did **3**. It was concluded that neither compound was an ion channel former and that in this dynamic milieu, Na^+ and K^+ ions failed to pass through the calix[4]arene's rather crowded interior (Figure 8.5).

In more recent work from the same laboratories, calix[4]arenes and calix[6]arenes were examined as both headgroups and central relays [67] in compounds having structural properties appropriate to channel function. When the calix[4]arene

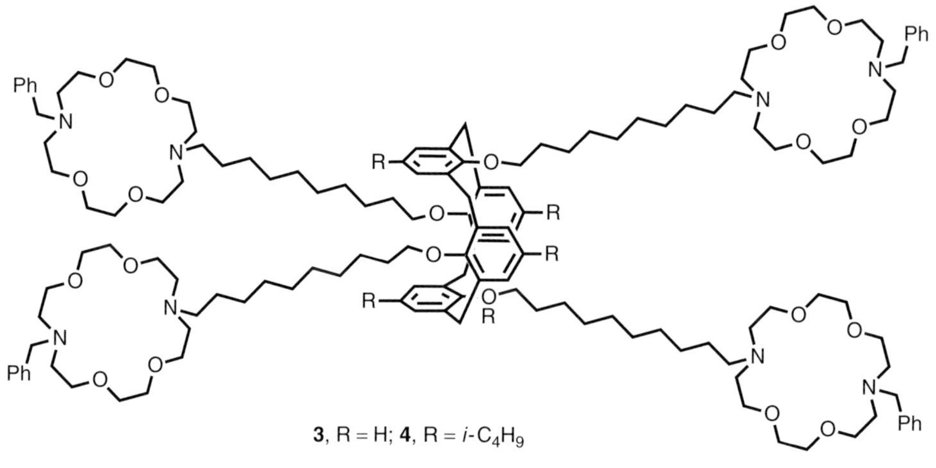

3, R = H; **4**, R = *i*-C_4H_9

Figure 8.5 A tetramacrocycle design incorporating calix[4]arene in the 1,3-alternate conformation as the central relay.

residue was present either as a central relay or as headgroups, only modest transport of either cations or anions was observed. When the calix[6]arene residue was incorporated as a headgroup, no ion transport could be detected. It seemed likely that the calix[6]arene was unable to pass through (flip-flop) the membrane in order to form a conductance pathway.

Davis and coworkers obtained a crystal structure of a calixarene (shown below as **5** in Figure 8.6) organizing in the presence of Cl^-. Two different structures showing a bound anion were observed [68]; one of these channellike motifs is illustrated below. Calixarene (**5**) was shown to function as a proton/chloride anion symporter in a lipid bilayer [69] in patch-clamp experiments, as well as in the human embryonic kidney (HEK) cells [70]. Interestingly, the methyl analog of calixarene (**5**) (in which a methyl group is replaced by a butyl chain) did not produce a crystalline Cl^-–arene complex [69] nor did it show activity when assayed using the above-mentioned experiments.

In order to investigate structure–activity relationships, six analogs of **5** were synthesized (shown in Figure 8.6 as **6**–**11**). The key structural elements remained, while the cyclic nature of the parent compound did not. The trimer **8** was found to be the most effective transporter. At similar concentrations, acyclic **8** was found to be orders of magnitude more efficient than the parent calixarene (**5**) at chloride transport [71]. Further study revealed that, in fact, the amide residues were responsible for anion complexation and the ester analog of **8** was inactive.

The critical residues in these systems were the amide groups, which are capable of hydrogen bond donation to the anion. For calixarene (**5**) to be an effective transporter, the cyclic molecule had to be flexible enough to adopt the cone conformation to allow for this key interaction. Later studies showed that when *t*-butyl groups were added to **5** (shown in Figure 8.6 as **12**), no transport activity was

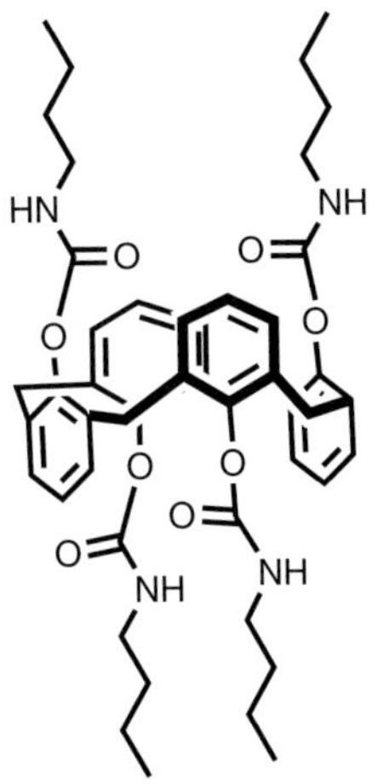

5

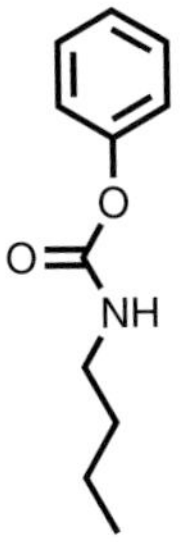

6

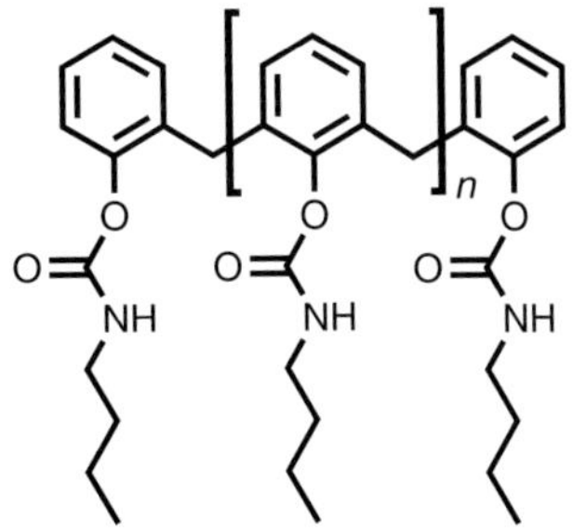

7, $n = 0$ 10, $n = 3$
8, $n = 1$ 11, $n = 4$
9, $n = 2$

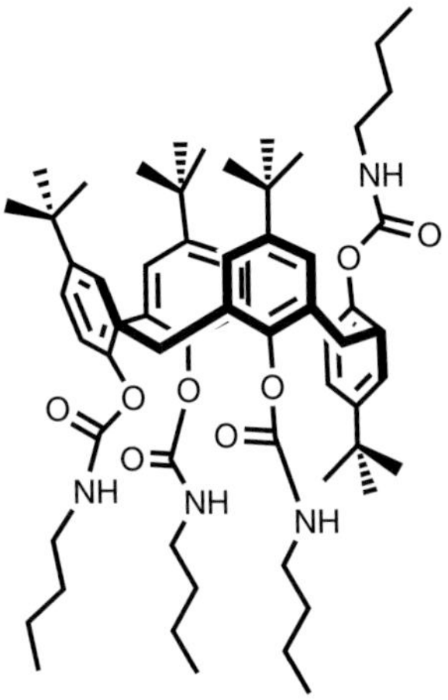

12

Figure 8.6 A calix[4]arene ion transporter and control compounds.

observed [72]. Solid-state structures suggest that the large *t*-butyl groups prevent **12** from adopting a "cone" structure, thereby obviating the necessary amide–anion interaction.

8.4.5
Oligophenylenes and π-Slides

Amphotericin B (AmB) is an antibiotic, nonpeptidic polyene macrocycle [73]. It has been referred to as a "*rigid rod*" and is thought to function by a "barrel stave" mechanism [74, 75]. Matile and coworkers attempted to mimic what they

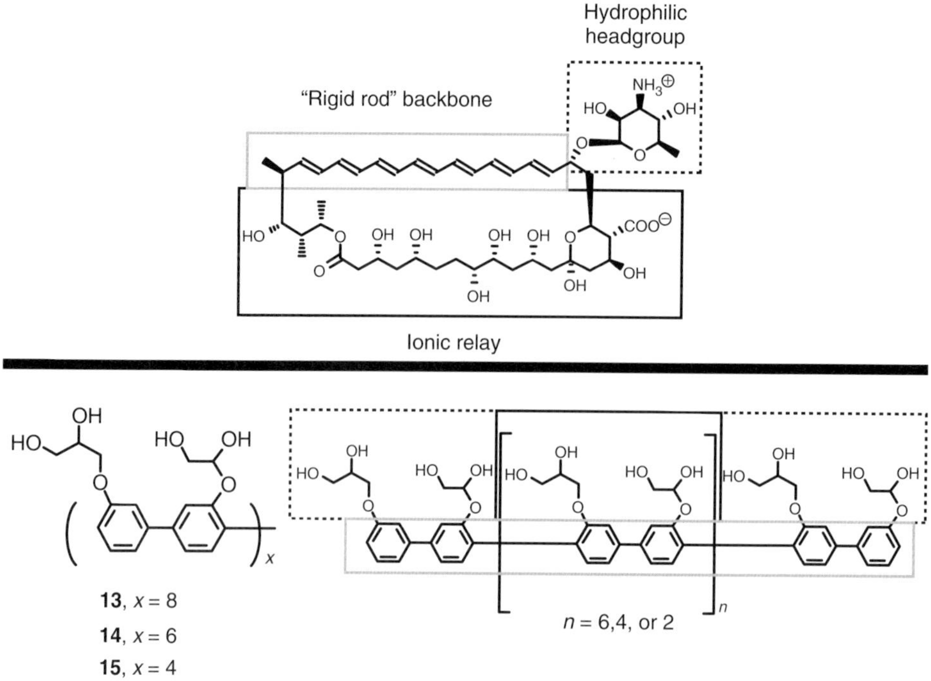

Figure 8.7 Structure of amphotericin B (top) and π-slide channel former that mimics its features.

recognized as the key structural elements of an AmB molecule, namely, (i) a hydrophilic element at the membrane/aqueous interface, (ii) a rigid polyene rod that would be stable in the hydrophobic slab of a membrane, and (iii) repeating hydrophilic residues that would function as an ionic relay system. These structural elements of AmB are highlighted in the top panel of Figure 8.7.

Matile's first rigid rod channel appeared in 1997 [73]. It is shown in the lower panel of Figure 8.7 as **13** and was conceptualized using the amphotericin model. The hydrophilic element at the interface and in the ionic relay was the same: repeating glycerol units. The rigid rod backbone was an octa(*p*-phenylene). It was estimated that the molecule would span 34 Å, the approximate thickness of a bilayer membrane. A hexaphenyl (**14**) and a tetraphenyl (**15**) compound were prepared as controls. Their lengths were estimated to be 17 and 26 Å, respectively. The shorter compound, **15**, was inactive, and hexaphenyl (**14**) was only ~30% as active in transporting ions as was **13**.

The channel versus carrier mechanisms may be tested by thickening the membrane by steroid addition. The change in membrane properties is expected to affect carrier transport more than channel function. Ion transport mediated by Matile's octa(*p*-phenyl) rigid rods showed little variation on steroid addition to the bilayer. It was inferred that the compound functions by a channel mechanism [76].

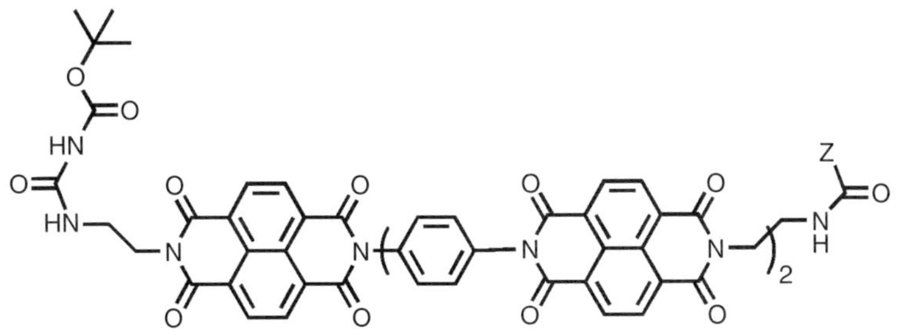

Figure 8.8 Second-generation anion-π slides. Compound **16**: Z = NHBoc; **17**: Z = $NH_3{}^+$TFA.

Matile recognized a problem in his rigid rod design. The hydrophilic residues in the hydrophobic slab of the membrane, while a proven effective ion relay, would obviously be incompatible with the surrounding alkyl chains of the bilayer's phospholipids. Assuming that the octa(*p*-phenyl) backbone inserted normal to the bilayer axis, the hydroxyl residues would reside orthogonal to that axis. A propylene spacer [77] was therefore added to the second generation of Matile's AmB-inspired channels. The new channels showed activity that increased by 100-fold. It remains unclear if the improved activity is due to a greater amount of bilayer insertion, a more stable and more efficient pore, or some combination thereof.

A third generation of Matile's oligophenylene channels took advantage of the π-acid/base properties of polyphenylene. Cation–π interactions are now well known and documented [78, 79]. Anion–π interactions are far less so and remain controversial [80]. Nevertheless, Matile's extended π-systems were already proved to shuttle Cl^- anions across a membrane. It is known that cation–π interactions are favored by increased π-basicity of the arene. Matile reasoned that electron-deficient arenes might function as an "anion–π-slide" [81]. Compounds **16** and **17** (Figure 8.8) were shown to be effective transporters of anions in the following selectivity order: $Cl^- > F^- > Br^- > I^-$. This selectivity order implies strong anion binding.

Two oligo-(*p*-phenylene)-*N*,*N*-naphthalenediimide (O-NDI) rod dimers of the compounds **16** and **17** were synthesized. They are shown in Figure 8.9 as **18** and **19** [82]. While their activity was poor, presumably because of low solubility and aggregation issues, they were reported to act as π-acidic semiconductors [83]. Triglutamate tails were added to increase solubility in the aqueous phase, and these compounds were tested and shown to be active chloride-selective channels. The anionic preference comports with the concept of an anion–π shuttle system.

8.4.6 Cholapods as Ion Transporters

Recent developments in electroneutral anion transporters include a family of compounds that have been called *cholapods* [84]. Cholapods are cholic acid derivatives and share the general structure shown in Figure 8.10.

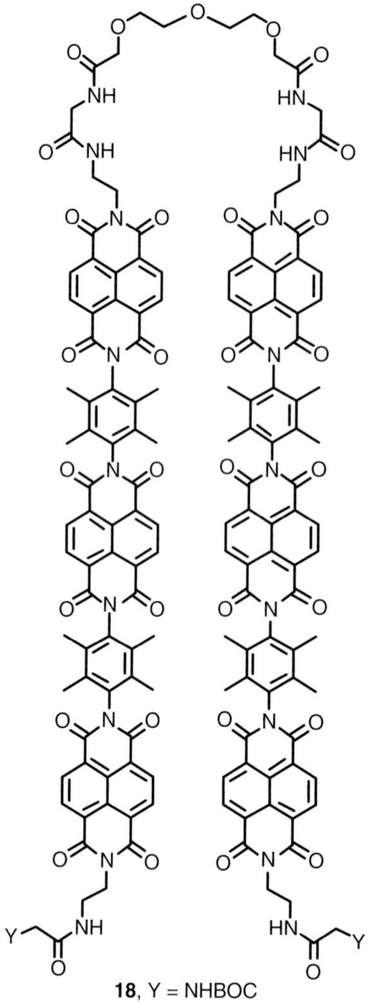

18, Y = NHBOC
19, Y = NH_2

Figure 8.9 Cation–π scaffolds: electron-deficient oligo-(*p*-phenylene)-*N*,*N*-naphthalenediimide (O-NDI) rods (**18**, **19**).

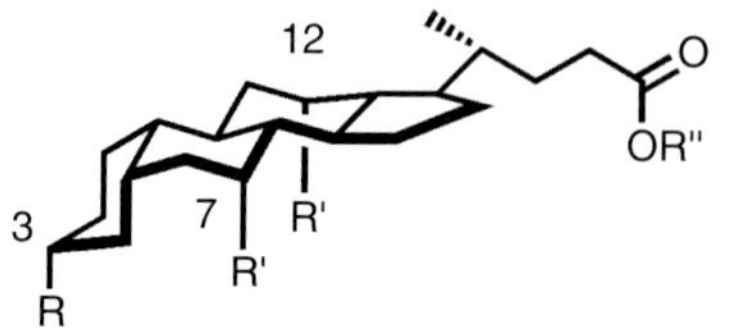

Figure 8.10 The cholapod backbone.

The substitutions at C-7 and C-12 are typically the same and provide rigid H-bond donation/anion recognition sites. The steroidal backbone and axial positions afford a preorganized anion-binding scaffold and inhibit intramolecular hydrogen bonding. The substituent C-3 either participates in anion recognition by providing an additional hydrogen bonding site(s) or is electron withdrawing to increase the acidity of the donors. Finally, R″ has been extended in chain length from methyl to eicosyl to increase lipophilicity and solubility in organic solvents.

Cholapods exhibit a high affinity for chloride and bromide. Measured K_A (M^{-1}) values were reported to be between 10^6 and 10^{11}. For each cholapod, the association constant for the binding of chloride was at least 1 power of 10 greater than that for bromide. These values have been obtained [85] using Cram's extraction method, by interface voltammetry [86], by fluorescence [85], by using an ion-selective electrode (ISE) [87], and by NMR titrations [88]. The strong binding was surprising because association constants for previously reported electroneutral synthetic anion binders were 7–8 orders of magnitude lower than for the cholapods when assayed using the same analytical techniques. The cholapods incorporating thiourea residues at C-7 and C-12 were found to have binding constants ranging from 10^8 to 10^9. The thiourea derivatives were superior mediators of Cl^- transport from vesicles as detected by both fluorescence and ISE methods. These compounds are presumed to function by a carrier mechanism.

8.4.7 Transport Mediated by Isophthalamides and Dipicolinamides

Smith and Lambert [89] have reviewed anion transport results related to cholapods and isophthalamide derivatives that transport ion pairs [90]. Smith and coworkers have also explored membrane-active compounds that mediate phospholipid translocation. Although "flip-flop" of membrane monomers is a well-recognized biochemical process, it formally involves transport across a membrane [91]. For a membrane monomer such as phosphatidic acid, this constitutes anion transport.

8.5 The Development of Amphiphilic Peptides as Anion Channels

Our laboratory had extensively explored the function of tris(macrocycle) and hydraphile channels as cation transporters [8]. Much of this work was under way when the Nobel Prize-winning solid-state structure of the KcsA channel appeared [23]. Shortly thereafter, we began to contemplate whether similar success was possible in the development of an anion channel. In thinking about a simple design, we considered primitive channels or pores. There is no fossil record of the earliest cells, and multiple theories have been presented to account for the emergence of life [92].

Notwithstanding these extensive and thoughtful conceptualizations of emergent life, we began our efforts with the presumption that an enclosed or encapsulated system must have existed at some very early stage of evolution. Such an enclosure was probably leaky at the outset, and both nutrients and waste products could pass through the tenuous external barrier. As this barrier improved, the need for regulation and the associated transporters emerged. The chemical details of such a system are obviously obscure. We therefore focused on the modern cell and used a phospholipid membrane monomer as a model.

8.5.1
The Bilayer Membrane

We often describe a membrane monomer as an amphiphile having a headgroup and a hydrophobic tail. In fact, phospholipids have three distinct polarity regimes. The headgroup contains two regimes: it is a phosphatidic acid or is an ammonium ion, an amino acid, or sugar terminated. The various terminal elements of the phospholipid are charged and extensively hydrated. The glyceryl ester regime is of intermediate polarity [93] and links these charged elements to the hydrocarbon chains that comprise the fatty acid tails. In a "typical" bilayer, the composite headgroups on opposite membrane surfaces are, together, similar in dimension to the insulator regime or "hydrocarbon slab" at the center. The latter is usually stated to be 30–35 Å thick. Of course, the actual dimension will depend at least on the various phospholipid monomers present, their headgroups and fatty acyl chains (whether components such as steroids and sphingolipids are present), and the extent of hydration.

Figure 8.11 shows the solid-state structure [Cambridge structural database (CSD) code: LAPETM10] of didecanoylphosphatidylethanolamine. The three polarity regimes are apparent from top to bottom. The cationic amino group protons are not visible, but this positively charged residue is attached via an ethylene unit to the phosphoryl unit. This, in turn, is esterified to one of glycerol's primary hydroxyl groups. The remaining 1° and the single 2° hydroxyl groups are esterified to dodecanoic acid residues. The undecyl chains are shorter than those that one typically encounters in membranes, but they adequately illustrate the nonpolar portion of the bilayer.

8.5.2
Initial Design Criteria for Synthetic Anion Transporters (SATs)

The notion was to mimic the phospholipid monomer structure in order to create a Cl^--selective pore. For the reasons described in more detail below, three elements were incorporated. These are (i) hydrocarbon tails, (ii) a midpolar regime mimic and linker, and (iii) a peptide headgroup. The last was envisioned as both the polar headgroup and the selectivity element. In principle, the peptide headgroup would either be cationic or anionic depending on whether the tail chains were linked to the C- or N-terminus of the peptide. For synthetic reasons, the hydrocarbon tails were linked to the peptide's N-terminus, leaving a free carboxyl group at the structure's opposite end. A decision was made to esterify the carboxyl to prevent ionization. It was later recognized, as discussed below, that the identity of the C-terminal residue played a role in the transport function.

Inspiration for the peptide was drawn from the sequences of C-peptide [94] and the putative conductance pore of the ClC protein chloride transporter [95].

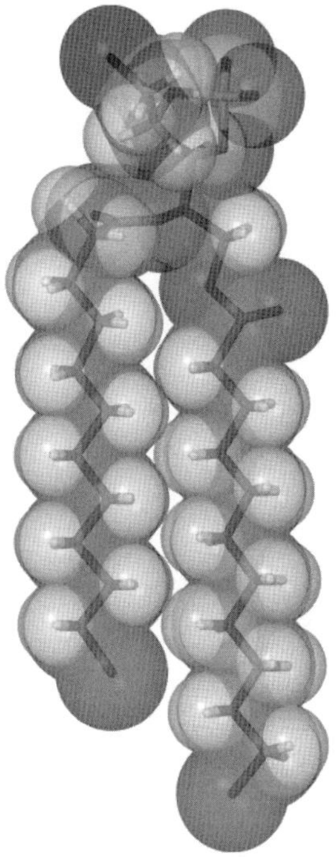

Figure 8.11 Solid-state structure of didodecanoylphosphatidylethanolamine from (CSD: LAPETM10), rendered in the stick metaphor overlayed by the van der Waals surfaces.

The simplest equivalent of the sequence Gly-Xxx-Xxx-Pro was Gly-Gly-Gly-Pro. Triglycine has the advantages of being commercially available and lacking any stereochemical issue. Intuition and an examination of molecular models suggested that a pore of suitable size to transport hydrated Cl^- could be formed from two V-shaped peptides. Proline would constitute a bend in any peptide sequence, so the initial goal became the peptide sequence $\sim$(Gly)$_3$-Pro-(Gly)$_3\sim$ (i.e., GGGPGGG). Molecular models suggested that two such sequences adjacent to and approximately reflecting each other would comprise an opening of 7–8 Å. The crystallographic diameter of chloride anion is reported to be 3.5 Å [20], but the hydrated diameter is calculated to be 6.5 Å [96].

8.5.3 Synthesis of the N-Terminal Anchor Module

One possible source of the hydrocarbon tails was natural lipids with a replaced headgroup. Cleavage of residues by using phospholipase or other enzymatic methods was considered but discarded in favor of a completely synthetic modular approach. Thus, the hydrocarbon tails and the midpolar regime would be incorporated in a single step by using a diamine and a diacid. Many dialkylamines are commercially available, and others may readily be prepared by the formation of an amide followed by reduction ($R^1{-}COOHd + H_2N{-}R^2 \rightarrow R^1{-}CO{-}NH{-}R^2 \rightarrow R^1{-}CH_2{-}NH{-}R^2$). The reaction of a dialkylamine and an anhydride is typically a clean and facile reaction. We found that the reaction of diglycolic anhydride with dioctadecylamine

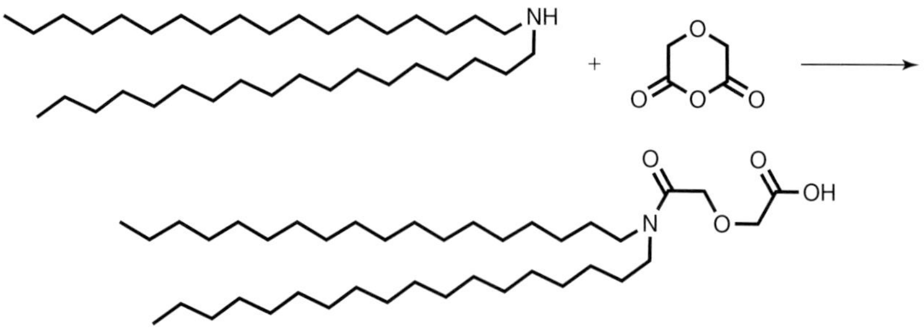

Figure 8.12 Synthesis of the anchor and linker modules of heptapeptide anion transporters.

in hot toluene in the absence of any catalyst required only solvent evaporation for workup [97] (Figure 8.12).

8.5.4 Preparation of the Heptapeptide

The C-terminal tripeptide was prepared as a benzyl ester (~GGG-OCH_2Ph). It was treated under standard peptide-coupling conditions (1-(3-dimethylaminopropyl)-3-ethylcarbodiimide hydrochloride) with $(C_{18}H_{37})_2NCOCH_2OCH_2COOH$ to give $(C_{18}H_{37})_2NCOCH_2OCH_2CO$–NH–Gly-Gly-Gly–$OCH_2Ph$. The ester group was removed by hydrogenolysis in EtOH. The fragment thus obtained comprised the anchor chains, the midpolar regime mimic, and three of the seven amino acids that would comprise the headgroup. The tetrapeptide required to complete the sequence was prepared by coupling H_2N–Gly-Gly-Gly–OCH_2Ph with *N*-Boc-Pro–OH to give Boc-Pro-Gly-Gly-Gly–OCH_2Ph. The Boc group was cleaved by treatment with HCl in dioxane. The two fragments were coupled using the same conditions noted above, with the resulting formation of $(C_{18}H_{37})_2NCOCH_2OCH_2CO$–NH–Gly-Gly-Gly-Pro-Gly-Gly-Gly–OCH_2Ph (**20**). The sequence is illustrated in Figure 8.13.

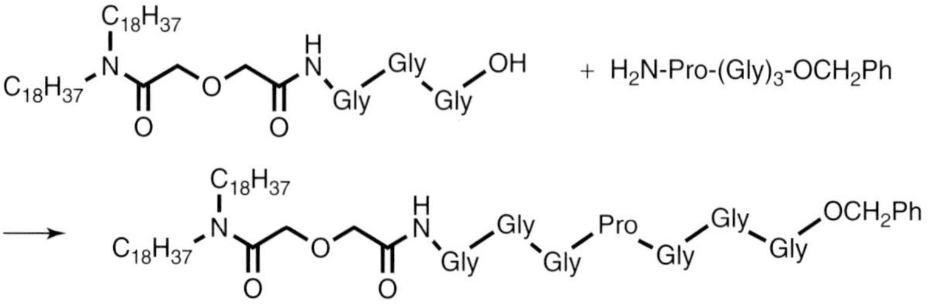

20

Figure 8.13 Coupling of anchor + linker and C-terminal tripeptides.

8.5.5
Initial Assessment of Ion Transport

Our initial success with synthetic anion transporters [98] was realized with the molecule $(C_{18}H_{37})_2NCOCH_2OCH_2CO$–$(Gly)_3$-Pro-$(Gly)_3$–$OCH_2Ph$ (**20**). We have used planar bilayer voltage clamp [99], ISE [100], and fluorescence methods [99, 101] to characterize these synthetic anion transporters, which we abbreviate as SATs. Release from liposomes mediated by the ionophore may be monitored in a variety of ways. If a chloride ISE is used, vesicles are prepared so that chloride is present in the liposomes but not in the surrounding buffer. When the ionophore is added, an increase in chloride ion concentration is detected as an electrical response in the bulk aqueous suspension. Ideally, this response is concentration dependent, as shown in Figure 8.14b. In the planar bilayer experiment, the ionophore is introduced to a membrane separating two aqueous ion reservoirs. A voltage is applied, and the response of the membrane, normally an insulator, is recorded. Channel activity is apparent in an open–close response. Because the channel (or pore) is of a particular size, the openings are uniform. When the channel is closed, no ions flow. This response is shown in Figure 8.14a.

Although several techniques were used to assay transport, the most critical was to obtain planar bilayer conductance (the "BLM" experiment) data. The first analysis was performed on $(C_{18}H_{37})_2N$–$COCH_2OCH_2CO$–$(Gly)_3$-Pro–$(Gly)_3$–OCH_2Ph (**20**) [99]. The data showed clear evidence for open–close behavior, indicating the formation of a conductance pore. Second, a plot of current as a function of voltage ("the I–V curve") showed that the pore or channel was selective for Cl^- over K^+ by at least 10-fold. Third, and unexpectedly, the conductance behavior showed voltage-dependent gating [99]. Thus, the open–close behavior altered as the voltage changed, such that the channels remained open all the time at more positive voltages and showed far fewer openings or none at all at lower or negative potentials.

Release of an anionic dye, such as carboxyfluorescein (CF), may be monitored by fluorescence methods. When liposomes are loaded with the dye, the high concentration causes self-quenching. If the ionophore mediates the passage of dye through the bilayer, the dye concentration in the external medium will be much lower than in the interior of the liposome. Quenching will no longer occur, and the fluorescence will be proportional to the dye release. This experiment is often called "*fluorescence dequenching*," and the observed absorbance will generally resemble the right panel traces of Figure 8.14. CF and Cl^- are clearly very different anions. Nevertheless, the results obtained in separate experiments for CF^- and Cl^- transport were consistent when the N-terminal (R^1) chains were either *n*-decyl or *n*-octadecyl.

Transport efficacy for $(C_{18}H_{37})_2N$–$COCH_2OCH_2CO$-$(Gly)_3$-Pro-$(Gly)_3$–OCH_2Ph (**20**) was initially assayed by measuring Cl^- release from phospholipid vesicles. In the presence of **20** ([**20**] = 24–154 μM), the anion was released in a concentration-dependent manner. Compound **21** [$(C_{18}H_{37})_2NCOCH_2OCH_2CO$–$(Gly)_3$-Leu-$(Gly)_3$–$OCH_2Ph$] was prepared as a control. Further, proline appears

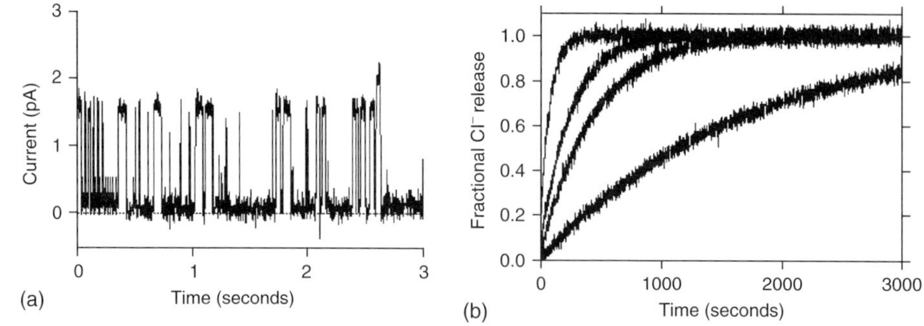

Figure 8.14 (a) Open–close behavior of a SAT molecule observed in a planar bilayer conductance experiment. (b) Concentration-dependent (increasing from bottom to top) fractional chloride release forms liposomes mediated by **20**.

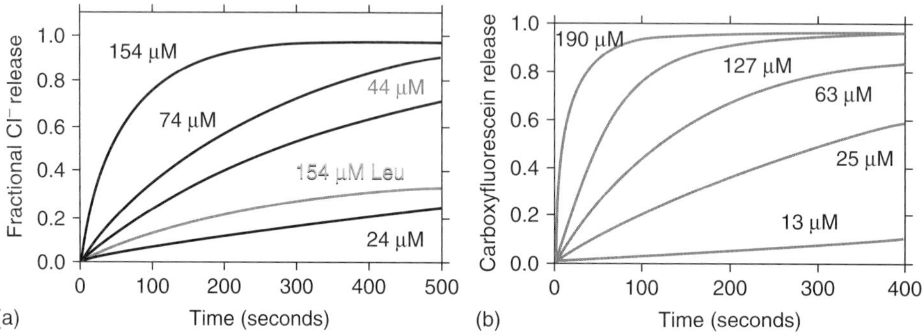

Figure 8.15 (a) Fractional chloride release mediated by **20** at concentrations of 24, 44, 74, and 154 μM and by **21** (Pro → Leu) at 154 μM. (b) Fractional carboxyfluorescein release mediated by **20** at concentrations of 13, 25, 63, 127, and 190 μM.

to be important in the ion-conducting peptide sequence. Transport of Cl^- by **21** at 154 μM was almost identical to that observed for **20** at 24 μM (~one-sixth the concentration). Leucine and proline have a similar number of carbon atoms, but proline is cyclic and leucine is open-chained. Moreover, proline typically enforces a bend in a peptide conformation, whereas leucine is more flexible. Additionally, the truncated SAT $(C_{18}H_{37})_2NCOCH_2OCH_2CO{-}(Gly)_3{-}OCH_2Ph$ failed to transport anions [102]. The left panel of Figure 8.15 shows this result.

The results shown in Figure 8.15 assess anion transport differently. Chloride release in the left panel was observed by ISE methods. At the right, the anion was CF, and its release from vesicles was observed by fluorescence dequenching. The use of the fluorescent dye lucigenin has recently been reported as a means to

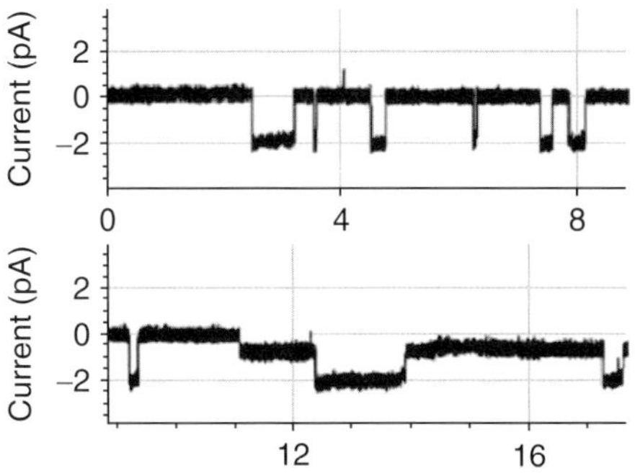

Figure 8.16 Typical raw data obtained in the planar bilayer conductance experiment. The abscissa is time, usually in seconds as shown here, and current is plotted on the ordinate, usually in picoamperes as shown here.

assay Cl^- transport [103]. Lucigenin within vesicles is fluorescent but undergoes collisional quenching when Cl^- is transported into the liposome.

A comparative study was undertaken to determine how transport assessments differed as a function of analytical method [104]. Seven heptapeptide derivatives of the general form $R^1_2N–COCH_2OCH_2CO–(Gly)_3$-Pro-$(Gly)_3–OR^2$ were prepared. Four of the compounds contained twin *n*-hexyl, *n*-decyl, *n*-tetradecyl, and *n*-octadecyl (R^1) groups at the N-terminus and a C-terminal (R^2) benzyl group. Three other compounds contained the C_6, C_{12}, and C_{18} alkyl substituents (R^1 groups), but the C-terminal ester was *n*-heptyl. Chloride ion release from or flux into vesicles was assessed for this family of compounds by Cl^- release (ISE), by CF dequenching, and by chloride quenching of lucigenin. The conclusion reached in this study was that the measurement of Cl^- transport by ISE or lucigenin quenching gave generally similar, although not identical, results.

Later studies involved a more detailed analysis of the planar bilayer behavior of several SAT molecules [105]. An example of the type of data obtained in these experiments is shown in Figure 8.16. The open–close behavior activity of the channel is recorded over time, indicated in seconds (0–16) on the abscissa in the two traces. When the channel opens, the current deflects from 0 to about −2 pA in the figure shown. It is consistently this value when the channel is open in the upper trace, but there is a brief period (near 12 s) in the lower trace when another conductance state is apparent. The conductance of the channel can be calculated from the applied potential and the observed current. The pore size can then be estimated from the conductance [106].

The BLM experiment is the definitive indication of channel activity. Alternatives noted above are to record the release of Cl^- or CF^- from liposomes. Detecting Cl^- by using an ISE is convenient. The advantage of studying CF release is that CF^- release from liposomes may be detected reliably over a wide concentration range [101]. When the concentration range studied is sufficient, typically 10-fold, application of the Hill equation [107] gives information about cooperativity. Thus, studies performed on a range of amphiphilic heptapeptides suggested that the aggregation state of the pore was ~2–3. This corresponds reasonably well with the original design, which called for two monomers to form a pore. The range of 2–3 suggests that larger pores also form.

8.6 Structural Variations in the SAT Modular Elements

Although the initial design considered the C-terminal ester as a protecting group, its effect on transport soon became apparent. Thus, we consider that the synthetic anion transporter molecules possess the four structural modules described above. These are (i) the N-terminal hydrocarbon tails, (ii) the diacid connector element, (iii) the peptide, and (iv) the C-terminal ester or amide residue. In the first SAT studied, $(C_{18}H_{37})_2N–COCH_2OCH_2CO–(Gly)_3$-Pro-$(Gly)_3–OCH_2Ph$ (**20**), the units were (i) dioctadecylamine, (ii) diglycolic acid, (iii) the peptide GGGPGGG, and (iv) benzyl ester. Systematic variation of the dialkyl chains from methyl to octadecyl showed that transport increased as the chains were shortened. This seemed counterintuitive. Further scrutiny showed that transport increased at the expense of selectivity, that is, both Cl^- and Na^+ ions were traversing the pore [102]. A range of dicarboxylic acids were examined, and when X in ~CO–X–CO~ had conformational restrictions, transport diminished [108]. Variations in the peptide sequence were significant and are discussed below.

8.6.1 Variations in the N-Terminal Anchor Chains

The dual hydrocarbon chains (R^1) that comprise the N-terminus of $(R^1)_2NCOCH_2OCH_2CO$-$(Aaa)_7$-OR^2 have been varied systematically. The twin N-terminal sidechains (R^1 in the structure above) studied were *n*-propyl, *n*-hexyl, *n*-octyl, *n*-decyl, *n*-dodecyl, *n*-tetradecyl, *n*-hexadecyl, and *n*-octadecyl. Fractional chloride release was determined (ISE) in 200 nm vesicles (0.31 mM) in the presence of each compound (65 μM). We expected the compound having the longest dialkyl chains to be the most effective in forming pores and thus in transporting Cl^-. As noted above, the results were quite different, as shown in the graph of Figure 8.17.

The most active ion transporters in the family $R^1_2N–COCH_2OCH_2CO–G_3PG_3–OCH_2Ph$ were those in which R^1 = octyl, decyl, or hexyl, in diminishing order of reactivity. The only shorter chain tested, propyl, showed low activity. The SATs having the longest chains studied, that is, C_{14}, C_{16}, and C_{18}, were very poorly active and were exceeded only slightly by C_{12}. Our initial dioctadecyl SAT, that is, **20**, was among the least active compounds studied. The release of CF was undertaken in an effort to compare the trend in ion release with anchor chain length. The experimental conditions differed from those used above, and the *n*-propyl and *n*-hexyl compounds were absent from the study. Nevertheless, nearly parallel results were obtained and are illustrated in the graph of Figure 8.18. In the CF release case, ion release diminishes in a nearly linear manner from *n*-octyl to *n*-octadecyl.

Further study of the bis(decyl) compound revealed that anion selectivity was essentially lost [102] in the shorter-chained compounds. Indeed, planar bilayer voltage clamp experiments showed that both Cl^- and Na^+ were transported by the C_{10} SAT. The conductance (at 89.7 μM) for **20** (C_{18} SAT) was 349 ± 69 pS compared

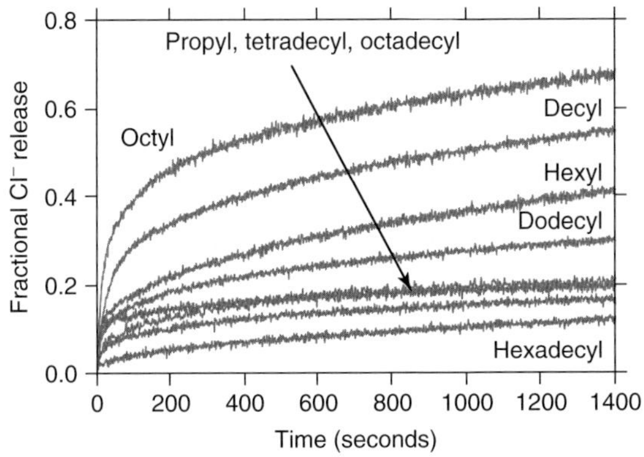

Figure 8.17 Fractional chloride ion release from vesicles for $R^1{}_2N\text{–}COCH_2OCH_2CO\text{–}G_3PG_3\text{–}OCH_2Ph$ as a function alkyl group R^1.

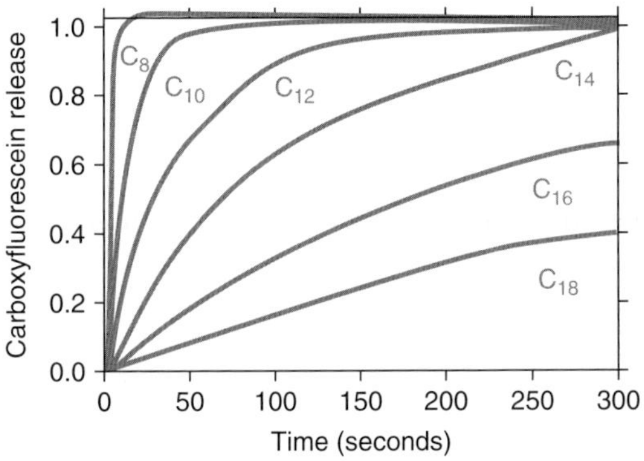

Figure 8.18 Carboxyfluorescein release from vesicles for $R^1{}_2N\text{–}COCH_2OCH_2CO\text{–}G_3PG_3\text{–}OCH_2Ph$ as a function alkyl group R^1.

to the C_{10} SAT (at 5.2 μM) of 573 ± 13 pS. Dextran sizing experiments (method [109]) showed that the apparent pore diameter was similar in both cases. The initial assumption that longer chains alone would form "better" pores is flawed. The formation of a conductance pathway must depend at least on the stability of the pore once formed, on how readily the monomers insert into the bilayer, and on how effectively they organize within it. In fact, the process must also involve the C-terminal residue, initially envisioned only as an element to prevent ionization.

8.6.2
Anchoring Effect of the C-Terminal Residue

In the initial design, the C-terminal ester was intended to protect the carboxyl group (i.e., prevent ionization) and reduce polarity. As above, a family of SAT molecules

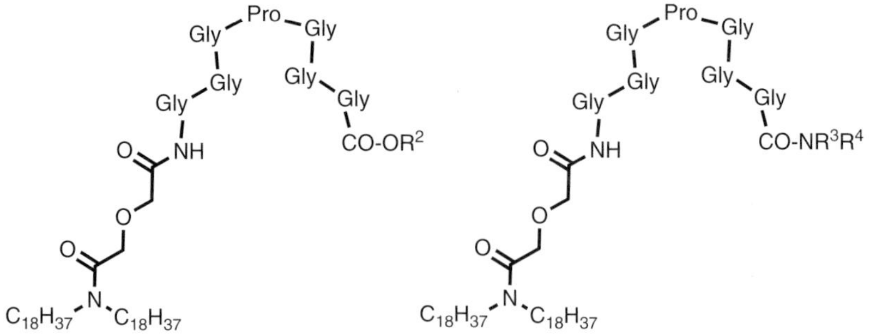

22, $R^2 = OCH_3$

23, $R^2 = OCH(CH_3)_2$

24, $R^2 = O(CH_2)_6CH_3$

25, $R^2 = O\text{-c-}C_6H_{11}$

26, $R^2 = O(CH_2)_9CH_3$

27, $R^2 = O(CH_2)_{17}CH_3$

28, $R^3 = R^4 = (CH_2)_6CH_3$

29, $R^3 = H$; $R^4 = (CH_2)_9CH_3$

30, $R^3 = R^4 = (CH_2)_9CH_3$

31, $R^3 = H$; $R^4 = (CH_2)_{17}CH_3$

32, $R^3 = R^4 = (CH_2)_{17}CH_3$

Figure 8.19 Structural variations at Y (**22–32**), the C-terminus of $(C_{18}H_{37})_2NCOCH_2OCH_2CO$–(Gly)$_3$-Pro-(Gly)$_3$–Y.

having the general structure $(C_{18}H_{37})_2NCOCH_2OCH_2CO$–(Gly)$_3$-Pro-(Gly)$_3$–OR^2 was prepared. The ester chains (OR^2) included OCH_2CH_3 (**22**), $OCH(CH_3)_2$ (**23**), $O(CH_2)_6CH_3$ (**24**), OCH_2–*c*-C_6H_{11} (**25**), $O(CH_2)_9CH_3$ (**26**), and $O(CH_2)_{17}CH_3$ (**27**). The ester residue was replaced in six cases by amides (**28–32**, $YR^3 = NR^4$). The C-terminal amides (NR^4) were $N[(CH_2)_6CH_3]_2$ (**28**), $NH(CH_2)_9CH_3$ (**29**), $N[(CH_2)_9CH_3]_2$ (**30**), $NH(CH_2)_{17}CH_3$ (**31**), and $N[(CH_2)_{17}CH_3]_2$ (**32**) (Figure 8.19).

As above, the results were surprising (shown in Figure 8.20) [103, 110]. The esters studied ranged from as short as propyl to as long as octadecyl. The ion release data revealed that by far, the most active compound in the family of esters, $(C_{18}H_{37})_2NCOCH_2OCH_2CO$–(Gly)$_3$-Pro-(Gly)$_3$–OR^2, was R^2 = *n*-heptyl (**24**). Normal alkyl ester residues both longer and shorter were significantly less active. Heptyl, cyclohexylmethyl, and benzyl all have seven carbon atoms, but the cyclic compounds proved to be far less effective mediators of Cl^- release from liposomes. The results with C-terminal amides showed a similar trend. Thus, the amides having the shortest (diheptyl, **28**) and longest (octadecyl **31**, **32**) chains were poorer transport mediators than didecylamide (**29**).

The role for which the C-terminal residue was initially envisioned was simply to prevent ionization of the carboxyl group. Benzyl was chosen so that it could be removed by hydrogenolysis without affecting the peptide chain or other modules. Clearly, however, its hydrophobicity permits it to insert into the bilayer and serve as a secondary anchor. We surmise that when the alkyl ester chain is short, it is a poor anchor and the headgroup conformation of the heptapeptide is conformationally

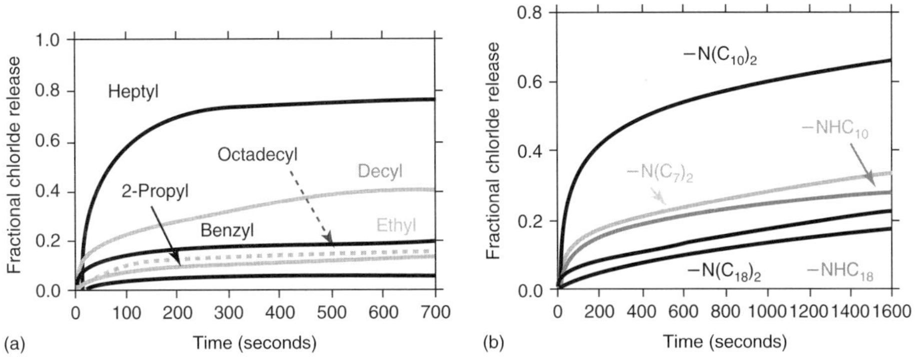

Figure 8.20 (a) Fractional chloride release mediated by **22–27** at concentrations of 65 μM. Release was measured using 200 nm phospholipid vesicles ((lipids) = 0.20 mM). (b) Fractional carboxyfluorescein release mediated by **28–32** at concentrations of 65 μM. Release was measured using 200 nm phospholipid vesicles ((lipids) = 0.20 mM).

flexible. When the C-terminal chain is long, this anchoring process should be better in the sense that embedding more deeply within the bilayer should afford greater conformational stability. This seems plausible, but it ignores the potential for aligning with the two other octadecyl chains at the SAT's N-terminus. If all three chains align tightly, mutually solvate each other, and largely exclude water, the molecule will be unable to form a conducting pathway. The *n*-heptyl ester seems to constitute a compromise in this respect. It has sufficient length to provide a stable C-terminal anchor, but it is does not interact strongly enough with the two octadecyl chains that it collapses the conductance pathway. Additional consideration of these questions is presented in Section 8.6.9.

The preceding discussion ignores the questions of association in the bulk phase and insertion into the bilayer. For the longer chained, N-terminal anchor chains, the possibility exists that aggregation will occur in the bulk phase. To the extent that the monomers aggregate, the proportion of them that insert from the aqueous suspension into the bilayer will be diminished.

8.6.3 Studies of Variations in the Peptide Module

We found in our initial studies [99] that when the fourth amino acid (Aaa) in $(C_{18}H_{37})_2NCOCH_2OCH_2CO$–$(Gly)_3$-Aaa-$(Gly)_3$–$OCH_2Ph$, was leucine rather than proline, Cl^- release from liposomes diminished by more than sixfold (see Section 8.5.5). We inferred from this that the bend in the peptide provided by proline was critical, as expected from the protein sequence information that initially inspired its use. This result begged the question of how sensitive the peptide sequence was to other substitutions in the fourth amino acid position within the heptapeptide.

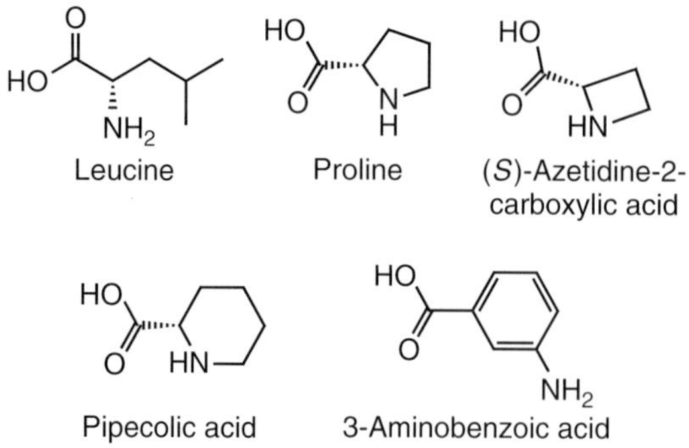

Figure 8.21 Structural variations in the fourth position of the SAT heptapeptide module.

8.6.3.1 Structural Variations in the Heptapeptide

It was found in early work that when the anchor chains of $(C_{18}H_{37})_2NCOCH_2OCH_2CO$–$(Gly)_3$-Pro-$(Gly)_3$–$OCH_2Ph$ were changed from *n*-octadecyl to *n*-decyl, ion transport was enhanced but at the expense of Cl^- selectivity. The greater dynamic range of ion release encouraged us to study the bis(decyl) derivatives in our assessment of amino acid variations. Thus, the compounds studied had the general structure $(C_{10}H_{21})_2NCOCH_2OCH_2CO$–$(Gly)_3$-Aaa-$(Gly)_3$–$OCH_2Ph$. The acids substituted for proline were leucine, (*S*)-azetidine-2-carboxylic acid, pipecolic acid, and 3-aminobenzoic acid. The relative stereochemistry was the same in all applicable cases. Measurement of fractional chloride release from liposomes was used to evaluate the structural changes. The results were somewhat surprising in that there was little variation among the SATs in the bis(decyl) series.

Comparisons among the G_3ProG_3 and G_3PipG_3 sequences having decyl or octadecyl chain lengths proved interesting. First, essentially identical ion release profiles were obtained for $(C_{18}H_{37})_2NCOCH_2OCH_2CO$–$G_3ProG_3$–$OCH_2Ph$ (**20**) and $(C_{10}H_{21})_2NCOCH_2OCH_2CO$–$G_3ProG_3$–$OCH_2Ph$ but only when the longer chained compound was present in a concentration ~400-fold greater than the shorter chained SAT (Figure 8.21). When $(C_{18}H_{37})_2NCOCH_2OCH_2CO$–$G_3ProG_3$–$OCH_2Ph$ was compared to $(C_{18}H_{37})_2NCOCH_2OCH_2CO$–$G_3PipG_3$–$OCH_2Ph$, the latter was found to be less active in mediating Cl^- release by a factor of 5–6.

Perhaps the most surprising results were obtained with SATs having anchor chains of unequal length. Two such structures (**33** and **34**) were prepared and compared to a symmetrical analog (**35**). Studies of fractional Cl^- release from vesicles mediated by **33–35** were conducted according to typical lab procedures. It was found that when the N-terminal anchor chains were ethyl, decyl (**34**), or didecyl (**35**), 50–60% of the available Cl^- was released during the 1500 s duration of the experiment (Figure 8.22). During this same interval, the ethyl, decyl-GGGLGGG derivative **33** showed quantitative Cl^- release. In fact, essentially complete Cl^- release was achieved by **33** in less than 200 s. A cooperativity study undertaken on **33** suggested that the active aggregation state was ~4 rather than the more typical two to three. Larger pores are expected to pass ions more readily, but no information is available on changes in selectivity that might result.

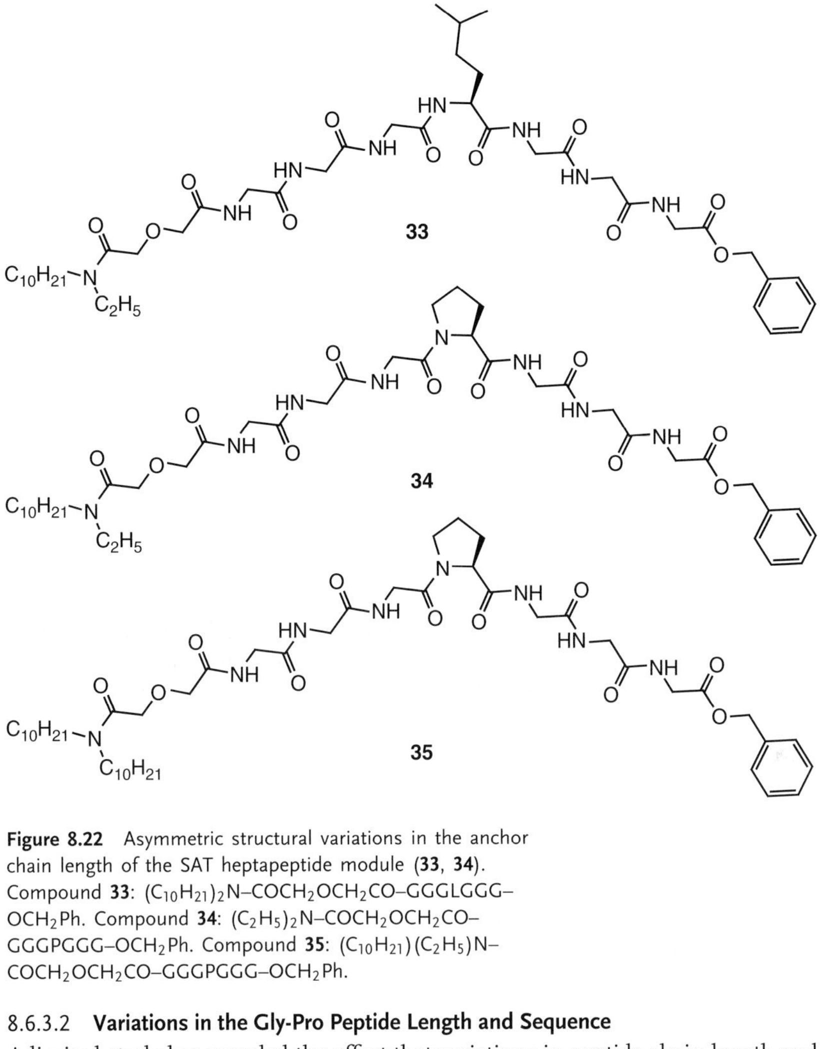

Figure 8.22 Asymmetric structural variations in the anchor chain length of the SAT heptapeptide module (**33**, **34**). Compound **33**: $(C_{10}H_{21})_2N–COCH_2OCH_2CO–GGGLGGG–OCH_2Ph$. Compound **34**: $(C_2H_5)_2N–COCH_2OCH_2CO–GGGPGGG–OCH_2Ph$. Compound **35**: $(C_{10}H_{21})(C_2H_5)N–COCH_2OCH_2CO–GGGPGGG–OCH_2Ph$.

8.6.3.2 **Variations in the Gly-Pro Peptide Length and Sequence**

A limited study has revealed the effect that variations in peptide chain length and proline placement have on Cl^- transport from vesicles [111]. Five sequences were explored: $\sim$(Gly)$_2$-Pro-(Gly)$_2\sim$ (**36**); $\sim$(Gly)$_2$-Pro-(Gly)$_4\sim$ (**37**); $\sim$(Gly)$_3$-Pro-(Gly)$_3\sim$ (**20**); $\sim$(Gly)$_4$-Pro-(Gly)$_2\sim$ (**38**); and $\sim$(Gly)$_4$-Pro-(Gly)$_4\sim$ (**39**). Fractional chloride release mediated by **36**, **37**, and **20** was poor and essentially indistinguishable. Compound **38** is the isomer of **37** and **20**, but its heptapeptide sequence places a tetraglycine fragment before the proline rather than diglycine or triglycine. Chloride ion release from vesicles mediated by the G_4PG_2 isomer is about twice that observed for either the G_2PG_4 or the G_3PG_3 SAT. The most dramatic improvement in chloride release was apparent when the G_4P sequence was retained but two additional glycine residues were added on the C-terminal side of the proline. This gave the peptide sequence G_4PG_4 (**38**). At an arbitrarily observed time point of

36
37
20
38
39

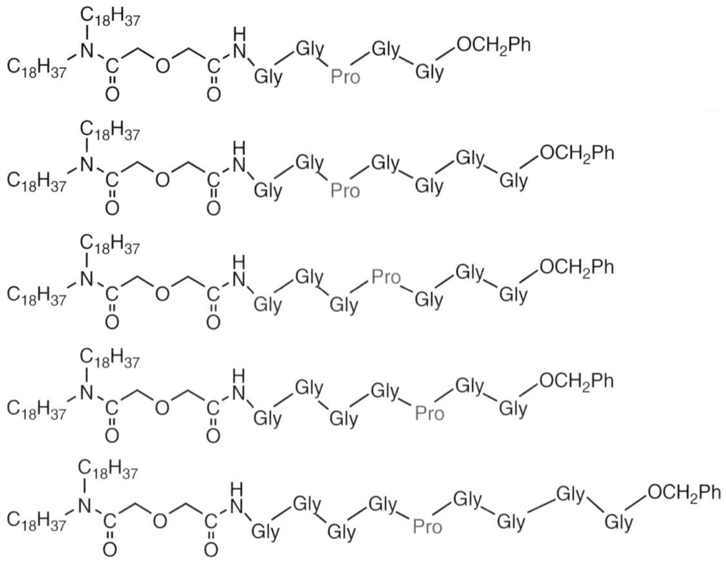

Figure 8.23 Structural and length variations in the $(Gly)_nPro(Gly)_m$ peptide module of SATs.

1200 s, Cl^- release mediated by G_4PG_4 (**38**) was 63% of theoretical value and for G4PG2 (**37**) it was only 34%. As noted, however, in both cases, the ion release was far greater than observed for **36**, **37**, or **20** (Figure 8.23).

Previous studies of heptapeptide SAT molecules suggested that the aggregation state for the active pore was two to three monomer units. A CF release study was undertaken to determine a Hill plot and to assess the aggregation state of the most active compound, nonapeptide (**39**). These experiments were performed over a CF concentration range of 0.05–9.9 μM or nearly 20-fold. The nearly linear slope observed for eight points had $r^2 = 0.97$ and a slope of 1.4. We interpreted this to mean that pores were forming from individual peptides as well as from dimers.

8.6.4
Variations in the Anchor Chain to Peptide Linker Module

The diacid linker module was designed to connect the N-terminal anchors with the peptide headgroup. Diglycolic acid was chosen in part because it is convenient to incorporate. A more important reason was the polarity correspondence it offered with a phospholipid membrane monomer. This is noted above in Sections 8.5.1–8.5.3. Figure 8.24 demonstrates this correspondence by comparing an ethanolamine-terminated diglycolamide with a phosphatidylethanolamine monomer (see also Figure 8.11). From the linker's terminus, the hexadecyl chains span 19–20 Å from the junction, denoted by an asterisk. The other distances, estimated from molecular models, are indicated numerically in Ångstrom units. The overall lengths from the charged headgroup to the omega-methyl

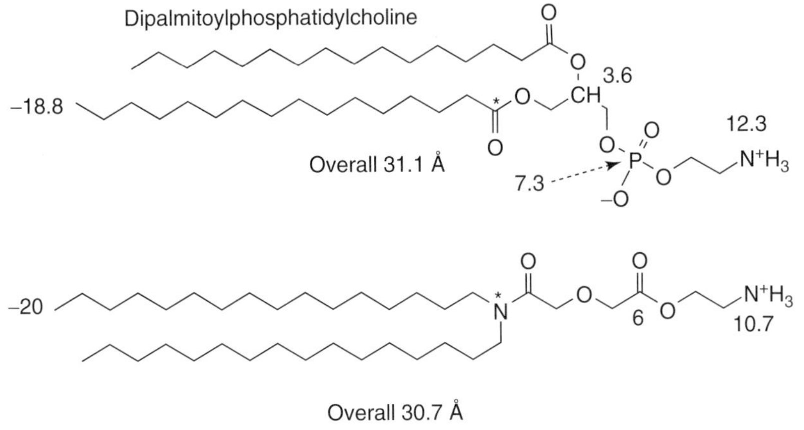

Figure 8.24 Comparison of SAT design with a phospholipid monomer.

group are nearly the same. The successful transport function of SAT molecules that incorporated a diglycolic acid module caused us to delay the examination of variations in this unit. As with all other modules, small modifications led to increased transport activity with no or only a modest cost in synthetic access.

This study required the preparation of nine amphiphilic heptapeptides of the form $(C_{18}H_{37})_2N$–Y–(Gly)$_3$-Pro-(Gly)$_3$–OCH_2Ph, in which Y represents a variation in the linker module. The diacids incorporated were succinic (**40**), glutaric (**41**), diglycolic (**20**), 3-thiaglutaric (**42**), *N*-methyliminodiglycine (**43**), isophthalic (**44**), and terephthalic acids (**45**). In addition, one amphiphilic peptide was constructed, which simply lacked a linker and one added an N-terminal glycine to the heptapeptide, formally incorporating acetic acid as a module. Their structures are, respectively, $(C_{18}H_{37})_2N$–(Gly)$_3$-Pro-(Gly)$_3$–OCH_2Ph (**46**) and $(C_{18}H_{37})_2N$–(Gly)$_4$-Pro-(Gly)$_3$–OCH_2Ph (**47**) [111].

Transport efficacy was assayed by detecting Cl^- release from vesicles prepared from a 7 : 3 w/w (2 : 1 mol/mol) mixture of 1,2-dioleoyl-*sn*-glycero-3-phosphocholine (DOPC) and 1,2-dioleoyl-*sn*-glycero-3-phosphate (DOPA). The HEPES buffer (pH 7) in which the liposomes were suspended contained KCl, and Cl^- was assayed by ISE methods [111]. The linker modules were evaluated by comparing the fractional chloride release (as a percentage of total chloride) at an arbitrary 1500 s time point. Release values ranged from 21 to 53%. The activity profile was as follows: succinic = glutaric > thiaglutaric > terephthalic > *N*-methyliminodiacetic = acetyl (G_4PG_3) > diglycolic > isophthalic = none (G_3PG_3) (Figure 8.25).

Our initial design for amphiphilic heptapeptides incorporated diglycolic acid because of its overall size, polarity, and synthetic convenience. The studies described above suggest that 3-thiaglutaric acid would be an equally effective choice, although what, if any, effect the sulfur atom may have on the system has not been investigated. Terephthalic acid was an effective midpolar regime mimic (isophthalic acid was

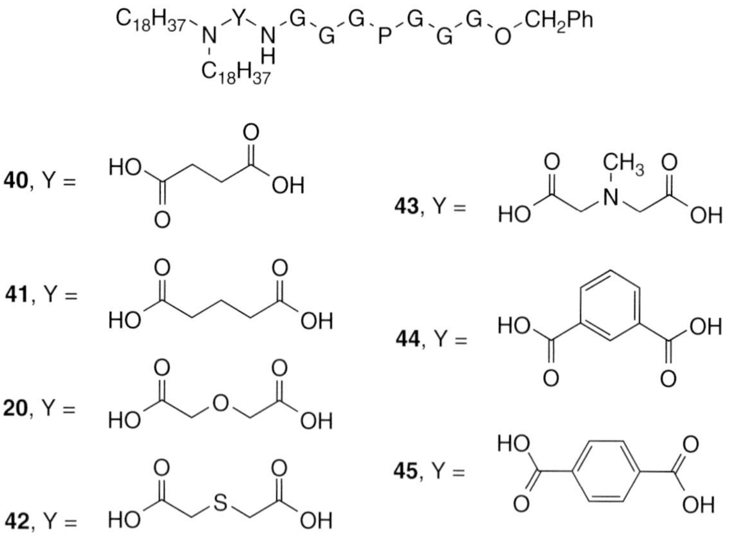

Figure 8.25 Variations in the midpolar/linker module of SATs.

not), but it does not form an intramolecular anhydride and is therefore less convenient to incorporate.

8.6.5
Covalent Linkage of SATs: *Pseudo*-Dimers

Several lines of evidence suggested that **20** and related SATs functioned as dimers to transport anions through a phospholipid bilayer. For example, Hill plots have a slope of ~2 corresponding to a dimeric aggregation state. Dextran block experiments suggested that the pore was 7–8 Å in diameter. This is consistent with a dimer structure as determined by molecular modeling. If the pore was dimeric, covalent linkage of two molecules of **20** should produce a compound having higher activity than **1** itself.

Two *pseudo*-dimers that were linked either at the C-terminus (**48**) of the molecule or the N-terminus of the heptapeptide (**49**) were designed. Limiting the connector in the latter case to the N-terminal end of the peptide was essential because linking the hydrocarbon anchor chains would be more difficult and likely to alter more dramatically the conformations available to the new structures. In the "N-linked twin" (**49**), covalent connection was established through the diglycolic acid residue, ~$COCH_2OCH_2CO$~. Replacement of the diglycolic ether oxygen by nitrogen would provide a convenient linking site as ~$COCH_2NYCH_2CO$~, where Y is the linking residue. An ethylene link was an obvious solution because ethylenediaminetetraacetic acid (EDTA) is well known and readily available. In practice, EDTA was converted into its bis(anhydride), which was treated with dioctadecylamine to give the dimer base having four anchor chains. Once the "base" was established, the twin peptide chains were elaborated much as had been done for the individual SAT molecules [112] (see structure **49**).

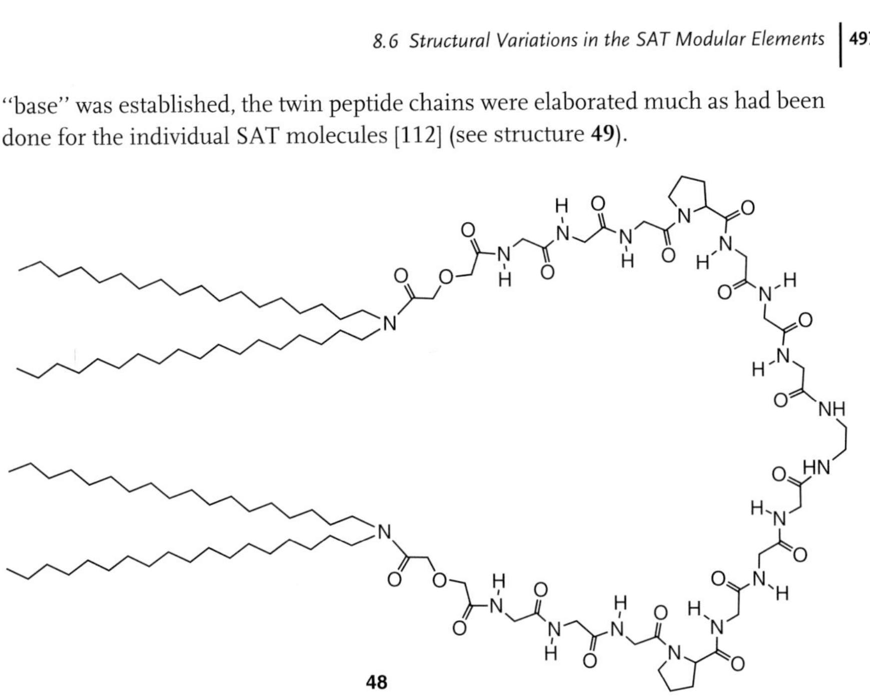

48

The preparation of **48**, the C-linked *pseudo*-dimer, was accomplished by preparing **20**, removing the ester function, and linking two molecules with ethylenediamine. In both **48** and **49**, nitrogen atoms replaced oxygens. The central link of **49** required the chains to be closer together in the lateral sense while the C-terminal connection in **48** produces a molecule with high flexibility and great length.

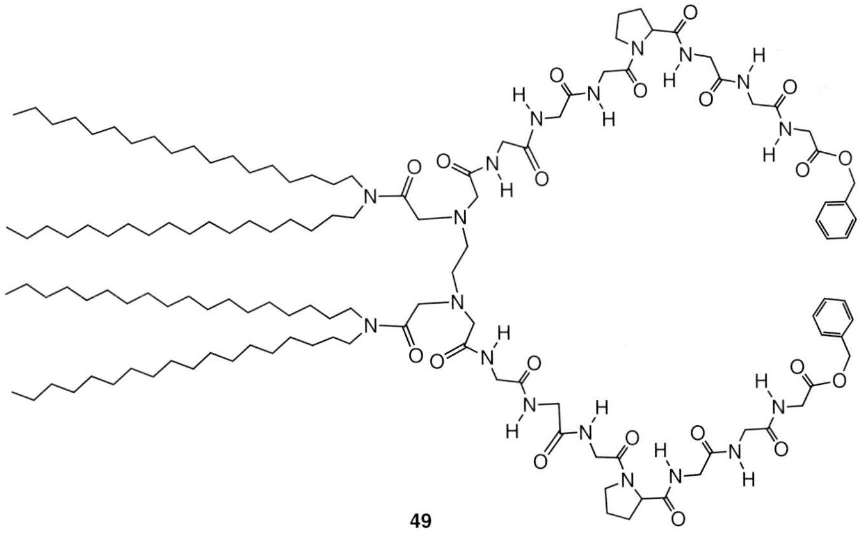

49

The critical test of function was accomplished by assessing anion release from vesicles by CF dequenching and by detecting Cl^- release by using an ISE. In the latter case, vesicles were prepared as usual from a 7 : 3 w/w (2 : 1 mol/mol) mixture of DOPC and DOPA. The internal buffer contained KCl, and the external buffer contained K_2SO_4. Chloride release was then recorded in separate experiments after addition of **20** (65 μM), **48** (32.5 μM), or **49** (32.5 μM). Ion transport mediated by N-twin **49** showed a somewhat steeper release curve than did the C-twin, **48**. In both cases, however, the fractional Cl^- release after 800 s was ~25%, whereas twice the concentration of monomer **20**, under otherwise identical conditions, released only ~15% of the total Cl^-.

The fact that half the concentration of **48** or **49** is more active than **20** does not confirm dimer formation by **20** within the bilayer. The data obtained from cooperativity studies suggested that **20** and its close relatives aggregate to the extent of two to three within the bilayer, and these results comport with those findings. Further, if dimer formation were inimical to function, **48** and **49** would transport Cl^- either poorly or not at all.

8.6.6
Chloride Binding by the Amphiphilic Heptapeptides

The selective transport of chloride requires recognition of this ion. The binding of Cl^- by a heptapeptide was determined by NMR methods in homogeneous $CDCl_3$ solution [110]. It was recognized that the long (typically octadecyl) hydrocarbon chains would enhance solubility in nonpolar solutions and potentially complicate the NMR spectra. It was not known with certainty whether the hydrocarbon chains would play a role in complexation. Even so, a decision was made to simplify the study by analyzing a heptapeptide having di-*n*-propyl, rather than di-*n*-octadecyl, N-terminal anchors. Subsequent studies confirmed that the hydrocarbon chain length was not an important variable [113]. The heptapeptide sequence contained six glycine residues whose chemical shifts required definition. It was therefore necessary to synthesize six SAT molecules. In each SAT, one of the glycine residues was deuterated. Study of these compounds permitted us to identify the signals contributed to the spectrum by each glycine and its associated NH proton.

Our initial studies revealed a binding constant between Bu_4NCl and **50** in $CDCl_3$ of ~1750. We sought to confirm this constant by NMR titration of other salts, such as Bu_4PCl and Ph_4PCl. The binding constants observed were significantly different, as shown in Table 8.1. In order to calibrate our effort, we undertook an NMR titration of the isophthalamide derivative (**51**) first prepared by Crabtree and coworkers, which has been extensively studied [114]. We confirmed the reported binding constant for the triarene with Bu_4NCl. We found that the binding of "Cl^-" was strongly cation dependent.

Table 8.1 Receptor–salt complexation in $CDCl_3$ determined by NMR.

Salt	Receptor	
	50	**51**
Bu_4NCl	1 750	2 500
Bu_3NCH_2PhCl	1 700	–
Bu_4PCl	510	–
Me_3NCH_2PhCl	$1\,750 \pm 750$	850
Ph_4PCl	$33\,500 \pm 1\,500$	$5\,000 \pm 100$

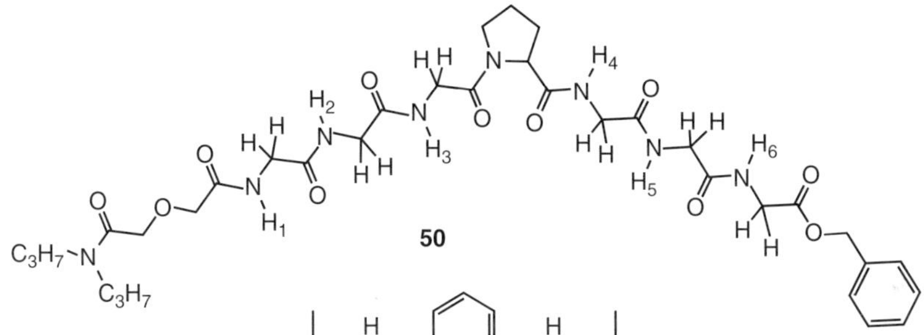

50

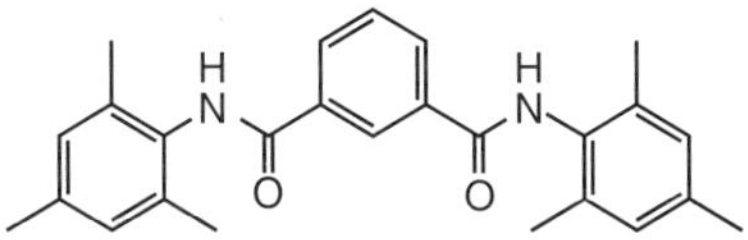

51

On the basis of the chemical shifts, it was apparent that the main contributors to anion binding were the amidc hydrogens located on the fifth and seventh amino acid residues (~GGGPGGG~, $^5G_{NH}$, $^7G_{NH}$). In principle, alterations in these amino acids will impact binding and recognition and, ultimately, transport. A structure determination of the **50**·Bu_4NCl complex was undertaken in collaboration with Tomich and coworkers. They found by one-dimensional solution-state NMR titration studies combined with computational molecular simulation studies that the peptide and salt interact as an ion pair. The amide H-bonding interactions with Cl^- were also confirmed [115].

8.6.7
The Effect on Transport of Charged Sidechains

It is thought that a glutamic acid residue plays a role in gating ion transport in the ClC transporter proteins [116]. We therefore prepared a family of structures incorporating a glutamic acid on the C-terminal side of proline. The general structure is $(C_{18}H_{37})_2N–COCH_2OCH_2CO–(Gly)_3$-Pro-$(Aaa)_3–O(CH_2)_6CH_3$, and the $(Aaa)_3$ sequence is EGG, GEG, or GGE. The latter is shown as **52**. In each case, the benzyl ester of the glutamic acid residue was obtained and studied as well.

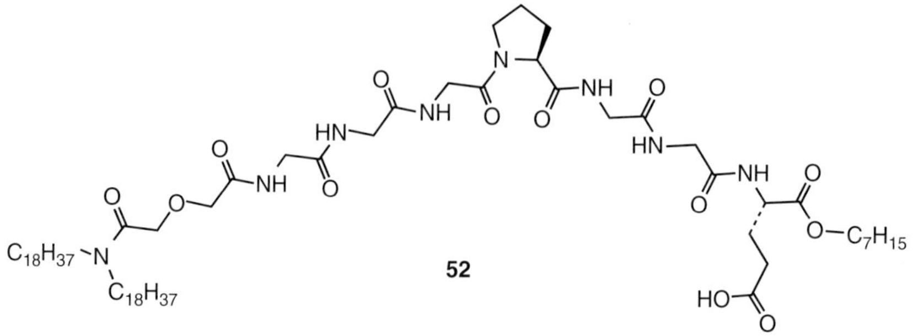

52

The best transport activity was observed for **52**, which has the heptapeptide sequence GGGPGGE. Chloride transport activity was higher for the glutamate benzyl esters than for the free carboxylate residue in all three cases. The influence of a carboxyl-containing side chain compared with its corresponding ester was greater for ^{5}Glx and ^{7}Glx than for ^{6}Glx. The amide hydrogens at Gly-5 and Gly-7 are those that were shown to be involved in Cl^- complexation [114, 117]. It is also interesting to note that negative ion electrospray mass spectrometry (ES-MS) showed that Cl^- was bound to the neutral ester glutamate SATs well but the analogs containing a free carboxyl group were weaker binders. It was also found that transport involving these SATs was selective for Cl^- over K^+ even when the carboxyl group was neither esterified nor ionizable. These results suggest that glutamate could function as a gating element in the natural protein.

8.6.8
Fluorescent Probes of SAT Structure and Function

Although the compounds described in this chapter are synthetic, the tools used by biophysicists to understand peptide and protein interactions are as valuable for them as for natural systems. Fluorescent probes are particularly useful because they report interactions at the very low concentrations at which these compounds function. Further, different types of fluorescent probes can be used to obtain different types of information about interactions of the peptides with each other as well as with the bilayer.

Four SAT derivatives that have well-characterized fluorescent residues attached [117] were prepared. These are illustrated as **53–56**. The SAT framework is identical to the initially designed **20** except that the C-terminal benzyl ester has been replaced by *n*-heptyl ester (see Section 8.6.2). In **53–55**, the seventh amino acid (glycine) has been substituted. In **53**, lysine replaced the C-terminal glycine; the heptapeptide sequence is GGGPGGK. Lysine has been substituted by nitrobenzodioxazole to form a nitrobenzodiazole "(NBD)" derivative. This fluorescent residue is useful as a probe of membrane interactions because its fluorescence is water quenched. Thus, its fluorescent intensity increases on insertion into a bilayer membrane. The NBD residue is also valuable in the sense that it is physically small. There is always a concern that the presence of a fluorescent residue will alter the property being probed, unless, of course, green fluorescent protein is used.

Pyrene is the fluorescent residue in **54**; it is linked as an ester to glutamic acid (GGGPGGE). The fluorescent response of pyrene is medium-dependent, but an important property is that proximate pyrenes form excimers. When two pyrene residues are close enough, an extended molecular orbital leads to a distinct redshift not apparent when pyrene is isolated.

Compound **55** incorporates a C-terminal tryptophan; it has the peptide sequence GGGPGGW. The indole side chain of tryptophan is a natural fluorescent probe that has been used extensively in biophysical analysis. Indole absorbs energy at about 280 nm and emits near 340 nm. The indole emission may be observed directly, and changes in its properties will reflect its environment. Of particular value, however, is that fluorescent residues such as pyrene (**54**) absorb energy at wavelengths near the indole emission. When the second fluorescent residue is near indole, energy may be transferred from indole to it. The result is described as fluorescence resonance energy transfer (FRET), and its application may reveal the proximity of various peptides.

The fourth compound illustrated (**56**) has the sequence GGGPKGG. A dimethylaminosulfonyl or dansyl group is attached to the lysine side chain. The dansyl group is valuable as a reporter of its environment as there is a redshift in its fluorescence emission in a nearly linear manner with increasing polarity. Further, it is excited near 340 nm and emits above 500 nm. Thus, it can be used as a partner for tryptophan (indole) in a FRET experiment. A plot of λ_{max} as a function of the polarity parameter E_T [118] showed that the dansyl group of **56** experienced an environment within the bilayer intermediate in polarity between dichloromethane and isopropyl alcohol [119]. We note that E_T values for CH_2Cl_2 and DMF are similar even though the reported dielectric constants differ significantly. Considering the dansyl group's position at the end of a lysine side chain, it seems reasonable that the fluorescent residue would be closer to the nonpolar than to the polar regimes of the bilayer.

The utilization of any of these compounds as probes of transport requires that the presence of a fluorescent residue does not impair function. Each of **53–56** was compared with the parent compound, $(C_{18}H_{37})_2N–COCH_2OCH_2CO–(Gly)_3\text{-Pro-}(Gly)_3–O(CH_2)_6CH_3$. Each of the fluorescent SAT molecules was at least as active as a chloride transporter as the parent. Compound **55**, which has a C-terminal tryptophan in the peptide sequence (i.e., GGGPGGW), was found to be significantly more active than the parent or the other fluorescent compounds. This may reflect a special membrane-anchoring effect of tryptophan that has been suggested for proteins [119] and previously explored in our laboratory [120] (Figure 8.26).

8.6.8.1 Aggregation in Aqueous Suspension and in the Bilayer

It is clear that a compound that mediates transport through a phospholipid bilayer must be present within it and must assume a conformation or aggregation state suitable to the task. In the studies summarized here, the SAT molecules are added to the aqueous suspension containing liposomes. The amphiphiles must insert into the bilayer and organize appropriately for function. The SAT monomers are

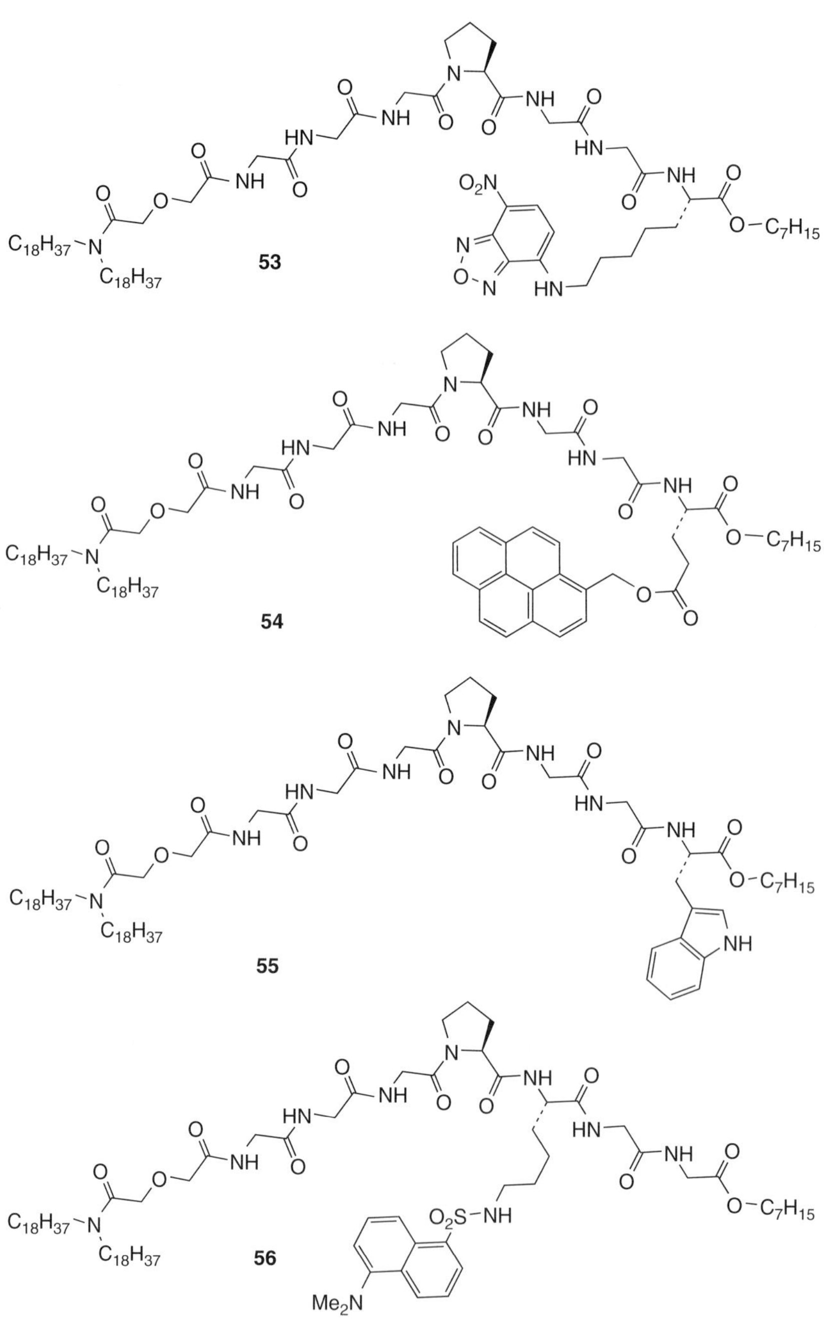

Figure 8.26 Synthetic anion transporters having fluorescent reporter tags.

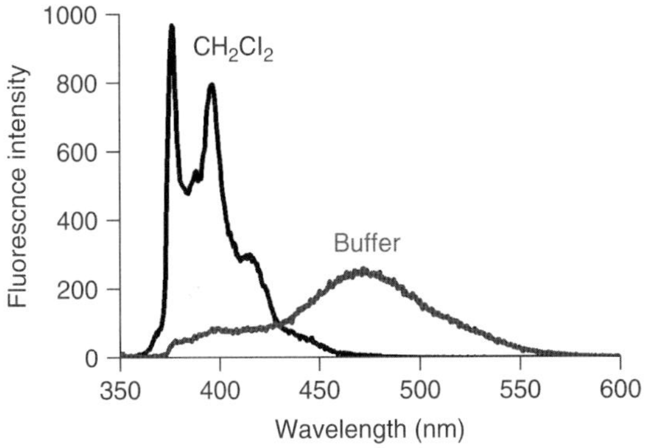

Figure 8.27 Fluorescence spectrum for pyrenyl-SAT **54** recorded in CH_2Cl_2 or aqueous buffer.

amphiphiles and may therefore aggregate in the aqueous buffer. To the extent this occurs, it alters the dynamics of membrane insertion.

When pyrene is present in dilute solution, excitation at 345 nm leads to a fluorescence emission spectrum having two major peaks at ~375 and 395 nm. This spectrum is attributed to the dispersed monomer that is not interacting with other pyrene molecules. When two pyrene residues are proximate, they associate into an excimer that exhibits a broad emission band centered at ~472 nm. When pyrenyl-SAT **H2** was dissolved in either CH_2Cl_2 or CH_3CH_2OH, the characteristic monomer spectrum (375, 395 nm) was observed. This was expected because **54** is soluble in these solvents and not expected to aggregate (Figure 8.27). When **54** was added to aqueous HEPES buffer, the broad (472 nm) excimer band was observed. From these observations, it was inferred that **54** aggregates in aqueous buffer although the extent of aggregation could not be quantified from these experiments [119].

In a separate experiment, **54** dissolved in minimal 2-propanol was injected into a buffer containing liposomes. The fluorescence emission spectrum of **54** was nearly identical whether pure DOPC or 7 : 3 DOPC:DOPA vesicles were present in the buffer (Figure 8.28). In either case, both monomers and excimers were apparent. In buffer alone, little, if any, monomer was apparent; the excimer (aggregates) dominated. We infer from this that on the bilayer surface or within the liposomal bilayer, separate monomers and aggregates are present [119]. Cooperativity studies suggested that ion conductance resulted from dimer formation, so these results seem reasonable. We presume that individual monomers do not form ion-conducting pores, but we lack direct evidence for this inference. The question of aggregate formation in the buffer, at or near the bilayer surface and within it, is obviously critical to understanding function. Various analytical methods have been used to address these questions, and the results of those studies follow.

8.6.8.2 Fluorescence Resonance Energy Transfer Studies

The proximity of SAT monomers in liposomes was further assessed by FRET studies using tryptophan-SAT (Tryp-SAT, **55**) and pyrenyl-SAT (Pyr-SAT, **54**) in

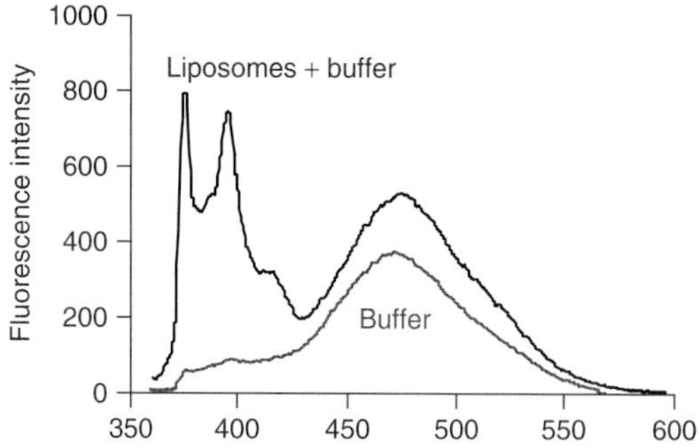

Figure 8.28 Fluorescence spectrum for pyrenyl-SAT **54** recorded in aqueous buffer or in the presence of liposomes suspended in aqueous solution.

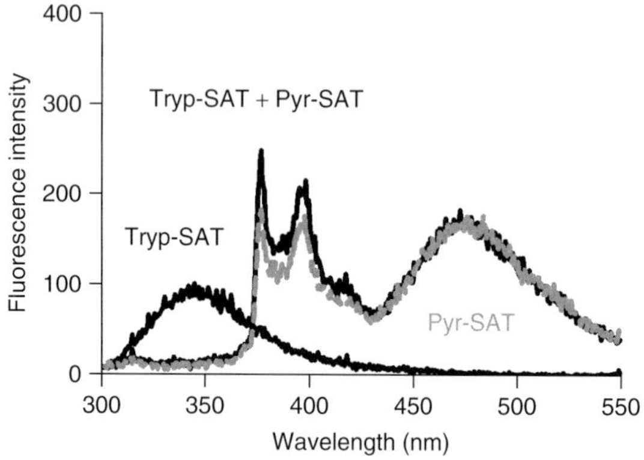

Figure 8.29 Fluorescence resonance energy transfer (FRET) experiments conducted with indole- and pyrene-substituted synthetic anion transporters.

combination. The tryptophan compound shows a single fluorescence emission with λ_{max} near 345 nm. The pyrenyl-SAT shows emission corresponding both to monomer and excimer in the presence of DOPC:DOPA vesicles, as shown in Figure 8.29. When both **54** and **55** are present in the liposomal suspension and the system is excited at ~280 nm, no emission is observed for Tryp-SAT **55**; its energy is transferred to Pyr-SAT **54**, which is present as before (see Figure 8.29), as both monomers and excimers [119].

8.6.8.3 **Insertion of SATs into the Bilayer**

The fluorescent NBD residue is valuable as a probe of the environment. It is highly fluorescent in nonpolar media but quenched by water. When NBD-SAT (**53**) was added to aqueous buffer, excitation at 465 nm produced almost no emission. When **53** was added to a suspension containing DOPC:DOPA vesicles that encapsulated KCl, ion release was observed along with concentration-dependent fluorescence emission at $\lambda_{max} = 565$ nm. The concentration-dependent ion release and fluorescence allowed us to estimate that ~35% of the available SAT partitioned into the liposomal bilayer [119].

8.6.8.4 Position of SATs in the Bilayer

The high fluorescence emission of NBD-SAT (**53**) when it is in the bilayer was used to determine the position of the amphiphile in the membrane. Fluorescence intensity increases as the SAT partitions into the bilayer, but it can be quenched by addition of $Na_2S_2O_4$ to the aqueous buffer. The increase in fluorescence emission as a function of added **53** was monitored in aqueous suspensions of DOPC:DOPA. In otherwise identical experiments, $Na_2S_2O_4$ was added after about 600 s. Essentially, immediate fluorescence quenching was observed over a 10-fold concentration range. We infer from the nearly complete quenching that the SAT present in the bilayer remains predominantly or exclusively in the membrane's outer leaflet [119].

8.6.9 Self-Assembly Studies of the Amphiphiles

The basic SAT structure that we have studied is $(C_{18}H_{37})_2N$–$COCH_2OCH_2CO$–$(Gly)_3$-Pro-$(Gly)_3$–OR. In the first example that was prepared and studied, R was benzyl [99]. Owing presumably to the twin octadecyl tails, the solubility of this compound in water was poor. A critical requirement in our studies is that a synthetic channel or pore former must be able to insert into the external membrane of a liposome or organism from the aqueous phase. We thus explored a variety of C-terminal groups ranging from methyl to octadecyl, branched and cyclic esters, and including variations such as dihydroxyethylamide (OR = $N(CH_2CH_2OH)_2$) [103].

We anticipated that when the ester was a long hydrocarbon chain (R = octadecyl), the formation of pores would be improved because of additional membrane anchoring by the third octadecyl group. In fact, this compound showed poor transport. We surmised that when R = $C_{18}H_{37}$, the three octadecyl chains align, solvate each other, and prohibit pore formation. When the ester chain is short (R = methyl or ethyl), there is insufficient "secondary anchoring" to improve pore formation and transport. In accord with this hypothesis, the *n*-heptyl group proved to be the most favorable of those tested for Cl^- transport. Detailed consideration of the problem suggested that multiple equilibria were possible in the aqueous phase, on the liposomal surface, and in the bilayer. The possible equilibria that can occur in the aqueous phase are as shown in the following scheme, from which it is clear that a range of equilibria and interactions are possible, only some of which lead to Cl^- transport. As noted above, only about 35% of the SAT monomers are ultimately inserted into the bilayer.

$$\mathrm{SAT_{solvent}} \xrightarrow{k_1} \mathrm{SAT_{aqueous}} \underset{k_{-2}}{\overset{k_2}{\rightleftharpoons}} (\mathrm{SAT_{aqueous}})_n$$

The Langmuir trough and Brewster angle microscope (BAM) were used to better understand the aggregation behavior of these systems [121]. The Langmuir trough is simply a "pan of water" on which a molecule of interest is spread as a monolayer. Movable, lateral bars permit the available surface area to be compressed. A transducer called a *Wilhelmy plate* measures the surface pressure as the lateral bars are moved. The pressure–area isotherm gives information about the surface area occupied by a molecule. This, in turn, gives insight into molecular organization on the water surface. The BAM uses a laser to visualize surface phenomena.

Figure 8.30 shows surface pressure–area isotherms for R_2N–$COCH_2OCH_2CO$–(Gly)$_3$-Pro-(Gly)$_3$–OCH_2Ph, in which R is hexyl, decyl, hexadecyl, or octadecyl (**20**). The SAT having twin hexyl anchors shows little organization or stability and fails to form an ordered surface. In contrast, the SAT having twin octadecyl chains is both organized and stable. The minimum area at collapse (hook atop the C_{18} trace) is ~40 Å^2. The cross-sectional area of a hydrocarbon chain is ~20 Å^2. The *n*-decyl- and *n*-hexadecyl-substituted compounds form monolayers whose stabilities are intermediate between those formed from the *n*-hexyl and *n*-octadecyl compounds. Unlike the *n*-hexyl compound, the isotherms obtained for *n*-decyl and *n*-hexadecyl have clearly defined areas at collapse, which suggests that these compounds organize into a monolayer at the air–water interface. However, these areas at collapse (61 A^2, C_{16}, and 70 A^2, C_{10}) occur at higher areas than expected for compounds with twin, normal alkyl chains (i.e., ~40 A^2). We concluded from this observation and the fact that the surface pressures at collapse are ~30 mN/m less than for the *n*-octadecyl compound that the monolayers formed from the *n*-decyl and *n*-hexadecyl compounds are less stable than those formed from the *n*-octadecyl-substituted compound.

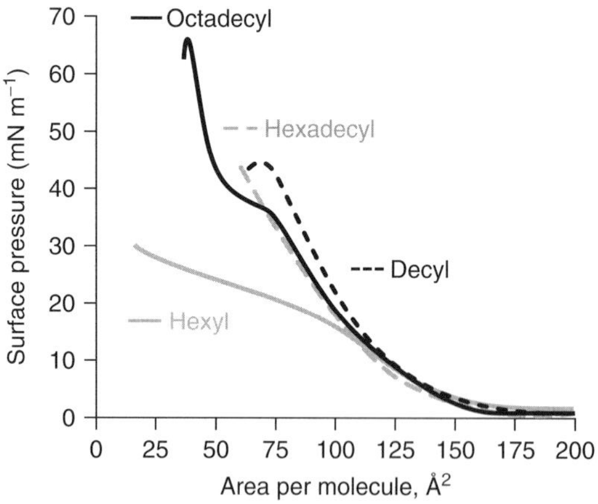

Figure 8.30 Surface pressure–molecular area isotherm data for $(R^1)_2N$–$COCH_2OCH_2$ CO–Gly-Gly-Gly-Pro-Gly-Gly-Gly–OCH_2Ph, where R^1 = –hexyl (–); –decyl (- -); –hexadecyl (– -); –octadecyl (**20**) (–).

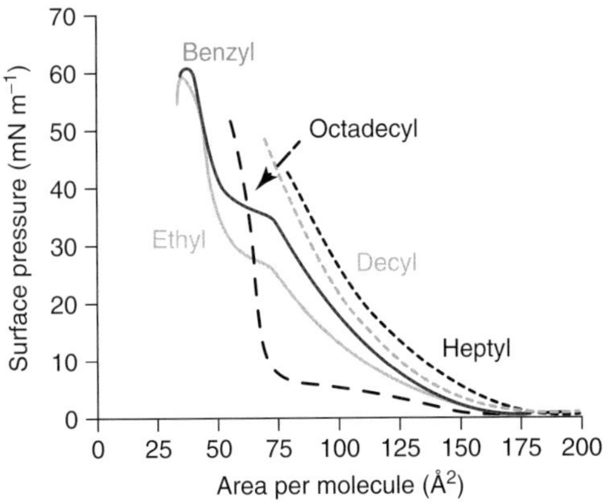

Figure 8.31 Surface pressure–molecular area isotherm data for $(C_{18}H_{37})_2N–COCH_2OCH_2CO–Gly-Gly-Gly-Pro-Gly-Gly-Gly–OR^2$, where R^2 = −ethyl (–); −benzyl (–); −octadecyl (— –); −hepyl (- -); −decyl (- -).

Dynamic light scattering (DLS) measurements showed that in aqueous suspension, $(C_{18}H_{37})_2N–COCH_2OCH_2CO–(Gly)_3$-Pro-$(Gly)_3–OCH_2Ph$ formed stable aggregates of approximate diameter 195 nm (~2000 Å). The precise shape of these aggregates in suspension is unknown: the instrument's software treats all aggregates as spheres. However, transmission electron micrographs obtained of these aggregates after solvent evaporation showed spherical clusters of aggregates.

When the C-terminal residue, R^2, of $(C_{18}H_{37})_2N–COCH_2OCH_2CO–(Gly)_3$-Pro-$(Gly)_3–OR^2$ was varied, where R^2 = ethyl, heptyl (**24**), decyl (**26**), octadecyl (**27**), and benzyl (**20**), the surface pressure–area isotherms shown in Figure 8.31 were obtained. The minimum area of a saturated alkyl chain is 20 Å^2, and the ethyl- and benzyl-substituted SATs collapse at the predicted value of ~40 Å^2, that is, 38 and 39 Å^2, respectively. As with the R^1-substituted compounds, intermittent stability is again apparent for both R^2 = heptyl (**24**) and decyl (**26**). These medium-chain substituted SATs collapse at higher areas than would be expected for an amphiphile having two normal chains (~40 Å^2), at ~75 Å^2. The monolayers formed from the ethyl- and benzyl (**20**)-substituted SATs are also more stable in that they are capable of withstanding more pressure exerted on them. Indeed, the collapse pressures of R^2 = ethyl and benzyl (**20**) are ~20 mN/m greater than those of R^2 = heptyl (**24**) and decyl (**26**).

In this study, a unique behavior was observed for R^2 = octadecyl (**27**). After the first transition, the octadecyl-substituted SAT enters a region in which the pressure increases only 7 mN/m over an 80 Å^2 surface area. During this *pseudo*-plateau region (~70–150 Å^2), Brewster angle microscopic images of solid domains that formed on the water surface (white areas shown in Figure 8.32) were obtained. At ~70 Å^2, the last transition point, these domains coalesce into a uniform monolayer, which was documented by BAM (Figure 8.32d). The coalescence of

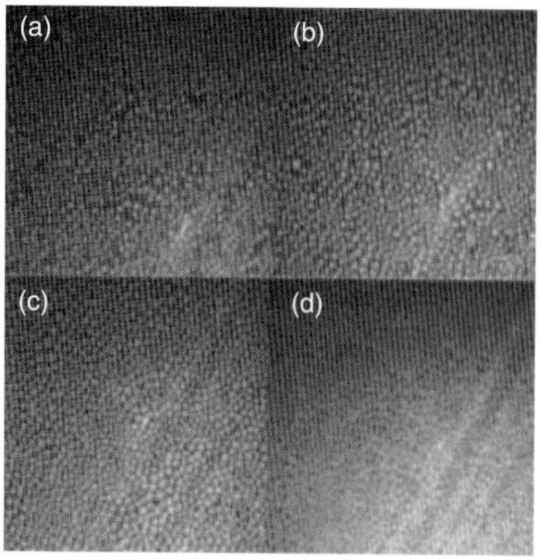

Figure 8.32 Brewster angle micrographs of $(C_{18}H_{37})_2N$–$COCH_2OCH_2CO$–Gly-Gly-Gly-Pro-Gly-Gly-Gly–$OC_{18}H_{37}$ (**27**) taken at (a) 121 Å^2, (b) 110 Å^2, (c) 78 Å^2, and (d) 73 Å^2. Image field of view is 2.0 mm × 2.0 mm.

visualized domains, which results in a featureless BAM image, is indicative of a homogeneous two-dimensional surface.

Note that the minimum area is ~60 Å^2 for $(C_{18}H_{37})_2N$–$COCH_2OCH_2CO$–Gly-Gly-Gly-Pro-Gly-Gly-Gly–$OC_{18}H_{37}$, which suggests that all three octadecyl chains (~20 Å^2 each) align. If the three chains pack tightly together in the bilayer, as we surmise they do, to form the ordered solid domains visualized with BAM, there would be no space for ions to pass through. Indeed, the tris(octadecyl) compound showed very poor Cl^- release from liposomes.

The findings with the Langmuir trough, the BAM, and DLS comport nicely. The SATs that form the most stable and/or most ordered monolayers also aggregate in the most stable and reproducible way in solution. These SATs are also the poorest Cl^- transporters. In contrast, the best Cl^- transporters form the least stable monolayers and do not aggregate in a way that is easily detected by DLS studies. Our findings show that the most effective Cl^- transporters in this series possess twin *n*-decyl chains at their N-terminus and either an *n*-heptyl or *n*-decyl chain at their C-terminus. An effective transporter must have enough hydrophobic anchoring so that it will interact and quickly insert into a bilayer. However, too much hydrophobicity causes an extensive aggregation (a large k_2), which, in turn, causes the amphiphile to precipitate. An effective ion transporter must also possess a secondary anchoring chain at its C-terminus. Functioning SATs in this family of transporters must aggregate at least as dimers within the membrane to form a conducting pore. SATs must therefore be capable of lateral diffusion or the insertion dynamics must be rapid in order for one SAT to "find" a second SAT. It appears that the medium alkyl chain lengths favor this ideal hydrophobic/hydrophilic balance.

8.6.10
The Biological Activity of Amphiphilic Peptides

An important challenge for developers of synthetic anion transporters is to produce a pharmaceutical that has efficacy for the treatment of cystic fibrosis [122]. This hereditary disease affects one in 2500 humans of northern

European descent. We have recently shown, by using an Ussing chamber, that $(C_{18}H_{37})_2N{-}COCH_2OCH_2CO{-}(Gly)_3$-Pro-$(Gly)_3{-}OCH_2Ph$ (**20**) alters Cl^- transport in vital, mammalian airway epithelial cells [123]. The study proved that $(C_{18}H_{37})_2N{-}COCH_2OCH_2CO{-}(Gly)_3$-**Pro**-$(Gly)_3{-}OCH_2Ph$ (**20**) functioned as a Cl^- transporter in this context, whereas the control compound, $(C_{18}H_{37})_2N{-}COCH_2OCH_2CO{-}(Gly)_3$-**Leu**-$(Gly)_3{-}OCH_2Ph$ (**21**), known to be sixfold less effective as a Cl^- transporter, was inactive. Admittedly, compound **20** has a molecular weight of 1168 Da, and it is a peptide. Still, introduction into the lungs could be effected by a simple inhaler such as used to administer steroids. The critical finding is that these compounds alter chloride transport in mammalian cells.

8.6.11 Nontransporter, Membrane-Active Compounds

The hydrophobic component of the SATs most studied by us derives from the basic structure $(C_{18}H_{37})_2N{-}COCH_2OCH_2COOH$ (**17**). We studied the effect on liposomes of **17**–**19** (shown above). Compound **17** formed stable vesicles of 100–150 nm diameter [124]. Aggregates could also be detected by sonication of **18** and **19**, but they were not stable and vesicle formation was accompanied by some precipitation. Compound **18** mediated Cl^- release from liposomes. Most remarkable, however, was the observation that when **18** was added to liposomes prepared from 2 : 1 DOPC/DOPA, dramatic, concentration-dependent size increases were observed for the vesicles. On the basis of fluorescence quenching and modeling, we concluded that **18** induces vesicular fusion. Such significant effects of simple amphiphiles have been noted previously by others [125] and clearly deserve additional study.

8.7 Conclusions

A broad range of approaches and much success have been realized in the design and characterization of synthetic ion channels. Even so, the effort is really in its infancy. Studies in this area hold forth the promise of better understanding of supramolecular interactions, of molecule membrane interactions, and ultimately the development of novel pharmaceuticals for pathologies involving impaired channel function.

Acknowledgments

The work described in this chapter that was done in the authors' laboratory was supported by the NIH through two awards: GM 36262 and GM 63190. This support is gratefully acknowledged.

References

1. (a) Degrado, W.F. (1988) Design of peptides and proteins. *Adv. Protein Chem.*, **39**, 51–124; (b) Lear, J.D., Wasserman, Z.R., and DeGrado, W.F. (1988) Synthetic amphiphilic peptide models for protein ion channels. *Science*, **240**, 1177–1181; (c) DeGrado, W.F., Wasserman, Z.R., and Lear, J.D. (1989) Protein design, a minimalist approach. *Science*, **243**, 622–628; (d) DeGrado, W.F. and Lear, J.D. (1990) Conformationally constrained alpha-helical peptide models for protein ion channels. *Biopolymers*, **29**, 205–213; (e) Åkerfeldt, K.S., Kim, R.M., Camac, D., Groves, J.T., Lear, J.D., and DeGrado, W.F. (1992) Tetraphilin: a four-helix proton channel built on a tetraphenylporphyrin framework. *J. Am. Chem. Soc.*, **114**, 9656–9657; (f) Åkerfeldt, K.S., Lear, J.D., Wasserman, Z.R., Chung, L.A., and DeGrado, W.F. (1993) Synthetic peptides as models for ion channel proteins. *Acc. Chem. Res.*, **26**, 191–197; (g) Betz, S.F., Bryson, J.W., and DeGrado, W.F. (1995) Native-like and structurally characterized designed alpha-helical bundles. *Curr. Opin. Struct. Biol.*, **5**, 457–463; (h) Bryson, J.W., Betz, S.F., Lu, H.S., Suich, D.J., Zhou, H.X., O'Neil, K.T., and DeGrado, W.F. (1995) Protein design: a hierarchic approach. *Science*, **270**, 935–941.
2. (a) Oiki, S., Danho, W., and Montal, M. (1988) Channel protein engineering: synthetic 22-mer peptide from the primary structure of the voltage-sensitive sodium channel forms ionic channels in lipid bilayers. *Proc. Natl. Acad. Sci. U.S.A.*, **85**, 2393–2397; (b) Montal, M., Montal, M.S., and Tomich, J.M. (1990) Synporins – synthetic proteins that emulate the pore structure of biological ionic channels. *Proc. Nat. Acad. Sci. U.S.A.*, **87**, 6929–6933; (c) Montal, M. (1990) in *Ion Channels*, vol. 2 (ed. T. Naranishi), Plenum, New York, pp. 1–31; (d) Grove, A., Tomich, J.M., and Montal, M. (1991) A molecular blueprint for the pore-forming structure of voltage-gated calcium channels. *Proc. Natl. Acad. Sci. U.S.A.*, **88**, 6418–6422; (e) Grove, A., Mutter, M., Rivier, J.E., and Montal, M. (1993) Template-assembled synthetic proteins designed to adopt a globular, four-helix bundle conformation from ionic channels in lipid bilayer. *J. Am. Chem. Soc.*, **115**, 5919–5924; (f) Oblatt-Montal, M., Buhler, L.K., Iwamoto, T., Tomich, J.M., and Montal, M. (1993) Synthetic peptides and four-helix bundle proteins as model systems for the pore-forming structure of channel proteins. I. Transmembrane segment M2 of the nicotinic cholinergic receptor channel is a key pore-lining structure. *J. Biol. Chem.*, **268**, 14601–14607; (g) Reddy, G.L., Iwamoto, T., Tomich, J.M., and Montal, M. (1993) Synthetic peptides and four-helix bundle proteins as model systems for the pore-forming structure of channel proteins. II. Transmembrane segment M2 of the brain glycine receptor is a plausible candidate for the pore-lining structure. *J. Biol. Chem.*, **268**, 14608–14615; (h) Iwamoto, T., Grove, A., Montal, M.O., Montal, M., and Tomich, J.M. (1994) Chemical synthesis and characterization of peptides and oligomeric proteins designed to form transmembrane ion channels. *Int. J. Pept. Protein Res.*, **43**, 597–607; (i) Montal, M. (1995) Design of molecular function: channels of communication. *Annu. Rev. Biophys. Biomol. Struct.*, **24**, 31–57; (j) Oblatt-Montal, M., Yamazaki, M., Nelson, R., and Montal, M. (1995) Formation of ion channels in lipid bilayers by a peptide with the predicted transmembrane sequence of botulinum neurotoxin A. *Protein Sci.*, **4**, 1490–1497.
3. (a) Jullien, L. and Lehn, J.-M. (1988) The 'chundle' approach to molecular channels. Synthesis of a macrocycle-based molecular bundle. *Tetrahedron Lett.*, **29**, 3803–3806; (b) Jullien, L. and Lehn, J.-M. (1992) An approach to channel type molecular structures. 1. Synthesis of

bouquet-shaped molecules based on an [18]-O6 polyether macrocycle. *J. Inclusion Phenom.*, **12**, 55–74.

4. Carmichael, V.E., Dutton, P.J., Fyles, T.M., James, T.D., Swan, J.A., and Zojaji, M. (1989) Biomimetic ion transport: a functional model of a unimolecular ion channel. *J. Am. Chem. Soc.*, **111**, 767–769.
5. Nakano, A., Xie, Q., Mallen, J.V., Echegoyen, L., and Gokel, G.W. (1990) Synthesis of a membrane-insertable, sodium cation conducting channel: kinetic analysis by dynamic sodium-23 NMR. *J. Am. Chem. Soc.*, **112**, 1287–1289.
6. (a) Canceill, J., Jullien, L., Lacombe, L., and Lehn, J.-M. (1992) Channel-type molecular structures. Synthesis of bouquet-shaped molecules based on a B-cyclodextrin core. *Helv. Chim. Acta*, **75**, 791–812; (b) Pregel, M.J., Jullien, L., and Lehn, J.-M. (1992) Toward artificial ion channels: transport of alkali metal ions across liposomal membranes by 'bouquet' molecules. *Angew. Chem. Int. Ed. Engl.*, **31**, 1637–1640.
7. (a) Dutton, P.J., Fyles, T.M., Suresh, V.V., Fronczek, F.R., and Gandour, R.D. (1993) Solid-state chemistry of polycarboxylate crown ether cation complexes: cooperative binding of water and metal ions by flexible chorands. *Can. J. Chem.*, **71**, 239–253; (b) Fyles, T.M., James, T.D., Pryhitka, A., and Zojaji, M. (1993) Assembly of ion channel mimics from a modular construction set. *J. Org. Chem.*, **58**, 7456–7468; (c) Fyles, T.M., James, T.D., and Kaye, K.C. (1993) Activities and modes of action of artificial ion channel mimics. *J. Am. Chem. Soc*, **115**, 12315–12321; (d) Fyles, T.M., Heberle, D., Van Straaten-Nijenhuis, W.F., and Zhou, X. (1995) Synthetic ion transporters in bilayer membranes. *Supramol. Chem.*, **6**, 71–77; (e) Fyles, T.M. and Van Straaten-Nijenhuis, W.F. (1996) in *Comprehensive Supramolecular Chemistry*, vol. 10 (eds D.N. Reinhoudt), Pergamon, Oxford, pp. 53–77; (f) Fyles, T.M., Loock, D., van Straaten-Nijenhuis, W.F., and Zhou, X. (1996) Pores formed by bis-macrocyclic bola-amphiphiles in vesicle and planar bilayer membranes. *J. Org. Chem.*, **61**, 8866–8874; (g) Fyles, T.M. (1997) Bilayer membranes and transporter models. *Curr. Opin. Chem. Biol.*, **1**, 497–505.
8. Gokel, G.W. (2000) Hydraphiles: design, synthesis, and analysis of a family of synthetic, cation-conducting channels. *Chem. Commun.*, 1–9.
9. (a) Voyer, N. and Robataille, M. (1995) A novel functional artificial ion channel. *J. Am. Chem. Soc.*, **117**, 6599–6600; (b) Meillon, J.-C. and Voyer, N. (1997) A synthetic transmembrane channel active in lipid bilayers. *Angew. Chem. Int Ed. Engl.*, **36**, 967–969.
10. (a) Gokel, G.W. and Murillo, O. (1996) Synthetic organic chemical models for transmembrane channels. *Acc. Chem. Res.*, **29**, 425–432; (b) Gokel, G.W. and Mukhopadhyay, A. (2001) Synthetic models of cation-conducting channels. *Chem. Soc. Rev.*, **30**, 274–286; (c) Sakai, N., Mareda, J., and Matile, S. (2005) Rigid-rod molecules in biomembrane models: from hydrogen-bonded chains to synthetic multifunctional pores. *Acc. Chem. Res.*, **38**, 79–87; (d) Sisson, A.L., Shah, M.R., Bhosale, S., and Matile, S. (2006) Synthetic ion channels and pores (2004–2005). *Chem. Soc. Rev.*, **35**, 1269–1286; (e) Fyles, T.M. (2007) Synthetic ion channels in bilayer membranes. *Chem. Soc. Rev.*, **36**, 335–347.
11. Bianchi, A., Bowman-James, K., and Garcia-España, E. (1997) *Supramolecular Chemistry of Anions*, Wiley-VCH Verlag GmbH, New York.
12. Sessler, J.L., Gale, P., and Cho, W.-S. (2006) *Anion Receptor Chemistry*, Royal Society of Chemistry, Cambridge.
13. Clarke, L.L., Grubb, B.R., Gabriel, S.E., Smithies, O., Koller, B.H., and Boucher, R.C. (1992) Defective epithelial chloride transport in a gene-targeted mouse model of cystic fibrosis. *Science*, **257**, 1125–1128.
14. Furstner, A. (2003) Chemistry and biology of roseophilin and the prodigiosin alkaloids: a survey of the last 2500 years. *Angew. Chem. Int. Ed. Engl.*, **42**, 3582–3603.

15. Lazaro, J. E., Nitcheu, J., Predicala, R. Z., Mangalindan, G. C., Nesslany, F., Marzin, D., Concepcion, G. P., and Diquet, B. (2002) Heptyl prodigiosin, a bacterial metabolite, is antimalarial in vivo and non-mutagenic in vitro. *J. Nat. Toxins*, **11**, 367–377.
16. Rapoport, K. and Holden, G. (1962) The synthesis of prodigiosin. *J. Am. Chem. Soc.*, **84**, 635–642.
17. Konno, H., Matsuya, H., Okamoto, M., Sato, T., Tanaka, Y., Yakoyama, K., Kataoka, T., Nagai, K., Wasserman, H., and Ohkuma, S. (1998) Prodigiosins uncouple mitochondrial and bacterial F-ATPases: evidence for their H^+/Cl^- symport activity. *J. Biochem.*, **124**, 547–556.
18. Manderville, R.A. (2001) Synthesis, proton-affinity and anti-cancer properties of the prodigiosin-group natural products. *Curr. Med. Chem.*, **1**, 195–218.
19. Davis, J.T., Gale, P.A., Okunola, O.A., Prados, P., Iglesias-Sánchez, J.C., Torroba, T., and Quesada, R. (2009) Using small molecules to facilitate exchange of bicarbonate and chloride anions across liposomal membranes. *Nat. Chem.*, **1**, 138–144.
20. (a) Shannon, R.D. (1976) Revised effective ionic radii and systematic studies of interatomic distances in halides and chalcogenides, *Acta Crystallogr., Sect. A*, **A32**, 751–767; (b) Darnell, J.E., Lodish, H., and Baltimore, D. (1986) *Molecular Cell Biology*, W. H. Freeman Co., p. 618 and 725.
21. (a) Kumpf, R.A. and Dougherty, D.A. (1993) A mechanism for ion selectivity in potassium channels: computational studies of cation-pi interactions. *Science*, **261**, 1708–1710; (b) Dougherty, D. (1996) Cation-pi interactions in chemistry and biology. A new view of benzene, Phe, Tyr, and Trp. *Science*, **271**, 163–168.
22. Heginbotham, L., Lu, Z., Abramson, T., and MacKinnon, R. (1994) Mutations in the K^+ channel signature sequence. *Biophys. J.*, **66**, 1061–1067.
23. Doyle, D.A., Cabral, J.M., Pfuetzner, R.A., Kuo, A., Gulbis, J.M., Cohen, S.L., Chait, B.T., and MacKinnon, R. (1998) The structure of the potassium channel: molecular basis of K+ conduction and selectivity. *Science*, **280**, 69–77.
24. Gokel, G.W. and Barkey, N. (2009) Transport of chloride ion through phospholipid bilayers mediated by synthetic ionophores. *New J. Chem.*, **33**, 947–963.
25. Liesegang, G.W., Vasquez, A., Purdie, N., and Eyring, E.M. (1977) Ultrasonic absorption kinetic studies of the complexation of aqueous Li^+, Na^+, Rb^+, Tl^+, Ag^+, $NH_4{}^+$, and Ca^{2+} by 18-crown-6. *J. Am. Chem. Soc.*, **99**, 3240.
26. Pedersen, C.J. (1967) *J. Am. Chem. Soc.*, **89**, 7017–7036.
27. Gokel, G.W. and Korzeniowski, S.H. (1982) *Macrocyclic Polyether Syntheses*, Springer-Verlag, Berlin, p. 410.
28. Izatt, R.M., Bradshaw, J.S., Nielsen, S.A., Lamb, J.D., Christensen, J.J., and Sen, D. (1985) Thermodynamic and kinetic data for cation-macrocycle interaction. *Chem. Rev.*, **85**, 271–339.
29. Izatt, R.M., Pawlak, K., Bradshaw, J.S., and Bruening, R.L. (1991) Thermodynamic and kinetic data for macrocycle interactions with cations and anions. *Chem. Rev.*, **91**, 1721–1785.
30. (a) Izatt, R.M., Bradshaw, J.S., Pawlak, K., Bruening, R.L., and Tarbet, B.J. (1992) Thermodynamic and kinetic data for macrocycle interaction with neutral molecules. *Chem. Rev.*, **92**, 1261–1354; (b) Izatt, R.M., Pawlak, K., and Bradshaw, J.S. (1995) Thermodynamic and kinetic data for macrocycle interaction with cations, anions, and neutral molecules. *Chem. Rev.*, **95**, 2529–2586.
31. (a) Dietrich, B., Fyles, T.M., Lehn, J.-M., Pease, L.G., and Fyles, D.L. (1978) Anion receptor molecules. Synthesis and some anion binding properties of macrocyclic guanidinium salts. *J. Chem. Soc. Chem. Commun.*, 934–936; (b) Lehn, J.M., Sonveaux, E., and Willard, A.K. (1978) Molecular recognition. Anion cryptates of a macrobicyclic receptor molecule for linear triatomic species. *J. Am. Chem. Soc.*, **100**, 4914–4916.

32. (a) Newcomb, M. and Blanda, M.T. (1988) Macrocycles containing tin. A small, exclusive host for the fluoride ion. *Tetrahedron Lett.*, **29**, 4261–4264; (b) Blanda, M.T. and Newcomb, M. (1989) Macrocycles containing tin. Ditopic, tricyclic, Lewis acidic hosts with four binding sites. *Tetrahedron Lett.*, **30**, 3501–3504; (c) Blanda, M.T., Horner, J.H., and Newcomb, M. (1989) Macrocycles containing tin. The preparation of macrobicyclic Lewis acidic hosts containing two tin atoms and tin-119 NMR studies of their chloride and bromide binding properties in solution. *J. Org. Chem.*, **54**, 4626–4636; (d) Newcomb, M., Horner, J.H., Blanda, M.T., and Squattrito, P.J. (1989) Macrocycles containing tin. Solid complexes of anions encrypted in macrobicyclic Lewis acidic hosts. *J. Am. Chem. Soc.*, **111**, 6294–6301.
33. (a) Schmidtchen, F.P. (1977) Inclusion of anions in macrotricyclic quaternary ammonium salts. *Angew. Chem.*, **89**, 751–752; (b) Schmidtchen, F.P. (1981) Macrocyclic quaternary ammonium salts. II. Formation of inclusion complexes with anions in solution. *Chem. Ber.*, **114**, 597–607; (c) Schmidtchen, F.P. (2006) Reflections on the construction of anion receptors. *Coord. Chem. Rev.*, **250** (23+24), 2918–2928.
34. Echavarren, A., Galan, A., De Mendoza, J., Salmeron, A., and Lehn, J.M. (1988) Anion-receptor molecules: synthesis of a chiral and functionalized binding subunit, a bicyclic guanidinium group derived from L- or D-asparagine. *Helv. Chim. Acta*, **71**, 685–693.
35. (a) Valiyaveettil, S., Engbersen, J.F.J., Verboom, W., and Reinhoudt, D.N. (1993) Synthesis and complexation studies of neutral anion receptors. *Angew. Chem. Int. Ed. Engl.*, **32**, 900–901; (b) Yang, X., Knobler, C.B., Zheng, Z., and Hawthorne, M.F. (1994) Host-guest chemistry of a new class of macrocyclic multidentate Lewis acids comprised of carborane-supported electrophilic mercury centers. *J. Am. Chem. Soc.*, **116**, 7142–7143; (c) Nishizawa, S., Buhlmann, P., Iwao, M., and Umezawa, Y. (1995) Anion recognition by urea and thiourea groups: remarkably simple neutral receptors for dihydrogen phosphate. *Tetrahedron Lett.*, **36**, 6483–6486; (d) Choi, K. and Hamilton, A.D. (2003) Rigid macrocyclic triamides as anion receptors: anion-dependent binding stoichiometries and 1H chemical shift changes. *J. Am. Chem. Soc.*, **125**, 10241–10249; (e) Choi, K. and Hamilton, A.D. (2003) Macrocyclic anion receptors based on directed hydrogen bonding interactions. *Coord. Chem. Rev.*, **240**, 101–110; (f) Bondy, C.R., Gale, P.A., and Loeb, S.J. (2004) Metal-organic anion receptors: arranging urea hydrogen-bond donors to encapsulate sulfate ions. *J. Am. Chem. Soc.*, **126**, 5030–5031; (g) Kato, R., Cui, Y.-Y., Nishizawa, S., Yokobori, T., and Teramae, N. (2004) Thiourea–isothiouronium conjugate for strong and selective binding of very hydrophilic H_2PO_4- anion at the 1,2-dichloroethane–water interface. *Tetrahedron Lett.*, **45**, 4273–4276.
36. Beer, P.D. and Gale, P.A. (2001) Anion recognition and sensing: the state of the art and future perspectives. *Angew. Chem. Int. Ed. Engl.*, **40**, 486–516.
37. Beer, P.D. and Hayes, E.J. (2003) Transition metal and organometallic anion complexation agents. *Coord. Chem. Rev.*, **240**, 167–189.
38. (a) Gale, P.A. (ed.) (2006) *Anion Coordination Chemistry II*, vol. 250, Elsevier B.V., Amsterdam, p. 327; (b) Gale, P.A. (2006) Structural and molecular recognition studies with acyclic anion receptors. *Acc. Chem. Res.*, **39**, 465–475.
39. (a) Bowman-James, K. and Alfred, W. (2005) Revisited: the coordination chemistry of anions. *Acc. Chem. Res.*, **38**, 671–678; (b) Kang, S.O., Begum, R.A., and Bowman-James, K. (2006) Amide-based ligands for anion coordination. *Angew. Chem. Int. Ed. Engl.*, **45**, 7882–7894.
40. (a) Kubik, S., Reyheller, C., and Stuewe, S. (2005) Recognition of anions by synthetic receptors in aqueous solution. *J. Inclusion Phenom. Macrocyclic Chem.*, **52**, 137–187; (b) Kubik,

S. (2009) Amino acid containing anion receptors. *Chem. Soc. Rev.*, **38**, 585–605.

41. Sessler, J.L., Gale, P., and Cho, W.-S. (2006) *Anion Receptor Chemistry*, Royal Society of Chemistry, Cambridge, p. 413.
42. Pressman, B.C., Harris, E.J., Jagger, W.S., and Johnson, J.H. (1967) Antibiotic-mediated transport of alkali ions across lipid barriers. *Proc. Natl. Acad. Sci. U.S.A.*, **58**, 1949–1956.
43. Eanes, E.D. and Costa, J.L. (1983) A ionophore-mediated calcium transport and calcium phosphate formation in Pressman cells. *Calcif. Tissue Int.*, **35**, 250–257.
44. Hernandez, J.C., Trafton, J.E., and Gokel, G.W. (1991) A direct comparison of extraction and homogeneous binding constants as predictors of efficacy in alkali metal cation transport. *Tetrahedron Lett.*, **32**, 6269–6272.
45. Szoka, F. and Papahadjopoulos, D. (1978) Procedure for preparation of liposomes with large internal aqueous space and high capture by reverse-phase evaporation. *Proc. Natl. Acad. Sci.*, **75**, 4194–4198.
46. Miller, C. (2006) ClC chloride channels viewed through a transporter lens. *Nature*, **440**, 484–489.
47. Accardi, A., Lobet, S., Williams, C., Miller, C., and Dutzler, R. (2006) Synergism between halide binding and proton transport in a CLC-type exchanger. *J. Mol. Biol.*, **362**, 691–699.
48. Matulef, K. and Maduke, M. (2007) The CLC 'chloride channel' family: revelations from prokaryotes. *Mol. Membr. Biol.*, **24**, 342–350.
49. Lisal, J. and Maduke, M. (2008) The ClC-0 chloride channel is a 'broken' Cl−/H+ antiporter. *Nat. Struct. Mol. Biol.*, **15**, 805–810.
50. Siskind, L.J., Fluss, S., Bui, M., and Colombini, M. (2005) Sphingosine forms channels in membranes that differ greatly from those formed by ceramide. *J. Bioenerg. Biomembr.*, **37**, 227–236.
51. Helsel, A.J., Brown, A.L., Yamato, K., Feng, W., Yuan, L., Clements, A.J., Harding, S.V., Szabo, G., Shao, Z., and Gong, B. (2008) Highly conducting transmembrane pores formed by aromatic oligoamide macrocycles. *J. Am. Chem. Soc.*, **130**, 15784–15785.
52. Schlieper, P. and De Robertis, E. (1977) Triton X-100 as a channel-forming substance in artificial lipid bilayer membranes. *Arch. Biochem. Biophys.*, **184**, 204–208.
53. (a) Wallace, D.P., Tomich, J.M., Iwamoto, T., Henderson, K., Grantham, J.J., and Sullivan, L.P. (1997) A synthetic peptide derived from glycine-gated Cl^- channel induces transepithelial Cl^- and fluid secretion. *Am. J. Physiol.*, **272**, C1672–C1679; (b) Tomich, J.M., Wallace, D., Henderson, K., Mitchell, K.E., Radke, G., Brandt, R., Ambler, C.A., Scott, A.J., Grantham, J., Sullivan, L., and Iwamoto, T. (1998) Aqueous solubilization of transmembrane peptide sequences with retention of membrane insertion and function. *Biophys. J.*, **74**, 256–267; (c) Mitchell, K.E., Iwamoto, T., Tomich, J., and Freeman, L.C. (2000) A synthetic peptide based on a glycine-gated chloride channel induces a novel chloride conductance in isolated epithelial cells. *Biochim. Biophys. Acta*, **1466**, 47–60; (d) Wallace, D.P., Tomich, J.M., Eppler, J.W., Iwamoto, T., Grantham, J.J., and Sullivan, L.P. (2000) A synthetic channel-forming peptide induces Cl^- secretion: modulation by Ca^{2+}-dependent K^+ channels. *Biochim. Biophys. Acta*, **1464**, 69–82; (e) Broughman, J.R., Mitchell, K.E., Sedlacek, R.L., Iwamoto, T., Tomich, J.M., and Schultz, B.D. (2001) NH_2-terminal modification of a channel-forming peptide increases capacity for epithelial anion secretion. *Am. J. Physiol.*, **280**, C451–C458; (f) Gao, L., Broughman, J.R., Iwamoto, T., Tomich, J.M., Venglarik, C.J., and Forman, H.J. (2001) Synthetic chloride channel restores glutathione secretion in cystic fibrosis airway epithelia. *Am. J. Physiol.*, **281**, L24–L30; (g) Broughman, J.R., Shank, L.P., Prakash, O., Schultz, B.D., Iwamoto, T., Tomich, J.M., and Mitchell, K.

(2002) Structural implications of placing cationic residues at either the NH_2- or COOH-terminus in a pore-forming synthetic peptide. *J. Membr. Biol.*, **190**, 93–103; (h) Broughman, J.R., Shank, L.P., Takeguchi, W., Schultz, B.D., Iwamoto, T., Mitchell, K.E., and Tomich, J.M. (2002) Distinct structural elements that direct solution aggregation and membrane assembly in the channel-forming peptide M2GlyR. *Biochemistry*, **41**, 7350–7358; (i) Broughman, J.R., Brandt, R.M., Hastings, C., Iwamoto, T., Tomich, J.M., and Schultz, B.D. (2004) Channel-forming peptide modulates transepithelial electrical conductance and solute permeability. *Am. J. Physiol. Cell Physiol.*, **286**, C1312–C1323; (j) Cook, G.A., Prakash, O., Zhang, K., Shank, L.P., Takeguchi, W.A., Robbins, A., Gong, Y.X., Iwamoto, T., Schultz, B.D., and Tomich, J.M. (2004) Activity and structural comparisons of solution associating and monomeric channel-forming peptides derived from the glycine receptor m2 segment. *Biophys. J.*, **86**, 1424–1435; (k) Iwamoto, T., You, M., Li, E., Spangler, J., Tomich, J.M., and Hristova, K. (2005) Synthesis and initial characterization of FGFR3 transmembrane domain: consequences of sequence modifications. *Biochim. Biophys. Acta*, **1668**, 240–247; (l) Cook, G.A., Pajewski, R., Aburi, M., Smith, P.E., Prakash, O., Tomich, J.M., and Gokel, G.W. (2006) NMR structure and dynamic studies of an anion-binding, channel-forming heptapeptide. *J. Am. Chem. Soc.*, **128**, 1633–1638; (m) Johnston, J.M., Cook, G.A., Tomich, J.M., and Sansom, M.S. (2006) Conformation and environment of channel-forming peptides: a simulation study. *Biophys J.*, **90**, 1855–1864.

54. Tabushi, I., Kuroda, Y., and Yokota, K. (1982) A,B,D,F-Tetrasubstituted *beta*-cyclodextrin as artificial channel compound. *Tetrahedron Lett.*, **23**, 4601–4604.

55. Canceill, J., Jullien, L., Lacombe, L., and Lehn, J.-M. (1992) Channel-type molecular structures. Synthesis of bouquet-shaped molecules based on a *beta*-cyclodextrin core. *Helv. Chim. Acta*, **75**, 791–812.

56. Madhavan, N., Robert, E.C., and Gin, M.S. (2005) A highly active anion selective aminocyclodextrinion channel. *Angew. Chem. Int. Ed. Engl.*, **44**, 7584–7587.

57. Riddell, F.G. and Hayer, M.K. (1985) The monensin-mediated transport of sodium ions through phospholipid bilayers studied by ^{23}Na-NMR spectroscopy. *Biochem. Biophys. Acta*, **817**, 313–317.

58. Murillo, O., Watanabe, S., Nakano, A., and Gokel, G.W. (1995) Synthetic models for transmembrane channels: structural variations that alter cation flux. *J. Am. Chem. Soc.*, **117**, 7665–7679.

59. Madhavan, N. and Gin, M.S. (2007) Increasing pH causes faster anion- and cation-transport rates through a synthetic ion channel. *ChemBioChem*, **8**, 1834–1840.

60. Borisenko, V., Burns, D.C., Zhang, Z., and Woolley, G.A. (2000) Optical switching of ion-dipole interactions in a gramicidin channel analogue. *J. Am. Chem. Soc.*, **122**, 6364–6370.

61. Woolley, G.A. (2005) Photocontrolling peptide alpha helices. *Acc. Chem. Res.*, **38**, 486–493.

62. Banghart, M., Borges, K., Isacoff, E., Trauner, D., and Kramer, R.H. (2004) Light-activated ion channels for remote control of neuronal firing. *Nat. Neurosci.*, **7**, 1381–1386.

63. Fortin, D.L., Banghart, M.R., Dunn, T.W., Borges, K., Wagenaar, D.A., Gaudry, Q., Karakossian, M.H., Otis, T.S., Kristan, W.B., Trauner, D., and Kramer, R.H. (2008) Photochemical control of endogenous ion channels and cellular excitability. *Nat. Methods*, **5**, 331–338.

64. Jog, P.V. and Gin, M.S. (2008) A light-gated synthetic ion channel. *Org. Lett.*, **10**, 3693–3696.

65. Murray, C.L., Shabany, H., and Gokel, G.W. (2000) The central 'relay' unit in hydraphile channels as a model for the water-and-ion 'capsule' of channel proteins. *Chem. Commun.*, 2371–2372.

66. de Mendoza, J., Cuevas, F., Prados, P., Meadows, E.S., and Gokel, G.W. (1998) A synthetic cation-transporting calix[4]arene derivative active in phospholipid bilayers. *Angew. Chem. Int. Ed. Engl.*, **37**, 1534–1537.
67. Iglesias-Sanchez, J.C., Wang, W., Ferdani, R., Prados, P., DeMendoza, J., and Gokel, G.W. (2008) Synthetic cation transporters incorporating crown ethers and calixarenes as headgroups and central relays: a comparison of sodium and chloride selectivity. *New J. Chem.*, **32**, 878–890.
68. Seganish, J.L., Santacroce, P.V., Salimian, K.J., Fettinger, J.C., Zavalij, P., and Davis, J.T. (2006) Regulating supramolecular function in membranes: calixarenes that enable or inhibit transmembrane Cl^- transport. *Angew. Chem. Int. Ed. Engl.*, **45**, 3334–3338.
69. Sidorov, V., Kotch, F.W., Kuebler, J.L., Lam, Y.F., and Davis, J.T. (2003) Chloride transport across lipid bilayers and transmembrane potential induction by an oligophenoxyacetamide. *J. Am. Chem. Soc.*, **125**, 2840–2841.
70. Sidorov, V., Kotch, F.W., Abdrakhmanova, G., Mizani, R., Fettinger, J.C., and Davis, J.T. (2002) Ion channel formation from a calix[4]arene amide that binds HCl. *J. Am. Chem. Soc.*, **124**, 2267–2278.
71. Seganish, J.L., Fettinger, J.C., and Davis, J.T. (2006) Facilitated chloride transport across phosphatidylcholine bilayers by an acyclic calixarene derivative: structure-function relationships. *Supramol. Chem.*, **18**, 257–264.
72. Okunola, O.A., Seganish, J.L., Salimian, K.J., Zavalij, P.Y., and Davis, J.T. (2007) Membrane-active calixarenes: toward 'gating' transmembrane anion transport. *Tetrahedron*, **63**, 10743–10750.
73. Sakai, N., Brennan, K.C., Weiss, L.A., and Matile, S. (1997) Toward biomimetic ion channels formed by rigid-rod molecules: length-dependent ion-transport activity of substituted oligo(*p*-phenylene)s. *J. Am. Chem. Soc.*, **119**, 8726.
74. Gennis, R.B. (1989) *Biomembranes: Molecular Structure and Function*, Springer-Verlag, New York.
75. Gruszecki, W.I., Gagos, M., Herec, M., and Kernen, P. (2003) Organization of antibiotic amphotericin B in model lipid membranes. A mini review. *Cell Mol. Biol. Lett.*, **8**, 161–170.
76. Weiss, L.A., Sakai, N., Ghebremariam, B., Ni, C., and Matile, S. (1997) Rigid rod-shaped polyols: functional nonpeptide models for transmembrane proton channels. *J. Am. Chem. Soc.*, **119**, 12142–12150.
77. Ni, C. and Matile, S. (1998) Side-chain hydrophobicity controls the activity of proton channel forming rigid rod-shaped polyols. *Chem. Commun.*, 755–756.
78. Gallivan, J.P. and Dougherty, D.A. (1999) Cation-pi interactions in structural biology. *Proc. Natl. Acad. Sci. U.S.A.*, **96**, 9459–9464.
79. Ma, J.C. and Dougherty, D.A. (1997) The cation-pi interaction. *Chem. Rev.*, **97**, 1303–1324.
80. Schottel, B.L., Chifotides, H.T., and Dunbar, K.R. (2008) Anion-pi interactions. *Chem. Soc. Rev.*, **37**, 68–83.
81. Gorteau, V., Bollot, G., Mareda, J., Perez-Velasco, A., and Matile, S. (2006) Rigid oligonaphthalenediimide rods as transmembrane anion-pi slides. *J. Am. Chem. Soc.*, **128**, 14788–14789.
82. Gorteau, V., Bollot, G., Mareda, J., and Matile, S. (2007) Rigid-rod anion-pi slides for multiion hopping across lipid bilayers. *Org. Biomol. Chem.*, **5**, 3000–3012.
83. Perez-Velasco, A., Gorteau, V., and Matile, S. (2008) Rigid oligoperylenediimide rods: anion-pi slides with photosynthetic activity. *Angew. Chem. Int. Ed. Engl*, **47**, 921–923.
84. Ayling, A.J., Perez-Payan, M.N., and Davis, A.P. (2001) New "cholapod" anionophores; high-affinity halide receptors derived from cholic acid. *J. Am. Chem. Soc.*, **123**, 12716–12717.
85. McNally, B.A., Koulov, A.V., Lambert, T.N., Smith, B.D., Joos, J.B., Sisson, A.L., Clare, J.P., Sgarlata, V., Judd, L.W., Magro, G., and Davis, A.P. (2008) Structure-activity relationships

in cholapod anion carriers: enhanced transmembrane chloride transport through substituent tuning. *Chemistry*, **14**, 9599–9606.

86. Dryfe, R.A., Hill, S.S., Davis, A.P., Joos, J.B., and Roberts, E.P. (2004) Electrochemical quantification of high-affinity halide binding by a steroid-based receptor. *Org. Biomol. Chem.*, **2**, 2716–2718.

87. Koulov, A.V., Lambert, T.N., Shukla, R., Jain, M., Boon, J.M., Smith, B.D., Li, H., Sheppard, D.N., Joos, J.B., Clare, J.P., and Davis, A.P. (2003) Chloride transport across vesicle and cell membranes by steroid-based receptors. *Angew. Chem. Int. Ed. Engl.*, **42**, 4931–4933.

88. Clare, J.P., Ayling, A.J., Joos, J.-B., Sisson, A.L., Magro, G., Perez-Payan, M.N., Lambert, T.N., Shukla, R., Smith, B.D., and Davis, A.P. (2005) Substrate discrimination by cholapod anion receptors: geometric effects and the "affinity-selectivity principle". *J. Am. Chem. Soc.*, **127**, 10739.

89. Smith, B.D. and Lambert, T.N. (2003) Molecular ferries: membrane carriers that promote phospholipid flip-flop and chloride transport. *Chem. Commun.*, 2261–2268.

90. Koulov, A.V., Mahoney, J.M., and Smith, B.D. (2003) Facilitated transport of sodium or potassium chloride across vesicle membranes using a ditopic salt-binding macrobicycle. *Org. Biomol. Chem.*, **1**, 27–29.

91. (a) Lambert, T.N., Boon, J.M., Smith, B.D., Perez-Payan, M.N., and Davis, A.P. (2002) Facilitated phospholipid flip-flop using synthetic steroid-derived translocases. *J. Am. Chem. Soc.*, **124**, 5276–5277; (b) DiVittorio, K.M., Lambert, T.N., and Smith, B.D. (2005) Steroid-derived phospholipid scramblases induce exposure of phosphatidylserine on the surface of red blood cells. *Bioorg. Med. Chem.*, **13**, 4485–4490.

92. (a) Miller, S.L. (1953) Production of amino acids under possible primitive earth conditions, *Science* **117**, 528–529; (b) Cairns-Smith, A.G. (1985) *Seven Clues to the Origin of Life: A Scientific Detective Story*, Canto/Cambridge University Press, Cambridge; (c) Dyson, F. (1999) *Origins of Life*, 2nd edn, Cambridge University Press, Cambridge; (d) Holm, N.G., Cairns-Smith, A.G., Daniel, R.M., Ferris, J.P., Hennet, R.J., Shock, E.L., Simoneit, B.R., and Yanagawa, H. (1992) Marine hydrothermal systems and the origin of life: future research. *Orig. Life Evol. Biosph.*, **22**, 181–242; (e) Ferris, J.P. and Ertem, G. (1993) Montmorillonite catalysis of RNA oligomer formation in aqueous solution. A model for the prebiotic formation of RNA. *J. Am. Chem. Soc.*, **115**, 12270–12275; (f) de Duve, C. (1993) The RNA world: before and after? *Gene*, **135**, 29–31; (g) Wachtershauser, G. (1994) Life in a ligand sphere, *Proc. Natl. Acad. Sci. U.S.A.*, **91**, 4283–4287; (h) Wachtershauser, G. (1998) in *The Molecular Origins of Life: Assembling Pieces of the Puzzle*, Brack, A. (ed.) Cambridge University Press, Cambridge, pp. 206–218; (i) Wachtershauser, G. (1998) in *Thermophiles: The Keys to the Molecular Evolution and the Origin of Life?* Wiegel, J., and Michael, A. W. W. (eds) CRC Press, Boynton Beach, FL, pp. 47–57; (j) Gesteland, R.F., Cech, T.R., and Atkins, J.F. (eds) (2005) in *The RNA World*, 3rd edn, Cold Spring Harbor Press, New York, p. 709; (k) Morigaki, K., Dallavalle, S., Walde, P., Colonna, S., and Luisi, P.L. (1997) Autopoietic self-reproduction of chiral fatty acid vesicles. *J. Am. Chem. Soc.*, **119**, 292–301.

93. Ohki, S. and Arnold, K. (1990) Surface dielectric constant, surface hydrophobicity, and membrane fusion. *J. Membrane Biol.*, **114**, 195–203.

94. Ido, Y., Vindigni, A., Chang, K., Stramm, L., Chance, R., Heath, W.F., DiMarchi, R.D., DiCera, E., and Williamson, J.R. (1997) Prevention of vascular and neural dysfunction in diabetic rats by C-peptide. *Science*, **277**, 563–566.

95. (a) Jentsch, T.J., Pusch, M., Rehfeldt, A., and Steinmeyer, K. (1993) The ClC family of voltage-gated chloride channels: structure and function. *Ann. NY Acad. Sci.*, **707**, 285–293; (b) Maduke, M., Miller, C., and Mindell, J.A. (2000) A decade of CLC chloride channels: structure, mechanism and many unsettled questions. *Annu. Rev. Biomol. Struct.*, **29**, 411–438; (c) Dutzler, R., Campbell, E.B., Cadene, M., Chait, B.T., and MacKinnon, R. (2002) X-ray structure of a ClC chloride channel at 3.0 Å reveals the molecular basis of anion selectivity. *Nature*, **415**, 287–294.

96. Zhou, J., Lu, X., Wang, Y., and Shi, J. (2002) Molecular dynamics study on ionic hydration. *Fluid Phase Equilib.*, **194–197**, 257–270.

97. Schlesinger, P.H., Ferdani, R., Pajewska, J., Pajewski, R., and Gokel, G.W. (2003) Replacing proline at the apex of heptapeptide-based chloride ion transporters alters their properties and their ionophoretic efficacy. *New J. Chem.*, **27**, 60–67.

98. Schlesinger, P.H., Ferdani, R., Liu, J., Pajewska, J., Pajewski, R., Saito, M., Shabany, H., and Gokel, G.W. (2002) SCMTR: a chloride-selective, membrane-anchored peptide channel that exhibits voltage gating. *J. Am. Chem. Soc*, **124**, 1848–1849.

99. Schlesinger, P.H., Djedovic, N.K., Ferdani, R., Pajewska, J., Pajewski, R., and Gokel, G.W. (2003) Anchor chain length alters the apparent mechanism of chloride channel function in SCMTR derivatives. *Chem. Commun.*, 308–309.

100. Djedovic, N., Ferdani, R., Harder, E., Pajewska, J., Pajewski, R., Schlesinger, P.H., and Gokel, G.W. (2003) The C-terminal ester of membrane anchored peptide ion channels affects anion transport. *Chem. Commun.*, 2862–2863.

101. Djedovic, N., Ferdani, R., Harder, E., Pajewska, J., Pajewski, R., Weber, M.E., Schlesinger, P.H., and Gokel, G.W. (2005) The C- and N-terminal residues of synthetic heptapeptide ion channels influence transport efficacy through phospholipid bilayers. *New J. Chem.*, **29**, 291–305.

102. Schlesinger, P.H., Ferdani, R., Pajewski, R., Pajewska, J., and Gokel, G.W. (2002) A hydrocarbon anchored peptide that forms a chloride-selective channel in liposomes. *Chem. Commun.*, 840–841.

103. (a) McNally, B.A., Koulov, A.V., Smith, B.D., Joos, J.B., and Davis, A.P. (2005) A fluorescent assay for chloride transport; identification of a synthetic anionophore with improved activity. *Chem. Commun.*, 1087–1089; (b) Seganish, J.L., Fettinger, J.C., and Davis, J.T. (2006) Facilitated chloride transport across phosphatidylcholine bilayers by an acyclic calixarene derivative: structure-function relationships. *Supramol. Chem.*, **18**, 257–264.

104. Ferdani, R., Li, R., Pajewski, R., Pajewska, J., Winter, R.K., and Gokel, G.W. (2007) Transport of chloride and carboxyfluorescein through phospholipid vesicle membranes by heptapeptide amphiphiles. *Org. Biomol. Chem.*, **5**, 2423–2432.

105. Ferdani, R. and Gokel, G.W. (2006) Planar bilayer studies reveal multiple conductance states for synthetic anion transporters. *Org. Biomol. Chem.*, **4**, 3746–3750.

106. Hille, B. (2001) *Ionic Channels of Excitable Membranes*, 3rd edn, Sinauer Associates, Sunderland, MA, p. 814.

107. Segel, I. (1975) *Enzyme Kinetics. Behavior and Analysis of Rapid Equilibrium and Steady-State Enzyme Systems*, John Wiley & Sons, Inc., New York (Wiley Classics Edition, 1993), pp. 371–375.

108. Pajewski, R., Pajewska, J., Li, R., Fowler, E.A., and Gokel, G.W. (2007) The effect of midpolar regime mimics on anion transport mediated by amphiphilic heptapeptides. *New J. Chem.*, **31**, 1960–1972.

109. Saito, M., Korsmeyer, S.J., and Schlesinger, P.H. (2000) BAX dependent cytochrome-c transport reconstituted in pure liposomes. *Nat. Cell Biol.*, **2**, 553–555.

110. Pajewski, R., Ferdani, R., Pajewska, J., Li, R., and Gokel, G.W. (2005) Cation dependence of chloride ion complexation by open-chained receptor molecules in chloroform solution. *J. Am. Chem. Soc.*, **126**, 18281–18295.

111. Ferdani, R., Pajewski, R., Pajewska, J., Schlesinger, P.H., and Gokel, G.W. (2006) Glycine position permutations and peptide length alterations change the aggregation state and efficacy of ion-conducting, pore-forming amphiphiles. *Chem. Commun.*, 439–441.
112. Pajewski, R., Ferdani, R., Pajewska, J., Djedovic, N., Schlesinger, P.H., and Gokel, G.W. (2005) Evidence for dimer formation by an amphiphilic heptapeptide that mediates chloride and carboxyfluorescein release from liposomes. *Org. Biomol. Chem.*, **3**, 619–625.
113. Pajewski, R., Ferdani, R., Schlesinger, P.H., and Gokel, G.W. (2004) Chloride complexation by heptapeptides: influence of C- and N-terminal sidechains and counterion. *Chem. Commun.*, 160–161.
114. Kavallieratos, K., Bertao, C.M., and Crabtree, R.H. (1999) Hydrogen bonding in anion recognition: a family of versatile, nonpreorganized neutral and acyclic receptors. *J. Org. Chem.*, **64**, 1675–1683.
115. Cook, G.A., Pajewski, R., Aburi, M., Smith, P.E., Prakash, O., Tomich, J.M., and Gokel, G.W. (2006) NMR structure and dynamic studies of an anion-binding, channel-forming heptapeptide. *J. Am. Chem. Soc.*, **128**, 1633–1638.
116. Dutzler, R., Campbell, E.B., and MacKinnon, R. (2003) Gating the selectivity filter in ClC chloride channels. *Science*, **300**, 108–112.
117. You, L. and Gokel, G.W. (2008) Fluorescent, synthetic amphiphilic heptapeptide anion transporters: evidence for self-assembly and membrane localization in liposomes. *Chemistry*, **14**, 5861–5870.
118. Reichardt, C. (1994) Solvatochromic dyes as solvent polarity indicators. *Chem. Rev.*, **94**, 2319–2358.
119. (a) Schiffer, M., Chang, C.-H., and Stevens, F.J. (1992) The functions of tryptophan residues in membrane proteins. *Protein Eng.*, **5**, 213–214; (b) Sansom, M.S., Bond, P.J., Deol, S.S., Grottesi, A., Haider, S., and Sands, Z.A. (2005) Molecular simulations and lipid-protein interactions: potassium channels and other membrane proteins. *Biochem. Soc. Trans.*, **33**, 916–920.
120. (a) Abel, E., Fedders, M.F., and Gokel, G.W. (1995) Vesicle formation from *N*-alkylindoles: implications for tryptophan-water interactions. *J. Am. Chem. Soc.*, **117**, 1265–1270; (b) Gokel, G.W. (2006) Indole, the aromatic element of tryptophan as a pi-donor and amphiphilic headgroup. *Int. Congr. Ser.*, **1304**, 1–14 (Interdisciplinary Conference on Tryptophan and Related Substances: Chemistry, Biology, and Medicine).
121. (a) Elliott, E.K., Daschbach, M.M., and Gokel, G.W. (2008) Aggregation behavior and dynamics of synthetic amphiphiles that self-assemble to anion transporters. *Chemistry*, **14**, 5871–5879; (b) Elliott, E.K., Stine, K.J., and Gokel, G.W. (2008) Air-water interfacial behavior of amphiphilic peptide analogs of synthetic chloride ion transporters. *J. Membr. Sci.*, **321**, 43–50.
122. Cuthbert, A.W. (2006) The prospects of pharmacotherapy for cystic fibrosis. *J. R. Soc. Med.*, **99** (Suppl. 46), 30–35.
123. Li, H., Sheppard, D.N., and Hug, M.J. (2004) Transepithelial electrical measurements with the Ussing chamber. *J. Cyst. Fibros.*, **3** (Suppl. 2), 123–126.
124. Pajewski, R., Djedovic, N., Harder, E., Ferdani, R., Schlesinger, P.H., and Gokel, G.W. (2005) Pore formation in and enlargement of phospholipid liposomes by synthetic models of ceramides and sphingomyelin. *Bioorg. Med. Chem.*, **13**, 29–37.
125. Siskind, L.J., Kolesnick, R.N., and Colombini, M. (2002) Ceramide channels increase the permeability of the mitochondrial outer membrane to small proteins. *J. Biol. Chem.*, **277**, 26796–26803.

9
Anion Sensing by Fluorescence Quenching or Revival

Valeria Amendola, Luigi Fabbrizzi, Maurizio Licchelli, and Angelo Taglietti

9.1
Introduction

The occurrence of a receptor–anion equilibrium in solution can be followed through a variety of procedures. The most direct and informative measurement is related to the ^{1}H NMR titration experiment. In fact, the receptor–anion interaction is in many cases based on hydrogen-bonding interactions, and the variation in the ^{1}H NMR pattern of the receptor on anion addition can provide essential pieces of information both on the stability of the receptor–anion complex(es) and on the geometrical–stereochemical aspects of the interaction. One drawback of this approach is that NMR spectra have to be taken at a relatively high concentration scale, rarely lower than 10^{-3} M, which prevents safe determination of binding constants higher than 10^3–10^4. Other convenient responses of anion addition to the receptor solution involve a change (i) of an electrode potential, (ii) of the color and the absortion spectrum, and (iii) of the fluorescence and the emission spectrum. As anions are in most cases redox inactive and colorless and do not fluoresce, the envisaged signal has to be provided by the receptor, which, for instance, should contain a redox-active, a chromogenic, or a fluorogenic subunit, respectively. Anion binding is expected to interfere with the mechanism of signal production, thus modifying signal energy and intensity and providing a convenient way to evaluate the stability of the receptor–anion complex(es). In these cases, we now have so-called electrochemical, colorimetric, and fluorescent *sensors*. Sensor, now a popular word, was introduced in the common language in the twentieth century and refers to "a device which detects or measures a physical property and records, indicates, or otherwise responds to it" [1]. Such a term has been more recently transferred into the molecular world. According to IUPAC, "a *chemical sensor* is a device that transforms chemical information, ranging from the concentration of a specific sample component to total composition analysis, into an analytically useful signal" [2]. Chemical sensors are typically classified on the basis of their physical response: optical (which includes absorbance and fluorescence), electrochemical, and so on. A SciFinder inspection in June 2011 gave nearly 500 articles for the topic "anion fluorescent sensor(s)," most of which

Anion Coordination Chemistry, First Edition. Edited by Kristin Bowman-James, Antonio Bianchi, and Enrique García-España.

referred to individual molecules, not assembled in nano- or macro-devices. Nearly 300 papers were associated with the topic "anion colorimetric sensors" (even if the chromogenic subunit present in the receptor is often also fluorogenic) and about 400 to "anion electrochemical sensors." The use of fluorescence as a signal of anion detection is related to several factors: (i) it is visual; (ii) it can be detected at rather low concentration levels, thus allowing the safe determination of binding constants as high as 10^7-10^8; (iii) instrumentation is not particularly expensive (more than 1 order of magnitude lower than that for NMR studies, for instance); and (iv) it can be controlled (switched on–off) through two well-defined mechanisms: electron transfer and energy transfer. Very importantly, from the point of view of the application, fluorescent receptors allow the detection in cells and tissues of anionic analytes, including amino acids, in real time and real space, using a fluorescence microscope.

This chapter is not intended to provide an exhaustive coverage of the literature on anion fluorescent sensors over the past 10 years (a period corresponding to the intense development of the topic), but it rather considers some significant and representative examples to illustrate the potential of the method. The state of the art of the investigation of anion recognition chemistry, including different experimental approaches, has been recently considered in a themed issue on the supramolecular chemistry of anionic species of *Chemical Society Reviews* [3].

9.2
Anion Recognition by Dynamic and Static Quenching of Fluorescence

The most direct use of fluorescence as an analytical signal involves the quenching of the emission of a fluorophore on addition of aliquots of the envisaged analyte.

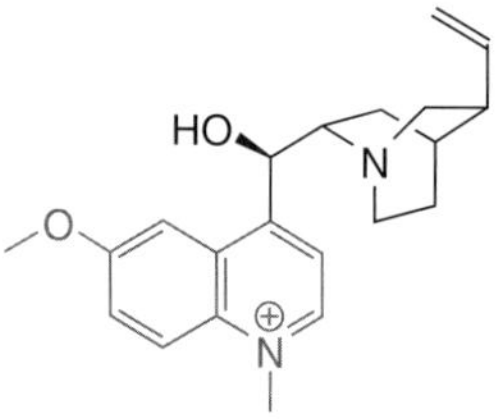

1

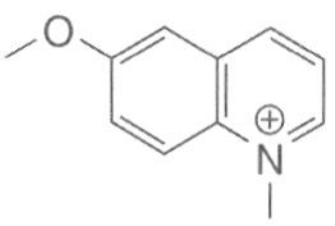

2

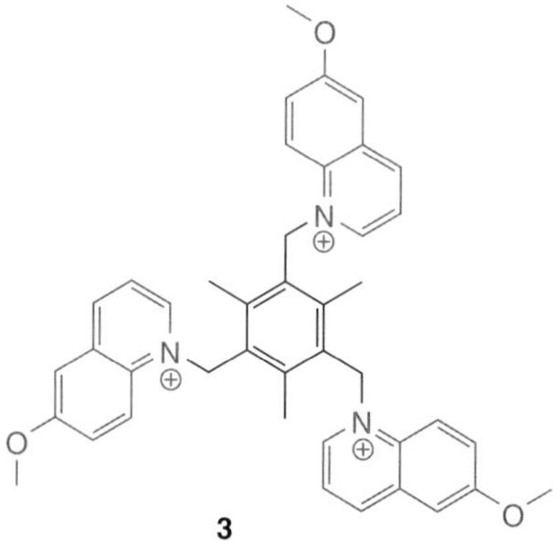

3

Quenching of fluorescence was first observed in 1869 by Stokes, when he noted that the light emission of quinine (**1**), in a solution of dilute sulfuric acid, was substantially reduced on addition of hydrochloric acid, in particular by Cl^- ions [4]. More recently, dyes based on the same structural motif of quinine, for example, the quinolinium subunit (e.g., 6-methoxy-1-methylquinolinium (**2**)), have been

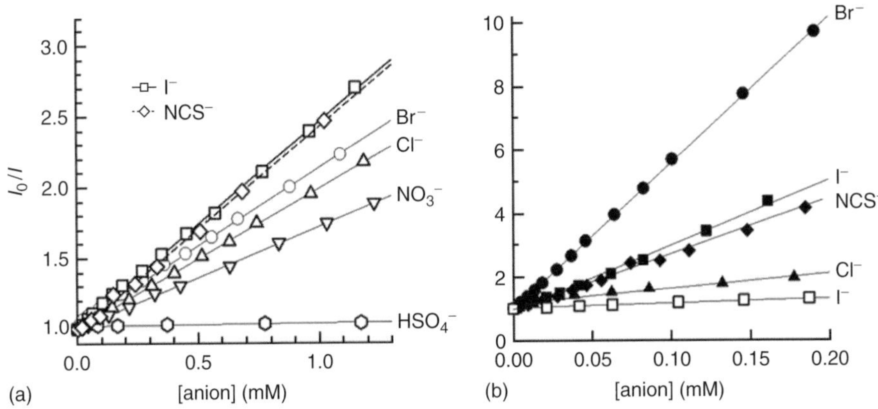

Figure 9.1 Stern–Volmer profiles obtained from the spectrofluorimetric titrations in an MeCN solution, at 25 °C, with some inorganic anions: (a) fluorescent probe **2** and (b) fluorescent receptor **3**. Open symbols: **2**; filled symbols: **3**; straight lines in Figure 9.1b, referring to fluorescent sensor **3**, do not fit data satisfactorily (filled symbols); nonlinear fitting is required (see text).

synthesized [5] and used as chloride fluorescent indicators in water for biological applications, in particular, for diagnosing cystic fibrosis, which is characterized by an unusually high Cl^- concentration in a patient's sweat and saliva [6]. In particular, on excitation of a solution of **2** at $\lambda = 330$ nm, a rather intense emission band develops at 440 nm. The photoexcited quinolinium fragment is quenched by the halide ion through an electron transfer mechanism, following a bimolecular collision [7], while the process is satisfactorily described by the Stern–Volmer equation:

$$\frac{I_0}{I} = 1 + k_q\tau_0[A] = 1 + K_{SV}[A] \tag{9.1}$$

where k_q is the quenching constant, τ_0 is the lifetime of the excited fluorophore and its product, $k_q \times \tau_0 = K_{SV}$ is the Stern–Volmer constant. Specifically, from the I_0/I linear plot it is possible to determine quantitatively the analyte A, over a wide range of concentration, from the measured fluorescence intensity I.

Quenching of **2** by a variety of anions was later investigated in an MeCN solution [8]. Figure 9.1a shows the Stern–Volmer plots obtained for titrations of an MeCN solution 10^{-6} M in **2** with halides (Cl^-, Br^-, I^-), pseudohalides (NCS^-), and oxoanions (NO_3^-, HSO_4^-), added as MeCN solutions of their tetrabutylammonium salts. Titration data fit well the linear equation (Eq. (9.1)), indicating the occurrence of a dynamic quenching process, that is, solely associated with the diffusional collisions between the anion and the excited fluorophore **2**. Pertinent values of K_{SV} are reported in Table 9.1.

It is observed that I^- and NCS^- exert the highest quenching effect on **2**, as expressed by K_{SV} values (visually, the slope of the I_0/I vs $[X^-]$ straight line), distinctly higher than Cl^-. In particular, the sequence of K_{SV} values parallels the sequence of the potentials of the oxidation half-reactions involving the anion:

Table 9.1 Parameters associated with the quenching of **2** and **3** in an MeCN solution by selected inorganic anions [7].

Anion	2/log K_{SV}[a]	3/log K_{SV}[b]	3/log K_{ass}[b]	3/log K_{SV}[c]	3/log K_{ass}[c]	ρ[c]
Cl^-	2.99 ± 0.03	–	–	3.16 ± 0.02	4.60 ± 0.06	0.71 ± 0.01
NO_3^-	2.86 ± 0.02	–	–	2.26 ± 0.02	3.87 ± 0.05	0.31 ± 0.01
HSO_4^-	1.27 ± 0.02	–	–	2.83 ± 0.03	3.78 ± 0.06	0.98 ± 0.04
Br^-	3.04 ± 0.04	–	4.55 ± 0.05	–	–	–
I^-	3.16 ± 0.04	4.09 ± 0.05	3.48 ± 0.01	–	–	–
NCS^-	3.16 ± 0.03	3.91 ± 0.05	3.40 ± 0.02	–	–	–

[a] Calculated by the fitting of the I_0/I versus [X] data with the Stern–Volmer equation (Eq. (9.1)).
[b] Calculated by the fitting of the I_0/I versus [anion] data with Eq. (9.2).
[c] Calculated by the fitting of the I_0/I versus [anion] data with Eq. (9.3).

$X^- \leftrightarrows X^\cdot + e^-$ and reflects anion tendencies to transfer an electron to the excited fluorophore.

Systems showing a linear Stern–Volmer behavior, thus displaying a purely dynamic quenching, may be useful under particular circumstances but have to be considered as rather primitive sensing devices for anions. In particular, sensing selectivity is solely related to the quenching properties of the anion (in most cases related to its reducing tendencies), excluding the more valuable geometrical effects, that is, size and shape complementarity with the receptor. Thus, since several years, interest has been developed on the design of fluorescent sensors capable of interacting firmly with the envisaged analyte (X^-), giving rise an adduct of 1 : 1 stoichiometry, characterized by a definite association constant K_{ass} (related to the equilibrium: $R + X^- \leftrightarrows [R.X]^-$). In order to form stable complexes, such receptors must contain multiple sites of interactions with the analyte, strategically positioned within their framework. From this perspective, system **3** was designed, in which three 6-methoxy-1-methylquinolinium fragments have been implanted on a mesityl platform [8]. The 1,3,5-trialkylbenzene scaffold has been frequently used in order to build up a variety of receptors [9]. In particular, system **3** possesses three symmetrically placed positive charges and may generate a cavity suitable for anion inclusion. On the other hand, molecular mechanics studies on system **3**/Cl^- gave the structure shown in Figure 9.2 [8]: in the model complex, the receptor is arranged in such a way that a well-defined cavity is generated and the chloride ion stays inside the bowl, close to the positively charged nitrogen atoms.

These considerations prompted us to investigate the capability of the tripodal receptor **3** to act as a fluorescent sensor for anions, and, for comparative purposes, we carried out spectrofluorimetric titration experiments under the same conditions as for the fluorescent probe **2** (MeCN solution, 25 °C). Figure 9.1b shows the Stern–Volmer plot obtained for the titration of **3** with anions and discloses two distinctive features: (i) the quenching effect exerted by anions on receptor **3**, as expressed by the slope of the roughly linear Stern–Volmer plots, is much more

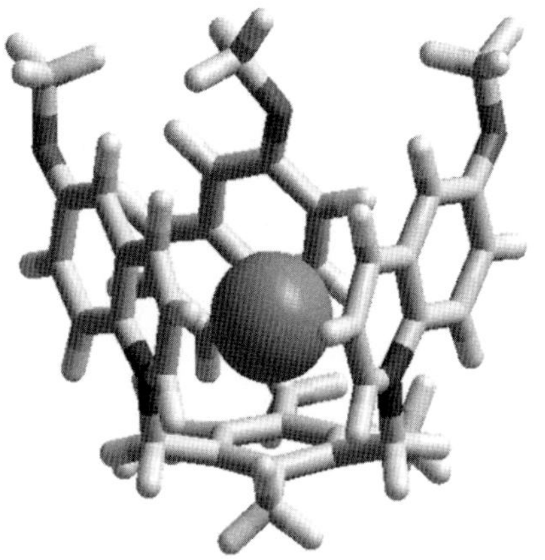

Figure 9.2 The structure of the $[\mathbf{3}\cdot Cl]^{2+}$ complex, as obtained through molecular mechanics calculations.

pronounced than observed for **2** (the plot pertinent to the system $\mathbf{2}/I^-$, empty squares, is shown in Figure 9.1b for comparison) and (ii) the sequence of slope values is different than for **2** and does not parallel the sequence of anion oxidation potentials. However, a closer inspection of the I_0/I versus $[X^-]$ profiles in the first steps of the titration experiment indicates that they are not perfectly linear and do not fit Eq. (9.1). Moreover, two distinct types of behavior are observed, depending on the more or less pronounced reducing tendencies of the anion.

1) Type **a**, reducing anions (I^-, Br^-, NCS^-): a parabolic behavior of the I_0/I versus [anion] plots is observed, with an upward curvature (Figure 9.3a). This is consistent with the simultaneous occurrence of both static and dynamic quenching. In particular, the observed plots fit the equation:

$$\frac{I_0}{I} = (1 + K_{SV}[A])(1 + K_{ass}[A]) \tag{9.2}$$

where K_{ass} is the equilibrium constant associated with the formation of the 1 : 1 receptor–anion complex $[\mathbf{3}\cdot X]^{2+}$. Pertinent values of K_{SV} and K_{ass} are reported in Table 9.1. Notice that the Br^- ion shows a linear, rather than a parabolic, curve: this indicates that, over the investigated quencher concentration, $1 + K_{SV}[Q] \approx 1$. As a consequence, the slope of the linear plot corresponds to the association constant K_{ass}.

2) Type **b**, poorly reducing anions (Cl^-, NO_3^-, HSO_4^-). A closer inspection of the titration data for chloride, over the lower concentration range, shows that I_0/I versus [anion] plots present a pronounced convex curvature (see Figure 9.3b). Such a behavior can be explained by assuming that the emission of the receptor–anion complex is not completely quenched by the included anion. In particular, if we define ρ as the residual fluorescence, the following equation can be derived:

$$\frac{I_0}{I} = \frac{(1 + K_{SV}[A])(1 + K_{ass}[A])}{1 + \rho K_{ass}[A]} \tag{9.3}$$

It is observed that spectrofluorimetric titration data fit well Eq. (9.3). In particular, for Cl^-, $\rho = 0.71$, that is, the chloride complex emits 71% of the fluorescence of the

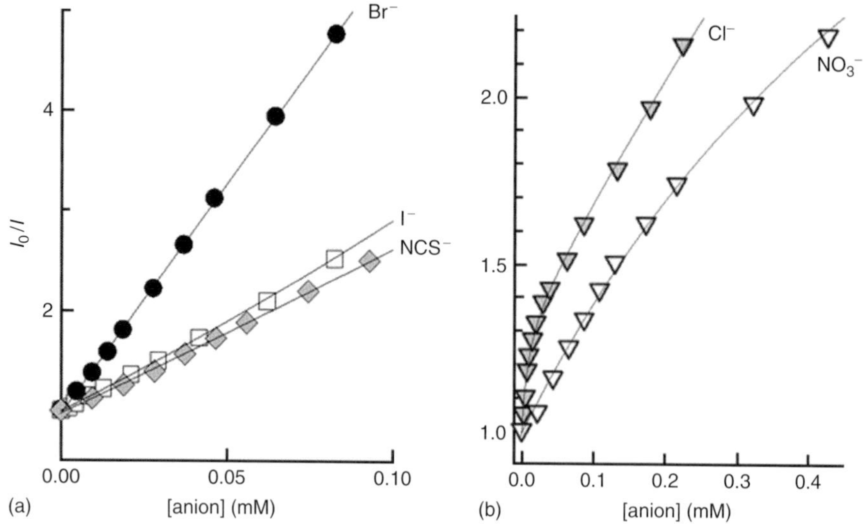

Figure 9.3 I_0/I profiles obtained from the spectrofluorimetric titrations of **3** in MeCN solution, at 25 °C: (a) plots show a moderate concave curvature and fit Eq. (9.2) and (b) plot shows a definite convex curvature and fit Eq. (9.3).

uncomplexed receptor, while the included nitrate ion exerts a more pronounced quenching of the receptor's fluorescence ($\rho = 0.31$). On the other hand, the effect exerted by the included HSO_4^- is very moderate ($\rho = 0.98$).

Thus, analysis of spectrofluorimetric titration data allows one to determine K_{ass} values (which assess receptor's selectivity toward anion) and to define the details of the quenching mechanism. The accuracy in the determination of K_{ass} values can be assessed by carrying out titration experiments on other observables, for example, absorbance (spectrophotometric titrations) or chemical shift (^{1}H NMR titrations). The system **3**/Br^- has been considered as an example. Figure 9.4a shows the absorption spectra recorded over the course of the titration of an MeCN solution 4×10^{-4} M in **3** with $[Bu_4N]Br$.

Definite spectral modifications are observed, following the addition of Br^-. The titration profile at 340 nm, shown in Figure 9.4b, clearly indicates the formation of a complex of 1 : 1 stoichiometry. Moreover, fitting of titration data over the 300–400 nm interval, using a nonlinear least-squares method, gave a log K value of 4.55 ± 0.05, in good agreement with that obtained by spectrofluorimetric studies.

Excellent agreement between log K values determined by both spectrofluorimetric and spectrophotometric methods was observed, as shown in Table 9.2. For chloride and bromide, log K values were determined also by ^{1}H NMR titration experiments, in a CD_3CN solution at 25 °C.

Figure 9.5a shows some representative ^{1}H NMR spectra taken over the course of the titration with chloride. The most remarkable modification refers to the H3 atom, which, on addition of Cl^-, undergoes a downfield shift, thus giving rise to a

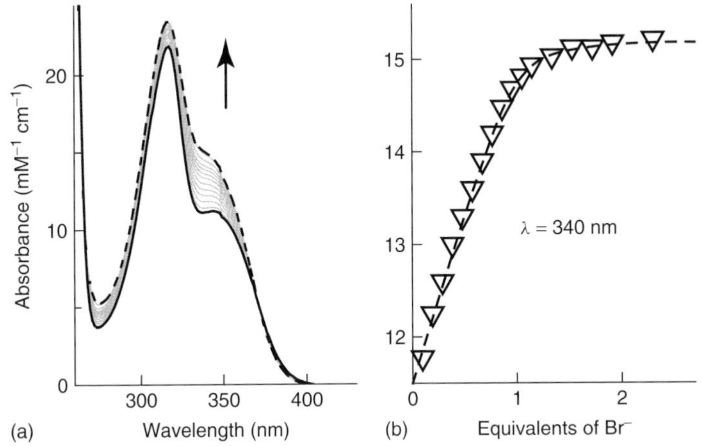

Figure 9.4 (a) Spectra recorded over the course of the titration of an MeCN solution 4×10^{-4} M in **3** with $[Bu_4N]Br$ and (b) titration profile at 340 nm (symbols); fitting for $\log K = 4.55 \pm 0.05$ (dashed line).

Table 9.2 Log K values for the equilibrium $\mathbf{3} + X^- \leftrightarrows [\mathbf{3}\cdot X]^{2+}$ in MeCN, determined by different methods, based on spectrofluorimetric, spectrophotometric, and ^1H NMR (CD_3CN) titration experiments, at 25 °C.

Anion	Emission	Absorbance	NMR
Cl^-	4.60 ± 0.06	4.55 ± 0.03	4.64 ± 0.06
Br^-	4.55 ± 0.05	4.42 ± 0.04	4.28 ± 0.05
I^-	3.48 ± 0.01	3.48 ± 0.01	–
NO_3^-	3.87 ± 0.05	3.91 ± 0.03	–
HSO_4^-	3.78 ± 0.06	3.79 ± 0.02	–
NCS^-	3.40 ± 0.02	3.40 ± 0.02	–
ClO_4^-	–	<2	–
$CF_3SO_3^-$	–	<2	–

saturation profile (see Figure 9.5b). The log K value (4.64 ± 0.06), calculated from ^1H NMR titration data (see Figure 9.5b), is in good agreement with the values obtained from spectrofluorimetric and spectrophotometric studies.

Figure 9.1b shows that the bowl-shaped fluorescent sensor **3** affords a much better fluorimetric response to anions (in terms of quenching of the emission) than the simple fluorophore **2**. Moreover, equilibrium studies carried out on different observables have shown that the tripodal receptor **3** exerts a rather moderate selectivity with respect to the anions investigated. For instance, in the case of halides, the stability decreases along the series $Cl^- > Br^- > I^-$, which reflects the decrease in charge density of the anion and its decreasing tendency to establish

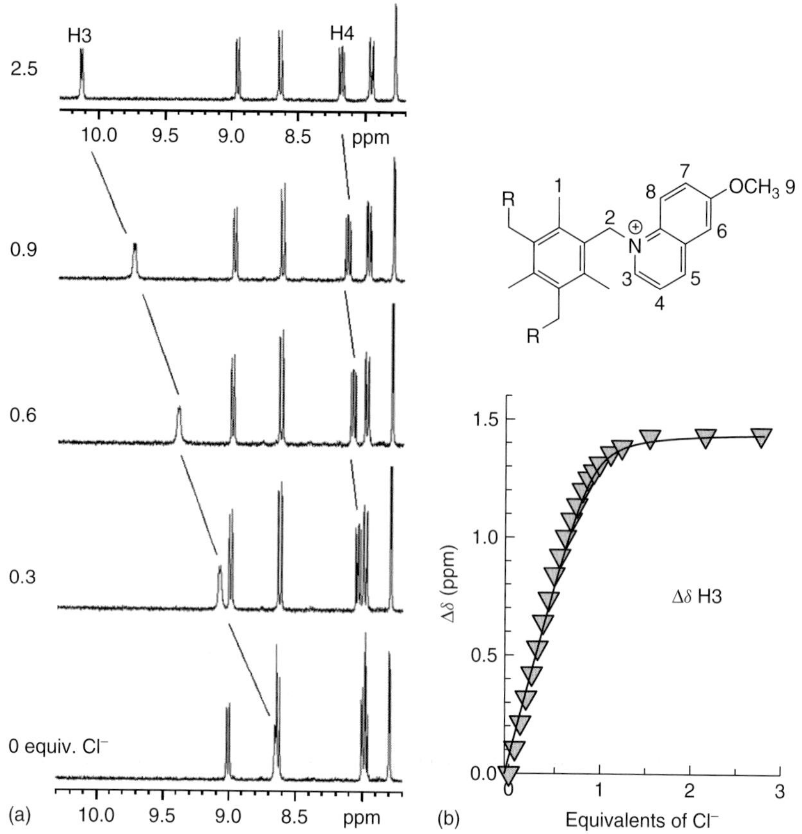

Figure 9.5 (a) ^{1}H NMR spectra of a solution of **3** containing varying amounts (equivalents) of chloride (25 °C, CD_3CN) and (b) titration profile based on the chemical shift of proton H3 (symbols); fitting for log $K = 4.64 \pm 0.06$ (dashed line).

electrostatic interactions. Studies on molecular recognition over the past decades have shown that selectivity typically results from the matching of the geometrical features of the receptor and of the substrate, in particular, in terms of size and shape complementarity. In this sense, the tripodal receptor **3**, due to its open nature and pronounced flexibility, can easily adapt its cavity to any geometrical requirement, thus disfavoring size (and shape) selectivity.

Figure 9.6 summarizes the behavior of **3** as a fluorescent sensor for halide and pseudohalide anions: selectivity (expressed by log K) smoothly decreases along the series $Cl^- > Br^- > I^- > NCS^-$, but the fluorimetric response (expressed in terms of quenching efficiency, I_0/I, in the presence of an anion excess) follows a different order: $Br^- \gg I^- > NCS^- > Cl^-$. The neat predominance of bromide reflects the successful combination of the relatively high association constant and of the reasonable reducing tendencies of the anion.

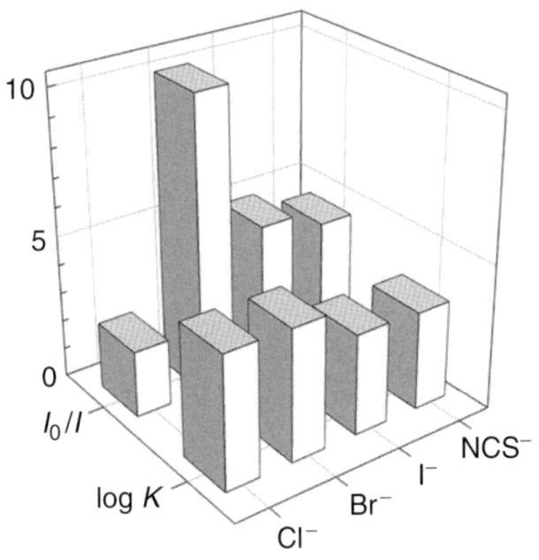

Figure 9.6 Binding affinity (log K) and signaling efficiency (I_0/I) in the interaction of receptor **3** with selected anions. I_0/I values refer to the experimental conditions of Figure 9.1b, with [anion] = 0.2 mM. The highest affinity (Cl^-) does not correspond to the highest quenching efficiency (Br^-).

In conclusion, the behavior of system **3** is fairly distant from perfection, which should entail the highest selectivity both in the binding affinity and in the fluorimetric response.

9.3
Fluorescent Sensors Based on Anthracene and on a Polyamine Framework

The first elaborate molecular sensor for anions providing a fluorimetric response was reported by Czarnik [10]. This system consisted of a branched tetramine subunit covalently linked, through a $-CH_2-$ spacer, to an anthracene fragment (L). Anthracene is a convenient and largely used fluorophore for three main reasons: (i) it displays a rather intense and well-structured emission band at the border between ultraviolet and visible regions (blue fluorescence), (ii) it is chemically resistant, and (iii) it provides two positions, C(9) and C(10) suitable for easy derivatization. Anthracene is intrinsically hydrophobic but, if functionalized with polar and/or electrically charged groups, may be soluble in water or in aqueous organic media. At a pH close to neutrality, 6.8, the anthrylamine L exists in the tri-protonated form, LH_3^{3+} (**4** in Scheme 9.1).

It has been suggested, on the basis of the spectrofluorimetric behavior, that the less basic amine group, unprotonated at pH = 6.8, is the benzylic one, because of the presence of the proximate electron-withdrawing anthracene fragment. In fact, anthracene fluorescence in the LH_3^{3+} species (**4**) is almost completely quenched because of electron transfer from the benzylamine nitrogen atom to the excited fluorogenic fragment. On addition of phosphate, the $[LH_3\cdots HPO_4]^+$ complex forms, in which an HPO_4^{2-} ion is wrapped by the polyammonium moiety of **4** and establishes four hydrogen-bonding interactions with the receptor. In particular, each one of the three anion oxygen atoms detaining a partial negative charge (formally two-thirds) receives an H–bond from an N–H fragment of an ammonium group of **4**. Then, the O–H fragment of HPO_4^{2-} donates an H–bond to the lone pair of the benzylic nitrogen atom (**5** in Scheme 9.1). However, the $[LH_3\cdots HPO_4]^+$

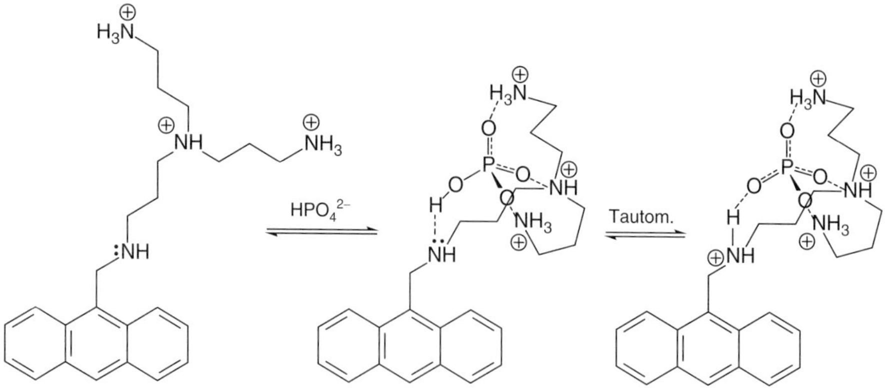

4
Low fluorescence

5

6
High fluorescence

Scheme 9.1 The interaction of phosphate with LH_3^{3+} at pH = 7. The $[LH_3 \cdots HPO_4]^+$ complex that forms can be conveniently represented by a tautomeric equilibrium between **5** and **6**.

complex could be better represented through a tautomeric equilibrium between **5** and **6**, in which the proton has been definitively transferred from the O–H fragment to the benzylamine nitrogen atom. In fact, the hydrogen bond interaction is currently viewed as a "frozen" partial transfer of a proton from the donor to the acceptor [11]. The state of the advancement of such an intracomplex acid–base reaction depends on the relative strength of the acid (H–bond donor) and of the base (H–bond acceptor). Therefore, restoration of anthracene fluorescence on complexation is due to the fact that the lone pair of the benzylic nitrogen atom becomes engaged in a hydrogen-bonding interaction, so that the electron transfer to the proximate anthracene subunit is thermodynamically prevented, which allows the fluorophore to display its natural emission. The sulfate ion under the same conditions induces only a small enhancement of the fluorescence of **4**: this seems to be due to the fact that at pH = 6.8 sulfate exists 100% as sulfate and does not possess any hydrogen atom to establish an H-bond interaction with the benzylic nitrogen atom of **4**.

Receptor **7**, in which two branched tetramine subunits have been implanted on an anthracene fragment, displays a more sophisticated function [12]. It is represented in Scheme 9.2 in its hexaprotonated form (the predominant species at pH = 7).

In a neutral solution, the fluorescence of the anthracene subunit is almost completely quenched in view of the occurrence of an intramolecular electron transfer from one of the benzylic nitrogen atoms. However, on addition of pyrophosphate, fluorescence is fully revived. This is due to the capability of each of the two –OH fragments of the pyrophosphate ion, $H_2P_2O_7^{2-}$, to donate one H-bond to each benzylic nitrogen atom of the receptor, to a more or less advanced extent, thus preventing the occurrence of the intramolecular electron transfer

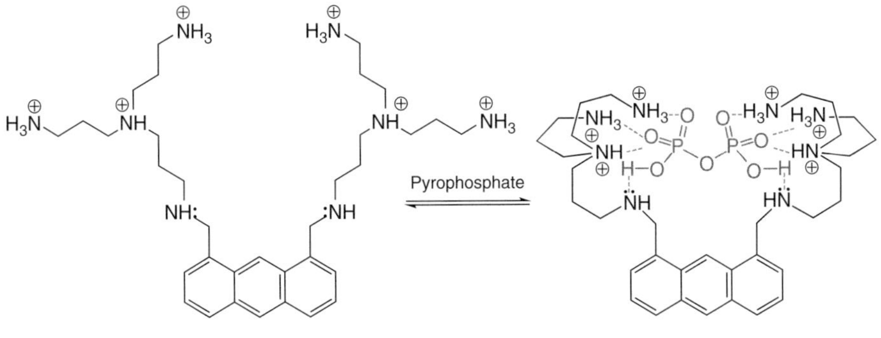

7
Low fluorescence

8
High fluorescence

Scheme 9.2 The interaction of pyrophosphate with **7** (LH_6^{6+}) at pH = 7 to give the $[LH_6 \cdots H_2P_2O_7]^{4+}$ complex **8**.

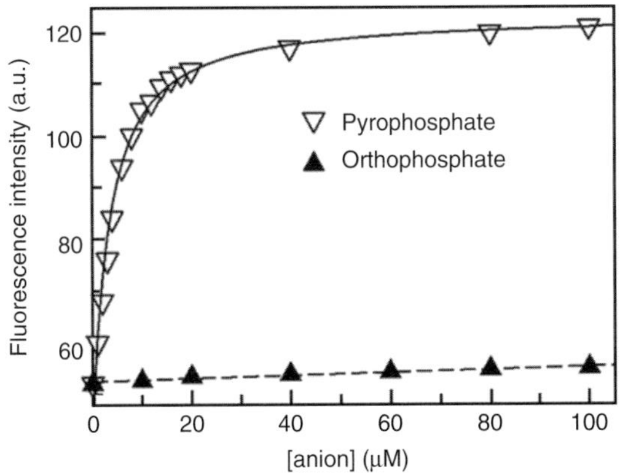

Figure 9.7 Fluorescence intensity profiles of the titrations of **7** at pH = 7 with pyrophosphate and orthophosphate.

processes. Figure 9.7 shows the profile of the fluorescence intensity on addition of pyrophosphate (open symbols).

On the other hand, addition of phosphate does not induce any fluorescence revival (filled symbols in Figure 9.7). In this connection, it should be considered that the first added phosphate ion goes to interact with one of the tetramine subunit, preventing the occurrence of the benzylic nitrogen-to-anthracene photoinduced electron transfer process. In contrast, the addition of the second phosphate ion is disfavored because of steric and electrostatic repulsive effects. Thus, the benzylic nitrogen atom-to-anthracene photoinduced electron transfer process keeps operating, and fluorescence remains quenched.

The anthrylamines (**4** and **7**) represent interesting examples of turn-on fluorescent sensors, in which a preliminary quenched emission is restored on interaction with the analyte. Turn-on systems are, in principle, more valuable than turn-off analogs in which recognition is signaled by fluorescence quenching. Such a preference is probably related to the experiment of fluorescence microscopy in which the activity of the envisaged analyte has to be revealed in real time and, in particular, in real space. In these circumstances, it seems easier to detect a spot switching on in the dark (e.g., an electrical torch in the Black Forest) than one switching off in the light (a lighted window in the night of Manhattan). Fluorescence turning on is based on an amine group, which behaves like a switch of the emission of a close fluorophore. Such an anthrylamine switch can be operated not only by anions, but also by metal ions (alkali [13] and group XII) [14].

There also exist turn-off polyamine-based receptors, in which the electron-donating tendencies of nitrogen atom(s) are inhibited by the coordination to a metal center. This is the case of a series of systems based on the tetramine tren. One of the first examples was the anthrylamine (**9**) (L) [15].

In MeOH, the fluorescence of the anthracene subunit is quenched by the typical photoinduced electron transfer from the proximate amine nitrogen atom. On addition of 1 equiv. of Zn^{II}. the complex $[Zn^{II}(\mathbf{9})]^{2+}$(= **10**) forms, which displays the full fluorescence of anthracene. In fact, the lone pair of the benzylamine nitrogen atom is engaged in the coordinative interaction with the Zn(II) ion and is no longer available for the electron transfer. In particular, the metal complex exhibits a trigonal bipyramidal geometry, with four coordination sites occupied by the amine groups, as sketched in Figure 9.8. A solvent molecule S resides in an axial position and can be replaced by a ligand displaying higher affinity, for example, by benzoate. The formation of the ternary complex $[Zn^{II}(\mathbf{9})(C_6H_5COO)]^+$ does not affect the emission of the anthracene subunit. However, on addition of a benzoate bearing an electron donor

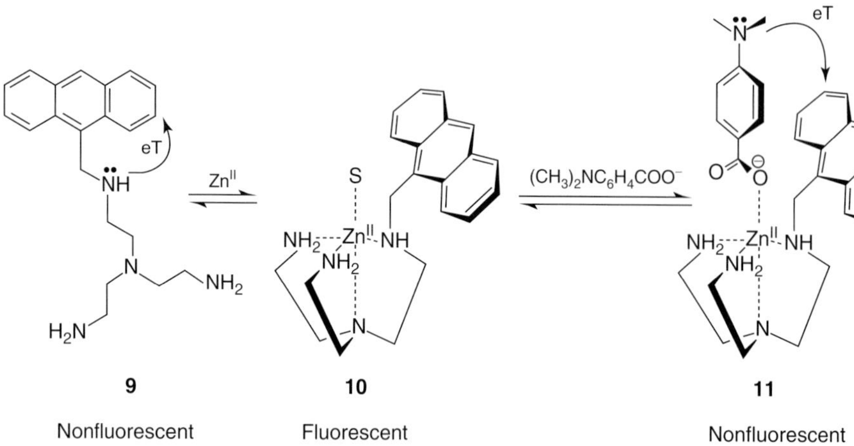

Figure 9.8 Consecutive equilibria involving (i) the formation of the Zn(II) tetramine complex **10** and (ii) the interaction with the anion 4-*N*,*N*-dimethylaminobenzoate in a methanolic solution to give complex **11** [15].

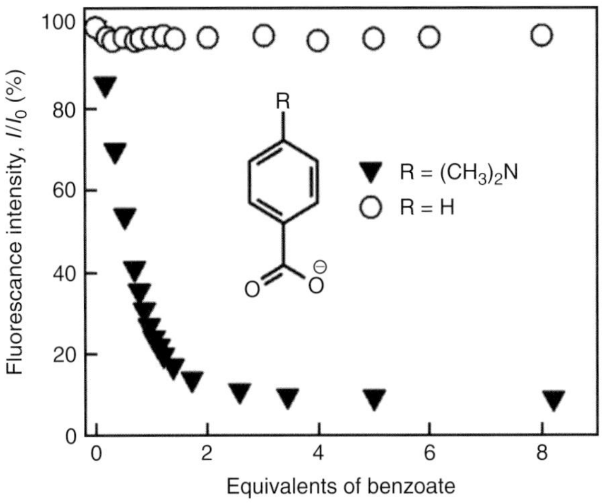

Figure 9.9 Profiles of the percent fluorescence emission of the anthracene subunit recorded over the course of titrations of the $[Zn(II)(\mathbf{9})]^{2+}$ receptor in a methanolic solution with benzoate (circles) and 4-*N*,*N*-dimethylaminobenzoate (triangles) [15].

substituent, for example, *N*,*N*-dimethylamine-4-benzoate, anthracene fluorescence is quenched. The titration profile in Figure 9.9 clearly indicates the formation of a receptor/anion adduct of 1 : 1 stoichiometry.

Moreover, on least-squares treatment of the titration profile, a log *K* value of 5.45 ± 0.03 was calculated for the equilibrium $[Zn^{II}(\mathbf{9})]^{2+} + X^- \leftrightarrows [Zn^{II}(\mathbf{9})X]^+$. Quenching of fluorescence within the receptor/anion complex has to be ascribed to the occurrence of a fast and effective electron transfer process from the *N*,*N*-dimethylaniline fragment to the excited anthracene subunit, as pictorially illustrated in Figure 9.8 (structural formula **11**). Thus, the analyte itself imposes the signal transduction mechanism. It was observed that quenching takes place in the presence of other benzoates containing an R substituent of either donor (e.g., ferrocene) or acceptor tendencies (e.g., the $-NO_2$ group of a 4-nitrophenyl fragment).

The experiment illustrated in Figures 9.8 and 9.9 demonstrates that metal polyamine receptors such as $[Zn^{II}(\mathbf{9})]^{2+}$ can act as sensors for anions displaying definite electron donor (or acceptor) tendencies, through an on/off signaling mechanism. Such an approach can be extended to anionic substrates of higher practical relevance than 4-substituted benzoates: for instance, amino acids. An example has been provided by the zinc(II) complex **12**, in which the tren ligand has been equipped with two anthracenyl and one benzyl substituents (see Figure 9.10) [16].

The complex is fluorescent and acts as a receptor for amino acids. In particular, the axially bound solvent molecule can be easily replaced by one oxygen atom of the carboxylate group of the amino acid, which confirms the high affinity of the $-COO^-$ group toward coordinatively unsaturated Zn^{II} polyamine complexes.

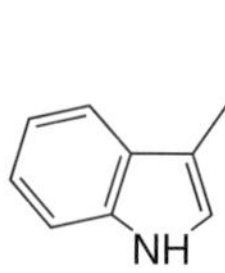

trp

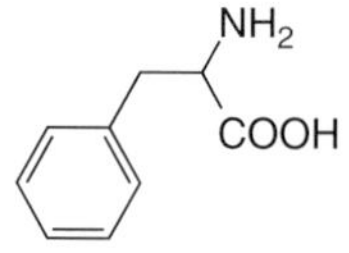

phe

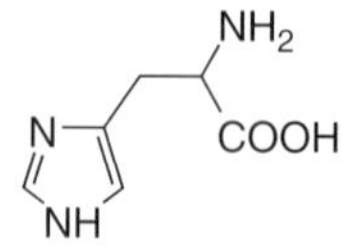

his

The interaction of a number of α-amino acids with receptor **12** was investigated in a mixture dioxane water (4 : 1 v/v), adjusted to pH = 6.8 with lutidine buffer. While all amino acids formed fairly stable complexes with **12**, as indicated by spectrophotometric titration experiments, only tryptophan (trp) was able to quench the fluorescence of the anthracene subunit (Figure 9.11).

Figure 9.11 shows the profile of the spectrofluorimetric titration of **12** with trp in a dioxane–water solution (4 : 1 v/v), buffered at pH 6.8. In particular, the normalized emission intensity at 415 nm is plotted versus the added equivalent of the amino acid. It is observed that binding of the carboxylate group of trp to the Zn^{II} center induces a definite fluorescence quenching. Nonlinear least-squares analysis of the titration profile indicated the formation of a 1 : 1 receptor–amino acid adduct with an equilibrium constant $\log K = 4.28 \pm 0.04$. Quenching was ascribed to the occurrence of an intracomplex electron transfer process from the indole moiety of trp to the excited fluorophore. Such a process is probably favored by the structural arrangement of the complex, as tentatively sketched in Figure 9.10. Other amino

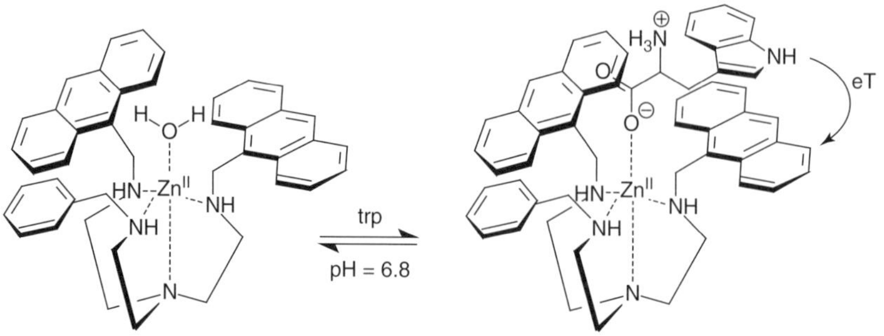

12
Fluorescent

13
Nonfluorescent

Figure 9.10 The interaction of a Zn^{II} tetramine receptor with tryptophan (trp). The fluorescence of the anthracene subunit(s) is quenched by an intracomplex electron transfer process from the indole substituent to the fluorophore [16].

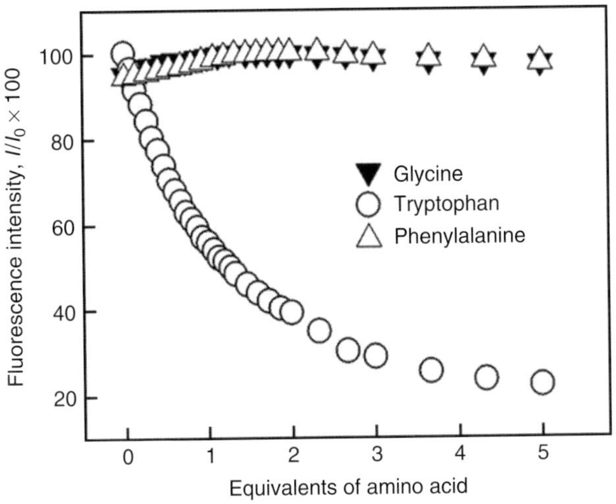

Figure 9.11 Profiles of the percent fluorescence emission of the anthracene subunit recorded over the course of titrations of the $[Zn^{II}(\mathbf{10})]^{2+}$ receptor in a dioxane/water solution (4 : 1 v/v) with tryptophan (trp, circles), glycine (gly, full triangles), and phenylalanine (phe, open triangles) [16].

acids (e.g., glycine, leucine, and alanine) do not induce any quenching (the titration profile for glycine, filled triangles, is shown in Figure 9.11). These amino acids form with receptor **12**, under the same conditions, complexes of moderate stability with log K values ≈ 3, as evaluated from spectrophotometric titration experiments (looking at modifications of $\pi-\pi^*$ bands of the organic backbone of the receptor). These values are distinctly smaller than that observed for trp (the log K value from spectrophotometric studies, 4.21 ± 0.02, is practically coincident with that obtained through spectrofluorimetry). Modeling studies have shown that, in the complex **13**, the indole residue and one anthracene moiety stay in parallel planes, at a distance of 3.5–3.6 Å, a situation favorable for the establishment of $\pi-\pi$ interactions. It is probably this additional contribution that is responsible for the extra stability of the receptor–trp complex. The amino acid phenylalanine, phe, displays distinctive behaviors: (i) it forms a stable complex with receptor **12** (log $K = 4.48 \pm 0.05$, spectrophotometrically determined) and (ii) it does not quench the fluorescence of the receptor (see open triangles in Figure 9.11). The high thermodynamic stability can be ascribed to the establishing of $\pi-\pi$ interaction involving the phenyl residue of the amino acid; photophysical inactivity should be related to the poor or nil redox tendencies of the phenyl substituent, which prevents any photoinduced electron transfer process.

A zinc(II) polyamine platform cannot ensure an extremely selective recognition of amino acids, as it is mainly based on the metal–carboxylate interaction, while any amino acid contains the carboxylate group. Selectivity could be obtained if the interaction with the receptor could involve the **R** substituent of the α-amino acid H_2N–CH(**R**)–COOH. At least one example exists and refers to histidine (HIS),

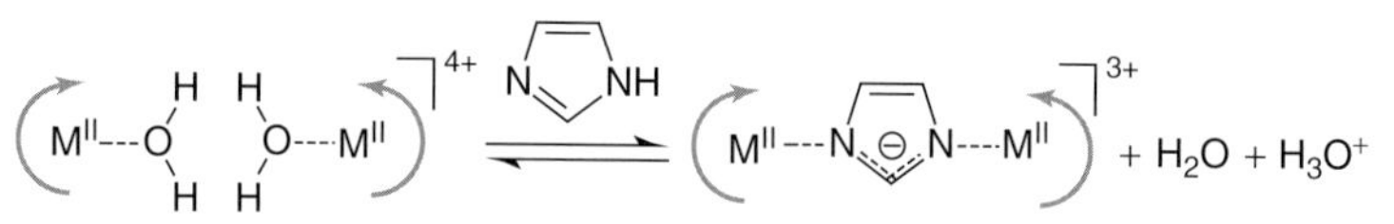

Figure 9.12 Deprotonation of imidazole with simultaneous ambidentate coordination of two metal ions (Cu(II), Zn(II)), pre-positioned within a multidentate ligand framework.

which contains an imidazole (ImH) residue. ImH, $pK_A = 14.5$, is a weaker acid than H_2O and does not deprotonate in water. However, in the presence of two divalent cations M^{II} (Cu^{II}, Zn^{II}) pre-positioned through the coordination by an appropriate ligating framework, the ImH molecule deprotonates, giving rise to the imidazolate ion Im^-, which bridges the two metal centers, according to a process described by equilibrium (Eq. (9.4)) and pictorially illustrated in Figure 9.12:

$$[M^{II}_2]^{4+} + ImH = [M^{II}_2(Im)M^{II}]^{3+} + H^+ \qquad (9.4)$$

The pK_A of ImH in the presence of the $[M^{II}_2]^{4+}$ receptor may vary between 5 and 9, depending on the nature of the cation and the coordinative and geometrical features of the ligand.

ImH uptake and deprotonation is made easy because of the contribution of different energy terms: (i) the energy of the M^{II}–N interactions (enthalpy), (ii) the energy related to the M^{II}–M^{II} electron pairing interaction through the electron-permeable imidazolato bridge (only for Cu^{II}, and (iii) the entropy advantage associated with the receptor's preorganization and rigidity.

All these features are presented by the homoditopic octamine ligand **14**, in which two tren subunits are covalently linked by a 9,10-anthracenyl spacer (see the structure in Figure 9.13) [17]. In a weakly alkaline solution, the bistren ligand **14** coordinates two Zn^{II} ions to give the dinuclear complex $[Zn^{II}_2(\mathbf{14})]^{4+}$(= **15**). Then, an added ImH molecule deprotonates and bridges the two Zn(II) ions, giving the $[Zn^{II}_2(\mathbf{14})(Im)]^{3+}$ complex(= **16**). The occurrence of the stepwise processes of Zn^{II} complexation and imidazolate inclusion, illustrated in Figure 9.13, is accompanied by drastic changes in the fluorescent emission: OFF (bistren ligand **14**), ON (dimetallic complex **15**), OFF again (imidazolate complex **16**). In particular, addition of ImH to an aqueous solution of the $[Zn^{II}_2(\mathbf{14})]^{4+}$ complex, buffered at pH = 9.6, quenches the anthracene fluorescence (Figure 9.14, open triangles).

Nonlinear fitting of I_F versus number of equivalents profile indicated the formation of the 1 : 1 receptor–analyte adduct, with a conditional equilibrium constant $\log K = 3.65 \pm 0.04$. Fluorescence quenching in the $[Zn^{II}_2(\mathbf{14})(Im)]^{3+}$ adduct was ascribed to the occurrence of an intracomplex electron transfer process from a π orbital of the electron-rich Im^- moiety to a π^* orbital of the photoexcited anthracene fragment. Notice that in Figure 9.13, the anthracene subunit and the imidazolate fragment have been sketched as lying in parallel planes, a situation that should allow the overlap of the appropriate π orbitals and favor the occurrence of the electron transfer process. Noteworthy, such a geometrical arrangement has been observed in the homologous dicopper(II) complex, $[Cu^{II}_2(\mathbf{14})(Im)]^{3+}$,

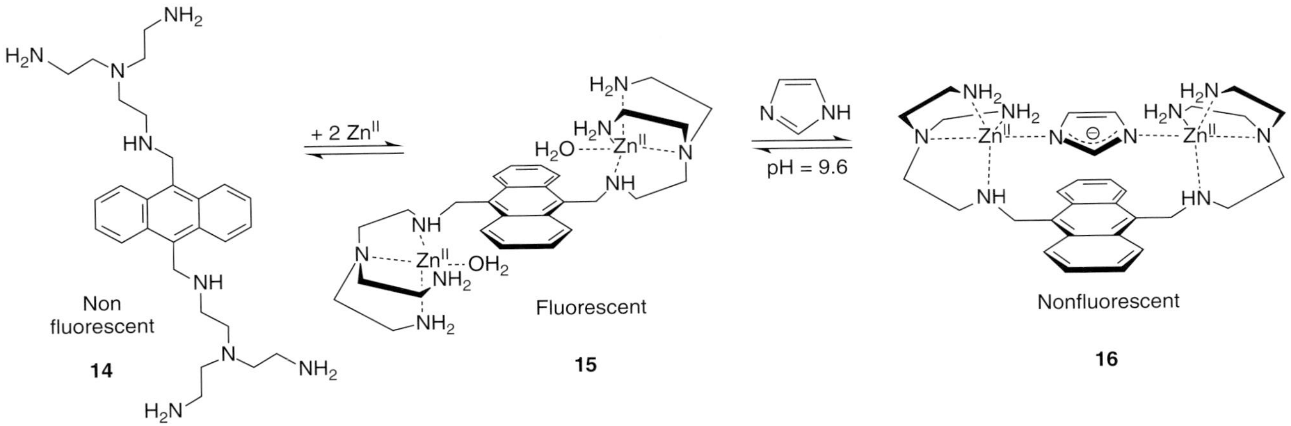

Figure 9.13 Recognition and sensing of the imidazole residue by a dinuclear zinc(II) complex. At pH = 9.6 imidazole deprotonates and bridges the two Zn^{II} centers of the dimetallic complex **15**. Then, in the ternary complex **16**, the electron-rich imidazolate moiety transfers one electron to the proximate anthracene subunit, quenching its emission [17].

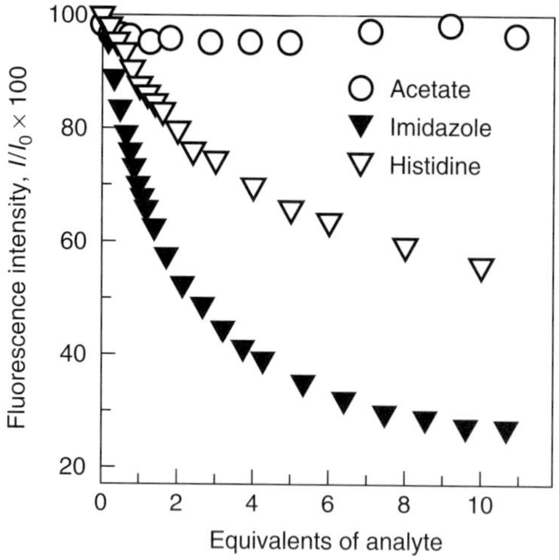

Figure 9.14 Profiles of the percent fluorescence emission of the anthracene subunit recorded over the course of titrations of the $[Zn^{II}_2(\mathbf{14})]^{4+}$ receptor in an aqueous solution buffered to pH = 9.6, with acetate, imidazole, and histidine [17].

whose molecular structure has been determined by X-ray diffraction studies [18]. Noticeably, titration with 1-methyl-ImH, which neither undergoes deprotonation nor bridges the two Zn^{II} ions, did not induce any modification of the anthracene fluorescent emission.

Titration of the $[Zn^{II}_2(\mathbf{14})]^{4+}$ complex with the ImH-containing amino acid histidine induced fluorescence quenching, but the titration profile was less steep (Figure 9.14, filled triangles) because of the lower value of the binding constant ($\log K = 2.92 \pm 0.01$): the lower stability of the complex may reflect the unfavorable contribution of the steric repulsions between the receptor and the amino acid functionality. Most interestingly, the titration profile is not modified when the solution contains even a large excess of any other amino acid. Such an absence of competitivity seems to be due to the fact that the carboxylate group of amino acids does not show pronounced bridging tendencies toward the two metal centers of the $[Zn^{II}(\mathbf{14})]^{4+}$ receptor. This has been confirmed by the observation that titration with the acetate ion does not alter at all the anthracene emission (see Figure 9.14, open circles).

9.4
Turning on Fluorescence with the Indicator Displacement Approach

The design of the fluorescent sensors described in the two previous sections is based on a "covalent" approach, that is, on the permanent linking of the receptor and of the fluorophore. Such a paradigm is often indicated by the fluorophore-spacer-receptor (FSR) [19]. Recognition and signaling of a substrate X^- by a covalently linked sensor is pictorially illustrated in Figure 9.15.

The synthesis of a covalently linked fluorescent sensor may involve a tedious multistep procedure. Moreover, unless the receptor contains an electron donor group acting as a switch, for example, the amine group close to the fluorogenic

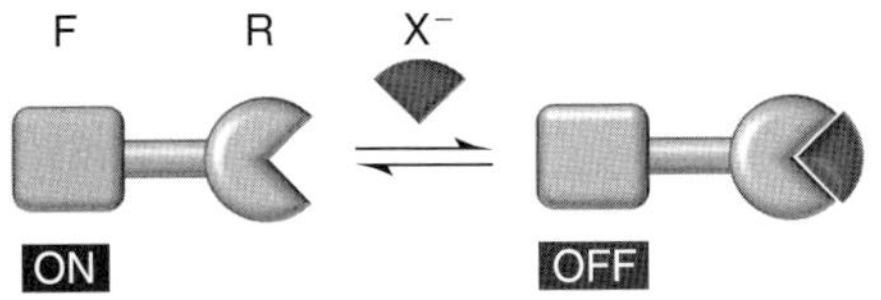

Figure 9.15 The "covalent" approach to the design of a fluorescent sensor for anions. The fluorophore subunit (**F**) and the receptor moiety (**R**) are covalently linked through a spacer. An anion **X**$^-$, displaying redox tendencies, quenches the fluorescence of F through a photoinduced electron transfer process. Anion recognition is signaled by a turning off of fluorescence.

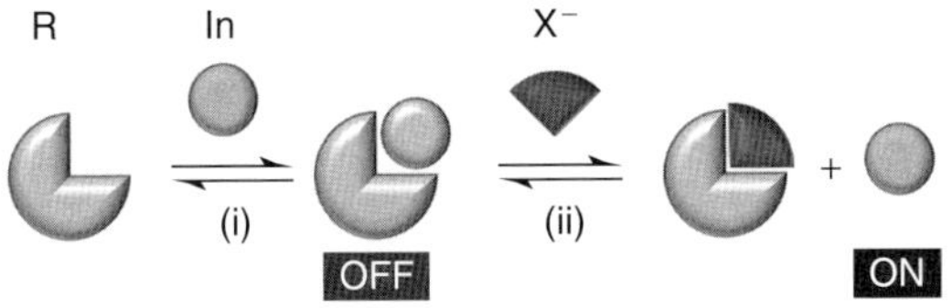

Figure 9.16 The "indicator displacement" approach to the design of a fluorescent sensor for anions. Process (i): the receptor **R** binds, not too strongly, the fluorescent indicator **In** and quenches its fluorescence; process (ii): the anion **X**$^-$, which possesses a specific affinity for **R**, displaces **In**, which, released to the solution, displays its full fluorescence. Anion recognition is signaled by a turning on of fluorescence.

subunit of system **4**, the signaling activity of the system relies on the redox properties of the anion: only anions capable of transfer/uptake of an electron to/from the excited fluorophore quench fluorescence and signal the occurrence of their recognition.

More recently, a new paradigm has been proposed for the design of optical sensors, which is based on indicators [20]. This approach first involved colorimetric indicators and spectrophotometric studies and has later been extended to the spectrofluorimetric determination of anions. It is pictorially illustrated in Figure 9.16.

According to the sketch in the figure, equilibrium (i), the receptor **R** first interacts with the fluorescent indicator **In**, which should have an anionic nature and, in particular, should possess a group suitable for the interaction with the binding site of the receptor (in general, a carboxylate group). As a stringent requirement, the receptor should be able to quench the emission of the bound indicator, thus making the pertinent complex [**R-In**] nonfluorescent. Then, in the second step, equilibrium (ii), the anion X^- is added, which, if possessing a higher affinity for the receptor, displaces the indicator. When released into the solution, the indicator displays its original fluorescence, signaling the occurrence of anion recognition. The indicator need not be present in stoichiometric amounts but at a concentration 10- to 1000-fold lower than the anion. Occurrence of the displacement process requires that the receptor–anion association constant K_X is higher than the receptor–indicator association constant K_{In}. Also, if K_X and K_{In} have comparable values, displacement can take place because of a mass effect. The mixture of the receptor and the indicator, suitable for optical titration experiments, has been defined "chemosensing ensemble" by Lavigne and Anslyn [21], probably inspired by the concerted musical performance of a duo.

One of the first examples of fluorescence sensing of anions according to the indicator displacement paradigm involved the dicopper(II) cryptate of the bistren ligand **17**, sketched in Figure 9.17 [22]. In the cryptate **18**, that is, $[Cu_2^{II}(\mathbf{17})]^{4+}$, each metal center exhibits a trigonal bipyramidal coordination geometry: the three secondary amine groups occupy the equatorial positions, while the tertiary amine nitrogen atom resides in one of the two axial positions. In an aqueous solution, such a position is available for the coordination by a water molecule or by an OH^- ion, depending on pH. At a neutral pH, the dominant species is $[Cu_2^{II}(\mathbf{17})(H_2O)(OH)]^{3+}$, in which one metal center binds a water molecule and the other, a hydroxide ion. Such interactions are labile, and the two water molecules can be replaced by anions displaying definite coordinating tendencies. Replacement is especially favored when the two donor atoms belong to an ambidentate anion of appropriate size, which will bridge the two Cu(II) centers. As an example, Figure 9.18 shows the crystal and molecular structure of an inclusion complex, in which receptor **18** has encapsulated the linear triatomic anion azide: $[Cu_2(\mathbf{17})(N_3)]^{3+}$ [23].

The affinity of receptor $[Cu_2(\mathbf{17})]^{4+}$ for anions was preliminarily investigated in an aqueous solution buffered at pH = 7 (HEPES 0.05 M), through spectrophotometric titration experiments. In particular, spectral changes in the range of anion-to-metal charge-transfer transitions obtained on anion addition to the dimetallic receptor were monitored, and the pertinent association constants determined. In particular, the conditional constants referred to the equilibrium (Eq. (9.5)):

$$[Cu_2^{II}(\mathbf{17})(H_2O)(OH)]^{3+} + X^- \rightleftarrows [Cu_2^{II}(\mathbf{17})(X)]^{3+} + H_2O + OH^- \qquad (9.5)$$

Log K values of such constants obtained for a variety of anions are plotted in Figure 9.19a versus the "bite length" of the anion.

Bite length is a geometrical parameter that corresponds to the distance between two consecutive donor atoms of the envisaged anion. As an example, in Figure 9.19b, the bite length for two selected anions, the Y-shaped HCO_3^- and the rodlike NCO^-, is indicated by a double-headed arrow. Figure 9.19a discloses a nice peak selectivity related to the matching of the geometrical features of the dimetallic receptor and of the anion. In particular, the most stable complex is formed with azide, because the N_3^- ion has the right bite length for placing its donor atoms in the positions desired by the Cu^{II} centers. Thus, its inclusion does not induce any conformational rearrangement of the cage framework (endothermic). On the other hand, anions with a smaller or a larger bite than N_3^- will form less stable inclusion complexes. The extent of such a destabilization is related to the extent of the deviation from the "ideal" value of azide: the larger the deviation the more endothermic the conformational rearrangement and the lower the thermodynamic stability of the association complex. HCO_3^- and NCO^- ions possess a bite length close to that of N_3^- and give inclusion complexes of slightly lower stability. The carbonate ion is relevant in many practical aspects, and a spectroscopic technique for its determination is highly required. Spectrophotometry cannot be recommended,

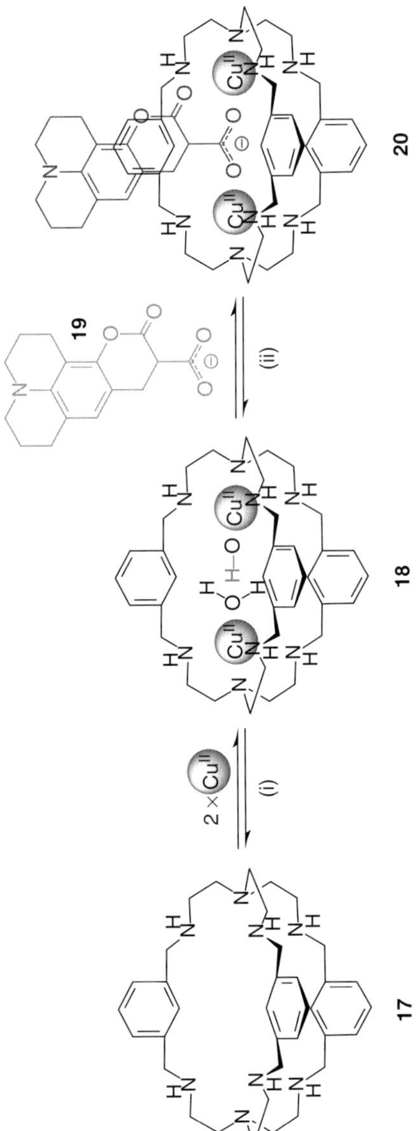

Figure 9.17 Cascade process in which, in an aqueous solution, the bistren cryptand **17** first incorporates two Cu(II) ions according to equilibrium (i). Then, the dicopper(II) cryptate **18** encapsulates the anionic form of the fluorescent dye coumarine (**19**), to give the inclusion complex **20**, according to equilibrium (ii). On inclusion into the receptor, the green fluorescence of coumarine is quenched.

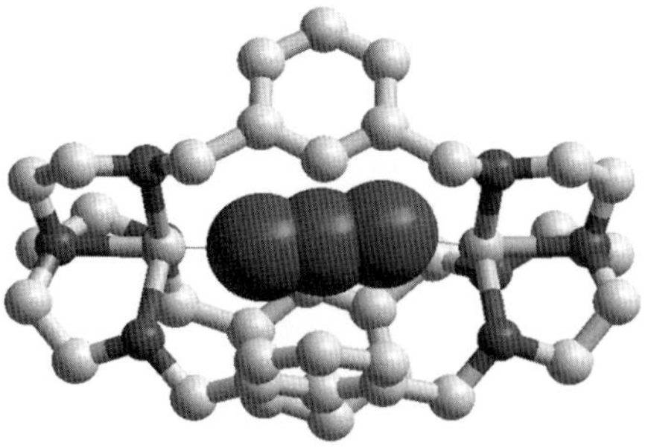

Figure 9.18 The molecular structure of the inclusion complex$[Cu^{II}_2(\mathbf{17})(N_3)]^{3+}$, in which an azide ion bridges two Cu(II) ions [23]. Hydrogen atoms have been omitted for clarity. Each metal center exhibits a compressed trigonal bipyramidal coordination geometry. A rather unusual alignment of seven atoms (N–Cu–N–N–N–Cu–N) is observed. Structure redrawn from data deposited at the Cambridge Crystallographic Data Center (TOGNUY).

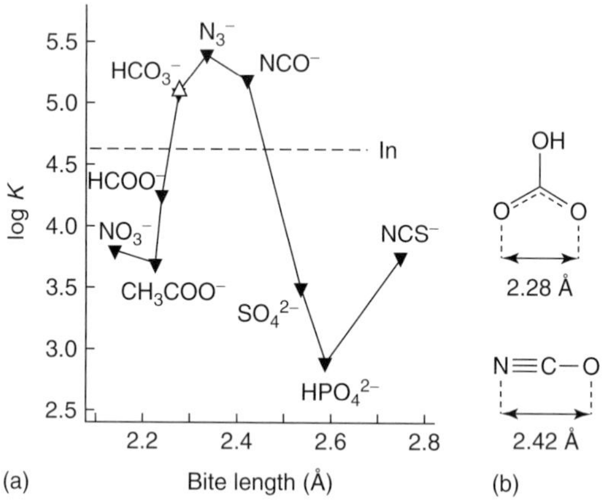

Figure 9.19 (a) Plot of log K for the equilibrium $[Cu^{II}_2(\mathbf{17})]^{4+} + X^- \leftrightarrows [Cu^{II}_2(\mathbf{17})(X)]^{3+}$ in an aqueous solution, buffered at pH = 7 (HEPES 0.05 M). Filled symbols refer to log K values determined spectrophotometrically; the open symbol (HCO_3^-) corresponds to log K determined spectrofluorimetrically. The dashed line corresponds to log K for the interaction of the receptor with the indicator coumarine 343: $[Cu^{II}_2(\mathbf{17})]^{4+} + In^- \leftrightarrows [Cu^{II}_2(\mathbf{17})(In)]^{3+}$ and (b) the bite length in HCO_3^- and NCO^- ions [22].

as HCO_3^- addition to a solution of $[Cu_2(\mathbf{17})]^{4+}$ induces a moderate change of spectral feature, while the color changes to pale blue to pale green (for a receptor's concentration of 5×10^{-4} M). Such considerations prompted us to develop an analytical procedure for carbonate determination based on receptor $[Cu_2(\mathbf{17})]^{4+}$ and operating through the fluorescent indicator paradigm. As an indicator we chose coumarine 343 (see **19** in Figure 9.17), which contains a carboxylate group and can therefore interact with receptor $[Cu_2(\mathbf{17})]^{4+}$, bridging the two Cu(II) centers. The occurrence of such an interaction was investigated by carrying out a spectrofluorimetric titration experiment.

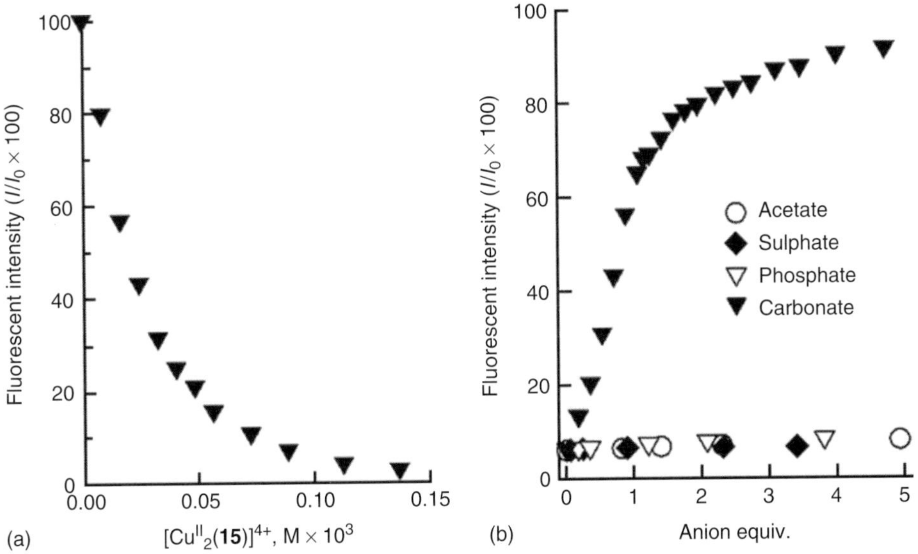

Figure 9.20 (a) Titration profile (487 nm) obtained over the course of the titration of solution 10^{-7} M of coumarine (**19**), buffered to pH 7 (HEPES 0.05 M), with a solution of receptor $[Cu_2(\mathbf{17})]^{4+}$. (b) Titration profiles (487 nm) obtained over the course of the titration with selected anions of a solution containing the "chemosensing ensemble" (2×10^{-4} M in $[Cu_2(\mathbf{17})]^{4+}$ and 10^{-7} M in coumarine) and buffered to pH 7 (HEPES 0.05 M) [22].

On titrating a degassed solution 10^{-7} M of coumarine, buffered to pH 7 (HEPES 0.05 M), with a solution of receptor $[Cu^{II}_2(\mathbf{17})]^{4+}$, complete quenching of the coumarine emission, centered at 487 nm, was observed. Nonlinear least-squares curve fitting of the titration profile, shown in Figure 9.20a, was consistent with the formation of an adduct of 1 : 1 stoichiometry, for which an association equilibrium constant $\log K_{ass} = 4.8 \pm 0.1$ was determined. A tentative structural arrangement of the receptor–indicator has been sketched in Figure 9.17 (**20**). Fluorescence quenching was ascribed to the occurrence of an intracomplex electronic energy transfer process involving the photoexcited coumarine fragment and one of carboxylate bound Cu^{II} centers.

Thus, the chemosensing ensemble was prepared: it consisted of an aqueous degassed solution, buffered to pH 7 with HEPES 0.05 M, 2×10^{-4} M in $[Cu^{II}_2(\mathbf{17})]^{4+}$, and 10^{-7} M in coumarine. Figure 9.20b shows the profiles corresponding to the titration of the chemosensing ensemble with a variety of anions (emission intensity at 487 nm vs equivalents of the envisaged anion). Only on addition of carbonate a full revival of the coumarine emission was observed, indicating successful competitive binding of HCO_3^- and displacement of the indicator from the host cavity. On the other hand, titration with a variety of ambidentate anions (acetate, formate, sulfate, phosphate, nitrate) induced only a slight fluorescence enhancement (some titration profiles are shown in Figure 9.20b).

Such a selective behavior can be fully accounted for on the basis of the diagram in Figure 9.19. The dashed horizontal line corresponds to the association constant, log K, of the receptor/coumarine complex. It stays well below the value of the HCO_3^- complex, which justifies the effective displacement of the indicator and full fluorescence revival. On the other hand, the other investigated anions have association constants distinctly lower than that of coumarine 343, they cannot compete successfully with the indicator, and do not induce any serious restoration of fluorescence. The described approach was proved to effectively detect carbonate over a substantial concentration range ($4 \times 10^{-4}/2 \times 10^{-1}$ M, corresponding to a 24/1222 mg l^{-1} of HCO_3^-), an interval covering, for instance, all the commercially available mineral waters. The fluorimetric procedure seems easier and more selective than that currently used to determine HCO_3^- in water, based on volumetric titration of alkalinity with standard hydrochloric or sulfuric acid [24], which does not discriminate HCO_3^- from other bases present in solution.

The presence of one or more Cu^{II} ions is essential in the design of a receptor operating under the regime of the fluorescent indicator displacement paradigm, for two main reasons: (i) Cu^{II} gives the most stable complexes among divalent 3d metal ions and is able to bind quite strongly the carboxylate group of any fluorescent dye and (ii) owing to the incomplete filling of the 3d level and a definite redox activity, Cu^{II} can effectively quench the proximate fluorophore through either an energy transfer or electron transfer process, respectively.

In this vein, a receptor containing two Cu^{II} centers has been used for the selective fluorimetric determination of the amino acid histidine following the indicator displacement approach [25], as an alternative to the ON/OFF Zn^{II}-containing fluorescent receptor described in the previous section. In the present case, the two metal centers were pre-positioned inside the hexamine macrocycle **21**, in which two triamine (diethylenetriamine, dien) subunits had been linked by 1,4-xylyl spacers.

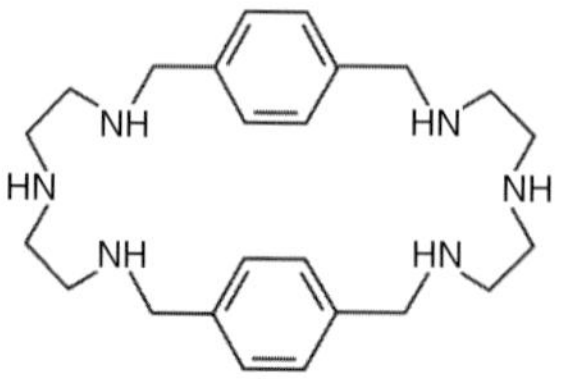

21

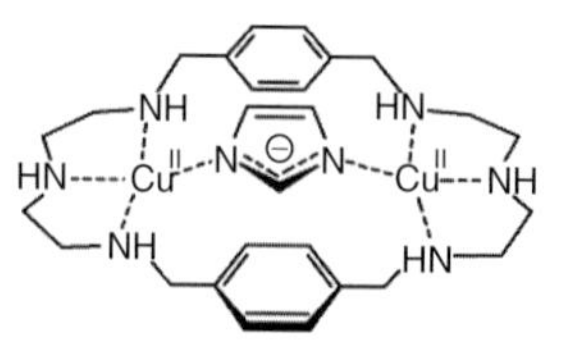

22

In the present case, each copper(II) center is coordinated by three secondary amine nitrogen atoms and needs a further ligand to complete its highly preferred coordinative arrangement: the square. Thus, the $[Cu^{II}_2(\mathbf{21})]^{4+}$ complex shows a

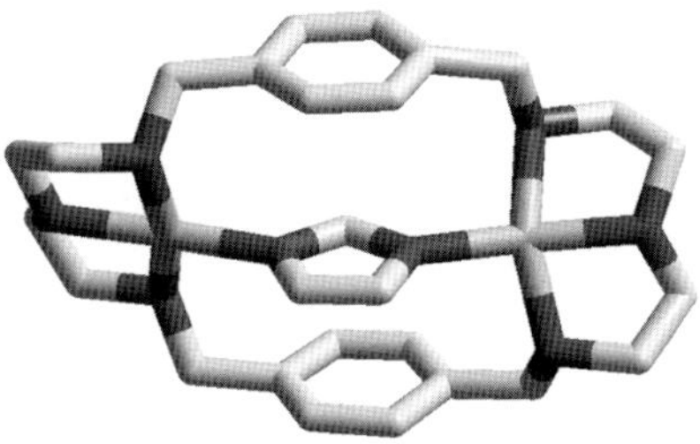

Figure 9.21 The crystal and molecular structure of the $[Cu^{II}_2(\mathbf{21})(Im)]^{3+}$ ternary complex [26]. Hydrogen atoms have been omitted for clarity. The system is further stabilized by π-stacking interactions between the imidazolate ion (donor) and the phenyl groups of the 1,4-xylyl spacers (acceptors). Structure redrawn from data deposited at the Cambridge Crystallographic Data Center (QUHLAG).

strong affinity toward ambidentate ligands, in particular the imidazolate ion. In fact, $[Cu^{II}_2(\mathbf{21})]^{4+}$ in water, at pH = 7, induces deprotonation of ImH and incorporates the imidazolate anion Im^-, to give the stable ternary complex **22**, $[Cu^{II}_2(\mathbf{21})(Im)]^{3+}$. Such a species has been isolated in the crystalline form and its molecular structure elucidated through X-ray diffraction studies (see Figure 9.21) [26].

It is shown that the imidazolate anion bridges the two copper(II) ions, completing a slightly distorted square coordinative arrangement of each metal center. The square is the coordination polygon preferred by Cu^{II}, which may account for the high stability of the $[Cu^{II}_2(\mathbf{21})(Im)]^{3+}$ complex. Moreover, two further favorable energy terms seem to contribute to the stability of the ternary complex: (i) the π-stacking interaction between the imidazolate ion and the phenyl ring of the two 1,4-xylyl spacers and (ii) the interaction between the unpaired electrons of the two Cu^{II} centers, mediated by the imidazolate bridge and responsible for a weak antiferromagnetic coupling.

Thus, system $[Cu^{II}_2(\mathbf{21})]^{4+}$ seemed an ideal receptor for the fluorimetric sensing of the ImH-containing amino acid histidine through the indicator displacement paradigm. An additional advantage was that $[Cu^{II}_2(\mathbf{21})]^{4+}$, because of the favorable energy terms outlined before, binds the imidazolate ion at pH = 7 and not in a distinctly alkaline solution such as dinuclear complexes of Zn^{II}. The fluorescent dye coumarine 343 (**19**) was tested as a possible partner of $Cu^{II}_2(\mathbf{21})]^{4+}$ in the chemosensing ensemble. The constant associated with the formation of the $[Cu^{II}_2(\mathbf{21})(In)]^{3+}$ complex was preliminarily determined: $\log K_{In} = 4.3$. At this point, the solution of the chemosensing ensemble was prepared: coumarine 343: 10^{-6} M, $[Cu^{II}_2(\mathbf{19})]^{4+}$: 2.5×10^{-4} M, pH = 7 (HEPES 0.05 M). Then, such a solution was titrated with histidine (HIS), and, in subsequent experiments, with some selected amino acids: glycine (GLY), alanine (ALA), phenylalanine.

Figure 9.22a shows the profiles obtained for the varying titration experiments, indicating how the emission band of the coumarine indicator develops on addition of the amino acid. Results seem rather poor from the point of view of the selectivity. In fact, the titration profiles obtained for HIS and GLY are coincident, showing no

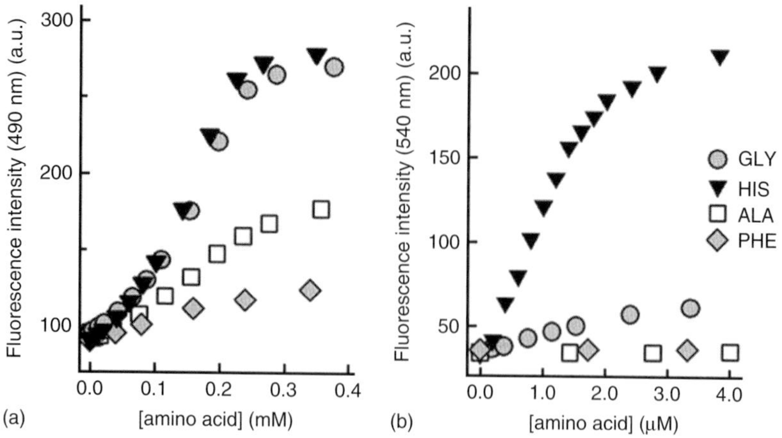

Figure 9.22 Profile of the spectrofluorimetric titrations with selected amino acids of (a) an aqueous solution containing coumarine 343 $\times 10^{-6}$ M, $[Cu^{II}_2(\mathbf{21})]^{4+}$ 2.5×10^{-4} M, buffered to pH = 7 (HEPES 0.05 M) and (b) eosine Y 10^{-6} M, $[Cu^{II}_2(\mathbf{21})]^{4+} = 2.4 \times 10^{-6}$ M [25].

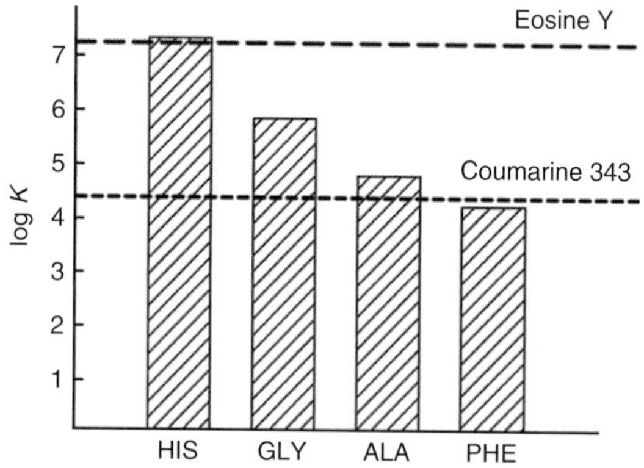

Figure 9.23 Log K_{ass} values associated with the equilibrium $[Cu^{II}_2(\mathbf{21})]^{4+} + AA \leftrightarrows [Cu^{II}_2(\mathbf{21})(AA)]^{4+}$ in aqueous solution at pH = 7 (bars, AA = aminoacid). Dashed lines indicate log K_{In} values for the equilibrium: $[Cu^{II}_2(\mathbf{21})]^{4+} + In^- \leftrightarrows [Cu^{II}_2(\mathbf{21})(In)]^{3+}$ (In^- = indicator) [24].

discrimination. However, these titration experiments were not useless, because, from the profiles in Figure 9.22a, the value of K_{in} for $[Cu^{II}_2(\mathbf{21})]^{4+}$ being known, it was possible to calculate, using a nonlinear least-squares procedure, the constant of the association equilibrium involving the dimetallic receptor and each amino acid. Pertinent log K_{ass} values are showed in the bar diagram in Figure 9.23.

It is observed that histidine displays the expected higher affinity for the dimetallic receptor $[Cu^{II}_2(\mathbf{21})]^{4+}$, forming, in particular, a complex whose association constant is more than 1 order of magnitude than that for glycine. However, the indicator coumarine 343 (**19**) forms with $[Cu^{II}_2(\mathbf{21})]^{4+}$ a complex whose stability is distinctly lower than that observed for the complexes of HIS and GLY and is therefore quantitatively displaced by *both* amino acids. Thus, in order to obtain discrimination, one should choose an indicator with a K_{In} definitely greater than K_{ass} of GLY but lower than K_{ass} of HIS. Eosine Y (**23**) was a good choice.

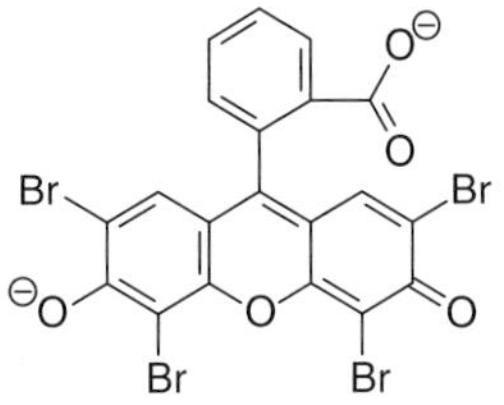

23

In fact, log K_{In}, determined by titrating with $[Cu^{II}_2(\mathbf{21})]^{4+}$ a solution of eosine Y buffered to pH = 7, was 7.2. Such a value is only slightly lower than that of log K_{ass} of HIS, but, in any case, the indicator displacement should be favored by a mass effect, due to the fact that, over the course of the titration, the concentration of the amino acid becomes higher and higher than that of the indicator (up to 1000-fold). In fact, titration experiments, illustrated by pertinent profiles in Figure 9.21b, disclosed a well-defined selective behavior in favor of HIS.

Ultimately, receptor $[Cu^{II}_2(\mathbf{21})]^{4+}$ prefers imidazolate with respect to the carboxylate group because it can establish stronger metal–ligand interactions. However, the same receptor, in the presence of bridging anions of the same coordinating tendencies, can exert geometrical selectivity. This occurs, for instance, in a classical issue of anion chemistry: the discrimination of orthophosphate (Pi) and pyrophosphate (PPi) [27]. Figure 9.24a shows the profiles obtained by titrating with Pi and PPi a neutral solution containing the chemosensing ensemble $\{[Cu^{II}_2(\mathbf{21})]^{4+} + \text{coumarine 343}\}$ [28].

However, such a system does not work well: PPi displaces effectively the indicator, producing a full revival of fluorescence, while Pi induces a less pronounced recovery of the emission, but still displays a competitive behavior. This state of affairs can be accounted for on the basis of the bar diagram shown in Figure 9.25.

The log K_{In} value for coumarine 343 (dashed line) is distinctly lower than log K_{ass} for PPi, which accounts for complete indicator displacement by pyrophosphate. However, log K_{In} is only slightly lower than log K_{ass} for Pi, and a significant displacement of the indicator takes place on excess addition of the anion. Thus, in order to achieve discrimination between PPi and Pi, a fluorescent indicator displaying a higher affinity for the receptor is required. Such an indicator can again be eosine Y. In fact, titration profiles shown in Figure 9.24b indicate the occurrence of an effective discrimination of PPi with respect to Pi, a behavior that can be satisfactorily interpreted on the basis of the bar diagram in Figure 9.25. Eosine Y forms with receptor $[Cu^{II}_2(\mathbf{21})]^{4+}$ a complex of stability comparable to that of PPi and much higher than that of Pi.

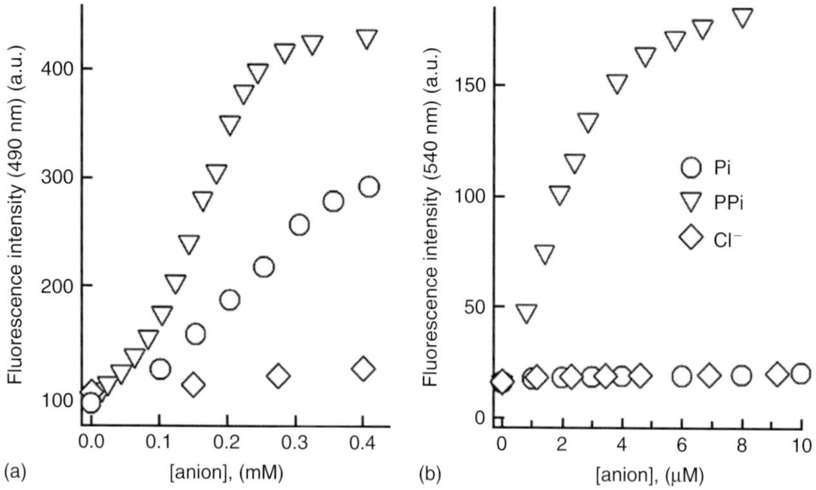

Figure 9.24 Titration of the chemosensing ensemble {$[Cu^{II}_2(\mathbf{21})]^{4+}$ + indicator(In)} in an aqueous solution buffered to pH = 7, with pyrophosphate (▽), orthophosphate (○), and chloride (◇); (a) In = coumarine 343 : 10^{-6} M, $[Cu^{II}_2(\mathbf{21})]^{4+} = 2.5 \times 10^{-4}$ M and (b) In = eosine Y : 10^{-6} M, $[Cu^{II}_2(\mathbf{21})]^{4+} = 2.4 \times 10^{-6}$ M. Nitrate and sulfate display the same behavior (no indicator displacement and no fluorescence revival) as chloride with both indicators [28].

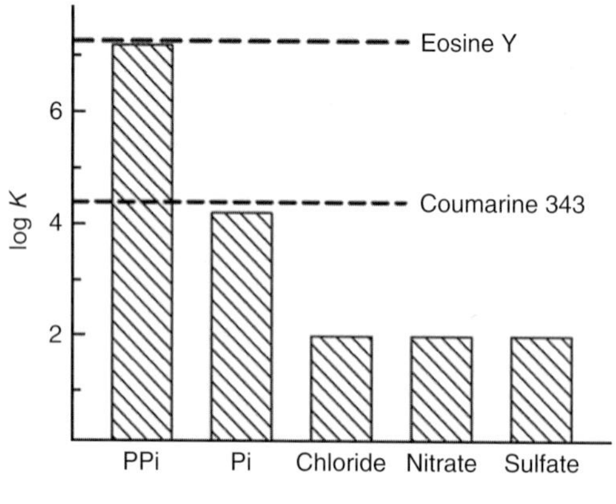

Figure 9.25 Log K_{ass} values associated with the equilibrium: $[Cu^{II}_2(\mathbf{21})]^{4+} + X^{n-} \leftrightarrows [Cu^{II}_2(\mathbf{21})(X)]^{(4-n)+}$ in aqueous solution at pH = 7 (bars, X^{n-} = anion). Dashed lines indicate log K_{In} values for the equilibrium: $[Cu^{II}_2(\mathbf{21})]^{4+} + In^- \leftrightarrows [Cu^{II}_2(\mathbf{21})(In)]^{3+}$ (In^- = indicator) [28].

Selectivity may derive from the capability of PPi to place its terminal oxygen atoms in the vacant coordinative sites of the two Cu^{II} centers, without inducing any serious structural modification of the hexamine macrocycle, relaxed to its minimum energy conformation. Such a conformation does not necessarily correspond to that observed in the $[Cu^{II}_2(\mathbf{21})(Im)]^{3+}$ complex in Figure 9.21 but can be similar to that

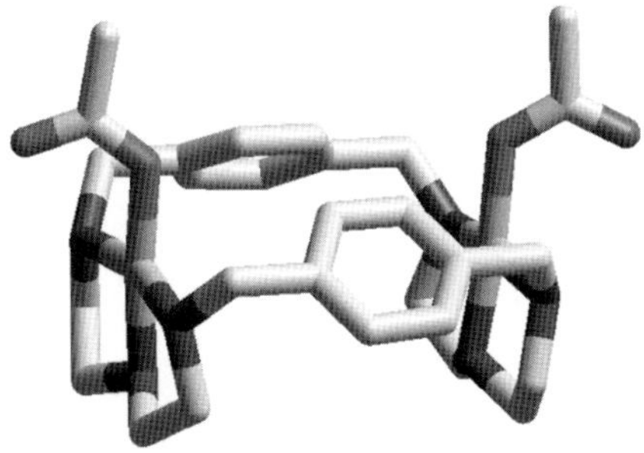

Figure 9.26 The crystal and molecular structure of the $[Cu^{II}_2(\mathbf{21})(CH_3COO)_2]^{2+}$ complex [29]. Hydrogen atoms have been omitted for clarity. Structure redrawn from data deposited at the Cambridge Crystallographic Data Center (KEBZAS).

assumed by the envisaged receptor in its complex with two acetate ions, as shown by the crystal and molecular structure displayed in Figure 9.26 [29].

On the other hand, bridging by the orthophosphate anion of the two Cu^{II} ions when the receptor is conformationally arranged as shown in Figure 9.26 should induce a preliminary pronounced structural reorganization, whose endothermicity is reflected in the much lower value of log K_{ass}. Chloride, nitrate, and sulfate ions experience an even higher difficulty in spanning the two metal ions of the $[Cu^{II}_2(\mathbf{21})]^{4+}$ receptor ($K_{ass} < 10^3$) and do not displace eosine Y and coumarine 343, even if added in a large excess.

The use of the indicator displacement paradigm seems quite convenient in designing analytical procedures for anion determination in solution, compared to the approach that involves the covalent linking of the fluorogenic subunit to the receptor's framework. The first evident advantage is that no synthetic efforts are required to link covalently the fluorophore to the envisaged receptor. Moreover, synthetic modifications on the receptor may change its binding tendencies and alter its genuine binding selectivity. Finally, when planning the synthesis of a fluorescent sensor using the covalent approach, one cannot predetermine the signal transduction mechanism (whether emission will be quenched or enhanced on anion interaction), an event that has to be experimentally verified. Conversely, the critical requirement to be fulfilled when setting up a fluorescent "chemosensing ensemble" is that the receptor must effectively quench the noncovalently bound fluorophore. Such an essential feature is provided by receptors containing as binding sites paramagnetic transition metal ions, which are photophysically active through electron transfer or energy transfer (Dexter type) mechanisms, but can be hardly achieved by receptors that do not contain metals. In this sense, dicopper(II) bistren cryptates such as $[Cu^{II}_2(\mathbf{17})]^{4+}$ are ideal candidates for the fluorescent sensing of a variety of anionic substrates. Further recent examples are shown in Scheme 9.3.

The dimetallic receptor $[Cu^{II}_2(\mathbf{25})]^{4+}$ includes dicarboxylates in water at pH = 7 and displays linear recognition selectivity [30]. Among phthalates, it shows a pronounced affinity for the 1,4-derivative (terephthalate), with respect to the 1,3-(isophthalate) and 1,2-(phthalate) positional isomers. The indicator used was rhodamine, which contains a 1,4-benzenedicarboxylate subunit. Among linear aliphatic dicarboxylates $^{-}OOC{-}(CH_2)_n{-}COO^{-}$, $[Cu^{II}_2(\mathbf{24})]^{4+}$ shows a definite preference for derivatives with $n = 3$ (glutarate) and $n = 4$ (adipate). Most interestingly,

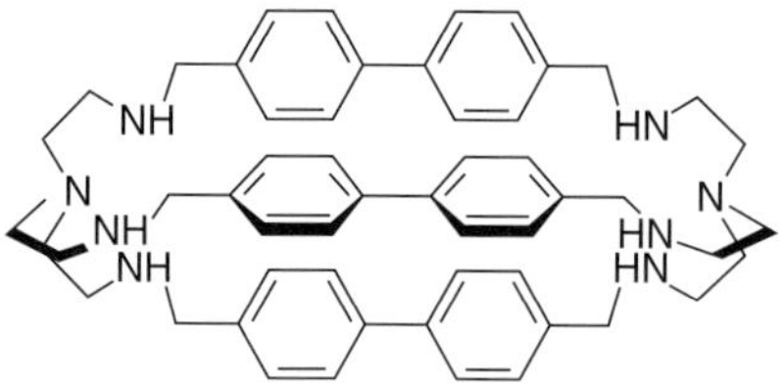

24

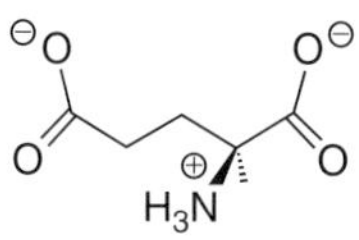

L-Glutamate

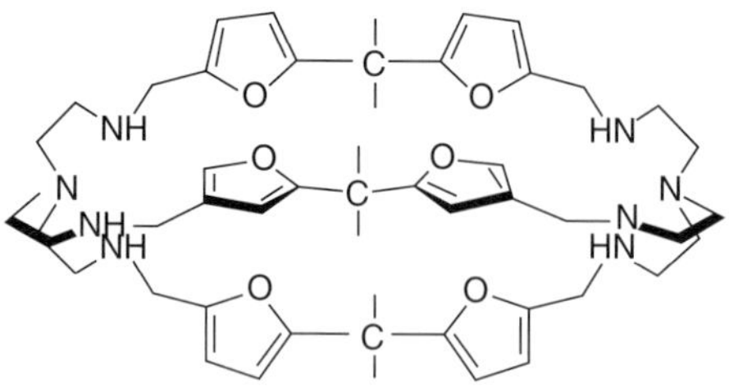

25

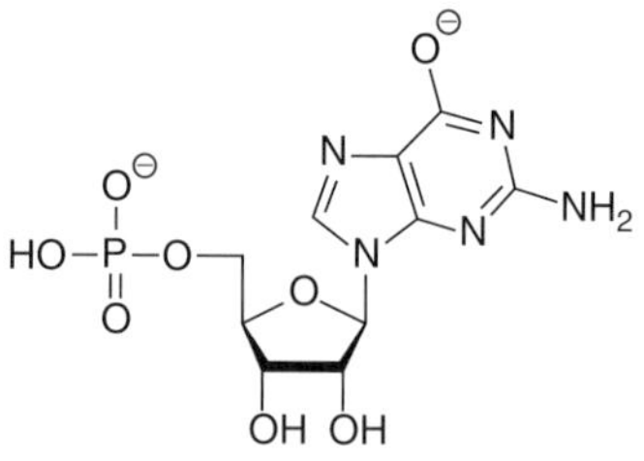

GMP

Scheme 9.3 Bistren cryptands, whose dicopper(II) complexes show selective affinity for L-glutamate (**24**) [30] and GMP (**25**) [31] in an aqueous neutral solution.

it recognizes L-glutamate in the presence of any other neurotransmitter containing $-NH_2$ and/or $-COO^-$ functionalities (including L-aspartate and GABA) [30].

Receptor $[Cu^{II}_2(\mathbf{25})]^{4+}$, showing the larger ellipsoidal cavity among investigated dimetallic bistren cryptates, is able to include, in an MeOH/water solution (50 : 50, v/v), buffered at pH, nucleoside monophosphates and displays selective affinity toward guanosine monophosphate (GMP) [31]. It has been hypothesized that GMP bridges the two Cu^{II} centers by two oxygen atoms: one from the phosphonate group, the other from the nucleobase in its enolate mesomeric form, as sketched in Scheme 9.3. Among the several tested indicators, 6-carboxyfluorescein gave the most satisfactory behavior.

9.4.1 Epilog

In this chapter, we have tried to describe an approach to the design of fluorescent sensors for anions. Such an approach reflects essentially the activity of this group in the field and how it developed over the past 15 years. A few selected examples have been presented and discussed to make the reader acquainted with the basic principles and experimental aspects of the fluorimetric investigations. If steady-state spectrofluorimetry is the main technique, the study of the receptor–anion interaction cannot be based solely on the determination of fluorescence spectra, whether the envisaged anion induces an enhancement of a decrease in the emission intensity of a given fluorophore. Selectivity is a thermodynamic property, and the affinity of the receptor for a given anion is expressed by the value of the equilibrium constant for the formation of the receptor–anion complex. Thus, detailed equilibrium studies should be carried out in order to fully characterize the system under investigation and to define the stoichiometry of the species present at equilibrium. Therefore, titration data (families of spectra, titration curves) should be carefully analyzed, for instance, by using the excellent packages available for data treatment, based on nonlinear least-squares fitting (the authors have used HyperQuad) [32]. In the presence of more complex equilibria, it may be useful to also carry out spectrophotometric titration experiments, as absorbance spectra typically show more rich and complex patterns, usually spread along a larger wavelength interval.

Chemists have a native tendency to explain reactivity (of any kind) on a structural basis, and in this particular field, a selective or specific interaction is interpreted on the basis of an especially favorable matching of the geometrical features of the receptor and the anion. Thus, availability of X-ray structures of the receptor and its anion complex favor understanding of the solution behavior. Most importantly, the knowledge of structural details can provide valuable suggestions for the design and synthesis of an anion receptor of improved selectivity, which, ultimately, remains the most fascinating and challenging aspect of chemical science.

References

1. Oxford Dictionaries Online-English Dictionary, *http://oxforddictionaries.com/definition/sensor* accessed on 24 June, 2011.
2. Hulanicki, A., Głab, S., and Ingman, F. (1991) *Pure Appl. Chem.*, **63**, 1247–1250.
3. Gale, P.A. and Gunnlaugsson, T. (Guest Editors) (2010) *Chem. Soc. Rev.*, **39**, 3581–4008 (themed issue on: Supramolecular chemistry of anionic species).
4. Stokes, G.G. (1869) *J. Chem. Soc.*, **22**, 174–181.
5. Geddes, C.D., Apperson, K., Karolin, J., and Birch, D.J.S. (2001) *Anal. Biochem.*, **293**, 60–66.
6. Warick, J.W., Huang, N.N., Waring, W.W., Cherian, A.G., Brown, I., Stejskal-Lorenz, E., Yeung, W.H., Duhon, G., Hill, J.G., and Strominger, D. (1986) *Clin. Chem.*, **32**, 850–853.
7. Jayaraman, S. and Verkman, A.S. (2000) *Biophys. Chem.*, **85**, 49–57.
8. Amendola, V., Fabbrizzi, L., and Monzani, E. (2004) *Chem. – Eur. J.*, **10**, 76–82.
9. (a) Wiskur, S.L., Ait-Haddou, H., Lavigne, J.J., and Anslyn, E.V. (2001) *Acc. Chem. Res.*, **34**, 963–972; (b) Abouderbala, L.O., Belcher, W.J., Boutelle, M.G., Cragg, P.J., Dhaliwal, J., Fabre, M., Steed, J.W., Turner, D.R., and Wallace, K.J. (2002) *Chem. Commun.*, 358–359; (c) Amendola, V., Boiocchi, M., Fabbrizzi, L., and Palchetti, A. (2005) *Chem. – Eur. J.*, **19**, 5648–5660; (d) Fabbrizzi, L., Foti, F., and Taglietti, A. (2005) *Org. Lett.*, **7**, 2603–2606; (e) Vacca, A., Nativi, C., Cacciarini, M., Pergoli, R., and Roelens, S. (2004) *J. Am. Chem. Soc.*, **126**, 16456–16465.
10. Huston, M.E., Akkaya, E.U., and Czarnik, A.W. (1989) *J. Am. Chem. Soc.*, **111**, 8735–8737.
11. Steiner, T. (2002) *Angew. Chem. Int. Ed.*, **41**, 48–76.
12. Vance, D.H. and Czarnik, A.W. (1994) *J. Am. Chem. Soc.*, **116**, 9397–9398.
13. Bissell, R.A., de Silva, A.P., Gunaratne, H.Q.N., Lynch, P.L.M., Maguire, G.E.M., McCoy, C.P., and Sandanayake, K.R.A.S. (1993) *Top. Curr. Chem.*, **168**, 223–264.
14. Czarnik, A.W. (1994) *Acc. Chem. Res.*, **27**, 302–308.

15. De Santis, G., Fabbrizzi, L., Licchelli, M., Poggi, A., and Taglietti, A. (1996) *Angew. Chem. Int. Ed. Engl.*, **35**, 202–204.
16. Fabbrizzi, L., Licchelli, M., Perotti, A., Poggi, A., Rabaioli, G., Sacchi, D., and Taglietti, A. (2001) *J. Chem. Soc., Perkin Trans. 2*, 2108–2113.
17. Fabbrizzi, L., Francese, G., Licchelli, M., Perotti, A., and Taglietti, A. (1997) *Chem. Commun.*, 581–582.
18. Fabbrizzi, L., Licchelli, M., and Taglietti, A. (2003) *Dalton Trans.*, 3471–3479.
19. Callan, J.F., de Silva, A.P., and Magri, D.C. (2005) *Tetrahedron*, **61**, 8551–8588.
20. (a) Wiskur, S.L., Ait-Haddou, H., Lavigne, J.J., and Anslyn, E.V. (2001) *Acc. Chem. Res.*, **34**, 963–972; (b) Nguyen, B.T. and Anslyn, E.V. (2006) *Coord. Chem. Rev.*, **250**, 3118–3127.
21. Lavigne, J.J. and Anslyn, E.V. (1999) *Angew. Chem. Int. Ed.*, **38**, 3666–3669.
22. Fabbrizzi, L., Leone, A., and Taglietti, A. (2001) *Angew. Chem. Int. Ed.*, **40**, 3066–3069.
23. Harding, C.J., Mabbs, F.E., MacInnes, E.J.L., McKee, V., and Nelson, J. (1996) *J. Chem. Soc., Dalton Trans.*, 3227–3230.
24. Standard Methods for the Examination of Water and Wastewater-CO_2 Carbon Dioxide, *http://standardmethods.org/store/productview.cfm?ProductID=179* (accessed on 24 June, 2011).
25. Ansa Hortalá, M., Fabbrizzi, L., Marcotte, N., Stomeo, F., and Taglietti, A. (2003) *J. Am. Chem. Soc.*, **125**, 20–21.
26. Zhu, H.-L., Hang, Q.-W., Zhao, J., Duan, C.-Y., Tang, W.-X., and Fu, D.-G. (1999) *Transition Met. Chem.*, **24**, 131–134.
27. For a recent review see: Kim, S.K., Lee, D.H., Hong, J.-I., and Yoon, J. (2009) *Acc. Chem. Res.*, **42**, 23–31.
28. Fabbrizzi, L., Marcotte, N., Stomeo, F., and Taglietti, A. (2002) *Angew. Chem. Int. Ed.*, **41**, 3811–3814.
29. Zhu, H.-L., Zheng, L.-M., Duan, C.-Y., Huang, X.-Y., Bu, W.-M., Wu, M.-F., and Tang, W.-X. (1999) *Polyhedron*, **17**, 3909–3917.
30. Boiocchi, M., Bonizzoni, M., Fabbrizzi, L., Piovani, G., and Taglietti, A. (2004) *Angew. Chem. Int. Ed.*, **43**, 3847–3852.
31. Amendola, V., Bergamaschi, G., Buttafava, A., Fabbrizzi, L., and Monzani, E. (2010) *J. Am. Chem. Soc.*, **132**, 147–156.
32. Gans, P., Sabatini, A., and Vacca, A. (1996) *Talanta*, **43**, 1739–1753, *http://www.hyperquad.co.uk/index.htm* (accessed on 24 June, 2011).

Index

Anion Coordination Chemistry, First Edition. Edited by Kristin Bowman-James, Antonio Bianchi, and Enrique García-España.